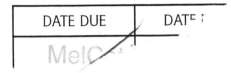

MERCURY POLLUTION
Integration and Synthesis

Edited by

Carl J. Watras, Ph.D.

Bureau of Research
Wisconsin Department of Natural Resources
Madison, Wisconsin
&
Center for Limnology
University of Wisconsin-Madison
Trout Lake Station
Boulder Junction, Wisconsin

and

John W. Huckabee, Ph.D.

Electric Power Research Institute
Palo Alto, California

LEWIS PUBLISHERS
Boca Raton   Ann Arbor   London   Tokyo

Library of Congress Cataloging-In-Publication Data

Mercury pollution : integration and synthesis / edited by Carl J.
   Watras and John W. Huckabee
      p.   cm.
   "Papers presented at the International Conference on Mercury as a
Global Pollutant held during June 1992 in Monterey, California"—Foreword
   Includes bibliographical references and index.
   ISBN 1–56670–066–3
   1. Mercury—Environmental aspects—Congresses. I. Watras, Carl J.
   II. Huckabee, John W. III. International Conference on Mercury as a Global Pollutant (1992 :
   Monterey, California)

TD196.M38M47   1994   94–15244
628.1′683—dc20   CIP

© 1994 by CRC Press, Inc.
Lewis Publishers is an imprint of CRC Press

No claim to original U.S. Government works
International Standard Book Number 1–56670–066–3
Library of Congress Card Number 94-15244
Printed in the United States of America 1 2 3 4 5 6 7 8 9 0
Printed on acid-free paper

FOREWORD

Today, concern about human impact on the environmental cycling of mercury (Hg) is widespread. Reports of contaminated fish stocks increased dramatically during the 1980s, particularly in freshwaters of the northern hemispheres.[1,2] There are serious public health ramifications, especially for developing fetuses and for populations that subsist on native fish. In 1989, the State of Michigan issued a blanket health advisory that covered all of Michigan's inland lakes. Pregnant women, nursing women, women who intend to have children, and all children age 15 and under were advised not to eat any largemouth bass (*Micropterus salmoides*), smallmouth bass (*Micropterus dolomieui*), walleye (*Stizostedion viteum*), northern pike (*Esox lucious*), or muskellunge (*Esox masquinongy*) caught in Michigan waters.[3] Findings in Wisconsin, Minnesota, Florida, Ontario, and Scandinavia have raised many questions about the sources and fates of mercury in freshwaters.

Mass balance studies now show that atmospheric transport and deposition is the dominant pathway delivering Hg to many of the world's lakes and oceans. Atmospheric Hg deposition, either direct or indirect via watersheds, is sufficient to account for almost all the Hg in many aquatic ecosystems. Mercury enrichment in surficial lake sediments suggests that atmospheric Hg deposition has increased by a factor of 3 to 5 since the industrial revolution,[4,5] but there is considerable uncertainty about the relative importance of short-range and long-range atmospheric transport. The tropospheric distribution of elemental Hg indicates long-range transport with atmospheric residence times of many months.[1] But it is other mercury species that rain out of the troposphere and/or accumulate in fish, and we are just beginning to understand the processes governing the distribution and deposition of these Hg species.

From a human perspective, methyl-Hg is the chemical species of greatest interest. Almost all of the Hg in fish is methylated and fish consumption is the major route for Hg uptake by humans. It is also the most toxic form of Hg in mammals. Methyl-Hg may arise via in situ methylation in the sediments and anoxic waters of both lakes and oceans.[6,7] Direct inputs of methyl-Hg from external sources may also be important in some lakes. Although methyl-Hg constitutes only a small percentage of the Hg in rain and snow, this may be sufficient to account for a significant fraction accumulated annually by gamefish.[8] In some lakes, atmospheric deposition may account for all the methyl-Hg cycling through the ecosystem annually.[9]

A substantial research effort is needed to identify the processes that generate atmospheric mercury species and control their transport and transformation in watersheds and in surface waters. While it may seem clear that human activity has perturbed the mercury cycle, it is not clear how this perturbation has interacted with other environmental factors to produce the fish contamination problem we now face. The relative importance of local, regional, hemispheric, and global phenomena remain to be assessed.

This volume comprises 56 papers presented at the *International Conference on Mercury as a Global Pollutant,* held during June 1992 in Monterey, California. During this conference, roughly 300 individuals representing 18 nations participated in a forum on environmental mercury. Sponsorship for the conference was provided by the Electric Power Research Institute (EPRI), the U.S. Environmental Protection Agency (EPA), the Swedish Power Board (Vattenfall), and Environment Canada. Sixty platform papers and fifty-two poster papers were presented in sessions dealing with bioaccumulation, methylation, human health, atmospheric cycling, terrestrial cycling, aquatic cycling, analytical chemistry, modeling, mitigation, and economics. It is hoped that the publication of these papers will facilitate the integration and synthesis of biogeochemical, biomedical, ecological, and sociological research.

We thank the authors of each chapter, and we also thank the reviewers listed below for their critical comments on the contributed papers.

Robert Ambrose, U.S. EPA
Thomas Atkeson, Florida DEP
Tamar Barkay, U.S. EPA
James P. Bennett, U.S. Park Service
Janina Benoit, University of Wisconsin
Gary N. Bigham, PTI Environmental Services
Nicolas S. Bloom, Frontier Geoscience
Drew Bodaly, Canada Dept. of Fish & Oceans
Gary Breece, Georgia Power Co.
Patricia Brewer, Tennessee Valley Authority
Thomas Clarkson, University of Rochester
Carolyn K. d'Almeida, U.S. EPA
Frank M. D'ltri, Michigan State University
Miriam Diamond, University of Toronto
Daniel R. Engstrom, University of Minnesota
R. Douglas Evans, Trent University
Russell Flegal, University of California
Robert Fleit, Fleit Research
Gary A. Gill, Texas A&M University
Cynthia C. Gilmour, BERL
Gary Glass, U.S. EPA
Reed Harris, Ontario Hydro
Harold Hemond, MIT
Marvin Hora, MN PCA
Robert Hudson, University of California
Tom Hutchinson, Trent University
J.D. Joslin, Tennessee Valley Authority
Gerald Keeler, University of Michigan
Ki Kim, Oak Ridge National Laboratory

David P. Krabbenhoft, U.S. Geological Survey
William M. Landing, Florida State University
Marcel Laperle, Hydro-Quebec
Leonard Levin, Electric Power Research Institute
Steven E. Lindberg, Oak Ridge National Laboratory
Robert P. Mason, MIT
David Michaud, Wisconsin Electric Power
Jerome Nriagu, Environment Canada
Susanne Padberg, Research Centre Jülich
Curtis D. Pollman, KBN Engin. & Appl. Sciences
Donald Porcella, Electric Power Research Institute
Ronald Rossmann, U.S. EPA
John Rudd, Canada Dept. Fish & Oceans
Pradeep Saxena, Electric Power Research Institute
Kent Schreiber, U.S. Fish and Wildlife Service
Abraham Silvers, Electric Power Research Institute
Jerry Stober, U.S. EPA
Edward B. Swain, Minn. Pollution Control Agency
George Taylor, Desert Research Institute
Joy Taylor, Michigan DNR
Huu V. Tra, Universite du Quebec
Ralph R. Turner, Oak Ridge National Laboratory
Grace M. Vandal, University of Connecticut
Matti Verta, Natl. Board of Waters, Finland
Jim Wiener, U.S. Fish and Wildlife Service
Ron Wyzga, Electric Power and Research Institute
Janice Yager, Electric Power and Research Institute
Edward Zillioux, Florida Power and Light

We especially thank and acknowledge the contributions of the six conference rapporteurs: Doctors Brian Wheatley, Ron Wyzga, Ken Bruland, Tom Hutchinson, Harry Hemond, and Mr. John Jansen. These rapporteurs provided a vital synthesis of the conference discussions at the final plenary session, and of equal importance, sparked many of the formative ideas for subsequent research.

REFERENCES

1. **Lindqvist, O., Ed.,** Mercury in the Swedish environment—recent research on causes, consequences and corrective measures, *Water Air Soil Pollut.,* 55, 1991.
2. **Watras, C.J. et al.,** Sources and fates of mercury and methylmercury in Wisconsin lakes, in *Mercury Pollution: Integration and Synthesis,* Watras, C.J. and Huckabee, J.W., Eds., Lewis Publishers, Chelsea, MI, 1994, chap. I.12.
3. **Michigan Department of Public Health,** *1989 Fish Consumption Advisory,* MDPH, Lansing, MI, 1989.
4. **Rada, R.G., Wiener, J.G., Winfrey, M.R., and Powell, D.E.,** Recent increases in atmospheric deposition of mercury to north-central Wisconsin lakes inferred from sediment analyses, *Arch. Environ. Contam. Toxicol.,* 18, 175, 1989.
5. **Swain, E.B., Engstrom, D.R., Brigham, M.E., Henning, T.A., and Brezonik, P.L.,** Increasing rates of atmospheric mercury deposition in midcontinental North America, *Science,* 257, 784, 1992.

6. **Gilmour, C.C. and Henry, E.A.,** Mercury methylation in aquatic systems affected by acid deposition, *Environ. Pollut.,* 71, 131, 1991.
7. **Winfrey, M.R. and Rudd, J.W.M.,** Environmental factors affecting the formation of methylmercury in low pH lakes, *Environ. Toxicol. Chem.,* 9, 853, 1990.
8. **Bloom, N.S. and Watras, C.J.,** Observations of methylmercury in precipitation, *Sci. Tot. Environ.,* 87/88 199, 1989.
9. **Hultberg, H., Iverfeldt, A. and Lee, Y.-H.,** Methylmercury input/output and accumulation in forested catchments and critical loads for lakes in southwestern Sweden, in *Mercury Pollution: Integration and Synthesis,* Watras, C.J. and Huckabee, J.W., Eds., Lewis Publishers, Chelsea, MI, 1994, chap III.3.

Carl J. Watras
Wisconsin Department of Natural Resources
John W. Huckabee
Electric Power Research Institute

THE EDITORS

Carl J. Watras, Ph.D., has been a Research Scientist with the Wisconsin Department of Natural Resources since 1984. He came to Wisconsin following doctoral training in Aquatic Ecology at the University of New Hampshire and postdoctoral research in Water Resources at the Massachusetts Institute of Technology. He has been actively involved in mercury research since 1986, serving as Chief Scientist and Project Director for two multi-institutional projects sponsored by EPRI on the biogeochemistry of mercury in freshwater ecosystems. Dr. Watras has served on several national and international mercury panels and advisory councils. He  co-chaired the *International Conference on Mercury as a Global Pollutant* in Monterey, California, 1993. Dr. Watras is also affiliated with the Center for Limnology, University of Wisconsin-Madison as an Associate Research Scientist. He works at the UW Trout Lake Research Station in northern Wisconsin.

John W. Huckabee, Ph.D., is Program Manager, Ecological Studies Program, Environment Division, at the Electric Power Research Institute (EPRI) in Palo Alto, California. At present, he is responsible for projects dealing with the ecological effects of the generation, transmission, and distribution of electric power. He has held his present position since 1985.

Previously, Dr. Huckabee was a Project Manager responsible for projects dealing with toxic substances, terrestrial and aquatic resources, and atmospheric deposition. He joined the Institute in 1979. Before joining the Institute, Dr. Huckabee  spent 8 years at the Oak Ridge National Laboratory, Oak Ridge, Tennessee, as a member of the Research Staff.

He received his Ph.D. in 1971 and his M.S. degree in 1965 from the University of Wyoming in the fields of zoology and physiology. He received a B.S. degree in biology from Sul Ros State College, Alpine, Texas in 1963.

Dr. Huckabee is a member of AAAS and the Ecological Society of America. He is the author of numerous papers on the cycling and effects of biologically active trace substances and radioisotopes.

CONTRIBUTORS

M. R. Anderson
Department of Fisheries and Oceans
Science Branch
St. John's, Newfoundland, Canada

A. W. Andren
University of Wisconsin
Madison, Wisconsin

Ilkka Aula
Department of Limnology and
  Environmental Protection
University of Helsinki
Helsinki, Finland

Franco Baldi
Università di Siena
Dipartimento di Biologia Ambientale
Siena, Italy

Corrado Barghigiani
Istituto di Biofisica CNR
Pisa, Italy

Frank W. Beaver
Energy and Environmental Research Center
University of North Dakota
Grand Forks, North Dakota

B. A. Bennett
Jacques Whitford Environment Limited
St. John's, Newfoundland, Canada

J. M. Benoit
Department of Marine Sciences
The University of Connecticut
Groton, Connecticut

Nicolas S. Bloom
Geosciences Inc.
Seattle, Washington

Gunnar Ch. Borg
Swedish Environmental Research
  Institute (IVL)
Göteborg, Sweden

A. Bortoli
Chemical Department, P.M.P.
Venice, Italy

Hannu Braunschweiler
Department of Limnology and
  Environmental Protection
University of Helsinki
Helsinki, Finland

Denis Brouard
Groupe Environnement Shooner inc.
Québec City, Québec, Canada

Raymond D. Butler
Energy and Environmental Research Center
University of North Dakota
Grand Forks, North Dakota

David S. Charlton
Energy and Environmental Research Center
University of North Dakota
Grand Forks, North Dakota

S. A. Claas
Bureau of Research
Wisconsin Department of Natural
  Resources
Boulder Junction, Wisconsin

Thomas W. Clarkson
Department of Environmental Medicine
University of Rochester School of
  Medicine
Rochester, New York

Dean Cocking
Department of Biology
James Madison University
Harrisonburg, Virginia

D. Cossa
Institut Français de Recherche pour
  l'Exploitation de la Mer
Nantes, France

A. W. H. Damman
Department of Ecology and Evolutionary
  Biology
The University of Connecticut
Groton, Connecticut

A. De Liso
Istituto di Biofiscia CNR
Pisa, Italy

Jean François Doyon
Groupe Environnement Shooner inc.
Québec City, Québec, Canada

Charles T. Driscoll
Department of Civil and Environmental
  Engineering
Syracuse University
Syracuse, New York

F. Dulac
CFR-LMCE
CNRS/CEA
Gif sur Yvette, France

H. Edner
Department of Physics
Lund Institute of Technology
Lund, Sweden

W. D. Ehmann
Department of Chemistry
University of Kentucky
Lexington, Kentucky

J. Elder
U.S. Geological Survey
Madison, Wisconsin

James M. Evans
Environmental and Health Research
Gas Research Institute
Chicago, Illinois

Romano Ferrara
Istituto di Biofiscia CNR
Pisa, Italy

Marco Filippelli
Laboratorio Chimico d'Igiene
  e Profilassi
La Spezia, Italy

W. F. Fitzgerald
Department of Marine Sciences
The University of Connecticut
Groton, Connecticut

R. P. Gambrell
Wetland Biogeochemistry Institute
Louisiana State University
Baton Rouge, Louisiana

Steven A. Gherini
Tetra Tech, Inc.
Lafayette, California

Steven P. Gloss
Wyoming Water Research Center
University of Wyoming
Laramie, Wyoming

Douglas L. Godbold
Forstbotanisches Institut
Universität Göttingen
Göttingen, Germany

D. F. Grigal
Department of Soil Science
University of Minnesota
St. Paul, Minnesota

Terry A. Haines
National Biological Survey
National Fisheries Contaminant Research
  Center
Field Research Station
Orono, Maine

Björn Hall
Department of Inorganic Chemistry
Chalmers University of Technology and
  University of Göteborg
Göteborg, Sweden

John A. Harju
Energy and Environmental Research Center
University of North Dakota
Grand Forks, North Dakota

DeVerle P. Harris
Department of Geoscience
Faculty of Science
The University of Arizona
Tucson, Arizona

Tacachi Hatanaka
Construçao e Comercio Camargo
  Correa s/a
UHE Tucuruí
Centro de Proteção Ambiental, Vila
  Permanente
Tucuruí-PA, Brazil

Robert Hayes
Department of Biology
James Madison University
Harrisonburg, Virginia

Kevin R. Henke
Energy and Environmental Research Center
University of North Dakota
Grand Forks, North Dakota

E. L. Hill
Lower Churchill Development Corporation
Environmental Affairs Department
St. John's, Newfoundland, Canada

P. S. Homann
Department of Forest Science
Oregon State University
Corvallis, Oregon

Marion E. Hoyer
The University of Michigan Air Quality
  Laboratory
Ann Arbor, Michigan

Xudong Huang
MIT Nuclear Reactor Laboratory
Cambridge, Massachusetts

John W. Huckabee
Electric Power Research Institute
Palo Alto, California

Robert J. M. Hudson
Institute of Marine Sciences
University of California
Santa Cruz, California
  and Tetra Tech, Inc.
Lafayette, California

Hans Hultberg
Swedish Environmental Research
  Institute (IVL)
Göteborg, Sweden

James P. Hurley
Bureau of Research
Wisconsin Department of Natural
  Resources
Monona, Wisconsin and
  Water Chemistry Program
University of Wisconsin
Madison, Wisconsin

Åke Iverfeldt
Swedish Environmental Research
  Institute (IVL)
Stockholm, Sweden

Charles H. Jagoe
Savannah River Ecology Laboratory
Aiken, South Carolina

Arne Jensen
Water Quality Institute
Horsholm, Denmark

Kjell Johansson
Swedish Environmental Protection Agency
Solna, Sweden

Anastácio Juras
Eletronorte
Centrais Elétricas do Norte do Brasil SA,
  EEA
Brasília-DF, Brazil

Ralph W. Karcher
New York State Department of
  Environmental Conservation
Gloversville, New York

E. J. Kasarskis
Department of Neurology
University of Kentucky
Lexington, Kentucky

Gerald J. Keeler
The University of Michigan Air Quality
  Laboratory
Ann Arbor, Michigan

Mary Lou King
Department of Biology
James Madison University
Harrisonburg, Virginia

Barbara A. Knuth
Human Dimensions Research Unit
Department of Natural Resources
Cornell University
Ithaca, New York

Victor T. Komov
Institute for Biology of Inland Waters
Academy of Sciences of Russia
Nekouz, Yaroslavl, Russia

D. Krabbenhoft
U.S. Geological Survey
Madison, Wisconsin

Walter A. Kretser
New York State Department of
  Environmental Conservation
Ray Brook, New York

Vit Kühnel
Energy and Environmental Research Cen
University of North Dakota
Grand Forks, North Dakota

Kestutis Kvietkus
Institute of Physics
Lithuanian Academy of Sciences
Vilnius, Lithuania

Carl H. Lamborg
The University of Michigan Air Quality
  Laboratory
Ann Arbor, Michigan

Brenda Lasorsa
Battelle Marine Sciences Laboratory
Sequim, Washington

Ying-Hua Lee
Swedish Environmental Research
  Institute (IVL)
Göteborg, Sweden

L. J. LeDrew
Newfoundland and Labrador Hydro
Environmental Services and Properties
  Department
St. John's, Newfoundland, Canada

Tuija Leino
Department of Limnology and
  Environmental Protection
University of Helsinki
Helsinki, Finland

Lian Liang
Brooks Rand, Ltd.
Seattle, Washington

S. E. Lindberg
Environmental Sciences Division
Oak Ridge National Laboratory
Oak Ridge, Tennessee

Oliver Lindqvist
Department of Inorganic Chemistry
Chalmers University of Technology and
  University of Göteborg
Göteborg, Sweden

Martin Lodenius
Department of Limnology and
  Environmental Protection
University of Helsinki
Helsinki, Finland

Elsemarie Lord
Swedish Environmental Research
  Institute (IVL)
Göteborg, Sweden

Kimmo Louekari
Institute of Occupational Health
Helsinki, Finland

Ismo Malin
Department of Limnology and
  Environmental Protection
University of Helsinki
Helsinki, Finland

V. Marin
Institute of Hygiene
University of Padua
Padua, Italy

W. R. Markesbery
Department of Neurology and Sanders-
  Brown Center on Aging
University of Kentucky
Lexington, Kentucky

B. E. Maserti
Istituto di Biofiscia CNR
Pisa, Italy

R. P. Mason
Department of Marine Sciences
The University of Connecticut
Groton, Connecticut
  and University of Maryland
Chesapeake Biological Laboratory
Solomons, Maryland

Edward J. Massaro
U.S. EPA
Health Effects Research Laboratory
Office of Research and Development
Perinatal Toxicology Branch (MD-67)
Development Toxicology Division
Research Triangle Park, North Carolina

T. Matilainen
Department of Limnology and
  Environmental Protection
University of Helsinki
Helsinki, Finland

Karl May
Research Centre (KFA) Jülich
Institut für Angewandte Physikalische
  Chemie (IPC)
Jülich, Germany

Markus Meili
Institute of Limnology
Uppsala University
Uppsala, Sweden

Tarja-Riitta Metsälä
Lammi Biological Station
University of Helsinki
Lammi, Finland

G. Moretti
Institute of Hygiene
University of Padua
Padua, Italy

K. A. Morrison
Wisconsin Department of Natural
  Resources
Boulder Junction, Wisconsin

Kenneth A. Morrison
Faculty of Applied Sciences
University of Sherbrooke
Sherbrooke, Québec, Canada

Arun B. Mukherjee
Department of Limnology and
  Environmental Protection
University of Helsinki
Helsinki, Finland

R. Munson
Tetra Tech, Inc.
Hadley, Massachusetts

John Munthe
Swedish Environmental Research
  Institute (IVL)
  and Department of Inorganic Chemistry
The University of Göteborg and Chalmers
  University of Technology
Göteborg, Sweden

E. A. Nater
Department of Soil Science
University of Minnesota
St. Paul, Minnesota

M. Niemi
Water and Environment Research Institute
National Board of Waters and the
  Environment
Helsinki, Finland

J. O'Donnell
Department of Marine Sciences
The University of Connecticut
Groton, Connecticut

Ilhan Olmez
MIT Nuclear Reactor Laboratory
Cambridge, Massachusetts

Per Östlund
Swedish Environmental Research
  Institute (IVL)
Stockholm, Sweden

J. G. Owens
Environmental Sciences Division
Oak Ridge National Laboratory
Oak Ridge, Tennessee

Susanne Padberg
Research Centre (KFA) Jülich
Institut für Angewandte Physikalische
  Chemie (IPC)
Jülich, Germany

Helena Parkman
Swedish Environmental Research
  Institute (IVL)
Stockholm, Sweden

Preeda Parkpian
Wetland Biogeochemistry Institute
Louisiana State University
Baton Rouge, Louisiana

W. H. Patrick, Jr.
Wetland Biogeochemistry Institute
Louisiana State University
Baton Rouge, Louisiana

E. L. Petticrew
Jacques Whitford Environment Limited
St. John's, Newfoundland, Canada

Donald B. Porcella
Electric Power Research Institute
Ecological Studies Program
Palo Alto, California

Petri Porvari
Department of Limnology
University of Helsinki and
  Water and Environment Research Institute
National Board of Waters and the
  Environment
Helsinki, Finland

D. Powell
University of Wisconsin
La Crosse, Wisconsin

R. Rada
University of Wisconsin
La Crosse, Wisconsin

P. Ragnarson
Department of Physics
Lund Institute of Technology
Lund, Sweden

Martti Rask
Finnish Game and Fisheries Research
  Institute
Evo State Fisheries and Aquaculture
  Research Station
Evo, Finland

Pat E. Rasmussen
Earth Sciences Department
University of Waterloo
Waterloo, Ontario, Canada

E. Ravazzolo
Institute of Hygiene
University of Padua
Padua, Italy

Michael Rieber
Department of Economics/Mining and
  Geological Engineering
College of Business and Public
  Administration
The University of Arizona
Tucson, Arizona

L. Rislove
University of Wisconsin
La Crosse, Wisconsin

Torquato Ristori
Istituto di Biofisica CNR
Pisa, Italy

Lisa Ritchie
Department of Biology
James Madison University
Harrisonburg, Virginia

D. W. Rodgers
Natural Sciences
Ontario Hydro Technologies
Toronto, Ontario, Canada

Jonas Sakalys
Institute of Physics
Lithuanian Academy of Sciences
Vilnius, Lithuania

Kalevi Salonen
Lammi Biological Station
University of Helsinki
Lammi, Finland

Matts-Ola Samuelsson
Swedish Environmental Research
  Institute (IVL)
Stockholm, Sweden

Peter Schager
Department of Inorganic Chemistry
Chalmers University of Technology and
  University of Göteborg
Göteborg, Sweden

Roger Schetagne
Environmental Department
Hydro-Québec
Montréal, Québec, Canada

Craig R. Schmit
Energy and Environmental Research Center
University of North Dakota
Grand Forks, North Dakota

Carl L. Schofield
Department of Natural Resources
Cornell University
Ithaca, New York

W. H. Schroeder
Atmospheric Environment Service
Environment Canada
Downsview, Ontario, Canada

D. A. Scruton
Department of Fisheries and Oceans
Science Branch
St. John's, Newfoundland, Canada

Howard A. Simonin
New York State Department of
  Environmental Conservation
Rome, New York

Daniel J. Stepan
Energy and Environmental Research Center
University of North Dakota
Grand Forks, North Dakota

Markus Stoeppler
Research Centre (KFA) Jülich
Institut für Angewandte Physikalische
  Chemie (IPC)
Jülich, Germany

W. J. Stratton
Department of Chemistry
Earlham College
Richmond, Indiana

D. Strömberg
Department of Inorganic Chemistry
Chalmers University of Technology and
  University of Göteborg
Göteborg, Sweden

S. Svanberg
Department of Physics
Lund Institute of Technology
Lund, Sweden

John Symula
New York State Department of
  Environmental Conservation
Rome, New York

F. Tan
Wetland Biogeochemistry Institute
Louisiana State University
Baton Rouge, Louisiana

Normand Thérien
Faculty of Applied Sciences
University of Sherbrooke
Sherbrooke, Québec, Canada

A. Uusi-Rauva
Faculty of Agriculture and Forestry
Instrument Centre
University of Helsinki
Helsinki, Finland

G. M. Vandal
The University of Connecticut
Groton, Connecticut

M. Verta
Water and Environment Research Institute
National Board of Waters and the
  Environment
Helsinki, Finland

E. Wallinder
Department of Physics
Lund Institute of Technology
Lund, Sweden

Shu Shen Wang
Guangxi Province Anti-Epidemic and
  Hygiene Station
Nanning, China

Carl J. Watras
Bureau of Research
Wisconsin Department of Natural
  Resources and
  Center for Limnology
University of Wisconsin
Trout Lake Station
Boulder Junction, Wisconsin

J. G. Wiener
U.S. Fish and Wildlife Service
La Crosse, Wisconsin

M. Winfrey
University of Wisconsin
La Crosse, Wisconsin

Z. F. Xiao
Department of Inorganic Chemistry
Chalmers University of Technology and
  University of Göteborg
Göteborg, Sweden

U. P. Williams
Department of Fisheries and Oceans
Science Branch
St. John's, Newfoundland, Canada

Zu Qiu Xie
Guangxi-Nanning Nan Hu
  Park
Nanning, China

CONTENTS

SECTION II: ATMOSPHERIC CYCLING, TRANSPORT, AND DEPOSITION

SECTION III: TERRESTRIAL AND WATERSHED PROCESSES

SECTION IV: BIOACCUMULATION

SECTION VIII: HUMAN HEALTH AND PUBLIC POLICY

SECTION I

Freshwater and Marine Ecosystems

Mercury in the Environment: Biogeochemistry

Donald B. Porcella

CONTENTS

ABSTRACT: Atmospheric deposition is a source of mercury to freshwater fish that previously has not been adequately assessed, because past analytical and sampling procedures caused inaccurate estimates of deposition and of emissions from natural and anthropogenic sources. More accurate studies during the last decade have begun to fill this gap. However, global circulation and the relative importance of regional and local sources of mercury need further quantification. Recently developed techniques have led to better estimates of global sources and sinks, but considerable work is needed on the estimation of anthropogenic and terrestrial sources. At a smaller scale—a Wisconsin, U.S. seepage lake—the annual mercury atmospheric deposition (about 10 $\mu g/m^2$/year) is greater than the mercury included in the total fish biomass of the lake. Atmospheric concentrations of gaseous mercury in Wisconsin are similar to those in Northern Hemisphere oceanic air masses. However, mercury deposition appears to be principally on particles containing oxidized mercury, Hg(II), produced directly from emissions or oxidized in raindrops. The fraction of total atmospheric input that was elemental mercury oxidized near the deposition site and the fraction that was particulate inorganic mercury transported from regional sources is largely unknown. Best estimates suggest that about 5% of the total mercury deposition within the lake system is transformed into methylmercury to be accumulated by fish. Estimates of methylmercury deposition from the atmosphere to the lake and net input to the sediments are both less than 1% of total deposition. At least 95% of mercury in whole, 1-year-old yellow perch (*Perca flavescens*) from these seepage lakes is methylmercury. The Mercury Cycling Model (MCM, Version 1.0) simulates the cycling of environmental mercury, including fish bioaccumulation of mercury. The model can assess risk as well as test hypotheses about processes.

I. INTRODUCTION

Substantial concentrations of methylmercury occur in both freshwater and marine fishes. Methylmercury is a neurotoxin and is the most toxic known chemical form of mercury. Biotic and abiotic methylation of inorganic mercury produces methylmercury which fish accumulate from water and their diet. Fish consumption by human populations raises health concerns, especially for fetuses and young children (Clarkson, 1990; Clarkson, 1994). Similarly, methylmercury can adversely affect developing neural tissue of mammals and birds that consume fish (e.g., Roelke et al., 1991). Nearly all mercury in fish flesh (>95%) occurs as methylmercury (Huckabee et al., 1979; Grieb et al., 1990; Bloom, 1992). Fish species that prey on other fish have higher concentrations of mercury, and the mercury concentration increases in fish as they age (Grieb et al., 1990). For humans and wildlife, the non-fish diet is relatively low in methylmercury, and fish consumption is the major contributor to mercury risk. For example, panthers (*Felis concolor*) that prey on mammals dependent on aquatic food sources appear to accumulate more mercury than panthers dependent on terrestrial food chains (Roelke et al., 1991).

Regulatory agencies focus on fish as the target organism to protect the health of humans and other sensitive organisms. For example, the U.S. Food and Drug Administration set an advisory standard of 1 ppm wet-weight in fish flesh, and the World Health Organization (WHO) recommends that the dose should not exceed 30 µg/day to protect adult humans. Many states use 0.5 ppm wet-weight in fish flesh to set fish-consumption advisories. Finding fish in a body of water that exceed established advisories leads health agencies to issue health warnings regarding sportfish consumption.

Mercury has been observed in a wide variety of environments, but attention has recently focused on regions having dilute, relatively unproductive waters. Many had assumed that the problem of mercury was solved by eliminating methylmercury discharges, methylmercury fungicides, and by reducing industrial mercury discharges. The discovery of high levels of mercury in fish in areas remote from human activities disproved that assumption, and implicated atmospheric deposition as a source (e.g., Rada et al., 1989).

This chapter contains a summary of recent information about the global cycle of mercury and its local cycle in a single lake. In addition, major uncertainties in our knowledge of the mercury cycle are highlighted. For atmospheric mercury behavior on a regional scale, the reader should see Lindqvist et al. (1991).

II. MERCURY BIOGEOCHEMISTRY AND RISK

The U.S. 1990 Clean Air Act (CAA) amendments require an assessment of health risk to humans and wildlife caused by mercury emissions. When accurate risk assessments indicate that little or no decrease in health risk will occur with expensive emission controls, alternative ways of generating power or producing goods, or more efficient methods of emission control may be necessary. For example, an important future development includes new generation techniques that promise higher energy efficiency along with significant reductions of mercury and other pollutant emissions at much lower costs than retrofitting controls onto older inefficient generating units (EPRI, 1990; EPRI, 1991).

Mercury risk assessment should include global, regional, and local scales because of mercury's significant atmospheric cycle. Mercury is transported in air, land, and water at all scales. In addition mercury undergoes transformations among chemical forms, including elemental mercury [Hg(0)], inorganic mercury [Hg(II)], and methylmercury (CH_3Hg^+). The chemical form affects transport in and between air, land, and water as well as chemical and biological behavior.

In the past, scientists inferred mercury's biogeochemical behavior by comparing fish mercury accumulation with water chemistry parameters and other limnological factors

(Rudd et al., 1983; Håkanson, 1980). Analytical problems forced this inference, because only sediments and biota contained sufficient concentrations for easy detection whereas sample contamination would have little effect. Recently, sample contamination problems have been overcome by development of ultraclean sampling and laboratory procedures along with more sensitive analytical techniques. These recent methods appear to provide accurate estimates of air and water concentrations of the different mercury forms (Fitzgerald and Gill, 1979; Slemr et al., 1985; Gill and Fitzgerald, 1987; Brosset, 1987; Bloom and Fitzgerald, 1988; Bloom, 1989; Fitzgerald and Watras, 1989; Gill and Bruland, 1990; Iverfeldt, 1990; Lindqvist et al., 1991; Fitzgerald et al., 1991; Porcella et al., 1992). Values of 1 to 2 ng/m^3 in air and <1 to 2 ng/L in water for ambient samples taken distant from point sources are within the range of concentrations that appear to be accurate. Concentrations in precipitation are considerably higher, requiring less sensitive techniques (5 to 25 ng/L) (Bloom and Watras, 1989; Fitzgerald et al., 1991). The ultraclean methods for measuring mercury in these concentration ranges allow more accurate characterization of mercury biogeochemistry.

A major application of mercury biogeochemical research is to contribute to an accurate risk assessment of mercury. Cost-effective risk management requires accurate data and models. To enable accurate risk assessments of mercury, we need to predict the relationship between human activities including emissions of mercury from major industrial sources, natural emissions, reemission from previously deposited mercury, mercury transport and transformation in air, land, and water, and the chemical and biological factors that lead to its accumulation in fish. Major questions include how methylmercury is formed, what factors control its bioavailability, and how it is accumulated. Fortunately, we have learned a great deal about mercury biogeochemistry, largely because of improved analytical and sampling procedures. Since 1988, many scientists have begun to apply ultraclean sampling and analysis methods that allow accurate data to be obtained, and thereby permit more accurate models of mercury biogeochemistry.

III. GLOBAL MERCURY

Mercury in the atmosphere tends to occur almost exclusively as Hg(0). Oxidized forms [Hg(II) and methylmercury] typically constitute less than 2% of the total concentration in air (Fitzgerald, 1986, 1989). However, virtually all of the deposition is in the oxidized forms (Fitzgerald et al., 1991). We have little information about temporal trends in the atmosphere and treat mercury as if it were at steady state, cycling through the atmosphere with about a 1-year residence time (Fitzgerald, 1986, 1989). For the mercury mass balance, the total mass of mercury in the atmosphere has been estimated at 5000 to 6000 metric tons (Fitzgerald, 1986, 1989; Slemr et al., 1985). Fitzgerald, (1986, 1989; also see Fitzgerald and Clarkson, 1991) provides extensive data that place mercury deposition at about 6000 metric tons per year.

Less extensive data on the major categories of input to the atmosphere exist. Fitzgerald (1986) estimated that about 2000 metric tons entered the atmosphere from the ocean surface, about the same amount from anthropogenic sources, and the remainder from land surfaces. He estimated that volcanoes were a minor source, based on relatively few samples. In round numbers, about one third of the mercury input to the atmosphere came from each of the three categories—ocean, land, and anthropogenic. More recent estimates (see Lindqvist et al., 1991) suggest that anthropogenic input could be as much as half of the total, reducing land evasion of mercury to one sixth. Although estimates of anthropogenic input have considerable variance, estimates of natural mercury evasion from the land are very inaccurate and are determined by difference.

Estimates of open ocean input of mercury to the atmosphere are the most credible (Fitzgerald, 1986, 1989). Open ocean atmospheric concentrations varied between 1 to 2 ng/m^3, with the Southern Hemisphere concentrations being about half the Northern Hemisphere levels. Fitzgerald (1986) suggested that the anthropogenic sources were focused in the Northern Hemisphere, leading to the observed difference in air levels because the atmosphere does not mix rapidly across the equator.

Various authors have estimated anthropogenic mercury emissions. Lindqvist et al. (1991) used literature data and calculations from Swedish results to estimate a mean of 4500 metric tons per year. Using the geometric mean of the range of mercury emissions compiled by Nriagu and Pacyna (1988), present-day worldwide fossil fuel combustion produces about 1500 metric tons per year (300 from electricity generation and 1200 from other industrial use), waste incineration produces about 600 metric tons per year, and smelting and wood combustion about 250 metric tons per year. (The geometric mean was used to decrease the effect of outliers; outliers tend to err on the high side when analytical contamination problems exist.) Lindqvist et al. (1991) estimate that diffuse anthropogenic sources add about 1000 metric tons per year. These sources account for approximately 3350 metric tons per year, or about 60% of the total influx into the atmosphere. However, actual emissions could vary by a factor of two. Incinerator emissions are extremely difficult to estimate because the number of incinerators is increasing, controls on emissions are rapidly changing along with the mercury content of the wastes being combusted, and accurate emissions data have been obtained only recently (Lindqvist et al., 1991). Estimates of emissions from fossil fuel combustion vary widely because the fuel can be quite clean or may contain substantial inclusions of contaminants such as minerals that include mercury. Also, Bloom (1992) has shown the need for clean techniques for measuring fuels and emissions. Using older fuel mercury concentrations, a recent study estimated that the U.S. fossil fuel contribution to the atmosphere was 81 metric tons per year, or about 1.5% of the total global atmospheric input (Neme, 1990).

A key issue concerns the form of mercury in the atmosphere. Oceanic emissions to the atmosphere are as Hg(0). Land and water emissions from reducing processes (whether biotic or abiotic) will largely be Hg(0). Gaseous mercury [Hg(0)] must be transformed to particulate oxidized mercury [Hg(II)] to contribute substantially to mercury deposition. Anthropogenic sources will largely provide oxidized forms that become associated with particles prior to or soon after emission, and these particles will have varying atmospheric residence times depending on particle size and wind speed. Understanding the particulate mercury cycle is the key to understanding the contribution of local and regional sources to the global cycle. The global cycle depicted in the previous paragraphs does not account for the effect of speciation on transport of mercury emitted from coal combustion, incineration, smelting, and other sources.

Two important assumptions underlie the global mercury mass balance. First, the system is assumed to be at steady state, i.e., concentrations and fluxes are not changing, and second, the global anthropogenic emissions estimates are assumed to be reasonable. To evaluate whether mercury levels in the atmosphere are at steady state, estimates of historical trends are needed. Some investigators have used sediment and peat cores to estimate present and preindustrial mercury deposition as a predictor of atmospheric concentrations. These measurements indicate that present-day mercury deposition is a factor of two to five times greater than preindustrial deposition measurements (Lindqvist et al., 1991).

The temporal pattern of deposition inferred from cores taken in northern Minnesota does not suggest a monotonic increase in atmospheric mercury concentrations (Benoit et al., 1994; Swain et al., 1992). Analysis of the variance of results from the Minnesota peat cores suggests that the peak deposition occurred prior to 1970. Previous uses of mercury may have introduced more mercury than present activities. In the Virginia City mines of Nevada more than 7000 metric tons of mercury were used in gold extraction prior to the turn of the century, and only about half has been accounted for by soil and water measurements (Cooper et al., 1985). More recently, atmospheric inputs of mercury could have occurred from other major uses including gold mining in the Amazon River basin (Nriagu et al., 1991). Other emissions resulting from use of mercury in the U.S. by battery manufacturers and as a fungicide in agriculture and in paints, and emissions from chlor-alkali plants, have declined in the last decade (Neme, 1990). In addition, Lindqvist et al. (1991) clearly show that Swedish mercury emissions to the atmosphere peaked prior to 1960. Assuming that many industrialized countries used mercury in a similar pattern, atmospheric mercury emissions could have peaked prior to the most recent decades. These estimations have been obtained only from local geographic areas, making it difficult to extrapolate to the global atmospheric mercury cycle, thus suggesting the need for additional assessments.

Fish mercury concentrations measured over different time intervals have been reviewed, and only one increasing trend has been reported (Swain and Helwig, 1989). As Swain and Helwig report, the data are too sparse to draw conclusions and are appropriate only for formulating hypotheses about mercury. As will be shown later in this book, other factors can confound such reported trends. Slemr and Langer (1992) provided data to show that mercury in the air had increased in mid-latitudes of the Atlantic Ocean. However, the evidence is equivocal because long-term continuous records at permanent stations are not available.

Like ambient measurements, anthropogenic emission estimates obtained in previous years suffer from inaccurate techniques (see Bloom, 1993). In fact, most measurements fail to speciate emissions—a critical piece of information because the global cycle will largely be unaffected by Hg(II) emissions unless Hg(II) is associated with very fine particles which can be transported long distances or, if deposited, Hg(II) is reemitted as Hg(0). Note that Fitzgerald et al. (1994) identify transport of fine particles as important to the atmospheric cycle, and consider that the global cycle is affected by Hg(II) because of reemission of deposited mercury. Land sources have been determined by difference, and therefore it is important to constrain them by independent estimates of evasion from land surfaces using accurate methods. Furthermore, one must assume that at least part of the oceanic and terrestrial inputs is the reemission of previously deposited mercury. These two factors—temporal variation in the emission of mercury and inaccurate estimates of flux from human activities and land evasion—make the global cycle of mercury and its contribution to risk from the consumption of fish uncertain.

The next smaller scale of interest is regional. Natural and man-made emissions into the atmosphere likely enter the global cycle if they are in the form of Hg(0), but deposit locally or regionally if they are oxidized [Hg(II)] (see Lindqvist et al., 1991). Regional analyses need to consider transport and transformation rates to perform an assessment. Swedish investigators have studied regional processes in considerable detail (Brosset, 1987; Iverfeldt, 1991; Munthe, 1991; Lindqvist et al., 1991), and they will not be discussed here.

IV. MERCURY CYCLING IN LAKES

With all the uncertainties of the global cycle, a focused study at a smaller scale—a single lake—can provide useful insights about mercury biogeochemistry. Biogeochemical studies of lakes that contain fish with high mercury levels (Grieb et al., 1990; Spry and Wiener, 1991), yet remote from point sources and mercury-containing geological strata, can delineate the role of atmospheric exchange of mercury. The investigators on the Mercury in Temperate Lakes project (MTL) (Watras et al., 1994) selected seepage lakes located in northern Wisconsin that were isolated from all but atmospheric sources to quantify mercury cycling and mass balances. MTL was coordinated with a NAPAP (National Acidic Precipitation Assessment Project) funded project assessing the effect of acidification on Little Rock Lake. Little Rock Lake was divided into two nearly equal parts, one of which was treated with acid and called the treatment basin and the other retained as the reference basin (see Watras and Frost, 1989; Wiener et al., 1990; Bloom et al., 1991; Wiener et al., in preparation). This paper summarizes a mass balance for the Little Rock Lake treatment basin (Watras et al., 1994). In addition, a mass balance for a reservoir with an abandoned mercury mine in the

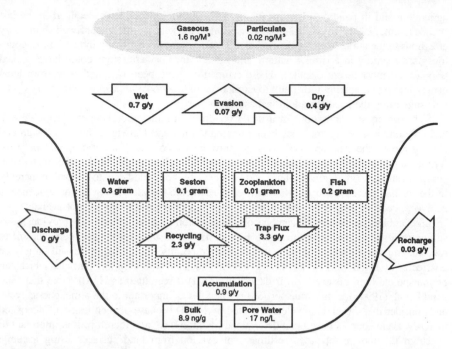

Figure 1 Mass balance of total mercury for Little Rock Lake treatment basin, showing that atmospheric sources can account for all mercury in fish. Three phases are shown: atmosphere, lake, and sediments.

watershed is discussed (Gill and Bruland, 1992). Also, implications of mercury biogeochemistry and the applications of a simulation model for mercury cycling are summarized.

A. MASS BALANCES

The Little Rock Lake treatment basin is 10 hectares in area, relatively shallow, and has a small hypolimnion which is lacking in the shallower reference basin (Watras and Frost, 1989). Measurements taken over a 3-year period gave the mass balance for total mercury in the treatment basin (Figure 1). In round numbers, the standard deviation of the annual mean fluxes and pools varied between 5 and 30%. The atmospheric mercury concentration of 1.6 ng/m^3 and wet deposition input of 0.7 g/year was typical of open ocean values in the northeastern Pacific and northwestern Atlantic (Fitzgerald et al., 1991; Slemr et al., 1985). Almost all (99%) of the atmospheric mercury was Hg(0), and particulate mercury accounted for almost all of the mercury input as wet (measured) and dry (calculated from particle deposition velocities; Fitzgerald et al., 1991) deposition. Inputs of mercury in surface water and ground water were nil, supporting the assumption of a system dominated by atmospheric inputs. Evasion of mercury as Hg(0) from the lake to the atmosphere was about 5% of the deposited input. Other MTL lakes range from 10 to 50% evasion of the total deposition (Watras et al., 1994).

Cycling of mercury occurred within the lake, with about as much mercury in the biota (the seston was mostly phytoplankton) as in the water column, an observation atypical of many microconstituents in lakes. Considerable resuspension and settling occurred, but the bulk of the deposited mercury entered the sediment pool.

Atmospheric deposition was the major source of mercury to the lake. Deposition was balanced by sedimentation plus evasion and a small flux to the ground water, and the annual atmospheric deposition of mercury was greater than the total mercury in the biotic pool.

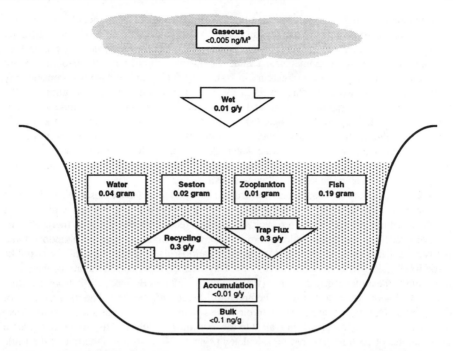

Figure 2 Mass balance of methylmercury for Little Rock Lake treatment basin, showing that in-lake methylation is a key process affecting mercury accumulation by fish. Three phases are shown: atmosphere, lake, and sediments.

Table 1 **Biomagnification of methylmercury in the aquatic food chain of Little Rock Lake (Treatment Basin).**

Compartment	Relative concentrations		
	Methyl-Hg	Non-methyl-Hg	Methyl-Hg (%)
Water	1	10	10
Phytoplankton	10^5	$10^{5.7}$	15
Zooplankton	$10^{5.5}$	$10^{5.9}$	30
Fish	$10^{6.5}$	10^5	95

Modified from Watras, C. J. and Bloom, N. S., *Limnol. Oceanogr.*, 37:1313–1318 (1992). With permission.

The mass balance for methylmercury differed greatly from that for total mercury (Figure 2). Standard deviations varied from 5 to 100% of the mean fluxes and pools (Watras et al., 1994). Methylmercury content varied more than total mercury, primarily due to the seasonal variation in water column concentrations. Although detected in the atmosphere, methylmercury accounted for only about 1% of the total mercury input to the lake. Fish were the dominant methylmercury pool in the water column. The measured sediment trap flux was substantial, but sediment methylmercury concentration was only about 1% of total mercury. Unlike total mercury, the net flux of methylmercury to and from the sediments appeared to be zero in this lake. The mass balance showed that essentially all methylmercury was formed within the lake ecosystem.

Biomagnification of methylmercury occurred in the Little Rock Lake treatment basin (Table 1). Two processes seemed to be involved in producing the high concentration of methylmercury in higher trophic levels: a higher affinity (defined as the net effect of uptake and depuration) for nonmethylmercury in organisms of lower trophic levels, and a higher affinity for methylmercury in fish. For 1-year-old yellow perch (*Perca flavescens*), 95% of the total mercury was methylmercury, a value consistent with other results (Bloom, 1992a). The overall percentage of methylmercury in young perch (1-year-old fish) was considerably higher than that in lower trophic levels in the treatment basin. The fish accumulated the methylmercury from both food sources and the water column, and the bioaccumulation factor, calculated from the ratio of total whole fish and aqueous methylmercury measurements, was $10^{6.5}$. This indicates that methylmercury was concentrated by a factor of 3 million times in whole fish relative to water in this dilute-water lake. These results emphasize the importance of methylation as a key process in mercury biogeochemistry. Furthermore, it suggests that aqueous methylmercury should correlate with fish mercury.

A less detailed mass balance was constructed for the Davis Creek Reservoir in northern California. The dilute-water, low pH Wisconsin lake provides a strong contrast with the more productive, high pH California reservoir which also receives more input of mercury. The reservoir receives substantial inflow from its watershed which contains an abandoned and partly remediated mercury mine and smelter (Gill and Bruland, 1992). Quantitatively unlike Little Rock Lake, the Hg(0) evasion from the Davis Creek Reservoir to the atmosphere is at least double the atmospheric deposition (calculated by Gill and Bruland, 1992, from literature values) and about equal to the watershed mercury loading to the lake. The median mass of dissolved reactive mercury in the reservoir is less than that of the particulate form. Areal deposition rates were assumed equivalent to the Wisconsin results (10 $\mu g/m^2$/year), but areal evasion rates of 25 $\mu g/m^2$/year were calculated as about 30 to 40 times the rate at the Little Rock Lake treatment basin (0.7 $\mu g/m^2$/year), supporting the arguments of Fitzgerald et al. (1994) that a higher pH favors production and subsequent evasion of elemental Hg. Outflow during the period of study removed about the same mass as was deposited, and assuming a steady state, the water column and sediment compartments would be in balance. Low rainfall due to drought occurred during the study, suggesting that additional mercury loading from the watershed would have occurred under more typical hydrologic conditions and account for increased input of mercury to sediments during high rainfall years.

B. BIOGEOCHEMICAL PROCESSES

Previous research had suggested that pH and DOC (dissolved organic carbon) affected methylmercury accumulation in fish (see reviews in Greib et al., 1990; Spry and Wiener, 1991), so seven lakes were selected for study in the MTL project which spanned a range of pH and DOC (Watras et al., 1994). If atmospheric deposition was the sole factor controlling fish mercury, the fish from these lakes would presumably have relatively similar mercury concentrations; in fact, the concentrations varied tenfold (Figure 3). It is not possible to explain the differences solely on the basis of pH and DOC, although pH is clearly involved (Wiener et al., in preparation; Hudson et al., 1994). Rada et al. (1993) showed that areal burdens of mercury in sediments varied strongly with pH and hypothesized that pH-related efflux of gaseous mercury [Hg(0)] from water to the atmosphere was partly responsible. Other chemical factors such as, chlorophyll α, sulfate, chloride, and calcium vary twofold (Watras et al., 1994) and appear to be at least partly responsible for some of the differences (Hudson et al., 1994). Lake morphometry varies considerably, but no clear relationship to methylmercury concentration exists.

Concentrations of methyl- and total mercury were correlated among the seven lakes (Figure 4), suggesting that total Hg(II) serves as the substrate for methylation (see analysis by Hudson et al., 1994). Three methods were used to bound an estimate of the amount of total mercury transformed to methylmercury: regression, mass balance, and modeling. The regression slope was heavily influenced by a single point (Russet Lake); accepting that, the slope suggests that about 20% of the total mercury was transformed to methylmercury

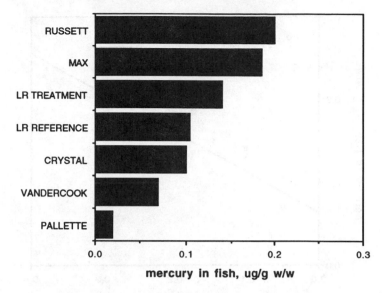

Figure 3 Order of magnitude differences in mercury content of 1-year-old yellow perch from seven northern Wisconsin lakes suggest other factors than atmospheric deposition can lead to differences in mercury accumulation.

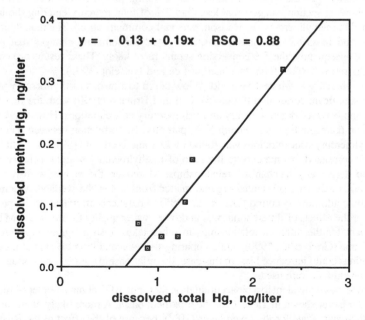

Figure 4 Dissolved methylmercury is highly correlated with total mercury in seven northern Wisconsin lake water columns.

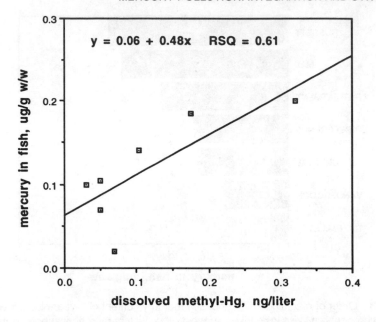

Figure 5   Mercury in 1-year-old yellow perch from seven northern Wisconsin lakes is highly correlated with methylmercury in water.

(Figure 4). The regression estimate was within a factor of two of the sediment trap measurements shown in the mass balances for the treatment basin (Figures 1 and 2), which suggest a little more than 10% transformation (0.3 g/year methylmercury recycled out of 2.3 g/year of total mercury). Simulations with the Mercury Cycling Model (Hudson et al., 1994, also see the following section) suggest that less than 5% of total mercury entering the lake was transformed to methylmercury (R. Hudson, personal communication). The model, unlike the regression and mass balance calculations, accounted for temporal changes that probably occurred. Consequently, the 5% conversion seems most likely. These results are consistent with similar ratios (0.02 to 0.06) tabulated in Lee and Iverfeldt (1991).

Relating mercury in whole 1-year-old yellow perch to aqueous methylmercury yielded a lower but significant correlation (Figure 5). Gill and Bruland (1990) were the first to show a relationship between organomercury and fish mercury concentrations. However, their data do not fit on the same line as in Figure 5. Apparently, no correlation between fish mercury and methylmercury concentrations was found by Lee and Iverfeldt (1991) for eight Swedish lakes; they correlated fish mercury to the ratio of methylmercury to total mercury, but it is difficult to derive any mechanistic relationships. Moreover, the eight Swedish lakes are drainage lakes receiving substantial organic matter from watershed wetlands where abiotic methylation could have occurred (Lee et al., 1985). Analytical differences, fish species differences, or the effects of different nutrient and other water quality factors in the MTL lakes may account for the different relationships in other lakes. Fish integrate exposure over a period of time (Grieb et al., 1990), and comparing fish concentrations to annual mean water concentrations could introduce bias. In this case, the relationship seemed appropriate because only 1-year-old fish were used.

The slope was equivalent to a bioaccumulation factor of $10^{5.7}$, about an order of magnitude less than the factor shown in Table 1 for the treatment basin. A more likely bioaccumulation factor is the mean value for the seven lakes ($10^{6.3}$), because of the effect of the Russet Lake value on the regression. Given the slopes in Figures 4 and 5, a bioaccumulation factor of

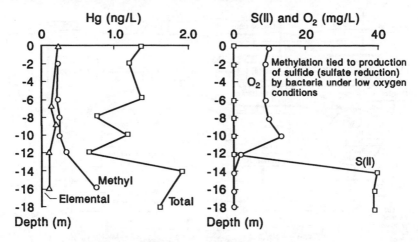

Figure 6 Late summer profiles in a stratified Wisconsin seepage lake support a linkage between sulfate reduction and net methylation of mercury. (Graph provided by courtesy of Steven Gherini, Tetra Tech, Inc.)

about 100,000 was calculated from the ratio of total fish mercury and total mercury in water. This factor may change with water quality, as illustrated by the attempt to compare with Gill and Bruland's (1990) data. Nutrient levels can substantially affect mercury concentrations in fish (e.g., Rudd et al., 1983), and model analyses performed by R. Hudson (personal communication) show substantial decreases in fish mercury concentration from increased primary production. Moreover, large fish can have a higher bioaccumulation because of age and food sources. Because of these results, the concept of a bioaccumulation factor does not seem applicable to mercury risk assessment over regions, but should be considered only on a site-specific basis. Furthermore, the results reemphasize the need to understand methylation processes.

As a first step in explaining the methylation process, concentration profiles in Pallette Lake (Figure 6) indicate that redox conditions and sulfur cycling affect methylation and mercury dynamics (data from Watras et al., 1994 and a graph provided by R. Hudson and S. Gherini, unpublished, 1992). Mercury species are shown in the left panel of Figure 6 and total sulfide (S(II)) and dissolved oxygen (O$_2$) are shown in the right panel. One key factor appears to be dissolved oxygen, which declined at a depth below 10 m to near anoxia at 14 meters. At these same depths, sulfide increased from near 0 to almost 40 mg/L due to sulfate reduction. Although total mercury [almost 90% Hg(II)] generally declined with depth for the mixed layer profile (above 10 m depth), there was a substantial increase in concentration to the range of 1.5 to 2 ng/L at 14 m. At 10 m, Hg(0) approached zero, suggesting that reduction of Hg(II) occurred only in the mixed layer. Although relatively constant in the mixed layer and constituting less than 10% of the total mercury, methylmercury concentration increased greatly below 10 m, accounting for about 30% of the total mercury in the anoxic layer and suggesting that methylation was markedly greater in the anoxic zone than the oxic zone. Methylmercury did not account for the entire increase observed in the total mercury concentrations. Although transport of particulate material from the mixed layer may account for some of the methylmercury produced and for the increased total mercury concentrations, the increased relative concentration of methylmercury suggests that net methylation increased in this layer. The results lend credence to the hypothesis that methylation is linked to sulfur cycling, perhaps via sulfate reduction (Compeau and Bartha, 1985; Gilmour and Henry, 1991).

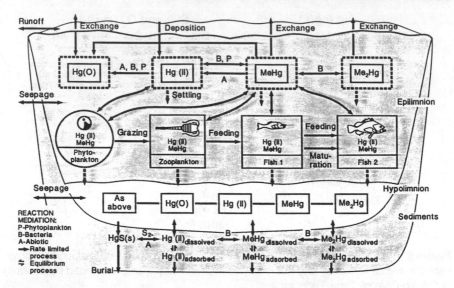

Figure 7   Schematic presentation of Mercury Cycling Model (MCM v. 1.0).

C. MERCURY CYCLING MODEL (MCM, VERSION 1.0)

To further examine the large variety of processes and factors affecting mercury methylation and bioaccumulation in fish, a simulation model incorporating our best understanding of mercury biogeochemistry was developed during the MTL project (Hudson et al., 1994). The Mercury Cycling Model (MCM) represents a theory of mercury cycling in lakes, and one of the most important uses of MCM will be the prediction of methylmercury in fish (recall that fish consumption causes the major exposure of target organisms to methylmercury).

As presently formulated, the MCM is bounded by the atmosphere, the lake margins, and a deep sediment layer (Figure 7). Reactions in the watershed and the atmosphere are not modeled; volumetric inflows and concentrations are measured to provide mercury input at the model boundary. MCM tracks all three major species of mercury [Hg(0), Hg(II), CH_3Hg^+] in three physical compartments: an upper mixed layer (epilimnion), a lower layer (hypolimnion), and the sediments. At one time dimethylmercury was considered as a possibly important chemical species, but so far it has been observed only at extremely low levels in marine environments (Mason and Fitzgerald, 1990). Four biotic compartments are defined as occurring in the two aquatic layers: phytoplankton, zooplankton, a forage fish population, and a piscivorous fish population. Although simulation output can include any variable, the target of interest is mercury concentrations in piscivorous fish. The MCM has a monthly time step and runs on Macintosh™ computers. Additional information about the model can be found in Hudson et al. (1994).

Four scenarios briefly illustrate how the MCM can predict methylmercury concentrations under different conditions (Figure 8). The graph shows how methylmercury in piscivorous fish responds over a 10-year period to changes of initial conditions in the Little Rock Lake treatment basin. The fish modeled in the treatment basin do not represent sport fish communities that include larger and older fish containing higher mercury concentrations (Grieb et al., 1990). The base case for the simulation shows that fish mercury varies seasonally around a mean concentration of 0.18 μg/g; the MCM calculates concentrations from seasonally variable predictions of methylmercury in the piscivorous fish compartment and the predicted biomass, producing each annual cycle. If atmospheric mercury deposition was reduced by 5%, a small change in fish methylmercury would begin to be observed at about the eighth year, reflecting the lag in response in organisms with long life cycles.

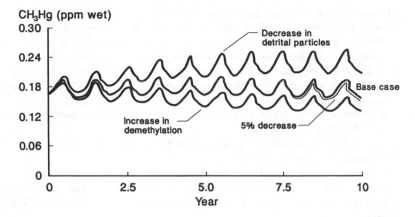

Figure 8   Illustration of Mercury Cycling Model (MCM v. 1.0) simulation of different scenarios compared to the base case (Little Rock Lake Reference Basin): effect of a decrease in detrital particles by a factor of 10; the effect of increased demethylation rate; the effect of a 5% decrease in atmospheric deposition.

The other two scenarios shown in Figure 8 describe conditions that affect the rate of production of methylmercury. Detrital particles include clay and organic materials that compete with methylating processes for Hg(II). If the detrital particles were to decrease by a factor of ten, more Hg(II) would be available for methylation causing an increase in fish bioaccumulation, as shown in the simulation. Analogously, an increase by a factor of two in the rate of demethylation leads to lower bioaccumulation in fish. In both cases, the simulated steady-state annual cycle occurs after year 10 due to the lag in response associated with organisms of a longer life span.

The illustration of model capabilities shows how risk assessment can be performed for different mitigation alternatives. As mercury inputs are changed with different mitigation alternatives and mercury controls, MCM can calculate mercury concentration in fish at a specific site, and thereby simulate exposures for humans and fish-eating wildlife. More generic and broadly applicable assessments can be made with regionally averaged or quartile range characteristics of lakes with regional deposition measurements and simulation results.

V.   CONCLUSIONS

1. Biogeochemical processes link mercury inputs to a lake with accumulation of mercury in fish.
2. Improved sampling and analytical procedures have improved our understanding of mercury biogeochemistry.
3. Globally, mercury is probably not at steady state in the atmosphere.
4. Fluxes to the atmosphere have different levels of uncertainty. Estimates of open ocean inputs to the atmosphere appear most certain, whereas anthropogenic inputs are much less certain. Anthropogenic emissions have not been accurately sampled on a global basis, and there is a need for accurate speciation of the mercury forms in atmospheric emissions. Little reliable data exist to estimate terrestrial mercury evasion into the atmosphere; consequently, this value is determined by difference.
5. The historical pattern of mercury deposition indicates that deposition in the most recent decades is about two to three times preindustrial emissions. However the peak deposition and anthropogenic emissions may have been greater than in the most recent decade. Better understanding of the multiple sources of historical mercury emissions will reduce uncertainty about the role of reemissions, e.g., from the ocean and land surfaces.

6. The mass balance of a single seepage lake in northern Wisconsin shows:
    a. Atmospheric deposition of mercury accounts for all the mercury in fish.
    b. Evasion and sedimentation are major sinks for mercury in lakes, but the transport to these sinks vary depending on lake water quality and mercury loading.
    c. A large fraction of the methylmercury is formed within the lake ecosystem.
    d. Net methylation is the key process affecting mercury in fish, where 95% of the mercury is methylmercury.
7. The amount of methylmercury accumulated by fish appears to be controlled by important factors other than just deposition.
8. In one northern Wisconsin seepage lake, about 5% of the total deposited mercury was accumulated as methylmercury by fish.
9. For seven northern Wisconsin lakes, fish mercury correlated with water mercury concentrations. However, the concept of bioaccumulation factors appears flawed when applied to a broad spectrum of lake types, adding uncertainty to risk assessment performed with such factors.
10. Field results support the contention that sulfate reducing bacteria are involved in methylation.
11. A simulation model, the Mercury Cycling Model (MCM), provides a way to assess hypotheses about mercury behavior. Also, it provides an excellent tool for mercury risk assessment.

ACKNOWLEDGMENTS

This paper borrows liberally from the work performed and reported elsewhere by William Fitzgerald, Carl Watras, Nicolas Bloom, Steven Gherini, Gary Gill, Robert Hudson, James Hurley, David Krabbenhoff, Anders Andren, Douglas Knauer, David Powell, Ronald Rada, James Wiener, and Michael Winfrey. I gratefully acknowledge their help along with the help of John Huckabee, Leonard Levin, and other reviewers. The work was funded by the Electric Power Research Institute and the Wisconsin Department of Natural Resources. Also, the U.S. Fish and Wildlife Service and the U.S. Geological Survey funded part of this research. In addition, cooperative research funded through NAPAP by the USEPA provided an experimental manipulation and important ancillary information on Little Rock Lake.

REFERENCES

Benoit, J. M., W. F. Fitzgerald, and A. W. H. Damman, 1994. Historical atmospheric mercury deposition in the mid-continental United States as recorded in an ombrotrophic peat bog, in: *Mercury Pollution: Integration and Synthesis,* Watras, C. J. and Huckabee, J. W., Eds., Lewis Publishers, Chelsea, MI, Chap. II.2.

Bloom, N., 1989. Determination of picogram levels of methylmercury by aqueous phase ethylation, followed by cryogenic gas chromatography with cold vapour atomic fluorescence detection, *Can. J. Fish. Aquat. Sci.,* 46:1131–1140.

Bloom, N. S., 1992. On the chemical form of mercury in edible fish and marine invertebrate tissue, *Can. J. Fish. Aquat. Sci.,* 49:1010–1017.

Bloom, N. S., 1993. Mercury speciation in fluegases: overcoming the analytical difficulties, in: *Managing Hazardous Air Pollutants: State of the Art,* Chow, W. and Connor, K. K., Eds., CRC Press, Boca Raton, FL, p. 148.

Bloom, N. S. and W. F. Fitzgerald, 1988. Determination of volatile mercury species at the picogram level by low-temperature gas chromatography with cold-vapor atomic fluorescence detection, *Anal. Chim. Acta,* 208:151–161.

Bloom, N. S. and C. J. Watras, 1989. Observations of methylmercury in precipitation, *Sci. Total Environ.,* 87/88:199–207.

Bloom, N. S., C. J. Watras, and J. P. Hurley, 1991. Impact of acidification on the methylmercury cycling of remote seepage lakes, *Water Air Soil Pollut.*, 56:477–491.

Brosset, C., 1987. The behavior of mercury in the physical environment, *Water Air Soil Pollut.*, 34:145–166.

Clarkson, T. W., 1990. Human health risks from methylmercury in fish, *Environ. Toxicol. Chem.*, 9:821–823.

Clarkson, T. W., 1994. The toxicology of mercury and its compounds, in: *Mercury Pollution: Integration and Synthesis*, Watras, C. J. and Huckabee, J. W., Eds., Lewis Publishers, Chelsea, MI, Chap. VIII.1.

Compeau, G. C. and R. Bartha, 1985. Sulfate-reducing bacteria: principal methylators of mercury in anoxic estuarine sediment, *Appl. Environ. Microbiol.*, 50:498–502.

Cooper, J. J., R. O. Thomas, and S. M. Reed, 1985. Total mercury in sediment water and fishes in the Carson River drainage, west-central Nevada, Nevada Division of Environmental Protection, Carson City, NV.

EPRI, 1990. Cool water coal gasification program. Final Report. EPRI GS-6806, Electric Power Research Institute, Palo Alto, CA.

EPRI, 1991. Analysis of alternative SO2 reduction strategies. Final Report. EPRI EN/GS-7132, Electric Power Research Institute, Palo Alto, CA.

Fitzgerald, W. F., 1986. In: The Role of Air-Sea Exchange in Geochemical Cycling. NATO Advanced Science Institutes Series, P. Buat-Menard, Ed., Reidel, Dordrecht, 363–408.

Fitzgerald, W. F., 1989. Atmospheric and oceanic cycling of mercury, In: *Chemical Oceanography*, Academic Press, New York, Chap. 57.

Fitzgerald, W. F. and T. W. Clarkson, 1991. Mercury and monomethylmercury: present and future concerns, *Environ. Health Persp.*, 96:159–166.

Fitzgerald, W. F. and G. A. Gill, 1979. Subnanogram determination of mercury by two-stage gold amalgamation and gas-phase detection applied to atmospheric analysis, *Anal. Chem.*, 51:1714–1720.

Fitzgerald, W. F., R. P. Mason, and G. M. Vandal, 1991. Atmospheric cycling and air-water exchange of mercury over mid-continental lacustrine regions, *Water Air Soil Pollut.*, 56:745–768.

Fitzgerald, W. F., R. P. Mason, G. M. Vandal, and F. Dulac. 1994. Air-water cycling of mercury in lakes, in: *Mercury Pollution: Integration and Synthesis*, Watras, C. J. and Huckabee, J. W., Eds., Lewis Publishers, Chelsea, MI, chap. II.3.

Fitzgerald, W. F., and C. J. Watras, 1989. Mercury in the surficial waters of rural Wisconsin lakes, *Sci. Total Environ.*, 87/88:223–232.

Gill, G. A. and K. W. Bruland, 1990. Mercury speciation in surface freshwater systems in California and other areas, *Environ. Sci. Technol.*, 24:1392–1400.

Gill, G. A. and K. W. Bruland, 1992. Mercury speciation and cycling in a seasonally anoxic freshwater systems: Davis Creek Reservoir. Final Report to EPRI Draft.

Gill, G. A. and W. F. Fitzgerald, 1987. Picomolar mercury measurements in seawater and other materials using stannous chloride reduction and two-stage gold amalgamation with gas phase detection, *Mar. Chem.*, 20:227–243.

Gilmour, C. C. and E. A. Henry, 1991. Mercury methylation in aquatic systems affected by acid deposition, *Environ. Pollut.*, 71:131–169.

Grieb, T. M., C. T. Driscoll, S. P. Gloss, C. L. Schofield, G. L. Bowie, and D. B. Porcella, 1990. Factors affecting mercury accumulation in fish in the Upper Michigan Peninsula, *Environ. Toxicol. Chem.*, 9:919–930.

Haakanson, L., 1980. The quantitative impact of pH, bioproduction and Hg-contamination on the Hg-content of fish (pike), *Environ. Pollut.*, B1:285–304.

Huckabee, J. W., J. W. Elwood, and S. G. Hildebrand, 1979. Accumulation of mercury in freshwater biota, in: J. O. Nriagu, Ed., *The Biogeochemistry of Mercury in the Environment*, Elsevier/North Holland, New York, 277–302.

Hudson, R. J. M., S. A. Gherini, C. J. Watras, and D. B. Porcella, 1994. Modeling the biogeo-chemical cycle of mercury in lakes: the mercury cycling model (MCM) and its application to the MTL study lakes, In: *Mercury Pollution: Integration and Synthesis*, Watras, C. J. and Huckabee, J. W., Eds., Lewis Publishers, Chelsea, MI, chap. V.1.

Iverfeldt, Å., 1990. Structural, Thermodynamic and Kinetic Studies of Mercury Compounds; Applications Within the Environmental Cycle, Ph.D. Thesis, Department of Inorganic Chemistry CTH, S-412 96, Goteborg, Sweden.

Iverfeldt, Å., 1991. Occurrence and turnover of atmospheric mercury over the Nordic countries, *Water Air Soil Pollut.*, 56:251–266.

Lee, Y.-H. and Å. Iverfeldt, 1991. Measurement of methylmercury and mercury in runoff, lake and rain waters, *Water Air Soil Pollut.*, 56:309–321.

Lee, Y. H., H. Hultberg, and I. Andersson, 1985. Catalytic effect of various metal ions on the methylation of mercury in the presence of humic substances, *Water Air Soil Pollut.*, 25:391–400.

Lindqvist, O., K. Johansson, M. Åstrup, A. Andersson, L. Bringmark, G. Hovsenius, Å. Iverfeldt, M. Mieli, and B. Timm, 1991. Mercury in the Swedish environment. Recent research on causes, consequences and corrective methods, *Water Air Soil Pollut.*, 55:i–261.

Mason, R. P. and W. F. Fitzgerald, 1990. Alkylmercury species in the equatorial Pacific, *Nature*, 347:457–459.

Munthe, J., 1991. The Redox Cycling of Mercury in the Atmosphere, Ph.D. Dissertation. Department of Inorganic Chemistry, University of Goteborg, Sweden.

Neme, C., 1991. Electric utilities and long-range transport of mercury and other toxic air pollutants, Center for Clean Air Policy, Washington, D.C., 126p.

Nriagu, J. O. and J. M. Pacyna, 1988. Quantitative assessment of worldwide contamination of air, water and soils by trace metals, *Nature*, 333:134–139.

Nriagu, J. O., W. C. Pfeiffer, O. Malm, C. M. M. de Souza, and G. Mierle, 1992. Mercury pollution in Brazil, *Nature*, 356:389.

Porcella, D. B., C. J. Watras, and N. S. Bloom, 1992. Mercury species in lake water, in: The Deposition and Fate of Trace Metals in our Environment, Verry, S. and S. J. Vermette, Eds., Gen. Tech. Rep. NC-150, St. Paul, MN, U.S. Dept. Agric., Forest Service, North Central Forest Exp. Station, pp. 127–138.

Rada, R. G., J. G. Wiener, M. R. Winfrey, and D. E. Powell, 1989. Recent increases in atmospheric deposition of mercury to north-central Wisconsin lakes inferred from sediment analyses, *Arch. Environ. Contam. Toxicol.*, 18:175–181.

Rada, R. G. and D. E. Powell, 1993. Whole-lake burdens and spatial distribution of mercury in surficial sediments in Wisconsin seepage lakes, *Can. J. Fish. Aquat. Sci.*, 50: in press.

Roelke, M. E., D. P. Schultz, C. F. Facemire, S. F. Sundlof, and H. E. Royals, 1991. Mercury contamination in Florida panthers. Report to the Florida Panther Interagency Committee, Florida Game and Fresh Water Fish Commission, Gainesville, FL 32601.

Rudd, J. W. M., M. A. Turner, A. Furutani, A. Swick, and B. E. Townsend, 1983. I. A synthesis of recent research with a view towards mercury amelioration, *Can. J. Fish. Aquat. Sci.*, 40:2206–2217.

Slemr, F. and E. Langer, 1992. Increase in global atmospheric concentrations of mercury inferred from measurements over the Atlantic Ocean, *Nature*, 355:434–437.

Slemr, F., W. Seiler, and G. Schuster, 1985. Distribution, speciation, and budget of atmospheric mercury, *J. Atmos. Chem.*, 3:407–434.

Spry, D. J. and J. G. Wiener, 1991. Metal bioavailability and toxicity to fish in low-alkalinity lakes: a critical review, *Environ. Pollut.*, 71:243–304.

Swain, E. B., D. R. Engstrom, M. E. Brigham, T. A. Henning, and P. L. Brezonik, 1992. Increasing rates of atmospheric mercury deposition in midcontinental North America, *Science*, 257:784–787.

Swain, E. B. and D. D. Helwig, 1989. Mercury in fish from northeastern Minnesota lakes. Historical trends, environmental correlates, and potential sources, *J. Minn. Acad. Sci.,* 55:103–110.

Watras, C. J. and N. S. Bloom, 1992. Mercury and methylmercury in individual zooplankton. Implications for bioaccumulation, *Limnol. Oceanogr.,* 37:1313–1318.

Watras, C. J. and T. M. Frost, 1989. Little Rock Lake (Wisconsin). Perspectives on an experimental ecosystem approach to seepage lake acidification, *Arch. Environ. Contam. Toxicol.,* 18:157–165.

Watras, C. J., N. S. Bloom, W. F. Fitzgerald, J. G. Wiener, R. Rada, R. J. M. Hudson, S. A. Gherini, and D. B. Porcella, 1994. Sources and fates of mercury and methylmercury in remote temperate lakes, in *Mercury Pollution: Integration and Synthesis,* Watras, C. J. and Huckabee, J. W., Eds., Lewis Publishers, Chelsea, MI, chap. I.12.

Wiener, J. G., W. F. Fitzgerald, C. J. Watras, and R. G. Rada, 1990. Partitioning and bioavailability of mercury in an experimentally acidified Wisconsin lake, *Environ. Toxicol. Chem.,* 9:909–918.

Wiener, J. G., D. E. Powell, and R. G. Rada, Mercury accumulation by fish in Wisconsin seepage lakes: relation to lake chemistry and acidification, in preparation.

Levels of Mercury in the Tucuruí Reservoir and its Surrounding Area in Pará, Brazil

*Ilkka Aula, Hannu Braunschweiler, Tuija Leino,
Ismo Malin, Petri Porvari, Tacachi Hatanaka,
Martin Lodenius, and Anastácio Juras*

CONTENTS

I. INTRODUCTION

The problems caused by the increasing loads of mercury from gold mining in the Amazon area are alarming.[1-6] The first large-scale dams with hydropower plants built in a rain forest area started to operate some years ago in Brazil. The state of their environment is now being surveyed and many of their impacts are yet to be studied.[7-8] The transportation and accumulation of environmental pollutants to the reservoirs have not been fully studied, especially in tropical areas.[9] Nevertheless, in water courses in which great dams have been built, problems with environmental pollutants like mercury can be worrisome.[10-12]

The aim of this study was to determine the levels of total mercury (abbreviated from here on as Hg) in a tropical reservoir and its surrounding area and to investigate how the

1–56670–066–3/94/$0.00+$.50
© 1994 Lewis Publishers

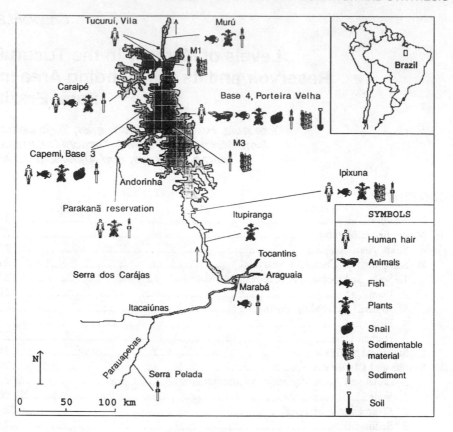

Figure 1   The location of the Tucuruí reservoir and the sampling areas in it and in its surroundings. The location of the study region in Brazil is indicated by a box in the small caption.

construction of the reservoir and how the most important source of Hg in Amazon, gold mining, affect these levels. The purpose was also to discover indicator species for monitoring the level of Hg. We studied the bioconcentration of Hg in the reservoir and its environments; whether the accumulation of Hg in fish will restrict the use of fish as food; and whether the levels of Hg will cause health problems for people of the region that consume the fish.

Another objective of this study was to estimate how much Hg is stored in different compartments of the reservoir and how this affects, or has affected, the present state of Hg pollution. At the same time, we gathered evidence of seasonal variation in Hg levels in sediments, sedimentable material, and plants and of differences in Hg levels in sediments, plants, fishes, and human hair in different areas of the Tucuruí reservoir and its surroundings.

II.   MATERIALS AND METHODS

A.   STUDY AREA

The following description is based on information supplied by Eletronorte, e.g., the summary report about the environmental conditions around Tucuruí.[13] The reservoir of Tucuruí is situated from 49° 20′ to 50° W and 3° 45′ to 5° S in the state of Pará, Brazil, (Figure 1). Its area at maximum normal water level is 2,430 km², of which about 25% (600 km²) was the former channel of the river Tocantins or its affluents, 1,180 km² covered by forest was inundated, and some 250 km^2 formed 600 islands. The area of Lake Caraipé is 270 km^2. It is the side basin formed to the west of the dam from waters of the River Caraipé and is almost totally separated from the main reservoir. The maximum change in the water level is 14 m. The whole reservoir has a mean volume of 45.8 × 10^9 m^3, a mean depth of 18.0 m, and a maximum depth of 105 m. The mean residence time of water in the reservoir is 46 d. The closing of the dam was finished in September 1984. The water had risen to the level of overflow by November 1984.

Geologically, the area of the reservoir belongs to a transition zone from crystalline to metamorphic rocks. The soils of the surrounding areas of the reservoir are mostly Podzoil Red and Yellow with minor areas of Latosol Red and Yellow. Podzoil Red and Yellow consist mostly of clay and hence are not very permeable and are of high consistency. Most of the land on the western side of the reservoir is deforested for cattle ranching. There has been some low production gold mining in the area between Marabá and Itupiranga. The next catchment area to the east of the reservoir belongs to the river Mojú.

Serra Pelada, one of the most famous gold mining areas in Brazil, is situated about 110 km to the southwest from the Tucuruí reservoir. Gold mining started there in 1979. At its peak in 1983, Serra Pelada produced more than a metric ton of gold a month and had a population of between 80,000 and 100,000.[14] Now there are only some 20,000 people in the area. Its production of gold between 1980 to 1986 was 44.5 t according to official statistics.[14] The amount of gold not registered is estimated to be about 90% and the amount of Hg released per kilogram of produced gold to be 0.8 kg.[3,15] Thus, the emissions of Hg from Serra Pelada have been about 360 t between 1980 and 1986.

Some 3,350 families totalling over 17,300 people lived at the area of the reservoir at the time before its filling. About two thirds of these lived in rural areas. The two larger cities in the area are Marabá (1985 population of 133,600) and Tucuruí (84,000).

The Tocantins-Araguaia basin, with its large catchment area extending from 46° to 55° W and 2° to 18° S, 758,000 km^2, drains into the Tucuruí reservoir. Of this 334,000 km^2 corresponds to the Tocantins, 382,000 km^2 to the river Araguaia (its main tributary), and 42,000 km^2 to the river Itacaiúnas (the principal affluent in the lower course). This southeasternmost Amazonian tributary integrates the landscape of the Central Plateau of Brazil (composed of seasonally dry shrub lands with 1,000 to 1,600 mm of rain per year and which covers 76% of the basin) to the hot, humid Amazonian rain forest (2,000 mm/year) in the lower course and the Itacaiúnas. Between them, the river crosses a transitional zone (1,600 to 1,800 mm/year) with pre-Amazonian environments.[16]

The Tocantins and the Araguaia are remarkably different. The 2,500-km long Tocantins is a channelized river with narrow riparian zones. Unlike the main course, the Araguaia is a floodplain river 2,115-km long. The Itacaiúnas is the largest tributary draining the typically Amazonian environment. It rises in Serra dos Carajás and flows over a steep slope to its mouth near Marabá. The affluents of the Itacaiúnas flow through the gold mining area of Serra Pelada.

The drainage basin has an average discharge of 11,000 m^3 s^{-1} almost equally shared by the Tocantins (40%) and the Araguaia (45%) with a small contribution of the Itacaiúnas (5%). The hydrologic regime is remarkably defined. The rising water period extends from October to April, with highwater peaks in February (the upper Tocantins) and March (middle and lower courses). On the Araguaia, floods are higher and one month delayed (March–April) due to overflowing at the floodplains. The maximum discharge to date has been 68,400 m^3 s^{-1}. Both rivers dry up from May to October, with a low water peak in September throughout the basin. The minimum discharge has been 1,511 m^3 s^{-1}. The Tocantins and its tributaries originally could be classified as clear, nutrient-poor, low ion and sediment load rivers, but recent heavy anthropogenic contributions are changing this pattern rather quickly.[16] The Tocantins carries an average of 200,000 t d^{-1} of solids. The water temperature in the reservoir is generally between 27 to 32° C. The prevailing winds in the area around Tucuruí come from between north and east.[17] The ichthyofauna of the Tocantins-Araguaia basin is poor compared with the total number of fish species in Amazonia: about 300 fish species have been found in the Tocantins.

B. SAMPLING AND CHEMICAL ANALYSIS

The organisms studied were aquatic macrophytes, fishes (main dietary fish species), predators at different levels of the trophic chains (mammals, reptiles), detrivores (snails), and humans. The Hg burden in the local population was estimated from hair samples. Non-biological samples were collected from soils, sedimentable material, and sediments. All samples were collected from between September 1990 and March 1991. Water samples were taken from the river and reservoir for basic water quality analysis as a part of the routine monitoring by the staff of Eletronorte.

The samples for Hg were analyzed mostly in the laboratory of the Environmental Research Center of Eletronorte at Tucuruí. The analytical method for total Hg is modified from that described by Nuorteva et al.[18] Hg concentrations of soil, sediment, plant, and hair samples were corrected for humidity by using the dry weight at 105° C. The accuracy of the analytical method was tested using parallel determinations and standard samples of dried fish muscle of two Hg levels, prepared by the National Board of Waters and the Environment, Finland.[19] Further parallel determinations were always made if the standard deviation was more than 10% of the average concentration of a single sample. Representative examples of all the various kinds of samples were taken to the Institute of Biophysics, Federal University of Rio de Janeiro, and to the University of Helsinki for calibration of the analysis.

According to the intercalibration and the standard addition method, the results of the Hg analyses were judged reliable. The recovery varied from 84.2 to 114%. The results of the same hair samples were generally 10% lower in the laboratory of Helsinki than the results from Tucuruí. Also the results of the analysis of the standard fish sample[19] of 6.9 ± 1.04 mg kg^{-1} of Hg were 10% lower in the laboratories in Rio de Janeiro and Helsinki than in Tucuruí.

1. Soils, Sediments, and Sedimentable Material

Samples of soils and some of the samples of sediments and sedimentable material were collected from Porteira Velha, a small, sheltered bay and its catchment area on the outskirts of the reservoir (Figure 1). The soils of the area are classified as Podzoil Red and Yellow. The properties of the soils of the catchment area surrounding the bay were studied by systematic sampling. Samples of sediments and sedimentable material were collected about every three weeks: sediments from two sites and sedimentable material from one site.

Sediment profiles were taken from other areas if the bottom was soft enough. Otherwise, as in the littoral zone and upstream, only surface sediment was taken (Figure 1). Nine cores and 21 surface sediment samples were taken. The samples were packed in plastic boxes and stored deep-frozen.

In addition to the trap in Porteira Velha there were three other traps for the collection of sedimentable material in the reservoir which were all in the old river channel. Each trap had four cylinders, 4.5 cm in diameter and 50 cm long, from which samples were analyzed separately. The traps were placed 1 m above the bottom. Sedimentable material was taken monthly from the traps.

The soil samples were stored air dried, the other samples were deep-frozen. The extraction of readily soluble cations by acidic 1 M ammonium acetate-EDTA solution (pH 4.65) and the analysis of soluble Al, Ca, Mg, and K was modified from that of Lakanen and Erviö[20] and Halonen et al.[21] The analysis of soluble Hg from the extractant was done according to the draft standard method.[22] Soil pH was measured with 0.01 M $CaCl_2$ after standing overnight and shaking, and the soil-solution volume ratio was 1:2.5.[23] The contents of organic matter in the soil, sedimentable matter, and sediment samples were measured as loss on ignition by ashing at 550° C for 2 h.[24]

2. Plants

Samples of floating plants—*Eichhornia crassipes, Salvinia auriculata,* and *Scirpus cubensis*—were collected from different areas of the reservoir and from up- and downstream of the reservoir during the dry and rainy seasons (Figure 1). In each area, samples of *Salvinia auriculata* and *Scirpus cubensis* were taken from ten plants and *Eichhornia crassipes* from one plant. The concentrations of Hg in *Eichhornia* and *Scirpus* were analyzed from roots and shoots and of *Salvinia* from submerged and floating leaves. The submerged and floating leaves of *Salvinia* have the same physiological role as the roots and shoots. The samples were washed with river water and stored in paper bags. In the laboratory the samples were dried at room temperature and then homogenized.

3. Animals (Excluding Fish)

Snails and their eggs were collected from Capemi and two places near Base 4 in January and February (Figure 1). After measuring the length of the shell, the samples were deep-frozen in plastic bags. The concentration of Hg was analyzed from the whole soft tissue of the homogenized snails. Caimans (*Paleosuchus* sp., an alligator-like reptile) and capybaras (*Hydrochoerus hydrochaeris,* a rodent) were shot during December and the eggs of the turtle (*Podocnemis unifilis*) were collected during September 1990 (Figure 1). Body weights, lengths, and sex of the caimans and capybaras were determined. The eggs of the turtle were in the early stage of development. Liver, muscle, and hair samples of capybaras, the liver and muscle samples of caimans, and combined white-yolk samples of eggs were prepared for Hg analysis. All the samples were stored at $-18°$ C in glass containers until analysis.

4. Fish

The fish were obtained either with a line and hook or with a net or bought from fishermen. The seven fish species studied were tucunaré (*Cichla temensis*), pescada (*Plagioscion squamosissimus*), mapará (*Hypophthalmus marginatus*), curimatã (*Prochilodus nigricans*), piau (*Anostomidae* spp.), piranha (*Serrasalmus* sp.), and piaba (*Brycon* sp.). The muscle Hg concentration was analyzed from a total of 230 fishes from the reservoir and the Tocantins upstream and downstream of the reservoir (Figure 1). Two pieces of white muscle were removed from below the dorsal fin for the determination of Hg. The whole muscle was removed from small fish. The samples were stored at $-18°$ C until analysis.

5. Human Hair

All together, 147 hair samples were collected from people living around the Tucuruí reservoir (Figure 1). Fifteen of these originated from the Parakanã Indian reservation, which was selected to be a control area as it is geographically closely situated to the reservoir, actually upstream of it on the river Andorinha, and there is no gold mining in this region. Information was obtained about the fish consumption habits of the individuals, i.e., how many times a week fish was eaten and which species of fish were consumed most. Also, the age, sex, and time lived in the area by the person was obtained and recorded. Some people from whom hair samples had been collected had to be left out from the study material due to their short stay in the area or unreliable fish consumption information. The remaining hair samples were from 125 adults who had lived an average of 8.9 years in the vicinity of the reservoir. Each hair sample weighed around 200 to 400 mg. Samples were stored in dry envelopes. The samples were washed as described by Kosta et al.[25] but otherwise they were analyzed in the same manner as the animal samples.

C. STATISTICAL METHODS

Statistical processing of the collected data was performed at the University of Helsinki. The statistical significance of differences in the concentrations of Hg in the different areas of the Tucuruí reservoir and its surroundings were tested using analysis of variance or covariance or the Kruskal-Wallis test and the test of Honestly Significant Difference (HSD). The seasonal variation of Hg levels in *Salvinia auriculata* was tested using the Student's *t*-test. Pearson's and Spearman's coefficients were used to calculate the correlations.

III. RESULTS AND DISCUSSION

A. SEASONAL AND AREAL VARIATION IN CONCENTRATIONS

1. Soils

The composition of the soils studied (Table 1) is typical for central Amazon forests.[26] The absence of anthropogenic Hg deposition can be seen from the increasing trend of Hg concentrations with the depth of the soil. The concentration in the topsoil is much lower than that found near Hg emission sources, and it is similar to the calculated background concentration.[25] Acid-extractable Hg concentrations of soils had the strongest correlation with base saturation or soluble Mg^{2+}, Ca^{2+}, and Al^{3+} (r = 0.66 . . . 0.51***, N = 42) (Table 1). Readily soluble Hg content in soils had the highest correlation with the cation exchange capacity (r = 0.51***, N = 42). On the other hand, the correlation of Hg concentrations (acid-extractable or soluble) with the content of organic matter was insignificant ($|r| < 0.23$, N = 42). This is opposite to observations from Nordic acid forest soils, but it may be due to dissolved organic components which can also make Hg mobile. Some of this mobile Hg leaches to water courses, especially during heavy rains. This leaching is enhanced by activities such as clear cutting and ditching.[27]

2. Sedimentable Material

In the sedimentable matter, the organic content and Hg concentration decreased from the dry to rainy season at the mouth of the reservoir at Ipixuna. Ipixuna is an erosional sediment area, so it was not reasonable to calculate the flux of Hg. In the middle of the reservoir, in M3, the amount of deposition of sedimentable material increased to over tenfold and the Hg concentration was almost halved from the dry to rainy season, but the flux of Hg increased to over fivefold during this period (Table 2). On the other hand, in Porteira Velha on the outskirts, the Hg flux pattern was quite different, being at all times lower and steadier than in the former river channel (Table 2).

Particle sizes in M1 were smaller than in M3 due to the longer distance from the mouth of the reservoir. In M1, Hg concentration was higher than in M3 at the same time, because finer particles can absorb more Hg due to their larger specific surface area. The correlation between the content of Hg and that of organic matter in sedimentable material was 0.78*** (N = 21). Although Hg concentrations in the sediments are low, the Hg flux in the former river bed is high compared to other studies. Rekolainen et al.[28] found a deposition of 25 to 50 $\mu g \ m^{-2} \ a^{-1}$ in Finnish natural forest lakes and 370 $\mu g \ m^{-2} \ a^{-1}$ in lakes situated downstream from large ditching areas. In the Pocone gold mining region, central Brazil, Lacerda et al.[29] estimated that there was a deposition rate for Hg of 90 to 120 $\mu g \ m^{-2} \ a^{-1}$ in two lakes.

3. Sediments

Most lakes have the highest levels of Hg in the surface layer of sediments due to recent anthropogenic contamination.[30-34] There were no similar patterns with depth in the sediments of the Tucuruí reservoir because it was possible to get only recent sediments as the reservoir is so young. Hg content in lake sediments is dependent mainly on the content of organic matter in the sediment.[32] This was also noticed in the Tucuruí reservoir (Table 3). The correlation between the content of Hg and that of organic matter in the surface sediment was 0.89*** (N = 24). The only place where there were differences was Serra Pelada, which was a consequence of gold mining.

Table 1 Concentrations of Hg (total or acid extractable, and readily soluble per dry weight basis) and other properties of the soils studied.

Soil layer	Total Hg mg kg⁻¹	Soluble Hg µg kg⁻¹	Soluble Hg (%)	Org. matter (%)	Density kg dm⁻³	pH 0.01 M CaCl₂	CEC meq/100 g	Base saturation %	Soluble Al³⁺ mg kg⁻¹	Soluble Ca²⁺ mg kg⁻¹	Soluble Mg²⁺ mg kg⁻¹	Soluble K⁺ mg kg⁻¹
Litter	0.084 (0.020) N=20 A	—	—	85 (5.5) N=18 A	—	—	—	—	—	—	—	—
Humus 0–2 cm	0.071 (0.014) N=22 B	0.5 (0.3) N=21 A	0.7 (0.45) N=21 A	7.0 (1.8) N=22 B	1.2 (0.10) N=21 A	3.38 (0.098) N=20 A	3.5 (0.81) N=21 A	21 (6.8) N=21 A	250 (69) N=21 A	48 (23) N=21 A	37 (14) N=21 A	65 (26) N=21 A
Yellow 2–15/50 cm	0.094 (0.018) N=22 A	0.4 (0.2) N=22 A	0.5 (0.25) N=22 A	4.8 (0.68) N=22 B	1.3 (0.092) N=22 B	3.53 (0.061) N=21 B	4.0 (1.2) N=21 A	6.5 (2.1) N=21 B	330 (105) N=21 B	8.1 (4.4) N=21 B	14 (8.1) N=21 B	35 (10) N=21 B
Red 50–110 cm	0.13 (0.007) N=4 C	0.3 (0.07) N=4 A	0.2 (0.061) N=4 B	5.4 (3.3) N=4 B	1.2 (0.043) N=4 AB	3.59 (0.072) N=4 B	4.7 (0.24) N=4 A	4.2 (1.0) N=4 B	400 (24) N=4 AB	6.9 (3.4) N=4 B	9.1 (3.0) N=4 C	33 (1.5) N=4 B

Note: Soluble Hg (%) is the percentage of soluble Hg from the acid extractable Hg. Cation exchange capacity (CEC) is expressed as the sum of exchangeable Al³⁺, H⁺, Ca²⁺, Mg²⁺, and K⁺. Base saturation is the proportion of soluble Ca²⁺, Mg²⁺, and K⁺ of CEC. The results are the means of the soil samples with standard deviations in parenthesis, N = the number of samples analyzed, and in each column, soil layers that differ from each other at the significance level of 0.05 are presented with different letters. A dash indicates "not analyzed". The yellow soil layer was mostly sampled down to 15 cm though it continued down to 40 to 60 cm.

Table 2   **Mean concentrations of Hg (mg kg⁻¹ dry weight) in sedimentable material, the monthly flux of Hg, the monthly sedimentation, content of organic matter, and water depth and pH at the sampling sites.**

Place/month	Hg (mg kg^{-1})	Flux of Hg (μg m^{-2} month^{-1})	Sedimentation (g m^{-2} month^{-1})	Org. matter (%)	Depth (m)	pH
M1, February	0.12	160	1,359	13	60	6.8
M3, December	0.14	38	276	—	24	7.7
M3, January	0.066	109	1,660	12	24	7.2
M3, February	0.078	238	3,049	12	24	6.9
Porteira Velha, December	0.17	38	225	17	10	6.5
Porteira Velha, January	0.24	36	148	—	11	6.7
Porteira Velha, February	0.30	43	145	—	12	7.1
Ipixuna, January	0.059	—	—	8.1	6	7.2

Note: Dashes indicate ''not analyzed''.

Table 3   **Mean concentrations of Hg (mg kg⁻¹ dry weight), their range and the content of organic matter (%) in the surface layer (0 to 5 cm) of sediments in different areas of the reservoir and its surroundings.**

Area	Hg mg kg^{-1}	Min-Max	Org. matter (%)	N	Depth (m)
Murú	0.037	0.022–0.047	6.0	3	3
M1	0.062	—	10.3	1	60
Caraipé	0.13	0.12–0.14	24.6	4	2–22
Capemi	0.10	—	46.5	1	12
Parakanã reservation[a]	nd	—	2.1	1	0.7
Parakanã reservation[b]	0.088	—	11.8	1	0.7
Base 4	0.093	0.089–0.096	24.8	2	13
Porteira Velha	0.13	0.047–0.21	14.9	9	2–10
M3	0.056	—	—	1	97
M3	0.087	0.086–0.088	11.9	2	40
Base 3	0.096	0.088–0.10	12.6	2	18
Ipixuna	0.054	nd–0.10	8.2	6	6–12
Marabá Tocantins	nd	—	0	1	3
Marabá Itacaiúnas	0.018	—	0.4	1	3
Serra Pelada[c]	0.18	0.15–0.22	8.0	2	0.25
Serra Pelada[d]	0.080	0.067–0.093	6.6	2	0.2

[a] River sediment.
[b] Floodplain.
[c] Downstream of the gold mining area.
[d] Upstream of the gold mining area, for other sites see Figure 1.
Note: Water depth at the sampling sites is indicated. N = the number of Hg and organic matter samples analyzed; nd = not determined; a dash indicates ''not analyzed''.

4. Plants

The concentrations of Hg in submerged and floating leaves of *Salvinia auriculata* were significantly higher than the corresponding concentrations in roots and shoots of *Scirpus cubensis* and *Eichhornia crassipes* ($p < 0.001$ for both parts; Table 4). In all the plant species studied the concentrations of Hg were higher in the roots and submerged leaves than in the shoots and floating leaves. The ratios between the concentrations of Hg in roots and shoots and in submerged and floating leaves were as follows: *Scirpus* 2.5:1, *Eichhornia* 2.1:1, *Salvinia* 1.6:1.

The accumulation of Hg occurred mainly in the roots and submerged leaves. There is a possibility that Hg forms a protein complex with a sulfhydryl group and is strongly bonded in roots and submerged leaves, preventing it from passing into the shoots and floating leaves.[35] Higher concentrations of Hg in roots and submerged leaves may be the result of surface adsorption.[36] Siegel et al.[37] have also found that aerial parts are capable of releasing Hg as a vapor.

The concentrations between the roots and shoots of *Scirpus cubensis* and the submerged and floating leaves of *Salvinia auriculata* were significantly correlated (*Scirpus* r = 0.56**, N = 15; *Salvinia* r = 0.72***, N = 36; *Eichhornia* r = 0.19, N = 19). Significant positive correlation between Hg concentrations in plant leaves and roots is a well-known phenomenon.[38–40] This might be attributed to the transport of Hg from roots or submerged leaves to shoots or floating leaves. These results indicate that all three species, especially *Salvinia auriculata,* are suitable for monitoring the levels of Hg in a tropical water ecosystem.

Seasonal variation of Hg levels was tested on *Salvinia auriculata.* The concentrations of Hg in submerged and floating leaves were higher in the dry season than in the rainy season (submerged leaves: $p = 0.0005$, floating leaves: $p = 0.006$; Table 4). This could be explained by competition with nutrients[41] or a growth dilution effect.[42] During the rainy season the levels of nutrients (nitrogen and phosphorus) in the surface water increase in all parts of the reservoir and its drainage area.[43] Increased levels of solid material occur only in the former river channel. Other limnological parameters, for example the levels of Cl, Fe, Ca, Mg, Na, K, pH, O_2, and temperature remain unchanged. These above-mentioned factors, especially the macronutrients, might explain the diminished levels of Hg on the outskirts of the reservoir during the rainy season. Potentially, the upstream Hg which is mainly from the gold mining area, binds to particulate matter[44] and therefore becomes unavailable to floating plants.

Variations in the concentration of Hg between different areas of the reservoir and its surrounding was observed both in the shoots and roots of *Eichhornia crassipes* and *Scirpus cubensis* and in submerged and floating leaves of *Salvinia auriculata* (Table 4). In the rainy season, differences between the areas, studied only with *Salvinia auriculata,* disappeared (Table 4).

The lowest concentrations were observed in the area of Caraipé; the highest in Base 4 and in the river Andorinha. These results, as well as results from other studies,[45,46] indicate that the type of vegetation in the drainage area and the natural fluctuation of water might affect the concentrations of Hg in water and therefore the concentrations in the floating plants studied. Also, the inundated vegetation cover of Lake Caraipé is different from other studied areas because only Caraipé and the area in front of the dam were deforested before the power plant started to operate.

5. Animals (Excluding Fish)

The Hg concentration in snails varied a lot within each area and there were no significant correlations between the mass or length of the snails and Hg concentration. Hg concentration was under the detectable limit for both older (white) and younger (red) eggs (Table 5).

Table 4 Mean concentrations of mercury (mg kg^{-1} dry weight) in floating plants in different areas of the reservoir and its surroundings during the dry and rainy season.

	Murú (downstream)	Caraípé	Capemi	Andorinha	Base 4	Base 3	Ipixuna	Itupiranga (upstream)	Mean of all samples
Salvinia									
Submerged (dry)	—	0.045 (0.006) A, N=3	0.16 (0.015) B C, N=3	0.20 (0.036) B D, N=3	0.27 (0.05) D, N=3	0.083 (0.013) A C, N=3	0.084 (0.014) A C, N=3	—	0.12 (0.038) N=36
Floating (dry)	—	0.016 (0.003) A, N=3	0.079 (0.005) A B, N=3	0.10 (0.047) A B, N=3	0.16 (0.067) B, N=3	0.074 (0.021) A B, N=3	0.063 (0.021) A, N=3	—	0.075 (0.043) N=36
Submerged (rainy)	—	—	0.080 (0.031) A, N=3	—	0.11 (0.010) A, N=3	0.13 (0.021) A, N=3	0.087 (0.010) A, N=3	—	—
Floating (rainy)	—	—	0.057 (0.021) A, N=3	—	0.066 (0.007) A, N=3	0.088 (0.006) A, N=3	0.092 (0.025) A, N=3	—	—
Scirpus									
Root (dry)	—	0.024 (0.001) A, N=3	0.072 (0.025) B, N=3	—	0.098 (0.02) B, N=3	0.095 (0.051) B, N=3	—	—	0.075 (0.038) N=15
Shoot (dry)	—	0.013 (0.003) A, N=3	0.034 (0.007) B, N=3	—	0.032 (0.009) B, N=3	0.033 (0.004) B, N=3	—	—	0.030 (0.011) N=15
Eichhornia									
Root (dry)	0.09 (0.008) A, N=3	—	0.067 (0.018) A B, N=3	—	0.045 (0.013) B, N=3	0.058 (0.004) A B, N=3	0.072 (0.018) A B, N=4	0.051 (0.017) A B, N=3	0.066 (0.020) N=19
Shoot (dry)	0.046 (0.009) A, N=3	—	0.041 (0.016) A, N=3	—	0.020 (0.002) B, N=3	0.041 (0.004) A, N=3	0.020 (0.004) B, N=3	0.020 (0.004) B, N=3	0.032 (0.015) N=19

Note: "Submerged" is submerged leaves, "floating" is floating leaves, "rainy" is rainy season, "dry" is dry season. Standard deviations are presented in parenthesis, N = number of samples. In each row, areas that differ from each other at the significance level of 0.05 are presented with different letters. A dash indicates "not sampled". "Means of samples" also contain data from other areas not indicated in the table.

Table 5   **Mean concentrations of mercury and their range in the animal samples.**

Sample	Hg (mg kg^{-1})	Min-Max
Snail (tissue)	0.057 (0.046) N=16	0.011–0.17
Snail (egg)	nd N=11	—
Capybara (meat)	0.015 (0.006) N=5	0.0066–0.026
Capybara (liver)	0.010 (0.002) N=5	0.0055–0.012
Capybara (hair)	0.16 (0.03) N=5	0.120–0.19
Turtle (egg)	0.012 (0.005) N=10	0.0072–0.023
Caiman (meat)	1.9 (0.9) N=5	1.2–3.6
Caiman (liver)	19 (6.1) N=5	11–30

Note: Concentrations are presented on a wet weight basis, except for hair of capybara on a dry weight basis. Standard deviations are presented in parenthesis, N = the number of samples analyzed, and "tissue" means the soft tissues of snail samples.

Table 6   **Mean concentration of Hg (mg kg^{-1} wet weight) in fish in different areas of the reservoir and in the up- and downstreams of the reservoir.**

	Murú	Caraipé	Capemi	Base 4	Base 3	Ipixuna	Marabá	All
Piranha	2.4	2.2	—	2.9	—	—	—	2.6
	(0.67)	(0.96)		(0.95)				(0.91)
	N=3	N=4		N=8				N=15
Tucunaré	1.1	1.2	0.99	1.2	1.3	1.0	—	1.1
	(0.62)	(1.39)	(0.35)	(0.81)	(0.53)	(0.51)		(0.81)
	N=7	N=15	N=9	N=12	N=5	N=5		N=53
Pescada	—	0.99	0.95	1.4	—	1.2	1.2	1.2
		(0.30)	(0.35)	(0.74)		(0.57)	(0.32)	(0.58)
		N=2	N=6	N=9		N=12	N=4	N=33
Mapará	0.43	0.42	—	—	—	0.25	0.48	0.41
	(0.22)	(0.15)				(0.06)	(0.16)	(0.17)
	N=7	N=11				N=4	N=5	N=27
Piau	0.16	—	—	0.30	—	0.25	0.20	0.22
	(0.08)			(0.26)		(0.13)	(0.09)	(0.15)
	N=15			N=7		N=9	N=5	N=36
Piaba	—	0.27	—	0.60	—	—	—	0.39
		(0.08)		(0.28)				(0.24)
		N=15		N=9				N=24
Curimatã	—	—	—	—	—	0.07	0.05	0.06
						(0.04)	(0.07)	(0.05)
						N=5	N=2	N=7

Note: Standard deviations are presented in parentheses and N = the number of samples. A dash indicates "not sampled".

The concentrations of Hg in the soft tissues of capybaras and in the eggs of turtles were low (Table 5) and can be considered nontoxic because the background Hg levels in herbivorous and reptiles rarely exceed 0.2 mg kg^{-1} in soft tissues or eggs.[47–49] Capybara is capable of excreting mercury to its hair, as observed by Wren[49] for other herbivorous species. The caimans, being a carnivorous species, have the highest Hg concentrations among the animals studied (Tables 5 and 6). Mercury is biomagnified effectively to their livers.

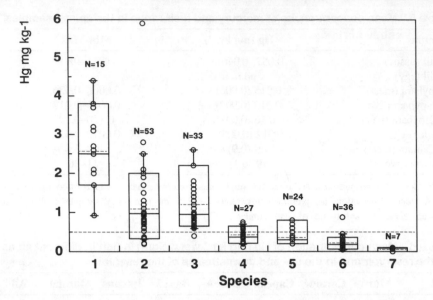

Figure 2 The Hg concentrations (mg kg^{-1} wet weight) in predatory fish: piranha (1), tucunaré (2), pescada (3); and in nonpredatory fish: mapará (4), piaba (5), piau (6), and curimatã (7). The samples are shown as circles, whiskers are drawn from each box end to the 5th and 95th percentiles, each box has a rectangle beginning at the 10th and ending at the 90th percentile, the solid line within the box represents the median and the dashed line within the box is the mean. The dashed line at 0.5 mg Hg kg^{-1} indicates the safety limit for fish consumption used in many countries.

6. Fish

The highest concentrations of Hg were measured in predatory fish. The herbivorous fish, curimatã, had the lowest concentration while planktivorous (mapará) and omnivorous (piaba and piau) fishes had intermediate concentrations (Table 6, Figure 2).

The lengths of the two predatory fishes, tucunaré and pescada, correlated significantly with Hg concentrations (r = 0.76***, N = 65; r = 0.67***, N = 33; respectively). Among tucunarés, the relation between weight and Hg concentration was linear in fishes with weights less than 2 kg (r = 0.75***, N = 65) and the same phenomenon was observed in pescadas (r = 0.66***, N = 33). The relation between length and Hg concentration was nonlinear in piranha, planktivorous, and omnivorous fish. They did not have any significant correlation between fish size and Hg concentration.

The high concentrations in predatory fish indicate severe Hg contamination in the reservoir. The Hg levels of predatory fish in the reservoir of Tucuruí are of the same level as measured in Hg-contaminated areas in Amazonia, but lower than in water courses contaminated by gold mining in other parts of the world.[50-52] The Hg concentrations are of the same level or higher than measured in reservoirs in Canada, Finland, and the U.S.[12,53-56]

Tucunaré is an excellent indicator of Hg pollution in fish of the Tucuruí reservoir. It is the most important dietary species and most of the catch is transported for sale to Belém, the capital of Pará state in the delta of the Amazon River.

Areal variations in Hg concentrations were tested with different species. Differences in Hg concentrations were tested with tucunaré for four areas (Murú, Caraipé, Capemi, and Base 4) and with pescada for three areas (Capemi, Base 4, and Ipixuna). Variations were tested by analysis of covariance, the length of fish being covariant. Only the Hg concentrations of tucunaré at Caraipé area differed from those at other areas (p <0.05).

Table 7   **The means and the variation (standard deviation and range) of the
Hg concentrations in the hair (mg kg⁻¹ dry weight), age, number of weekly fish
meals, and the number of people in each group of adults in the different areas
of surrounding the reservoir.**

	N	Age	Meals	Hg (mg kg^{-1})	Min–Max	SD
Main reservoir	45	28	14	65	6–241	58
Tucuruí	16	21	14	37	11–64	21
Caraipé	11	32	13	31	14–74	20
Mojú	8	27	10	11	2.3–19	6.1
Parakanã	12	32	2	8.5	3.3–12	2.8
Ipixuna	18	32	14	10	1.2–34	9.7
Vila	15	33	1	11	0.9–37	9.8

The standardized tucunarés of Caraipé had less Hg than the other three areas. No significant differences were observed in the Hg concentrations of mapará in the Murú, Caraipé, and Marabá areas.

An explanation for the areal differences in the Hg concentrations of tucunaré could be the location of Caraipé separated from the old river channel. This is valid if most of the Hg comes within plankton into the food chains. Another reason could be that the drainage area of Caraipé is clear cut. Therefore, just a little organic material (mainly humus) is left in the topsoils of the drainage area of Caraipé and consequently probably less Hg flushes from the soils.

7. Human Hair

Mean Hg concentration in the hair of people fishing in the reservoir was 47 mg kg⁻¹ (variation 4–241), the standard deviation being 10.2 and the average number of weekly fish meals being 14. Parakanã Indians, as well as employees of the power company living in Vila, ate fish less often than people fishing in the reservoir. The Hg concentrations in both groups were clearly lower than that for the fishermen (Table 7, Figure 3).

The concentration of Hg in the hair samples correlated with the number of weekly fish meals amongst the reservoir fishermen who ate predatory fish ($r = 0.75***$, $N = 48$). Hg concentrations in people who ate other kinds of fish did not significantly correlate with the number of weekly fish meals. The number of women and children was small in this study, but inside the family they tended to have lower Hg concentrations than men.

The Hg concentrations in people who consume predatory fish were compared in the following areas: the mouth of the reservoir (Ipixuna), the main reservoir, the Caraipé side reservoir, and the downstream city of Tucuruí (Table 7, Figure 3). The amount of fish eaten in a week did not significantly differ between these areas ($p = 0.7$). The Hg concentrations differed statistically ($p < 0.001$). Only the mouth of the reservoir differed from the other parts. This can partly be explained by the fact that at the mouth of the reservoir the most commonly eaten fish is herbivorous, but it was forbidden to fish for it during the period of the study and therefore people did not admit that they were eating that fish.

As we consider fish to be the main route of Hg into the human body, it is possible to count the amount of fish needed to be eaten each day for the Hg concentration in hair to rise to a certain level. For a 60-kg man to reach a hair Hg concentration of 50 mg kg⁻¹ he needs to ingest 0.3 mg Hg daily, according to a Japanese study of 765 people.[57] Swedish experts have obtained similar results.[58,59] To ingest this amount of Hg from fish containing 1 mg Hg per kilogram (e.g., tucunaré in Tucuruí), one has to eat 330 g of fish daily. Fishermen in this study generally ate two daily meals of fish. On the other hand, it will be difficult to restrict

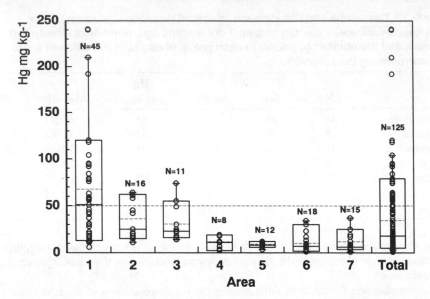

Figure 3　The Hg concentrations (mg kg^{-1} dry weight) in the hair in different areas: main reservoir (1), Tucuruí (2), Caraipé (3), Mojú (4), Parakanã (5), Ipixuna (6), Vila (7), and in all areas (total). The samples are shown as circles, whiskers are drawn from box end to the 5th and 95th percentiles, each box has a rectangle beginning at the 10th and ending at the 90th percentile, the solid line within the box represents the median and the dashed line within the box is the mean. The dashed line at 50 mg Hg kg^{-1} indicates the hair concentration above which adult people may have a risk of neurological damage.[64]

the use of fish as alimentation in the area because fish is the main protein source, especially in the fishermen's families, and cannot be substitute with meat for economical reasons. The population should, however, eat fish other than predatory fish or small predatory fish.

The Hg concentration in human hair in this study are much higher than in studies performed in other reservoirs. In Finland, the concentrations in human hair in the northern reservoirs have been between 0.02 to 35 mg kg^{-1}, the mean in the Lokka reservoir has been 4.4 mg kg^{-1}, and in the Porttipahta reservoir the mean was 5.9 mg kg^{-1}.[60]

In South America, values also have generally been much lower than in this study. Amongst Yamomami Indians Hg concentrations in hair were between 0.3 to 1.4 mg kg^{-1} (mean = 1.0 mg kg^{-1}).[61] The Japanese immigrants in South America had hair Hg values as high as 40 mg kg^{-1} but the mean was 4.5 mg kg^{-1} in the northeastern Bolivian Amazon.[62] In Brazil the highest value was 9.2 mg kg^{-1} (mean = 2.2 mg kg^{-1}) amongst the men in that same study.

In studies made in the gold mining area of Carajás in the state of Pará, Fernandes et al.[4] found Hg concentrations of 0.25 to 15.7 mg kg^{-1} in hair samples. In the Cumaru gold mining area, also situated in Pará, the Hg concentrations in the hair samples were between 1.5 and 13.7 mg kg^{-1} (mean 5.1 mg kg^{-1}) and in the area of Cachoeiro the values were between 2.0 to 69.0 mg kg^{-1} (mean 11.5 mg kg^{-1}), respectively.[63] These values are clearly lower than the results of our study and could be explained by the smaller role of fish in daily alimentation in gold mining areas, where people generally can afford to eat more meat.

It has been estimated that adult people with 50 mg kg^{-1} Hg concentration in hair, corresponding to a daily intake of 0.2 mg Hg, would have a low risk of neurological damage.[64] However, the fetus is more sensitive, and psychomotor retardation has become detectable with maternal hair concentrations as low as 20 mg kg^{-1} during pregnancy.[64,65] Hair concentrations above these levels were common in this study.

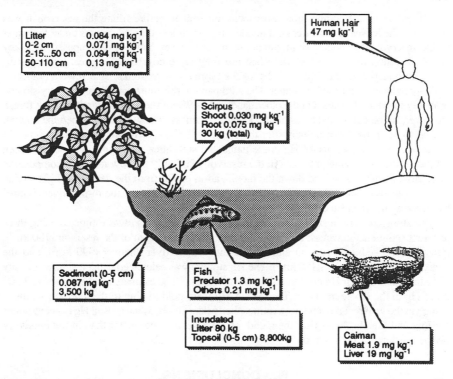

Figure 4 The mean concentrations and storages of Hg in different compartments of the reservoir and its surrounding area. The storage of Hg in the whole *Scirpus* biomass of the reservoir is calculated from biomass (kg m^{-2} dry weight) multiplied by the concentration of the plant (mg kg^{-1} dry weight) and the area covered by *Scirpus* (km^2). The storage of Hg in surface sediments is calculated from Hg concentrations weighted with the proportional area which each site represents.

B. STORAGE OF MERCURY

The estimation of biomass of the studied plants is based on work by Meyer,[66] and the estimate of the area covered up by each plant is *Eichhornia crassipes,* 5%, *Salvinia auriculata,* 20%, and *Scirpus cubensis,* 70%. The area covered by the studied plants is 655 km^2.[67] The amount of Hg stored in *Salvinia* is 1.4 kg and in *Eichhornia,* 1.0 kg.

The average amount of fine litter in a seasonal evergreen rain forest in November has been estimated to be 6,050 kg ha^{-1} (dry weight)[68] and the amount of leaves 8,647 kg ha^{-1} (dry weight).[69] Thus, the storage of Hg inundated with litter in the forested and deforested areas (1,580 km^2) was 80 kg and that inundated with leaves was 110 kg. The storage of Hg in the inundated humus layer was 1,500 kg m^{-1} of soil layer. In the yellow soil layer (2 to 15 cm deep) the storage should be 1,900 kg cm^{-1}. Thus the storage of Hg in the 0 to 5 cm layer inundated soil with litter should be about 8,800 kg. This storage (Figure 4) is about the same size as that calculated for the topsoils of Nordic countries where inundated soils have caused mobilization of mercury in reservoirs.[27] These rough estimations do not take account of the potential volatilization of Hg from deforested soils nor the possible differences in the Hg concentration of latosol types of soil. However, this is about one third to one sixth of the annual load of Hg to the environment from the Serra Pelada gold mining area during the years 1980 to 1986.

According to a decomposition experiment undertaken before filling the reservoir it was noted that the leaves of local trees did not decompose in less than 22 months under the water in the reservoir. In another similar experiment, the leaves had not decomposed with any visual alteration in 5 years, probably because they were covered by a great load of sedimentable material. In the forests 57.2% of the biomass were trunks, 36.3% were branches, 3.3% were leaves, and 3.2% was litter. The hard parts of this biomass decompose very slowly, probably in 40 to 50 years. On the other hand, only 49% of the area is inundated with forest. However, in the deforested areas of 400 km^2 the biomass of new leaves reached the amount it used to be in the three months after the clearing.[13] Thus, there are great uncertainties in the estimates of the storage of Hg in the inundated soil, litter, and biomass which has been subject to the mobilizing processes in the reservoir. Probably the low rate of decomposition of organic matter has slowed down the mobilization of Hg from the decomposing material. The high loads of solids has further reduced this mobilization near the former river channel by covering the inundated soils and decomposing organic matter.

The storage of Hg in surface sediments (0 to 5 cm; Figure 4) was estimated using three different sedimentation areas: erosion sediment area at the mouth of the reservoir (100 km^2), transportation sediment area in and around the former river channel (990 km^2), and the outskirts of the reservoir (1,180 km^2). The Hg storage was calculated using estimated density of sediments and their Hg concentrations in each area.[70] The estimate for the storage is 3,500 kg of Hg in the whole reservoir. The load of Hg from gold mining to the reservoir seems to come via the watercourses and not as atmospheric deposition, as the topsoil Hg concentrations are low and the Hg fluxes near the mouth of the reservoir are higher than in the middle or on the outskirts of the reservoir.

IV. CONCLUSIONS

A promising indicator species for monitoring Hg concentration in water of tropical reservoirs is *Salvinia auriculata*. The predatory fish species, tucunaré, is the most usable indicator species for monitoring both the accumulation Hg in food chains and the level of Hg in dietary fish in Amazonian reservoirs.

According to the results of this study, the hair Hg concentrations amongst the fishermen in Tucuruí reservoir and its surrounding area are high enough to cause health effects. Methyl mercury poisoning threatens thousands of people in the area.

The main source of Hg in the reservoir is gold mining upstream of the reservoir. On the outskirts of the reservoir the effect of inundated soil and vegetation on the levels of Hg can be seen. Even in the deforested separated area of Caraipé, there is an 0.7 mg Hg kg^{-1} in a standardized tucunaré of 0.7 kg in weight, which is over the safety limit for dietary fish. This is probably due to the Hg mobilized mainly from soil.

Because of these alarming results the levels of mercury in other Amazonian and tropical reservoirs should be studied.

ACKNOWLEDGMENTS

This study has been partly financed by FINNIDA and the Academy of Finland. Of the substudies, Ilkka Aula has been responsible for the plants and other animals apart from fish, Hannu Braunschweiler undertook the study in the drainage area of Porteira Velha, Tuija Leino did the human hair studies, Ismo Malin studied the snails, sedimentable material, and sediments (other than Porteira Velha), and Petri Porvari did the fish study. We would like to express our warmest thanks for all the help and advice to Paulo Edgar Dias Almeida, Wilze Frey Casanova, the other staff of Base 4, and the environmental laboratory in Tucuruí, Dr. Olaf Malm from the Federal University of Rio de Janeiro, and the Water and Environment Research Institute, Finland. We also wish to thank Donald Smart for correcting the English.

REFERENCES

1. Mallas, J. and Benedicto, N., Mercury and gold mining in the Brazilian Amazon, *Ambio*, 15, 248, 1986.
2. Martinelli, L., Ferreira, J., Forsberg, B., and Victoria, R., Mercury contamination in the Amazon: a gold rush consequence, *Ambio*, 17, 252, 1988.
3. Pfeiffer, W. C. and de Lacerda, L. D., Mercury inputs into the Amazon regions, Brazil, *Envir. Technol. Lett.*, 9, 325, 1988.
4. Fernandes, R., Guimarães, A. F., Bidone, E. D., Lacerda, L. D., and Pfeiffer, W. C., Monitoramento do mercúrio na area do projeto Carajás da Companhia Vale do Rio Doce (CVRD), Estado do Pará, Brasil. Presented at Control of pollution by mercury in the Amazon: New technologies and environmental education, Belém, Brazil, December 4 to 6, 1989, 1.
5. Lacerda, L. D., Pfeiffer, W. C., Ott, A. T., and Silveira, E. G., Mercury contamination in the Madeira river, Amazon—mercury inputs to the environment, *Biotropica*, 21, 91, 1989.
6. Malm, O., Pfeiffer, W., Souza, C., and Reuther, R., Mercury pollution due to gold mining in the Madeira River basin, Brazil, *Ambio*, 19, 11, 1990.
7. Barrow, C. J., The environmental impacts of the Tucuruí dam on the middle and lower Tocantins River basin, Brazil, *Regul. Rivers*, 1, 49, 1987.
8. Hildyard, N., The social and environmental effects of large dams, *Ecologist*, 14, 1, 1989.
9. Saleh, M., Saleh, M., Fouda, M., Saleh, M., Lattif, M., and Wilson, B., Inorganic pollution of the man-made lakes of Wadi El-Raiyan and its impact on aquaculture and wildlife of the surrounding Egyptian Desert, *Arch. Environ. Contam. Toxicol.*, 17, 391, 1988.
10. Potter, L., Kidd, D., and Standiford, D., Mercury in Lake Powell. Bioamplification of mercury in man made reservoir, *Environ. Sci. Technol.*, 9, 41, 1975.
11. Alfthan, G., Järvinen, O., Pikkarainen, J., and Verta, M., Mercury and artificial lakes in northern Finland. Possible ecological and health consequences, *Nordic Counc. Arct. Med. Res. Rep.*, 35, 77, 1983.
12. Bodaly, R., Hecky, R., and Fudge, R., Increases in fish mercury levels in lakes flooded by the Churchill River diversion, Northern Manitoba, *Can. J. Fish. Aquat. Sci.*, 41, 682, 1984.
13. Eletrobrás and Eletronorte Companies, *Livro Branco, Sobre o meio ambiente na usina hidrelétrica Tucuruí*, 1st ed., Brasília, DF, 1986, 1.
14. Cleary, D., *Anatomy of the Amazon Gold Rush*, Macmillan, Oxford, 1990, 245.
15. Salati, E., The climatology and hydrology of Amazonia, in *Key Environments, Amazonia*, Prance, G. T. and Lovejoy, T. E., Eds., Pergamon Press, Oxford, 1985, chap. 2.
16. Ferreira, R. C. H. and Appel, L. E., Mercury: Sources and uses in Brazil. 1st Int. Symp. Environmental Studies on Tropical Humid Forest, Manaus, Brazil, Oct. 1990, 1.
17. Ribeiro, M. C. L. B., Petrere, M., and Juras, A. A., Fisheries ecology and management prospects on the Tocantins–Araguaia river basin: The Southeasternmost Amazonian Frontier, ASIH Symp. "Fish Ecology in Latin America", University of Texas, 1993, 1.
18. Nuorteva, P., Autio, S., Lehtonen, J., Lepistö, A., Ojala, S., Seppänen, A., Tulisalo, E., Veide, P., Viipuri, J., and Willamo, R., Levels of iron, aluminium, zinc, cadmium and mercury in plants growing in the surroundings of an acidified and a non-acidified lake in Espoo, Finland, *Ann. Bot. Fennici*, 23, 333, 1986.
19. Verta, M., Mercury in Finnish forest lakes and reservoirs. Anthropogenic contribution to the load and accumulation in fish, *Publications of the Water and Environment Research Institute*, National Board of Waters and the Environment, Finland, 6, 1, 1990.
20. Lakanen, E. and Erviö, R., A comparison of eight extractants for the determination of plant available micronutrients in soils, *Acta Agric. Fenn.*, 123, 223, 1971.
21. Halonen, O., Tulkki, H., and Derome, J., Nutrient analysis methods, *Metsäntutkimuslaitoksen tiedonantoja*, 121, ISSN 0358–4283, Valtion Painatuskeskus, Helsinki, 1983 (in Finnish).

22. INSTA-VH 93., Determination of mercury in water, sludge and sediment by flameless atomic absorption spectrometry, digestion with nitric acid, draft proposal, Scientific Advisory Committee of the National Board of Waters, Working Group for Water Analysis, Finland, December 1986, 12 (in Finnish).

23. Conyers, M. K. and Davey, B. G., Observations of some routine methods for soil pH determination, *Soil Sci.,* 145, 29, 1988.

24. SFS3008, Determination of total residue and total fixed residue in water, sludge and sediment, 1990–12–03, 1 (in Finnish).

25. Kosta, L., Horvat, M., Byrne, A. R., and Stegnar, P., Determination of methylmercury, total mercury and selenium in human hair by gas liquid chromatography and cold vapour atomic absorption spectrophotometry, presented at the Joint WHO/FAO/UNEP Meet. Mediterranean Health Related Environmental Quality Criteria, Bled, September 12 to 16, 1989, 13.

26. Salati, E. and Vose, P. B., Amazon basin: a system in equilibrium, *Science,* 225, 129, 1984.

27. Lindqvist, O., Ed. Mercury in the Swedish environment, mercury in terrestrial systems, *Water Air Soil Pollut.,* 55, 73, 1991.

28. Rekolainen, S., Verta, M., and Liehu, A., The effect of airborne mercury and peatland drainage on sediment mercury content in some Finnish forest lakes, *Publications of the Water Research Institute,* National Board of Waters, Finland, 65, 11, 1986.

29. Lacerda, L. D., Salomons, W., Pfeiffer, W. C., and Bastos, W. R., Mercury distribution in sediment profiles from lakes of the high Pantanal, Mato Grosso State, Brazil, *Biogeochemistry,* 14, 91, 1991.

30. Dietrich, P. G. and Boyge, P., Mercury and zinc in silts and interstitial waters of Gdansk Bay in the Baltic Sea, *Oceanol. Acad. Sci. U.S.S.R.,* 23, 195, 1983.

31. Rudd, J. W. M., Turner, M. A., Furutani, A., Swick, A. L., and Townsend, B. E., The English-Wabigoon River system. I. A synthesis of recent research with a view towards mercury amelioration, *Can. J. Fish. Aquat. Sci.,* 40, 2206, 1983.

32. Björklund, I., Borg, H., and Johansson, K., Mercury in Swedish lakes—its regional distribution and causes, *Ambio,* 13, 118, 1984.

33. Meger, S. A., Polluted precipitation and the geochronology of mercury deposition in lake sediment of Northern Minnesota, *Water Air Soil Pollut.,* 30, 411, 1986.

34. Evans, R. D., Sources of mercury contamination in the sediments of Small Headwater Lakes in South-Central Ontario, Canada, *Arch. Environ. Contam. Toxicol.,* 15, 505, 1986.

35. Sigel, H., *Metal Ion in Biological System,* Marcel Dekker, New York, 1973, 2.

36. Togunaga, T., Furata, N., and Morimoto, M., Accumulation of cadmium in *Eichhornia crassipes* Solms, *J. Hyg. Chem.,* 22, 234, 1976.

37. Siegel, S. M., Siegel, B. Z., Barghiani, C., Aratani, C., Penny, C., and Penny, D., A contribution to the environmental biology of mercury accumulation in plants, *Water Air Soil Pollut.,* 33, 65, 1987.

38. Window, H. L. and Kendall, D. R., Accumulation and biotransformation of mercury in coastal and marine biota, in *The Biogeochemistry of Mercury in the Environment,* Nriagu, J. O., Ed., Elsevier/North-Holland, New York, 1979, 303.

39. Breteler, R. J., Valiela, J., and Teal, J. M., Bioavailability of mercury in several north eastern USA Spartina ecosystems, *Estuarine Coastal Shelf Sci.,* 12, 155, 1981.

40. Lyngby, J. E. and Brix, H., Monitoring of heavy metal contamination in the Limfjord, Denmark, using biological indicators and sediments, *Sci. Total Environ.,* 64, 239, 1987.

41. Crowder, A., Acidification, metals and macrophytes, *Environ. Pollut.,* 1, 171, 1991.

42. Huckabee, J. W., Elwood, J. W., and Hildebrand, S. G., Accumulation of mercury in freshwater biota, in *The Biogeochemistry of Mercury in the Environment,* Nriagu, J. O., Ed., Elsevier/North-Holland, New York, 1979, 277.

43. Uhe Tucuruí, Plano de utilização, do reservatorio, Programa de qualidade da água, Relatório técnico de qualidade da água (Junho/1986 a Junho/1987), TUC-10–26871-RE, Eletronorte, 1988, 1.

44. Benes, P. and Havlík, B., Speciation of mercury in natural waters, in *The Biogeochemistry of Mercury in the Environment*, Nriagu, J. O., Ed., Elsevier/North-Holland, New York, 1979, 175.

45. Lee, Y. H. and Hultberg, H., Methylmercury in some Swedish surface waters, *Environ. Toxicol. Chem.*, 9, 833, 1990.

46. Meili, M., Iverfeldt, Å., and Håkanson, L., Mercury in the surface water of Swedish forest lakes—concentrations, speciation and controlling factors, *Water Air Soil Pollut.*, 56, 439, 1991.

47. Dustman, E. H., Stickel, L. F., and Elder, J. B., Mercury in wild animals, Lake St. Clair, 1970, in *Environmental Mercury Contamination*, Hartung, R. H. and Dinman, B. D., Eds., Ann Arbor Science Publishers, MI, 1972, 46.

48. Terhivuo, J., Lodenius, M., Nuorteva, P., and Tulisalo, E., Mercury content of common frogs (*Rana temporaria* L.) and common toads (*Bufo bufo* L.) collected in southern Finland, *Ann. Zool. Fennici*, 21, 41, 1984.

49. Wren, C. D., A review of metal accumulation and toxicity in wild mammals. I. Mercury, *Environ. Res.*, 40, 210, 1986.

50. Walter, C., June, F., and Brown, H., Mercury in fish, sediments, and water in Lake Oahe, South Dakota, *J. Water Pollut. Control Fed.*, 45, 2203, 1973.

51. Moore J. and Sutherland D., Mercury concentration in fish inhabiting two polluted lakes in Northern Canada, *Water Res.*, 14, 903, 1980.

52. Bycroft, B., Coller, B., Deacon, G., Coleman, D., and Lake, P., Mercury contamination of the Lederderg River, Victoria, Australia, from an abandoned gold field, *Environ. Pollut., Ser. A.*, 28, 135, 1982.

53. Abernathy, A. and Cumbie, P., Mercury accumulation by largemouth bass (*Micropterus salmoides*) in recently impounded reservoirs, *Bull. Environ. Contam. Toxicol.*, 17, 595, 1977.

54. Cox, J., Carnahan, J., DiNunzio, J., McCoy, J., and Meister, J., Source of mercury in fish in new impoundments, *Bull. Environ. Contam. Toxicol.*, 23, 779, 1979.

55. Lodenius, M., Seppänen, A., and Herranen, M., Accumulation of mercury in fish and man from reservoirs in Northern Finland, *Water Air Soil Pollut.*, 19, 237, 1983.

56. Verta, M., Rekolainen, S., and Kinnunen, K., Causes of increased fish mercury levels in Finnish reservoirs, *Publications of the Water Research Institute*, National Board of Waters, Finland, 65, 44, 1986.

57. Kojima, K. and Araka, T., Normal mercury levels in food in Japan. (Data of the polluted area are excluded), *Stencils*, Tokyo, 1, 1972.

58. Birke, G., Johnels, A. G., Plantin, L. O., Sjöstrand, B., and Westermark, T., Mercury poisoning through eating fish?, *Läkartidning*, 64, 3628, 1967.

59. Swedish National Institute for Public Health, Methylmercury in fish, a toxicologic epidemiologic evaluations of risks, *Nordisk Hygienisk Tidskrift*, Suppl. 4, 75, 1971 (in Swedish).

60. Lodenius, M. and Seppänen, A., Hair mercury contents and fish eating habits of people living near a Finnish man-made lake, *Chemosphere*, 11, 755, 1982.

61. Hecker, L. H., Allen, H. E., Dinman, B. D., and Neel, J. V., Heavy metal levels in acculturated and unacculturated populations, *Arch. Environ. Health*, 29, 181, 1974.

62. Tsugane, S., The mercury content of hair of Japanese immigrants in various locations in South America, *Sci. Total Environ.*, 63, 69, 1987.

63. Couto, R. C., Camara, V. M., and Sabroza, P. C., Intoxicação mercurial: resuldados preliminares em duas areas garimpeiras no Estado do Pará, *Pará Desenvolvimento*, 23, 63, 1988.

64. WHO, Methylmercury, *Environmental Health Criteria 101,* World Health Organization, Geneva, 1990.

65. Clarkson, T. W., Human health risks from methylmercury in fish, *Environ. Toxicol. Chem.,* 9, 957, 1990.

66. Meyer, M., Atividades do programa macrófitas aquáticas referentes a outumbro 1988-setembro 1989, Relatório, unpublished data, 1989 (in Portuguese).

67. Abdon, M. M. and Meyer, M., Variacão temporal de áreas ocupadas por macrófitas aquáticas no reservatório de Tucuruí através de dados do satélite Landsat/TM, presented at VI Simpósio Brasileiro de Sensoriamento Remoto, Manaus, June 24 to 29, 1990, 1.

68. Klinge, H., Preliminary data on nutrient release from decomposing leaf litter in a neotropical rain forest, *Amazoniana,* 6, 193, 1977.

69. Uhl, C. and Jordan, C. F., Succession and nutrient dynamics following forest cutting and burning in Amazonia, *Ecology,* 65, 1476, 1984.

70. Axelsson, V. and Håkanson, L., The relation between mercury distribution and sedimentological environment in lake Ekoln. Part 1. Purpose and methods of analysis, *UNGI Rapport 11,* University of Uppsala, Dept. of Physical Geography, Sweden, 1971. 1.

The Distribution of Mercury in a Mediterranean Area

Corrado Barghigiani and Torquato Ristori

CONTENTS

ABSTRACT: This project is aimed at studying the mercury distribution in an area of the northern Tyrrhenian sea affected by a geochemical anomaly. The importance of the contamination sources, the transfer of mercury from these sources to the terrestrial and marine ecosystems, and the exchange of mercury among the abiotic and biotic environmental components were also studied.

I. INTRODUCTION

The Mediterranean Basin covers only 1% of the earth's surface, but comprises about 65% of the world's cinnabar deposits. The Mt. Amiata area, located in central Italy, is an important part of the Mediterranean cinnabar anomaly. The cinnabar deposits of Mt. Amiata were first worked by the Etruscans. Closed later by the Romans, they were rediscovered in 1868. In this area mining activity reached peak production in 1969, was drastically reduced in the 1970s, and ceased in 1980.[1]

Due to the geological anomaly, and above all to the mining activities, the mercury levels on Mt. Amiata were found to be high in both the abiotic and biotic environmental components,[1] and though the mining and smelting activities have ceased these environmental mercury levels still seem to be high.[2-5]

The mercury contamination extends from Amiata to the northern Tyrrhenian sea, where it affects both sediments and marine organisms.[6] The aim of this study is to assess mercury distribution in both biotic and abiotic environmental components of the marine and terrestrial environment of this area and to determine the major sources of contamination.

1–56670–066–3/94/$0.00+$.50

II. EXPERIMENTS

The metal concentrations were determined in air, surface (5 cm), and deep (15 to 20 cm) soil; in vegetation in the terrestrial environment (Mt. Amiata area), and in sediments and marine organisms in the marine environment (northern Tyrrhenian sea). The samplings were performed from 1985 to 1990.

A. STUDY AREAS

Different sampling stations were chosen on Mt. Amiata, characterized by the presence of mercury contamination sources of various nature. The studied stations, shown in Figure 1A, were station A, the old mine and smelting plant of Abbadia S. Salvatore; B and C, located 3 km southwest and 7 km northwest of the Abbadia mine, respectively; D, near Abbadia, on spoil banks of roasted cinnabar; E, near the thermal spring of Bagni S. Filippo; and F and G, near the geothermal power plants of Bagnore and Piancastagnaio, respectively.

The study area of the northern Tyrrhenian sea was that shown in Figure 1B.

B. TERRESTRIAL VEGETATION AND MARINE SPECIES ANALYZED

The terrestrial species studied were *Pinus nigra* and *Abies alba* (conifers), *Cytisus scoparius* (broom), and *Parmelia sulcata* and *Xanthoria parietina* (lichens). Pine, spruce, and broom were chosen because they were found to be good Hg biomonitors of mining areas,[3,5,7] and lichens because they are important for atmospheric Hg biomonitoring.[8] For both species of conifers, needles of the same age were analyzed since Hg content increases with needle age.[5,7] For spruce, 7-year-old needles were analyzed as suggested by the literature;[5] for pine, 1-year-old needles were examined.

The marine species studied were *Citharus linguatula*, *Lepidorhombus boscii*, and *Solea vulgaris* (benthic flatfish); *Nephrops norvegicus* and *Parapenaeus longirostris* (benthic crustaceans); *Eledone cirrhosa* (benthic cephalopod); and *Merluccius merluccius* and *Trisopterus minutus capelanus* (pelagic fish). These species were chosen because in the study area they are very important both commercially and for their abundance.[9]

C. MERCURY ANALYSIS

Organism, soil, and sediment samples were digested with nitric acid in a closed system under pressure at 120° C for 6 h and analyzed by atomic absorption spectrometry on cold vapor. The metal concentrations were expressed as dry weight. Organic Hg was determined by the method of Capelli et al.[10] The analytical procedures were tested using Certified Reference Materials 1572 (citrus leaves: 0.08 ± 0.02 µg g^{-1} Hg) from the National Bureau of Standards (U.S.), BRC (calcareous loam soil: 56.8×10^{-3} µg g^{-1} Hg) from the Community Bureau of Reference of Brussels (Belgium), and DORM-1 (dogfish muscle: 0.798 ± 0.074 µg g^{-1} Hg) from the National Research Council of Canada.

D. AIR COLLECTION AND DETERMINATION

Atmospheric mercury was collected on gold traps connected to battery-powered portable pumps. The traps consisted of gold sheets inserted into quartz tubes (2-mm internal diameter) with a filter (Millipore®, 0.45 µm pore size) at the entrance. Mercury was stripped from the gold amalgam by heating the trap to 600° C with a resistance, then transported with a nitrogen flow of 1 L min^{-1} into the measuring cell, and analyzed by atomic absorption spectrometry. The calibration of the traps was done in the laboratory with Hg^0 vapor by injecting known amounts of volatile elemental mercury sucked from a bottle containing a drop of liquid mercury. The Hg^0 concentration of the mercury-saturated air in the bottle was calculated by treating Hg^0 as an ideal gas.

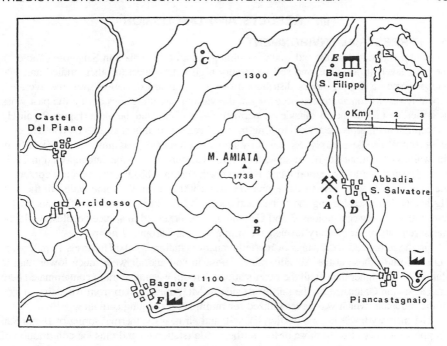

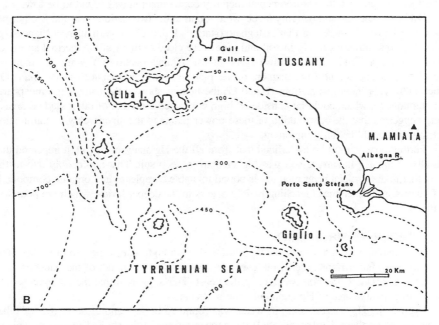

Figure 1   Study areas of Mt. Amiata, with sampling stations (A), and of the Tyrrhenian sea (B).

III. RESULTS AND DISCUSSION

A. TERRESTRIAL ENVIRONMENT

At station A, located on the old smelting-mining plant of Abbadia San Salvatore (Table 1), mercury levels in air, pine, and soil were very high with respect to other studied areas[2,5,7] and presented a heterogeneous distribution (see the standard deviation from the average), probably due to contamination occurring at the various sampling sites during the past work activities. Just a few meters outside the perimetrical walls around the plant (Figure 2), air Hg concentrations sharply decreased to 7 ng m^{-3} and remained almost constant. Also, Hg contents in pine needles sharply decreased by about one order of magnitude just 200 m from the walls and continued to decrease with distance. Coming back to the data reported in Table 1, it appeared that at the station located 7 km northwest of Abbadia, the mercury content in the pine needles was close to those pine needles 1300 m from the mine walls. In the soil, Hg levels decreased leaving the mine (stations B and C). Furthermore, a difference in Hg concentration between stations B and C could also be observed in spruce needles, while the measurements of air mercury levels gave similar values—around 7 ng m^{-3}.

On the reforested spoil banks of roasted cinnabar (station D), the Hg levels in pine were high—close to those at the Abbadia mine. Those in soil and air were much lower than at station A, but higher than at all the other stations. Broom was much more contaminated here than at station C, located on the northwest side of Mt. Amiata, 7 km from Abbadia—a side of the mountain which was never utilized for mining and smelting activities.

The mercury levels in pine (station E), soil, and air near the thermal spring of Bagni San Filippo were very close to those of the northern side (station C) and thus the contribution of this thermal spring to the environmental mercury contamination was found to be low.

Different pine and soil mercury levels were found at the two stations near the geothermal power plants of Bagnore and Piancastagnaio (stations F and G). In particular, at Piancastagnaio the concentrations of Hg in soil and pine were close to those at the thermal spring of Bagni S. Filippo and on the northwest side of Mt. Amiata (station C), while the mercury levels in pine and soil near the Bagnore power plant were higher than at Piancastagnaio. On the contrary, at these two stations (F and G) the lichens showed the same Hg atmospheric contamination, which could be considered high compared with that of other studied areas,[8] and suggested that the contribution of these power plants to the air mercury contamination of this area could be fairly important.

Furthermore, it must be underlined that from all the Hg analyses of soil, it appeared that at all stations Hg in surface soil was higher than in deep soil. This is probably due to the fact that in the surficial layer mercury is bound to stable complexes and organic compounds of humus, and hence metal leaching mainly occurs in the deeper layers, which are poorer in humus.

B. MARINE ENVIRONMENT

Hg concentrations in marine sediments (Figure 3) were high with respect to other Mediterranean areas. Furthermore, the metal levels decreased from the mouth of the Albegna river going west offshore and northwest along the coast. This suggested that the Albegna should be an important source of Hg contamination in this area.

Even if the Hg concentration in water was found to be low and close to that of other areas of the northern Tyrrhenian sea,[11] the concentrations in the studied marine organisms (Table 2), both pelagic and benthic (except sole), were found to be high considering that the maximum limit accepted by the European Economic Community (EC) for edible parts of marine organisms[12] is 0.7 μg g^{-1}. Indeed, since the average dry weight of an analyzed sample was 20% of the fresh one, 50% of the studied species had an average Hg content over the EC limit, and the others were close to the limit except for sole, which presented a low Hg concentration, probably due to its feeding behavior.[13] Furthermore, in fish, organic mercury was found to be 80–86% of the total, and in *Eledone* 50 to 90% (in crustaceans it was not

Table 1　**Mercury concentrations in vegetation and soil (μg g^{-1} d.w.) and in the air (ng m^{-3}) at the different sampling stations.**

	Hg	No.
Abbadia mine		
Pine needles	2.01 ± 1.46	16
Soil	645.0 ± 197.0	9
Air	491.0 ± 248.0	6
3 km southwest of Abbadia		
Spruce needles	0.41 ± 0.07	5
Soil		
Surface	1.01 ± 0.30	5
Deep	0.34 ± 0.11	5
Air	5.40 ± 1.20	3
7 km northwest of Abbadia		
Spruce needles	0.24 ± 0.04	5
Pine needles	0.08 ± 0.02	5
Broom twigs	0.10 ± 0.08	5
Soil		
Surface	0.81 ± 0.25	5
Deep	0.25 ± 0.08	5
Air	8.80 ± 3.40	3
Spoil banks of roasted cinnabar near Abbadia		
Broom twigs	0.98 ± 0.18	9
Pine needles	1.50 ± 0.17	6
Spruce needles	1.04 ± 0.09	3
Soil		
Surface	85.0 ± 18.0	5
Deep	5.75 ± 1.05	5
Air	21.92 ± 4.12	5
Thermal spring of Bagni S. Filippo		
Pine needles	0.13 ± 0.04	10
Lichens	0.17 ± 0.03	10
Soil		
Surface	0.71 ± 0.25	10
Deep	0.19 ± 0.05	10
Air	5.90 ± 1.60	4
Geothermal power plant of Bagnore		
Pine needles	0.39 ± 0.06	5
Lichens	2.33 ± 0.76	5
Soil		
Surface	16.51 ± 5.12	5
Deep	1.32 ± 0.81	5
Geothermal power plant of Piancastagnaio		
Pine needles	0.11 ± 0.06	5
Lichens	2.11 ± 0.80	5
Soil		
Surface	1.98 ± 0.60	5
Deep	0.18 ± 0.09	5

Note: Average values ± S.D. are given; No. is the number of samples.

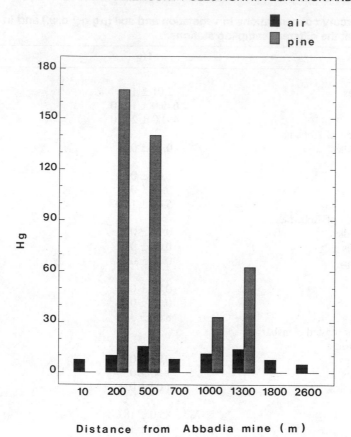

Figure 2   Mercury concentrations in air (ng m^{-3}) and pine needles (ng g^{-1}) at various distances from the mine of Abbadia S. Salvatore.

determined), while in sediments, although biological and/or chemical methylation may occur,[14,15] organic mercury was found to be only 1 to 2% of the total.

IV.   CONCLUSIONS

From the results it appears that, in general, the Hg distribution in the air, vegetation, and soil of Mt. Amiata is heterogeneous, and depends on the nature of and the distance from the different Hg sources. Air mercury concentration is very high (about 500 ng m^{-3}) near the mining-smelting plant of Abbadia S. Salvatore, then sharply decreases and reaches quite homogeneous levels for all the area (7 ng m^{-3}), except on the spoil banks of roasted cinnabar where it is about 20 ng m^{-3}. From the data on pine (Figure 4), which among the studied species in this area is the most widespread, it seems that the environmental Hg levels could be distributed in the following order of decreasing importance: Abbadia mine-smelting plant, roasted cinnabar banks, geothermal power plants, thermal springs, and the northern side.

Concerning the marine environment, sediments, due to their high mercury levels, seem to represent the most important Hg reserve pool and are the main mercury contamination source for marine organisms. Indeed, all studied organisms (except sole) were found to be very contaminated.

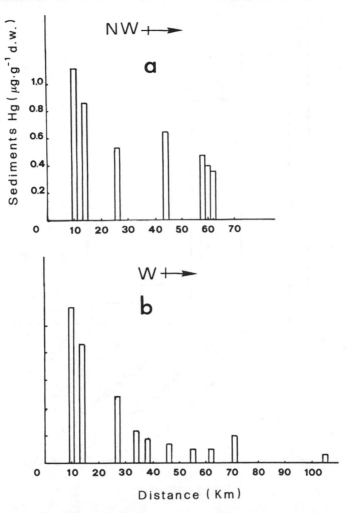

Figure 3   Mercury concentrations in the marine sediments of the study area shown in Figure 1B, going northwest along the coast (a), and west offshore (b) from the mouth of the Albegna river, which is represented by the origin (0) of the axes.

Table 2   **Mercury concentration in marine organisms.**

Organism	Hg (μg g^{-1} d.w.)	Length range (cm)	Sample no.
M. merluccius	2.01 ± 1.51	25–35	21
T.m. capelanus	5.05 ± 2.16	15–25	13
E. cirrhosa	3.64 ± 1.36	8.5–10.5	82
N. norvegicus	6.81 ± 4.20	3.5–4.5	25
P. longirostris	3.14 ± 1.07	2–3	47
L. boscii	2.83 ± 1.13	16.5–18.5	22
C. linguatula	5.40 ± 2.61	14.5–22.0	22
S. vulgaris	0.68 ± 0.32	19.0–26.5	10

Note: Average values ± S.D. are given.

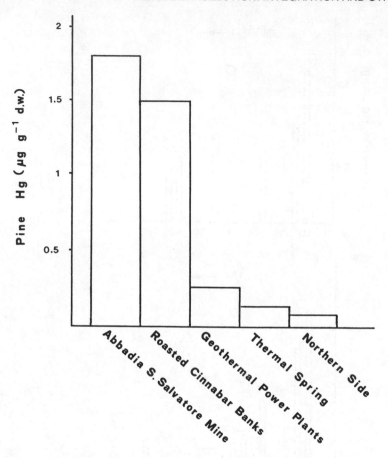

Figure 4   Average Hg concentrations in pine needles from the sites studied.

REFERENCES

1. Bombace, M.A., Cigna Rossi, L., Clemente, G.F., Zuccaro Labellarte, G., Allegrini, M., Lanzola, L., and Gatti, L., Ricerca ecologica sulle zone mercurifere del Monte Amiata, *Ig. Sanita Pubblica,* 29, 191, 1983.
2. Breder, R. and Flucth, R., Mercury levels in the atmosphere of various regions and locations in Italy, *Sci. Total Environ.,* 40, 231, 1984.
3. Barghigiani, C. and Bargagli, R., Mercury uptake by plants in a mining area, in *Current Perspectives in Environmental Biogeochemistry,* Giovannozzi-Sermanni G. and Nannipieri P., Eds., CNR-IPRA Rome, 1987, chap. 3.
4. Ferrara, R., Masrerti B., and Breder, R., Mercury in abiotic and biotic compartments of an area affected by a geochemical anomaly (Mt. Amiata, Italy), *Water Air Soil Pollut.,* 56, 219, 1991.
5. Barghigiani, C. and Bauleo, R., Mining area environmental mercury assessment using *Abies alba, Bull. Environ. Contam. Toxicol.,* 49, 31, 1992.
6. UNEP/FAO/WHO, Assessment of the state of pollution of the Mediterranean sea by mercury and mercury compounds, *MAP Tech. Rep. Ser.* No 18, 1987.
7. Barghigiani, C., Ristori, T., and Bauleo, R., Pinus as an atmospheric Hg biomonitor, *Environ. Technol.,* 12, 935, 1991.

8. Bargagli, R. and Barghigiani, C., Lichen biomonitoring of mercury emission and deposition in mining, geothermal and volcanic areas of Italy, *Environ. Monit. Assess.*, 16, 265, 1991.

9. Barghigiani, C. and De Ranieri, S., Mercury content in different size classes of important edible species of the northern Tyrrhenian sea, *Mar. Pollut. Bull.*, 22, 406, 1992.

10. Capelli, R., Fezia, C., Franchi, A., and Zanicchi, G., Extraction of methylmercury from fish and its determination by atomic absorption spectroscopy, *Analyst*, 104, 1197, 1977.

11. Barghigiani, C., Ferrara, R., Seritti, A., Petrosino, A., Masoni, A., and Morelli, E., Determination of reactive, total and particulate mercury in the coastal water of Tuscany (Italy) by atomic fluorescence spectrometry, in *Proc. 5th Journée Etud. Pollut. C.I.E.S.M. 1980*, Cagliari, 1981, 124.

12. EEC, Objectif de qualitè rejets industriels. Conseil des Ministres, G.V. Mo L74/49, March 17, 1984.

13. Pellegrini, D. and Barghigiani, C., Feeding behavior and mercury content in *Solea vulgaris* and *Lepidorhombus boscii* of the Northern Tyrrhenian Sea, *Mar. Pollut. Bull.*, 20, 443, 1988.

14. Jensen, S. and Jernelov, A., Biological methylation of mercury in acquatic organisms, *Nature*, 223, 753, 1969.

15. Jewett, K.L. and Brinkman, F.E., Transmethylation of heavy metal ions in water, *Div. Environ. Chem. Am. Chem. Soc.*, 14, 218, 1974.

Distribution and Speciation of Mercury in the Water and Fish of Nan Hu (South Lake), Guangxi Province, People's Republic of China

Nicolas S. Bloom, Lian Liang, Zu Qiu Xie, and Shu Shen Wang

CONTENTS

ABSTRACT: Total and methylmercury concentrations were measured in the water and fish of Nan Hu (South Lake), a small (1.0 km^2), shallow ($z_{max} = 2$ m) impoundment adjoining the city of Nanning. The lake, which appears to be highly eutrophic, is used as a discharge point for municipal sewage and industrial effluents. In addition, to promote production of fish for consumption by the local residents, large quantities of fish-food pellets are added to the lake throughout the year. In September, 1991, duplicate surface water samples from seven sites within the lake, the municipal sewage outfall, and a multi-use industrial discharge channel were collected using ultraclean handling techniques. Four specimens each of five important food-fish species [*Cyprinus carpio* (common carp), *Hypophthalmichthus molitrix* (silver carp), *Tilapia nolotica* (Nile tilapia), *Labeo rohita* (shark minnow), and *Aristichthys nobilis*] were also collected for analysis. Total mercury concentrations in the lake water varied from 3.42 ng/L^{-1} near the eastern (rural) end to 11.52 ng/L^{-1} at the western (urban) end. Methylmercury values were much lower, and less variable across the lake, averaging 0.100 ± 0.021 ng/L^{-1} for all samples. The mercury concentration in fish tissue was remarkably low, averaging only 0.0121 ± 0.0071 µg g^{-1} Hg (wet weight, skin-off axial muscle) for 3-year-old, 1-kg fish. The fraction of the total mercury in the methylated form was 91.9 $\pm 8.9\%$, (n = 21).

I. INTRODUCTION

Many researchers have recently reported mercury levels in excess of WHO consumption guidelines (0.5 µg g^{-1}) in the freshwater fish of Northern Hemispheric lakes, even in lakes remotely located from direct mercury sources.[1-4] Some evidence suggests that fish tissue mercury concentrations are increasing over time, perhaps due to increased atmospheric loading.[5]

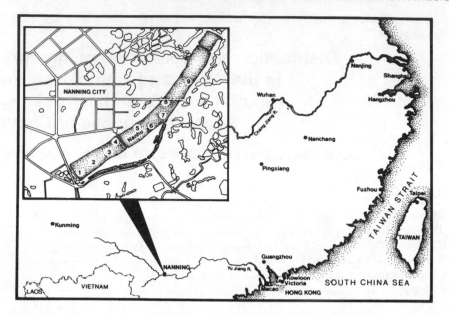

Figure 1 Map showing location of Nan Hu in relation to city of Nanning, People's Republic of China. Numbers in lake correspond to water sampling locations discussed in text.

In the People's Republic of China, a substantial portion of protein in the human diet is provided by fish from government-stocked public lakes. These lakes are typically amended throughout the year with large quantities of fish-food pellets containing growth hormones. The lakes are stocked with omnivorous fish such as carp and tilapia, which are fast-growing species and thrive in the turbid, eutrophic water.

In the summer of 1991, we examined the potential for mercury-related human health risk posed by the consumption of fish from one such lake (Nan Hu). In this study we document the rather surprising results obtained from Nan Hu, which is located in southeast Nanning (population 500,000), in Guangxi province (Figure 1). Nan Hu, apparently formed from a cut-off oxbow of the river Yu Jiang, is long and narrow (4 km × 0.25 km), with a mean depth of only 1 m. The lake is well stocked with several species of food fish, supporting a local fishery that has yielded an average of 360 tonnes/year^{-1} over the past 5 years. To maintain this level of production, the lake is amended with approximately 125 tonnes/year^{-1} of dry food pellets composed largely of grain and soy by-products (Table 1). Nan Hu receives water at the southwestern end from the Nanning municipal sewage treatment facility, and near the center from two small streams that carry the discharge of several light manufacturing industries. Much of the shoreline is in parkland and the lake is extensively used for recreation. Typical chemical and physical parameters for Nan Hu are shown in Table 1.

II. METHODS

Twenty-one fish, all judged by the sampling crew to be approximately 3 years of age, were collected in September of 1991. Within 4 h the fish were weighed and filleted, with 10 to 20 g of axial muscle tissue being dissected out and placed into acid-cleaned Teflon® vials. All handling of the tissues was under trace-metal clean conditions,[6,7] including the use of clean-room gloves, and stainless steel dissecting implements. Samples were collected from

Table 1   **Lake characteristics and chemical parameters regarding the Nan Hu impoundment.**

Lake and water chemistry parameters

Area: 1.0 km^2	Maximum depth: 2 m
Mean depth: $\sim$1 m	Mean pH: 7.1
Hardness: 69.6 mg/L^{-1} as CaCO$_3$	Chemical oxidation demand: 11.1 mg/L^{-1} O$_2$
Inorganic nitrogen: 1.7 mg/L^{-1}	Total phosphate: 12 mg/L^{-1}

Recent fish data

Hg$_{(tot)}$: 0.0177 ± 0.0145 µg g^{-1}
CH$_3$Hg: 0.0178 ± 0.0058 µg g^{-1}

Fish feeding program

Amount added: 125 tonnes/year^{-1}

Recent harvests (tonnes/year^{-1})

1987: 356
1988: 370
1989: 357
1990: 360

Food pellet composition

Wheat shell: 45%
Defatted soybean: 16%
Defatted rapeseed: 10%
Fish meal: 4%
Corn: 25%

Data from the Guanaxi Province Anti-Epidemic and Hygiene Station Environmental Monitoring Group.

4 to 5 individuals each of commonly eaten species [*Cyprinus carpio* (common carp), *Hypophthalmichthus molitri* (silver carp), *Aristichthys nobilis, Tilapia nolotica* (Nile tilapia), and *Labeo rohita* (shark minnow)].

Water samples were collected by immersing an acid-cleaned 125 mL Teflon® bottle from the bow of a forward-moving boat. Clean-room gloves were used to hold the bottle, and were changed between samples. All samples were frozen until shipment to the laboratory (Seattle, WA), and arrived partially thawed, but still at 0° C with significant ice present. Water samples were immediately acidified to 1% with low mercury HCl (<5 pg Hg/mL), and allowed to sit overnight prior to analysis. Tissues were refrozen until analysis.

Tissue samples were analyzed for total mercury by hot acid digestion (7:3 HNO$_3$/H$_2$SO$_4$), followed by SnCl$_2$ reduction, dual amalgamation, and cold vapor atomic fluorescence spectrometry (CVAFS).[6] Total mercury (Hg$_t$) in water was analyzed similarly, following cold oxidation with bromine monochloride.[8] Tissue methyl mercury (CH$_3$Hg) concentrations were assayed from alkaline (1:3 KOH/methanol) digestates by aqueous phase ethylation, cryogenic GC separation, and CVAFS detection.[9] Methylmercury in water samples was preextracted into methylene chloride and then back extracted into deionized water prior to analysis by the ethylation technique.[9]

III. RESULTS AND DISCUSSION

Analytical variability, as indicated by the quadruplicate water analysis at lake site 3 (Table 2), was about 5 to 10% for these analyses. The method detection limits, based upon 2 SD of the reagent blanks, were about 0.0001 µg g^{-1} (wet weight) for both Hg$_t$ and CH$_3$Hg in fish tissue. In water, the measured detection limits were 0.012 ng/L^{-1} for Hg$_t$, and 0.016 ng/L^{-1} for CH$_3$Hg. As indicated in Table 3, measurements of NRCC certified fish tissue (DORM-1) were close to the central value for both total and methylmercury.

Total mercury in the lake water varied from a maximum of 11.52 ng/L^{-1} at the urbanized (southwestern) end, to a minimum of 3.42 ng/L^{-1} near the rural (northeastern) end of the lake (Table 2). Within the main body of the lake, concentrations were quite similar, at 6.38 ± 1.40 ng/L^{-1} (n = 5 sites). Total mercury concentrations in water from the sewage outfall (16.1 ng/L^{-1}) and the industrial discharge stream (10.2 ng/L^{-1}) were somewhat elevated

Table 2   **Total mercury and methylmercury in unfiltered surface water samples from Nan Hu (September, 1991).**

Sample map ID	Water location and type	Blank and yield corrected mercury concentrations, ng/L^{-1} as Hg	
		Total	Methyl
1	West (urban) end	11.53	0.114
2	West central	7.37	0.078
3	West central	5.77 ± 0.30 (n=4)	0.105 ± 0.010 (n=4)
4	Nanning sewage outfall	16.09	<0.008
5	North central	6.22	0.123
6	South central	8.07	0.069
7	South central	4.47	0.075
8	Industrial outfall	10.15	0.115
9	Northeastern end (rural)	3.42	0.116

Note: All results are the mean of duplicate samples, except #3, which is the mean of quadruplicate samples.

Table 3   **Fish muscle tissue mercury concentrations from Nan Hu (collected September, 1991).**

Species	Fresh weight (g)	Mercury conc. (µg g^{-1} wet wt.)		
		Total	Methyl	% Methyl
Cyprinus carpio				
Mean (n=4)	925	0.0065	0.0051	80.0
S.D.	124	0.0022	0.0016	5.3
Hypophthalmichthus molitrix				
Mean (n=5)	713	0.0120	0.0112	94.6
S.D.	85	0.0019	0.0014	7.1
Aristichthys nobilis				
Mean (n=4)	1263	0.0133	0.0132	98.8
S.D.	75	0.0034	0.0034	0.9
Tilapia nolotica				
Mean (n=4)	351	0.0064	0.0056	88.6
S.D.	82	0.0019	0.0014	6.9
Labeo rohita				
Mean (n=4)	1488	0.0223	0.0215	97.3
S.D.	311	0.0093	0.0090	8.5
Fish food pellets (n=1)	—	0.0071	0.0018	25.4
NRCC DORM-1				
Mean (n=3)	—	0.841	0.716	—
S.D.	—	0.013	0.051	—
Certified	—	0.796	0.732	—
S.D.	—	0.079	0.069	—

compared to the average lake water, but much lower than similar discharges in North America.[8,10] Methylmercury accounted for about 1.5% of the total, and was relatively consistent over the entire length of the lake (0.100 ± 0.021 ng/L^{-1}, n = 7 sites). The exception was the sewage outfall, where no methylmercury was detected (<0.016 ng/L^{-1}), possibly an effect of a wastewater treatment process such as chlorination, which could oxidize CH_3Hg to Hg(II). Overall, these values are somewhat higher than those typically reported for pristine lakes (1–3 ng/L^{-1} total mercury and 0.02–0.1 ng/L^{-1} methylmercury),[4,7,11] but still considerably lower than reported for mercury-contaminated waters.[10,12]

The mercury concentration in axial muscle was remarkably low, averaging only 0.0121 ± 0.0071 µg/g (total Hg, wet weight basis) for fish approximately 3 years old and 1 kg in mass. This is more than an order of magnitude lower than values reported for similar benthic fishes found in natural environments (i.e., *C. carpio*, 0.17 to 0.39 µg g^{-1}; *L. rohita*, 0.17 to 0.20 µg g^{-1})[13] and up to two orders lower than attained by free-swimming piscivores in remote North American lakes [i.e., *Stizostedion vitreum* (walleye), 0.12 to 1.74 µg g^{-1}].[2,5] These results are quite similar to recent data obtained by the Guangxi Anti-Epidemic and Hygiene Station (Table 1), although direct comparison cannot be made as the Chinese data set was not divided by individual fish species.

The fraction of methylmercury for all fish analyzed averaged $91.9 \pm 8.9\%$ of the total Hg present. A significant positive correlation between mean fish mass and both total mercury concentration and percent methylmercury was observed. Inorganic mercury concentration was not correlated with fish size or species, but was rather randomly distributed with a mean of 0.0008 ± 0.0009 µg g^{-1}. These observations suggest that the inorganic mercury is merely residual low-level sampling or laboratory contamination of the samples, the percentage of which is only significant due to the very low levels of total mercury in the samples. This result is supported by recent work describing the nature of analytically introduced variability in fish tissue mercury speciation data.[6]

The anomalously low mercury levels in the fish are an unexpected finding, especially given the relatively high average concentrations of mercury in the water. The lake is circumneutral and eutrophic, two factors which favor lower levels of methylmercury in fish.[2–4] Another probable factor giving low fish tissue mercury concentrations is biomass dilution, caused by rapid growth and an artificial, low-mercury diet made largely from grain products (see Table 1 for gross composition, Table 3 for mercury speciation). These food pellets are added to the lake at a rate of approximately twice the annual fish production of the lake (dry weight basis). The presence of an abundance of artificial food also diminishes the effects of bioaccumulation by predation, normally the dominant route of mercury uptake in fish.[2–4]

The dissolved and particulate mercury levels were not determined, but it was visually apparent that the lakewater was very turbid due to the presence of organic particles. Thus, another contributing factor to the low tissue concentrations may be a sequestering of the methylmercury in forms that are unavailable for direct uptake.

The unexpectedly low values for tissue mercury concentrations from the fish of Nan Hu are contrary to observations in Wisconsin of a near constant relationship between water column and fish tissue mercury concentrations.[11] This suggests that there is still much to be learned with regard to the subtle interplay between biology and geochemistry, which ultimately determines the degree of biological accumulation of mercury in different ecosystems. The continued study of extreme cases such as Nan Hu can provide rigorous and informative tests of the general applicability of mercury biogeochemistry models now being developed and parameterized on a relatively narrow North American database.

ACKNOWLEDGMENTS

We wish to thank Dr. James G. Wiener (National Fisheries Research Center, Lacrosse, WI), for his many useful comments in reviewing this manuscript. This work was conducted largely with the internal support of Brooks-Rand, Ltd., Seattle, WA. Additional support, as well as the co-use of analytical equipment came from the Electric Power Research Institute, Palo Alto, CA, under research contract RP2020–10.

REFERENCES

1. Bjorklund, I., Borg, H., and Johansson, K., Mercury in Swedish lakes—its regional distribution and causes, *Ambio*, 13:118, 1984.
2. Grieb, T.M., Driscoll, C.T., Gloss, S.P., Schofield, C.L., Bowie, G.L., and Porcella, D.B., Factors affecting mercury accumulation in fish in the upper Michigan peninsula, *Environ. Toxicol. Chem.*, 9:919, 1990.
3. Jackson, T.A., Biological and environmental control of mercury accumulation by fish in lakes and reservoirs of northern Manitoba, Canada, *Can J. Fish Aquat. Sci.*, 48:2449, 1991.
4. Wiener, J.G. and Stokes, P.M., Enhanced bioaccumulation of mercury, cadmium, and lead in low-alkalinity waters: an emerging regional environmental problem, *Environ. Toxicol. Chem.* 9:821, 1991.
5. Swain, E.B., Engtrom, D.R., Brigham, M.E., Henning, T.A., and Brezonik, P.L., Increasing rates of atmospheric mercury deposition in mid-continental North America, *Science*, 257:784, 1992.
6. Bloom, N.S., On the methylmercury content of fish and marine invertebrates, *Can. J. Fish Aquat. Sci.*, 49:1010, 1992.
7. Fitzgerald, W.F. and Watras, C.J., Mercury in the surficial waters of rural Wisconsin lakes, *Sci. Total Environ.*, 87/88:223, 1989.
8. Bloom, N.S. and Crecelius, E.A., Determination of mercury in seawater at subnanogram per liter levels, *Mar. Chem.*, 14:49, 1983.
9. Bloom, N.S., Determination of picogram levels of methylmercury by aqueous phase ethylation, followed by cryogenic gas chromatography with cold vapor atomic fluorescence detection, *Can. J. Fish Aquat. Sci.*, 46:1131, 1989.
10. Bloom, N.S. and Effler, S.W., Seasonal variability in the mercury speciation of Onondaga Lake (New York), *Water Air Soil Pollut.*, 53:251, 1990.
11. Bloom, N.S., Watras, C.J., and Hurley, J.P., Impact of acidification on the methylmercury cycling of remote seepage lakes, *Water Air Soil Pollut.*, 56:477, 1991.
12. Jackson, T.A., Parks, J.W., Jones, R.N., Woychuk, J.A., Sutton, J.A., and Hollinger, J.D., Dissolved and suspended mercury in the Wabigoon river (Ontario, Canada): seasonal and regional variations, *Hydrobiologia*, 92:473, 1982.
13. Mitra, S., *Mercury in the Ecosystem*, Transtech Publications, Aedermannsdorf, Switzerland, 1986, pp. 182–189.

Chemical Speciation of Mercury in a Meromictic Lake

D. Cossa, R. P. Mason, and W. F. Fitzgerald

CONTENTS

ABSTRACT: The chemical speciation and partitioning of mercury in Pavin Lake, France was studied in August 1990. This small meromictic crater lake is 92 m deep. A seasonal thermocline was present, as well as the permanent well-defined chemocline around 60 m. Bottom waters are permanently anoxic. Total dissolved Hg [$(Hg_T)_D$] was highest in the surface waters (9.1 pM), lowest near 40 m, and there was a subsurface peak just below the chemocline (6 pM). Dissolved reactive Hg [$(Hg_R)_D$] was highest within the subsurface productive zone (3.9 pM at 20 m) and decreased with depth. Dissolved monomethylmercury (MMHg$_D$), undetectable in the first 20 m, increased to 140 fM at the chemocline. The average concentration in the anoxic zone was 144 ± 30 fM (n=5). No dimethylmercury (DMHg) was detected (<25 fM). Maxima in total particulate Hg (Hg$_P$) and particulate MMHg (MMHg$_P$), 5.2 pM and 242 fM, respectively, were found in the upper reaches of the chemocline, where a nepheloid layer was observed. Mercury enters the lake principally via atmospheric deposition and is transported downward by large particles and accumulates at the chemocline. The peak in MMHg$_P$ at the interface could be the result of in situ production, or could result from the accumulation of sinking particulate matter. The low concentrations of MMHg$_D$ in the oxic region, and the higher concentrations below the chemocline, are consistent with other aquatic studies and indicate that MMHg$_D$ is formed within these low oxygen waters. The results of this study demonstrate that, even though the Hg$_T$ concentrations are relatively low in this lake and the pH is relatively high (>6), MMHg is formed. Furthermore, low oxygen conditions seem to promote the formation of MMHg$_D$.

1–56670–066–3/94/$0.00 + $.50
© 1994 Lewis Publishers

I. INTRODUCTION

Elevated mercury levels, frequently exceeding public health guidelines (1 μg g^{-1}) have been found in fish from lakes far from urban and industrial influences.[1] More than 80% of the mercury in fish is present as monomethylmercury (MMHg), which is not produced by the living animal tissue[2,3] but rather is accumulated from water and food. Mass balance calculations for total mercury in a remote Wisconsin lake show that MMHg is formed primarily by in-lake processes.[4] Thus it is important that we better understand the biogeochemical cycling of mercury, particularly the cycle of its methylated forms, within natural waters.

The results of studies in northern Wisconsin,[4] Sweden,[3] and elsewhere[7] suggest that low pH and oxygen concentrations are the principal water characteristics associated with mercury methylation in lakes.[5] This has been confirmed by field observations: MMHg levels increase seasonally when waters become oxygen-depleted in an acidified lake.[6]

We selected a mountain lake remote from anthropogenic mercury sources, and which is characterized by the presence of a permanently anoxic monimolimnion and a relatively high pH,[8] for further study of the relationship between oxygen, pH, and MMHg production. Results demonstrate that even though the total mercury concentrations are relatively low and the pH is high, MMHg is formed within the lake and formation is promoted by low oxygen conditions.

II. METHODS

A. STUDY SITE

Pavin Lake (45°55' N and 2°54'E) is located at 1197 m altitude in the Mont-Dore range, part of the Massif Central (France). Its hydrology and geochemistry has been described by Martin.[8] This crater lake was formed 3500 years ago and represents the most recent volcanic activity in the Massif Central. It is quite circular with a mean diameter of 750 m (area: 0.44 Km^2) and relatively deep (92 m) (Figure 1). Its drainage basin is very small (0.8 Km^2) and covered by beeches and conifers. The limnic ecosystem is characterized by the predominance of diatoms.[9]

The water structure of Pavin Lake consists of two stratified layers. The mixolimnion, affected by winter mixing, is oxygenated and the monimolimnion is permanently anoxic. The upper layer receives 58 L s^{-1} via rainfall, runoff, and surficial springs, while underlacustrine freshwater springs have been estimated to supply 40 L s^{-1}, making the bottom water heavily mineralized.[8] There is an estimated exchange between the mixolimnion and the monimolimnion of 5 L s^{-1}. Calculation of the mean residence times in the two compartments shows that there is a slower renewal of the upper layer (6.1 years) compared to the deep waters (1.7 years), the meromixis of the lake being maintained by a chemocline around 60 m.

B. SAMPLING

Water was collected using ultraclean metal-free techniques between the 7th and 11th of August 1990 from a plastic boat. Two Teflon® 500 mL Mercos® samplers,[10] and one or two 5 L Teflon®-coated Go-Flo® bottles were filled for each of the 10 depths sampled (10, 20, 30, 40, 50, 60, 65, 70, 80, and 85 m). Additional surface samples were taken by hand using arm-length polyethylene gloves. The samplers were set on a PVC-coated stainless steel hydrowire equipped with a resin-coated lead weight and PVC-coated messengers. Water from one Mercos® sampler was used on board for stripping dimethylmercury (DMHg) with Hg-free Ar (trace Hg was removed by passing the gas over gold-coated sand), while the other sampler was used for field measurements of conductivity, temperature, and pH immediately after sampling.

The Go-Flo® bottles were used for collecting waters for the other parameters. After subsampling for O$_2$, on-line filtrations for particulate and dissolved mercury species were

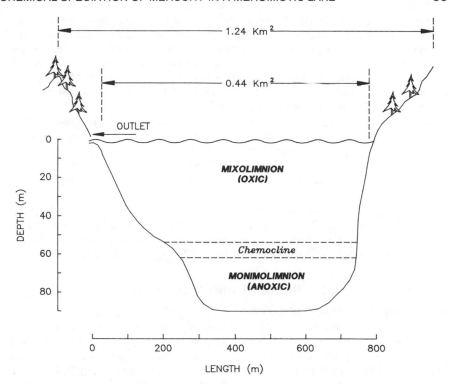

Figure 1 Schematic representation of the Pavin crater lake. (Adapted from Martin, J.M., *Chemical Processes in Lakes,* Stuum, W., Ed., John Wiley & Sons, New York, 1985, With permission.)

performed on board by pressurizing (2.5 bar) the Go-Flo® bottles with Hg-free Ar and filtering through quartz fiber filters (QM® from Mililipore which had been acid washed, rinsed, and heated overnight at 600° C) held in polypropylene filter holders. This method allowed the collection of water samples from the anoxic layer without any visible precipitation of hydroxides. Samples were also collected on-line using pre-cleaned Nuclepore® membranes (0.4 μm) for dissolved Fe, Mn, and suspended particulate matter measurements. Dissolved Fe and Mn were determined by flame atomic absorption spectrometry, and oxygen by the standard Winkler method.

Waters sampled for dissolved monomethylmercury ($MMHg_D$) and dissolved reactive mercury [$(Hg_R)_D$] determinations were stored at 4° C in tightly sealed PFA Teflon® bottles wrapped in polyethylene bags until analysis, which was performed within 8 hours of collection. Samples collected for total dissolved mercury ($Hg_T)_D$ were acidified in the field with concentrated HCl (Suprapur®, Merck) to yield a 0.5% acid solution.

Teflon® bottles were washed prior to use with a concentrated HCl/HNO_3 mixture, using the method described by Gill and Fitzgerald.[11] Conventional plasticware was leached for one week in 50% HNO_3, rinsed with MilliQ® water and leached again for one week in 10% HNO_3 at 50° C for another week; they were finally abundantly rinsed with MilliQ® water.

C. ANALYSIS

Mercury speciation has been divided in six fractions: total dissolved mercury ([$Hg_T]_D$), dissolved "reactive" or "easily reducible" mercury ([$Hg_R]_D$), particulate mercury (Hg_P), dissolved monomethylmercury ([$MMHg]_D$), particulate monomethylmercury ($MMHg_P$), and dimethylmercury (DMHg). All Hg species were detected by atomic fluorescence spectroscopy (AFS) after transformation to Hg^0, using a Merlin® atomic fluorescence monitor (PSA Instrument) or a University of Connecticut Analyzer.[12] $[Hg_T]_D$ was measured after reduction by $NaBH_4$ and Au amalgamation, following the method described by Gill and Bruland.[13] $[Hg_R]_D$ was also obtained by gold amalgamation, but after reduction with $SnCl_2$.[11] Particulate Hg was determined after an HCl/HNO_3 digestion in PFA Teflon® bombs and subsequent reduction by $SnCl_2$ without preconcentration.[14]

Dissolved methylated Hg species were determined by cryogenic chromatography and AFS quantification. DMHg was sparged from an unamended sample while an ethylation derivatization step was required to convert $MMHg_D$ to a volatile form prior to degassing.[15] $MMHg_P$ was analyzed similarly after solubilization with 5 mL of 30% KOH and pH adjustment to 5. The stripping gas and carrier gas used for cryogenic chromatography were Hg-free Ar and He.

Overall procedural blanks were 95 ± 20 pg, 30 ± 5 pg, and 5 ± 2 pg for $[Hg_T]_D$, $[Hg_R]_D$, and $MMHg_D$, respectively. The corresponding detection limits (DL: three times the standard deviation of the blank expressed per unit sample volume analyzed) were 600, 300, and 25 fM. Other DLs were 300 fM for Hg_P, 10 fM for $MMHg_P$, and 25 fM for DMHg.

III. RESULTS AND DISCUSSION

A. CHEMICAL CHARACTERISTICS

Conductivity and O_2 distributions (Figure 2) were typical of this meromictic lake.[8] At the chemocline, between 60 and 70 m, the conductivity jumped from 84 to 676 µS. The O_2 concentration was supersaturated in the first 30 m (up to 175% near the surface) and decreased continuously with depth to zero at 65 m. Waters below 65 m were anoxic. The sharp redox change was also visible in the dissolved Fe and Mn profiles (Figure 3), with concentration increasing from undetectable (<0.4 µM for Fe and <0.1 µM for Mn) in the mixolimnion up to 338 µM and 29 µM, respectively, at 60 m. The pH was quite high in the upper layer of the mixolimnion, with values higher than 8, but decreased to 6.6 near 60 m and remained constant in the anoxic zone.

Temperature showed, in addition to the summer thermocline (a 13° C gradient in the first 20 m below the surface), a slight increase below the minimum value found at 50 m. This increase has been observed previously and reflects hot spring activities at the bottom of this lake.[8]

B. TOTAL AND REACTIVE MERCURY

Total dissolved Hg in the Pavin Lake waters ranged from 2.8 to 9.1 pM with a mean concentration of 5.1 ± 2.2 pM, while $(Hg_R)_D$ concentrations ranged from 0.7 to 3.9 pM with a mean of 1.6 ± 1.0 pM (Table 1). Few data are available from other lakes for comparison. Fitzgerald and Watras[16] found 4.7 to 9.7 pM for Hg_T in unfiltered waters from Wisconsin lakes, while Gill and Bruland[13] measured 2.1 to 63 pM Hg_T in filtered samples from pristine alpine lakes and contaminated systems from California. The Hg_R concentrations are similar to those found in the Wisconsin lakes (0.7 to 2.9 pM)[6] and compare with the lowest levels measured by Gill and Bruland[13] in Californian freshwaters. Thus, Pavin Lake can be categorized in the pristine lakes group on the basis of its total and reactive dissolved mercury concentration ranges (Table 2).

The highest $(Hg_T)_D$ concentration occurred in the surface waters (9.1 pM; Figure 4) and the concentration decreased to a minimum near 40 m (2.8 pM) before increasing again to a subsurface maximum just below the chemocline (6 pM). Total dissolved Hg then decreased down to 3.3 pM near the bottom of the lake. The vertical distribution of $(Hg_R)_D$ (Figure 4) peaked just below the thermocline, within the subsurface productive zone (3.9 pM at 20 m),

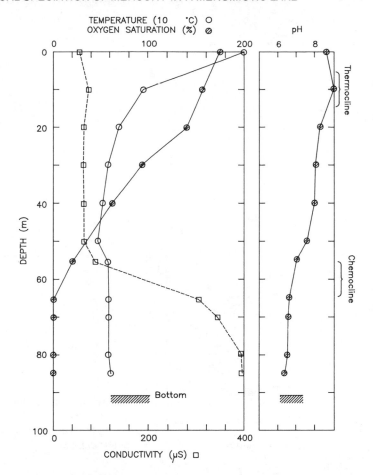

Figure 2 Distribution of temperature, dissolved oxygen, and pH during the August 1990 sampling of Pavin Lake.

but immediately decreased with depth and remained low in the anoxic waters (1.2 ± 0.5 pM; n = 4). Particulate mercury was low in the mixolimnion (1.0 to 2.3 pM) and higher in the anoxic monimolimnion (3.8 to 5.2 pM) (Figure 5). The maximum Hg$_P$, in the upper reaches of the chemocline, coincides with the nepheloid layer (Figure 6), while a minimum (1.4 pM) was found where the suspended matter concentration was lowest (50 m) (Figures 5 and 6). The peak in Hg$_P$ also coincided with the peak in particulate Fe. These observations are similar to those seen by Mason et al.[7] in the Pettaquamscutt Estuary, Rhode Island, and by the studies in the northern Wisconsin lakes.[4] The results suggest that the cycling and distribution of dissolved and particulate Hg is closely linked to that of Fe, and is strongly influenced by the redox chemistry of the system, as demonstrated by a recent study of coastal marine sediments.[17]

Vertical profile of (Hg$_T$)$_D$ (Figure 4) suggests that mercury enters the lake principally via atmospheric deposition, mostly as dissolved species (<30% Hg$_P$). The data also suggest that mercury is transported downward by large particles and accumulates at the chemocline, within the nepheloid layer. In the oxic waters, Hg$_P$ averages $29 \pm 11\%$ (n = 5) of the Hg$_T$, while Hg$_P$ is a larger fraction of the total in the bottom waters ($50 \pm 6\%$; n = 4).

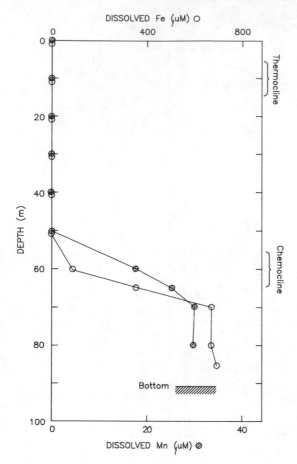

Figure 3 Distribution of dissolved iron and manganese during the August 1990 sampling of Pavin Lake.

Table 1 **Mercury species distributions in the water column of Pavin Lake.**

Depth (m)	$(Hg_T)_D$ (pM)	$(Hg_R)_D$ (pM)	$MMHg_D$ (fM)	Hg_P (pM)	$MMHg_P$ (fM)
0	9.12	—	<25	—	191
10	8.52	2.74	<25	2.29	130
20	4.43	3.85	<25	1.40	51
30	3.94	0.72	63	1.74	47
40	2.80	0.80	84	2.39	33
50	5.18	1.16	140	1.05	135
60	6.03	1.32	140	5.19	242
65	4.88	0.62	140	3.79	70
70	—	1.02	130	3.79	116
80	3.29	1.85	195	3.89	65
85	3.34	—	116	4.24	88

Note: $(Hg_T)_D$: Total dissolved mercury; $MMHg_D$: Dissolved monomethylmercury; $(Hg_R)_D$: Reactive dissolved mercury; $MMHg_P$: Particulate monomethylmercury; and Hg_P: Total particulate mercury.

Table 2 **Proportion of the various mercury species in the water column of Pavin Lake.**

Depth (m)	$(Hg_R)_D/(Hg_T)_D$ (%)	$MMHg_D/(Hg_T)_D$ (%)	$MMHg_P/Hg_P$ (%)
0	—	<0.3	—
10	32	<0.3	5.7
20	87	<0.6	3.6
30	18	1.6	2.7
40	29	3.0	1.4
50	22	2.7	12.9
60	22	2.3	4.7
65	13	2.9	1.8
70	—	5.9	3.1
80	56	3.5	1.7
85	—	—	2.1

Note: $(Hg_T)_D$: Total dissolved mercury; $MMHg_D$: Dissolved monomethylmercury; $(Hg_R)_D$: Reactive dissolved mercury; $MMMHg_P$: Particulate monomethylmercury; and Hg_P: Total particulate mercury.

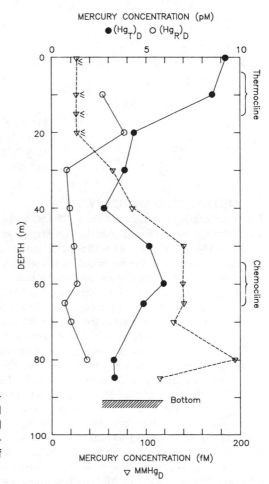

Figure 4 Distribution of total dissolved mercury $(Hg_T)_D$, dissolved reactive $(Hg_R)_D$, and dissolved monomethylmercury $MMHg_D$, during the August 1990 sampling of Pavin Lake.

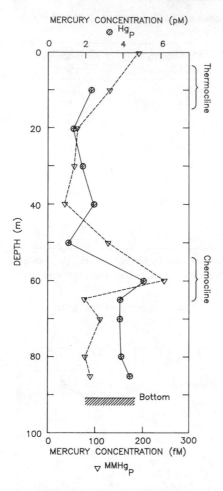

Figure 5 Distribution of total particulate mercury (Hg_P) and particulate monomethylmercury ($MMHg_P$) during the August 1990 sampling of Pavin Lake.

C. METHYLATED MERCURY

Dissolved MMHg was undetectable (<25 fM) in the first 20 m, then increased steadily to 140 fM at the chemocline (Figure 4). The average concentration of $MMHg_D$ in the anoxic zone was 144 ± 30 fM (n = 5), with a maximum concentration at 80 m (195 fM). Particulate MMHg was highest in the nepheloid layer and was minimum in the low suspended load zone (40 m), having a distribution very similar to that of Hg_P (Figures 5 and 6). No evidence of DMHg was found (<25 fM) in this lake. Total MMHg ($MMHg_D$ + $MMHg_P$) concentrations in Pavin Lake waters are comparable to the lower range of the concentrations measured in other lakes: <0.5 to 2.0 pM in three Swedish lakes,[16] and 0.2 to 50 pM in five lakes from Wisconsin and New York states.[6,18,19] The latter authors also reported higher MMHg concentrations in anoxic waters. However, in the permanently stratified Pettaquamscutt Estuary, the highest concentrations were found at the redox interface.[7]

The peak in $MMHg_P$ at the interface could be the result of in situ production, or could result from the accumulation of sinking particulate matter [$MMHg_P$ averages 90 ± 68 fM (n = 5) in the upper 40 m]. Particulate methylmercury averages $3.5 \pm 2\%$ of the Hg_P in the oxic region, $2.8 \pm 1.3\%$ in the waters below 60 m, and has a maximum relative concentration (13%) at 50 m, suggesting formation of the $MMHg_P$ at this depth. However, the peak in %$MMHg_P/Hg_P$ could also result from diffusion of $MMHg_D$ across the chemocline, followed by scavenging by particulate formed due to iron/manganese precipitation at the oxic/anoxic interface. The maximum in $MMHg_P/Hg_P$ coincides with the Mn_P maximum (Figures 5

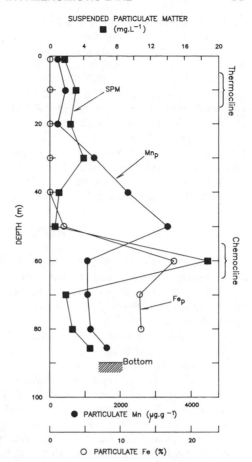

Figure 6 Distribution of suspended particulate matter, particulate iron, and particulate manganese during the August 1990 sampling of Pavin Lake.

and 6). This correspondence suggests that the sorption of MMHg occurs in conjunction with Mn precipitation. $MMHg_D$ likely diffuses upward from the higher concentration in the anoxic waters into the oxic zone, where it is scavenged by Mn precipitation. However, the maximum $MMHg_P$ concentration was found at the depth of the maximum Fe_P, suggesting that Fe precipitation is also scavenging Hg species from the water column. Thus, it seems likely that the redox cycling of Fe and Mn play an important role in the scavenging and recycling of Hg species at the chemocline.

The low concentrations of $MMHg_D$ in the oxic region, and the higher concentrations below the chemocline, are consistent with other studies in freshwater lakes in Wisconsin,[4,6,19] estuarine (Pettaquamscutt Estuary, Rhode Island, U.S.),[7] and open ocean studies (the equatorial Pacific[20] and Alboran Sea, Mediterranean[21]), and suggest that methylated Hg is formed within the low oxygen waters. While there is no information available concerning the bacterial community in the lake and its relationship to MMHg formation, the results of this study corroborate the other studies[7,19–21] and provide further evidence that the concentration and distribution of MMHg depend not only on the biological processes, but are also influenced by the redox cycling of Fe and Mn at the oxic/anoxic interface.

IV. CONCLUSIONS

This study demonstrates that, even though the Hg_T concentrations are relatively low in this lake (average 7.6 ± 2.2 pM for the entire water column), and the pH is relatively high (>6),

MMHg is formed. Furthermore, low oxygen conditions seem to promote the formation of methylated Hg. Again, these conclusions are in agreement with findings from marine environments [20,21] where relatively low amounts of methylated Hg (<1 pM) were found in the low oxygen regions. To date, the highest MMHg concentrations in remote aquatic systems have been found in the low pH lakes (pH <6) in northern Wisconsin[4,19] and in the lakes of Sweden.[3] These results, and the low MMHg concentrations in the high pH Pavin Lake, confirm the ubiquity of the inverse pH-MMHg relationship already observed in the water column and at the sediment-water interface.[5]

Thus, the overall cycling of Hg in Pavin Lake is similar to what has been observed in other systems.[3,4,7,16,19–21] Methylation occurs primarily within the low oxygen region of the lake, but the final distribution of MMHg species depends both on the site of formation and on the redox cycling of Fe and Mn at the oxic/anoxic interface. This study further demonstrates that Hg is methylated in both remote and impacted aquatic systems.

REFERENCES

1. Björklund, I., Borg, H., and Johansson, K., Mercury in Swedish Lakes—regional distribution and causes, *Ambio,* 13, 118, 1984.
2. Huckabee, J.W., Elwood, J.W., and Hildebrand, S.G., Accumulation of mercury in freshwater biota, in *The Biogeochemistry of Mercury,* Nriagu, J.O., Ed., Elsevier/North Holland, Amsterdam, 1979, 277.
3. Lindqvist, O., Mercury in the Swedish environment, *Water Air Soil Pollut.,* 55, 1, 1990.
4. Watras, C. J., Mercury in Temperate Lakes, 1990 Annual Report, Electric Power Research Institute, Palo Alto, CA, 1991.
5. Winfrey, M.R. and Rudd, J.W.M., Environmental factors affecting the formation of methylmercury in low pH lakes, *Environ. Toxicol. Chem.,* 9, 853, 1990.
6. Bloom, N.S. and Watras, C.J., Seasonal and vertical variability in the mercury speciation of an acidified Wisconsin lake, in *Heavy Metals in the Environment,* Vernet, J.-P., Ed., Proc. 7th Int. Conf., CEP Consultants, Edinburgh, 1989.
7. Mason, R.P., Fitzgerald, W.F., Hurley, J., Hanson, A.K., Donaghay, P.L., and Sieburth, J.M., Mercury biogeochemical cycling in a stratified estuary, *Limnol. Oceanogr.,* 38, 1227, 1993.
8. Martin, J.-M., The Pavin Crater Lake, in *Chemical Processes in Lakes,* Stuum, W., Ed., John Wiley & Sons, New York, 1985, 169.
9. Romagoux, J.-C., Caractéristiques du microphytobenthos d'un lac volcanique méromictique (Lac Pavin, France), *Int. Revue Ges. Hydrobiol.,* 65, 849, 1980.
10. Freiman, P., Schmidt, D., and Schomakere, K., Mercos—A simple Teflon sampler for ultra-trace metal analysis in seawater, *Mar. Chem.,* 14, 43, 1983.
11. Gill, G.A. and Fitzgerald, W.F., Picomolar mercury measurement in seawater and other materials using stannous chloride reduction and two-stage gold amalgamation with gas phase detection, *Mar. Chem.,* 20, 227, 1987.
12. Bloom, N.S. and Fitzgerald, W.F., Determination of volatile mercury species at the picogram level by low temperature gas chromatography with cold-vapor atomic fluorescence detector, *Anal. Chim. Acta,* 208, 151, 1988.
13. Gill, G.A. and Bruland, K.W., Mercury speciation in surface freshwater systems in California and other areas, *Environ. Sci. Technol.,* 24, 1392, 1990.
14. Cossa, D. and Martin, J.-M., Mercury in the Rhône delta and adjacent marine areas, *Mar. Chem.,* 36, 291, 1991.
15. Bloom, N.S., Determination of picogram level of methylmercury by aqueous phase ethylation, followed by cryogenic gas chromatography with atomic fluorescence detection, *Can. J. Fish. Aquat. Sci.,* 46, 1131, 1989.

16. Lee, Y.-H. and Hultberg, H., Methylmercury in some Swedish surface waters, *Environ. Toxicol. Chem.,* 9, 833, 1990.
17. Gobeil, C. and Cossa, D., Mercury in sediments and sediment pore waters in the Laurentian trough, *Can. J. Fish. Aquat. Sci.,* 50, 1794, 1993.
18. Hurley, J.P., Watras, C.J., and Bloom, N.S., Mercury cycling in a northern Wisconsin seepage lake. The role of particulate matter in vertical transport, *Water Air Soil Pollut.,* 56, 543, 1991.
19. Bloom, N.S., Watras, C.J., and Hurley, J.P., Impact of acidification on the methylmercury cycle of remote seepage lakes, *Water Air Soil Pollut.,* 56, 477, 1991.
20. Mason, R.P. and Fitzgerald, W.F., Alkylmercury species in equatorial Pacific, *Nature, London,* 347, 457, 1990.
21. Cossa, D. and Martin, J.-M., Dimethylmercury in the sub-thermocline region of the Alboran Sea and Strait of Gibraltar, in *Proc. European River Ocean System (EROS-2000) Fourth Workshop on the Western Mediterranean Sea,* Plymouth, UK, Sep. 28–Oct. 2, 1992, Martin, J.-M. and Barth, H., Eds., *Water Pollut. Res. Rep. CEC,* Brussels, 30, 139, 1993.

Distribution and Flux of Particulate Mercury in Four Stratified Seepage Lakes

James P. Hurley, Carl J. Watras, and Nicolas S. Bloom

CONTENTS

ABSTRACT: The influence of particle-mediated transport in regulating mercury (Hg) cycling in four northern Wisconsin seepage lakes was examined by analyzing total Hg (Hg_T) and monomethyl Hg (MeHg) in suspended and sedimenting material during 1990. Annual mean suspended particulate Hg_T content ranged from 0.12 to 0.68 $\mu g\ g^{-1}$ while mean MeHg content ranged from 0.047 to 0.16 μg^{-1}. Among the study lakes, mean sedimenting flux rates for the ice-free period ranged from 126 to 305 ng $m^{-2}d^{-1}$ for Hg_T and from 2.2 to 34 ng $m^{-2}d^{-1}$ for MeHg, suggesting greater variability in MeHg than Hg_T fluxes. In all lakes, MeHg content of suspended particulates increased during summer stratification. Similarly, settling particles collected in sediment traps exhibited MeHg increases during this period. Upon settling, particles became depleted in MeHg relative to Hg_T, emphasizing differences in cycling between these forms of Hg. Trends in particle MeHg content among lakes were also consistent with observed trends in yellow perch tissue Hg content.[1]

I. INTRODUCTION

Recent evidence from mercury cycling studies in seepage lakes of north-central Wisconsin[2] has shown that by using ultraclean methods for field sampling and sample preparation,[3] coupled with low-level (picomolar) analytical techniques,[4-6] patterns of seasonal and spatial variability in Hg cycling in the water column of lakes can be observed. For instance, in stratified lakes that develop anoxia in the hypolimnion during summer stratification, large increases of both total Hg (Hg_T) and monomethyl Hg (MeHg) are evident at depth.[7-9]

Similarly, analyses of volatile Hg (Hg^0) in lakes at femtomolar levels reveal strong seasonal response.[10] By applying these new methodologies to this and similar studies in other lakes[11] a better understanding of Hg cycling in lakes is beginning to emerge.

Current knowledge of the processes controlling MeHg formation and accumulation in fish of low pH and low alkalinity lakes has been reviewed by Spry and Wiener[13] and Winfrey and Rudd.[14] In Wisconsin, a survey of 1,103 walleyes (*Stizostedion vitreum vitreum Mitchill*) from 219 lakes has shown that a large proportion of legal angling size fish from low acid-neutralizing capacity (ANC) lakes exceeded the state health advisory level of 0.5 $\mu g\ g^{-1}$ Hg.[15] More detailed sampling and analyses of yellow perch (*Perca flavescens*) from seven MTL study lakes has shown similar increases in Hg fish content with decreasing pH.[1] Since Hg present in fish tissue is predominantly in the methyl form,[16,17] it becomes increasingly important to better understand transport and uptake mechanisms of MeHg which lead to eventual bioconcentration.

Particle phase uptake can be an important vector for contaminant transport in the environment.[18] Uptake of trace metals by phytoplankton may represent the initial stages of trophic transport by providing a food source for higher levels such as zooplankton or planktivorous fish. In two basins of the experimentally acidified Little Rock Lake, Watras and Bloom[19] found that the proportion of Hg_T in the MeHg form increased from 5 to 12% in lake water, 13 to 31% in lake seston, 29 to 91% in zooplankton, to >98% in yellow perch. Furthermore, Hg_T in zooplankton of the acidified basin is comprised of a greater proportion of MeHg than their counterparts in the unacidified reference basin. These results suggest both an effect of acidification and the importance of trophic transfer and biocencentration on Hg cycling.

In this study, we evaluate the seasonal dependence of sestonic content and particulate deposition of Hg_T and MeHg in four stratified seepage lakes of northern Wisconsin. In examining particulate distribution and flux from the water column, we attempt to identify lake conditions or events which may regulate Hg cycling in lakes. Details of trends in dissolved Hg_T and MeHg in the study lakes are given elsewhere.[20]

II. METHODOLOGY

A. FIELD SITES

The four lakes chosen for this study were chosen as a subset of lakes from the Mercury in Temperate Lakes (MTL) project.[1] The original design of MTL included lakes along both a pH and dissolved organic carbon (DOC) gradient. Three lakes (Max, Vandercook, Little Rock Reference) are not included in this discussion because their shallow character (<7 m) and lack of thermal stratification made them unsuitable for sediment trap studies. Our study lakes are located in noncalcareous glacial outwash in the Northern Highlands region of north-central Wisconsin. All of the lakes are seepage lakes that receive hydrologic inputs from rainfall and subsurface flow. Physical and chemical characteristics are summarized in Table 1.

B. WATER COLUMN SAMPLING

The water column of each lake was sampled five times from March through November, 1990. Sampling was designed to incorporate the following lake conditions: maximum ice cover (March); spring mixis (after ice-out, April); mid-stratification (July); late stratification; and, fall mixis. Sampling dates may vary by up to two months among lakes due to differences in the breakdown of thermal stratification in the fall. In shallow Little Rock Treatment Basin (10 m), the water column became well mixed in mid-September. Russet Lake, with a similar maximum depth, is well protected from prevailing winds and did not fully mix until November, as did the deeper Pallette and Crystal Lakes.

Particulate Hg_T, which includes both organic and inorganic Hg, and monomethyl Hg were measured at numerous depths in the water column. Water samples were pumped from depth

Table 1 **General physical and chemical characteristics of study lakes.**

Lake	Surface area (ha)	Z_{max} (m)	Z_{mean} (m)	pH	ANC	DOC	Anoxic hypolimnion
Little Rock treatment	9.8	10.3	3.8	5.0	−4.5	1.1	Yes
Pallette	70	18.2	9.6	7.0	127	5.3	Yes
Russet	19	11.0	2.9	5.8	27	6.8	Yes
Crystal	36	20.4	10.4	6.3	25	1.7	No

Note: Acid neutralizing capacity (ANC) in units of µeq/L and dissolved organic carbon (DOC) in mg/L.[20]

using acid-cleaned ¼″ O.D. Teflon® tubing. Particulate Hg was collected by pumping lake water through precombusted (12 h at 550° C to remove Hg, then stored in a Teflon® jar) quartz fiber filters (Whatman #1851, nominal pore size 0.8 µm) housed in acid-cleaned Teflon® filter holders. After sampling, filters were frozen in their filter holders and sent to the laboratory packed in dry ice. Particulate mass was determined by filtering known volumes of lake water through preweighed 0.8-µm Nuclepore® filters. All mass concentration values are expressed on a dry weight basis.

C. SEDIMENTATION

Sediment traps[21] were used to estimate the downward flux of particles toward the sediment-water interface. These were acid-cleaned acrylic and Teflon® sediment traps, 90 cm tall, with a 4:1 aspect ratio.[22] At the bottom of the collection area, the trap is funneled into a 250 ml Teflon® bottle, designed to reduce the effects of "swimmers"[23] in traps. Traps were positioned in the hypolimnion at the maximum lake depth, with the bottom of the trap collection area 0.5 m above the sediment surface. Duplicate nonpoisoned traps were set for two-week intervals, staggered throughout the ice-free period of 1990 in the four study lakes. Samples from October were lost in Crystal Lake. Collected material was sieved through a 253-µm precleaned Nitex sieve to remove larger zooplankton and invertebrates from settling phytoplankton and detrital material.

Aliquots of trap material were transferred to precleaned Teflon® vials and shipped to the laboratory where they were filtered onto Whatman quartz fiber filters for Hg analyses. Additional aliquots were taken for mass estimates and pigment analyses. The remainder of the sample was freeze-dried and used for determination of total Fe, Mn, and S.

D. ANALYSES

Particulate Hg_T was determined by wet oxidation of samples, reduction with $SnCl_2$, volatilization, and purging onto a gold-coated sand trap. The sample was then desorbed off the trap at 450° C and carried to a cold-vapor atomic fluorescence detector with Hg-free carrier gas.[5,24] Particulate MeHg was determined on samples predigested in 25% KOH in methanol, followed by aqueous phase ethylation, cryogenic gas chromatographic separation, and cold vapor atomic fluorescence detection.[6]

To estimate the influence of biogenic material to the settling particulate flux, we measured the pigment content of sediment trap particulates. High performance liquid chromatographic (HPLC) techniques[25,26] enabled the separation of major chlorophyll *a* derivatives (chlorophyll *a*, chlorophyllide *a*, pheophorbide *a*, pheophytin *a*—summed here as total *a* phorbin) and bacteriochlorophylls *d* and *e*. In three of the study lakes which develop strong anoxia (Little Rock, Pallette, and Russet Lake), phototrophic sulfur bacteria[27] are major components of hypolimnetic suspended particulates.[26,28] This technique does not account for nonphotosynthetic bacterial sedimentation.

Particulate Fe, Mn, and S were determined on digested (HCl, HNO$_3$) sediment trap material. Analyses were conducted by Inductively Coupled Plasma Atomic Emission Spectrophotometry (ICP-AES) at the University of Wisconsin Soil and Plant Analysis Laboratory.

III. RESULTS AND DISCUSSION

A. SUSPENDED PARTICULATE MATTER

Mean volume-weighted particle mass (mg L^{-1}) and mass-weighted mean Hg$_T$ and MeHg content (μg g^{-1}) of suspended particulates were measured five times during 1990. Volume-weighted particle mass accounts for hypsometric differences among study lakes. Mass weighted mean concentrations for a given lake were used to allow for normalization of concentrations based on the total mass of particulate matter. Mean lake particle mass (volume-weighted) was calculated based on hypsometry (based on the volume of water that the sample represented). Mean particle content was then calculated by multiplying each observed particle concentration (μg g^{-1}) by the hypsometrically weighted particle masses to determine the mean for the entire water column. If one were to view particulate matter solely on a ng L^{-1} basis, increases in Hg could be reflective of merely increased numbers of particles present. Instead, mass-weighted particle content (μg g^{-1}) was used to view seasonal differences in Hg content based on the number of particles in the water column on selected dates. Particulate Hg$_T$ and MeHg burdens on a ng L^{-1} basis can be calculated by multiplying the particle content (μg g^{-1}) by the suspended particle mass (mg L^{-1}) in Figure 1.

In all four lakes, strong seasonal responses in particulate matter and corresponding Hg particle content, and major differences in Hg particle content among the lakes were observed. Mean particle mass was generally lowest under ice cover in March and increased throughout the ice-free period (about April 15 to November) with highest particle masses in most lakes near fall overturn (Figure 1). In Little Rock and Russet, however, lowest particle masses were observed following fall overturn, which may in part be due to increased grazing of plankton by zooplankton. In all of the study lakes, suspended particle masses generally reflected biogenically derived particles in the water column (strongly correlated with chlorophyll a levels),[20] with the exception of fall overturn where resuspended sediment may be important. Mean particle mass ranged from 0.3 to 3.2 mg L^{-1} (Table 2). At individual depths, particle masses can exceed 30 mg L^{-1}, especially in the hypolimnion of Little Rock Treatment Basin.[20]

Hg$_T$ levels of particulates suspended in the water column in all of the study lakes ranged from 0.05 to 1.10 μg g^{-1} Hg$_T$, with the highest annual mean values in Russet Lake (0.68 μg g^{-1}) and the lowest in Pallette Lake (0.12 μg g^{-1}) (Table 2). MeHg levels ranged from 0.019 to 0.371 μg g^{-1}. Similar to Hg$_T$ particle content, the lowest MeHg annual mean was observed in Pallette Lake (0.047 μg g^{-1}) and highest levels in Russet Lake (0.16 μg g^{-1}). In all of the study lakes, the highest particle MeHg content was observed under ice cover when particle levels were generally low. While Hg$_T$ particle content tends to decrease during the summer with increasing particle loads (with the exception of Little Rock), MeHg particle content increased in all lakes during summer stratification. If dissolved MeHg formation is occurring at increased rates during stratification, it appears to be reflected in the increasing MeHg content of particles.

B. SEDIMENTATION OF PARTICLES

Positioning of sediment traps 0.5 m above the sediment-water interface in the hypolimnion was designed so that particles collected in traps would be assumed similar to those particles deposited at the sediment-water interface. We acknowledge that additional degradation may take place in the water column below this depth, especially for particles smaller than 1 μm, but Stoke's Law calculations[29] suggest minimal retention of nonliving plankton remains in the water column below the trap depth. Calculated settling rates ranged from 2 to 15 cm d^{-1}

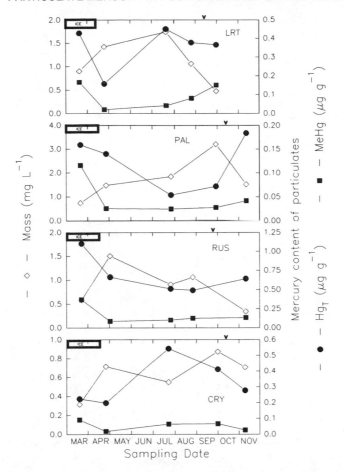

Figure 1   Total suspended particulate mass, Hg_T, and MeHg content of particles in the four study lakes. LRT = Little Rock Lake, PAL = Pallette Lake, RUS = Russet Lake, and CRY = Crystal Lake. V on upper axis denotes approximate date of fall overturn.

Table 2   **Suspended particulate matter in study lakes.**

Lake		Mass (mg L⁻¹)	Hg_T content (µg g⁻¹)	MeHg content (µg g⁻¹)
Little Rock treatment	Range	0.47–1.74	0.16–0.45	0.022–0.166
	Mean	0.89	0.36	0.092
Pallette	Range	0.74–3.2	0.054–0.18	0.025–0.12
	Mean	1.75	0.12	0.047
Russet	Range	0.34–1.50	0.49–1.10	0.085–0.37
	Mean	0.88	0.68	0.16
Crystal	Range	0.31–0.87	0.20–0.54	0.019–0.091
	Mean	0.63	0.33	0.053

Note: Total mass, Hg_T, and MeHg content, March through November, 1990.

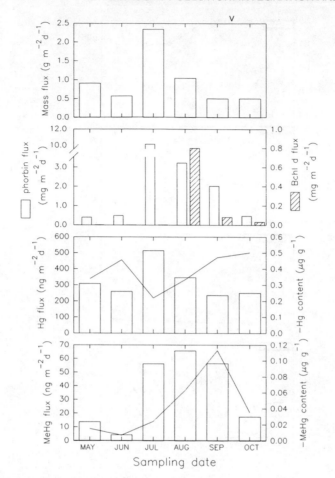

Figure 2 Mass flux, total Hg flux, and total Hg particle content, and methyl Hg and methyl Hg particle content determined from sedimentation traps. A: Little Rock Treatment Basin; B: Pallette Lake; C: Russet Lake; D: Crystal Lake; V on upper axis denotes approximate date of fall overturn.

for the smallest phototrophic bacteria (ca 1 μm radius) to 20 cm d^{-1} to ≥2 m d^{-1} for chrysophytes and cryptophytes (calculated as ca 7 μm diameter and accounting for shape factors). These results are conservative compared to sinking estimates of planktonic remains of about 50 cm d^{-1} to >10 m d^{-1} in a nearby lake.[30]

Mass deposition rates in Pallette, Russet, and Crystal Lake were similar (0.17 to 0.87 g m^{-2}d^{-1}—Figure 2; Table 3) while deposition rates in Little Rock Treatment were about two to three times greater (0.50 to 2.30 g m^{-2}d^{-1}). Our annual mass flux estimate for Little Rock Lake is similar to the rate estimated by Baker et al.[31] from 1985 through 1986. Of particular interest in Little Rock was an event in July which accounted for large increases in mass flux—three times greater than normal deposition rates in Little Rock or any of the other study lakes. Based on phorbin (next section) and carotenoid pigment analysis,[28] we attribute this event to deposition of green algae. A similar mass deposition event in Little Rock was also noted in July 1989.[8] In other lakes, a general trend of highest mass deposition at overturn periods and lowest deposition during late stratification was observed.

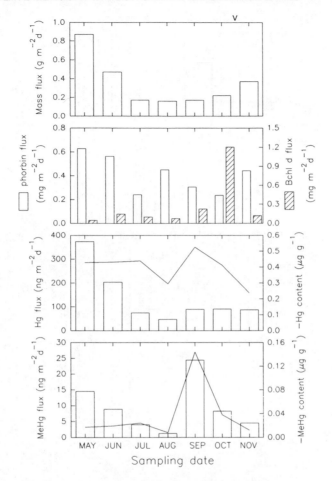

Figure 2 Continued.

Table 3 **Mass, Hg$_T$, methyl-Hg deposition rates, and particle Hg content of sediment trap material.**

Lake		Mass flux (g m^{-2}d^{-1})	Hg$_T$ flux (ng m^{-2}d^{-1})	Methyl Hg flux (ng m^{-2}d^{-1})	Hg$_T$ content (µg g^{-1})	Me-Hg content (µg g^{-1})
Little Rock	Range	0.50–2.30	240–510	4.2–66	0.22–0.50	0.01–0.11
treatment	Mean	0.89	305	34	0.41	0.044
Pallette	Range	0.17–0.87	47–350	1.2–24	0.24–0.52	0.008–0.14
	Mean	0.35	138	9.4	0.40	0.037
Russet	Range	0.21–0.49	110–540	7.0–57	0.36–1.10	0.023–0.12
	Mean	0.34	250	17	0.69	0.044
Crystal	Range	0.12–0.45	55–250	1.0–3.0	0.25–0.69	0.004–0.011
	Mean	0.31	126	2.2	0.42	0.008

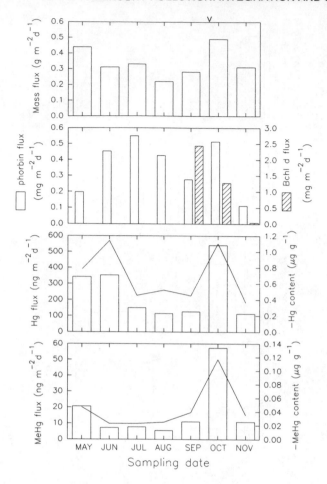

Figure 2   Continued.

Total a phorbin and bacteriochlorophyll analysis of sediment trap material was used to identify the type of planktonic material associated with deposition events. Differences in pigment fluxes among lakes are reflective of the magnitude of plankton blooms and sedimentation of plankton from the water column. Additionally, location of the plankton bloom to the sediment trap may affect pigment content,[28] as blooms which occur deep in the water column are more effectively preserved. Surface bloom sedimentation should result in greater pigment degradation during settling, therefore the influences of surface blooms may be underestimated by pigment fluxes. In our study lakes which develop strong anoxic hypolimnia and contain sufficient levels of sulfide, phototrophic sulfur bacteria are present. Therefore during stratification, settling phototrophic bacteria can be an important fraction of the mass flux in three of the four study lakes (Figure 2). However, on an annual flux basis, deposition of phytoplankton to the sediment surface is more important than phototrophic bacterial deposition in these lakes.

Hg_T deposition rates ranged from 47 to 540 ng m^{-2}d^{-1} for individual trapping periods. The lowest individual Hg_T flux was observed in Pallette Lake in August (also the lowest mass flux period), while the highest Hg_T flux was seen in Little Rock Lake and associated with the large deposition event in July. Mean Hg_T deposition rates ranged from 126 to 305 ng m^{-2}d^{-1}, with the lowest means in Crystal Lake and highest in Little Rock Lake.

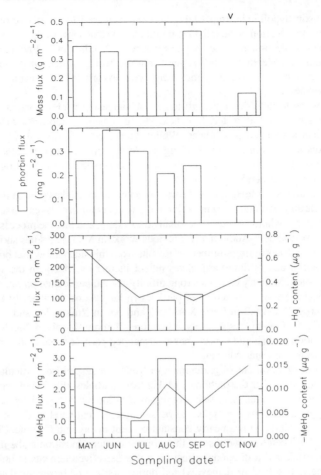

Figure 2   Continued.

Particle Hg_T content of sediment trap particulates (annual means ranged from 0.40 to 0.69 µg g^{-1} Hg_T) exhibited less variability than flux rates (annual mean fluxes were 126 to 305 ng m^{-2}d^{-1}) among lakes. Individual event Hg_T content ranged from 0.22 to 1.10 µg g^{-1} Hg_T. Mean Hg_T content ranged from 0.40 to 0.69 µg g^{-1}, with the lowest levels in Pallette Lake and highest in Russet Lake, consistent with trends in suspended particulates. Greater variability existed in suspended particulates than in trap material, yet the ranges of Hg_T content are similar.

In contrast to Hg_T, MeHg flux and particle content exhibited strong seasonal response. Flux rates ranged from 1.0 to 66 ng m^{-2}d^{-1}, with greater flux rates in all lakes observed during late stratification and at fall overturn. Similarly, trap particulate MeHg levels reach their highest values in late summer, with the exception of Crystal Lake. While MeHg content of trap particles in Little Rock, Pallette, and Russet Lakes typically exceed 0.02 µg g^{-1} (with a high of 0.14 in Little Rock) MeHg levels in Crystal lake trap material were consistently below 0.011 µg g^{-1} MeHg.

Particulate Hg_T and MeHg content and did not follow similar trends during stratification in all of the lakes. Although the maximum Hg_T flux in Little Rock Lake occurred during July, this was not true for MeHg deposition. However, as MeHg content in suspended particles increased in late summer, similar trends were noted in trap particulates. In lakes that developed anoxic hypolimnia, trap MeHg content increased four- to fivefold from early summer to late stratification in the fall. One might assume that some of the increases in MeHg particle content in the fall may be due to resuspended sediments. However, preliminary analyses of sediment MeHg content[32] suggest that the MeHg content of surficial sediments (0 to 2 cm) is quite low (<10 ng g^{-1}), thus increased trap particle MeHg cannot be accounted for by resuspension.

It is interesting to note that although trends of increasing MeHg content of suspended particulate are similar to trap material, particle MeHg content of trap particulates is substantially lower than suspended particulates. While mean levels of suspended particulates for individual lakes range from 0.053 to 0.16 µg g^{-1} MeHg, mean trap particulates only range from 0.008 to 0.044 µg g^{-1}. This contrasts with the observation of increasing Hg_T content in trap material over suspended particulates.

These observations of MeHg flux and particle content among lakes present four interesting hypotheses regarding particulate Hg behavior in the water column of lakes. First, other studies have shown that methylation of Hg is seasonally dependent, and that the process is stimulated during warmer periods.[33,34] Additionally, previous work on MTL lakes has shown increased dissolved MeHg in lakes during summer.[7,20] Results from this study show that both suspended particulate material and settling particulates reflect increasing MeHg in the water column during stratification. Mercury exhibits strong affinity for the particulate phase. Annual mean log Kd values (Kd [L kg^{-1}] = (conc. of particles [g kg^{-1}]/conc. in water [g L^{-1}]) for Hg_T and MeHg were 5.8 and 6.2 in Little Rock, 5.6 and 5.9 in Russet, 5.2 and 6.2 in Pallette, and 6.0 and 6.5 in Crystal, respectively. Therefore, if more dissolved MeHg is available during stratification, one should expect greater suspended particle content and greater vertical MeHg fluxes from the water column.

Secondly, our observations of enrichment of MeHg in suspended particulates over trap material may suggest either dissolution of MeHg from particles, dilution of trap material by other mineral or organic phases, or selective removal of methyl-rich particles before settling. Sediment trap material does not appear to be significantly enriched in Fe or Mn (Hurley, J. P., unpublished data) and has nearly constant content of organic C content in traps (310 to 490 mg g^{-1} C). Therefore, we assume that organic matter deposition is the major fraction of the trap material, and a dilutional effect by redox-sensitive elements is unlikely. Either MeHg is being redissolved or demeythylated during settling or bioaccumulated (i.e., zooplankton grazing).

Thirdly, our observations of MeHg content of seston among lakes is consistent with trends observed by Wiener et al.[1] for Hg content of yellow perch. In a comparison of lakes across a DOC gradient, yellow perch from Crystal Lake (low DOC) ranged from about 50 to 270 ng g^{-1} Hg_T, while yellow perch from Russet Lake ranged from about 100 to 460 ng g^{-1} Hg_T. Additionally, a comparison was made of lakes across a pH gradient. Yellow perch from Pallette Lake (high pH) had a mean Hg_T content of 30 ng g^{-1} while perch from Little Rock Treatment had a mean Hg_T content of 130 ng g^{-1}. If all of the Hg in perch is in the methyl form, and our suspended particulates are reflective of plankton, these results suggest increasing MeHg content as one proceeds to higher trophic levels. Other lakes were included in the Wiener et al.[1] comparison, but are not discussed here because of the lack of thermal stratification and the unsuitability for sediment trapping. Regardless, data from this study suggest that increases in fish Hg content are related to increased levels of MeHg in particulate matter (i.e., plankton) in lakes, further emphasizing the potentially important role of trophic transfer in MeHg biomagnification.[19]

Finally, results from sediment trap studies identify both similarities and dissimilarities between Hg_T and MeHg sedimentation in the study lakes. Both Hg_T and MeHg fluxes were regressed against Fe, Mn, S, *a* phorbin, and bacteriochlorophyll fluxes (Figure 3). Due to the abnormal flux in Little Rock Treatment in July, this date was regarded as an outlier and disregarded from the regressions. Both Fe and Mn were chosen because midsummer profiles

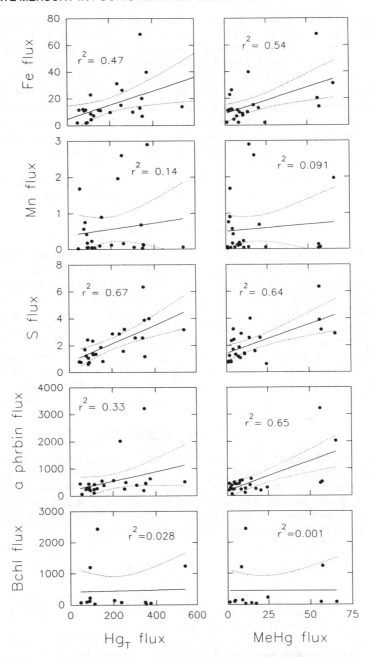

Figure 3 Regressions of Fe, Mn, S, *a* phorbin, and bacteriochlorophyll flux vs. Hg$_T$ and MeHg fluxes to sediment traps. Fe, Mn, and S fluxes are in mg m^{-2}d^{-1}, *a* phorbin and Bchl fluxes mg m^{-2}d^{-1}, and Hg$_T$ and MeHg fluxes are ng m^{-2}d^{-1}.

suggest substantial redox cycling in most of the study lakes during late stratification.[8,20] The strongest correlations were between Fe and Hg_T, Fe and MeHg, S and Hg_T, S and MeHg, and *a* phorbin and MeHg. Unfortunately, it is difficult to separate whether MeHg is associated directly with either Fe or S. Methyl Hg enrichment of particles occurs at the same time as Fe release and precipitation and total S deposition may be abiotic through FeS precipitation or biotic through deposition of phototrophic sulfur bacterial remains. Perhaps a division between biotic and abiotic sulfur would substantiate whether this relationship, in fact, is valid.

The greater differences between Hg species (Hg_T vs. MeHg) appears to be the correlation to total phorbin. The stronger correlation between this indicator of total phytoplankton and MeHg suggests that partitioning of MeHg to biota is stronger than nonmethylated forms of Hg. This finding is also supported by our log Kd data and from results of zooplankton analyses from Little Rock Lake.[19] Additional research, enabling physical separation between inorganic and organic sedimentary phases would better substantiate these results.

IV. CONCLUSIONS

Particulate transport of Hg in the water column of lakes represents an important pathway in biogeochemical Hg cycling. Uptake of dissolved Hg by the particulate phase represents the potential for uptake through the food chain while also representing an effective mechanism transport toward the sediments where it may be effectively buried. Analysis of Hg content of particulates both suspended in the water column and collected during settling exhibit strong seasonal responses. MeHg content of particles in all lakes progressively increased as stratification developed over the summer. Suspended particulate MeHg content follows similar increases in yellow perch among the study lakes.[1] These results stress the importance of both the seasonal effects of methylation and the implications of trophic transfer of MeHg. These results further emphasize differences in MeHg and nonmethyl Hg cycling,[7,19] in that MeHg appears to be more strongly correlated with settling biogenic particles in sediment traps.

ACKNOWLEDGMENTS

We thank S. Class, T. Hoffman, and K. Morrison for assistance in field sampling and sample processing. G. Lee constructed the sediment traps used in this study. We also thank the staff at the University of Wisconsin Trout Lake Station for providing the laboratory base for this work. J. Wiener provided information on fish Hg content. C. Babiarz reviewed an earlier version of this manuscript. This research was supported by the Electric Power Research Institute and the Wisconsin Department of Natural Resources.

REFERENCES

1. Wiener, J.G., Powell, D.E., and Rada, R.G., Mercury accumulation by fish in Wisconsin seepage lakes: relation to lake chemistry and acidification. Final report to Electric Power Research Institute (RP2020–10), Palo Alto, CA, 1992.
2. Watras, C.J., Andren, A.W., Bloom, N.S., Fitzgerald, W.F., Hurley, J.P., Krabbenhoft, D.P., Rada, R.G., and Wiener, J. G., Mercury in temperate lakes: a mechanistic field study, *Verh. Int. Verein. Limnol.*, 24, 2199, 1991.
3. Patterson, C.C. and Settle, D.M., The reduction of orders of magnitude errors in lead analyses of biological materials and natural waters by evaluating and controlling the extent and sources of industrial lead contamination introduced during sample collection, handling and analysis, in: *Accuracy in Trace Analysis: Sampling, Sample Handling, and Analysis*, P.D. LaFleur, Ed., U.S. National Bureau of Standards Special Publication 422, pp. 321, 1976.

4. Fitzgerald, W.F. and Gill, G.A., Subnanogram determination of mercury by two-stage gold amalgamation and gas-phase detection applied to atmospheric analysis, *Anal. Chem.*, 51, 1714, 1979.
5. Bloom N.S. and Fitzgerald, W.F., Determination of volatile mercury species at the picogram level by low temperature gas chromatography with cold vapor atomic fluorescence detection, *Anal. Chim. Acta*, 208, 151, 1988.
6. Bloom, N.S., Determination of picogram levels of methylmercury by aqueous phase ethylation, followed by cryogenic gas chromatography with cold vapor atomic fluorescence detection, *Can. J. Fish. Aquat. Sci.*, 46, 1131, 1989.
7. Bloom, N.S. and Watras, C.J., Seasonal and vertical variability in the mercury speciation of an acidified Wisconsin lake, in: *Heavy Metals in the Environment: International Conference*, Geneva, Vol. 2, J.P. Vernet, Ed., CEP Consultants, Publishers, Edinburgh, U.K., 1989.
8. Bloom, N., Watras, C.J., and Hurley, J.P., Impact of acidification on methylmercury cycling in remote seepage lakes, *Water Air Soil Pollut.*, 56, 477, 1991.
9. Hurley, J.P., Watras, C.J., and Bloom, N.S., Mercury cycling in a northern Wisconsin seepage lake. The role of particulate matter in vertical transport, *Water Air Soil Pollut.*, 56, 543, 1991.
10. Vandal, G.M., Mason, R.P., and Fitzgerald, W.F., Cycling of volatile mercury in temperate lakes, *Water Air Soil Pollut.*, 56, 791, 1991.
11. Gill, G.A. and Bruland, K.W., Mercury speciation in surface freshwater systems in California and other areas, *Environ. Sci. Technol.*, 24, 1392, 1990.
12. Bloom, N.S. and Effler, S.W., Seasonal variability in the mercury speciation of Onondaga Lake (New York), *Water Air Soil Pollut.*, 53, 251, 1990.
13. Spry, D.J. and Wiener, J.G., Metal bioavailability and toxicity to fish in low-alkalinity lakes: a critical review, *Environ. Pollut.*, 71, 243, 1991.
14. Winfrey, M.R. and Rudd, J.W.M., Environmental factors affecting the formation of methylmercury in low pH lakes, *Environ. Toxicol. Chem.*, 9, 853, 1990.
15. Lathrop, R.C., Rasmussen, P.W., and Knauer, D.R., Walley mercury concentrations in Wisconsin lakes, *Water Air Soil Pollut.*, 56, 295, 1991.
16. Grieb, T.M., Driscoll, C.T., Gloss, S.P., Schofield, C.L., Bowie, G.L., and Porcella, D.B., Factors affecting mercury accumulation in fish in the upper Michigan peninsula, *Environ. Toxicol. Chem.*, 9, 919, 1990.
17. Bloom, N.S., On the chemical form of mercury in fish and marine invertebrates, *Can. J. Fish. Aquat. Sci.*, 49, 1010, 1992.
18. Sigg, L., Metal transfer mechanisms in lakes: the role of particulate matter, in: *Chemical Processes in Lakes*, W. Stumm, Ed., John Wiley & Sons, New York, p. 283, 1985.
19. Watras, C.J. and Bloom, N.S., Mercury and methylmercury in individual zooplankton. Implications for foodweb transfer and bioaccumulation, *Limnol. Oceanogr.*, 37, 1313, 1992.
20. Watras, C.J. et al. Mercury in Temperate Lakes, 1990 Annual Report. Electric Power Research Institute, Palo Alto, CA, 1991.
21. Gardner, W.D., Field assessment of sediment traps, *J. Mar. Res.*, 38, 41, 1980.
22. Shafer, M.M., Biogeochemistry and Cycling of Water Column Particulates in Southern Lake Michigan, Ph.D. Thesis, University of Wisconsin-Madison, 428 p., 1988.
23. Lee, C., Hedges, J.I., Wakeham, S.G., and Zhu, N., Effectiveness of various treatments in retarding microbial activity in sediment trap material and their effects on the collection of swimmers, *Limnol. Oceanogr.*, 37, 117, 1992.
24. Bloom, N.S. and Crecelius, E.A., Determination of mercury in seawater at sub-nanogram per liter levels, *Mar. Chem.*, 14, 49, 1988.
25. Mantoura, R.F.C. and Llewellyn, C.A., The rapid determination of algal chlorophyll and carotenoid pigments and their breakdown products in natural waters by reverse-phase high performance liquid chromatography, *Anal. Chim. Acta*, 151, 297, 1983.

26. Hurley, J.P. and Watras, C.J., Analysis of bacteriochlorophylls in lake waters by HPLC and in-line diode array detection, *Limnol. Oceanogr.*, 36, 307, 1991.
27. Madigan, M.T., Microbiology, physiology, and ecology of phototrophic bacteria, in: *Biology of Anaerobic Microorganisms*, A.J.B. Zennder, Ed., John Wiley & Sons, New York, 1988.
28. Hurley, J.P. and Garrison, P.J., Composition and sedimentation of aquatic pigments associated with deep plankton in lakes, *Can. J. Fish. Aquat. Sci.*, 50, 2723, 1993.
29. Lerman, A., *Geochemical Processes: Water and Sediment Environments*, John Wiley & Sons, New York, 1979.
30. Leavitt, P.R. and Carpenter, S.R., Regulation of pigment sedimentation by photooxidation and herbivore grazing, *Can. J. Fish. Aquat. Sci.*, 47, 1166, 1990.
31. Baker, L.A., Tacconi, J.E., and Brezonik, P.L., Role of sedimentation in regulating major ion composition in a softwater, seepage lake in northern Wisconsin, *Verh. Int. Verein, Limnol.*, 23, 346, 1988.
32. Elder, J., unpublished data, 1991.
33. Matilainen, T., Verta, M., Niemi, M., and Uusi-Rauva, A., Specific rates of net methylmercury production in lake sediments, *Water Air Soil Pollut.*, 56, 595, 1991.
34. Korthals, E.T. and Winfrey, M.R., Seasonal and spatial variations in mercury methylation and demythylation in an oligotrophic lake, *Appl. Environ. Microbiol.*, 53, 2397, 1987.

Elemental Mercury Cycling within the Mixed Layer of the Equatorial Pacific Ocean

R. P. Mason, J. O'Donnell, and W. F. Fitzgerald

CONTENTS

ABSTRACT: Atmospheric deposition is the main source of mercury to the open ocean while gas evasion and particle sinking are the major pathways for removal of mercury from surface waters. Evasion rates for mercury at the sea surface in the equatorial Pacific Ocean were estimated from measurements of mixed layer dissolved gaseous mercury obtained during a cruise in January through February 1990 aboard the NOAA *Malcolm Baldrige*. The average flux of 500 ± 340 pmol m^{-2} day^{-1} was similar to previous estimates for this region. Preliminary flux calculations for the mixed layer of the equatorial Pacific Ocean suggested that high evasion rates of Hg^0 could only be supported by the supply of Hg_R via upwelling, and subsequent formation of Hg^0 within the mixed layer. A numerical model was developed to test these observations. The steady-state, one-dimensional advection-diffusion-reaction model took into account vertical eddy diffusion, particulate uptake and release of dissolved mercury, particle sinking, and upwelling. Model simulations confirmed that the system is severely substrate limited, with maximal sustainable evasion rates of less than 200 pmol m^{-2} day^{-1} in the absence of upwelling. Additionally, the model results suggested that upwelling rates of 80 to 160 m $year^{-1}$ would be sufficient to support an evasion rate of 500 pmol m^{-2} day^{-1}, with physical conditions and Hg concentrations similar to those found during the cruise. This work has provided further evidence that the supply of labile inorganic mercury limits Hg^0 formation (and methylation) in the aquatic environment.

I. INTRODUCTION

Recent investigations of the biogeochemical cycling of mercury (Hg) have demonstrated the importance of dissolved gaseous Hg (DGHg) in the cycling of Hg at the air/water

SAMPLING STATIONS 1990 MB—90—01—RITS CRUISE

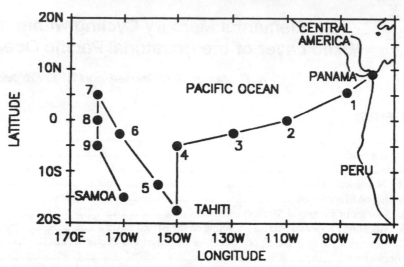

Figure 1 The cruise track of the *Malcolm Baldrige* during the cruise in January and February 1990 in the equatorial Pacific Ocean. The station locations are indicated. (Adapted from Mason, R.P. and Fitzgerald, W.F., *Nature,* 347, 457–459, 1990. Copyright © 1990 Macmillan Magazines Ltd. With permission.)

boundary.[1-5] Furthermore, it is now apparent that elemental Hg (Hg^0) is the principle component of the DGHg in the mixed layer of the ocean,[3,6] in freshwater lakes,[7] and in estuaries,[8] with dimethylmercury (DMHg) concentrations being below the detection limit of 5 fM for all these studies. Dimethylmercury has, however, been found in the low oxygen waters of the equatorial Pacific Ocean.[6] Evasion to the atmosphere of mixed layer Hg^0 and the downward particulate Hg flux are the principal sinks for Hg in open ocean waters.[9] Little is known, however, about the sources of Hg^0, or the mechanisms of Hg^0 formation in open ocean waters.

Biological processes in the mixed layer and low oxygen region play an important role in the cycling of Hg in the upper ocean of the equatorial Pacific.[5,9,10] The permanent stratification and resultant high stability of the thermocline effectively limit diffusional exchange between the surface mixed layer and the sub-thermocline waters.[9] High biological production is maintained by the upwelling of nutrients,[11] and particle sinking from the euphotic zone is the dominant exchange process between the mixed layer and the deeper waters.[12-14] The principal external source of Hg to the equatorial Pacific Ocean is precipitation.[9,10,15] Biological uptake in the mixed layer and particle sinking and dissolution in the sub-thermocline waters are the major routes for the transfer of Hg from the mixed layer to the deeper waters.[9]

The open ocean cycling of Hg was investigated during a cruise in the equatorial Pacific Ocean in January and February 1990 aboard the NOAA ship *Malcolm Baldrige*.[3] Nine stations were occupied (Figure 1) and samples were collected to a depth of 1000 m and analyzed, on board, for reactive Hg (Hg$_R$); DGHg (Hg^0 and DMHg); dissolved monomethylmercury (MMHg); and in some instances for particulate Hg. The data obtained on the distribution and speciation of Hg in the upper ocean of the equatorial Pacific allowed an assessment of the principal sources and sinks of Hg from this region, and of the primary reactions of Hg$_R$ within this zone.[9] The Hg$_R$ determination of open ocean seawater, adjusted for any DGHg present, has been shown to be a suitable measure of the labile Hg fraction (Hg substrate) available for methylation and Hg^0 formation.[6] Furthermore, the rate of supply of Hg_R substrate is likely to limit the formation of methylated Hg species and Hg^0 in ocean waters. Investigations in the open ocean,[6,9] in an estuary,[8] and in freshwater lakes[4] have demonstrated that the substrate hypothesis provides a constructive physiochemical interpretation of Hg biogeochemistry in aquatic systems.

This chapter details both box model and numerical simulations of the Hg cycle in the mixed layer of the equatorial Pacific Ocean that focus on assessing the validity of the substrate hypothesis. A companion paper concentrates on Hg cycling in the low oxygen region of the equatorial Pacific.[9] The modeling results suggest that formation of Hg^0 within the mixed layer is required to support the evasion rates estimated for the equatorial Pacific, and that the Hg^0 concentration and flux is limited by the rate of supply of Hg_R substrate. Moreover, Hg_R supply via upwelling is the principal source of mixed layer Hg_R and direct reduction of Hg_R is the principal source of mixed layer Hg^0.

II. METHODS

The analytical methods have been discussed in detail elsewhere.[3,15-18] Samples were collected using techniques designed to avoid trace metal contamination[16,19] from nine stations within the equatorial Pacific region (Figure 1). All stations except Station 5 were between 5° N and 5° S; five stations were south of the equator; two were on the equator, and two were north of the equator. Samples were analyzed on board for Hg_R, DGHg (Hg^0 and DMHg), MMHg, and particulate Hg. Analytical techniques relied on purging and trapping of volatile Hg species.[16] Dissolved gaseous Hg species were purged from solution directly, Hg_R species were converted into Hg^0 by $SnCl_2$ reduction[20] and MMHg was converted into methylethylmercury by derivitization using tetraethylaborate[18] prior to degassing. For the determination of methylated Hg species the volatile Hg was trapped on Carbotrap®, a carbon absorbent, while gold-coated sand was used for the DGHg and Hg_R methods. Methylated Hg species were separated and identified using cryogenic gas chromatography with quantification by cold vapor atomic fluorescence spectroscopy.[17] Procedural blanks were consistent over the duration of the cruise (DMHg <5 fM, Hg_R 0.58 ± 0.16 pM, MMHg 40 ± 17 fM, DGHg 145 ± 45 fM).[16]

III. THE NUMERICAL MODEL

A steady-state, one-dimensional advection-diffusion-reaction model was developed to test the hypothesis that the rate of Hg^0 evasion is determined by the rate of supply of labile Hg, and that equatorial upwelling is the principal source of mixed layer Hg_R. Three Hg species were considered in the model: labile ionic Hg [$Hg_R - Hg^0$; designated as Hg(II) in the model], Hg^0, and particulate Hg (Hg_s) (Figure 2). As DMHg concentrations in the mixed layer of the equatorial Pacific were below the detection limit, it is assumed that all the DGHg is Hg^0. The choice of Hg species and interactions included in the model was based on the box model calculations outlined in Section IV. Scaling calculations showed that methylation and demethylation were not important reactions within the mixed layer of the equatorial Pacific. The mass balance estimates for the mixed layer indicate that the system is controlled by vertical diffusion (10^{-3} to 10^{-4} m^2 s^{-1}),[31,32] particle sinking (50 to 150 m day^{-1}, 0.58 to 1.7×10^{-3} m s^{-1}),[33] and upwelling (50 to 150 m year^{-1}; 1.6 to 4.8×10^{-6} m s^{-1}).[34]

The sinking particle concentration (Hg_s) was estimated using the measured suspended particle concentration ($Hg_p = 10$ pg/L) and the relationship of Clegg and Whitfield[33] [$Hg_s/Hg_p = 0.025$]. A distinction is made in the model between large, rapidly sinking particulate matter (Hg_s) and smaller, suspended matter that does not sink (Hg_p). It is thought that Go-Flo® bottles sample Hg_p, while sediment traps sample Hg_s.[33,35] The interactions between Hg species include the conversion of Hg(II) to Hg^0, the aggregation and sinking of Hg_s, and the disintegration and dissolution of Hg_s. The three resultant second-order

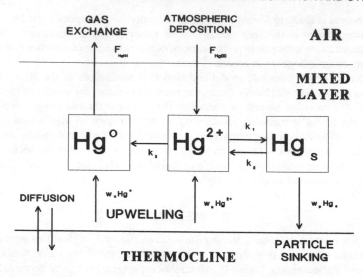

Figure 2   Diagram showing the interactions between labile Hg substrate [Hg(II)], elemental mercury (Hg⁰) and particulate mercury (Hg$_s$) considered in the model of mercury biogeochemical cycling in the mixed layer of the equatorial Pacific Ocean. The fluxes for these mercury species due to particulate sinking (w_p = particulate sinking velocity in m/s), upwelling (w_u = upwelling velocity in m/s), gas evasion ($F_{Hg(0)}$, pmol/m² day), and wet deposition ($F_{Hg(II)}$, pmol/m² day) are also illustrated.

differential equations require six boundary conditions. The surface fluxes of the three species and the concentrations at the thermocline were the boundary conditions chosen. In this simplified simulation the associated processes of uptake and release of dissolved species [Hg(II) ↔ Hg$_p$] were incorporated with the suspended particle-sinking particle interactions (Hg$_p$ ↔ Hg$_s$) into overall reaction coefficients: k_1 for Hg(II) ↔ Hg$_s$, and k_2 for Hg$_s$ ↔ Hg(II). These coefficients were derived from the data of Clegg and Whitfield[33] and the measured Hg concentrations (Hg$_R$ = 1 pM; Hg$_p$ = 50 fM; Hg$_R$/Hg$_p$ = 20). Both k_1 and k_2 were depth dependent functions having a maximum near the surface (10 to 20 m).

The derived differential equations are

$$d/dz[d\{D.Hg(II)\}/dz] - d\{w_u.Hg(II)\}/dz - (k_1 + k_3)\{Hg(II)\} + k_2\{Hg_s\} = 0 \qquad (1)$$

$$d/dz[d\{D.Hg_s\}/dz] - w_p.d\{Hg_s\}/dz + k_1\{Hg(II)\} - k_2\{Hg_s\} = 0 \qquad (2)$$

$$d/dz[d\{D.Hg^0\}/dz] - d\{w_u.Hg^0\}/dz + k_3\{Hg(II)\} = 0 \qquad (3)$$

where D is the diffusion coefficient, taken to be an exponential function; w_u the upwelling velocity, which is linear and depth dependent; w_p the particle sinking velocity; k_1 the particle scavenging coefficient; k_2 the particle dissolution coefficient; and k_3 the reaction rate constant for conversion of Hg(II) to Hg⁰ (Figure 2). The flux due to the upwelling of particulate matter is relatively small compared to the fluxes of the other transport processes and is not included in the model equations.[3]

The equations were solved numerically using the scheme outlined in Chapter 16 of *Numerical Recipes*[36]—a "two point boundary value problem" which was solved using the "shooting method". The boundary conditions were chosen such that the fluxes of each mercury species were set at the sea surface, while concentrations were specified at the bottom boundary. Boundary conditions were representative of the actual concentrations measured during the 1990 equatorial Pacific Ocean cruise, and of the fluxes estimated from the actual data. The flux of Hg(II) (depositional input) at the surface and the bottom concentration of Hg^0 were not changed to simplify interpretation of the results. The total flux of Hg_s at the surface was set at zero as rain was estimated to not contribute a significant amount of sinking particulate mercury to the mixed layer.[3]

It was necessary to stipulate that the reaction coefficients were zero when concentrations become zero or negative. Without this stipulation it was found that solutions were obtained that, while they conformed to the boundary conditions, were scientifically unreasonable. This artifact arose because a negative concentration changes an expression in the equations from a source to a sink of a particular species, due to the sign change (and vice versa). This resulted in instabilities in the solution. Under these conditions, the total flux was not conserved. To ensure the accurate performance of the model, the total flux at each depth was calculated as a test of the numerical scheme. In all the runs discussed in this chapter, the total flux was conserved.

The flux of Hg^0 at the surface was set at 500 pmol m^{-2} day^{-1}, the average value estimated from the data obtained during the equatorial Pacific Ocean cruise. The model, however, produced negative concentrations if the supply of Hg^0 did not match the evasional flux. In these instances, the Hg^0 flux was decreased until the concentration of Hg^0 became positive. This allowed a prediction of the maximum supportable flux under the chosen conditions. Thus, the Hg^0 flux was only changed when the model produced physically unreasonable results, i.e., negative concentrations. This manipulation provided the best illustration of the effect of upwelling on the production and subsequent evasion of Hg^0.

IV. RESULTS AND DISCUSSION

A. SOURCES OF ELEMENTAL MERCURY

Mixed layer concentrations of Hg^0 were between 40 and 325 fM. Concentrations were highest at Stations 1 (325 fM), 2 (180 fM), 8 (360 fM), and 9 (210 fM). Stations 1 and 2 were in the eastern equatorial Pacific, Stations 2 and 8 were on the equator (Figure 1), and Stations 8 and 9 were in a region of upwelling, based on acoustic doppler current profiles.[3,21] These relationships suggest that higher Hg^0 concentrations are associated with higher primary productivity, upwelling, or both. The surface waters were saturated with respect to Hg^0 (400 to 4000% saturated)[9] and thus Hg^0 is lost from the sea surface by gas exchange. Elemental Hg evasion rates, estimated from the mixed layer concentrations with an average gas transfer coefficient of 4 m day^{-1}, ranged from 160 to 1440 pmol m^{-2} day^{-1} (504 ± 340 pmol m^{-2} day^{-1}). These fluxes are similar to those determined previously for the equatorial Pacific Ocean.[5]

Vertical diffusion and advection of Hg^0 would not supply sufficient Hg^0 to the mixed layer to support these evasion rates at steady state. The estimated average diffusional flux of Hg^0 into the mixed layer from the thermocline for all stations is less than 5 pmol m^{-2} day^{-1} (about 30% or less of the Hg_R flux, D = 5 × 10^{-6} m^2 s^{-1}).[9] The upwelling flux of Hg^0 is about 80 pmol m^{-2} day^{-1} (upwelling at 100 m year^{-1}, 300 fM Hg^0). Thus upwelling and diffusion could not supply sufficient Hg^0 to the mixed layer to support even the lowest measured evasional flux of 160 pmol m^{-2} day^{-1}. In situ formation of Hg^0 within the mixed layer is therefore required to maintain the estimated evasional fluxes. In addition, a supply of Hg substrate to the mixed layer sufficient to sustain the formation of Hg^0 is needed.

Two possible sources of Hg^0 in the mixed layer are the direct production of Hg^0 from Hg_R and demethylation of MMHg. The maximum flux of MMHg into the mixed layer, less than 30 pmol m^{-2} day^{-1} (including both diffusion and upwelling),[9] is comparatively small and would provide insufficient MMHg to supply Hg^0 by demethylation. In situ methylation within the water column could supply additional MMHg. It is possible to estimate the overall rate of formation of MMHg in the mixed layer by using the estimate of the export of particulate Hg (Hg$_s$) from the euphotic zone (180 pmol m^{-2} day^{-1}),[9] and the observation that only 10 to 20% of the total production in the mixed layer is exported.[12,14,22] Based on the relationship between particulate MMHg and Hg_s (part.$MMHg/Hg_s$ = 0.01),[23] the total production of MMHg in the mixed layer is estimated as 18 to 36 pmol m^{-2} day^{-1} (<0.05% day^{-1}, assuming a Hg_R concentration of 1 pM and a 100 m mixed layer). This methylation rate is comparable to estimated rates of methylation in freshwater, derived from radiotracer experiments (0.01 to 0.3% day^{-1}).[24] Subsequent demethylation could supply Hg^0 at a comparable or lower rate. This low estimated methylation rate (<40 pmol m^{-2} day^{-1}) suggests that demethylation would not contribute significantly to the Hg^0 pool in the mixed layer, and therefore Hg^0 must be formed principally by direct reduction of Hg_R within the mixed layer. The demethylation rates calculated for the low oxygen region of the equatorial Pacific (10^{-8} to 10^{-9} s^{-1}; 0.1–0.01% day^{-1})[9] are similar to the demethylation rate estimated for the mixed layer. Thus, while demethylation likely occurs throughout the water column, production of Hg^0 by reduction of Hg_R, either biotically or abiotically, is the main route for Hg^0 formation within the mixed layer. Similar conclusions have been derived from investigations in the lakes of northern Wisconsin.[25]

The in situ Hg_R reduction rate in the mixed layer required to sustain the surface flux is about 6×10^{-8} s^{-1} (0.5% day^{-1}) if the production of Hg^0 occurs throughout the mixed layer. The conversion rate required would be higher; however, if the Hg_R distribution is supply limited, or if only a fraction of the mixed layer is involved in Hg^0 production. Rates of abiotic conversion of ionic Hg to Hg^0 as high as 2.5×10^{-6} s^{-1} have been measured under laboratory conditions. Alberts et al.[26] found the conversion to be pH dependent in the presence of humic acid in distilled water at room temperature, and studies with Wisconsin lakewater spiked with radiolabeled Hg showed that Hg^0 production increased with increasing pH.[27] These experiments indicate that abiotic conversion of ionic Hg could sustain the flux of Hg^0 at the sea surface in the equatorial Pacific.

The estimated average diffusional rate of supply of Hg_R to the mixed layer from the thermocline is 15 pmol m^{-2} day^{-1}.[9] This diffusional flux cannot provide the necessary Hg_R for the formation of Hg^0. However, upwelling of thermocline waters could supply additional Hg_R. The estimated delivery of Hg_R due to upwelling is 820 pmol m^{-2} day^{-1}, taking 3 pM as a representative thermocline Hg_R concentration and using an upwelling rate of 100 m $year^{-1}$.[28,29] The rate of supply of Hg_R via rain is estimated as 74 to 99 pmol m^{-2} day^{-1} (precipitation at 3 to 4 m $year^{-1}$),[30] and Hg_R of 9 pM;[15] analogous to previous estimates.[1,2] Thus, the inputs of Hg_R to the mixed layer via deposition and upwelling, a maximum of 10^3 pmol m^{-2} day^{-1}, are of the same order as the average evasional flux (500 ± 340 pmol m^{-2} day^{-1}). These calculations indicate that the evasion of Hg^0 at the sea surface is directly coupled to the rate of supply of Hg substrate to the mixed layer and that Hg^0 is formed principally by direct reduction of Hg_R, supplied primarily by upwelling and atmospheric deposition.

B. MODELING ELEMENTAL MERCURY CYCLING IN THE MIXED LAYER

The observations and conclusions of Section IV.A were studied further using a numerical simulation of the Hg cycle within the mixed layer of the equatorial Pacific. Table 1 summarizes the fluxes associated with these processes in the mixed layer and Figure 2 depicts the principal interactions, derived from the flux estimates and box model calculations, that are included in the model. The associated fluxes are of the same order, and therefore all of these processes must be included in any model of the biogeochemical interactions of Hg in the mixed layer of the equatorial Pacific Ocean.

The model was first used to assess whether the estimated average evasion rate of 500 pmol m^{-2} day^{-1} could be supported in the absence of upwelling. The following boundary conditions were chosen: at 125 m, Hg(II) 2 pM, Hg^0 0.2 pM, Hg_s 1 fM; Hg^0 formation rate (k_3) 2×10^{-7} s^{-1}; particle-dissolved interaction coefficients 3.2×10^{-7} s^{-1} (k_1 and k_2). As

Table 1   **Calculated fluxes for the predominant processes affecting the concentration and speciation of mercury in the mixed layer of the equatorial Pacific Ocean.**

Process	Parameters	Flux (pmol m^{-2} day^{-1})
Diffusion	$D = 10^{-3}$–10^{-4} m^2/s; dc/dz for Hg_R about 5 pmol/m^4	40–430
Particle sinking	$w_p = 0.6$–1.7×10^{-3} m/s; sinking particles 0.2 pg/L	50–150
Upwelling	$w_u = 1.6$–3.2×10^{-6} m/s; Hg_R 3pM within the thermocline	410–830
Reaction ($Hg^{2+} \leftrightarrow Hg^0$)	$k = 6 \times 10^{-7}$ s^{-1}; L = 150 m; Hg_R 1 pM in the mixed layer	500
Rain	Mason et al. 1992[15]	75–100

discussed in Section III, the chosen Hg^0 flux of 500 pmol m^{-2} day^{-1} could not be sustained under these chosen conditions or any reasonable conditions where upwelling was not included. The boundary condition was then changed to ascertain what flux could be supported. The highest sustainable Hg^0 evasion rate in the absence of upwelling was less than 200 pmol m^{-2} day^{-1} (curve 1, Figure 3a). For all the remaining simulations, the flux was kept at 500 pmol m^{-2} day^{-1}.

A flux at the surface of 500 pmol m^{-2} day^{-1} of Hg^0 is maintained with upwelling at 125 m of 252 m $year^{-1}$ (curve 2, Figure 3a). The upwelling rate increases linearly with depth in the model, and thus the average upwelling rate under these conditions is 126 m $year^{-1}$. This upwelling rate is comparable to values estimated based on both physical and chemical measurements in the equatorial Pacific.[28,29,34] At higher thermocline Hg(II) concentrations, the same flux is maintained at lower upwelling rates (curve 3, Figure 3a). Doubling the Hg(II) concentration from 2 to 4 pM leads to a Hg^0 flux of 500 pmol m^{-2} day^{-1} with an upwelling rate at 125 m of 63 m $year^{-1}$ (a factor of four change). These results indicate that the higher estimated evasion rates (>1000 pmol m^{-2} day^{-1}) could be supported in regions of active upwelling, where thermocline concentrations are somewhat elevated compared to the mixed layer. The average thermocline Hg(II) concentration ($Hg_R - Hg^0$) during the cruise (3.3 ± 1.7 pM) is comparable to the higher value used in the numerical model, and is elevated relative to the mixed layer (1.1 ± 0.5 pM).[3]

The thermocline is the principal source region of Hg(II) and gas evasion at the sea surface is the sink for Hg^0. Thus, the Hg(II) concentration profiles increase with depth if Hg(II) is removed by conversion to Hg^0 faster than it is supplied, and because the rate of supply decreases as the upwelling rate decreases surfaceward. In addition, the Hg^0 concentration reaches a mid-depth maximum if Hg^0 formation is constant with depth as there is less formation in the top portion of the mixed layer due to substrate limitation, and supply of Hg^0 via diffusion is slower than the gas evasion rate. Thus, decreasing the Hg^0 formation rate from 2×10^{-7} s^{-1} to 10^{-7} s^{-1} (curves 2 and 4, Figure 3a) leads to a substantial reduction in the surface Hg^0 concentration (0.19 to 0.02 fM), and a 20% decline in the maximum concentration, with a concurrent doubling in the Hg(II) concentration throughout the water column (curves 2 and 4, Figure 3b).

Dissolved-particle interactions also affect the availability of Hg(II). These interactions were investigated in the absence of upwelling under the following conditions: Hg^0 evasion rate 167 pmol m^{-2} day^{-1}; Hg(II) 2 pM; Hg_s 0.2 fM; k_2 0.05×10^{-7} s^{-1}; other conditions the same as Figure 3. As the rate of Hg uptake and removal by particles decreases (i.e., as

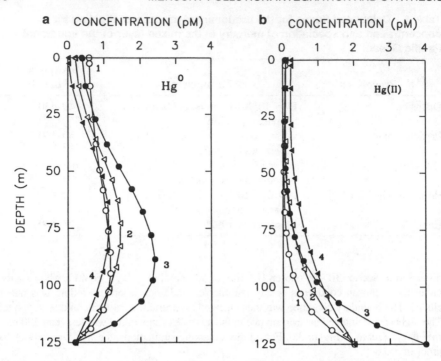

Figure 3 Model simulations illustrating the effect of upwelling on mercury concentrations in the mixed layer. (a) Elemental mercury concentrations. (b) Ionic mercury concentrations. Boundary conditions: at 125 m, Hg_s 1 fM, Hg^0 0.2 pM, Hg(II) 2 pM (curves 1, 2, 4) 4 pM (curve 3); depositional flux of Hg(II) 75 pmol m^{-2} day^{-1}; no overall Hg_s surface flux. Hg^0 formation rate 2×10^{-7} s^{-1} (curves 1–3), 10^{-7} s^{-1} (curve 4); particulate-dissolved interactions 3.2×10^{-7} s^{-1}. Evasional Hg^0 flux 167 pmol m^{-2} day^{-1} for curve 1, 500 pmol m^2 day^{-1} for curves 2–4. Upwelling rates: no upwelling (curve 1), 252 m year^{-1} at 125 m (curve 2 and 4), 63 m year^{-1} (curve 3).

k_1 changes from 0.1 to 0.05×10^{-7} s^{-1}; Figure 4), the Hg(II) supply to the mixed layer increases, resulting in an increase in Hg^0 concentration at the surface from 0 to 180 fM. Thus the rate of removal of Hg(II) from the mixed layer via particulate scavenging has an important influence on the rate of Hg^0 production. The model results suggest that Hg^0 formation would be hindered under conditions where Hg is rapidly sequestered by particulate and removed from the mixed layer.

The above simulations assume that Hg^0 formation is constant throughout the mixed layer. The rate constant for Hg^0 formation (k_3) could be a depth dependent function (e.g., if it is a photochemically induced process). An exponential equation $\{k_3 \text{ (scaled)} = \alpha[1 + 10 \exp(-\beta/z)];$ reaction scale $= 10^{-7}$ s$^{-1}\}$ was used to investigate the effect of the form of k_3 on the Hg(II) and Hg^0 profiles. Little difference can be ascertained when α is large as the system is strongly substrate limited as most of the Hg(II) is converted to Hg^0 close to the thermocline. However, as the reaction rate decreases (i.e., as α decreases), the effect of the rate of decrease in reaction rate with depth (i.e., in β) is more pronounced. With a depth dependent function ($\alpha = 0.75$, $\beta = 1$) the concentrations of Hg(II) are above 0.1 pM throughout the mixed layer, and substrate limitation is less severe. With a constant reaction coefficient throughout the mixed layer the concentration of Hg(II) is much smaller than the concentrations of Hg_R measured during the cruise (average 1.1 ± 0.5 pM),[3] while the concentrations predicted by the model are closer to the measured concentrations if the reaction coefficient is smaller at depth than at the surface.

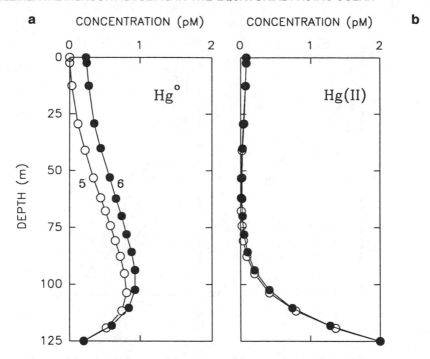

Figure 4 Model simulations illustrating the effect of particulate scavenging of mercury on the amount of elemental mercury formed. (a) Elemental mercury concentrations. (b) Ionic mercury concentrations. Hg^0 evasion at 167 pmol m^{-2} day^{-1}; Hg(II) 2 pM, Hg$_s$ 0.2 fM at 125 m; scavenging rate 0.1 × 10^{-7} s^{-1} (curve 5), 0.05 × 10^{-7} s^{-1} (curve 6); other conditions as in Figure 3.

Finally, the model was used to determine what conditions were necessary to produce model simulations with concentrations similar to the concentrations measured during the cruise in the equatorial Pacific. The measured concentrations of Hg(II) {Hg$_R$ − DGHg}[3] for the mixed layer (30 to 50 m) of the equatorial Pacific Ocean ranged from 1.7 pM (Station 4) to 0.22 pM (Station 2) and averaged 1.1 ± 0.5 pM, while the concentrations of Hg^0 ranged from 40 to 360 fM and averaged 148 ± 100 fM. The average Hg(II) thermocline concentration was 3.3 ± 1.7 pM. The data from Station 2 was used to test the ability of the model to produce profiles similar to those predicted from the limited data set available. Station 2 was chosen as the profiles obtained at this station were typical of all the stations; the station was on the equator; and surface samples were obtained.[3] The estimated surface, mid-depth (40 m), and thermocline concentrations of Hg^0 and Hg(II) and the flux of Hg^0 at the surface were used to define the boundary conditions. The winds were relatively light (average 12 knots) in the vicinity of Station 2 and therefore a transfer velocity of 3 m day^{-1} was appropriate.[37] The estimated flux was 510 pmol m^{-2} day^{-1} (170 fM Hg^0 at the surface).

An initial set of runs showed that an upwelling rate of at least 80 m year^{-1} was required to provide sufficient substrate under the constraint of water column Hg^0 concentrations similar to those measured at Station 2 and elsewhere (Figure 5). If the production rate of Hg^0 is constant with depth, the resultant profiles for Hg^0 show a distinct mid-depth maximum (Figure 5; curves 2 and 6). A reaction coefficient of 0.5 × 10^{-7} s^{-1} (curve 2) results in a reasonable Hg^0 profile. The effect of changing the formulation of the reaction coefficient on the Hg(II) profile is also shown. The curves 1, 2, and 3 (Figure 5a) provide reasonable agreement with the surface concentration, and curve 1 more closely matches the overall data. Curves 1 and

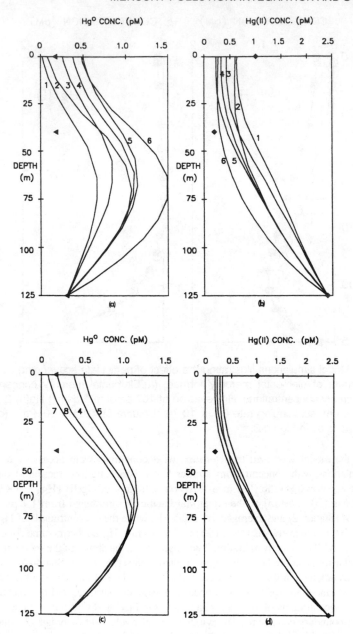

Figure 5 Model derived distributions for Hg^0 (Figure 5a) and $Hg(II)$ (Figure 5b). The concentrations of Hg^0 (triangles) and $Hg(II)$ (diamonds) from Station 2 are plotted for comparison on each graph. Hg^0 evasion rate is 500 pmol m^{-2} day^{-1}. Thermocline concentrations of Hg^0 and $Hg(II)$, respectively, are 300 fM and 2.4 pM. All other boundary conditions are the same as in Figure 4. The reaction coefficient for Hg^0 formation, k_3, is constant with depth for curve 2 (0.5×10^{-7} s^{-1}) and curve 6 (1×10^{-7} s^{-1}). For all the other curves, $k_3 = [\alpha(1 + 10 \exp(-\beta z)] \times 10^{-7} s^{-1}$. Curve 1: $\alpha = 0.2$, $\beta = 2$; curve 3: $\alpha = 0.2$, $\beta = 1$; curves 4 and 7: $\alpha = 0.5$; $\beta = 4$; curves 5 and 8: $\alpha = 0.5$; $\beta = 2$.

3 were generated using a depth dependent reaction coefficient, and differ only in the rate of decrease with depth (i.e., in β).

These model simulations show that it is possible to choose conditions such that the model derived concentrations are similar to the measured values (Figure 5). The correspondence between the measured and model derived concentrations of Hg^0 and Hg_R suggest that the rate of Hg^0 formation is a depth dependent function. High surface evasion rates can only be maintained under steady-state conditions by upwelling of thermocline waters of higher Hg concentration. In the absence of upwelling, sustainable evasion rates are less than 200 pmol m^{-2} day^{-1} and the system is strongly substrate limited.

The model suggests that there should be a measurable gradient in the concentration of Hg^0 and $Hg(II)$ in the mixed layer as reaction rates are substantially faster than substrate supply rates from the thermocline. To date, there has been no detailed profiling of either Hg_R or Hg^0 concentrations within the mixed layer of any oceanic system. Future work should endeavor to establish the vertical gradient of these species in the mixed layer to examine the conclusions derived from the modeling studies. Detailed information on the distribution and concentration of Hg^0 and $Hg(II)$ in the mixed layer could help ascertain whether the Hg^0 production reaction is a function of depth, and could further elucidate the sources of $Hg(II)$ to the mixed layer. An investigation of a region other than the equatorial Pacific would further test the conclusion that Hg^0 formation is limited by substrate supply in areas where upwelling is absent.

C. MERCURY CYCLING IN THE UPPER OCEAN OF THE EQUATORIAL PACIFIC

The Hg_R upwelled into the mixed layer of the equatorial Pacific Ocean is supplied by meridional advection of thermocline waters, and various studies have demonstrated that the upwelled water is derived principally from thermocline waters which are advected horizontally toward the equator from subtropical regions.[28,29,34] In addition, upwelled water is advected away from the equator within the upper reaches of the mixed layer. Thus, Hg^0 evasion at stations away from the equator could be sustained by Hg_R which is upwelled at the equator and subsequently advected to higher latitudes. Horizontal advection would move a parcel of water, upwelled at the equator, to a latitude of 5° within 60 days at 0.1 m s^{-1}. At an initial rate of loss of Hg^0 of 500 pmol m^{-2} day^{-1} (loss rate 5% day^{-1}) the Hg^0 from the top 100 m of the water column would be depleted in about 70 days (i.e., in 5 half-lives) if no formation was occurring (dC/dt = 500 pmol m^{-2} day^{-1} and C = 10^4 pmol m^{-2}). In addition, Hg_R supplied by equatorial upwelling would not all be converted to Hg^0 within 60 days at a rate of 0.5% day^{-1}, nor would it be removed via particle sinking processes (180 pmol m^{-2} day^{-1}; 0.2% day^{-1}). Evasion of Hg^0 at latitudes away from the equator could therefore be maintained by the Hg_R supply at the equator.

High Hg^0 concentrations were found at the equatorial stations and at Station 9 (5° S 180°). Meridional velocity profiles obtained during the second leg of the cruise[21] indicate that upwelling was occurring in the region 0–1° S, 180° during the cruise, and this could account for the higher Hg^0 at Stations 8 and 9 in response to substrate supply via upwelling. These results suggest that the rate of supply of reactive Hg substrate to the mixed layer is the mechanism controlling the rate of Hg^0 formation, and that Hg^0 formation in the equatorial region is driven principally by Hg_R supply via equatorial upwelling.

The meridional circulation pattern and its resultant effect on Hg biogeochemical cycling in the mixed layer of the equatorial Pacific suggests that atmospheric Hg deposition to subtropical regions,[1,3] which is larger than that of the equatorial regions,[15] is driving Hg^0 formation and evasion from the equatorial Pacific. Particle sinking from the mixed layer and Hg^0 evasion are the principal sinks for mixed layer Hg, and atmospheric deposition is the major external Hg flux (Figure 6a). Upwelling provides nutrients which sustain the high productivity, and a higher resultant particle flux, and thus Hg flux to deeper waters is higher

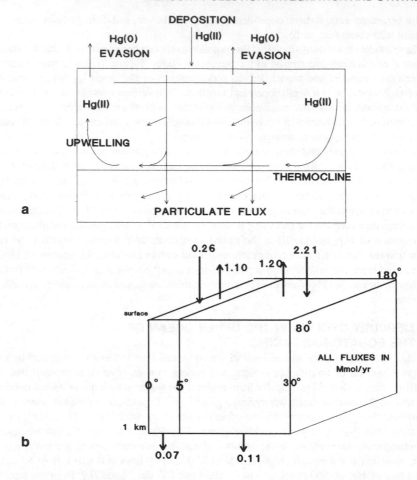

Figure 6 (a) A diagrammatic representation of the overall circulation pattern and mercury transport for the mixed layer and thermocline of the North Pacific showing the major transports, sources, and sinks of mercury to the region. The circulation pattern in the South Pacific would be similar. (b) A mass balance box model of the sources and sinks of mercury to the upper ocean of the north Pacific, based on data from this work, Kim[2] and Fitzgerald.[1] All fluxes are in Mmol year^{-1}.

in regions of enhanced productivity. In the equatorial region, evasion exceeds atmospheric input, but at higher latitudes deposition is greater than evasion, as portrayed in Figure 6a.

The estimated sources and sinks of Hg from the upper ocean (surface to 1000 m) of the Pacific Ocean (80° W to 180° W) between the equator and the tropics (30°) are depicted in Figure 6b. These flux estimates rely on data from the North Pacific Ocean and the meridional circulation within the Northern Hemisphere. The calculations for the southern Pacific would be similar. Average evasional losses are 500 pmol m^{-2} day^{-1} for the equatorial zone (0° to 5°; 6.05 × 10^{12} m^2). Depositional fluxes were calculated using values for total Hg in rain of 15 pM for the equatorial region[15] and 50 pM for the tropical zone (area 30.25 × 10^{12} m^2)[1] and using average rainfalls of 3 m year^{-1} (equatorial) and 1.5 m year^{-1} (tropical).[30] Particulate fluxes at 1000 m were included in the calculation of Hg losses from the upper ocean but were a minor component of the total Hg loss (30 pmol m^{-2} day^{-1} for the equatorial zone; 10 pmol m^{-2} day^{-1} for the tropics). These fluxes were calculated using the relative flux at

1000 m found by Knauer and Martin,[14] estimated average productivity for these regions,[38] and Hg/carbon ratios for particulate matter.[3] The evasional flux in the tropical Pacific (108 pmol m^{-2} day^{-1}) was calculated by difference, to balance the box model. This flux is equivalent to a mixed layer concentration for Hg^0 of 47 fM (K = 5 m/day). Kim[2] found a similar concentration of 26 fM for this region. The integrated fluxes (Mmol year^{-1} for the two regions are given in Figure 6b.

The overall evasional loss was estimated as 2.3 Mmol year^{-1} for the region between the equator and 30° N, and 80° W to 180° W; a value comparable to previous estimates.[1,2] The depositional flux to the region was estimated to be 2.5 Mmol year^{-1}. The overall mass balance calculation (Figure 6b) indicates that atmospheric deposition at higher latitudes could sustain the estimated evasional losses of Hg from the surface ocean in the equatorial region. In addition, the importance of the meridional transport of tropical surface water and subsequent upwelling in the biogeochemical cycling of Hg in the open ocean is demonstrated. Furthermore, the flux estimates illustrate the rapid cycling of Hg in the mixed layer of the ocean. Only a small proportion of the input (0.18 Mmol year^{-1} or 7%) is removed to the deeper waters, the principal region where Hg methylation occurs within the equatorial Pacific Ocean.[9] Thus, Hg^0 formation and evasion removes Hg from the ocean that might otherwise be transported to the low oxygen region where methylation occurs, and any factors that reduce the rate of Hg^0 production or limit the availability of Hg for Hg^0 formation (such as particle scavenging) could result in an increase in Hg methylation. While little is currently known about the mechanisms of Hg^0 formation, the results of this study indicate that it is an important area for further research, as Hg^0 formation has a profound effect on the biogeochemical cycling of Hg.

V. CONCLUSIONS

The modeling results strengthen the hypothesis that the rate of supply of labile inorganic Hg is the determining factor in the production of Hg^0 in the mixed layer of the equatorial Pacific. As the rate of Hg^0 formation in the equatorial Pacific is closely linked to the supply of Hg(II) by upwelling, lower conversion rates would be expected in the absence of upwelling. While more work is needed to further test the model formulation and assumptions, there is good agreement between the model results and the available data concerning Hg^0 evasion from oceanic regions (this study and Kim[2]). In the open ocean, and in particular in the equatorial Pacific Ocean, Hg^0 formation and gas evasion remove a substantial fraction of the Hg(II) input to the mixed layer. Estimated Hg^0 evasion (500 pmol m^{-2} day^{-1}) is a factor of three greater than particulate scavenging removal (180 pmol m^{-2} day^{-1}), and is an order of magnitude greater than estimated methylation within the mixed layer.

The results of the equatorial Pacific Ocean (and the Wisconsin lake studies[39]) demonstrate the importance of particulate matter in transporting Hg from the mixed layer to low oxygen subthermocline waters where methylation is enhanced, and future research should focus on the particulate phase as its relatively rapid sinking makes it an important transport process for Hg in aquatic systems. As Hg^0 formation is an important mechanism in aquatic systems for removing Hg(II) substrate that might otherwise be methylated, future work should also investigate the mechanisms involved in Hg^0 production and should aim to identify the biogeochemical factors that enhance the formation of Hg^0.

ACKNOWLEDGMENTS

We thank the captain, officers, crew, and survey technicians of the NOAA *Malcolm Baldrige* for their assistance. In particular, we thank G. Harvey for inviting us to participate in the cruise; Jim McElroy and others who helped in sample collection; Lloyd Moore for nutrient analysis, and Robert Hopkins for CTD deployment and dissolved O$_2$ analysis. We thank our colleagues at the Marine Sciences Institute for their help, especially Grace Vandal and Jane Knox. This research was supported by the Research Foundation of the University of Connecticut and formed part of Robert Mason's Ph.D. Thesis. This is contribution number 249 of the Marine Sciences Institute, University of Connecticut.

REFERENCES

1. Fitzgerald, W.F., Atmospheric and oceanic cycling of mercury, In: *Chemical Oceanography,* Vol. 10, Riley, J.P. and Chester, R., Eds., Academic Press, London, 1989, 151.

2. Kim, J.K., Volatilization and Efflux of Mercury from Biologically Productive Ocean Regions, Ph.D. Thesis, University of Connecticut, Storrs, 1987.

3. Mason, R.P., The Chemistry of Mercury in the Equatorial Pacific Ocean, Ph.D. Thesis, University of Connecticut, Storrs, 1991.

4. Fitzgerald, W.F., Vandal, G.V., and Mason, R.P., Atmospheric cycling and air-water exchange of mercury over mid-continental lacustrine regions, *Water Air Soil Pollut.,* 56, 745, 1991.

5. Kim, J.K. and Fitzgerald, W.F., An equatorial Pacific source of atmospheric mercury, *Science,* 231, 1131, 1986.

6. Mason, R.P. and Fitzgerald, W.F., Alkylmercury species in the equatorial Pacific, *Nature,* 347, 457, 1990.

7. Vandal, G.M., Fitzgerald, W.F., and Mason, R.P., Cycling of volatile mercury in temperate lakes, *Water Air Soil Pollut.,* 56, 791, 1991.

8. Mason, R.P., Fitzgerald, W.F., Hurley, J.P., Hanson, A.K., Jr., Donaghay, P.L., and Sieburth, J.M., Mercury biogeochemical cycling in a stratified estuary, *Limnol. Oceanogr.,* 38, 1227, 1993.

9. Mason, R.P. and Fitzgerald, W.F., The distribution and biogeochemical cycling of mercury in the equatorial Pacific Ocean, *Deep-Sea Res.,* 40, 1897, 1993.

10. Gill, G.A. and Fitzgerald, W.F., Mercury in the surface waters of the open ocean, *Global Biogeochem. Cycles,* 3, 199–212, 1987a.

11. Chavez, F.P. and Barber, R.T., An estimate of new production in the equatorial Pacific, *Deep-Sea Res.,* 34, 1229, 1987.

12. Eppley, R.W. and Petersen, B.J., Particulate organic matter flux and planktonic new production in the deep ocean, *Nature,* 282, 677, 1979.

13. Murray, J.W., Downs, J.N., Strom, S., Wei, C.-L., and Jannasch, H.W., Nutrient assimilation, export production and ^{234}Th scavenging in the eastern equatorial Pacific, *Deep-Sea Res.,* 36, 1471, 1989.

14. Knauer, G.A. and Martin, J.H., Primary production and carbon-nitrogen fluxes in the upper 1500 m of the northeast Pacific, *Limnol. Oceanogr.,* 26, 181, 1981.

15. Mason, R.P., Fitzgerald, W.F., and Vandal, G.M., The sources and composition of mercury in Pacific Ocean rain, *J. Atmos. Chem.,* 14, 489, 1992.

16. Mason, R.P. and Fitzgerald, W.F., Mercury speciation in open ocean waters, *Water Air Soil Pollut.,* 56, 779, 1991.

17. Bloom, N.S., Determination of picogram levels of methylmercury by aqueous phase ethylation, followed by cryogenic gas chromatography with atomic fluorescence detection, *Can. J. Fish. Aquat. Sci.,* 46, 1131, 1989.

18. Bloom, N.S. and Fitzgerald, W.F., Determination of volatile mercury species at the picogram level by low temperature gas chromatography with cold-vapour atomic fluorescence detector, *Anal. Chim. Acta,* 208, 151, 1988.

19. Gill, G.A. and Fitzgerald, W.F., Mercury sampling of open ocean waters at the picomolar level, *Deep-Sea Res.,* 32, 287, 1985.

20. Gill, G.A. and Fitzgerald, W.F., Picomolar mercury measurement in seawater and other materials using stannous chloride reduction and two-stage gold amalgamation with gas phase detection, *Mar. Chem.,* 20, 227, 1987b.

21. Moore, personal communication.

22. Karl, D.M. and Knauer, G.A., Vertical distribution, transport and exchange of carbon in the northeast Pacific Ocean: evidence for multiple zones of biological activity, *Deep-Sea Res.,* 31, 221, 1984.

23. Topping, G. and Davies, I.M., Methylmercury production in the marine water column, *Nature,* 290, 243, 1981.

24. Gilmour, C.G. and Henry, E.A., Mercury methylation in aquatic systems affected by acid deposition, *Environ. Pollut.,* 71, 131, 1991.

25. Fitzgerald, W.F., Vandal, G.M., Mason, R.P., and Dulac, F., Air-water cycling of mercury in lakes, in: *Mercury Pollution: Integration and Synthesis,* Watras, C.J. and Huckabee, J.W., Eds., Lewis Publishers, Chelsea, MI, 1994, chap. II.3.

26. Alberts, J.J., Schindler, J.E., Miller, R.W., and Nutter, D.E., Elemental mercury evolution mediated by humic acid, *Science,* 184, 895, 1974.

27. Winfrey, M.R. and Rudd, J.W.M., Environmental factors affecting the formation of methylmercury in low pH lakes, *Environ. Toxicol. Chem.,* 9, 853, 1990.

28. Quay, P.D., Stuiver, M., and Broecker, W.S., Upwelling rates for the equatorial Pacific Ocean derived from the bomb ^{14}C distribution, *J. Mar. Res.,* 41, 769, 1983.

29. Wyrtki, K., An estimate of equatorial upwelling in the Pacific, *J. Phys. Oceanogr.,* 5, 1205, 1981.

30. Hastenrath, S., Ocean circulation, and Regional circulation systems, in: *Climate and Circulation of the Tropics,* D. Reidel Publishers, Dordrecht, 1985, Chap. 4 and 6.

31. Gregg, M.C., Diapycnal mixing in the thermocline: a review, *J.G.R.,* 92(C5), 5249, 1987.

32. Peters, H., Gregg, M.C., and Toole, J.M., On the parameterization of equatorial turbulence, *J.G.R.,* 93(C2), 1199, 1988.

33. Clegg, S.L. and Whitfield, M., A generalized model for the scavenging of trace metals in the open ocean. I. Particle cycling, *Deep-Sea Res.,* 37A, 809, 1990.

34. Wyrtki, K. and Kilonsky, B., Mean water and current structure during the Hawaii-Tahiti shuttle experiment, *J. Phys. Oceanogr.,* 14, 242, 1984.

35. Murnane, R.J., Sarmiento, J.L., and Bacon, M.P., Thorium isotopes, particle cycling models and inverse calculations of model rate constants, *J.G.R.,* 95(C9), 16195, 1990.

36. Press, W.H., Flannery, B.P., Teukolsky, S.A., and Vetterling, W.T., *Numerical Recipes,* Cambridge University Press, Cambridge, 1986.

37. Watson, J.A., Upstill-Goddard, R.C., and Liss, P.S., Air-sea exchange in rough and stormy seas measured by a dual-tracer technique, *Nature,* 349, 145, 1990.

38. Berger, W.H., Smetacek, V.S., and Wefer, G., Appendix: Global Maps of Ocean Productivity, in: *Productivity of the Oceans: Present and Past,* Berger et al., Eds., John Wiley & Sons, New York, 1989, 429.

39. Hurley, J.P., Watras, C.J., and Bloom, N.S., Mercury cycling in northern Wisconsin seepage lakes: the role of particulate matter in vertical transport, *Water Air Soil Pollut.,* 56, 543, 1991.

Aqueous and Biotic Mercury Concentrations in Boreal Lakes: Model Predictions and Observations

Markus Meili

CONTENTS

I. INTRODUCTION

The large variation of Hg concentrations in lacustrine fish has been the concern of many studies. Statistical relationships have been found with a variety of chemical and geomorphological variables.[1-4] When searching for functional relationships, however, attention should be focused on biological variables and processes because of the strong interaction of Hg with organic matter in natural systems.[6]

Two of the key factors determining the Hg concentration in large animals are the flux control of the Hg concentration in the environment and the transfer of Hg from abiotic compartments into the low levels of the food web.[5] These processes are addressed here by a synthesis of: a biogeochemical concept of mercury cycling in the boreal forest zone,[6] model predictions based on water quality data,[7] and validations using observed Hg concentrations in surface waters[8] and fish.[9]

II. Hg IN LAKE WATERS: QUANTIFYING EXPOSURE

The uptake of Hg by lacustrine organisms is directly or indirectly related to the concentration and bioavailability of Hg in the lake water. Concentrations of total Hg in the surface water of boreal forest lakes are strongly related to the concentration of humic matter originating from surrounding forest soils.[8,10,11] Accordingly, Hg concentrations are correlated to water color (Figure 1), which is a good measure of allochthonous humic organic carbon in boreal lakes and streams.[7] The intercept of the relationship suggests that in clearwater lakes, concentrations of total Hg in surface water are around 1 ng L^{-1}, which agrees with other studies.[12] In humic lakes, however, mean concentrations can reach values above 5 ng L^{-1}.[8]

Thus far, aqueous Hg concentrations have only been measured in a few of the hundreds of thousands of boreal lakes. Variations within lakes are large, and the effort required to obtain reliable mean values is considerable, especially due to the necessary clean sampling protocols.[8] However, a conceptually derived model can be used to predict Hg concentrations in boreal lake waters.[6] Hg input to forest lakes is dominated either by input from the watershed or by direct deposition on the lake surface.[13] Accordingly, concentrations of total Hg in the surface water can be calculated as the sum of the contributions from each of these fluxes.[6]

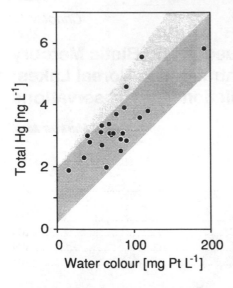

Figure 1 Relationship between total Hg concentration (oxidation with BrCl followed by reduction with $SnCl_2$) and water color in the surface water of 23 Swedish forest lakes. Observed summer mean values (n = 3) and predicted range (shaded area) in small head-water lakes with a mixed surface layer of 2 to 5 m, situated in forested areas with an annual Hg wet deposition of 7 to 25 μg m^{-2} and a surface runoff with a color/orgC ratio of 10 gPt (gC)$^{-1}$ and a Hg/orgC ratio of 0.25 μg g^{-1} (or up to 0.35 μg g^{-1} in areas situated 10 to 100 km from major atmospheric point sources of Hg,[6] lightly shaded area).

The primary source of most Hg in remote watersheds is atmospheric deposition. Hg concentrations in precipitation are likely to be related to overall air pollution and acidification.[6,14] Indeed, a combination of data from recent surveys over remote regions of northern Europe shows a strong relationship between Hg concentrations and pH in precipitation (Figure 2). Mean Hg concentrations during 1985 through 1989 were proportional to the proton concentration, with a Hg/H$^+$ ratio around 380 μg g^{-1}. In Sweden, this corresponds to a Hg/S ratio of 15 μg g^{-1}, which is similar to the ratio found in coal.[6] Mean proton concentrations were obtained from regular national surveys of precipitation pH,[6] and Hg data from a Nordic monitoring network[15] which included 7 remote stations situated at a distance >100 km from major Hg emission sources and large cities and with a mean precipitation pH ≥4.3.

The contribution from direct deposition to the Hg concentration in surficial lake waters can be predicted from the amount and contamination of precipitation, the depth of the mixed surface water layer, and the residence time of Hg in this layer, thereby assuming that net dry deposition (deposition minus volatilization) is negligible over lake surfaces.[6] The resulting estimate for most lakes varies between 0.3 and 2 ng L^{-1} (Figure 1) and is thus in agreement with field measurements in rural seepage lakes.[12] The contribution from the watershed can be estimated from the concentration of humic substances (converted from the water color in the lake) and by the Hg contamination of stream humic matter, which in many remote areas of the Swedish forest region is around 0.25 μg (g C)$^{-1}$.[6] By adding this terrestrial Hg contribution to the atmospheric contribution, mean concentrations of total Hg in the surface water of Swedish forest lakes can be conveniently estimated with similar confidence as in purely statistical models, and without lake-specific flux measurements although using a mechanistic approach.[6]

III. Hg IN FOOD WEBS: QUANTIFYING BIODILUTION

The uptake of Hg by biota is a product of environmental exposure and susceptibility. In the Swedish forest region, atmospheric Hg deposition varies about threefold[13] (Figure 2), whereas Hg concentrations in fish vary about tenfold between lakes in this area,[1] and about fourfold within small areas with similar deposition.[9] This large variability in susceptibility can likely be ascribed to biotic factors in watersheds and lakes, given the strong interaction of Hg with organic matter in boreal systems.[6] This is in line with the mercury/biomass (Hg/B) concept,[6]

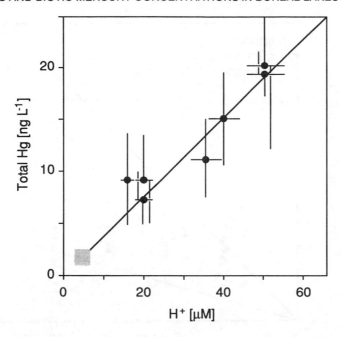

Figure 2 Relationship between concentrations of total Hg (oxidation with BrCl followed by reduction with $SnCl_2$) and H^+ in precipitation over remote areas of northern Europe. The shaded area shows the range of natural background concentrations.[6] The line describes a proportional weight relationship of $Hg = 0.00038 \cdot H^+$. Error bars denote 95% confidence limits of mean Hg concentrations ($n = 22$ to 200) and ± 0.05 pH units of mean proton concentrations. Hg data from Iverfeldt[15] and H^+ data from national surveys.[6]

according to which the biogeochemical cycling of Hg is controlled by the cycling of biogenic matter (see previous section).

Accordingly, concentrations of Hg and transitions from one compartment to another are dependent on the abundance and type of organic matrices. Hg concentrations in fish are well correlated with the concentrations in other animals from the same lakes.[6,9] On the other hand, Hg concentrations in fish observed in the study lakes show no correlation with Hg concentrations in lake sediments[9] which are often used as pollution indicators. This apparent "uncoupling" of biotic and abiotic compartments within systems indicates that most of the variability in food web contamination is induced at the transition of Hg from the abiotic to the biotic phase.[5] The Hg/B approach suggests that Hg concentrations in biota are proportional to the aqueous Hg concentration and, as a result of the high affinity of Hg for proteins, inversely proportional to the biomass at the base of the food web by which the available Hg is shared (biodilution).[6] Accordingly, biotic Hg concentration should be related to the ratio of Hg and autochthonous carbon (AOC) in the water column. A convenient measure for the standing stock of organic matter produced within lakes is the concentration of available nutrients.[7] In boreal lakes, the concentration of AOC (mg L^{-1}) can be estimated from phosphorus concentration ($\mu g\ L^{-1}$) and water color (mgPt L^{-1}) as $AOC = 0.35 \cdot (P - 0.05\ WC)$.[7] Such estimates, which are supported by stable carbon isotope ratios,[16] can be used to quantify biodilution which has so far mainly been used as an abstract term. Quantification is possible even for humic lakes,[7] where it is difficult to selectively measure AOC because it is operationally impossible to separate organic matter from different sources, and because of the intense recycling and mixing in the microbial food web.[16]

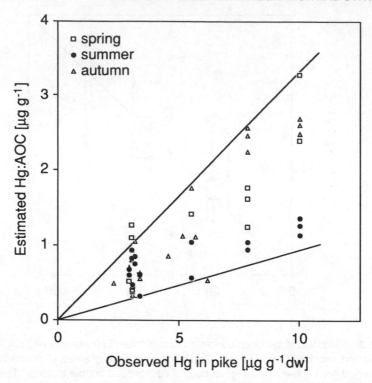

Figure 3 Estimated ratios of total Hg to autochthonous organic carbon (Hg/AOC) during different seasons in 16 Swedish forest lakes, vs. observed concentrations of total ($\approx$ methylated) mercury in the flesh of pike (*Esox lucius* L.) given as normalized mean values in spring at a somatic body weight of 1 kg and a sex ratio of 1:1. Lines show proportionality coefficients of 3 and 10.

The estimated Hg/AOC ratios are poorly correlated with the estimated Hg concentrations in lake water, despite the expected autocorrelation. Because of the large natural variability in AOC, Hg/AOC ratios vary about tenfold between lakes and seasons (typically 0.3 to 3 μg g^{-1}), whereas estimated aqueous Hg concentrations only vary about fourfold including seasonal changes (typically 1.5 to 5.5 ng L^{-1}). This relationship agrees with the above-mentioned field studies showing that the variability of Hg concentrations in lacustrine biota is much larger than the variability in Hg exposure. These findings suggest that biodilution alone may explain much of the variability in Hg concentration of lacustrine biota.

The applicability of this Hg/B approach can be tested by comparing Hg/AOC ratios with observed Hg concentrations in fish from the same lakes, e.g., in pike (*Esox lucius*). As predicted, estimated Hg/AOC ratios are not only well correlated, but are also proportional to observed Hg concentration in pike (Figure 3), and thus to the Hg concentrations in other animals from the same lakes.[9] Logarithmic lake mean values show correlation coefficients of 0.77 for all seasonal means (n = 28, p <0.0001); 0.77 for annual means (n = 13, p = 0.002); 0.74 in summer (n = 8, p <0.05); 0.86 in spring (n = 7, p = 0.01); and 0.80 in fall (n = 13, p = 0.001). The Hg/B concept has previously been validated with summer data from another, larger set of boreal lakes.[6]

IV. LAKE ACIDITY: AN OVEREMPHASIZED FACTOR?

Many survey studies have shown a correlation between Hg concentrations in lacustrine biota and lake water acidity (not to be confused with acidification).[2,3,17-19] In contrast, residuals of predictions based on the Hg/B approach in this study show no relationship with lake water pH covering over two orders of magnitude in acidity. Apparently, the Hg/B approach accounts for natural variations in acidity. This can be explained by the well-known correlation of lake productivity with lake water pH among temperate lakes, as was also found in this study ($p = 0.001$, Figure 4a), and is supported by the strong relationship of Hg/AOC ratios with the observed surface water pH (Figure 4b).

Consequently, high Hg levels in biota from acidic lakes may be a result of low productivity rather than of a direct influence of pH on the Hg cycle. While evidence is overwhelming that most Hg in natural systems is bound to some type of organic matter, be it in dissolved, detrital, or live phase,[4-6,8-13,17] experimental and empirical studies of the influence of pH on the Hg cycle provide contradictory results.[19,20] Figure 4b and many survey data[2,17,18] show about a doubling of fish Hg per unit pH. However, experimental acidification or liming of lakes rarely have such dramatic effects.[18,21] This implies that neither direct (e.g., chemical) nor indirect (e.g., biological) effects of changes in lake water pH can account for its strong empirical relationship with fish Hg among temperate lakes. Statistically significant correlations of acidity variables with biotic Hg concentrations may result from collinearity with both concentration and biodilution of Hg,[4,14,18,19] given the correlations between lake water pH, precipitation pH, and Hg loading to lakes and their catchment (Figure 2), and between weathering rate, buffering capacity, and nutrient concentration (Figure 4a), as well as between water color and pH resulting from the natural acidity of humic substances.[17]

If acidity accounted for the large variations in concentrations of fish Hg, which is mostly methylated, either the transformation of inorganic into methylmercury (MeHg) or the affinity of MeHg to organic matter and living organisms should be highly variable and correlated to pH. In contrast, the proportionality of biotic Hg concentrations and aqueous Hg/AOC ratios implicitly presupposes that the mean proportion of methylated Hg in total aqueous Hg (MeHg/TotHg) is similar among lakes during the growing season. Available data from two recent studies,[22,23] in fact, suggest that annual mean proportions rarely vary more than twofold between remote lakes. In a Swedish study, the mean proportions of MeHg/TotHg in lake surface water were around 4%, and not correlated to pH.[22] In four of five sites from a Wisconsin study based on different analytical methods, annual mean proportions were around 10% and unexpectedly showed a weak positive relationship with pH, while one acidic lake had an unusually high value of 26%.[23] The values of the second study are volume weighted and may not represent the productive environment, as they comprise winter samples and include hypolimnetic waters which are potentially enriched in MeHg.[12] Neither of the two studies gives support to an elevated net methylation in acidic waters.

If acidity increased the bioavailability of formed MeHg, pH would be correlated to *in situ* mean distribution coefficients (particulate MeHg/dissolved MeHg) or bioconcentration factors (fish MeHg/dissolved MeHg). However, no such relationship could be found in a study covering three orders of magnitude in acidity.[23] In the same study, bioconcentration factors showed a much larger variability than distribution coefficients, again suggesting that biological processes are responsible for the between-lake variation in food web contamination with Hg.

In conclusion, several lines of evidence suggest that Hg concentrations in lacustrine biota are largely regulated by the turnover of organic matter, particularly by the pelagic bioproductivity determining biodilution, whereas lake water acidity appears to be of minor causal importance. Severe acidification may, however, indirectly contribute to increased Hg levels in biota within a given lake, if biodilution is affected via a decrease in bioproduction.

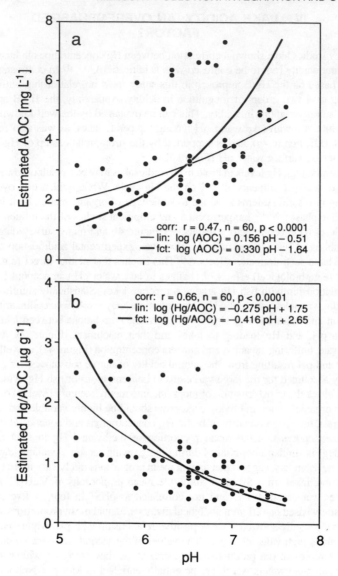

Figure 4 Relationship of (a) estimated concentrations of autochthonous organic carbon (AOC), and (b) estimated ratios of total Hg to AOC (Hg/AOC) with observed surface water pH in 16 Swedish forest lakes during different periods of the ice-free season. Correlation data as well as the solutions of simple linear (thin line) and functional[24] (bold line) regressions are given.

ACKNOWLEDGMENTS

The comments provided by two anonymous referees are gratefully acknowledged.

REFERENCES

1. Håkanson, L., Nilsson, Å., and Andersson, T., Mercury in fish in Swedish lakes, *Environ. Pollut.*, 49, 145, 1988.
2. Lathrop, R. C., Rasmussen, P. W., and Knauer, D. R., Mercury concentrations in walleyes from Wisconsin (USA) lakes, *Water Air Soil Pollut.*, 56, 295, 1991.
3. Wren, C. D., Scheider, W. A., Wales, D. L., Muncaster, B. W., and Gray, I. M., Relation between mercury concentrations in walleye (*Stizostedion vitreum vitreum*) and northern pike (*Esox lucius*) in Ontario lakes and influence of environmental factors, *Can. J. Fish. Aquat. Sci.*, 48, 132, 1991.
4. Verta, M., Mercury in Finnish forest lakes and reservoirs: anthropogenic contribution to the load and accumulation in fish, National Board of Waters and the Environment, Helsinki, *Publ. Water Environ. Res. Inst.*, 6, 33, 1990.
5. Meili, M., Mercury in boreal forest lake ecosystems, *Acta Univ. Ups.*, 336, 36, 1991.
6. Meili, M., The coupling of mercury and organic matter in the biogeochemical cycle— towards a mechanistic model for the boreal forest zone, *Water Air Soil Pollut.*, 56, 333, 1991.
7. Meili, M., Sources, concentrations and characteristics of organic matter in softwater lakes and streams of the Swedish forest region, *Hydrobiologia*, 229, 23, 1992.
8. Meili, M., Iverfeldt, Å., and Håkanson, L., Mercury in the surface water of Swedish forest lakes: concentrations, speciation and controlling factors, *Water Air Soil Pollut.*, 56, 439, 1991.
9. Meili, M., Mercury in forest lake ecosystems—bioavailability, bioaccumulation and biomagnification, *Water Air Soil Pollut.*, 55, 131, 1991.
10. Johansson, K. and Iverfeldt, Å., Factors influencing the run off of mercury from small watersheds in Sweden, *Verh. Int. Verein. Limnol.*, 24, 2200, 1991.
11. Mierle, G. and Ingram, R., The role of humic substances in the mobilization of mercury from watersheds, *Water Air Soil Pollut.*, 56, 349, 1991.
12. Hurley, J. P., Watras, C. J., and Bloom, N. S., Mercury cycling in a northern Wisconsin seepage lake: the role of particulate matter in vertical transport, *Water Air Soil Pollut.*, 56, 543, 1991.
13. Meili, M., Fluxes, pools and turnover of mercury in Swedish forest lakes, *Water Air Soil Pollut.*, 56, 719, 1991.
14. Håkanson, L., Nilsson, Å., and Andersson, T., Mercury in fish in Swedish lakes—linkages to domestic and European sources of emission, *Water Air Soil Pollut.*, 50, 171, 1990.
15. Iverfeldt, Å., Occurrence and turnover of atmospheric mercury over the Nordic countries, *Water Air Soil Pollut.*, 56, 251, 1991.
16. Meili, M., Fry, B., Kling, G. W., Bell, R. T., and Ahlgren, I., The utilization of humic substances in pelagic food webs: evidence from natural stable isotopes (^{13}C, ^{15}N) in boreal forest lakes, *Abst. ASLO 92 Aqu. Sci. Meet.*, 1992, and in preparation.
17. Mannio, J., Verta, M., Kortelainen, P., and Rekolainen, S., The effect of water quality on the mercury concentration of northern pike (*Esox lucius* L.) in Finnish forest lakes and reservoirs, National Board of Waters and the Environment, Helsinki, *Publ. Water Environ. Res. Inst.*, 65, 32, 1986.
18. Håkanson, L., Measures to reduce mercury in fish, *Water Air Soil Pollut.*, 55, 193, 1991.
19. Richman, L. A., Wren, C. D., and Stokes, P. M., Facts and fallacies concerning mercury uptake by fish in acid-stressed lakes. *Water Air Soil Pollut.*, 37, 465, 1988.

20. Winfrey, M. R. and Rudd, J. M. W., Environmental factors affecting the formation of methylmercury in low pH lakes, *Environ. Toxicol. Chem.,* 9, 853, 1990.
21. Wiener, J. G., Fitzgerald, W. F., Watras, C. J., and Rada, R. G., Partitioning and bioavailability of mercury in an experimentally acidified Wisconsin lake, *Environ. Toxicol. Chem.,* 9, 909, 1990.
22. Lee, Y. H. and Iverfeldt, Å., Methylmercury in run-off, lake and rain waters, *Water Air Soil Pollut.,* 56, 309, 1991.
23. Bloom, N. S., Watras, C. J., and Hurley, J. P., Impact of acidification on the methylmercury cycle of remote seepage lakes, *Water Air Soil Pollut.,* 56, 477, 1991.
24. Ricker, W. E., Linear regressions in fishery research, *J. Fish. Res. Board Can.,* 30, 409, 1973.

Methylmercury in a Permanently Stratified Fiord

Helena Parkman, Per Östlund, Matts-Ola Samuelsson, and Åke Iverfeldt

CONTENTS

ABSTRACT: In order to identify possible mechanisms for mercury methylation, sample collection and in situ incubation experiments were carried out in a Norwegian fiord providing differentiated environmental factors. The fiord is permanently stratified by two water masses with different salinities. The bottom water is chemically reduced and a redoxcline is formed in meter-scale between the water masses at about 20 m water depth. Different species of microorganisms are distributed over the redoxcline. Measurements of mercury in the water were made on water sampled on two occasions, summer (September 1991) and fall (November 1990), in filtered (0.2 μm) and unfiltered water at and on both sides of the redoxcline. Total mercury data from only the fall sampling are presented and were not found to vary significantly with depth. The methylmercury concentrations at and below the redoxcline were much larger than above the redoxcline, with higher levels in summer. The very small differences between filtered and unfiltered samples suggest that both inorganic and methylmercury, to a large extent, are either associated with colloids or are dissolved species. Results from in situ incubations, with additions of mercury to filtered and unfiltered water samples, suggest that mercury can be methylated without the presence of intact bacteria. In contrast, demethylation seems to take place primarily in the presence of bacteria.

    In order to explain the large differences in methylated mercury activities on both sides of the redoxcline, bacterial and exoenzyme activities were measured. The microbial activity was comparatively larger above the redoxcline, and the exoenzyme activity was larger below the redoxcline.

    Our results suggest that the amount of measured methylmercury is the net result of two processes—accidental methylation of Hg(II) by exoproducts in the chemical surrounding of bacterial cells, and active demethylation by the cells.

I. INTRODUCTION

Mercury (Hg) methylation and demethylation is reported to take place in many different matrices by a large number of processes, both biotic[1,2] and abiotic.[3,4] In aquatic systems, microbial processes in mainly anoxic sediments[5-7] (sulfate reducing bacteria[8]) are commonly held responsible. The mechanisms or reasons for these processes are poorly understood. Natural systems are complex and processes are difficult to distinguish from each other. Hence, results from mechanism studies in such systems are often difficult to interpret. On the other hand, results from simplified systems like pure culture experiments are of uncertain relevance to processes in nature. Consequently, a system with marked chemical and physical gradients offering a selective pressure on biological systems, resulting in markedly differentiated biological species, would be of great value for studies on Hg speciation. Such a system is found in the Norwegian fiord, Framvaren.

A. THE FRAMVAREN FIORD

Framvaren is a narrow fiord in the southernmost part of Norway. The catchment area is 31 km^2 and comprises several small lakes,[9] and mountains descending steeply into the fiord. The area is sparsely inhabited (population: 100). The precipitation is about 1000 mm year^{-1} and the overall freshwater input through five small streams has been estimated to ~1 to 2 m^3 sec.$^{-1}$ (annual average).[9] Framvaren is composed of two basins with volumes of 3.3 × 10^8 and 0.6 × 10^8 m^3, respectively. The total surface area of the fiord is 5.8 km^2 and the maximum depth is 183 m.[9] The water exchange between Framvaren and the Atlantic ocean (Skagerack) is hampered by shallow sills and channels of which the innermost, which connects Framvaren with the Helvikfjord, has a threshold depth of about 2 m. The water exchange over this sill is driven by the fluctuating sea-level difference between the two basins.[10] Owing to the steep morphometry, the shallow inlet, and the low rate of sea water inflow and exchange, two water masses with different salinities and, hence, densities are formed in Framvaren. The fiord is almost permanently stratified and a total turnover was reported only once during this century.[9] The water down to about 20 m depth is oxygenated, has a salinity of ~1%, and a pH that varies between 7 and 8.1 (1965 through 1985).[11] Below 20 m, the water is anoxic and a redoxcline is formed, the salinity is ~2%, and the pH is between 6.8 and 7.1.[11] The water masses are slowly mixed by diffusive forces and therefore large chemical gradients are formed and maintained, both over the redoxcline and in the total water body. For instance, the concentrations of H_2S ranges from 0 at the redoxcline to 8 mM in the deep water, and PO_4 increases constantly with depth to ~0.1 mM. Most metals either have a maximum at the redoxcline or increase successively down to a depth of about 100 m. In the bottom water, below 100 m, changes in salinity and chemistry are small.[9]

Exchange between the water masses is governed by diffusion and sedimentation. At 20 m depth, the density differences between the water masses causes sedimenting particles (organic and minerogenic matter) to accumulate. Most of the degradation of organic matter in the water mass takes place at the redoxcline.[12] The accumulation of particulate matter causes a light transmission minimum.[9] The redox shift causes remobilization of metals and other substances in settling Fe and Mn oxyhydroxides that are reduced and dissolved when crossing the redoxcline. However, metals are precipitated as sulfides in the sulfide-rich water below the redoxcline. Maximum amounts of particulate and total Fe, as well as total Pb, Mn, Co, and dissolved gaseous and dissolved "reactive" Hg have been observed at this boundary layer.[11,13,14] These conditions across the redoxcline cause a firm selective pressure resulting in a vertical zonation of different species of microorganisms. Similar zonation, extended over a couple of meters in Framvaren, is usually found within a few millimeters in oxic/anoxic sediments. The Framvaren system therefore combines advantages offered in natural systems with those in laboratory systems.

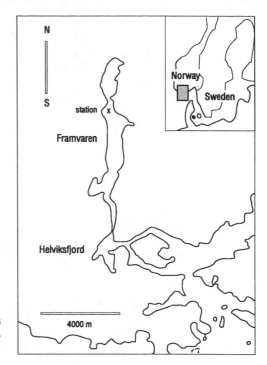

Figure 1 Map of Framvaren and its location in southern Norway. The sampling station is indicated by **X**.

B. OBJECTIVES

The main objectives were to study under natural conditions

- In situ levels of total and methylmercury which can be related to specific chemical environments
- In situ levels of total and methylmercury which can be related to specific microbial processes
- Methylation/demethylation capacities and rates by in situ incubations in specific environments with specific microbial processes

II. METHODS

All plastics and glass used were carefully washed, including leaching procedures in alkaline detergent as well as strong and weak acid, followed by an extensive rinsing with Milli-Q® water. The borosilicate incubation bottles used in November were autoclaved with 7 M HNO_3 (120° C, 1 h) before careful rinsing with Milli-Q® water. All clean material was kept in double plastic bags, and double plastic gloves were always used when handling open bottles.

Framvaren was visited November 2 through 6, 1990 and September 2 through 5, 1991. A buoy connected to a concrete anchor via a nylon line was set out in the northern part of the outer basin, and the redoxcline was localized (Figure 1); the next day, water samples were taken in a vertical profile. The vertical distances between the samples were small (50 cm) across the redoxcline. An acid-cleaned GoFlo sampler was used in November, and the water was poured into plastic (PE) bottles immediately after sampling. In September, the water was pumped into borosilicate glass bottles via acid-cleaned Teflon® tubings, using a peristaltic pump. Temperature and Eh are measured in separate samples, using an alcohol thermometer and Pt vs. calomel electrodes, respectively.

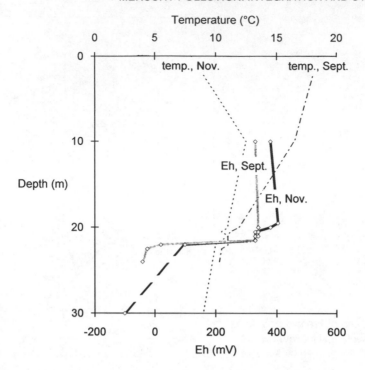

Figure 2   Temperature and redox potential (Eh) in Framvaren, November 3, 1990 and September 3, 1991.

In November, handling of the samples was carried out directly on the shore on a cart covered with plastic. In September, the samples were brought to a temporary laboratory. The water samples were divided into several 120-mL borosilicate bottles that were filled to the brim and capped with polypropylene screw caps. Some of the samples were first gently pressfiltered (0.2 MPa, 0.2 μm, individually packed sterile S&S membrane filter) in an N_2 atmosphere. Samples for measurements of in situ concentrations of methylmercury (MeHg) and total mercury (TotHg) were preserved with HCl (1 mL 6 M Merck Suprapur® in each 120-mL sample).

In November, an in situ incubation experiment was carried out. Mercury was added (1 mL, 5 μg L^{-1} $Hg(NO_3)_2$, Merck p.a.) to two bottles with filtered water and two bottles with unfiltered water from each depth, resulting in a concentration of 40 ng L^{-1} Hg above the natural level. The bottles were placed in netbags and incubated in situ at their original depth for 1 and 3 days. The incubations were stopped and the samples preserved with HCl.

In September, the bacterial activity was measured using the [³H]thymidine incorporation method.[15] Ten milliliters of water were incubated in the dark with [³H]thymidine for 100 min at 19° C, and the incubation was stopped by the addition of formaldehyde (2%) in phosphate buffer. The [³H] activity was measured with a liquid scintillation spectrometer. The [³H] activities in unfiltered samples were adjusted for the [³H] activity in filtered samples by subtraction.

The protease (exoenzyme) activity was measured by incubation of water samples with Hide Powder Azur (HPA, 1 mg mL^{-1}), a substance which is protolyzed by exoenzymes. A colored complex was formed that was proportional to the enzyme activity, and spectrophotometrically measured at 595 nm.[16]

TotHg was determined after oxidation with BrCl in 6 M HCl followed by prereduction in 12% $NH_2OH·HCl$ solution and reduction with $SnCl_2$ before being volatilized. Mercury

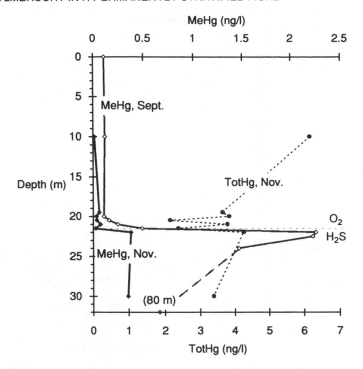

Figure 3 Depth profiles of total mercury (TotHg) (November 3, 1990) and methylmercury (MeHg) (November 3, 1990 and September 3, 1991) in Framvaren. The location of the redoxcline is indicated by a dotted line.

was determined by atomic fluorescence spectrometry after preconcentration by double amalgamation.[17]

MeHg was determined by aqueous phase ethylation, followed by cryogenic gas chromatography with cold vapor atomic fluorescence detection.[18]

III. RESULTS

Above 20 m depth the water temperatures were higher in late summer 1991 (September) than in fall 1990 (November), while below 20 m only small differences were found in the temperature—in the range of 10 to 12° C (Figure 2). On both occasions the redox potential (Eh) was >300 mV throughout the oxic part of the water mass (Figure 2). The redoxcline was found between 21.5 and 22 m depth, where Eh dropped to close to 0 mV. Below this depth the water samples had an odor of H_2S.

The TotHg concentrations (November 1990) ranged between 2.14 and 6.15 ng L^{-1} and the highest value was found at the 10-m depth (Figure 3). The in situ concentrations of MeHg (November) above the redox cline were 0.01 and 0.07 ng L^{-1}, while the concentrations at and below the cline were in the range of 0.35 to 0.38 ng L^{-1} (Figure 2). In September, the MeHg concentrations ranged between 0.11 to 0.57 ng L^{-1} and 1.47 to 2.23 ng L^{-1} above and below the cline, respectively. On both sampling occasions the highest concentrations were found at 22 m.

The in situ MeHg concentrations in filtered water (only September) were similar to those in unfiltered water (Figure 4). A minimum (~70%) in the relative amount of MeHg in filtered relative to MeHg in unfiltered water was found at 22 m. This minimum coincided with a

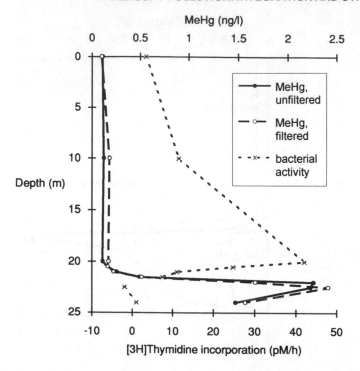

Figure 4 Depth profiles of methylmercury concentrations (MeHg) in unfiltered and filtered water and bacterial activity ([³H]thymidine incorporation) in Framvaren, September 3, 1991.

maximum in particulate matter, semiqualitatively measured as the number of filters needed for filtering 1.2 L of water. The water samples and the filters were distinctly colored red between 20 and 22.5 m, while the filters from 24 and 30 m were grayish.

The bacterial activity ([³H]thymidine incorporation) in September at the surface was 3.5 pM h^{-1}, was highest at 20 m (42 pM h^{-1}), and was drastically lower (0 to 1 pM h^{-1}) below the redoxcline (Figure 4) (the 22-m sample was accidentally lost). The activities in the filtered samples were equivalent to 2.5 to 5 pM h^{-1} [³H]thymidine incorporation. These values were subtracted from the activity for the unfiltered waters.

The HPA activity analyses executed immediately after sampling did not result in any absorbance, owing to too short an incubation period. On returning to the laboratory, HPA powder was added to water samples from various depths and left for 4 days. There may be some doubt about the validity of this analysis and that is why only the measured absorbancy values (595 nm) are presented here (means of two samples). The values for unfiltered waters were 0.003 (21 m), 0.007 (22 m), and 0.112 (24 m), while the absorbancies for filtered waters were 0.004, 0.001, and 0.008, respectively.

The results from the incubation experiment performed in situ in November, where $Hg(NO_3)_2$ was added to filtered and unfiltered water resulting in a level of 40 ng L^{-1} over the natural concentrations, are shown in Figure 5. The concentrations of MeHg after 3 days of incubation were usually higher than after 1 day, and the concentrations in filtered samples were usually higher than in the unfiltered samples. In the filtered waters, at most depths, the concentrations after incubation were higher than the initial MeHg concentrations, and the largest increases (3 to 4 times the initial) took place at 30 m. In the unfiltered waters, incubation usually resulted in a decrease of MeHg. However, after 3 days at 30 m, MeHg was as high as in the filtered sample (4 times the initial concentration).

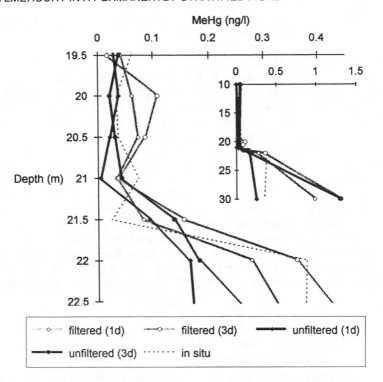

Figure 5 Results of in situ incubations in Framvaren, November 3 through 6, 1990. Hg(NO$_3$)$_2$ to a concentration of 40 ng L^{-1} over the natural was added to bottles with filtered and unfiltered water. Methylmercury (MeHg) in the bottles close to the redoxcline after 1 and 3 days of incubation is shown in the blowup. The inserted graph shows the results from the total profile. In situ MeHg concentrations in unfiltered water at day 0 are also shown.

Due to analytical disturbances, TotHg values from September are not reported. The reason for these disturbances is not clear, but is tentatively related to the Framvaren water matrix. This matrix effect was not present in the MeHg analyses of the samples collected in September.

IV. DISCUSSION

The TotHg concentrations in the upper 30 m of the water column in Framvaren (2 to 6 ng L^{-1}, measured in November) agreed with the total mercury concentrations in surface waters of Swedish forest lakes.[19] This is also to be expected as a consequence of the common Hg source: atmospheric deposition. Considerably varying concentrations in "reactive and non-reactive" Hg have previously been measured in Framvaren.[14] In September 1983, the concentrations above the redoxcline were similar to the total concentrations found in the present study (November 1990). Below 22 m, the concentrations were ~10 ng L^{-1} and a maximum (30 ng L^{-1}) was found in association with the redoxcline. The concentrations in February 1985 were much lower, 0.2 to 1.6 ng L^{-1}, throughout the water column. Such large seasonal variations cannot be explained by variations with time in the deposition of Hg in the area (35 µg m^{-2} year $^{-1}$)[20] and may at least partly be ascribed to a failure in the analysis of TotHg in the samples collected near the redoxcline in 1985, which was also reported.[14] The same effect cannot be ruled out completely in the present study. The TotHg material from 1983 was derived after a somewhat different pretreatment procedure, which may have been more adequate for the samples near and below the redoxcline.

A. MERCURY SPECIATION

Hg speciation in freshwater is reported to be highly dependent on the input from the catchment area, and Hg is suggested to be associated to organic matter.[19] The relatively small catchment area and the meager soils around Framvaren suggest that direct deposition could be comparably significant for the input and speciation of Hg in the fiord. Both ''reactive and nonreactive'' Hg as well as MeHg in Framvaren (this study) exist almost exclusively in dissolved forms[14] (''reactive + nonreactive'' Hg <0.4, MeHg <0.2 μm), except at one depth just below the redoxcline (22 m), where the dissolved made up ~70% of total MeHg. A large proportion (sometimes >50%) of TotHg in the oxic water is dissolved gaseous Hg (DGM), which indicates effluxes of Hg to the atmosphere.[14] Considering the negligible differences in Hg between filtered and unfiltered water, the MeHg and ''reactive + nonreactive'' Hg (excluding DGM) in the upper water mass is probably bound to small (<0.2 μm) particulates or is present as dissolved complexes. Below the redoxcline, $HgS(s)$ is expected to form. However, the lack of difference between filtered and unfiltered samples may be a consequence of the high concentrations of S^{2-} (0 to 8 mM)[21] which increases the otherwise extremely low solubility of $HgS(s)$. For example, the following equilibria may occur;[22]

$$HgS(s) = Hg^{2+} + S^{2-} \qquad \qquad logK_{sp} = -53 \qquad \qquad (1)$$

$$HgS(s) + H^+ = Hg^{2+} + H_2S \qquad logK_{sp} = -30.8 \qquad (2)$$

$$HgS(s) + H_2S(g) = Hg(SH)_2 \qquad logK_{sp} = -6.2 \qquad (3)$$

$$HgS(s) + S^{2-} = HgS_2^{2-} \qquad \qquad logK_{sp} = -1.5 \qquad (4)$$

This suggests that inorganic Hg may also exist as HgS_2^{2-} or $Hg(SH)_2$ in the intermediate to slightly acid water. MeHg below the redoxcline is probably complexed to sulfides or low molecular organic matter such as CH_3HgSH or CH_3HgSR.[23]

B. MICROBIAL PROCESSES

The maximum bacterial activity in Framvaren (42 pM incorporated [^{3}H]thymidine h^{-1}) was found in the oxidizing water at 20 m (~1 m above the redoxcline) where the microbial flora probably consists of a mixture of different aerobic heterotrophic bacteria.[24] In comparison, incorporation rates of up to 4 pM h^{-1} was measured in the nutrient-rich Rhône River plume (in January),[25] indicating a rather high bacterial activity over the redoxcline in Framvaren. No correlation was found between MeHg concentrations and bacterial activity, although the maximum in MeHg concentration at 22 m depth coincided with a maximum in particle density (>0.2 μm, semiquantitatively measured), which visually consisted of purple and green (sulfur) bacteria. According to the [^{3}H]thymidine incorporation method, bacterial activity below the redoxcline was very low, which could mean that the bacteria observed by eye were inactive. However, there have been suggestions that the [^{3}H]thymidine incorporation method drastically underestimates bacterial production in systems where sulfate reduction is quantitatively important.[26] Our semiquantitative observations may support this statement.

Previous reports on bacterial biomass (measured as adenosine triphosphate) and bacterial chlorophyll in Framvaren showed maxima in bacterial biomass at the redoxcline, with *Chromatium* sp. above *Chlorobium* sp., but also showed active biomass several meters below the cline where no chlorophyll was detected.[24,27] In our study, the water at the redoxcline was red, while the color at 22 m was a more brownish-red, indicating a mixture of red and green sulfur bacteria. At this depth, MeHg concentrations were highest and a minimum in the relative amount of dissolved MeHg was also found (~70%). This agrees with results from Little Rock Lake, Wisconsin, U.S. where a 'dramatic' peak in particulate MeHg (80% of

TotHg) was found at the redoxcline interface, a layer primarily containing phototrophic green sulfur bacteria,[28] indicating that these bacteria either produce MeHg or are exposed to comparatively large amounts of MeHg bound to accumulated particulates.

The incubation experiments show a comparatively high demethylation capacity at 22 m, indicating that the high in situ concentrations of MeHg at this level is a consequence of accumulation of MeHg-bearing matter, rather than an in situ production. Genes coding for Hg resistance have been located in several types of bacteria, and the cell-bound detoxification process involves both cleavage of carbon-Hg bonds as well as reduction of Hg^{2+} to volatile elemental Hg.[29] Therefore, addition of inorganic Hg may lead to a decrease in MeHg, owing to an activation of the detoxification system. This is further supported by the fact that maximum in DGM have been measured at a corresponding depth.[14]

The incubation experiments also resulted in some trends, valid for most depths in the profile, although the variability is considerable. Addition of inorganic Hg resulted, at least initially, in more MeHg in samples without cells (>0.2 µm), compared to samples where cells were present. In comparison with in situ concentrations measured before the incubations, MeHg increased in cell-free waters after addition of $Hg(NO_3)_2$ while the immediate effect in waters with cells was a decrease in MeHg. After 3 days of incubation with cells, MeHg concentrations had increased again. The variability in these trends is probably partly due to the zonation of the water and, consequently, the dominating bacteria in the layers close to the redoxcline. In conclusion, during both oxidizing and reducing conditions, methylation of Hg takes place without the presence of bacteria, and the addition of inorganic Hg may stimulate demethylation of MeHg when bacteria are present. However, since we did not incubate samples without addition of Hg, our results cannot be taken as proof for such a stimulating effect. At 30 m the water was grayish, indicating the occurrence of particulate sulfides and the dominance of unpigmented bacteria, probably sulfate reducers. In samples from this depth, methylation during incubation was also very high in the presence of cells (after 3 days), although an initial demethylation was detected. It has been reported that selective inhibition of sulfate reducers causes a drastic reduction in methylation rates.[8] Our results suggests that activity from sulfate reducers may be of importance for the methylation of Hg, but that cells do not have to be present.

These heterotrophic bacteria are capable of decomposing organic matter, a process that can be catalyzed by exoenzymes. We propose that the activity of these exoenzymes, present in unfiltered and filtered water, by 'accident' also catalyzes the formation of MeHg. The results from the exoenzyme activity measurements in September may support this. Much higher activity was observed in waters from 24 m (were sulfate reducing bacteria dominated) relative to in waters from 21 and 22 m. Other "abiotic" processes have been proposed for methylation of Hg, e.g., methylation catalyzed by metal ions (Fe^{2+}, Fe^{3+}) in the presence of fulvic acids.[3] Iron concentrations up to 200 nM/L have been measured in Framvaren,[11] however, the high sulfide concentrations below the redoxcline and the low solubility of FeS suggest that the Fe is unreactive.

C. SEASONAL VARIATIONS

The seasonal variations observed for MeHg in Framvaren is probably explained by favorable physical conditions (temperature and light) for the organisms in late summer, in combination with a higher input of nutritive organic matter. Data on sedimentation rates in Framvaren 1983 through 1984 showed that sedimentation of organic matter at 20 and 40 m depth was highest during July through September.[12]

Throughout the investigated part of the water column in Framvaren, MeHg concentrations, on an average, were five times higher in September than in November. On both occasions, MeHg concentrations increased steeply between 21.5 and 22 m and this increase coincided with the most drastic drop in Eh and also with the first smell of H_2S in the samples. In September, a distinct maximum in MeHg was discovered just below the redoxcline. Owing to the sparse sampling below the cline in November, an undetected maximum could have existed. On an average, the percentage of MeHg relative to TotHg was ~1% above the redoxcline while it was ~10% below the cline (November). The percentages and concentrations of MeHg reported from occasionally stratified freshwater systems are usually higher than those found in this study. For instance, Bloom et al.[28] found an increase of MeHg in the bottom water of Little Rock Lake during summer stratification, with maximum values in late summer of 3 to 10 ng L^{-1}. From measurements in several lakes, the same authors report that MeHg usually made up 7 to 25% of the TotHg in the oxic epilimnetic waters, and 50 to 90% of the total in the anoxic hypolimnion. Parks et al[30] found ~10% MeHg in oxic compared to ~30% MeHg in the anoxic hypolimnetic waters of Clay Lake. On the other hand, in the Pacific Ocean high percentages of MeHg (up to 25% compared to "reactive" Hg, were measured at depths which did not coincide with the depths of minimum oxygen concentrations.[31]

V. CONCLUSIONS

The results from Framvaren support several earlier findings from laboratory experiments. Methylation as well as demethylation of Hg in natural systems takes place both in oxic and anoxic conditions, but net MeHg occurrence is favored in anoxic environments. Demethylation of Hg appears to be mediated by microbes in an intracellular detoxification process, which is possessed by different types of bacteria, and may be activated at elevated concentrations of inorganic as well as organic Hg. Demethylation may be of importance in the seasonal cycling of TotHg owing to the formation of volatile Hg(0). Methylation, on the other hand, appears to be an extracellular process, probably enhanced by bacterial exoenzyme activity in environments where microbial decomposition of organic matter is important. Such an environment is found below the redoxcline in Framvaren, where the accumulated organic material can be utilized by the dominating heterotrophic sulfate reducers. The sulfate reducers have also the ability to reduce and demethylate toxic Hg, but in anoxic environments the MeHg formed is probably complexed with sulfide, decreasing the toxicity of the MeHg and allowing a maintenance of high MeHg concentrations. The methylation of Hg probably takes place "by accident" and not as a means of detoxification, which previously had been suggested.[2,32]

ACKNOWLEDGMENTS

We express our gratitude to Dr. Y.-H. Lee and Ms. E. Lord for the mercury analyses, and to Professor R. Hallberg for improving the manuscript.

REFERENCES

1. Berman, M. and Bartha, R., Levels of chemical versus biological methylation of mercury in sediments, *Bull. Environ. Contam. Toxicol.*, 36, 401, 1986.
2. Pan-Hou, H. S. and Imura, N., Physiological role of mercury-methylation in *Clostridium cochlearium* T-2C, *Bull. Environ. Contam. Toxicol.*, 29, 290, 1982.
3. Lee, Y.-H., Hultberg, H., and Andersson, I., Catalytic effect of various metal ions on the methylation of mercury in the presence of humic substances, *Water Air Soil Pollut.*, 25, 391, 1985.
4. Weber, J. H., Reisinger, K., and Stoeppler, M., Methylation of mercury (II) by fulvic acid, *Environ. Technol. Lett.*, 6, 203, 1985.
5. Craig, P. J. and Moreton, P. A., Total mercury, methylmercury and sulphide levels in British estuarine sediments. III, *Water Res.*, 9, 1111, 1986.

6. Korthals, E. T. and Winfrey, M. R., Seasonal and spatial variations in mercury methylation and demethylation in an oligotrophic lake, *Appl. Environ. Microbiol.*, 53, 2397, 1987.

7. Furutani, A. and Rudd, J. W. M., Measurements of mercury methylation in lake water and sediment samples, *Appl. Environ. Microbiol.*, 40, 770, 1980.

8. Gilmour, C. C. and Henry, E. A., Mercury methylation in aquatic systems affected by acid deposition, *Environ. Pollut.*, 71, 131, 1991.

9. Skei, J. M., Framvaren—environmental setting, *Mar. Chem.*, 23, 209, 1988.

10. Stigebrandt, A. and Molvaer, J., On the water exchange in Framvaren, *Mar. Chem.*, 23, 219, 1988.

11. Skei, J. M., Formation of framboidal iron sulfide in the water of a permanently anoxic fiord, *Mar. Chem.*, 23, 345, 1988.

12. Naes, K., Skei, J. M., and Wassman, P., Total particulate and organic fluxes in anoxic Framvaren waters, *Mar. Chem.*, 23, 257, 1988.

13. Haraldsson, C. and Westerlund, S., Trace metals in the water columns of the Black Sea and Framvaren Fjord, *Mar. Chem.*, 23, 417, 1988.

14. Iverfeldt, Å., Mercury in the Norwegian Fjord Framvaren, *Mar. Chem.*, 23, 441, 1988.

15. Fuhrman, J. A. and Azam, F., Bacterioplankton secondary production estimates for coastal waters of British Columbia, Antarctica and California, *Appl. Environ. Microbiol.*, 39, 1085, 1980.

16. Albertson, N. H., Nyström, T., and Kjelleberg, S., Exoprotease activity of two marine bacteria during starvation, *Appl. Environ. Microbiol.*, 56, 218, 1990.

17. Bloom, N. and Fitzgerald, W. F., Determination of volatile mercury species at the picogram level by low-temperature gas chromatography with cold-vapour atomic fluorescence detection, *Anal. Chim. Acta*, 208, 151, 1988.

18. Bloom, N., Determination of picogram levels of methylmercury by aqueous phase ethylation, followed by cryogenic gas chromatography with cold vapour atomic fluorescence detection, *Can. J. Fish. Aquat. Sci.*, 46, 1131, 1989.

19. Meili, M., Iverfeldt, Å., and Håkanson, L., Mercury in the surface water of Swedish forest lakes—concentrations, speciation and controlling factors, *Water Air Soil Pollut.*, 56, 439, 1991.

20. Steinnes, E. and Andersson, E. M., Atmospheric deposition of mercury in Norway: temporal and spatial trends, *Water Air Soil Pollut.*, 56, 391, 1991.

21. Anderson, L. G., Dyrssen, D., and Hall, P. O. J., On the sulphur chemistry of a super-anoxic fjord, Framvaren, South Norway, *Mar. Chem.*, 23, 283, 1988.

22. Fergusson, J. E., *The Heavy Elements: Chemistry, Environmental Impact and Health Effects*, Pergamon Press, Oxford, 1990, chap. 3.

23. Dyrssen, D. and Wedborg, M., The sulphur-mercury (II) system in natural waters, *Water Air Soil Pollut.*, 56, 507, 1991.

24. Ormerod, K. S., Distribution of some non-phototrophic bacteria and active biomass (ATP) in the permanently anoxic fjord Framvaren, *Mar. Chem.*, 23, 243, 1988.

25. Kirchman, D., Soto, Y., Van Wambeck, F., and Bianchi, M., Bacterial production in the Rhône River plume: effect of mixing on relationships among microbial assemblages, *Mar. Ecol. Prog. Ser.*, 53, 267, 1989.

26. Gilmour, C. C., Leavitt, M. E., and Shiaris, M. P., Evidence against incorporation of exogenous thymidine by sulfate-reducing bacteria, *Limnol. Oceanogr.*, 35, 1401, 1990.

27. Sorensen, K., The distribution and biomass of phytoplankton and phototrophic bacteria in Framvaren, a permanently anoxic fjord in Norway, *Mar. Chem.*, 23, 229, 1988.

28. Bloom, N. S., Watras, C. J., and Hurley, J. P., Impact of acidification on the methylmercury cycle of remote seepage lakes, *Water Air Soil Pollut.*, 56, 477, 1991.

29. Brown, N. L., Bacterial resistance to mercury—*reductio ad absurdum?*, *Trends Biochem. Sci.*, 41, 400, 1985.

30. Parks, J. W., Lutz, A., and Sutton, J. A., Water column methylmercury in the Wabigoon/ English river-lake system: factors controlling concentrations, speciation, and net production, *Can. J. Fish. Aquat. Sci.*, 46, 2184, 1989.

31. Mason, R. P. and Fitzgerald, W. F., Alkylmercury species in the equatorial pacific, *Nature*, 347, 457, 1990.

32. Wood, J. M. and Wang, H.-K., Microbial resistance to heavy metals, *Environ. Sci. Technol.*, 17, 542A, 1983.

Methylmercury Sources in Boreal Lake Ecosystems

M. Verta, T. Matilainen, P. Porvari, M. Niemi, A. Uusi-Rauva, and N. S. Bloom

CONTENTS

ABSTRACT: Rates of potential methylation and demethylation were determined several times during 1990 and 1991 from submerged soil, littoral and profundal sediments (0 to 2 cm), lake water, and sedimenting seston in three small forest lakes in southern Finland. In addition, the rates of potential methylation and demethylation were measured from different kinds of surface soils. Total mercury (TotHg) and methylmercury (MeHg) concentrations were monitored in precipitation, runoff, lake water, and in sedimenting seston. Mercury methylation was observed in all environments studied, but showed differences of several orders of magnitude in potential. The order of methylation rates were sediment > peat > soil ≈ water ≈ sedimenting seston. Only small differences were found in potential demethylation rates between compartments, and high rates were recorded for long periods during the ice-free season. In both sediments and submerged soils, methylation was higher in anoxic conditions than under oxygen. High MeHg levels were measured in runoff water, but especially in the anoxic hypolimnion and in sedimenting particulate material. Sedimenting particles are thought to be an important MeHg source in the hypolimnion. Terrestrial catchments and submerged soils were the main MeHg source in the drainage lakes, whereas both atmospheric MeHg and in-lake MeHg production were more important in the seepage lake studied.

I. INTRODUCTION

The problem of increased methylmercury levels in biota in remote lakes in Scandinavia, the northern U.S., and Canada has gained wide attention in recent years.[1-3] Both seepage lakes and drainage lakes situated in the Precambrian shield are affected. There are indications that, in general, higher levels of mercury in fish are found in drainage lakes than in seepage lakes.[4] Large catchment-to-lake-area ratios also seem to be an indicator of high MeHg concentrations in fish.[5,6]

Whether the increase of fish methylmercury levels is caused by an increased load of atmospheric mercury or other environmental changes caused by increased acidification of soils and lakes, such as increased sulfur deposition followed by an increased sulfate reduction or other mechanisms, has been discussed by several authors.[7-9] It is widely agreed that not a single factor, but rather the combined effects of many processes are involved, and that different processes may be of importance in different type of lakes.

According to a statistical survey of 1000 lakes, the median lake surface area in southern and central Finland is 0.07 km^2.[10] Drainage lakes are most common and represent 69% of all lakes. Seepage lakes are of minor importance (10%). The median catchment area is 1.6 km^2 and the median catchment/lake ratio is 20.

The key process in the cycling of mercury leading to its bioaccumulation in fish is the methylation of inorganic mercury (Hg^{2+}) to monomethyl mercury (CH_3Hg^+). Monomethyl mercury is produced in the terrestrial environment,[11] in the water column,[12] in lake sediments,[13] and in the intestines[14] and the slime layer of fish.[15]

During a 3-year period (1989 through 1991) the potential of mercury methylation and methylmercury demethylation was studied in some of the compartments of three boreal forest lake ecosystems in southern Finland. The main objective was to identify those compartments that may be of importance as methylmercury sources for different types of lakes. The potential rates of methylation and demethylation were compared to methylmercury levels in the water and in the runoff of the study lakes. The lakes investigated represent both a headwater (seepage) lake and drainage lakes with different water qualities and high levels of methylmercury in fish. This chapter summarizes some of the major results of the study.

II. MATERIAL AND METHODS

A. STUDY AREA

The three lakes, Lake Hakojärvi (HAKO), Lake Keskinen Hakojärvi (KEHA), and Lake Iso Valkjärvi (IVA), chosen for this study, are situated in a sparsely populated area in southern Finland, some 30 km from the nearest municipality and a few kilometers apart from each other (Figure 1). The area receives one of the highest depositions of acidic compounds in Finland,[16] and according to sediment studies the atmospheric mercury load in the area has increased by a factor of about three during the last century.[17,18]

The lake KEHA watershed is the main inlet of lake HAKO and makes up 65% of its catchment. Both catchments are vegetated with Scots pine (*Pinus sylvestris* L.) and Norway spruce (*Picea abies* L.). In about 30% of the overall catchment area the morainic soil material is partly covered with shallow peatlands (<1 m) (Figure 1). Most of the peatlands were drained for forestry in the late 1960s. Extensive forestry management, such as clearcutting and forest fertilization, has been carried out since then.

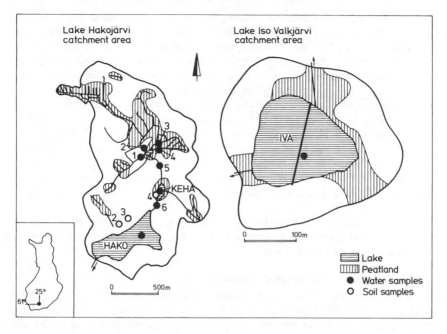

Figure 1 Location of study catchments. Open circles indicate sites of soil methylation studies. Numbers refer to the text (Site 1 outside the catchment). Closed circles indicate sites of water sampling (Sites 1–4 = ditches, site 5 = lake KEHA inlet, site 6 = lake HAKO inlet).

Table 1 **Morphometric characteristics and mean water quality in runoff of the four ditches and in the epilimnion of the study lakes.**

	Runoff	Lake KEHA	Lake HAKO	Lake IVA	Finland median[a]
Lake area (km²)		0.013	0.17	0.042	0.07
Catchment area (km²)		1.3	2.0	0.17	1.6
Catchment/lake ratio		100	12	4.0	20
TotHg (ng L⁻¹)	4.7	7.2	3.2	1.3	—
MeHg (ng L⁻¹)	0.66	2.2	0.43	0.12	—
TOC (mg L⁻¹)	22	22	12	6.0	14
pH	5.6	5.5	6.3	5.5	6.1
ANC (μeq L⁻¹)	49	78	86	8	69
CA + Mg (μeq L⁻¹)	340	230	200	71	211
SO₄ (μeq L⁻¹)	210	160	160	46	90

[a] The median water quality in southern Finland is estimated from a statistical survey of 1000 lakes.[10]

Lake KEHA was dammed by beavers in fall 1989 and consequently the water level has risen by 0.5 to 0.7 m, depending on the season. This increase resulted in a flooding of the terrestrial soils around the lake and enlarged the lake surface by some 30%. The water in the ditches of the catchment is very humic and poorly buffered, but has a relatively high pH (Table 1). The water quality of lake HAKO is close to the median found from a statistical survey of 1000 lakes (Table 1).

The fish assemblage of the lakes consists of roach (*Rutilus rutilus* L.), perch (*Perca fluviatilis* L.), northern pike (*Esox lucius* L.), burbot (*Lota lota* L.), and eel (*Anguilla anguilla*

L.). Lake HAKO was fished intensively during spring 1984 and 1985 to lower the high mercury levels in the top predator, northern pike (1.07 µg g⁻¹ in a 1-kg "standard" pike). A total catch of 29.5 kg ha⁻¹ was removed, representing about 50% of the total biomass. Mercury concentrations in burbot, large pike, and small roach had decreased by 1987 and 1988, but increased thereafter.[19] In 1991 mercury concentrations in the other species, except burbot, had reached the same level as before the intensive fishing.

The catchment of lake IVA (Figure 1) consists of sand and gravel of glacial fluvial origin and is vegetated by Scots pine. The lake has no visible inlet and derives most of its water from rain or groundwater. However, the increase in the water color observed in recent years indicates the leaching of water through upper soil layers. The two ditched outlets are dry during most of the year. The lake is acidified to a pH of about 5.2 during spring turnover and has lost its roach population. Perch, northern pike, and stocked whitefish (*Coregonus lavaretus* L.) are the only species left. This lake also has high concentrations of mercury in fish—1.13 µg g⁻¹ in pike.[20]

The lake was divided into two sub-basins by a plastic curtain in April 1991. One of the halves was limed in May 1991, whereas the other was left as an untreated control. The results presented here are for the untreated part.

B. COLLECTION OF MERCURY SAMPLES

During all sampling ultraclean handling techniques were employed. This included the use of polyethylene gloves, double bagging of samples, and the use of preanalyzed reagents. After rigorous acid cleaning of the sample bottles, a 1% HCl solution was kept in the bottles until sampling.

Rainwater samples were collected from the study area using an acid-cleaned glass collection funnel with direct discharge into an acid-cleaned Teflon® bottle. Depending on the rainfall intensity, sampling duration varied from 2 to 10 h. Samples were frozen prior to analysis. One snow sample (March 23, 1992) was collected from the Lake HAKO catchment. The sample was taken from a snow profile in a clearcut area into an acid-cleaned 5 L polyethene vessel by scraping snow from the wall of a trench dug into the snowpack (0.4 m deep). The sample was melted at 5° C and transferred into an acid-cleaned Teflon® bottle. One snow meltwater sample (March 23, 1992) was collected from the ice of Lake IVA straight into a Teflon® bottle. At the time of sampling, all the snow on the ice cover and most of the snow on the lake shore had melted.

Runoff water was collected from four of the six ditches located in the Lake KEHA catchment. The first sampling was carried out in September 1990 and was continued during the frost-free period from April to October 1991. The samples were collected directly into Teflon® bottles. During winter time no discharge occurred in the ditches.

Lakewater samples were collected from the deepest part of each lake, using a carefully washed Teflon®-coated water sampler that had been kept in 5% HCl for 24 h prior to sampling. The first set of samples (September 1990) was frozen within 4 h of collection before being sent for analysis.

Sedimenting seston was collected from Lake IVA and Lake HAKO using five parallel Plexiglass® sedimenting traps (diameter 5 cm, depth 50 cm) that were placed 1 m above the sediment surface at the deepest points of the lakes during the ice-free periods in 1990 and 1991. In Lake HAKO an epilimnetic trap (5 m) was used as well. Traps were sampled at about 1 to 2-month intervals. The sedimenting material was siphoned carefully from the bottom of the traps by a 100-mL acid-washed plastic syringe and transferred to Teflon® bottles for analysis of TotHg and MeHg.

The samples (except those frozen) were kept at <5° C for less than 24 h prior to courier shipment to Seattle, WA for analysis. Generally, the samples were filtered and preserved with 10 mL l⁻¹ HCl within 72 h of collection. In several cases, however, when the samples were delayed an additional 2 to 5 days in customs, the samples were simply acidified with HCl, and no dissolved/particulate partitioning was undertaken.

Although more thorough investigations are needed, experience (Bloom, N.S., unpublished data) indicates that fresh lakewater samples may be stored at least 2 to 4 weeks, unpreserved in Teflon® bottles (0 to 5° C, in the dark), with no significant change in speciation and/or concentration of Hg. Similar experiments indicate that the dissolved/particulate fraction is somewhat less stable, but even after 1 week under the above conditions changes are less than ±20%. Stability of all forms and species seems to be enhanced in highly colored waters.

C. ANALYSIS OF TOTAL AND METHYLMERCURY

Prior to analysis for total Hg, an aliquot of water (50 to 100 mL) is cold-oxidized for >1 h with 0.002 N BrCl[21] in order to convert bound and organic forms to labile inorganic Hg. The sample is prereduced with NH_2OH to neutralize free halogens, which can damage gold traps and interfere with the analysis. The sample is then reduced with $SnCl_2$ to covert Hg(II) to Hg^0, which is purged out with N_2 onto gold-coated sand traps.[22] The trapped mercury is then quantified using the two-stage gold amalgamation procedure, with cold vapor atomic fluorescence (CVAFS) detection.[23] The detection limit for this method is about 0.05 ng L^{-1} total Hg.

Methylmercury (MeHg) is analyzed from a separate aliquot following extraction from a KCl/HCl matrix with CH_2Cl_2, and then back extracted into deionized water.[24,25] The deionized water extract is then analyzed by aqueous phase ethylation, Carbotrap™ (Supleco Inc., Bellefonte, PA) precollection, and cryogenic GC separation of the methylethyl-Hg formed. The mercury is then quantified by CVAFS as Hg^0, following on-line pyrolytic breakdown of the organo-Hg species. The detection limit is a function of the MeHg content of the CH_2Cl_2, and varied in this study from 0.005 to 0.015 ng L^{-1} as Hg. All results were corrected for the mean extraction efficiency obtained from spiked samples (typically 70 to 90%).

D. POTENTIAL METHYLATION AND DEMETHYLATION RATES IN SEDIMENTS

Potential rates of mercury ($^{203}HgCl_2$) methylation and methylmercury ($^{14}CH_3HgI$) demethylation were determined in samples of surface (0 to 2 cm) sediments from littoral and profundal sites of each lake. In Lake KEHA all lake sediments collected were from under the oxycline; for littoral sediments inundated soils were collected.

Sediment samples were collected with a corer (Züllig, inner diameter, 60 mm) 4 to 6 times during open water. The intact surface layer (0 to 2 cm) was siphoned carefully into a 100 mL plastic syringe under the overlying water. The inundated soil samples were taken into syringes by hand under water. The samples were kept dark and cool during transport and processed within 24 h of collection.

The well-mixed sediment samples were transferred in 20 mL portion into 125 mL bottles, one blank and two replicates, for both anaerobic (under N_2 gas) and aerobic determination of methylation and demethylation rates. In cases of an anaerobic hypolimnion, only anaerobic incubation results are presented. When oxygen was present in the water above the sediment the results of both aerobic and anaerobic incubations are shown. In several cases, particularly in inundated soils, anoxic conditions most probably existed within the soils or sediments regardless of low oxygen concentrations in the water. This was clearly visible as black strings in the sediment cores or noted as hydrogen sulfide odor in the soils.

The mercury methylation activity was determined by incubating 20 mL samples with $^{203}HgCl_2$ and methylmercury demethylation with $^{14}CH_3HgI$ in the dark at in situ temperature for 24 h and then terminated with 4 N HCl. The incubation doses ranged from 32.8 to 73.3 kBq per 2.5 µg Hg in methylation assays and from 0.78 to 0.99 kBq per 0.33 µg Hg in demethylation assays. The blanks were killed with 2mL of 4 N HCl prior to the isotope additions.

The formed methylmercury ($Ch_3^{203}Hg^+$) was determined using a modification of the method developed by Furutani and Rudd[12] and the products of demethylation ($^{14}CO_2$ and

$^{14}CH_4$) by a method developed by Ramlal et al.[26] Radioactivity was measured with a liquid scintillation counter. A more detailed description of the methods is presented by Matilainen et al.[27] Rates given are expressed as percent of the added methylated or demethylated mercury per day.

E. POTENTIAL METHYLATION AND DEMETHYLATION RATES IN SOILS

Mercury methylation, methylmercury demethylation, and soil respiration were monitored in the upper humic layers of four sampling stations during the frost-free period in 1991. Three of the stations are located in the Lake HAKO catchment and the fourth in close vicinity. All sites (see Figure 1) represented soils and vegetation typical for the catchment:

Site	Soil type	Vegetation
1	Sand and gravel	Scots pine, *Calluna* sp.
2	Podsol	Norway spruce, *Myrtillus* sp.
3	Podsol	Norway spruce, *Myrtillus* sp., clearcut
4	Podsol	Norway spruce, *Sphaghnum* spp.

Ten core samples were taken five times at each sampling station from an area of 10 m^2 between June and October. The samples were sieved and visible plant material removed. Subsamples of 2 g of soil as dry weight were adjusted to a water-holding capacity of 60%, except the *Sphaghnum* sample which was adjusted to 90% capacity. The rates of potential methylation and demethylation were determined by incubating the samples for 24 h at $+14°$ C after labeling with $^{203}HgCl_2$ and $^{14}CH_3HgI$. Each test consisted of duplicate samples and a HCl-treated blank, and in some cases an additional formaldehyde-treated blank. The incubation dose ranged from 22.6 to 131 kBq per 5µg Hg in methylation assays and from 0.82 to 0.99 kBq per 0.33 µg Hg in demethylation assays.

F. POTENTIAL METHYLATION AND DEMETHYLATION RATES IN WATER AND SESTON

Parallel samples to those analyzed for TotHg and MeHg were taken with a Ruttner-type sampler and also measured for methylation and demethylation rates. Activities were determined from two replicates and one acid-killed blank after incubation with $^{203}HgCl_2$ and $^{14}CH_3HgI$. A 100 mL sample for methylation and a 50 mL sample for demethylation were used. The incubation doses ranged from 21.7 to 73.3 kBq per 2.5 µg Hg in methylation assays and from 0.40 to 0.48 kBq per 0.165 µg Hg in demethylation assays.

Samples from oxygen-free hypolimnia were kept anaerobic during sampling and transport. For the incubations, the subsamples were transferred and the isotope and HCl additions were made under a stream of N_2 gas and, for comparison, in two cases under ambient air. During anaerobic incubations nitrogen was left in the head space of the incubation bottle to maintain anaerobic conditions. The two parallel anaerobic/aerobic incubations did not reveal any differences in potential methylation or demethylation rates.

For processing potential methylation/demethylation in seston the sedimenting material was siphoned carefully from the bottom of the traps with a 100 mL acid-washed plastic syringe and transferred to glass bottles. Anoxic samples were kept in syringes during transport and processed under N_2 gas.

III. RESULTS AND DISCUSSION

A. MeHg IN RUNOFF AND PRECIPITATION

The MeHg concentrations in runoff (0.19 to 2.0 ng L^{-1}, Table 2) were higher than those measured in runoff from small catchments in Sweden (0.04 to 0.73 ng L^{-1}).[28-30] The

Table 2 **Median total MeHg concentrations, potential Hg methylation and MeHg demethylation rates in runoff water and in the study lakes.**

	n	MeHg (ng L⁻¹)		n	Methylation (% d⁻¹ × 10⁻³)		n	Demethylation (% d⁻¹)	
Runoff	24	0.51	(0.19–2.0)	24	3.1	(0–24)	20	0.53	(0.04–3.0)
KEHA									
(inlet)	4	0.50	(0.28–0.66)	5	4.8	(3.6–12)	5	2.2	(0.60–4.0)
(epil.)	3	2.3	(0.38–3.8)	4	12	(2.4–82)	4	2.6	(1.9–6.1)
(hypol.)	5	3.7	(1.9–6.0)	3	3.1	(3.0–7.2)	3	0.98	(0.98–1.3)
HAKO									
(inlet)	7	2.0	(0.31–7.3)	6	4.2	(1.2–9.6)	5	2.1	(0.1–2.7)
(epil.)	11	0.40	(0.13–0.70)	14	6.7	(2.4–280)	11	1.4	(0.1–2.7)
(hypol.)	8	0.50	(0.29–0.59)	7	2.4	(1.0–100)	5	0.91	(0.02–2.3)
IVA									
(epil.)	6	0.11	(0.05–0.17)	6	13	(2.4–290)	6	1.5	(0.56–4.2)
(hypol.)	7	0.19	(0.07–3.0)	7	9.6	(3.0–140)	7	1.8	(0.47–4.0)

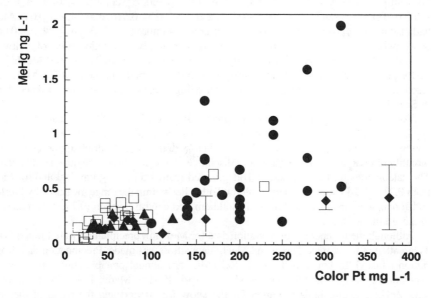

Figure 2 MeHg concentrations in runoff in three different studies in Scandinavia plotted against water color. Closed circle = present study, open square = Lee and Iverfeldt, 1991, diamond = Westling, 1991.

difference between our data and that of Lee and Hultberg[28] and Lee and Iverfeldt[29] was tested ($p < 0.001$, Student's t-test), and it is obvious between our data and that of Westling[30] (0.08 to 0.73 ng L⁻¹) as well (Figure 2).

The catchments in those studies were different in character. While large areas of our catchments are characterized by thin peatlands on top of the till soils (Figure 1), Westling[30] studied both drained and undrained thick peatlands. Lee and Hultberg[28] and Lee and Iverfeldt,[29] on the other hand, studied catchments on till soils with less than 20% peatlands (with one exception).

Concentrations of MeHg in runoff were usually lowest in spring, but did not show clear differences between summer and fall. Although the water quality in the ditches did not differ

Table 3　TotHg and MeHg concentrations (ng L^{-1}) in rainwater and snow meltwater collected from the study area 1991 through 1992.

	Hg_{tot}	$Hg_{diss.}$	$MeHg_{tot.}$	$MeHg_{diss.}$
Rainwater				
June 1991	2.94	—	—	—
Aug. 1991	1.23	—	—	—
Oct. 1991	1.76	0.45	0.093	0.093
Nov. 1991	2.43	0.75	0.058	0.057
Snow meltwater				
Mar. 1992				
IVA ice	11.09	5.09	0.295	0.287
HAKO clearcut	5.90	nd	0.136	0.044

Note: nd = Not detected.

much, significant correlations were observed between water quality and MeHg concentrations, in particular with TotHg (r = 0.457, n = 21, p <0.05), color (r = 0.496, n = 24, p <0.05), TOC (r = 0.474, n = 20, p <0.05), and ANC (r = 0.516, n = 24, p <0.01). MeHg had also a strong positive correlation with water temperature (r = 0.633, n = 24, p <0.001), possibly reflecting the methylation activity in soils or simply the dilution effect of meltwater.

Only a few speciations of dissolved/particulate phases of both TotHg and MeHg were performed from runoff water. Most MeHg was in the dissolved phase (>90%). MeHg represented from 3.6 to 35% of TotHg; the latter being somewhat higher than in other studies in Scandinavia.[28-30]

TotHg and MeHg runoff from the catchment of Lake HAKO were calculated using discharge data from a similar catchment nearby (distance 50 km, area 1.49 km^2) with continuous flow recording. Three different calculation methods (daily discharge × conc., mean monthly discharge × conc, mean yearly discharge × mean conc.) gave quite similar results. The calculations for the transport of TotHg varied from 1.5 to 1.7 g km^{-2}a^{-1} and for MeHg from 0.21 to 0.24 g km^{-2}a^{-1}. The result for TotHg lies within the range measured by Iverfeldt and Johansson[31] in small catchments in southern and central Sweden (0.7 to 6.1, mostly 2.0 to 3.3 g km^{-2}a^{-1}).

Although the number of concentration data for MeHg in precipitation were limited (Table 3), an attempt to estimate TotHg and MeHg deposition was made. Precipitation data from a station situated 25 km from Lake HAKO (610 mm annual precipitation) were used and deposition was calculated for the period from April 1991 to March 1992. The mean of the rain MeHg concentrations was used for the snow-free period and the mean of the snow meltwater concentration for the period from December 1991 to March 1992. The estimated wet deposition of TotHg was 2.6 g km^{-2}a^{-1} and the wet deposition of MeHg 0.07 g km^{-2}a^{-1}. This suggests that MeHg in precipitation cannot explain the MeHg transport from the catchments. MeHg deposition was about one third of MeHg runoff, indicating net methylmercury production in the catchments.

Our estimate for TotHg deposition in the area is notably lower than that measured by Iverfeldt[32] in central Finland (Tikkakoski) during 1987 through 1989 (11 g km^{-2}a^{-1}, mean precipitation 720 mm), but of the same order estimated by Rekolainen et al.[33] in 1983 through 1984 from the snowpack in southern and central Finland. The 1987–1989 study suggests that due to limited data we have underestimated the deposition of TotHg, and probably MeHg as well. Data on sulfate deposition (bulk) in the same area, on the other hand, show a remarkable (roughly 40%) decrease in deposition during the period between these two studies.[34,35] A strong positive correlation between Hg and sulfate has been found in

Table 4 **Medians and ranges of potential Hg methylation and MeHg demethylation rates (% d^{-1}) in different environmental compartments during ice- and frost-free seasons in 1990 through 1991.**

		\multicolumn{3}{c}{Methylation}			\multicolumn{3}{c}{Demethylation}		
		n	Median	Range	n	Median	Range
Peat	(O$_2$)	5	0.031	(0.018–0.15)	5	11	(8.3–14)
Humus layer	(O$_2$)	15	0.007	(0.002–0.029)	15	7.6	(3.9–23)
Inundated soil	(O$_2$)	4	0.097	(0.003–0.69)	4	3.6	(0.3–8.3)
	(N$_2$)	4	0.92	(0.42–1.2)	4	8.9	(3.9–14)
Littoral sediments	(O$_2$)	7	0.080	(0.005–0.50)	7	7.0	(0.7–16)
	(N$_2$)	7	0.59	(0.14–1.3)	7	14	(4.5–23)
Profundal	(O$_2$)	10	0.061	(0.023–0.79)	10	6.3	(2.2–19)
sediments	(N$_2$)	10	0.41	(0.17–1.2)	10	11	(5.4–19)
Sediment traps	(O$_2$)	9	0.002	(0.0–0.02)	9	4.4	(0.59–13)
	(N$_2$)	3	0.007	(0.005–0.010)	3	8.2	(6.6–9.4)
Epilimnion	(O$_2$)	17	0.012	(0.002–0.29)	15	1.6	(0.56–6.1)
Runoff	(O$_2$)	24	0.003	(0.0–0.024)	20	0.53	(0.04–3.0)

Note: O$_2$ = incubation in ambient air; N$_2$ = incubation under N$_2$ gas.

precipitation over the Nordic countries,[32] and indicates that the Hg deposition may have decreased as well.

B. POTENTIAL METHYLATION AND DEMETHYLATION IN SOILS AND RUNOFF

Potential methylation and demethylation rates in runoff were low–on the order of 0 to 0.02% for methylation and 0.04 to 3.0% for demethylation per day (Table 2). Correlation analysis with water quality data showed a strong positive correlation for demethylation with water temperature ($r = 0.753$, $n = 19$, $p < 0.001$), ANC ($r = 0.610$, $n = 19$, $p < 0.01$), and pH ($r = 0.576$, $n = 19$, $p < 0.01$). These correlations might reflect the biotic nature of demethylation. No significant correlation of the methylation rate with runoff water quality was found.

The same order of potential methylation/demethylation rates as in runoff (0.002 to 0.029% d^{-1} and 3.9 to 23% d^{-1}, respectively) were measured in the top soil (humus) layers of the study sites (Table 4). During the entire study period a clearly higher rate of methylation was recorded in peat soils than in other soils (0.02 to 0.15% d^{-1}, Table 4).

The net methylmercury production in soils is indicated by several findings. First, the lowest MeHg concentrations in runoff (0.19 to 0.40 ng L^{-1}) were measured in April 1991, at a time when snow meltwater accounted for a large fraction in the runoff. Second, positive correlations of MeHg with humic substances (color, TOC) originating from soils support the hypothesis of a significant net methylation in soils. And third, potential methylation/demethylation rates in soils were comparable to those measured in runoff and in lake water, and in some cases (peat) of the order measured in surficial sediments.

These findings support the hypothesis that the net methylation of mercury in the soil, particularly in peat surface layers, is an important process determining the MeHg levels in the runoff of the catchments. Net methylation in the ditch sediments, followed by leaching of MeHg, cannot be excluded, and may affect MeHg levels in runoff. The relatively short

KEHA 22.4. 1991

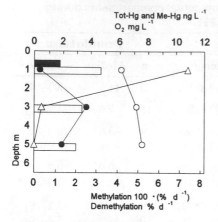

KEHA 13.8. 1991

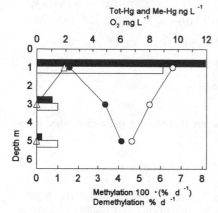

Figure 3 TotHg, MeHg and potential methylation/demethylation rates in lake KEHA during winter and summer stagnation in 1991. Open triangle = O_2, open circle = TotHg, closed circle = MeHg, black bars = methylation, open bars = demethylation.

residence time in the ditches compared to that in the soils suggests the soils are the primary source.

C. MeHg IN LAKES
1. Lake Keskinen Hakojärvi (KEHA)

During winter stratification Lake KEHA reached total anoxia from 4 m down, had spring meromixis, and at the end of summer was anoxic and sulfide rich below 2 m depth. At 1 m depth an oxygen concentration of only 1.9 mg L^{-1} was measured in August (Figure 3).

The water MeHg concentrations showed dramatic peaks in Lake KEHA compared to the ditches, to the inlet, and to the other lakes (Table 2, Figure 3), both in the epilimnion and in the hypolimnion. Extremely high MeHg concentrations were recorded, averaging 2.3 ng L^{-1} in the epilimnion and reaching up to 6.0 ng L^{-1} during summer stratification in the hypolimnion (Figure 3). At its maximum, MeHg represented 89% of TotHg.

The potential methylation rates in epilimnetic water during summer did not differ from those in other lakes (Table 2, t-test; $p > 0.05$). Instead, potential demethylation showed a small but significantly ($p < 0.05$) higher rate in Lake KEHA than in Lake HAKO, probably due to higher bacterial activity, as reflected in O_2 conditions.

In the oxygen-free hypolimnion practically no potential methylation activity was observed in March (Figure 3). Both in the epilimnion and in the hypolimnion of Lake KEHA an order of magnitude lower methylation potential was measured in August compared to that in Lake

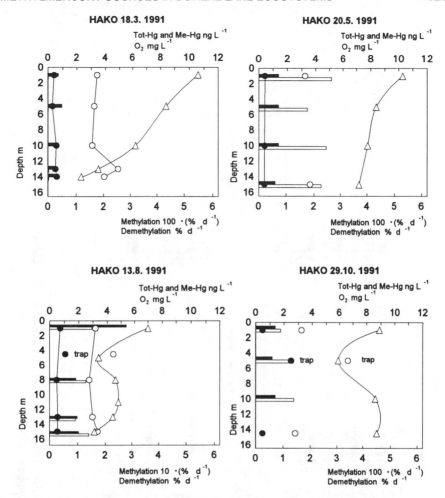

Figure 4   TotHg, MeHg and potential methylation/demethylation rates in lake HAKO during winter and summer stagnations and spring and fall mixis in 1991. (Note the different scaling in methylation in August.) Symbols as in Figure 3.

HAKO downstream (Figures 3 and 4, note the scale). Again, very low methylation potential was measured in the hypolimnion of Lake KEHA. At the same time, high MeHg concentrations were measured in the same layers.

These results do not indicate net production of MeHg in anoxic layers of the water. However, it should be noted that the sampling was performed with a Ruttner-type sampler and from a few depths only. If methylation in the water phase is caused by bacteria capable of producing MeHg in restricted environmental conditions, e.g., in oxic/anoxic boundaries, we probably have missed these layers. Thus, we cannot exclude the possibility of a water phase methylation to explain the MeHg concentrations in the hypolimnion.

Despite lower temperatures, the potential demethylation in the water phase showed rates as high in March as during summer stratification (Figure 3), probably due to an intensive heterotrophic bacterial activity and/or high concentrations of substrate (MeHg). This is in strong contrast to the other two lakes, where an order of magnitude lower demethylation potential was recorded during winter (Figures 4 and 5).

130

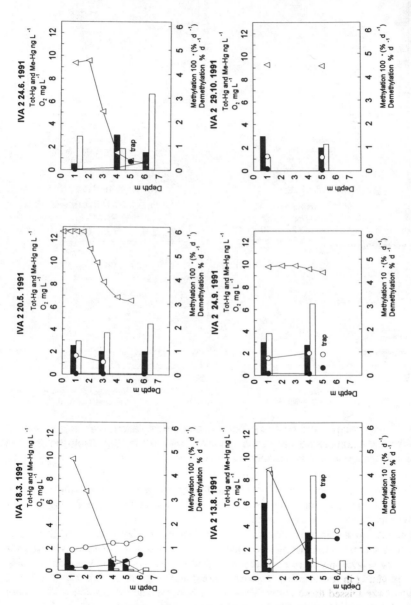

Figure 5 TotHg, MeHg and potential methylation/demethylation rates in water of lake IVA during 1991. (Note the different scaling in methylation in August through September.) Symbols as in Figure 3.

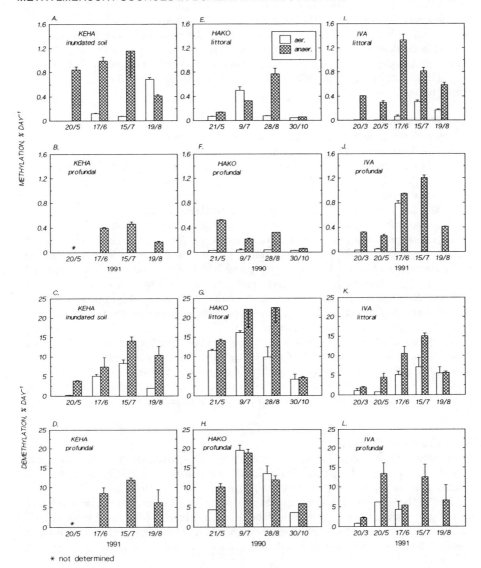

* not determined

Figure 6 The development of potential methylation and demethylation rates in the littoral and profundal sediments (0 to 2 cm) of the study lakes. For the anoxic hypolimnion only incubations under N_2 gas are shown.

In the inundated soil material the rate of potential methylation (per 20 ml of wet material) was somewhat higher than in profundal sediments of Lake KEHA and in sediments of Lake HAKO downstream (Figures 6A, B, E, F, anaerobic incubation). However, the difference was not significant. Instead, when calculated per dry weight, the submerged soils showed significantly higher rates (4.95 vs. 0.75% d^{-1}, t-test, $p < 0.05$, anaerobic incubation). The difference between these two calculations results from a much higher dry weight of surface sediments compared to soils in incubation.

The potential demethylation rate in inundated soils in Lake KEHA during summer was only about half of that in Lake HAKO (Figures 6C, G, $p < 0.05$), which again indicates much higher net methylmercury production in soils under water compared to lake sediments.

The reason for the high MeHg concentrations in the water of Lake KEHA presumably is the flooding of the terrestrial soils around the lake and subsequent high net MeHg production in the flooded soils. The data are in accordance with findings reported, e.g., Bodaly et al.[36] and Jackson[37] from man-made reservoirs in Canada. Furthermore, the soil methylation data indicate that MeHg already methylated in surface soils is dissolved to the water phase during flooding.

2. Lake Hakojärvi (HAKO)

During both winter and summer stagnations Lake HAKO remained oxic, although a clear oxygen depletion was observed. Low oxygen concentrations were also evident in the epilimnion in August, resulting from the very low oxygen concentrations in the inlet (KEHA outlet: 1.3 mg L^{-1}). TotHg, but not MeHg, showed weak peaks in the hypolimnion during winter stratification (Figure 4). The MeHg level remained almost constant from March to October and was notably lower than in Lake KEHA or in the inlet (Table 2).

Water collected from the bottom of a sediment trap at a depth of 5 m in August and in October contained about twice as much TotHg and two to five times more MeHg (up to 2.5 ng L^{-1} in October) as the water outside the trap (Figure 4). The dissolved MeHg concentration in the trap was 1.12 ng L^{-1} in October, represented 45% of the total MeHg, and exceeded the level of dissolved MeHg in the epilimnion (0.36 ng L^{-1}) by 200%. At the time of sampling, the traps had been in the lake for 4 weeks (July through August) and 10 weeks (August through October), had reached anoxis, and had sulfide odor on the bottom, although they were situated in an oxic epilimnion.

The potential methylation rates were very low in water samples from March, May, and October, but an order of magnitude higher in August (Figure 4, note the scale). In contrast, potential demethylation was at the same order of magnitude in all seasons, except in March under ice. The oxygen-containing hypolimnion usually showed as high a methylation and demethylation potential as the epilimnion, a behavior different from that in the anoxic lake KEHA hypolimnion.

3. Lake Iso Valkjärvi (IVA)

Of all the study lakes, Lake IVA was most frequently sampled, namely in March, May, June, August, September, and October. During both winter and summer stagnations the lake reached total anoxis at the sediment surface (Figure 5).

The MeHg behavior of this seepage lake showed some differences compared to Lakes KEHA and HAKO. With the exception of the winter stagnation, MeHg levels in the epilimnion were clearly lower than in the other two lakes. In March and during June through August MeHg maxima were recorded in the anoxic hypolimnion (max. 2.95 ng L^{-1}, Figure 5).

At spring overturn the MeHg level in water dropped to 20% of the original in the epilimnion and to less than 10% in the hypolimnion: from 0.26–1.47 to 0.06–0.07 ng L^{-1} (Figure 5). When the summer stagnation developed, the MeHg concentration increased in both the epilimnion and in the hypolimnion. During fall mixis the concentrations decreased again. However, in late October, shortly before permanent ice cover, the lake water contained nearly twice as much MeHg as in May (0.09 to 0.15 ng L^{-1}). The behavior was somewhat similar to that observed in the treatment and reference basins of acidified Little Rock Lake, Wisconsin.[27]

Since the sediment traps were placed only in May, the influence of sedimenting particles on scavenging water MeHg levels during spring mixis is uncertain. However, results presented by Bloom et al.[27] and Hurley et al.[38] from Little Rock Lake indicate such a process. The demethylation potential in water (Figure 5), and also in profundal sediments (Figure 6L), had already reached the same level in May as was recorded later during summer and may also have affected the MeHg decrease. The early rise of potential demethylation in May is similar to that found in Lake Hako which, however, had no MeHg decrease in water during spring. In Lake Hako the input of MeHg from Lake Keha during spring may confuse the picture.

The total MeHg concentration in a sediment trap in August, after 7 weeks collection, was 6.60 ng L^{-1}. Six weeks later (in September) an order of magnitude lower concentration, 0.66 ng L^{-1}, was recorded (Figure 5). The dissolved and particulate MeHg was analyzed in September, at a time when fall mixis had started. Particulate MeHg represented about 60% of the total MeHg in the water phase (0.094 and 0.152 ng L^{-1}, respectively). In the trap material particulate MeHg was more abundant, both in concentration and in fraction of total MeHg, representing 80% of the MeHg (0.54 vs. 0.66 ng L^{-1}). The particulate-associated MeHg flux to sediments was not measured.

The sedimenting seston on the bottom of the traps, as well as on the sediment surface, was bright-green during the ice-free season. In contrast to the other lakes, the whole water mass was penetrated by light. Nearly all the material in the traps presumably was algae and/ or photoautotrophic green bacteria. The organic matter content varied from 71 to 85% in the traps (data from summer 1990), and resuspension of old sediments was most probably of minor importance.

The algal biomass both in the epilimnion (1 to 2 m) and in the hypolimnion (4 to 5 m) during summer 1991 consisted mainly of *Gonyostomum semen* (Ehr.) Diesing (Raphidophyceae), which accounted for 40 to 90% of the biomass. The other abundant algae were flagellated Cryptophyceae, Chrysophyceae, and Chlorophyceae.[39] In early August, purple sulfur bacteria of the genus *Chromatium* were present at a quantity of about 90 cells per milliliter in the hypolimnion (4 to 5 m). At the same time, three times as much chlorophyll was present in the hypolimnion than in the epilimnion, regardless of a higher phytoplankton biomass in the latter. Comparing the absorbances at two different wavelengths (665 nm/654 nm), a ratio of 1.8 was measured in the epilimnion and 0.6 in the hypolimnion, indicating that photoautotrophic bacteria formed most of the chlorophyllous pigments in the hypolimnion.[39]

Very low, if any, potential methylation was measured in the hypolimnic samples in August (Figure 5), a time of maximum MeHg concentration. The methylation potential of the sedimenting seston collected from the traps was also low. However, the incubations in both water and in trap material were performed in the dark in the laboratory, and do not allow us to draw conclusions of potential methylation by photoautotrophic bacteria. The only incubations in light were performed in June in situ at a time of oxic hypolimnion, and showed a somewhat higher methylation potential (10 to 30%) in light than in the dark.[40]

Potential methylation rates were somewhat higher and demethylation rates somewhat lower in Lake IVA profundal sediments than in Lake HAKO (Figure 6). This was true for both aerobic and anaerobic incubations. Comparing aerobic M/D (methylation/demethylation) in Lake HAKO with anaerobic M/D in Lake IVA shows a 20 times higher net methylation potential in the latter (0.004 vs. 0.083). This is consistent with the findings of other studies[41] as well. Thus, the net methylation of Hg in the anaerobic sediments of Lake IVA may be at least one reason for the differences in MeHg concentrations in the hypolimnion of these two lakes.

The results indicate that sedimenting particles (mainly algae) and anaerobic sediments are important MeHg sources in the hypolimnion of Lake IVA. Net methylation of Hg is evident in the water phase as well. The sampling/analyzing methods used do not allow us to draw any conclusion on the MeHg production by photoautotrophic bacteria present in the anaerobic hypolimnion at layers of maximum MeHg concentrations.

D. THE ROLE OF METHYLATION/DEMETHYLATION IN LAKE CATCHMENTS AND THE STUDY LAKES

The calculated annual MeHg flux from direct wet deposition and snowmelt on the lake surface compared to catchment derived MeHg was less than 0.5% for Lake KEHA and 1 to

2% for Lake HAKO. The atmospheric derived MeHg on the whole catchment of Lake KEHA was estimated as 100 mg a^{-1}, but may be an underestimate due to few data on wet deposition. The MeHg runoff from the catchment was 270 mg a^{-1}, indicating a net methylmercury production in the catchment. During the study period Lake KEHA was clearly a source of MeHg (270 mg a^{-1} input vs. 840 mg a^{-1} output at the outlet), whereas Lake HAKO acted as a sink for MeHg (1000 mg a^{-1} input vs. 250 mg a^{-1} output).

Due to the small drainage area, atmospheric MeHg input and in-lake MeHg production may be of much greater importance in Lake IVA than in the Lakes HAKO and KEHA. Lake IVA had the lowest MeHg concentration in epilimnetic water and a clearly different seasonal fluctuation pattern compared to the other two lakes. The methylation potential and M/D ratio in littoral and profundal sediments and in the water phase of lake IVA were at least as high or higher than in the other two lakes.

A comparison among the three lakes suggests that the reasons for differences in the hypolimnion MeHg concentrations may be differences in methylation, but especially differences in MeHg demethylation. The anoxic conditions in the hypolimnion seem to reduce both methylation and demethylation in the water phase, but stimulate methylation in sediments. The significance of photoautotrophic bacteria as MeHg producers remains open. Methylmercury production by photoautotrophs cannot, however, explain MeHg concentrations in the very dark hypolimnion of Lake KEHA (Secchi disk transparency <1 m) during the summer or in Lakes KEHA and IVA during winter.

The methylation potential in water and sedimenting particles is most probably very low compared to that in sediments. In contrast, the demethylation potential may be of the same order in all compartments studied. This may lead to a more efficient demethylation in the water and at the sediment surface of aerobic hypolimnion than in anaerobic conditions.

It is worth noting that, in contrast to methylation, demethylation in the water phase and at the sediment surface showed 'high' rates during spring and fall. Thus, in-lake demethylation may be efficient for longer periods during ice-free seasons than methylation. The only lake with an aerobic hypolimnion during both winter and summer stagnations was Lake HAKO, with a net loss of MeHg. These findings furthermore emphasize the importance of demethylation.

IV. CONCLUSIONS

The results of this study indicate that Hg methylation in soils, littoral and profundal sediments, and in epilimnetic water may be of significance as a MeHg source for lakes. The fact that the volume of lake water and the volume of catchment soils are considerably greater than that of lake sediments emphasizes their importance not only as methylation sites, but also as sites of MeHg demethylation. The results support the hypothesis that high fish MeHg concentrations, commonly found in drainage lakes in Finland, are caused by MeHg production in soils. In seepage lakes, atmospheric input and in-lake production of MeHg are of much greater importance than in drainage lakes. Particulate sedimentation and demethylation of MeHg greatly affect the MeHg cycle in lakes. Methylation and demethylation activities in lakes may be primarily controlled by oxygen conditions.

ACKNOWLEDGMENTS

The study was financed by the Finnish Research Foundation for Natural Resources, the Ministry of the Environment, and the National Board of Waters and the Environment. We thank Ms. Petra Tallberg, Mr. Asko Särkelä, and Mr. Ismo Malin for their valuable assistance in laboratory work and sampling. We also thank M.Sc. Marko Järvinen for the water quality and biological data of Lake IVA.

REFERENCES

1. Lindqvist, O., Jernelöv, A., Johansson, K., and Rodhe, H., Mercury in the Swedish Environment. Global and Local Sources, Swedish Environment Protection Board, Report PM 1816, 1984.
2. Verta, M., Mercury in Finnish forest lakes and reservoirs: anthropogenic contribution to the load and accumulation in fish, in: *Publications of the Water and Environment Institute*, National Board of Water and the Environment, Finland, 6, 1990.
3. Wiener, J.W. and Stokes, P.M., Enchanged bioaccumulation of mercury, cadmium and lead in low-alkalinity waters: an emerging regional environmental problem, *Environ. Toxicol. Chem.*, 9, 821, 1990.
4. Grieb, T.M., Driscoll, C.T., Gloss, S.P., Scofield, C.L., Bowie, G.L., and Porcella, D.P., Factors affecting mercury accumulation in fish in the upper Michigan Peninsula, *Environ. Toxicol. Chem.*, 9, 919, 1990.
5. Suns, K., Curry, C., and Russel, D., The effects of water quality and morphometric parameters on mercury uptake by yearling yellow perch. Ontario Ministry of the Environment, Rexdale, Ontario, Technical Report LTS, 80–1, 1980.
6. Verta, M., Rekolainen, S., Mannio, J., and Surma-Aho, K., The origin and level of mercury in Finnish forest lakes, *Publication of the Water Research Institute*, National Board of Waters, Finland, 65, 21, 1986.
7. Richman, L.A., Wren, C.D., and Stokes, P.M., Facts and fallacies concerning mercury uptake by fish in acid stressed lakes, *Water Air Soil Pollut.*, 37, 465, 1988.
8. Winfrey, M.R. and Rudd, J.W.M., Environmental factors affecting the formation of the methylmercury in low pH lakes, *Environ. Toxicol. Chem.*, 9, 853, 1990.
9. Gilmour, G. C. and Henry, E.A., Mercury methylation in aquatic systems affected by acid deposition, *Environ. Pollut.*, 71, 131, 1991.
10. Forsius, M., Kämäri, J., Kortelainen, P., Mannio, J., Verta, M., and Kinnunen, K., Statistical lake survey in Finland. Regional estimates on lake acidification, in: Kauppi, P., K. Kenttämies, and P. Anttila, Eds., *Acidification in Finland*, Springer-Verlag, Heidelberg, 761, 1990.
11. Rogers, R.D., Abiotic methylation of mercury in soils, *J. Environ. Qual.*, 6, 463, 1977.
12. Furutani, A. and Rudd, J.W.M., Measurement of mercury methylation in lake water and sediment samples, *Appl. Environ. Microbiol.*, 40, 770, 1980.
13. Jensen, S. and Jernelöv, A., Biological methylation of mercury in aquatic organisms, *Nature*, 223, 753, 1969.
14. Rudd, J.W.M., Furutani, A., and Turner, M.A., Mercury methylation by fish intestinal contents, *Appl. Environ. Microbiol.*, 40, 777, 1980.
15. Jernelöv, A., Mercury and food chains, in: Hartung, R. and B.D. Dinman, Eds., *Environmental Mercury Contamination*, Ann Arbor Science Publishers, Ann Arbor, MI, 174, 1972.
16. Järvinen, O. and Vänni, T., Bulk deposition chemistry in Finland, in: Kauppi, P., K. Kenttämies, and P. Anttila, Eds., *Acidification in Finland*, Springer-Verlag, Heidelberg, 151, 1990.
17. Rekolainen, S., Verta, M., and Leihu, A., The effect of airborne mercury and peatland drainage on sediment mercury contents in some Finnish forest lakes, *Publications of the Water Research Institute*, National Board of Waters, Finland, 65, 11, 1986.
18. Verta, M., Tolonen, K., and Simola, H., History of heavy metal pollution in Finland as recorded by lake sediments, *Sci. Total Environ.*, 87/88, 1, 1989.
19. Verta, M., Changes in fish Mercury concentration in an intensively fished lake, *Can. J. Fish. Aquat. Sci.*, 47, 1888, 1990.
20. Rask, M. and Metsälä, T.-R., Mercury concentrations in northern pike, *Esox lucius* L., in small lakes of Evo area, southern Finland, *Water Air Soil Pollut.*, 56, 369, 1991.

21. Bloom, N.S. and Crecelius, E.A., Determination of mercury in seawater at subnanogram per liter levels, *Mar. Chem.*, 14, 49, 1983.

22. Fitzgerald, W.F. and Gill, G.A., Subnanogram determination of mercury by two-stage gold amalgamation and gas phase detection applied to atmospheric analysis, *Anal. Chem.*, 51, 1714, 1979.

23. Bloom, N.S. and Fitzgerald, W.F., Determination of volatile mercury species at the picogram level by low-temperature gas chromatography with cold-vapor atomic fluorescence detection, *Anal. Chim. Acta*, 208, 151, 1988.

24. Bloom, N.S., Determination of picogram levels of methylmercury by aqueous phase ethylation, followed by cryogenic gas chromatography with cold vapor atomic fluorescence detection, *Can J. Fish. Aquat. Sci.*, 46, 1131, 1989.

25. Bloom, N.S., Watras, C.J., and Hurley, J.P., Impact of acidification on the methylmercury cycle of remote seepage lakes, *Water Air Soil Pollut.*, 56, 477, 1991.

26. Ramlal, P.S., Rudd, J.W.M., and Hecky, R.E., Methods for measuring specific rates of mercury methylation and degradation and their use in determining factors controlling net rates of mercury methylation, *Appl. Environ. Microbiol.*, 51, 110, 1986.

27. Matilainen, T., Verta, M., Niemi, M., and Uusi-Rauva, A., Specific rates of net methylmercury production in lake sediments, *Water Air Soil Pollut.*, 56, 595, 1991.

28. Lee, Y.H. and Hultberg, H., Methylmercury in some Swedish surface waters, *Environ. Toxicol. Chem.*, 9, 833, 1990.

29. Lee, Y.H. and Iverfeldt, Å., Measurement of methylmercury and mercury in run-off, lake and rain waters, *Water Air Soil Pollut.*, 56, 309, 1991.

30. Westling, O., Mercury in runoff from drained and undrained peatlands in Sweden, *Water Air Soil Pollut.*, 56, 419, 1991.

31. Iverfeldt, Å. and Johansson, K., Mercury in run-off water from small watersheds, *Verh. Int. Verein. Limnol.*, 23, 1626, 1988.

32. Iverfeldt, Å., Occurrence and turnover of atmospheric mercury over the Nordic countries, *Water Air Soil Pollut.*, 56, 251, 1991.

33. Rekolainen, S., Verta, M., and Järvinen, O., Mercury in snow cover and rainfall in Finland 1983–1984, *Publications of the Water Research Institute,* National Board of Waters, Finland, 65, 3, 1986.

34. Järvinen, O. and Vänni, T., Bulk deposition chemistry in Finland in 1988, National Board of Waters and the Environment Finland, Report No. 235 (in Finnish), 1990.

35. Järvinen, O. and Vänni, T., Bulk deposition chemistry in Finland in 1990, National Board of Waters and the Environment Finland, Report No. 378 (in Finnish), 1992.

36. Bodaly, R.A., Hecky, R.E., and Ramlal, P.S., Mercury availability, mobilization and methylation in the Churchill River Diversion area. in: Technical Appendices to the Summary Report, Canada-Manitoba Agreement on the Study and Monitoring of Mercury in the Churchill River Diversion, Vol. 1, chap 3, Governments of Canada and Manitoba, Winnipeg, 1987.

37. Jackson, T.A., Biological and environmental control of mercury accumulation by fish in lakes and reservoirs in northern Manitoba, Canada, *Can. J. Fish. Aquat. Sci.*, 48, 2449, 1991.

38. Hurley, J.P., Watras, C.J., and Bloom, N.S., Mercury cycling in a Northern Wisconsin seepage lake. The role of particulate matter in vertical transport, *Water Air Soil Pollut.*, 56, 543, 1991.

39. Järvinen, M., personal communication, 1992.

40. Matilainen, T. and Verta, M., unpublished data, 1991.

41. Regnell, O. and Tunlid, A., Laboratory study of speciation of mercury in lake sediment and water under aerobic and anaerobic conditions, *Appl. Environ. Microbiol.*, 57, 789, 1991.

The Vertical Distribution of Mercury Species in Wisconsin Lakes: Accumulation in Plankton Layers

Carl J. Watras and Nicolas S. Bloom

CONTENTS

ABSTRACT: The aquatic cycling of highly bioreactive trace elements such as Hg is potentially influenced by layers of physiologically active plankton which develop in the thermocline and hypolimnia of lakes. Here we examine the distribution of waterborne mercury species (Hg and methyl-Hg in dissolved and particulate forms) with respect to the vertical distribution of plankton layers in three quite different Wisconsin lakes. We also examine corresponding distributions of waterborne Fe, Mn, dissolved oxygen, organic carbon, sulfate, and sulfide. To resolve fine-scale spatial features of the plankton layers, the watercolumn was mapped using computer-based, bio-optical techniques with an absolute resolving power of about 2 cm.

Mercury accumulated in all the plankton layers we observed, reaching maximum concentrations of 45 ng Hg/L (compared to 1 ng/L in the epilimnion), but there were distinct differences in the distributions of Hg and methyl-Hg. Methyl-Hg accumulated primarily in plankton layers below the oxic/anoxic (O/A) boundary, reaching 12 ng/L in anoxic layers (>100-fold higher than [methyl-Hg] in the epilimnion). Spatially and seasonally, [Hg] tended to track the concentration of chlorophyll-*a* (eukaryotic phytoplankton and cyanobacteria); [methyl-Hg] tracked the distribution of bacteriochlorophylls (phototrophic sulfur bacteria). Maxima of bacteriochlorophyll and methyl-Hg both occurred near the sulfide/sulfate transition zone below the oxic/anoxic boundary. Unlike Fe and Mn, the concentration of Hg and methyl-Hg tended to decrease in the region between the lowest plankton layers and the sediment surface. While the orthograde distributions of Fe and Mn indicate redox-driven diffusion from sediments into the overlying watercolumn, more complex processes appear to determine the observed vertical distribution of mercury species.

Given the strong affinity of Hg for biotic particles (log Kd = 5 to 6), either Hg scavenging/settling from above or Hg diffusion/sorption to plankton layers could account for the elevated [Hg] observed in all plankton layers. Both mechanisms could be involved if Hg was released from settling detritus and then resorbed by cells within the layers. Since [methyl-Hg] maxima were confined to anoxic plankton layers, either diffusion/sorption from below (i.e., sediments) or de novo production of methyl-Hg within these layers is implied. Our data tend to support the hypothesis that methyl-Hg is formed within anoxic layers via conversion of Hg derived from the upper watercolumn. We explore this hypothesis and its implications.

I. INTRODUCTION

With growing reports of methyl-Hg contamination in freshwater fisheries, there is an increasing need for accurate data on the concentration and distribution of waterborne mercury species in lakes and streams. Although the transformation of inorganic mercury to methylated Hg species is known to occur in freshwater, estuarine, and marine sediments,[1–3] there is also evidence of methylmercury production in the watercolumn[4–6] dating back to the calculations of Topping and Davies[7] which indicated that the annual production of methylmercury in the water and sediments of the world oceans was similar. There remains substantial uncertainty about:

1. The relative importance of sedimentary vs. watercolumn methylation;
2. The identity of methylating microbes;
3. Rates of methyl-Hg formation and destruction in natural ecosystems; and
4. The environmental dependencies of these rates.

These uncertainties stem, in part, from a lack of reliable data on the distribution of waterborne Hg species.

Recent analytical advances now permit the determination of Hg and alkyl-Hg species in natural waters with high precision and reproducibility.[8,9] Using trace metal clean sampling techniques, we have determined the ambient concentration and distribution of Hg species in pristine lake water, rain, air, sediments, porewaters, and the biota of several Wisconsin lakes.[10–18] In the surface waters of these lakes, [HgT] generally ranged from about 0.5 to 5 ng Hg/L and [methyl-Hg] ranged from 0.05 to 0.5 ng methyl-Hg/L. However, much higher concentrations of both Hg and methyl-Hg have been observed at depth in some lakes, particularly in metalimnia and hypolimnia where dense populations of plankton accumulate in discrete layers.[12]

Plankton layers typically comprise either phytoplankton or bacterioplankton along with assemblages of protozoans and small metazoans.[19] Guerrero and Mas[20] coined the acronym MPM (multilayered planktonic microbial communities) to describe plankton layers and they documented similarities between laminated microbial communities in the watercolumn and those observed in benthic mats. Gasol and Pedros-Alio[21] have demonstrated that plankton layers form via in situ growth rather than the passive accumulation of settling cells.

Table 1 **Characteristics of the three study lakes in north-central Wisconsin.**

Lake	Type	Surface area (ha)	Zmax (m)	pH[a]	Ref.
Pallette	Seepage, dimictic	70	18.2	7.3	33
Little Rock[b]	Seepage, dimictic	9.8	10.3	5.0	26, 34
Mary	Drained, meromictic	1.2	21.7	5.9	31, 35, 36

[a] Epilimnetic.
[b] Treatment Basin.

Although relationships between the physiological activity of plankton layers and the cycling of carbon and sulfur have been studied,[22] their roles in trace element cycling remain largely unknown. Our early work on mercury distributions in Wisconsin lakes suggested that plankton layers were potentially important sites for biologically driven mercury transformations.[12]

Here, we report data on the distribution of Hg and alkyl-Hg species relative to the distribution of plankton layers and other physicochemical gradients in the watercolumn of three small lakes in northern Wisconsin. These data were obtained using bio-optical instrumentation to map the fine-scale, vertical distribution of planktonic communities[23,24] combined with the "clean" analytical and sampling techniques used for low-level mercury determinations. We also examine hypotheses relating plankton layers and sulfur cycling to methyl-Hg formation.

II. METHODS

A. STUDY SITES

The three study sites are small, remote lakes located in the Northern Highlands Lake District of Wisconsin, near the Trout Lake Biological Station (46°N, 89°W). This region is heavily forested and sparsely populated. The area contains hundreds of seepage and drainage lakes situated in thick glacial tills and outwash sands at an elevation of about 500 m.[25] The three study lakes were selected because their stratification regimes result in very different patterns of plankton layering. In Little Rock Lake, plankton layers form close to the sediment/water interface. In Pallette Lake, they form at mid-water depths; and in Mary Lake they form in the upper waters well above the sediments. By comparing Hg accumulation in layers that form at such different locations, we hope to evaluate potential Hg sources and accumulation mechanisms.

The lakes were also chosen because they represent both holomictic and meromictic waters. Little Rock Lake and Pallette Lake are dimictic, clearwater seepage lakes that stratify thermally soon after ice-out in spring (April/May) and remain stratified until the autumn mixis (September/October). Mary Lake is dystrophic (color 100 PCU) and meromictic. Additional data for these lakes are listed on Table 1.

B. BIO-OPTICAL PROFILES

A computer-based, bio-optical profiling unit was used to measure temperature, dissolved oxygen, depth, optical density, right-angle light scattering, in vivo chlorophyll-*a* fluorescence, and downwelling irradiance at 1 to 2 cm depth intervals throughout the watercolumn of each lake. The sensor housing was lowered at 1 cm s^{-1} using an electric winch deployed from a stabilized 4-m boat (Boston Whaler). The profiler design was based on prototypes described by Watras and Baker[23] and Steenbergen et al.[24] Temperature was measured with a Thermometrics FP14 thermistor; dissolved oxygen with a continuously stirred, temperature-compensated, Martek polarographic electrode; depth with a Druck PTX 160/D depth transmitter; optical density with a 25-cm, folded-beam, Martek XMS transmissometer; right-angle scatter with a Turner Designs Model 10 field nephelometer; in vivo fluorescence with a Turner Designs Model 10–005R field fluorometer with the standard Chl-a filter set supplied by Turner Designs; and downwelling irradiance with a Licor underwater PAR cosine collector. Water was pumped through the deck-mounted fluorometer and the nephelometer using a submersible pump. The pump and all other sensors were aligned with the transmissometer beam at depth. All data were logged on a portable computer, corrected for pumping rate, and displayed real-time during descent.

C. MERCURY ANALYSIS

Water samples for Hg analysis were collected by extending a 1 m acrylic outrigger to the sensor housing of the bio-optical profiler and then pumping water to the surface through rigorously cleaned and blanked tubing using a nonmetallic pump. The sample intake port was aligned with the transmissometer beam. Water samples were collected on deck into precleaned, Teflon® bottles that were hermetically sealed and double-bagged for transport. Hg-clean technique was followed throughout the process of sample collection and handling.[10]

Total Hg, (HgT), was determined on unfiltered samples after BrCl wet-oxidation by $SnCl_2$ reduction, dual gold-column amalgamation, followed by thermal desorption and atomic fluorescence detection.[8] Dissolved Hg, (HgD), was determined the same way on lakewater samples filtered through Hg-free, 0.4 µm, 47 mm disposable filtration units. Particulate Hg (HgP) was calculated by difference. Operational detection limits for these Hg species were about 0.05 ng Hg/L in a 100 mL sample.

Total monomethylmercury, (methyl-HgT), was determined by extracting unfiltered samples from an HCl/KCl matrix into CH_2Cl_2, then back-extracting into pure water by solvent evaporation. The aqueous phase was ethylated with $NaB(C_2H_5)_4$ and the volatile ethyl analogs were quantified via pre-trapping by gas purging onto graphitized carbon, cryogenic GC separation, pyrolysis to Hg^0, followed by atomic fluorescence detection.[9] Dissolved methyl-Hg (methyl-HgD) was determined the same way on filtered samples. Particulate methyl-Hg (methyl-HgP) was determined by difference. Detection limits for these species were about 0.01 ng/L as Hg in a 50 mL sample.

C. ANCILLARY CHEMISTRY

In addition to the in situ oxygen probe, dissolved oxygen was analyzed at selected depths using the Pomeroy-Kirschner modification of the Winkler titration. Sulfate was determined by suppressed-column ion chromatography. Dissolved sulfide was determined after SAOBII addition using an Orion model 94–16BN silver/sulfide electrode and an Orion double-junction reference electrode (detection limit circa 100 nM).[27] Fe and Mn were determined on unfiltered samples using ICP. Dissolved organic carbon (DOC) was determined on 0.4 µm filtered samples by wet oxidation with sodium persulfate followed by heating (200° C) and non-dispersive CO_2 detection. Photosynthetic pigments were determined by reverse-phase HPLC, following the method of Hurley and Watras[28] for bacteriochlorophylls. Suspended particulate mass (SPM) was determined on oven-dried, preweighed filters.

III. RESULTS

A. Hg AND METHYL-Hg
1. Little Rock Lake

The development of two deep plankton layers in Little Rock Lake began soon after ice-out and the layers reached peak density in mid-August (Figures 1 and 2). The location of the layers changed as the thermocline deepened and anoxia developed in the hypolimnion. The upper layer comprised Chl-a-containing phytoplankton and the lower layer comprised BChl-d-containing bacterioplankton (Figure 9). The optical density trace went off-scale in August due to intense scattering associated with the small particles (Figure 2).

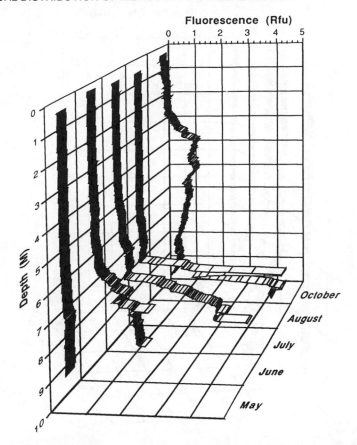

Figure 1 Seasonal development of plankton layers in the thermocline and hypolimnion of Little Rock Lake, as measured by in vivo chlorophyll fluorescence with a spatial resolution of 2 cm (RFU, relative fluorescence units). Profiles were similar at multiple stations.

Seasonal changes in [HgT] paralleled changes in [Chl-a] within the upper plankton layer (Figure 3). At peak stratification, [HgT] reached roughly 45 ng/L in this region. This represents an increase of 40- to 50-fold over the average epilimnetic [HgT] of about 1 ng/L.

Near the O/A boundary, there were pronounced changes in the chemical speciation and physical form of waterborne mercury (Figure 4). As noted above, maximum concentrations of total Hg occurred within the upper plankton layer which spanned the O/A boundary and most (94%) of the HgT was particulate (Figure 4a). Both [HgT] and [HgP] decreased from this point to the sediment surface. The dissolved Hg fraction (HgD) was relatively constant (8.95 to 8.97 ng/L) in the three bottom samples from 9.0 to 9.5 M.

Maximum concentrations of methyl-Hg occurred within the lower (anoxic) plankton layer (Figure 4b). [Methyl-Hg] reached 11.6 ng/l in this layer, over 100 times greater than epilimnetic concentrations. As observed with [HgP], [methyl-HgP] was highest near the plankton layer peak, where it comprised >50% of the methyl-HgT. Concentrations of methyl-HgT and methyl-HgD decreased between the layer and the sediment surface.

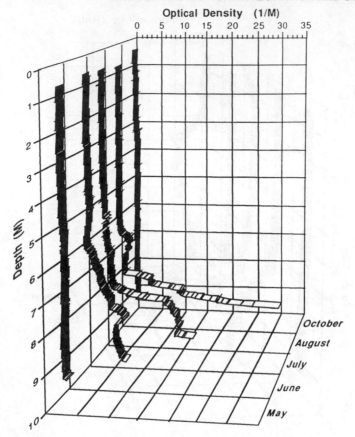

Figure 2   Seasonal development of plankton layers in Little Rock Lake as measured by transmissometry.

2.   Pallette Lake

In Pallette Lake we also observed two major plankton layers, but they were broader, more widely spaced, and located further from sediments than those in Little Rock Lake (Figure 5). The upper layer was associated with a dissolved oxygen maximum indicating oxygenic photosynthesis in the metalimnion. Microscopic inspection indicated that this upper plankton layer was dominated by small, filamentous cyanobacteria. The lower (anoxic) plankton layer was sharper and it was rich in Bchl-e, characteristic of brown-green (Chlorobiaceae) phototrophic sulfur bacteria (PSB). Unlike Little Rock Lake, where the PSB layer was close to the sediment surface, these PSB were 4–6 m above the bottom.

[HgT] peaked within both the oxic and anoxic plankton layers of Pallette Lake (Figure 5a). Near the top of both plankton layers, [HgP] constituted up to 96% of the [HgP] (Figure 5b). Concentrations of dissolved mercury were highest just below the particulate Hg maxima, but [HgD], [HgP], and [HgT] all decreased toward sediments.

As in Little Rock Lake, maximal [methyl-Hg] occurred in anoxic water just below the second [Hg] maximum (Figure 6). Methyl-Hg constituted roughly 30% of the HgT in the lower layer and about half of the methyl-Hg was in the particulate form. Due to analytical difficulties, we were not able to resolve the dissolved/particulate fractionation above or below the layers; but, as observed with [Hg], concentrations of methyl-HgT decreased toward the sediments.

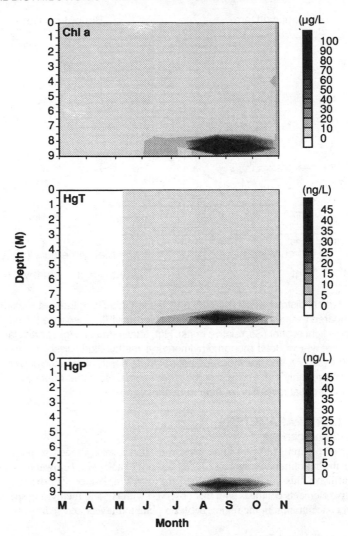

Figure 3 Seasonal and spatial development of chlorophyll-*a*, total mercury (HgT), and particulate mercury (HgP) maxima in Little Rock Lake.

3. Mary Lake

Mary Lake is meromictic and during August the thermocline was very close to the surface of the lake. The mixed layer was confined to the upper 1 m (Figure 7). Dissolved oxygen concentrations decreased rapidly below the narrow mixed layer and the O/A boundary was situated at about 2 m depth, roughly 17 m above the profundal sediments. A distinct plankton layer formed in this region.

Methyl-Hg concentrations were greatest in the plankton layer at 2 m (Figure 7a). At this depth, roughly 40% of the HgT was methylated (Figure 7b). In surface waters and at depth, less than 10% of the Hg was methylated. A secondary methyl-Hg maximum was located near the top of the monimolimnion. Below 10 m, the concentration of methyl-Hg was very low and decreased with depth to the sediment surface.

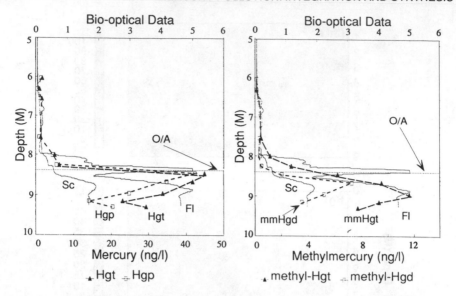

Figure 4 Vertical distribution of mercury species in Little Rock Lake at maximum stratification compared to cm-scale profiles of in vivo chlorophyll fluorescence (Fl, relative units) and right-angle light scatter (Sc, relative units). Hgt, total mercury; Hgp, particulate mercury (>0.4 μm); methyl-Hgt, total monomethylmercury; methyl-Hgd, dissolved monomethylmercury (<0.4 μm). Dotted line indicates approximate depth of the oxic/anoxic boundary (O/A boundary) roughly 1 m above profundal sediments. Note that depth scale begins at 5 m to improve spatial resolution of near sediment zone.

B. ANCILLARY PARAMETERS
1. Iron and Manganese
Distributions of Fe and Mn in all three lakes indicated a strong redox response to anoxic conditions in the hypolimnion (Figure 8). Increases in [Fe] and [Mn] below the O/A boundary suggested diffusive flux from sediments into the overlying water. The behavior of these redox-sensitive elements contrasts with that observed with Hg and methyl-Hg species, which decreased in concentration in the region between plankton layers and sediments.

2. Chlorophyll-*a* and Bacteriochlorophylls
Rather than following [Fe] and [Mn], the distribution of mercury species more closely followed the distribution of photosynthetic pigments in these lakes (Figure 9). In general, maxima of [Chl-a] and [Hg] tended to co-occur, as did maxima of [Bchl] and [methyl-Hg].

3. Dissolved Oxygen, Organic Carbon, Sulfate and Sulfide
Increased dissolved oxygen near peaks in Chl-a indicated active oxygenic photosynthesis in all the lakes (Figure 9). All profiles were made near midday on sunny afternoons, when one would expect high photosynthetic rates in nutrient-replete algae. DOC also tended to increase in these regions, perhaps as the result of leaking photosynthate.

In all cases, the methyl-Hg maximum occurred in anoxic water near the sulfate/sulfide transition zone. The vertical distributions of SO_4 and total sulfide were similar in all three lakes, although absolute concentrations varied (Figure 9). [SO_4], for example, was consistently higher in Little Rock Lake due to experimental additions during the acidification years.[26] Within the thermocline, [SO_4] assumed an S-shaped distribution indicating alternating

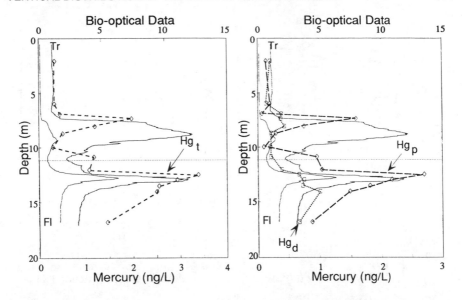

Figure 5 Vertical distribution of Hg species in Pallette Lake relative to centimeter-scale profiles of transmissivity (Tr = optical density m^{-1}), and in vivo chlorophyll fluorescence (Fl, relative units). Hg$_t$, total mercury; Hg$_p$ particulate mercury (>0.4 µm); Hg$_d$, dissolved mercury (<0.04 µm). Dotted line indicate O/A boundary roughly 5.5 m above profundal sediments.

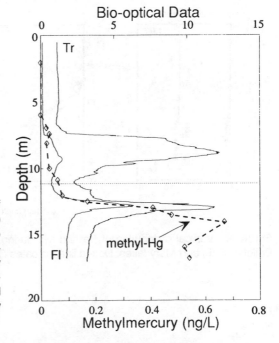

Figure 6 Vertical distribution of monomethylmercury (methyl-Hg) in Pallette Lake, WI, in relation to centimeter-scale profiles of transmissometry (Tr = optical density m^{-1}), and in vivo chlorophyll fluorescence (Fl, relative units). Dotted line shows O/A boundary roughly 5.5 m above sediments.

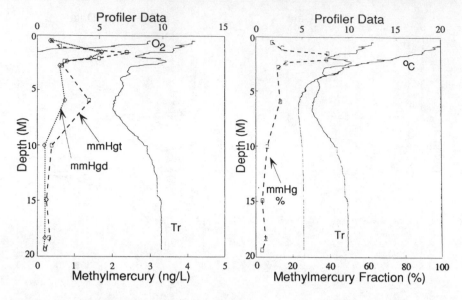

Figure 7 Vertical distribution of methyl-Hg species in meromictic Mary Lake relative to cm-scale profiles of dissolved oxygen (O_2, mg/L), transmissometry (Tr = optical density m^{-1}), and temperature (°C). Methyl-Hg fraction: % methyl-Hg = ([methyl-Hgt]/[Hgt] × 100).

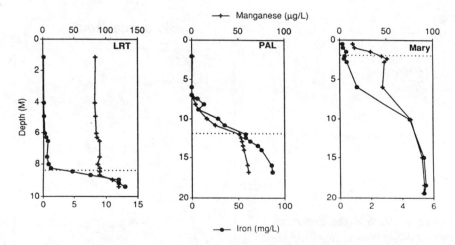

Figure 8 Vertical distribution of Fe and Mn during stratification in Little Rock (LRT), Pallette (PAL) and Mary lakes. Dotted line indicates O/A boundary.

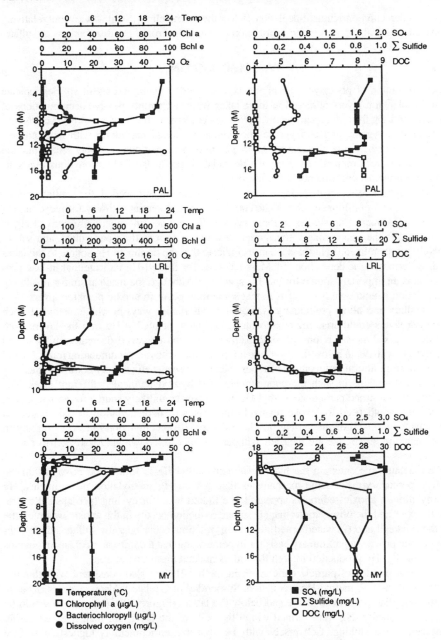

Figure 9 Vertical distribution of chlorophyllous pigments (chlorophyll-*a* and bacteriochlorophylls), dissolved organic carbon (DOC), dissolved oxygen (O₂), temperature, sulfate, and total sulfide during stratification in Little Rock (LRT), Pallette (PAL), and Mary (MY) Lakes.

zones of depletion-enrichment-depletion. Below the thermocline, $[SO_4]$ was low but relatively constant. Vertical changes in the distribution of sulfide mirrored the distributions of sulfate.

IV. DISCUSSION

The distribution of photosynthetic pigments, dissolved oxygen, and sulfur species indicate that the plankton layers observed in these lakes were not simply passive accumulations of material settling from the epilimnion. Microscopy confirmed that they comprised active populations of oxygenic and anoxygenic photosynthetic organisms, along with heterotrophic bacteria, protozoans, and metazoans. Although epilimnetic particles probably settled into the layers and were remineralized there, the degradation products of chlorophyllous pigments were not principal constituents of the layers.[28]

Although more intensive process-oriented investigations are needed, the settling and re-mineralization of epilimnetic particulate could provide a potential pathway for recycling Hg at depth. Sediment trap studies in these lakes indicate that atmospherically derived Hg is scavenged in the upper waters, transported downward with settling particulate, and subsequently remineralized.[17] Hg mass balances indicate that the dominant Hg source to the lakes is direct atmospheric deposition.[10,29] Mass balances for methyl-Hg indicate that in-lake production is an important source for this Hg species.[29] Although the mechanism for in situ Hg methylation is unknown, we speculate that some may occur in anoxic plankton layers.

Ancillary data allow preliminary evaluation of alternative ways in which methyl-Hg-rich plankton layers could arise in anoxic waters. Diffusion of methyl-Hg into the layers from below (i.e., sediments) is one possible pathway, but the observed decreases in dissolved methyl-Hg with depth below the layers are not consistent with this accumulation route. Lateral diffusion from littoral sediments is also possible, but it seems an unlikely common cause for similar distributions in such morphometrically and hydrodynamically different lakes. The settling of preformed particulate methyl-Hg also seems an unlikely source for anoxic layers, since one would need to invoke relatively high rates of demethylation in oxic waters to account for the lack of methyl-Hg-rich layers there. If demethylation did occur rapidly in oxic plankton layers, we are then left without a source for downward transport to anoxic layers.

By default, we entertain the hypothesis that methyl-Hg is formed de novo within the layers, perhaps via the same mechanisms that govern Hg methylation in sediments. Hg methylation in anoxic sediments appears to be linked to sulfur cycling (perhaps sulfate reduction),[3,30] and the observations that (1) sulfate reduction occurs in the anoxic watercolumn of these lakes[31] and (2) waterborne [methyl-Hg] maxima occur near the sulfate/sulfide transition zone provide a preliminary basis for hypothesizing that a common mechanism governs in situ methyl-Hg production in both the sediments and the watercolumn.

Vertical profiles of particle enrichment in methyl-Hg are also consistent with this hypothesis of in-layer methyl-Hg production. Suspended particulate in deep layers is richer in methyl-Hg than particulate above and below the layers (Figure 10). This pattern would be expected if methyl-Hg were formed within the cells of the layer and then released during senescence and sinking. Conversely with Hg, particle enrichment is highest in surface waters—as expected for an element scavenged in the epilimnion and remineralized at depth (Figure 10).

We have calculated hypothetical rates of in situ methylation within anoxic plankton layers based on a rough, seasonal time-course of methyl-Hg build-up (Table 2). Interestingly, the calculated rates of methyl-Hg production in layers are remarkably consistent (50–100 ng/m^2d) even though maximum concentrations vary almost 100-fold between lakes. Furthermore, these calculated rates are similar to estimates of methylation rates in other ecosystems, such as an oxic estuarine watercolumn (140 to 1400 ng/m^2d)[4] and flow-through, sediment-

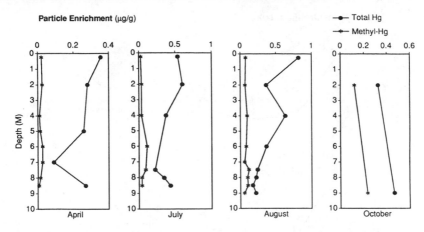

Figure 10   Seasonal patterns of mercury enrichment (μg/g, dry weight) in suspended particulate (>0.8 μm) of Little Rock Lake. Data collected the year prior to bio-optical profiling, hence the coarse spatial resolution.

Table 2   Estimated net rate of methylmercury production in anoxic plankton layers of Wisconsin lakes.

Lake	Methyl-Hg accumulated (ng)	Surface area of layer (m²)	Time (d)	Net flux (ng/m²/d)
LRT	4.0×10^7	8.0×10^3	100	50
PAL	3.2×10^9	3.5×10^5	100	92
Mary	7.0×10^7	1.3×10^4	100	54

Note: LRT: Little Rock Lake—treatment basin; PAL: Pallette Lake.

water microcosms (10 to 40 ng/m²d).[37] For Little Rock Lake, the calculated methyl-Hg flux is sufficient to account for about two thirds of the methyl-Hg annually accumulated by fish (assuming annual fish production of 30%). This would also account for a significant fraction of the methyl-Hg estimated to be cycling through all of the biota annually.[29] Thus, the planktonic production of methyl-Hg is potentially an important input to the lake. However, we emphasize that direct estimates of methylation are needed to confirm these calculations.

Previous work on the composition and physiological activity of microbial communities near O/A boundaries in sediments, benthic mats, and the watercolumn of lakes[20,32] indicates that some biogeochemical processes may be facultatively benthic or planktonic, depending on the location of the O/A boundary and the physical structure of the watercolumn. Since the pioneering work of Jernelov and colleagues,[38,39] most research on methylation in aquatic environments has focused on sediments.[2,3] If benthic microbial consortia tended to migrate from sediments and spread out into anoxic water, as illustrated on Figure 11, then important methylation sites might move with them. If so, anoxic microbial layers could serve as useful model systems for direct methylation studies, eliminating some of the complications involved in studying more tightly packed sediment communities.[40]

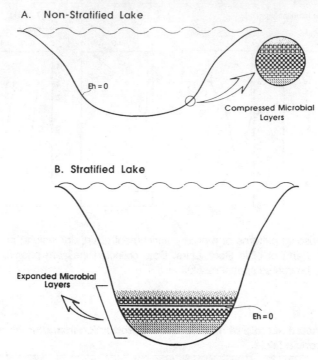

Figure 11   Theoretical comparison of microbial distributions in surficial sediments and in a layered plankton community, relative to the location of the O/A boundary. (Adapted from Guerrero, R. and Mas, J., *Microbial Mats: Physiological Ecology of Benthic Microbial Communities,* Cohen, V. and Rosenberg, E., Eds., American Society for Microbiology, Washington DC, 1989, and Sieburth, J. McN., *Microbes in the Sea*, Sleigh, M.A., Ed., Horwood, Chichester, England, 1987.)

ACKNOWLEDGMENTS

We thank Ken Morrison, Steve Claas, Tris Hoffman, Jim Hurley, and Lian Liang for assistance in the field and laboratory. Discussions with John Sieburth, Cindy Gilmour, Richard Bartha, Norbert Pfenning, John Stoltz, and Matti Verta helped to shape this manuscript. Research was supported by the Electric Power Research Institute (RP2020–10) and the Wisconsin Department of Natural Resources. The is a contribution from the Trout Lake Field Station, Center for Limnology, University of Wisconsin–Madison.

REFERENCES

1.   Berman, M. and Bartha, R., Levels of chemical versus biological methylation of mercury in sediments, *Bull. Environ. Contam. Toxicol.,* 36, 401, 1986.
2.   Winfrey, M.R. and Rudd, J.W.M., Environmental factors affecting the formation of methylmercury in low pH lakes, *Environ. Toxicol. Chem.,* 9, 853, 1990.
3.   Gilmour, C.C. and Henry, E.A., Mercury methylation in aquatic systems affected by acid deposition, *Environ. Pollut.,* 71, 131, 1991.
4.   Ramlal, P.S., Rudd, J.W.M., Furitani, A., and Xun, L., The effect of pH on methyl mercury production and decomposition in lake sediment, *Can. J. Fish. Aquat. Sci.,* 42, 685, 1985.
5.   Xun, L. Campbell, N.E.R., and Rudd, J.W.M., Measurements of specific rates of methylmercury production in the watercolumn and surface sediments of acidified and circumneutral lakes, *Can. J. Fish Aquat. Sci.,* 44, 750, 1987.

6. Steffan, R.J., Korthals, E.T., and Winfrey, M.R., Effects of acidification on mercury methylation, demethylation and volatilization in sediments from an acid susceptible lake, *Appl. Environ. Microbiol.*, 54, 2003, 1988.

7. Topping, G. and Davies, I.M., Methylmercury production in the marine watercolumn, *Nature*, 290, 243, 1981.

8. Bloom, N.S. and Fitzgerald, W.F., Determination of volatile mercury species at the picogram level by low temperature gas chromatography with cold vapor atomic fluorescence detection, *Anal. Chim Acta.*, 208, 151, 1988.

9. Bloom, N.S., Determination of picogram levels of methylmercury by aqueous phase ethylation, followed by cryogenic gas chromatography with cold vapor atomic fluorescence detection, *Can. J. Fish. Aquat. Sci.*, 46, 1131, 1989.

10. Fitzgerald, W.F. and Watras, C.J., Mercury in the surficial waters of rural Wisconsin lakes, *Sci. Total Environ.*, 87/88, 223, 1989.

11. Bloom, N.S. and Watras, C.J., Observations of methylmercury in precipitation, *Sci. Total Environ.*, 87/88, 199, 1989.

12. Bloom, N.S. and Watras, C.J., Seasonal and vertical variability in the mercury speciation of an acidified Wisconsin lake, in: *Heavy Metals in the Environment: Proceedings of the Seventh International Conference*, Vol. 2, Vernet, J.-P., Ed., CEP Consultants, Edinburgh, U.K., 1989, 349.

13. Watras, C.J., Bloom, N.S., Fitzgerald, W.F., Hurley, J.P., Krabbenhoft, D.P., Rada R.G., and Wiener, J.G., Mercury in temperate lakes: a mechanistic field study, *Verh. Int. Verein. Limnol.*, 24, 2199, 1991.

14. Wiener, J.G., Fitzgerald, W.F., Watras, C.J., and Rada, R.G., Partitioning and bioavailability of mercury in an experimentally acidified Wisconsin lake, *Environ. Toxicol. Chem.*, 9, 909, 1990.

15. Bloom, N.S., Watras, C.J., and Hurley, J.P., Impact of acidification on the methylmercury cycling of remote seepage lakes, *Water Air Soil Pollut.*, 56, 477, 1991.

16. Fitzgerald, W.F., Mason, R.P., and Vandal, G.M., Atmospheric cycling and airwater exchange of mercury over mid-continental lacustrine regions, *Water Air Soil Pollut.*, 56, 745, 1991.

17. Hurley, J.P., Watras, C.J., and Bloom, N.S., Mercury cycling in a northern Wisconsin seepage lake: the role of particulate matter in vertical transport, *Water Air Soil Pollut.*, 56, 543, 1991.

18. Watras, C.J. and Bloom, N.S., Mercury and methylmercury in individual zooplankton: implications for bioaccumulation, *Limnol. Oceanogr.*, 37, 1313, 1991.

19. Lindholm, T., *Ecology of Photosynthetic Procaryotes with Special Reference to Meromictic Lakes and Coastal Lagoons*, Abo Academy Press, Abo, Finland, 1987.

20. Guerrero, R. and Mas, J., Multilayered microbial communities in aquatic ecosystems: growth and loss factors, in: *Microbial Mats: Physiological Ecology of Benthic Microgial Communities*, Cohen, Y. and Rosenberg, E., Eds., American Society for Microbiology, Washington, D.C., 1989, chap. 4.

21. Gasol, J.M. and Pedros-Alio, C., On the origin of deep algal maxima: the case of Lake Ciso, *Verh. Int. Verein. Limnol.*, 24, 1024, 1991.

22. van Gemerden, H., Montesinos, E., Mas, J., and Guerrero, R., Diel cycle of metabolism of phototrophic purple sulfur bacteria in Lake Ciso (Spain), *Limnol. Oceanogr.*, 30, 932, 1985.

23. Watras, C.J. and Baker, A.L., The spectral distribution of downwelling light in northern Wisconsin lakes, *Arch. Hydrobiol.*, 112, 481, 1988.

24. Steenbergen, C.L.M., Korthals, H.J., Baker, A.L., and Watras, C.J., Microscale vertical distribution of algal and bacterial plankton in Lake Vechten (The Netherlands), *FEMS Microbiol. Ecol.*, 62, 209, 1989.

25. Magnuson, J.J., Bowser, C.J., and Kratz, T.K., Long-term ecological research (LTER) on north temperate lakes of the United States, *Verh. Int. Verein. Limnol.*, 22, 533, 1984.

26. Watras, C.J. and Frost, T.M., Little Rock Lake: perspectives on an experimental ecosystem approach to seepage lake acidification, *Arch. Environ. Contam. Toxicol.*, 18, 157, 1989.

27. van Gemerden, Competition between purple sulfur bacteria and green sulfur bacteria: role of sulfur, sulfide and polysulfides, in: *Ecology of Photosynthetic Procaryotes with Special Reference to Meromictic Lakes and Coastal Lagoons,* Lindholm, T., Ed., Abo Academy Press, Abo, Finland, 1987, 13.

28. Hurley, J.P. and Watras, C.J., Identification of bacteriochlorophylls in lakes using reverse-phase HPLC, *Limnol. Oceanogr.*, 36, 307, 1991.

29. Watras, C.J., Bloom, N.S., Hudson, R.J.M., Gherini, S., Wiener, J.G., Fitzgerald, W.F., and Porcella, D.B., Sources and fates of mercury and methylmercury in Wisconsin lakes, in: *Mercury Pollution: Integration and Synthesis,* Watras, C.J. and Huckabee, J.W., Eds., Lewis Publishers, Chelsea, MI, 1994, chap I.12.

30. Compeau, G.C. and Bartha, R., Sulfate-reducing bacteria: principal methylators of mercury in anoxic marine sediment, *Appl. Environ. Microbiol.*, 50, 261, 1987.

31. Parkin, T.B. and Brock, T.D., The role of phototrophic bacteria in the sulfur cycle of a meromictic lake, *Limnol. Oceanogr.* 26, 880, 1981.

32. Sieburth, J. McN., Contrary habitats for redox-specific processes: methanogenesis in oxic waters and oxidation in anoxic waters, in *Microbes in the Sea,* Sleigh, M.A., Ed., Horwood, Chichester, England, 1987.

33. Engle, S.S., Utilization of Food and Space by Cisco, Yellow Perch and Introduced Coho Salmon, With Notes on Other Species, in Pallette Lake, WI, Ph.D. Dissertation, University of Wisconsin–Madison, 1972.

34. Brezonik, P.L., Baker, L.A., Eaton, J.R., Frost, T.M., Garrison, P., Kratz, T.K., Magnuson, J.J., Rose, W.J., Shepard, B.K., Swenson, W.A., Watras, C.J., and Webster, K.E., The acidification of Little Rock Lake, *Water Air Soil Pollut.*, 31, 115, 1986.

35. Weimer, W.C. and Lee, G.F., Some considerations of the chemical limnology of mer-omictic Mary Lake, *Limnol. Oceanogr.*, 18, 414, 1973.

36. Hutchinson, G.E., *A Treatise on Limnology,* Vol. 1, John Wiley & Sons, New York, 1957.

37. Wright, D.R. and Hamilton, R.D., Release of methylmercury from sediments: effects of mercury concentration, low temperature and nutrient addition, *Can. J. Fish. Aquat. Sci.*, 39, 1459, 1982.

38. Jernelov, A., Release of methylmercury from sediment with layers containing inorganic mercury at different depths, *Limnol. Oceanogr.*, 15, 958, 1970.

39. Fagerstrom, T. and Jernelov, A., Formation of methylmercury from pure mercuric sulfide in anaerobic organic sediment, *Water Res.*, 5, 121, 1971.

40. Sweerts, J.-P.R.A., Oxygen Consumption Processes, Mineralization and Nitrogen Cycling at the Sediment-Water Interface of North Temperate Lakes, Ph.D. Thesis, Rijksuniversiteit Groningen, Holland, 1990.

Sources and Fates of Mercury and Methylmercury in Wisconsin Lakes

C. J. Watras, N. S. Bloom, R. J. M. Hudson, S. Gherini, R. Munson,
S. A. Claas, K. A. Morrison, J. Hurley, J. G. Wiener, W. F. Fitzgerald,
R. Mason, G. Vandal, D. Powell, R. Rada, L. Rislov, M. Winfrey,
J. Elder, D. Krabbenhoft, A. W. Andren, C. Babiarz,
D. B. Porcella, and J. W. Huckabee

CONTENTS

ABSTRACT: The mercury cycle in seven northern Wisconsin seepage lakes was characterized by high atmospheric influx, removal by sedimentation and evasive efflux, and by in-lake transformation of Hg to biologically sequestered methyl-Hg species. Direct depositional Hg loading from the atmosphere to lakes was roughly 10 $\mu g/m^2$/year, with rain and snow the principal delivery vectors. Annual atmospheric Hg deposition exceeded estimated fish bioaccumulation by a factor of roughly 10. The atmospheric Hg influx was roughly balanced by losses to sediments and the return of volatile Hg^0 to the atmosphere. The relative importance of sedimentation and gaseous evasion as Hg loss terms varied from lake to lake, with sedimentation/evasion ratios ranging from 9:1 to 1:1 in the seven lakes studied. Residence times for Hg varied from roughly 125 to 300 days in these lakes.

Methyl-Hg in these lakes also had an atmospheric source, estimated to be roughly 1% of the total Hg inputs. Although the direct atmospheric deposition and sediment accumulation of methyl-Hg roughly balanced, the atmospheric influx of methyl-Hg was much lower than annual rates of methyl-Hg bioaccumulation. Unless the recycling efficiency of methyl-Hg was extraordinarily high, in situ production was an important source of methyl-Hg species. Most of the methyl-Hg in these lakes was stored in fish tissue. Assuming fish production of 30%/year, the annual bioaccumulation of methyl-Hg exceeded sediment accumulation by a factor of 6 to 7. No dimethyl-Hg has been observed in any Wisconsin lake.

The distribution of Hg species in the study lakes was characterized by very dilute pools that varied seasonally and spatially. Waterborne Hg species had concentrations in the picomolar to femtomolar range, with parts per million to parts per billion concentrations in sediment and organisms. Average waterborne Hg and methyl-Hg concentrations correlated negatively with lakewater pH and positively with DOC. Seasonal cycles involved decreasing concentrations under ice cover, followed by build-up during summer. Epilimnetic concentrations ranged from 1 to 3 ng/L Hg and 0.05 to 0.5 ng/L methyl-Hg. Higher mercury concentrations were observed at depth in stratified lakes (Hg >45 ng/L and methyl-Hg >10 ng/L) and Hg maxima were observed near microbial layers in the watercolumn. In anoxic, sulfidic plankton layers, $>50\%$ of the Hg may be in the methyl-Hg form (vs. 5 to 15% in the epilimnion).

Methyl-Hg was biomagnified in the foodchain of Little Rock Lake, but there was evidence that nonmethyl-Hg species became more dilute at higher trophic levels. The bioconcentration factor for methyl-Hg increased by threefold for each trophic level, approaching 10^7 in fish. The Hg in fish was almost all methylated ($>95\%$), while the Hg in sediments was primarily nonmethyl-Hg ($>97\%$). Since most methyl-Hg in the study lakes appeared to be sequestered by fish biomass, fish contamination could be significantly enhanced by small increases in net rates of methyl-Hg production, recycling, or loading.

I. INTRODUCTION

Reports of Hg-contaminated freshwater fisheries in North America have steadily increased during the past decade (Figure 1). While the issuance of fish consumption advisories reflects both awareness and contamination, there are concerns about disruption of the natural Hg cycle on local, regional, and global scales. Unfortunately, risk assessment, source attribution, and remediation have been hindered by a lack of reliable data on Hg in aquatic environments. Laboratory contamination has compromised much of the data on waterborne Hg.[1-4]

With new analytical techniques[5,6] and the use of "clean" sampling protocols, a more reliable and comprehensive database on environmental Hg is emerging. Mass balances for lakes and watersheds in Wisconsin, Canada, and Sweden indicate that atmospheric deposition is the principal source of Hg.[1,7,8] In Wisconsin, the direct deposition of airborne Hg to precipitation-dominated lakes is sufficient to account for annual sediment and fish accumulation.[1,9] In Swedish and Canadian drainage lakes, watershed Hg inputs are clearly important,[7,10] but ultimately the principal Hg source for remote freshwaters in nonmercuriferous regions appears to be atmospheric deposition, either directly to the water surface or indirectly via export from atmospherically enriched, shallow soil horizons.[11-13]

Although waterborne mercury concentrations in unpolluted surface waters typically range from about 2 to 20 pM (0.5 to 4 ng/L),[1,2,8] mercury is biomagnified to such a high degree that contaminated fish stocks occur even in very remote, northern waters.[8,14,15] While geographical gradients in Scandinavia indicate a link between Hg depositional rates and mercury in fish,[8] factors such as pH, DOC, and trophic structure further influence bioaccumulation.[14,16] Since methyl-Hg seems to biomagnify most strongly in aquatic foodchains, and since almost all of the Hg in fish is methylated,[15,16] such mitigating factors could operate by regulating net rates of methyl-Hg formation in the ecosystem.[18,19]

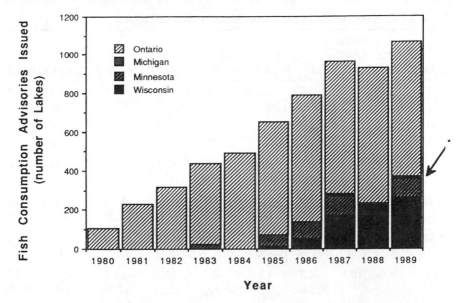

Figure 1 Number of fish consumption advisories issued for inland lakes in the Great Lakes region of North America during the past decade (1980–1990). Although Michigan issued a blanket advisory during 1989 restricting consumption of fish in all 11,000 lakes within the state, this advisory is counted only once (*). Data obtained from: Jim Amrhein, Wisconsin DNR, Madison; Charles Cox, Ontario Min. Environ., Toronto; Ed Swain, Minnesota PCA, St. Paul; and John Filpus, Michigan DPH, Lansing.

To further our understanding of the mercury cycle in freshwaters and provide a framework for ongoing process-oriented studies, we present here the results of investigations on seven Wisconsin lakes which span gradients of pH and DOC.[20] We report the concentration and distribution of Hg species in air, precipitation, water, sediments, and biota over seasonal and spatial scales, and we construct comparative mass balances for these lakes. We describe relationships between waterborne Hg species and water quality parameters and we examine differences between lakes in input:output budgets and bioaccumulation patterns. A conceptual model of Hg cycling in Wisconsin lakes is presented. The results of a dynamic modeling study are presented in another chapter in this book.[21]

II. METHODS

A. STUDY SITES

Seven lakes in north-central Wisconsin (circa 46°N and 89°W) were sampled routinely between August 1988 and January 1992. This region of Wisconsin is sparsely populated (6.6 people/km^2) and largely covered with second-growth Great Lakes forest and wetlands. Many of the small lakes in this area can be considered remote from direct anthropogenic influence, with no dwellings or other permanent structures on the shorelines. The study lakes were chosen to span gradients of pH and DOC (Table 1), variables which purportedly influence Hg bioaccumulation.[8,14,16] They are all seepage lakes (no permanent surface water inflow or outflow), and most of the lakes were hydrologically mounded above groundwater levels during the study period (except for Pallette Lake and, perhaps, Russett Lake[42]). Being isolated from terrestrial watersheds, the lakes were considered model ecosystems for examining atmospheric interactions and in-lake processes.

Table 1   **Characteristics of the seven Wisconsin study lakes.**

Lake[a]	Area (ha)	Mean depth (m)	pH	DOC (mg/L)	SPM[b] (mg/L)	SO_4^{2-} (μeq/L)	Ca^{2+} (μeq/L)	Cl^- (μeq/L)	Chl a (μg/L)
LRT	9.8	3.8	4.9 (0.15)	1.5 (0.53)	1.0 (0.39)	164.8 (24.0)	79.6 (8.1)	6.0 (1.8)	2.0 (1.8)
LRR	8.1	3.1	6.0 (0.30)	3.3 (0.50)	1.0 (0.45)	64.8 (7.1)	47.2 (4.6)	7.8 (0.7)	2.0 (1.3)
VAN	43.6	4.7	6.3 (0.37)	4.3 (0.65)	1.7 (1.00)	80.4 (4.8)	67.3 (4.6)	7.6 (0.9)	1.6 (1.2)
PAL	70.0	9.6	7.2 (.79)	5.4 (0.30)	1.4 (0.82)	56.4 (8.5)	111.2 (4.2)	7.0 (0.8)	2.8 (2.8)
MAX	9.2	2.9	5.4 (0.22)	3.3 (0.98)	0.6 (0.37)	113.0 (14.1)	59.0 (5.1)	7.5 (2.0)	1.2 (0.7)
CRY	36.0	10.4	6.3 (0.20)	1.8 (0.23)	0.5 (0.19)	72.5 (2.2)	57.2 (2.3)	13.1 (1.0)	1.2 (0.8)
RUS	19.0	3.9	5.8 (0.26)	6.8 (0.20)	0.8 (0.38)	71.9 (5.2)	59.3 (5.8)	5.9 (0.4)	2.8 (3.0)

[a] LRT (Little Rock Lake—treatment basin); LRR (Little Rock Lake—Reference basin); VAN (Vandercook Lake); PAL (Pallette Lake); MAX (Max Lake); CRY (Crystal Lake); RUS (Russett Lake).
[b] SPM: Suspended particulate matter >0.8 μM.
Note: Water quality data are epilimnetic averages for the study period. Standard deviations are in parenthesis.

B. WATERBORNE Hg SPECIES

Water column samples were collected at roughly 4- to 8-week intervals, targeting events such as peak stratification, overturn, and ice-out. Trace-metal clean protocols based on oceanographic techniques[22] were adapted for sampling mercury in small lakes.[1] Technicians wore lint-free suits, hoods and shoulder-length plastic gloves to reduce the possibility of particulate Hg contamination. Rigorously acid-cleaned sample bottles and equipment (soaked 12 h in hot concentrated HNO_3; boiled 12 h in Hg-free water), sealed in a double layer of plastic bags, were stored in a clean laboratory and removed only for sample collection and transport. Airborne Hg concentrations in the clean lab were about 2 ng/m³.

At the sampling site, field personnel positioned themselves so that wind could not transport contaminants from their equipment or clothing into samples. From a thoroughly cleaned nonmetallic boat or a hole in the ice, multiple samples were collected through the water column at the region of maximum depth in each lake. Typically, three samples were collected from each depth. A peristaltic pump threaded with a 15-cm length of silicone tubing was used to pull water through ½'' O.D. Teflon® tubing held vertical by a weighted Teflon® torpedo. Whole water samples were dispensed directly into 500 mL Teflon® bottles. Filtered samples were pumped through pre-ashed (12 h at 550° C) 47 mM diameter quartz fiber filters (Whatman #1851; nominal pore size 0.8 M) housed in Teflon® holders. The caps of all sample bottles were tightly wrenched after collection to prevent Hg^0 influx during storage.

Filtered and unfiltered aliquots were collected for total Hg, total particulate Hg, total dissolved Hg, total methyl-Hg, particulate methyl-Hg, and dissolved methyl-Hg determinations. Unfiltered samples for Hgt determinations were preserved with 10 mL 6 N HCl or HNO_3 at the time of collection. Filtered water for methyl-Hg determinations and seston

Table 2 **Detection limits and related parameters for waterborne Hg species.**

Mercury species	Sample size (mL)	Detection limit (ng/L)	Typical pristine oligotrophic waters (ng/L)
Total	100	0.15	1.0
Elemental	4000	0.006	0.02
Particulate total	1000	0.05	0.3
Total methyl	50	0.01	0.13
Dimethyl	4000	0.0006	<0.0001
Particulate methyl	1000	0.005	0.05

samples (on filters in their Teflon® holders) were preserved by freezing within 6 h of collection. Frozen samples were shipped overnight on dry ice to the analytical laboratory. Approximately 10% of the samples collected were replicates.

Recently, we have sampled using a submersible, nonmetallic pump attached to rigorously cleaned, Silastic® tubing. Samples were collected from depth with flow rates of 5 to 6 L/min directly into cleaned Teflon® bottles. Filtration was performed in the clean laboratory shortly after collection, using cleaned and blanked disposable filtration units (0.45 μm).[23]

Analytical protocols for waterborne Hg species followed Bloom and Fitzgerald[5] and Bloom.[6] All sample handling and analyses were conducted in a dedicated clean laboratory. Typical sample sizes and detection limits are summarized in Table 2. Absolute detection limits for all Hg species were less than 1 pg Hg.

For total Hg analysis, 100 mL aliquots of lakewater were wet-oxidized using BrCl. Following oxidation, samples were prereduced with NH_2OH/HCl, then further reduced with $SnCl_2$ and purged from solution with nitrogen onto gold traps. Trapped Hg was thermally desorbed into inert carrier gas for analysis by cold-vapor atomic fluorescence spectroscopy (CVAFS) detection. Particulate and dissolved Hg species were defined operationally. Particulate (total) mercury was defined as that fraction recovered after chemical digestion of the in-line quartz fiber filters, followed by the total mercury protocol described above. Dissolved Hg was computed as the difference between total and particulate Hg.

Methyl-Hg determinations were based on the cryogenic GC separation of volatile ethyl analogues.[6] Dissolved methyl-Hg was determined by extracting filtered samples from an HCl/KCl matrix into CH_2Cl_2, then back-extracting into pure water by solvent evaporation. The aqueous phase was ethylated with $NaB(C_2H_5)_4$ and the volatile ethyl analogue was separated from other species using cryogenic GC separation with CVAFS detection. Particulate methyl-Hg was defined as that fraction recovered after chemical digestion (KOH/MeOH) of the in-line quartz fiber filter, followed by the dissolved methyl-Hg protocol. Total methyl-Hg was computed as the sum of dissolved and particulate species.

With the adoption of in-lab filtration protocols mentioned above, the dissolved/particulate fractionation of Hg and alkyl-Hg species are determined in a slightly different way (particulate = total − dissolved).[23]

Dissolved gaseous Hg (DGM) determinations followed Vandal et al.[24] Water samples were collected using an 8-L Teflon®-coated Go-Flo bottle (General Oceanics) which had been thoroughly cleaned and blanked in the laboratory. The sampling protocol was based on Gill and Fitzgerald.[25] The bottle was lowered on a clean Kevlar® line by hand and triggered at depth with a clean Teflon® messenger. Analysis was made in a clean laboratory within 3 h of collection. DGM measurements were made by sparging a 4 L sample with Hg-free argon and trapping on either Au or Carbotrap™ columns. Detection limits using atomic fluorescence detection were DGM, 1 pg/L; dimethyl-Hg, 0.6 pg/L.[24]

C. ATMOSPHERIC Hg SPECIES

Atmospheric gaseous and particulate Hg samples were collected routinely from a sampling platform 2 m above the lake surface.[26] Total gaseous Hg (TGM) was trapped on gold columns, while speciation samples were collected using Carbotrap™ columns.[5] Particulate Hg (PM) was collected using quartz wool plugs. Sample volumes for TGM and speciation determinations were between 0.5 and 2 m^3, while larger volumes were needed for particulate measurements (2 to 10 m^3). Detection limits were TGM, 0.15 ng/m^3; MMHg and PM, 0.005 ng/m^3.

Rain and snow samples were collected on an event basis. The rain collector was a glass or Teflon® funnel deployed 2 m above the surface of one of the study lakes. The Teflon® funnel was housed in an acrylic box with a removable lid. The funnel and housing were constructed so that rain entering the funnel contacted only rigorously cleaned Teflon® parts. Prior to each collection, the funnel was rinsed and blanked with clean, acidified water. The funnel was open to the atmosphere only during the rinse and the rain event. Snow samples were collected in clean, 1 L Teflon® jars by scooping snow from the lake surface <12 h after a snowfall. The collector, wearing long plastic gloves, moved forward into the wind scooping the snow from areas free of visible terrestrial inputs (i.e., twigs, needles, etc.). The jars were wrenched tight, double bagged, and stored frozen until analysis in a clean laboratory. Analytical techniques were similar to those used for lakewater samples.

D. BIOTIC Hg SPECIES

Zooplankton samples were collected using a nonmetallic plankton net and then hand picked by species for analysis.[15] After each vertical tow, the net contents were backwashed with surface lakewater into Hg-clean, Teflon® jars which were hermetically sealed and double-bagged in clean, ziplocked plastic bags. Clean protocols were observed throughout sample collection and handling. Within 1 h of collection, individual zooplankton of a given size-class and species were transferred live into small Hg-free, Teflon® vials with rigorously cleaned pipets. Groups of 5 to 20 individuals were immediately frozen in each vial in a small drop of lakewater. Procedural blanks consisted of just the frozen drop of lakewater without zooplankton. For HgT determinations, the zooplankton were first digested in a 200-μL aliquot of 7:2 HNO_3/H_2SO_4 at 70° C for 2 h, and then the entire sample was reduced with $SnCl_2$ and preconcentrated using dual gold amalgamation. For methyl-Hg determinations, zooplankton were digested in 200 μL of 25% KOH/methanol for 2 h at 70° C followed by aqueous phase ethylation, cryogenic GC separation, and CVAFS detection.

Fish (1-year-old yellow perch, *Perca flavescens*) were collected using fyke nets or minnow traps.[9] Whole fish were then frozen, lyophilized to constant weight, and ground to a fine powder. Subsamples were digested for 14 h at 220° C in 12 mL of 36 N H_2SO_4 and 3 mL of 16 N HNO_3 and analyzed for total Hg by cold vapor AAS following $SnCl_2$ reduction (modified from Wiener et al.[9]). Precautions were taken to prevent Hg contamination during storage, preparation, and analysis of samples. The accuracy of Hg determinations was verified by analysis of standard reference materials, spiked samples, replicate samples, and procedural blanks with each batch of fish samples. Method limits of detection were 4 ng and limits of quantification were 13 ng for fish. Protocols for determining methyl-Hg in fish are described by Bloom.[17]

E. SEDIMENT Hg

For total Hg determinations in surficial sediment, samples were collected using a diver-operated PVC core sampler (22 cm I.D.) designed to collect the top 5 cm of sediment with minimal disturbance and compaction.[27] Analytical methods for determining total Hg in surficial sediments are described by Rada et al.[27] Samples for methyl-Hg analysis were collected from the top 1 cm of profundal cores from Little Rock Lake. Fresh (wet) sediment was

Table 3   **Mean epilimnetic concentrations of Hg species in the seven Wisconsin study lakes.**

Lake	HgT (ng/L)	Methyl-Hg (ng/L)	Dimethyl-Hg (pg/L)	Hg^0 (pg/L)	%S^a
LRT	1.11	0.091	ND^b	17	300
LRR	1.13	0.075	ND	27	500
VAN	0.90	0.088	ND	56	1490
PAL	0.79	0.083	ND	128	3350
MAX	1.35	0.261	ND	36	950
CRY	0.72	0.046	ND	157	4100
RUS	2.10	0.327	ND	84	2250

^{a} %S = Saturation of volatile Hg^0 with respect to atmospheric concentrations.
^{b} ND: Not detected, ≤0.6 pg/L.

leached with 25% KOH in methanol and the leachate was analyzed for methyl-Hg following protocols described above for biotic samples.

II.   RESULTS AND DISCUSSION

A.   WATERBORNE Hg: VARIABILITY BETWEEN LAKES

Typical epilimnetic concentrations of the Hg species measured in these lakes are shown on Table 3. Annual average [Hg] in all the study lakes was very low (picomolar or parts per trillion) and tended to vary within a narrow range (Ca. 1 to 3 ng/L). Methyl-Hg concentrations were lower by about an order of magnitude (ranging from roughly 0.05 to 0.3 ng/L). No alkyl-Hg species other than monomethylmercury have been observed in any of the lakes at a detection limit of about 3 fM. Note that surface waters in all the lakes were supersaturated with respect to Hg^0 during summer.

The mercury concentrations observed in these lakes are similar to those reported for open ocean sites,[28] and much lower than most of the freshwater data reported in North America before 1989. Clean sampling and novel analytical protocols are primarily responsible for the differences between the old and new freshwater data. Sample contamination has been shown to account for errors of up to 500% in historical data.[1] Even in recent studies, waterborne Hg detection limits often exceed the concentrations typical of many remote northern lakes.[29,30]

Both [Hg] and [methyl-Hg] tended to increase in the study lakes with increases in both [H$^+$] and DOC (Figures 2 and 3). However, because the number of lakes in this study was small, it is not possible to quantify the nature of the pH and DOC dependencies with reasonable confidence.

Most of the waterborne Hg and methyl-Hg tended to be in the dissolved phase (Figures 2 and 3). Although the dissolved/particulate fractionation was highly variable seasonally and spatially, 60 to 90% of the Hg and 40 to 70% of the methyl-Hg was in the operationally defined dissolved pool (>0.8 μm). The average mass of suspended particulate matter (SPM) during midsummer varied from about 0.5 to 3.0 mg/L between lakes. Most of the SPM in these lakes was living and dead plankton, rather than mineralogic particulate.

B.   WATERBORNE Hg: SEASONAL VARIABILITY

Although average waterborne [Hg] and [methyl-Hg] varied within a rather narrow annual range, there were detectable seasonal changes. A 2-year time-series for Little Rock lake shows that waterborne Hg concentrations tended to be high during summer and low during winter

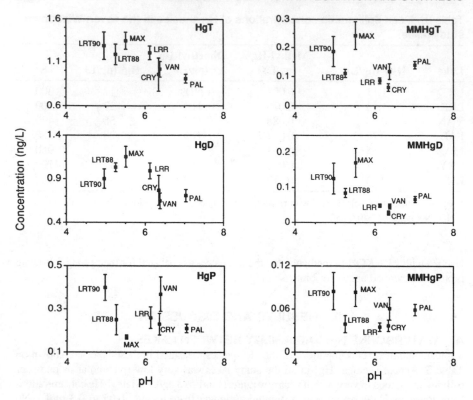

Figure 2 Waterborne Hg or methyl-Hg relative to pH in lakes with similar DOC. Data are annual average, volume-weighted concentrations of total Hg (HgT), dissolved Hg (HgD), particulate Hg (HgP), total methyl-Hg (MMHgT), dissolved methyl-Hg (MMHgD), and particulate methyl-Hg (MMHgP). Error bars indicate 2 SD.

(Figure 4). Winter declines in waterborne [Hg] are consistent with the observation that atmospheric inputs are the major source of Hg to these lakes.[1] Under ice cover the lakes are effectively sealed from this external source. During summer, both waterborne [Hg] and [methyl-Hg] tended to increase, reflecting the cumulative effects of increased atmospheric Hg deposition and, perhaps, increased methylation activity due to warmer temperatures.[18] Following autumn mixis, declines in [Hg] and [methyl-Hg] corresponded to the fall plankton bloom and a period of high sedimentation. During winter, we have observed the gradual decay of an Hg-enriched layer near the sediment surface.[31] We hypothesize that this decay results from the gradual settling of Hg-rich particulate into profundal sediments.

C. WATERBORNE Hg: SPATIAL VARIABILITY

During periods when the watercolumn was well mixed, [Hg] and [methyl-Hg] were relatively uniform from the lake surface to the sediments. However, during stratification there were striking discontinuities in the distribution of Hg species (Figure 5). [Hg] maxima were observed in the metalimnion and hypolimnion—regions where we also observed the development of deep plankton layers (Figure 5A). In these deep plankton layers, most of the Hg was particulate (ca. 80%) and Hg concentrations were up to 10 times higher than those observed

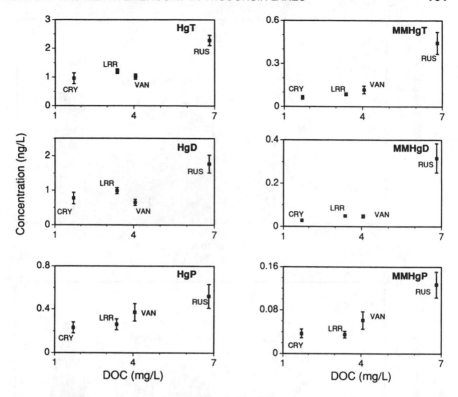

Figure 3 Waterborne Hg and methyl-Hg relative to DOC in lakes of similar pH. Data are annual average, volume-weighted concentrations of total Hg (HgT), dissolved Hg (HgD), particulate Hg (HgP), total methyl-Hg (MMHgT), dissolved methyl-Hg (MMHgD), and particulate methyl-Hg (MMHgP). Error bars indicate 2 SD.

in the epilimnion. In anoxic hypolimnetic plankton layers, [methyl-Hg] increased to levels approaching 100 times those in epilimnetic waters. Similar mercury profiles have been observed in all of the stratified lakes studied,[23] but the absolute [Hg] and [methyl-Hg] associated with plankton layers varied widely between lakes (compare Figures 5A and 5B).

The origin of Hg maxima in the watercolumn of stratified lakes remains uncertain, but their development suggests that freshly settled particulate may be an important Hg source. We hypothesize that atmospherically derived Hg may be scavenged by biogenic particles in upper waters and delivered to depth on settling particulate. Upon reaching layers of high microbial activity, decomposition or dissolution reactions may transfer or transform the suspended Hg burden.

In anoxic/sulfidic regions of the watercolumn, complexation with dissolved sulfide, polysulfide, or thiols may increase the dissolved Hg pools relative to the particulate fractions. Partitioning data tend to support this hypothesis. For example, during peak stratification in Little Rock Lake, partition coefficients for Hg and methyl-Hg decrease with depth below the deepest plankton layers (Figure 6). Lowered Kd at depth in Little Rock Lake could also be a function of high concentrations of particulate matter in the 8.5 to 9.5 m depth range.

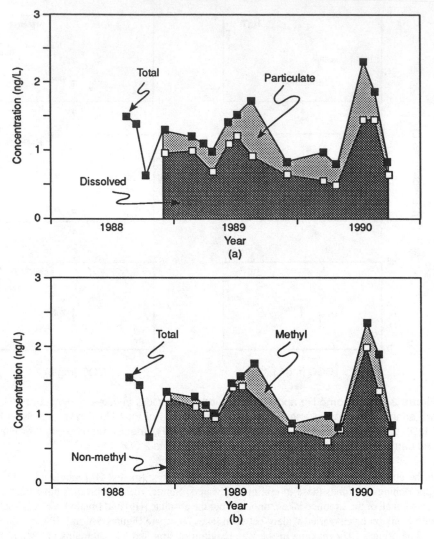

Figure 4   Seasonal changes in the concentration of waterborne mercury species in the treatment basin of Little Rock Lake. Data are volume-weighted, whole-basin averages.

D. MASS BALANCE FOR TOTAL Hg IN LITTLE ROCK LAKE

The mass balance for mercury in Little Rock Lake (Figure 7a) has evolved as studies progressed from 1988 to 1992.[1,9,26] During this period, airborne Hg was the dominant source to the lake, and sediments were the dominant sink. Inputs from the atmosphere were roughly balanced by sediment accumulation. Data from sediment traps indicated recycling of Hg within the hypolimnion.[31]

In the atmosphere above the lake, the annual average concentrations of total gaseous and particulate mercury (TGM, TPM) were 1.6 ± 0.4 ng/m^3 and 0.02 ± 0.02 ng/m^3, respectively.

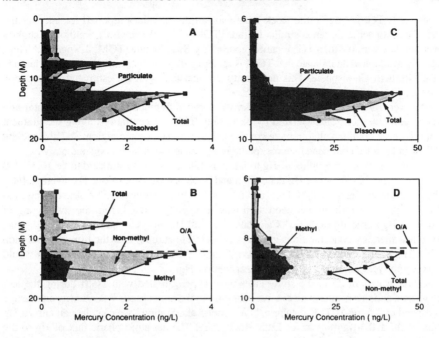

Figure 5 Vertical distribution of waterborne mercury species in Pallette Lake (A, B) and the treatment basin of Little Rock Lake (C,D).

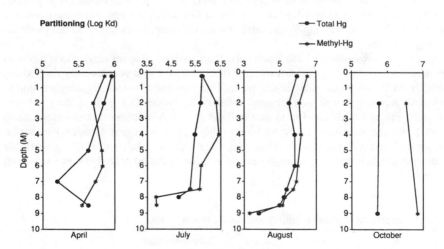

Figure 6 Particle-water partition coefficients for Hg and methyl-Hg in Little Rock Lake. Kd = log (Cp/Cw), where Cp and Cw are particulate and dissolved concentrations, respectively. CP >0.8 μm (dry wt. basis).

Atmospheric TGM above Little Rock Lake was similar to values reported for air over the North Central Pacific Ocean at similar latitude (TGM: 1.7 ± 0.15 ng/m^3),[32] although slightly lower and less variable than TGM values reported for Scandinavia (TGM: 2.5 ng/m^3).[8] Thus, presently available data indicate that TGM is rather uniformly distributed for a given latitude in the Northern Hemisphere. This uniformity is consistent with the estimated atmospheric residence time of Hg^0 (6 to 24 months).[8,32]

However, for TPM there are comparatively large differences between continental and oceanic concentrations characterized by decreasing TPM concentrations along the gradient from urbanized continental, to remote continental, to remote oceanic regions.[26] This gradient in TPM indicates a continental source, perhaps associated with anthropogenic activity.

The gross atmospheric influx of Hg to lakes in this region was estimated to be 10.3 ± 5.0 μg/m^2/year: roughly 70% wet Hg deposition and 30% dry Hg deposition. The average [Hg] in rain and snow was 10.5 ± 4.8 ng/L and 6.0 ± 0.9 ng/L, respectively.[26] A scavenging ratio of 437 suggests that Hg in wet deposition was principally derived from the scavenging of particulate Hg in the atmosphere.[26] Calculated leaf-fall Hg inputs to the lake were negligible, due to low mercury concentrations in leaves collected in autumn near the time of abscission (Table 4). Given leaf-litter inputs of 14 g/m^2/year,[43] the annual leaf-fall contribution would be 0.6 μg/m^2/year Hg and 0.003 μg/m^2/year methyl-Hg.

Surface waters of all of the study lakes were supersaturated with TGM during the ice-free season, and this results in an evasive back-flux to the atmosphere.[24] Assuming an average windspeed of 2.5 m/s (LTER database, UW-Trout Lake Station), we calculate an annual Hg efflux of 0.7 ± 0.3 μg/m^2/year for Little Rock Lake. The net atmospheric flux of Hg to the lake, then, is about 9.6 μg/m^2/year.

Removal of Hg from the watercolumn of Little Rock Lake occurred mainly via sedimentation. Losses to groundwater were calculated at 0.3 μg/m^2/year, assuming that groundwater recharge constituted about 47% of the hydrologic losses from the lake[46] and that [Hg] in out-seeping water was the same as lakewater [Hg] (i.e., 1 ng/L). Combined with gaseous evasion (0.7 μg/m^2/year) total Hg losses to the airshed and watershed constitute roughly 1 μg/m^2/year. The rest (9.3 μg/m^2/year, calculated by difference) was carried to sediments on settling particulate.

Our Hg mass balance for Little Rock Lake, then, indicates that net atmospheric inputs to the lake were roughly 1 g Hg annually and that most of this atmospherically derived Hg (roughly 0.9 g Hg) accumulated in sediments. This estimate of annual Hg accumulation in sediments agrees well with the measured sediment Hg burden (1.1 g Hg in the top 1 mm sediment, Figure 7) calculated from data in Rada et al.[27] A sediment accumulation rate of about 1 mm/year seems reasonable for lakes of this type in the region. Gaseous Hg evasion and groundwater recharge constitute small Hg losses for this lake (roughly 0.1 g/year), but as described below, the relative importance of evasion in the other study lakes varies with pH and DOC.

Table 4   **Mercury in leaves falling into Little Rock Lake.**

Leaf type	Concentration (μg/g, dw)	
	Total Hg	**Methyl-Hg**
Red maple	0.046	0.00025
Red oak	0.034	0.00035
White pine	0.048	0.00007
Paper birch	0.040	0.00019
Bigtooth aspen	0.048	0.00020

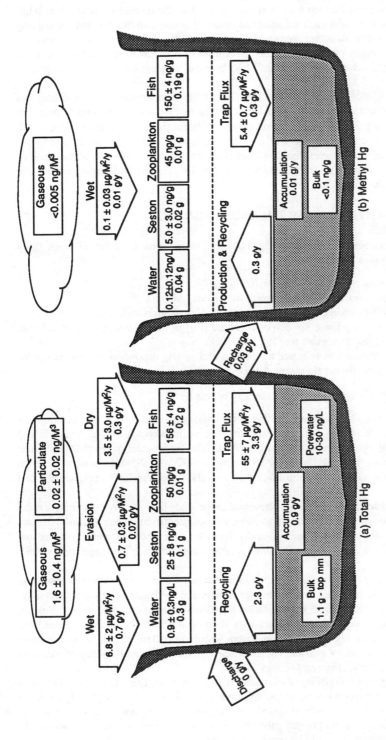

Figure 7 Mass balances for Hg and methyl-Hg in the treatment basin of Little Rock Lake. All concentrations are in ng/L or ng/g wet weight. Error terms are ±1 SD. Annual sediment Hg accumulation calculated by difference between measured inputs and outputs. Data for bulk sediment Hg calculated from Reference 27 and profundal porewater Hg from Reference 40. Annual methyl-Hg accumulation in sediments is based on the observation that methyl-Hg constituted roughly 1% of HgT in surficial profundal sediment. Dashed line represents the O/A boundary.

Biotic Hg pools in Little Rock Lake were estimated as the product of biomass × [Hg]. Phytoplankton biomass was assumed equal to suspended seston, and thus included both living and dead cells. Seston mass was measured directly on each sampling date by filtration and weighing. The biomass of crustacean zooplankton was estimated from a vertical series of collections using a Schindler-Patalis trap.[44] Fish biomass was estimated from mark and re-capture studies.[45]

The data for biotic Hg indicated that Hg concentrations increased along the continuum from phytoplankton (seston) to zooplankton to young fish by a factor of two to three per trophic level (Figure 7a). Given the biomass distribution for this lake, the pool of Hg in fish (roughly 0.2 g) was about double that in plankton (0.11 g). The greatest change in Hg concentration occurred between the dissolved phase (roughly 0.9 ng/kg) and the sestonic phase (25 μg/kg, wet weight)—a factor of roughly 3×10^4. The largest Hg reservoir in Little Rock Lake was the sediment.

E. COMPARATIVE Hg MASS BALANCES FOR THE MTL LAKES

Comparative input:output budgets for Hg (Figure 8) were constructed for all seven study lakes using the following criteria:

1. Local atmospheric Hg deposition = 10.3 μg/m²/year
2. Ground water discharge = zero, except for Pallette Lake
3. Ground water recharge varied from roughly 18–50% of hydrologic outputs
4. Hg evasion was a linear function of elemental [Hg] in surface water, extrapolated from the annual rates determined for Little Rock Lake
5. Hg sedimentation was estimated by difference: S = (Hg deposition + groundwater Hg input) − (groundwater Hg output + Hg evasion)

Note that groundwater inputs to Russett Lake may be non-zero.[42]

The resulting mass balances indicate that direct atmospheric deposition was the dominant Hg input for all the study lakes. However, the relative importance of sedimentation and volatile evasion as Hg loss terms varied from lake to lake. Sedimentation tended to dominate losses in lakes with high [H⁺] and low DOC (Figures 9B and 9D). Hg evasion assumed greater importance as a loss term in lakes with low [H⁺] and high DOC (Figures 9A and 9C).

The measured sediment Hg burdens observed in the study lakes[27] provide an independent check on our estimates of Hg loss via evasion and sedimentation. When sediment Hg accumulation rates are calculated both ways (S_1, from the difference between Hg influx and Hg evasion (Figure 9); or S_2, (from measured sediment Hg burdens),[27] the agreement is quite good (Figure 10).

There is evidence that the relative importance of evasion and sedimentation as loss terms may be related to the residence time of atmospherically derived Hg in the upper watercolumn (Figure 11). When evasion and sedimentation are plotted against epilimnetic Hg residence time, we see two clusters of lakes. Lakes with short epilimnetic Hg residence times have relatively high rates of Hg sedimentation. Lakes with long epilimnetic Hg residence times have relatively high rates of evasion. Hg residence times are not simply surrogate measures for mean depth or SPM (Figures 11C and 11D).

Hypothetically, the probability of elemental Hg evasion from these lakes would increase with increased Hg residence time in the epilimnion. Provided that elemental Hg does not engage in coordination reactions or absorb strongly to natural organic matter, as indicated by its weak partitioning into organic solvents, all the Hg⁰ formed as a result of longer Hg residence in upper waters should be free to volatilize.

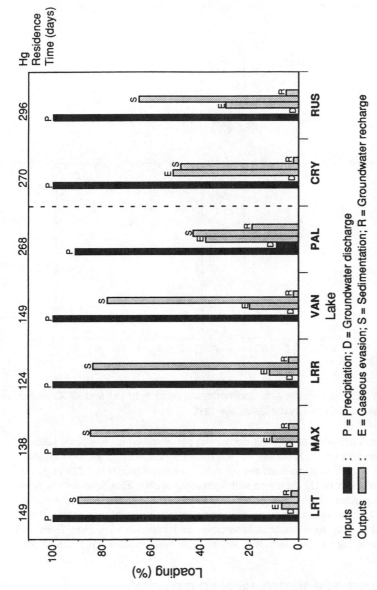

Figure 8 Input:output budgets for Hg in the seven Wisconsin study lakes. Sedimentation = (deposition + ground-water discharge) − (evasion + groundwater recharge). Residence time = {mean areal epilimnetic [Hg] (μg/m^2)} /{mean areal Hg influx (μg/m^2/d)}. Groundwater data for Pallette Lake from Reference 41. LRT: Little Rock Lake—treatment basin; MAX: Max Lake; LRR: Little Rock Lake—reference basin; VAN: Vandercook Lake; PAL: Pallette Lake; CRY: Crystal Lake; RUS: Russett Lake.

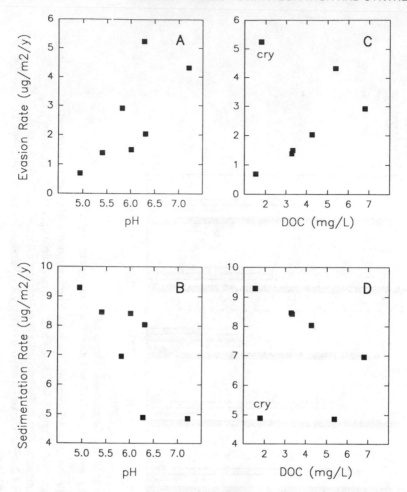

Figure 9 Hg removal via evasion and sedimentation relative to pH and DOC in the Wisconsin study lakes; CRY = Crystal Lake, see text.

The increased sedimentation of Hg with decreasing pH observed in Wisconsin lakes is consistent with observations made during the acidification of Lake 223 in the Canadian ELA.[33,34] Although acidification increased the solubility of several metals in L223 (e.g., Fe, Zn, Mn, Co), the solubility of Hg decreased with increasing acidity. This behavior has been attributed to the preferential binding of transition metals to acid-soluble oxides and the preferential binding of Hg to acid-insoluble organic substances, such as settling plankton.[33] Although low pH lakes may have faster sedimentation and lower evasion rates from the epilimnion, waterborne Hg concentrations may still increase with decreasing pH on a whole-lake basis (Figure 2).

F. MASS BALANCE FOR METHYL-Hg IN LITTLE ROCK LAKE

This mass balance indicates that atmospheric inputs of methyl-Hg were about 1% of the HgT inputs (Figure 7b). We estimate that the methyl-Hg influx was roughly balanced by loss to sediments, based on the observation that the methyl-Hg concentration in surficial profundal sediments of Little Rock Lake was about 1% of the total Hg concentration (0.01 g/year,

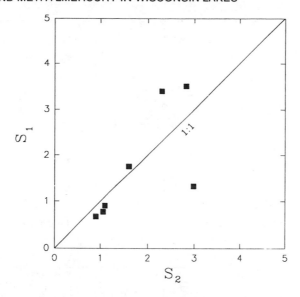

Figure 10 Independent estimates of annual Hg sedimentation in the seven Wisconsin study lakes. S_1 = (deposition + groundwater discharge) − (evasion + groundwater recharge), see Figure 9. S_2 estimated from measured sediment Hg burdens (μg/ha/mm)[26] assuming 1 mm of sediment accumulation per year.

Figure 7b). Sediment traps indicated a downward flux for methyl-Hg that was roughly 30 times higher than the airborne inputs or sediment accumulation. Although more extensive sediment data (seasonal and spatial) are needed to confirm our estimate of annual methyl-Hg accumulation rates, these data indicate substantial production and/or recycling of methyl-Hg within the lake.

In the atmosphere, <0.5% of the TGM was methyl-Hg. Waterborne methyl-Hg constituted roughly 10% of the total waterborne Hg in the lake. Across lakes, the average methyl-Hg fraction varied from about 5–20% of waterborne HgT. The methyl-Hg fraction increased with movement up the food chain, and in this lake most of the Hg in zooplankton and in fish was methyl-Hg (Table 5). Comparisons across lakes indicate that the methyl-Hg fraction in zooplankton varied substantially (ca. 30 to 90%), but in all fish assayed most (>95%) of the Hg has been methylated.[15,17]

Fish constituted the largest methyl-Hg reservoir in the watercolumn of Little Rock Lake. Assuming that our estimates for methyl-Hg in sediment are correct, fish biomass contained about 75% of the methyl-Hg in the entire lake, including freshly deposited sediments (i.e., a 1-year accumulation). Assuming 30% turnover of fish biomass annually, the net bioaccumulation of methyl-Hg in fishes (60 mg/year) exceeded our estimate of annual sediment accumulation (9 mg/year) by a factor of 6 to 7. Although our dataset is limited, the sediment burden of methyl-Hg observed for Little Rock Lake (about 1% of the total Hg) was within the range reported for sediments in other aquatic systems (<0.1–3%).[35]

Since methyl-Hg inputs and outputs were small relative to biotic concentrations and to the measured sediment trap flux, in situ production appears to be a major source of methyl-Hg to the system. Nevertheless, precipitation inputs of methyl-Hg can constitute a significant fraction of the amount accumulated annually by fish.[36] Highly efficient recycling (i.e., low demethylation) would increase the importance of external input fluxes. In any case, the observed distributions imply rapid transport of methyl-Hg into the foodchain.

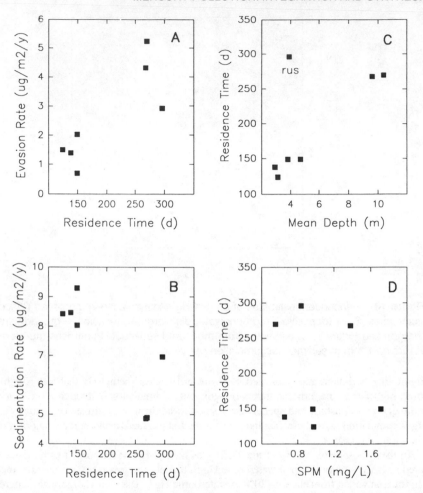

Figure 11 Rates of Hg evasion (A) and sedimentation (B) relative to Hg residence time in surface waters of the seven Wisconsin study lakes. Also, relation between Hg residence time and (C) mean depth or (D) suspended particulate mass (SPM). Hg residence times were computed as in Figure 8.

Table 5 Relative concentration of Hg species in phytoplankton, zooplankton, and fish in Little Rock Lake, all on a wet weight basis.

Compartment	Relative concentration (wet weight basis)		%Methyl-Hg
	Methyl-Hg	Non-methyl-Hg	
Water	1	10	10
Phytoplankton	$10^{5.2}$	$10^{5.4}$	35
Zooplankton	$10^{5.7}$	$10^{5.3}$	70
Young perch	$10^{6.0}$	$10^{4.7}$	95

Modified from Watras, C.J. and Bloom, N.S., *Limnol. Oceanogr.*, 37, 1313, 1992.

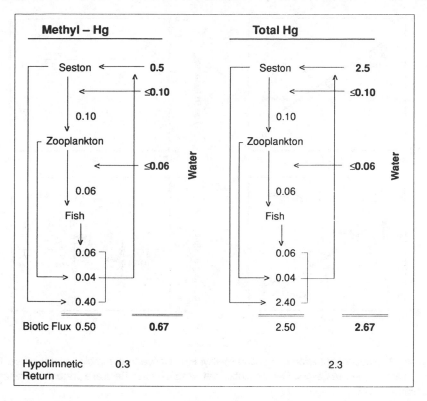

Figure 12 Biotic cycling of Hg and methyl-Hg in Little Rock Lake. All rates are g/year. The total biotic flux was computed two ways for each Hg species: (1) assuming all uptake is from water (bold text); (2) assuming all uptake is trophic transfer (standard text). Return from hypolimnion = sediment trap flux − sediment accumulation. (See text for details.)

G. BIOACCUMULATION AND BIOCONCENTRATION OF Hg AND METHYL-Hg

Biotic turnover rates for Hg species in the lake can be estimated using: (1) measured mercury concentrations in water, seston (phytoplankton), zooplankton, and fish; (2) production/biomass (P/B) ratios from the literature for each biotic compartment; and (3) some simplified assumptions about transfer between compartments (Figure 12). For fish, we use a P/B ratio of 0.3/year; for crustacean zooplankton, 10/year (ca. 0.1/d for 100 d);[37] and for phytoplankton, 25/year (0.25/d for 100 d).[38] We assume that the degree of trophic transfer falls between two extremes: (1) 100% direct uptake from water (no trophic transfer), and (2) 100% uptake from the lower trophic level (all trophic transfer). For these extreme cases, we find that 2.5 to 2.7 g Hg cycles through the biota per year. By comparison, 0.5 to 0.7 g methyl-Hg cycles through the biota per year. These biotic fluxes are very close to the turnover fluxes estimated from hypolimnetic sediment trap data: 2.3 g Hg recycled annually; 0.3 g methyl-Hg produced and/ or recycled annually (Figure 7).

Bioconcentration factors (Bf) and partitioning coefficients (Kd) measure the tendency of solutes to form solid phase species (biotic and abiotic) in aquatic environments. Computation is similar:

$$Bf = Cb/Cw; \ Kd = Cp/Cw$$

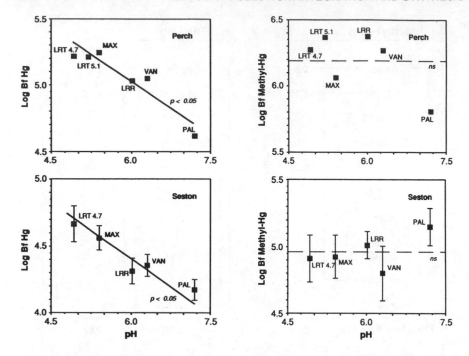

Figure 13   Bioconcentration of Hg and methyl-Hg relative to pH in clearwater lakes. Bf = log (Cb/Cw), where Cb and Cw are biotic (wet weight basis). Data are annual epilimnetic averages ±1 SE.

where Cb, Cp, and Cw are concentrations in biota, particulate matter, and water, respectively, (all mass:mass). Since most of the particulate matter in our study lakes was biogenic, we report Bf for seston and for fish—both on a wet weight basis—rather than computing a separate Kd for suspended particulate.

Bioconcentration factors for Hg and methyl-Hg were related to pH in different ways (Figure 13). For both seston and young perch, the Bf (Hg) tended to increase with decreasing pH. This observation is consistent with data from other studies showing increased Hg in fishes from acidic lakes.[8,14,15] Over the pH range 5 to 7, young perch bioconcentrated Hg 3 to 5 times more than bulk seston. In contrast to Hg, bioconcentration factors for methyl-Hg were relatively constant over the pH range 5 to 7. Across this pH range, then, [methyl-Hg] in biota and in water increased at similar rates. For bulk seston, the Bf (methyl-Hg) averaged about 5.0, while for young perch it was about an order of magnitude higher. This observation implies that methyl-Hg is bioaccumulated in proportion to supply and that bioaccumulation is greater at higher trophic levels.[15]

The relationships between bioconcentration factors and DOC shows some evidence of an effect working opposite to that of pH (Figure 14). The Bf (Hg) remained relatively constant in both seston and fish across the DOC range 1.5 to 6.5 mg/L. The Bf (methyl-Hg) decreased significantly with increasing DOC in the case of perch ($p < 0.05$), and there was a similar trend in seston (albeit not statistically significant). Absolute concentrations of Hg and methyl-Hg in both organisms and water are highest in the darkwater lake (Russette Lake, Figures 2 and 3), but a disproportionately higher fraction apparently remains in the waterborne phase.

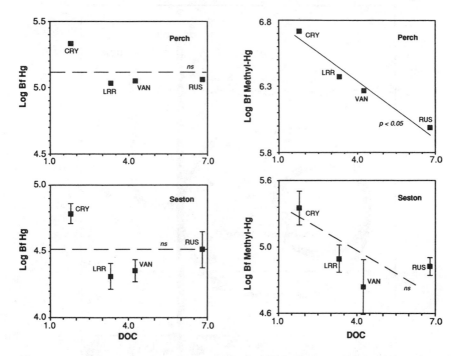

Figure 14 Bioconcentration of Hg and methyl-Hg relative to dissolved organic carbon (DOC) concentration in lakes of similar pH. Bf as in Figure 13. Data are annual epilimnetic averages ±1 SE.

H. A CONCEPTUAL MODEL OF SOURCES AND FATES

At least six processes have an important effect on the concentration and distribution of mercury in the lakes of northern Wisconsin: atmospheric deposition, gaseous evasion, sedimentation, methylation, bioaccumulation, and demethylation (Figure 15). Our studies of Wisconsin seepage lakes, and Scandinavian studies of drainage lakes,[8] indicate that Hg in aquatic ecosystems is ultimately derived from atmospheric deposition. Within lakes, atmospheric Hg inputs follow three main pathways: (1) particle scavenging and transport toward sediments; (2) conversion to elemental Hg and subsequent evasion; or (3) methylation and subsequent bioaccumulation and/or demethylation. These processes may compete for the available Hg and the outcome of this competition seems to be a rather complex function of environmental variables, such as pH and DOC.

Sedimentation appears to be an effective removal mechanism, since Hg is apparently buried below the profundal sediment/water interface and not remobilized by either the cation exchange or redox reactions that liberate other metals such as Fe.[31] However, sedimentation also transports Hg to methylation sites in sediments or in the watercolumn below the O/A boundary.[23,39] Thus, factors which favor Hg-particle binding, such as decreased pH, may enhance Hg burial and methyl-Hg formation simultaneously.

While methylation seems to occur in anoxic regions of the lake,[18,35] the supersaturation of Hg^0 in surface waters and undersaturation of Hg^0 in deep anoxic waters indicates that reactions governing the evasive Hg flux occur in the oxic watercolumn,[24] but the spatial separation of the evasion and methylation pathways does not rule out competitive interaction

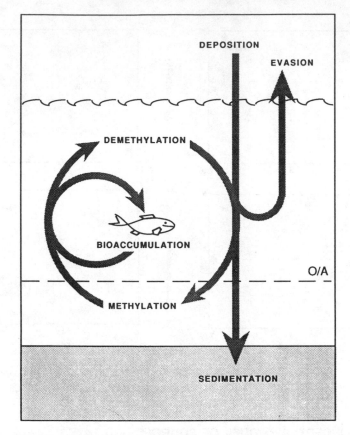

Figure 15 Conceptual model of the principal Hg cycling pathways in remote Wisconsin seepage lakes. (O/A) denotes the oxic/anoxic boundary, which may occur above or below the sediment/water interface.

between these processes. In fact, our data indicate that variables which tend to increase Hg residence in the upper water enhance evasive flux at the expense of downward transport to burial or methylation zones. However, it does not necessarily follow that competitive interaction limits rates of methylation. Biotic methylation is also a function of microbial activity. Thus, either the size and physiological status of microbial populations or the availability of Hg substrate may constrain methyl-Hg formation.

The distribution of methyl-Hg in the Wisconsin study lakes indicates that it is effectively bioaccumulated and biologically sequestered. However, we have no data which pertain to the process of demethylation and its importance in the cycling of alkyl-Hg. Since waterborne and sediment pools of methyl-Hg appear to be small, it must either be rapidly shunted back into organisms or demethylated when released via excretion or death. Understanding the bioaccumulation of Hg in these remote lakes clearly will require knowing the environmental dependencies of factors which regulate the formation, destruction, and trophic transfer of methyl-Hg.

ACKNOWLEDGMENTS

We thank Emily Greenberg, Steve Greb, Ann Grant, Lian Liang, David Sloey, Kent Hatch, D. William Fitzgerald, and Tris Hoffman for their contributions in the field and laboratory.

Research was funded by the Electric Power Research Institute (RP2020–10) and the Wisconsin Department of Natural Resources. This is a contribution from the Trout Lake Research Station, University of Wisconsin–Madison.

REFERENCES

1. Fitzgerald, W.F. and Watras, C.J., Mercury in the surficial waters of rural Wisconsin lakes, *Sci. Total Environ.*, 87/88, 223, 1989.
2. Gill, G.A. and Bruland, K.W., Mercury speciation in surface freshwaters systems in California and other areas, *Environ. Sci. Technol.*, 24, 1392, 1990.
3. Porcella, D., Watras, C.J., and Bloom, N.S., Mercury species in: lakewater, in Deposition and Fate of Trace Metals in our Environment, Verry, E.S. and Vermette, S.J., Eds., Gen. Tech. Rpt. MC-150, USDA-Forest Service, St. Paul, MN, 1992.
4. Watras, C.J., Bloom, N.S., Crecelius, E.A., Cutter, G.A., and Gill, G.A., Quantification of mercury, selenium and arsenic in aquatic environments, in: *Proceedings: 1990 International Conference on Measuring Waterborne Trace Substances,* Chow, W. et al., Eds., EPRI, Palo Alto, CA, 1992.
5. Bloom, N.S. and Fitzgerald, W.F., Determination of volatile mercury species at the picogram level by low temperature gas chromatography with cold vapor atomic fluorescence detection, *Anal. Chim Acta,* 208, 151, 1988.
6. Bloom, N.S., Determination of picogram levels of methylmercury by aqueous phase ethylation, followed by cryogenic gas chromatography with cold vapor atomic fluorescence detection, *Can. J. Fish. Aquat. Sci.*, 46, 1131, 1989.
7. Mierle, G., Aqueous inputs of mercury to precambrian shield lakes in Ontario, *Environ. Toxicol. Chem.*, 9, 843, 1990.
8. Lindqvist, O. et al., Mercury in the Swedish environment: recent research on causes, consequences and corrective methods, *Water Air Soil Pollut.*, 55, 1, 1991.
9. Wiener, J.G., Fitzgerald, W.F., Watras, C.J., and Rada, R.G., Partitioning and bioavailability of mercury in an experimentally acidified Wisconsin lake, *Environ. Toxicol. Chem.*, 9, 909, 1990.
10. Iverfeldt, Å. and Johansson, K., Mercury in run-off from small watersheds, *Verh. Int. Verein. Limnol.*, 23, 1626, 1988.
11. Meili, M., Mercury in Boreal Forest Lake Ecosystems, Ph.D. dissertation, Uppsala University, Sweden, 1991.
12. Mierle, G. and Ingram, R., The role of humic substances in the mobilization of mercury from watersheds, *Water Air Soil Pollut.*, 56, 349, 1991.
13. Padberg, S., Quecksilber im terrestrischen okosystem, Ph.D. dissertation, Universitat Tubingen, Germany, 1991.
14. Spry, D.J. and Wiener, J.G., Metal bioavailability and toxicity to fish in low alkalinity lakes: a critical review, *Environ. Pollut.*, 71, 243, 1991.
15. Watras, C.J. and Bloom, N.S., Mercury and methylmercury in individual zooplankton: implications for bioaccumulation, *Limnol. Oceanogr.*, 37, 1313, 1992.
16. Lathrop, R.C., Rasmussen, P.W., and Knauer, K.K., Walleye mercury concentrations in Wisconsin lakes, *Water Air Soil Pollut.*, 56, 295, 1991.
17. Bloom, N.S., On the methylmercury content of fish, *Can. J. Fish. Aquat. Sci.*, 49, 1992.
18. Winfrey, M.R. and Rudd, J.W.M., Environmental factors affecting the formation of methylmercury in low pH lakes, *Environ. Toxicol. Chem.* 9, 853, 1990.
19. Bloom, N.S., Watras, C.J., and Hurley, J.P., Impact of acidification on the methylmercury cycling of remote seepage lakes, *Water Air Soil Pollut.*, 56, 477, 1991.
20. Watras, C.J., Bloom, N.S., Fitzgerald, W.F., Hurley, J.P., Krabbenhoft, D.P., Rada, R.G., and Wiener, J.G., Mercury in temperate lakes: a mechanistic field study, *Verh. Int. Verein. Limnol.*, 24, 2199, 1991.

21. Hudson, R.J.M., Gherini, S.J., Watras, C.J., and Porcella, D.B., Modelling the biogeo-chemical cycle of mercury in lakes: the Mercury Cycling Model (MCM) and its application to the MTL study lakes, in: *Mercury Pollution: Integration and Synthesis,* Watras, C.J. and Huckabee, J.W. Eds., Lewis Publishers, Chelsea, MI, 1994, chap V.1.

22. Patterson, C. and Settle, D., The reduction of orders of magnitude errors in lead analyses of biological materials and natural waters by evaluating and controlling the extent and sources of industrial lead contamination introduced during sample collecting, handling and analysis, in: U.S. National Bureau of Standards Special Publication No. 422: Accuracy in Trace Analysis—Sampling, Sample Handling and Analysis, Washington, D.C., 1976, 321.

23. Watras, C.J. and Bloom, N.S., The vertical distribution of mercury and methylmercury in Wisconsin lakes: accumulation in plankton layers, in: *Mercury Pollution: Integration and Synthesis,* Watras, C.J. and Huckabee, J.W., Eds., Lewis Publishers, Chelsea, MI, 1994, chap I.11.

24. Vandal, G.M., Mason, R.P., and Fitzgerald, W.F., Cycling of volatile mercury in temperate lakes, *Water Air Soil Pollut.,* 56, 791, 1991.

25. Gill, G.A. and Fitzgerald, W.F., Mercury sampling of open ocean waters at the picomolar level, *Deep-Sea Res.,* 32, 287, 1985.

26. Fitzgerald, W.F., Mason, R.P., and Vandal, G.M., Atmospheric cycling and air-water exchange of mercury over mid-continental lacustrine regions, *Water Air Soil Pollut.,* 56, 745, 1991.

27. Rada, R.G., Powell, D.E., and Wiener, J.G., Whole-lake burdens and spatial distribution of mercury in surficial sediments in Wisconsin seepage lakes, *Can. J. Fish. Aquat. Sci.,* 50, 865, 1993.

28. Gill, G.A. and Fitzgerald, W.F., Mercury in surface waters of the open ocean, *Global Biogeochem. Cycles,* 1, 199, 1987.

29. Glass, G.E., Sorensen, J.A., Schmidt, K.W., and Rapp, G.R., New source identification of mercury contamination in the Great Lakes, *Environ. Sci. Technol.,* 24, 1059, 1990.

30. Sorensen, J.A., Glass, G.E., Schmidt, K.W., Huber, J.K., and Rapp, G.R., Airborne mercury deposition and watershed characteristics in relation to mercury concentrations in water, sediments, plankton and fish of eighty Minnesota lakes, *Environ. Sci. Technol.,* 24, 1716, 1990.

31. Hurley, J.P., Watras, C.J., and Bloom, N.S., Distribution and flux of particulate mercury in temperate seepage lakes, *Water Air Soil Pollut.,* 56, 543, 1991.

32. Fitzgerald, W.F.,Atmospheric and oceanic cycling of mercury, in: Duce, R.A., Riley, J.P., and Chester, R., Eds., *Chemical Oceanography: Vol. 10 SEAREX, The Sea/Air Exchange Program,* Academic Press, London, 1989, 151.

33. Jackson, T.A., Kipphut, G., Hesslein, R.H., and Schindler, D.W., Experimental study of trace metal chemistry in soft-water lakes at different pH levels, *Can. J. Fish Aquat. Sci.,* 37, 387, 1980.

34. Schindler, D.W., Hesslein, R.H., and Wagemann, R., Effects of acidification on mobilization of heavy metals and radionuclides from the sediments of a freshwater lake, *Can. J. Fish. Aquat. Sci.,* 37, 373, 1990.

35. Gilmour, C.C. and Henry, E.A., Mercury methylation in aquatic systems affected by acid deposition, *Environ. Pollut.,* 71, 131, 1991.

36. Bloom, N.S. and Watras, C.J., Observations of methylmercury in precipitation, *Sci. Total Environ.,* 87/88, 199, 1989.

37. Winberg, G.G., Patalas, K., Wright, J.C., Hillbricht-Ilkowska, A., Cooper, W.E., and Mann, K.H., Methods for calculating productivity, in: Edmondson, W.T. and Winberg, G.G., Eds., *A Manual on the Methods for the Assessment of Secondary Productivity in Freshwaters,* Blackwell, Scientific, Oxford, 1971, 296.

38. Harris, G.P., *Phytoplankton Ecology: Structure, Function and Fluctuation,* Chapman and Hall, London, 1986.

39. Bloom, N.S. and Watras, C.J., Seasonal and vertical variability in the mercury speciation of an acidified Wisconsin lake, in: Vernet, J.-P., Ed., *Heavy Metals in the Environment,* Vol. 2, Proceedings of the Seventh International Conference, CEP Consultants, Edinburgh, U.K., 1989, 349.

40. Hurley, J.P., Krabbenhoft, D.P., Babiarz, C.L., and Andren, A.W. Cycling processes of mercury across sediment/water interfaces in seepage lakes, in: Baker, L.A., Ed., *Environmental Chemistry of Lakes and Reservoirs,* American Chemical Society, Washington, D.C., (in press).

41. Krabbenhoft, D.P. and Babiarz, C.L., The role of groundwater transport in aquatic mercury cycling, *Water Res.,* 28, 3119, 1992.

42. Krabbenhoft, D.P, Personal communication.

43. Shelly, B.C.L., Personal communication.

44. Frost, T., Unpublished data.

45. Swenson, W., Personal communication.

46. Rose, W., Personal communication.

SECTION II

Atmospheric Cycling, Transport, and Deposition

Atmospheric Cycling of Mercury: An Overview

Oliver Lindqvist

CONTENTS

ABSTRACT: During the last decade it has become evident that mercury emissions to air are dispersed within the regional scale. This means that much of the emitted mercury is deposited within 1,000 to 2,000 km from the sources. Recent data indicate that the global background of atmospheric mercury is increasing by about 1% per year. It has been estimated that the mercury deposition over Scandinavia must be decreased by about 80% to reach a balanced situation. Due to the serious environmental damage caused by mercury, it is thus essential that mercury emissions are substantially decreased in most countries.

I. INTRODUCTION

Mercury has been well-known as a pollutant for several decades. As early as in the 1950s, it was established that emissions of mercury to the environment could have serious effects on human health. Toxic levels of methylmercury in fish have been reported in a large number of water bodies throughout the world. Such problems have often arisen as a result of direct emissions from various types of industrial activities.

During the last decade a new pattern has emerged with regard to mercury pollution, particularly in North America and the Nordic countries. Fish, mainly from nutrient-poor lakes, have often been found to contain high concentrations of mercury. Elevated concentrations have also been found in fish from marine areas. These effects cannot be linked to individual emissions of mercury, but are instead due to more widespread air pollution.

This review on atmospheric mercury will mainly focus on three questions:

- Regional deposition of mercury
- Emissions of mercury to the atmosphere
- Increasing levels of mercury in the atmosphere and in other parts of the environment, illustrating the global impact of mercury

1–56670–066–3/94/$0.00 + $.50
© 1994 Lewis Publishers

II. MERCURY AS A REGIONAL POLLUTANT

During the Conference on Mercury as an Environmental Pollutant held in Gävle, Sweden in 1990, many regional aspects of mercury pollution were raised. Due to the poisonous effects of mercury and the tendency of methylmercury to bioaccumulate in fish, much of the previous and current work has concentrated on local effects in lakes and rivers. However, during the 1980s it was recognized that mercury was transported long distances in the atmosphere and thus may be deposited within 1,000 to 2,000 km from the source areas. This means that lakes in relatively remote areas with no direct sources of mercury pollution are also often contaminated. The relatively high stability of elemental mercury in the atmosphere[1] means that emitted mercury will also contribute to a global background, hence influencing the deposition at every location.

The regional transport of mercury has been illustrated in several recent works, for example over Europe by Iverfeld,[2] and over the U.S. by Glass et al.[3] Trajectory studies clearly show high deposition values when the air masses have passed polluted areas within the last 72 hours or so, while values close to the background can be obtained for other wind directions. At the Rörvik station in southwestern Sweden, large deposition values occur in connection with air masses from Central Europe, while low values may be obtained if, for example, the wind direction is from the Norwegian Sea.

From available emission and deposition data measured in Sweden, a local mercury budget over this country has been calculated.[4] The picture that emerges is that, of a total estimated emission (point, diffuse, and natural sources) of 9,000 kg Hg in Sweden, only about 1,000 kg Hg is deposited in Sweden. However, the total deposition over Sweden is estimated to be 12,000 kg Hg, which means that more than 90% of the mercury arrives via long distance transport from other European countries, and from the global background of atmospheric mercury. A chemical model of part of the atmospheric cycling of mercury is presented by Munthe.[5]

III. MERCURY EMISSIONS TO THE ATMOSPHERE

To understand the regional deposition pattern of mercury, one very important task for the next few years is to perform good regional emission inventories. Münch and Pacyna[6] have started such investigations regarding point sources in Europe, and similar inventories have also been started in Canada and the U.S. To be able to construct a good model of the atmospheric cycling of mercury, the emission inventories must be refined and extended to include diffuse and natural sources. Also, other areas, for example in Asia, should be surveyed, in order to understand the situation there and also to get a better understanding of the global distribution of mercury.

There has been a drastic decrease of mercury emissions to the atmosphere in Sweden during the last 20 years. However, there are no clear indications that the deposition has decreased, which would mean that the emissions in other parts of Europe—at least until lately—have increased. The main sources of mercury emissions to air are industrial processes like the chlorine alkali process, waste incineration, coal combustion, and metal production. In certain parts of the world, emission from natural gas combustion may also occur.

An important issue when defining mercury emissions is the distribution between elemental mercury and divalent mercury compounds in the process. Divalent mercury compounds may exist in the gaseous phase, or they may be adsorbed on particulate matter, like fly ash in a combustion process. If no device for cleaning the flue gases is installed, emissions of divalent mercury compounds are considered worse than elemental mercury, since these compounds have a high tendency to deposit within the region. On the other hand, mercury cleaning systems in the flue gases often only remove the divalent forms, e.g., in filter adsorption processes or in wet scrubbers, where mercury may be precipitated as different sulfur compounds.

Table 1   **Estimates of fluxes of Hg to and from the global atmosphere.**

Process	Lindqvist[11] et al., 1984	Fitzgerald[11] 1986	Lindqvist[4] et al., 1991
Anthropogenic emissions	2–10	2	4.5 (3–6)
Natural emissions	<15	3–4	3 (2–9)
Total present emissions	2–17	5–6	7.5 (5–15)
Wet deposition	2–10	—	5 (4–6)
Dry deposition	<7	—	2.5 (1–9)
Total present deposition	2–17	5–6	7.5 (5–15)

Table 2   **Total gaseous mercury in air (ng m^{-3}) measured over the Atlantic indicating increasing levels in the global background.**

Year	Northern Atlantic	Southern Atlantic
1977–1979	1.91 (20)	1.27 (16)
1980	2.09 (10)	1.45 (08)
1990	2.25 (12)	1.50 (16)
Increase/year	1.5 ± 0.2%	1.2 ± 0.2%

From Slemr, F. and Langer, E., *Nature,* 355, 434–437, 1992. With permission.

Today, it is technically possible to clean up most kinds of point source mercury emissions. Due to the increasing effects of mercury in the environment, it is considered important that governments and environmental agencies perform detailed inventories of mercury emissions and regulate the emissions based on existing clean-up techniques.

Emissions of mercury from natural systems should also be better mapped, so that the scale of anthropogenic emissions may be judged correctly. It is also probable that emissions from natural systems today are higher than during preindustrial times, since accumulated depositions may cause reemission to air. A summary of estimated global emission and deposition of mercury is given in Table 1.

IV.   MERCURY AS A GLOBAL POLLUTANT

The regional connections between sources and deposition are evident. Another question of importance is whether the global background of mercury is changing. There are several indications that the mercury load in the atmosphere is increasing. Slemr and Langer[7] have compared concentrations of atmospheric mercury measured over the Atlantic during late 1970, 1980, and 1990. Their data are summarized in Table 2, and indicate a yearly increase of about 1% over the Southern Hemisphere, and slightly more over the Northern Hemisphere. The higher values over the Northern Hemisphere may be interpreted so that atmospheric mercury has a significant contribution from anthropogenic sources. In combination with other data, this information has also been used to estimate the residence time of mercury in the atmosphere to somewhere between 6 months and 2 years.[1]

In a series of measurements of total mercury in air over the Scandinavian countries, Iverfeldt[8] finds increasing average values for the 4 years 1986 through 1989. Even if the increasing tendency is clear within Iverfeldt's data, one must be aware that the time series is relatively short and that meteorological differences between the years may influence the data. The day-to-day variations within the Scandinavian data are much wider than in the data collected over the Atlantic[7] since the influence from anthropogenic sources is much greater in the former set.

A third indication of increasing load of mercury in the environment is obtained by studying the mercury budget over a lake drainage area. Based on mercury concentrations measured in deposition, in soils, in lake and ground water, and in run-off water before and after the lake, it was possible to calculate a yearly increase of about 1% of mercury bound to the top layer of the forest soil.[9]

A typical budget for a lake of 1 km^2 with a 10 km^2 drainage area in southern Sweden includes a wet deposition of 200 g Hg over 10 km^2. In the budget calculation we have estimated the dry deposition to be 50% of the wet deposition, resulting in a total deposition of 300 g Hg during 1 year. However, measurements have shown that no more than about 50 g Hg is leached from the top soil layer with the run-off water in the typical drainage area. Since the total load of mercury in the humic moory layer is about 25,000 g on 10 km^2, this means that the amount of accumulated mercury is increasing by about 1% per year. From such budget calculations, Johansson et al.[9] have shown that a reduction of about 80% from present atmospheric deposition has to be obtained to reach a balanced input/output budget. If this situation is typical for other forest systems in the world is not known, but the available data fit with the tendency that mercury is presently increasing by approximately 1% per year in the global environment.

By comparing the accumulated amounts of mercury in lake sediments and forest top soils with what is believed to be the natural levels, one may conclude that the load today is between 2 and 7 times higher than during preindustrial times in Sweden, the lower factor referring to the northern and the higher to the southern parts of the country (Johansson et al., 1991). This distribution reflects the fact that mercury is long-distance transported from south to north in a similar way as is measured for sulfur deposition. However, there are large local variations in the mercury amount in the moory layer which are clearly correlated to former large emission sources in Sweden. The available data on mercury concentrations in ground and run-off water indicate that the leakage time is about 100 years before the moory layer in a contaminated area is restored to natural values, if no more mercury deposition occurs. This is because mercury compounds bind so strongly to sulfur atoms in the humic acids.

V. CONCLUSIONS

The regional effects of mercury emissions have been demonstrated in forested land on the Northern Hemisphere. The connections between emission sources and effects within the region are evident. There is still a need for improved emission inventories in most countries. In order to improve budget calculations, there is also a need for improved understanding of deposition processes, especially dry deposition.

Data indicating that the global background of atmospheric mercury is increasing by about 1% per annum have been presented. There are also indications, at least in Scandinavia, that the mercury accumulation in forest top soil is also increasing by about 1% per year. These increased concentrations are probably related to increased bioaccumulation in lake systems.

Due to the negative environmental effects of mercury documented in several regions in the world, and since the environmental concentrations are increasing continuously, it is desirable that mercury emissions should be regulated and substantially decreased in most countries.

REFERENCES

1. Lindqvist, O. and Rodhe, H., Atmospheric mercury—a review, *Tellus,* 37B, 136–159, 1985.
2. Iverfeldt, Å., Occurrence and turnover of atmospheric mercury over the Nordic countries, *Water Air Soil Pollut.,* 56, 251–265, 1991.

3. Glass, G.E., et al., Mercury deposition and sources for the upper Great Lakes region, *Water Air Soil Pollut.,* 56, 235–249, 1991.

4. Lindqvist, O., et al., Mercury in the Swedish environment—recent research on causes, consequences and corrective methods, *Water Air Soil Pollut.,* 55, 1–261, 1991.

5. Munthe, J., The atmospheric chemistry of mercury—kinetic studies of redox reaction, in: *Mercury Pollution: Integration and Synthesis,* Watras, C.J. and Huckabee, J.W., Eds., Lewis Publishers, Chelsea, MI, 1994, chap. I.9.

6. Pacyna, J.M. and Münch, J., Anthropogenic mercury emission in Europe, *Water Air Soil Pollut.,* 56, 51–61, 1991.

7. Slemr, F. and Langer, E., Increase in global atmospheric concentrations of mercury inferred from measurements over the Atlantic Ocean, *Nature,* 355, 434–437, 1992.

8. Iverfeldt, Å., Atmospheric mercury over the Nordic countries, *Water Air Soil Pollut.,* 55, 33–47, 1991.

9. Johansson, K. et al., Mercury in Swedish forest soils and waters—assessment of critical load, *Water Air Soil Pollut.,* 56, 267–281, 1991.

10. Lindqvist, O., Jernelöv, A., Johansson, K., and Rodhe, H., Mercury in the Swedish Environment. Global and Local Sources, Nat. Swedish Env. Prot. Board, SNV-PM-1816, 1–105, 1984.

11. Fitzgerald, W.F., Cycling of mercury between the atmosphere and oceans, *NATO ASI Ser.,* C185, 363–408, 1986.

Historical Atmospheric Mercury Deposition in the Mid-Continental U.S. as Recorded in an Ombrotrophic Peat Bog

J. M. Benoit, W. F. Fitzgerald, and A. W. H. Damman

CONTENTS

I. INTRODUCTION

Atmospheric deposition represents a major source of mercury to freshwater[1] and marine environments.[2] This toxic trace metal accumulates in biota, and elevated concentrations may be present in both freshwater and marine organisms, particularly large predators.[3] In the U.S., Canada, and Scandinavia, health officials have issued advisories to limit human consumption of certain fishes due to Hg contamination.[4,5]

Global mass balances estimate that 30 to 60% of atmospheric Hg is from anthropogenic sources,[6,7] which implies a 1.5- to 3-fold increase in mercury in the atmosphere since industrialization. Enrichment of surficial sediments in lakes in central North America provide evidence of a two to threefold increase in anthropogenic loadings of mercury after 1875 through 1900.[8,9] Observed recent increases in mercury in lake sediments and freshwater fishes have been attributed to atmospheric deposition.[10-12] In order to assess the magnitude of the effect of human-related activities on global Hg cycling, it is necessary to establish background (preindustrial) atmospheric deposition and changes in atmospheric fluxes over time.

Ombrotrophic bogs receive all moisture and nutrients from the atmosphere. Hg entering in wet and dry deposition is adsorbed by the surface layer of moss, and retained in the peat layers as the bog continues to grow vertically, producing a record of changes in atmospheric deposition over time. Geochemical monitoring of atmospheric pollution in ombrotrophic peat bogs has been carried out for a variety of substances including acid deposition,[13] lead,[14,15] mercury,[16] organic contaminants,[17] and other metals.[18–20] Reviews on the use of peatlands as depositional archives are given by Glooschenko[21] and Livett.[22] Several assumptions underlie the use of ombrotrophic bogs as geochemical archives: (1) that peat cores can be accurately dated, (2) that there is no postdepositional migration of mercury, and (3) that bogs are spatially homogenous collectors of deposition.

Rapidly changing accretion rates that occur due to decomposition tend to confound attempts to accurately date peat. Postdepositional migration of several elements has been observed,[23,24] although the high adsorptive capacity of mosses suggests immobility of Hg in peat.[25–27] Previous depositional studies of Hg in ombrotrophic bogs[16,28] have measured distributions in only one core per sampling location, therefore they do not assess spatial variability. In the following discussion and analysis of the use of peat bogs as geochemical archives we concentrate on the validity of the dating scheme, evidence for mobility or retention of Hg, and the degree of horizontal homogeneity in recorded deposition.

II. MATERIALS AND METHODS

A. EXPERIMENTAL DESIGN

Arlberg bog, the peatland chosen for the study, is located in St. Louis County, MN about 20 miles west of Duluth. We entered the bog via a snowmobile trail that crosses State Road 8 exactly 5.5 miles west of Culver, and selected an ombrotrophic portion based on topography and vegetational characteristics. The sampling site is sheltered from direct dust inputs from the road by a dense black spruce and larch forest. We visited the site (46° 56'N, 92° 41'W) in October 1990 and 1991. Trace-metal-free laboratory protocols were employed during all stages of collection and analysis.

Hummocks were preferred to hollows for coring because a larger portion of peat lies above the high water table, and element retention is more efficient. Trace metal distributions were established on three cores (IB, IIB, IIIB) collected on the first sampling occasion. Separate cores (IA, IIA, IIIA) for bulk density were also taken from each of the three sampled hummocks. Pb-210 chronologies were determined in the trace metal cores and verified with the Cs-137 horizon and a pollen marker.

An additional three hummocks were sampled during the second visit. Cores IVB, VB, and VIB were collected for total excess Pb-210 to ascertain whether the average hummock inventory of this radioisotope was the same as that expected from atmospheric deposition, i.e., to determine if Pb-210 was lost from the hummocks. Three cores (IVA, VA, VIA) were taken for bulk density at this time to better estimate the variability associated with this measurement.

On both occasions additional cores (IC-VIC) were taken from the sampled hummocks with a coring device and sectioning technique that minimized artificial compaction of the peat. The bulk densities of these sections were compared to the others to determine where compression was the greatest in the trace metal cores.

B. CORING AND SECTIONING

The majority of the cores were collected with devices designed and built at the University of Connecticut in Groton. These corers were stainless steel tubes with sharpened teeth and removable handles. The "noncompressing" corer from UCONN in Storrs tapered to a diameter of 11.3 cm at the cutting edge. The teeth of the Storrs corer were specially constructed to cut through the roots of shrubs without artificially compacting the sample. A plastic sleeve inside the tube facilitated removal of the peat cylinder with little disturbance. Both types of corers were inserted by hand with very little downward pressure and a gradual twisting motion to shear roots that penetrated the peat.

All cores were sectioned in the field. After sampling, the Groton corers were attached vertically to an extruding device. A Teflon® piston on a threaded rod was gradually pushed through the cylinder and sections were cut from the top with SS scissors or saws. While this device allowed for fine sectioning (2.5 cm increments to 10 cm and 1 cm increments below), water was squeezed out of the core during the procedure. The Storrs cores were emptied completely intact onto a metal tray and sectioned every 5 cm with a SS serrated breadknife.

C. SAMPLE PREPARATION AND ANALYSIS

Bulk density sections from cores IA-VIA were oven-dried to constant weight at 70° C. The sections were weighed before and after drying. The sample volume was 120 to 300 mL. Bulk density sections from the Storrs corer were similarly processed, but sample volumes were 500 mL or more. The results from these cores are more reliable because there was less compression during collection and sectioning, and larger volumes of material were used. The trace metal sections were divided in a trace-metal-free clean laboratory and leaves, sticks, and large roots were removed. Subsections were dried on hotplates in acid-cleaned Pyrex® petri dishes under a stream of ultrapure argon at 60° C. Pb-210 sections were oven-dried at 70° C. The dried peat was ground for 2 min in a Black and Decker Handy Chopper®, which produced a fairly homogeneous powder.

To prepare the dried, ground peat for metal analyses, 100 mg portions were microwave digested in Savillex® bombs with concentrated nitric acid for Hg or in Parr® bombs with HF/HNO_3/HCl for Al, Mn, and Fe. Digestates were diluted with sub-boiling distilled water (QH_2O), and filtered through 45 μm polyprophylene filters.

Mercury was determined in the digestates by atomic fluorescence[29] using a modification of the cold vapor AAS method of Gill and Fitzgerald.[30] The detection limit for the analysis was 10 ng/g. Repeated analyses were performed on one of the peat sections and a standard reference material (NIST SRM 1575 Pine Needles). Overall precision of the method was 10% for both materials.

The other metal (Al, Mn, and Fe) analyses were performed at the Environmental Research Institute in Storrs, CT by inductivity coupled plasma AAS. Digested standard reference material, duplicate peat samples, and dissolved standards were included with unlabelled digestates to assure quality control. The detection limits were as follows: Al—10 μg/g, Mn—5 μg/g, and Fe—10 μg/g. The overall precision was Al—10%, Mn—5%, and Fe—12%.

Pb-210 activity (dpm/g) was determined on 300–500 mg samples via alpha counting of the daughter Po-210. Samples were spiked with 2.5 dpm Po-208 and refluxed in nitric acid (≤80° C) for about 5 h. The solutions were brought to near dryness, and taken up in HCl, then plated onto silver planchettes as outlined by Flynn.[31] Supported Pb-210 was taken as the low, constant activity that was reached deep in all of the cores. Ages of the sections were estimated from excess Pb-210 using the constant rate of supply (CRS) method, which is the most reliable method when sediment accumulation rates are not constant over time.[32] The residence time of Pb-210 in the atmosphere is only 5 to 30 days,[33] and it takes 2 years for Po-210 to grow into equilibrium with Pb-210. When surface sections were less than 2 years old, the measured Pb-210 was corrected to the equilibrium activity.

Subsamples of the sections used for Pb-210 dating were gamma counted in a GeLi well detector at Lamont Doherty Geological Observatory in Palisades, NY to determine Cs-137 activity. All samples were packed to 3.5 cm in identical plastic tubes and had similar counting geometries. The detector was not specifically calibrated for peat, therefore the activities (cpm/g) give relative, not absolute, measurements.

Additional validation of the dating scheme came from pollen analysis performed at the Limnological Institute in Minneapolis, MN. Specifically, the section with concurrent decline in white pine (*Pinus strobus*) and increase in ragweed (*Ambrosia spp.*) pollen was identified in cores IA and IIIA. This horizon marks the time period of major disturbance in the area from white pine harvest and land clearing.

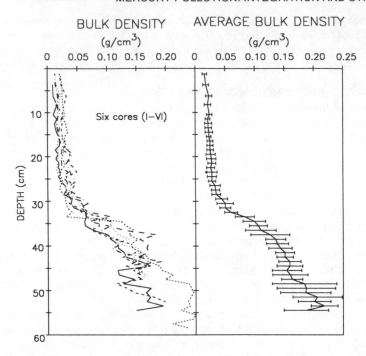

Figure 1 Individual bulk density distributions from hummocks I, II, III, IV, V, and VI, and the average bulk density for five of the cores. The bulk density for core IA is not included in the average because material was lost during sectioning. The error bars represent one standard deviation of the mean.

III. RESULTS AND DISCUSSION

A. BULK DENSITY

Bulk density profiles for cores IA-VIA are presented in Figure 1, along with the average and standard deviation for all six cores. These cores were collected and extruded in the same manner and in the same hummocks as those for trace metals, so the average bulk density of "A" cores should be representative of "B" cores. During sectioning of core IA, material was lost, and the measured bulk densities are systematically lower than the others, therefore it is not included in the average. The uncertainty associated with these estimates (one standard deviation of the mean) ranged from 10 to 39% with an average of 22%. This uncertainty is a function of both the in situ spatial variability in bulk density and analytical error associated with the sampling method.

The results from cores IA-VIA are compared to the "noncompressed" bulk densities in Table 1. The average uncertainty associated with these estimates is also 22%. The ratio of the average bulk density for the "compressed" to "noncompressed" cores provides an index of the relative magnitude of compression in each depth increment. This ratio is >1 at all depths, and it is largest between 35 and 45 cm. Loss of porewater during core extrusion probably caused the extreme compression (>5 times) in this region.

B. EXCESS Pb-210

Excess Pb-210 distributions from cores IB, IIB, and IIIB are shown in Figure 2. The activity in core IB and IIIB declines more or less steadily toward zero below about 15 cm, whereas in core IIB excess Pb-210 is still present at relatively high activities between 40 and 50 cm. The presence of large amounts of excess Pb-210 deep in the core cannot be explained by

Table 1   **Bulk density of "uncompressed" cores collected with the Storrs coring device.**

Section depth (cm)	Bulk density (g/cm³)								Ratio
	IC	IIC	IIIC	IVC	VC	VIC	Average		
0–5	0.012	0.015	0.013	0.014	0.008	0.014	0.013	0.002	1.26
5–10	0.013	0.013	0.014	0.012	0.007	0.019	0.013	0.003	1.55
10–15	0.011	0.011	0.018	0.019	0.011	0.014	0.014	0.003	1.46
15–20	0.009	0.010	0.012	0.016	0.013	0.017	0.013	0.003	1.90
20–25	0.012	0.014	0.017	0.019	0.012	0.017	0.015	0.003	1.75
25–30	0.021	0.014	0.015	0.017	0.016	0.014	0.016	0.003	2.16
30–35	0.021	0.010	0.019	0.016	0.018	0.024	0.018	0.004	3.55
35–40	0.023	0.019	0.023	0.019	0.027	0.024	0.023	0.003	5.24
40–45	0.027	0.023	0.034	0.022	0.031	0.026	0.027	0.004	5.56
45–50	0.037	0.042	0.030	0.034	0.074	0.041	0.043	0.014	3.85
50–55	0.070	0.029	0.049	0.063	0.072	0.053	0.056	0.014	3.58

Note: The Roman numeral refers to the hummock from which the core was taken. The ratio is the average bulk density for "compressed" cores (IIA-VIA) divided by the average for the "uncompressed" cores (IC-VIC). The reported uncertainties are one standard deviation of the mean.

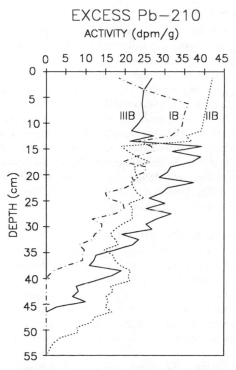

Figure 2   Excess Pb-210 distribution in cores IB, IIB, and IIIB. The distribution in core IIB is anomolous because there is excess PB-210 present at depth (below 40 cm) that could not have been supplied by atmospheric deposition, the rate of which has been constant over the past several centuries.

Table 2 **Total excess Pb-210 inventories.**

	Core				
	IB	IIIB	IVB	VB	VIB
Excess	16.6	23.9	16.0	15.6	16.2
Pb-210 (pCi/cm)	±3.9	±5.4	±3.6	±4.1	±3.7

Note: The depth of excess Pb-210 penetration was adjusted to 50 cm for cores I, III, and V.

atmospheric deposition, which has been constant over the last several hundred years. The deep excess Pb-210 may be caused by an abundance of small, living roots which supply "hot" lead to this depth. Whatever the controlling mechanism, the anomalous distribution in core II makes Pb-210 dating of it unreliable.

The remaining five cores can be depth correlated to account for differential compression by scaling the depth of Pb-210 penetration to 50 cm. The ratio in Table 1 indicates the relative magnitude of the adjustment to depth and to average bulk density in each 5 cm increment. Adjusted depths and bulk densities will be used in further analysis and discussion. Core inventories are given in Table 2. The average is 17.6 ± 4.1 pCi/cm^2, which corresponds to an atmospheric deposition rate for the past 250 years of 0.53 ± 0.12 pCi/cm$^2 \cdot$ year. This inventory agrees with the average inventory from atmospheric deposition, 14.8 ± 3.7 pCi/cm^2, and the deposition rate is within the range measured in North America, 0.34 to 0.68 pCi/cm$^2 \cdot$ year.[34]

C. GEOCHRONOLOGY

Pb-210 dating schemes for cores IB and IIIB are shown in Figure 3 along with Cs-137 distributions. The peak is associated with maximum fallout from bomb testing in 1963.[35] A sub-peak associated with intense fallout in 1959 cannot be distinguished as it has been obliterated by diffusion. Agreement between the Pb-210 geochronology and the location of the Cs-137 peak is quite good for both cores; the peak corresponds to the years 1963–1969 in core IB and 1967 through 1975 in core IIIB.

The pollen horizon indicated in the figure refers to the section in which a concurrent decline in white pine pollen and increase in ambrosia pollen was observed. The white pine industry in the Duluth area was most intense between 1890 and 1905, and evidence of major disturbance should appear in the sedimentary record between those dates.[36] In core IB this horizon corresponds to the period 1898 to 1921 and in core IIIB to the period 1918 to 1926. The marker appears a little later than expected in core IIIB but, within the error associated with the Pb-210 chronology, the dates agree.

D. OBSERVED METAL DISTRIBUTIONS

In order to understand the observed metal distributions in the cores, one must consider the structure and development of rainfed bogs. Ombrotrophic peat consists of two distinct layers: a surface layer (the acrotelm) which is biologically active, is variably aerated, and has high hydraulic conductivity; and a lower layer (the catotelm) which has low biological activity, is permanently anaerobic, and has very low hydraulic conductivity.[37] The water level is maintained by a balance between supply from the atmosphere and drainage through the acrotelm, so that the water table is perched above the saturated and highly impermeable catotelm.[24]

Seasonal changes in the amount and kind of precipitation cause water table fluctuations (of about 5 cm) over the course of a year, while accumulation of decomposed peat produces a net upward movement of the water table (on the order of millimeters per year) over longer time scales. Therefore, just above the catatelm there is a zone of water table fluctuation between the low and high water levels of a given annual cycle. At the study site in October,

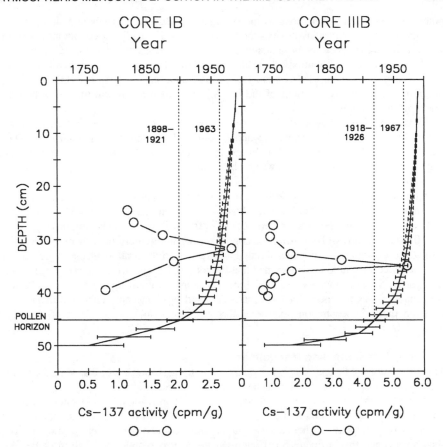

Figure 3 Geochronologies for cores IB and IIIB. The solid curve with error bars represents the year at the bottom of each section as a function of depth based on Pb-210 data. The Cs-137 activity peak is shown with open circles, and the year corresponding to the peak is indicated. The horizontal line marks the section in which the pollen horizon was identified (see text for details), and the corresponding dates at the bottom and top of the section are given.

it is likely that the recent lowest water levels occurred in the previous spring while the bog surface was frozen, that snowmelt supplied a pulse of water, and that the water level dropped gradually over the summer. On the sampling occasions the water table was probably somewhere between the highest and lowest levels for that year.

Elements in rain or melting snow are scavenged by living moss as they enter an ombrotrophic bog. Excess water drains laterally through the acrotelm in the zone of water table fluctuation. Moss has a high binding capacity for metals[26,27,38] and can effectively trap elements in the upper layers of peat. As decomposition proceeds, solid phase metal concentrations increase. Protons generated during decomposition contribute to the low pH of bog porewaters, thereby increasing the solubility of many metal compounds. In addition, the enhanced proton activity can displace elements from ligand exchange sites. Inorganic precipitation can immobilize dissolved metals, and anionic species that form highly stable precipitates will compete with organic ligands for metal ions.

The oxidation state of solutes is controlled by pe (the negative log of the electron concentration), which changes from positive to negative across the water table. The redox potential of peat porewaters controls the speciation of dissolved metals and sulfur. Metals may precipitate as oxides, hydroxides, or sulfates under the oxidizing conditions above the water table. In the permanently saturated zone, reduced metals may precipitate as sulfides.

We modeled the distributions of metals (Fe, Al, Mn, Hg) in peat by predicting the dominant species at constant pH over a range of redox conditions using the MINEQL$^+$ software program.[39] The relative locations of the Fe, Al, and Mn peaks reveal the redox potential gradients in our cores.

Grigal and Nord[40] measured pH in Arlberg bog porewater and found that the mean was 3.9 in the top 25 cm, 3.7 between 25 and 50 cm, and 3.8 between 50 and 75 cm. These results indicate that the pH does not change across the zone of water table fluctuation, and that the porewater pH in our cores was probably about 3.8 for the entire length. We will use this constant pH in our model.

Since we did not measure dissolved concentrations, it was necessary to obtain order of magnitude estimates from the literature. Porewater concentrations of S (as SO_4^{-2} and H_2S) and Fe in Marcell Bog porewaters were about $3 \cdot 10^{-5} M$,[41] and Al concentrations were about $1 \cdot 10^{-5} M$.[42] The dissolved metals in bogwaters are in equilibrium with inorganic solids through precipitation and organics through adsorption, so the total available is greater than that measured in solution. We assumed that the total S, Fe, and Al in the porewater system (designated S_T, Fe_T, and Al_T) were 10 times the concentrations reported above. The Mn_T concentration was assumed to be the same as Al_T, and Hg_T concentration was estimated from its solid phase concentration relative to that of iron: $[Mn_T] = 1 \cdot 10^{-4} M$, $[Hg_T] = 4 \cdot 10^{-9} M$.

1.   Iron, Aluminum, and Manganese

The distributions of Fe and Al are characterized by strong subsurface peaks, while the highest concentrations of Mn are found near the hummock surface (Figure 4).

According to our speciation calculations, the transition from sulfate to sulfide occurs at pe 1 in bog porewaters, and reduced S is mostly in the form of hydrogen sulfide (H_2S) at pH 3.8. Iron precipitates as hematite ($Fe_2O_{3(s)}$) above pe 4 and pyrite ($FeS_{(s)}$) below pe 1, as shown schematically in Figure 5. At intermediate pe it is mobile in the form of Fe^{+2}. Al precipitates as jarbanite ($ALSO_4OH_{(s)}$) above pe 1, below it is dissolved primarily in the form of Al^{+3}. The distribution of Mn appears to be influenced by uptake by vascular plants and leaching of detritus. The mechanism is similar to that described for N, P, and K by Damman.[24] The observed Mn distributions can also be explained by the reduction of Mn^{+4} at a higher pe than the reduction of Fe^{+3}. Oxidation of Mn^{+2} to Mn^{+4} and precipitation as manganite (MnO_4) occurs above pe 15. Removal of Mn from solution sets up a concentration gradient in the porewater, and diffusion can account for upward migration of Mn^{+2} toward the zone of precipitation near the hummock surface.

The observed distributions of these metals in cores IB and IIIB are consistent with a pe gradient ranging from about 18 at the living moss surface to 4 at the base of the Fe peak. The pe then drops to 1 at the base of the Al peak, and continues to decline gradually below. The most rapid transition in pe occurs across the top of the water table, with sulfate (SO_4^{-2}) predominating in the top several centimeters of saturated peat and H_2S below. When the water table rises after snow melt, dissolved Fe and Al are exposed to more highly oxidizing conditions and they precipitate. As the water recedes over the course of the summer, the solids are left behind forming peaks in the zone of water table fluctuation. At the time of sampling (October) the Al peak should be below the surface of the water table, and the Fe peak near the level of the most recent HWT.

Pyrite becomes highly soluble below a pe of -4. If dissolution of FeS were occurring in the peat, a secondary peak in solid phase iron would form due to diffusion of dissolved iron

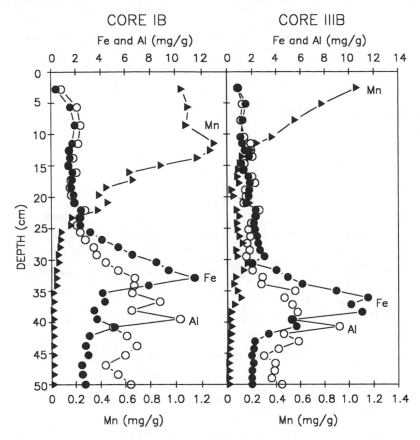

Figure 4   Iron, aluminum, and manganese distributions in cores IB and IIIB.

from below. Since a secondary maximum is not observed below the Al peak (see Figure 4), it is unlikely that the pe ever becomes lower than about −3 within the sampled depths.

2.   Mercury

A peak below the Al maximum appears in the Hg profiles for both cores IB and IIIB (Figure 6). Since it is located in the reducing zone (Figure 5), this peak may indicate a horizon where cinnabar (HgS) precipitates, underlain by a region of enhanced HgS solubility, with diffusion of dissolved Hg toward the zone of precipitation. Speciation calculations for Hg indicate that below a pe of −5 cinnabar is solubilized due to enhanced formation of $Hg(SH)_{2(aq)}$. However, the lower pe limit of −3, which is evidenced by the iron distributions, prevents mobility of Hg by this mechanism. According to the proximity of the Hg peak to the Al peak, the Hg peak occurs at a pe of about 0, and we can assume that Hg is largely immobile in peat after it is deposited.

An alternative explanation for the Hg peak is increased deposition after the turn of the century. As illustrated in Figure 7, peat accumulation rate has a sigmoidal shape as a function of depth. With a constant atmospheric flux, Hg concentration tends to increase with depth from the surface, first slowly and then more rapidly. The theoretical Hg concentration profile in Figure 7 illustrates that lower Hg deposition at depths below the second drop in mass accumulation produces a distinct peak similar to that observed in the cores. Peat Hg concentrations reflect changes in the rate of atmospheric input and can be used to estimate deposition rates.

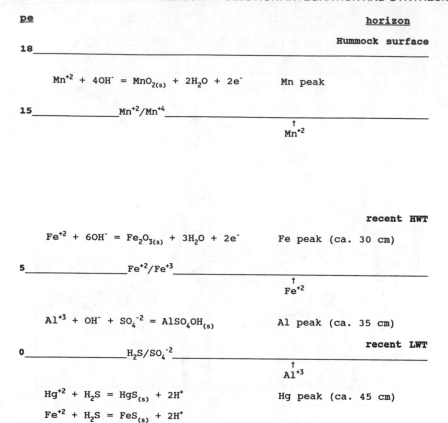

Figure 5 Schematic representation of the pe gradient in the hummocks and the important diagenetic reactions at various depths. The structure of the gradient and water table levels are inferred from speciation calculations and the observed positions of solid phase Fe, Al, and Mn peaks.

Table 3 **Mercury deposition in cores IB and IIIB from 1750 to present.**

Core	Mercury deposition ($\mu g/m^2 \cdot$ year)						
	1750–1900	1900–1935	1935–1950	1950–1960	1960–1970	1970–1980	1980–1991
IB	8.6 ±3.0	23.9 ±8.3	46.6 ±16.2	42.2 ±14.7	28.2 ±9.8	22.4 ±7.8	16.8 ±5.8
IIIB	5.4 ±1.7	29.0 ±9.0	38.3 ±11.8	71.1 ±22.0	45.6 ±14.3	30.8 ±9.5	29.2 ±9.0

E. ATMOSPHERIC MERCURY DEPOSITION

Since Hg is immobile in peat, we were able to estimate atmospheric deposition as a function of the Hg concentration (ng/g dry wt), the bulk density (g dry wt/cm^3) and the accretion rate (cm/year). The accretion rates were estimated from the Pb-210 ages at the bottom of each section (Figure 3).

Table 3 shows the Hg deposition rates for cores IB and IIIB integrated over the indicated time periods. A similar trend is evident in both cores, i.e., an increase in deposition after

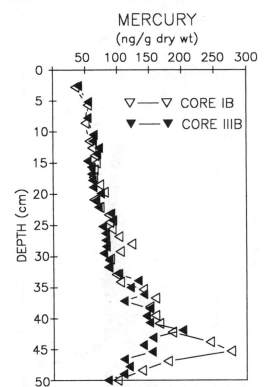

Figure 6 Mercury distribution in cores IB and IIIB.

1900, maximum deposition between 1935 and 1960, and a slight decrease after that. Due to the magnitude of the analytical error associated with the estimates, and the differences between the two cores, these trends have a high degree of associated uncertainty. However, it is noteworthy that apparent increases in deposition occurred as early as 1900 to 1935 relative to the previous 150 years. The mean pre-1900 deposition rate from the two cores is 7.0 µg/ m^2 · year.

The sectioning of cores IVB-VIB did not allow such fine-time resolution, but comparison with cores IB and IIIB illustrates the spatial variability of deposition within the bog. Figure 8 depicts deposition in all five cores over four time periods. A one-way ANOVA on log-transformed deposition data, followed by a Bonferroni multiple analysis ($\alpha = 0.05$), indicates that the mean deposition prior to 1935 is significantly different from later means. There is some suggestion of a decrease in deposition during the last decade compared to the mid-19th century, but the difference is not statistically significant. Atmospheric fluxes estimated over the period 1750 through 1935 include a low pre-1900 component and a higher 1900 through 1935 component, so they overestimate preindustrial fluxes.

The mean recent atmospheric deposition rate in Arlberg bog is higher than estimated from precipitation collections in north-central Wisconsin at a site about 150 miles away: 24.5 ± 7.9 µg/m^2 · year, vs. 11.4 ± 3.8 µg/m^2 · year.[43] There are several explanations for this difference. The Wisconsin estimate may be low due to the inability to sample every rain event. On the other hand, the bog estimate may be high due to direct uptake of gaseous mercury by living moss. The bias would carry through to all depths in the peat, so deposition would be overestimated for all dates. A third, most likely, explanation is that atmospheric mercury deposition varies regionally, and that the deposition at Arlberg bog is higher than at the more pristine site in Wisconsin. Regional sources of Hg to the atmosphere (e.g., Duluth, MN) affect inputs to Arlberg bog, and global or hemispheric changes in emissions should not be invoked to explain the observed changes.

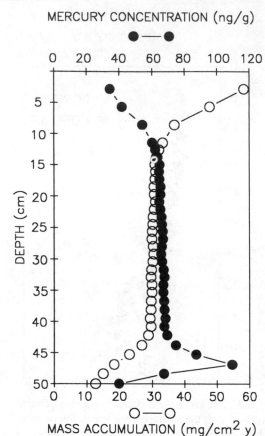

Figure 7 The peat accumulation rate in core IB (open circles) and the mercury concentration profile (solid circles) that results from an increase in atmospheric deposition from 5 μg/cm² · year between 48 to 50 cm to 10 μg/cm² · year between 46 to 48 cm to 20 μg/cm² · year above.

Pheiffer-Madson[16] reported an increase in peat-derived atmospheric Hg deposition after about 1860 in two Danish ombrotrophic bogs located 250 km apart, and attributed the trend to increased mercury emissions from industry, burning of fuel, or degassing from agricultural soils. Recent (1970s) deposition estimates were as high as 200 μg/m² · year, and pre-1860 estimates about an order of magnitude lower. The author noted that the more northern site generally exhibited higher rates, a point which underscores the regional nature of atmospheric mercury deposition. Similarly, Groet[44] observed highest concentrations of heavy metals in bryophytes collected in more highly industrialized portions of southern New England relative to northern sites, and a decreasing moss concentration gradient in Pb, Cu, Cr, and Ni with distance from New York City. Maps of regional differences in metal deposition in Scandinavia that are based on concentrations in moss carpets show distinct pattern of lowest rates in northern Norway and highest rates in southern Sweden.[45,46] The Hg contents of raw humus in Sweden show a similar trend.[47]

As mentioned above, land clearing in the Duluth area began as early as 1890. Removal of forests may have led to an increase of mercury evasion from exposed soils after this time. Increased coal combustion and industrialization in the region probably caused the increase in Hg deposition after 1935. If the slight decline during the last decade is real, it reflects a change in emissions on a regional scale.

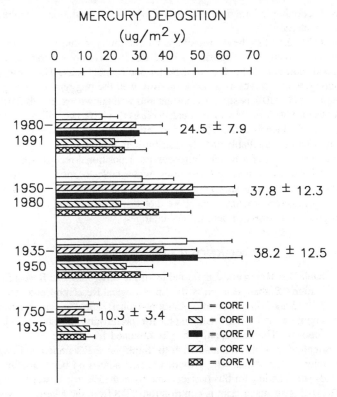

Figure 8   Bar graph depicting measured Hg deposition rates in cores IB, IIIB, IVB, VB, and VIB. The mean of the five cores and the mean error is given for each time period.

IV.   CONCLUSIONS

Al, Fe, and Mn are highly mobile in ombrotrophic bogs, and their vertical distributions can be attributed to redox conditions and water table fluctuations. Hg is much less mobile, and its concentration gradient with depth is best explained by changes in atmospheric deposition over time. Total inventories of Pb-210 and 250-year inventories of Hg indicate that Arlberg bog is a horizontally homogeneous collector of deposition over long periods of time. Horizontal variability was observed in estimates of deposition integrated over shorter time scales, but this variability was within the uncertainty associated with the estimates (at the 98% confidence level).

Mean atmospheric Hg deposition at the site shows a significant increase after 1935— from 10.3 ± 3.4 to 24.5 ± 7.9 µg/m^2 · year. The increase actually began as early as 1900, probably due to increased land clearing, coal burning, and industrialization in the area. There is indication of a slight decrease after 1980, but the difference is not statistically significant. Ombrotrophic bogs provide a powerful tool for monitoring changes in atmospheric mercury deposition. However, deposition recorded in peat bogs is sensitive to regional inputs, and study sites must be chosen carefully if they are to show larger-scale trends. Similarly, depositional information obtained for industrialized areas should not be extrapolated to more pristine areas.

The use of an average bulk density from different hummocks contributes a great deal to the associated uncertainty of deposition estimates. Measurement of bulk density and trace metal concentrations on subsections of a single core per hummock could help reduce this uncertainty. A large surface area corer is required to obtain a sufficient volume of material for analysis. Sectioning in larger increments will cut down on the time resolution, but will provide more material.

It is essential to establish the degree of spatial variability within a bog before attempting to assess temporal changes in deposition. Total Pb-210 inventories from several locations provide a good indicator of the degree of horizontal homogeneity. The corresponding Pb-210 deposition rate also serves as a check on how well the bog retains atmospheric inputs. Since atmospheric Pb-210 deposition is constant and well known, excess Pb-210 distributions and inventories provide a useful way to identify cores that have been disturbed (for example, by bioturbation). Such cores can not be used to estimate atmospheric Hg deposition because dating schemes will be unreliable and Hg distributions altered.

Vertical redistribution of a metal obliterates its depositional record. Ancillary measurements, e.g., redox-sensitive metal distributions, water table location, pH and pe gradients, and porewater sulfur concentrations supply information about peat post-depositional processes. These parameters provide clues about the degree of mobility of mercury and allow for an appropriate assessment of depositional changes.

ACKNOWLEDGMENTS

The authors thank Don Hobro and Liz Fumal from the Environmental Research Institute for metal determinations, Barbara Hansen of the Limnological Research Center for pollen analysis, Steve Chillrud and Mike Zibello of Lamont-Doherty Geological Observatory for Cs-137 gamma counting, and Kristin Chaloupka for performing digestions and platings for Pb-210 measurement. The high quality results obtained from those individuals made the present investigation possible. We also wish to thank Steve Eisenreich and Eville Gorham of the University of Minnesota, who provided expert advice on bog sampling and helped locate the study site. Funding for this project came from the Electric Power Research Institute Grant #RP202012. This manuscript is contribution #248 from the Marine Sciences Institute of the University of Connecticut.

REFERENCES

1. Wiener, J.G., Fitzgerald, W.F., Watras, C.J., and Rada, R.G., Partitioning and bioavailability of mercury in an experimentally acidified Wisconsin lake, *Environ. Toxicol. Chem.*, 9, 909, 1990.
2. Fitzgerald, W.F. and Gill, G.A., Depositional fluxes of mercury to the oceans, in *Int. Conf. Heavy Metals in the Environment*, Athens, 1, 79, 1985.
3. Wiener, J.G., Comparative analyses of fish populations in naturally acidic and circumneutral lakes in northern Wisconsin, FWS/OBS-80/40.16, U.S. Fish and Wildlife Service, Kearnegsville, WI, 1983.
4. Wisconsin Department of Natural Resources and Wisconsin Division of Health (WDNR/WDH), Health advisory for people who eat sport fish from Wisconsin waters, PUBL-IE-019, Madison, WI, 1988.
5. Wiener, J.G. and Stokes, P.M., Enhanced bioaccumulation of mercury, cadmium and lead in low-alkalinity waters—an emerging regional environmental problem, *Envrion. Toxicol. Chem.*, 9, 821, 1990.
6. Fitzgerald, W.F., Cycling of mercury between the atmosphere and oceans, in: *The Role of Air-Sea Exchange in Geochemical Cycling*, P. Buat-Ménard, Ed., D. Riedel Publishing, Dordrecht, 1986, 363–408.
7. Nriagu, J.O., A global assessment of natural sources of atmospheric trace metals, *Nature*, 338, 47, 1989.
8. Johnson, M.G., Culp, L.R., and George, S.E., Temporal and spatial trends in metal loadings to the Turkey Lakes, Ontario, Canada, *Can. J. Fish. Aquat. Sci.*, 43, 754, 1986.

9.   Meger, S.A. Polluted precipitation and geochronology of mercury deposition in lake sediments of northern Minnesota, *Water Air Soil Pollut.*, 30, 411, 1986.

10. Evans, R.D., Sources of mercury contamination in the sediments of small headwater lakes in south-central Ontario, Canada, *Arch. Environ. Contam. Toxicol.*, 15, 505, 1986.

11. Johnson, M.G., Trace element loadings to sediments of fourteen Ontario Lakes and correlations with concentrations in fish, *Can. J. Fish. Squat. Sci.*, 44, 3, 1987.

12. Rada, R.G., Powell, D.E., Wiener, J.G., and Winfrey, M.R., Recent increases in atmospheric deposition of mercury to north-central Wisconsin lakes inferred from sediment analyses, *Arch. Environ. Contam. Toxicol.*, 18, 175, 1989.

13. Schell, W.R., Sanchez, A.L., and Granlund, C., New data from peat bogs may give a historical perspective on acid deposition, *Water Air Soil Pollut.*, 30, 393, 1986.

14. Lee, J.A. and Tallis, J.H., Regional and historical aspects of lead pollution in Britain, *Nature*, 245, 216, 1973.

15. Hemond, H.F., Biogeochemistry of Thoreau's Bog, Concord, Massachusetts, *Ecol. Monogr.*, 50, 507, 1980.

16. Pheiffer-Madsen, P., Peat bog records of atmospheric mercury deposition, *Science*, 293, 127, 1981.

17. Rapaport, R.A. and Eisenreich, S.J., Atmospheric deposition of toxaphene to eastern North American derived from peat accumulation, *Atmos. Environ.*, 20, 2367, 1986.

18. Pakarinen, P. and Gorham, E., Mineral element composition of *Sphagnum fuscum* peats collected from Minnesota, Manitoba, and Ontario, in *Proc. Int. Peat Symp.*, Bemidji State University, Bemidji, MN, 1983, 417.

19. Glooschenko, W.A., Holloway, L., and Arafat, N., The use of mires in monitoring the atmospheric deposition of heavy metals, *Aquat. Bot.*, 25, 179, 1986.

20. Norton, S.A., The stratigraphic record of atmospheric loadings of metals at ombrotrophic Big Heath Bog, Mount Desert Island, Maine, USA, in: *Effects of Atmospheric Pollutants on Forests, Wetlands and Agricultural Ecosystems*, T.C. Hutchinson and K.M. Meema, Eds., Springer-Verlag, Berlin, 1987, 561.

21. Glooschenko, W.A., Monitoring the atmospheric deposition of metals by use of bog vegetation and peat profiles, *Adv. Environ. Sci. Technol.*, 17, 507, 1986.

22. Livett, E.A., Geochemical monitoring of atmospheric heavy-metal pollution—theory and applications, *Adv. Ecol. Res.*, 18, 65, 1988.

23. Damman, A.W.H., Distribution and movement of elements in ombrotrophic peat bogs, *Oikos*, 30, 480, 1978.

24. Damman, A.W.H., Hydrology, development, and biogeochemistry of ombrogenous peat bogs with special reference to nutrient relocation in a western Newfoundland bog, *Can. J. Bot.*, 64, 384, 1986.

25. Gorham, E., A comparison of the lower and higher plants as accumulators of radioactive fallout, *Can. J. Bot.*, 37, 327, 1959.

26. Rühling, A. and Tyler, G., Sorption and retention of heavy metals in the woodland moss *Hylocomium splendens* (Hedw.) Br. et Sch., *Okos*, 21, 92, 1970.

27. Huckabee, J.W., Mosses: sensitive indicators of airborne mercury pollution, *Atmos. Environ.*, 7, 749, 1973.

28. Jensen, A., Historical deposition rates of mercury in Scandinavia estimated by dating and measurement of mercury in cores of peat bogs, *Water Air Soil Pollut.*, 56, 769, 1991.

29. Bloom, N. and Fitzgerald, W.F., Determination of volatile mercury species at the nanogram level by low temperature gas chromatography with an ultrasensitive cold-vapor atomic fluorescence detector, *Anal. Chim. Acta*, 51, 1714, 1988.

30. Gill, G.A. and Fitzgerald, W.F., Picomolar mercury measurements in seawater and other materials using stannous chloride reduction and two-stage gold amalgamation with gas phase detection, *Mar. Chem.*, 20, 227, 1987.

31. Flynn, W.W., The determination of low levels of polonium-210 in environmental materials, *Anal. Chim. Acta,* 43, 221, 1968.
32. Oldfield, F., Appleby, P.G., Cambray, R.S., Eakins, J.D., Barber, K.E., Battarbee, R.W., Pearson, G.R., and Williams, J.M., Pb-210, Cs-137, and Pu-239 profiles in ombrotrophic peat, *Oikos,* 33, 40, 1979.
33. Robbins, J.A., Geochemical and geophysical applications of radioactive lead, in: *The Biogeochemistry of Lead in the Environment,* Vol. 1, J.O. Nriagu, Ed., Elsevier, New York, 1978, 285.
34. Urban, N.R., Eisenreich, S.J., Grigal, D.F., and Schurr, K.T., Mobility and diagenesis of Pb and Pb-210 in peat, *Geochim. Cosmochim. Acta,* 54, 3329, 1990.
35. Charles, M.J. and Hites, R.A., Sediments as archives of environmental pollution trends, in: *Sources and Fates of Aquatic Pollutants,* American Chemical Society, Washington, D.C., 365, 1987.
36. Larsen, A.M., *History of the White Pine Industry in Minnesota,* University of Minnesota Press, Minneapolis, 1949.
37. Ingram, H.A.P., Soil layers in mires, function and terminology, *J. Soil Sci.,* 29, 224, 1978.
38. Clymo, R.S., Ion exchange in *Spagnum* and its relation to bog ecology, *Ann. Bot.,* 27, 309, 1963.
39. Schecher, W.D. and McAvoy, D.C., MINEQL⁺ Version 2.1: A Chemical Equilibrium Program for Personal Computers, The Proctor and Gamble Co., Cincinnati, OH, 1991.
40. Grigal, D.F. and Nord, W.S., Investigations of heavy metal in Minnesota peatlands, Report to Minn. Dept. of Natural Resources, Div. of Minerals, St. Paul, 1983.
41. Urban, N.R., Eisenreich, S.J., and Grigal, D.F., Sulfur cycling in a forested *Sphagnum* bog in northern Minnesota, *Biogeochemistry,* 7, 81, 1989.
42. Helmer, E.H., Urban, N.R., and Kisenreich, S.J., Aluminum geochemistry in peatland waters, *Biogeochemistry,* 9, 247, 1990.
43. Fitzgerald, W.F., Mason, R.P., Vandal, G.M., and Dulac, F., Air-water cycling of mercury in lakes, Int. Conf. Mercury as a Global Pollutant, Monterey, CA, May 31-June 4, 1992.
44. Groet, S.S., Regional and local variations in heavy metal concentrations of bryophytes in the northeastern United States, *Oikos,* 27, 445, 1976.
45. Rūhling, A. and Tyler, G., Regional differences in the deposition of heavy metals over Scandinavia, *J. Appl. Ecol.,* 8, 497, 1971.
46. Rūhling, A. and Tyler, G., Heavy metal deposition in Scandinavia, *Water Soil Air Pollut.,* 2, 445, 1973.
47. Lindqvist, O., Johansson, K., Aastrup, M., Andersson, A., Bringmark, L., Hovsenius, G., Häkanson, L., Iverfeldt, A., Meili, M., and Timm, B., Mercury in the Swedish environment, recent research on causes, consequences and corrective measures, *Water Air Soil Pollut.,* 55, 49, 1991.

Air-Water Cycling of Mercury in Lakes

W. F. Fitzgerald, R. P. Mason, G. M. Vandal, and F. Dulac

CONTENTS

ABSTRACT: Global atmospheric chemical cycling of Hg and exchange at air-water interfaces are preeminent processes affecting the mobilization of Hg at the earth's surface. The principal transfer of Hg from the terrestrial to the marine environment occurs by eolian mechanisms. Atmospheric Hg deposition is also the primary input to many fresh water systems. As part of our long-term studies of mercury in the environment, we are investigating the tropospheric cycling, deposition, and air-water exchange of Hg in the mid-continental lacustrine environs of north-central Wisconsin. This work is part of the Mercury in Temperate Lakes (MTL) Program, which is a multidisciplinary examination into the reactions governing the aquatic biogeochemistry of Hg in temperate regions (1988–1991). The studies are continuing in the Mercury Accumulation, Pathways, and Processes (MAPP) Program. Recent analytical developments and trace-metal-free methodologies are utilized in the determination of total Hg (Hg_T), reactive Hg (Hg_R), inorganic Hg [Hg(II)], elemental Hg (Hg^0), and alkylated Hg species [monomethyl-Hg (MMHg); dimethyl-Hg (DMHg)] at the picomolar to femtomolar level in air, water, and precipitation. Preliminary results were reported at the first international meeting on Mercury as an Environmental Pollutant, Gävle, Sweden, 1990. The full papers appear in the conference volume.

Briefly, we reported that the total atmospheric Hg deposition (Hg_T) of circa 10 $\mu g\ m^{-2}\ year^{-1}$ (ca. 66% wet) observed during the first year (October 1988/October 1989) readily accounts for the total mass of Hg in fish, water, and accumulating in the sediments of Little Rock Lake, a temperate seepage lake. The mass balance was well constrained and suggested that small increases in atmospheric Hg loading could lead directly to elevated levels in the fish stock. Atmospheric deposition of MMHg, however, appeared insufficient to account for the amounts of MMHg observed in biota; thereby, indicating the need for an in-lake synthesis of MMHg. Preliminary estimates of scavenging ratios suggested that the Hg concentrations in rain were due to scavenging of atmospheric particulate Hg. Gaseous Hg in the atmosphere and lake water was principally Hg^0, and the evasional fluxes of Hg^0 were significant. Moreover, we postulated as part of the reactive Hg(II) substrate hypothesis that in-lake production and efflux of Hg^0 could provide a potential buffering and/or amelioration role in aqueous systems. We hypothesized that in-lake biological and chemical production processes for Hg^0 and MMHg compete with one another for the reactive Hg substrate which we suggest is labile Hg (II) species.

Here, we are presenting data from further investigations. In general, the results strengthened our hypotheses regarding: (1) the preeminence of the atmospheric particulate Hg cycle in delivering Hg to aquatic systems, (2) the prominence of Hg^0 in the processes controlling the behavior and fate of Hg in natural waters, and (3) the Hg (II) substrate biogeochemistry as a unifying physicochemical paradigm. In particular, while the Hg_T deposition at about 13 $\mu g\ m^{-2}\ year^{-1}$ (ca. 69% wet) found during October 1989/October 1990 was similar to the initial results, the Hg_R fraction [Hg(II) substrate] was much larger (circa 3 times). Correspondingly, substantially higher in-lakes values of Hg^0 supersaturation were observed in August 1990 compared to August 1989, and the estimated average evasional fluxes of Hg^0 were about three times greater and quite significant. A coupling between the production of Hg^0 and the supply/availability of Hg_R is consistent with the Hg(II) substrate hypothesis. We have found a similar linkage in the marine environment. At present, the principal sources of the high levels of Hg^0 supersaturation in surface waters are not known. In a preliminary model, however, we demonstrate that abiotic reduction of Hg_R can provide a sufficient source and important pathway for the in-lake production of Hg^0.

I. INTRODUCTION

Atmospheric processes such as mobilization, chemical oxidation, and deposition play preeminent roles in the geochemical cycling of Hg. Mass balance models indicate that atmospheric deposition is the predominant source of Hg to lacustrine systems and the open ocean (see for example, Gill and Fitzgerald, 1987; Fitzgerald et al., 1991; Mierle, 1990; Lindqvist et al., 1991). It is also clear that in situ production and water-air transfer of elemental Hg (Hg^0) exerts a major influence on the behavior and fate of Hg in the environment. Marine studies demonstrated that in situ synthesis of volatile Hg, which is principally Hg^0 in the mixed layer (Kim and Fitzgerald, 1986; Mason and Fitzgerald, 1991) and its subsequent evasion at the water-air interface were major features of the global Hg cycle (Fitzgerald et al., 1984; Fitzgerald, 1989; Iverfeldt, 1988). Most recently, fresh water investigations by Vandal et al., 1991, in Wisconsin and Xiao et al., 1991, in Sweden have shown a similar and important in-lake Hg^0 cycle which yields significant Hg^0 fluxes to the atmosphere. In

1990, we proposed a reactive Hg(II) substrate hypothesis to provide a consistent physico-chemical explanation for the speciation of Hg in natural waters and a unifying view of the biogeochemical cycling of Hg in aquatic systems (Fitzgerald et al., 1991). We postulated that in-lake biological and chemical production processes for Hg^0 and monomethyl-Hg (MMHg) compete for the reactive Hg substrate which we suggested was labile Hg (II) species. We also hypothesized that once Hg^0 is produced in the aqueous phase, it is unreactive and eventually lost from the system by evasion. Thus, lakes with limnological conditions favoring Hg^0 production would be less likely to have elevated levels of Hg in the fish stock, and would show smaller accumulations of Hg in sediment. The physicochemical availability of the substrate would be affected by inorganic and organic chemical components within a lake. For example, Vandal et al., 1991, found a strong direct relationship between pH and degree of saturation for Hg^0. Aspects of the Hg(II) substrate hypothesis were offered by Fagerström and Jernelöv in 1972, and Winfrey and Rudd, 1990, outlined a similar biogeo-chemical Hg cycle for freshwater lakes.

Here, we present new information from a continuing examination of the atmospheric Hg cycle and its influence on the behavior and fate of Hg in mid-continental seepage lakes. These investigations were part of the Mercury in Temperate Lakes Program (MTL) in north-central Wisconsin, U.S. (1988–1991). Previous work appears in Fitzgerald et al., 1991; Vandal et al., 1991; and Mason et al., 1992. Our research is continuing in the Mercury Accumulation, Pathways, and Processes (MAPP) Program. Major effort has been devoted to the chemical speciation of Hg in air and in deposition and its effect on the in-lake cycling and air-water exchange of Hg. Particular emphasis has been given to exploring potential linkages between the chemical forms of atmospherically derived Hg and the aquatic biogeo-chemistry of Hg, especially the Hg^0 cycle. We found substantially higher degrees of Hg^0 supersaturation during August 1990 investigations of the study lakes as compared to earlier results (Vandal et al., 1991). Moreover, the estimated average evasional fluxes of Hg^0 in the August 1990 study were quite significant and about three times greater than August 1989; and this increase coincided with a three times increase in the atmospheric input of reactive Hg (Hg_R). The total Hg (Hg_T) input, however, was approximately the same in August of both years. This apparent coupling between the production of Hg^0 and the supply/availability of Hg_R is consistent with the Hg(II) substrate hypothesis. Moreover, a similar correlation is evident in the marine environment (Mason et al., 1992). Although we have not yet identified the principal sources of the high levels of Hg^0 supersaturation in surface waters, we dem-onstrate using a preliminary model that abiotic reduction of Hg_R can provide a sufficient source and important pathway for the in-lake production of Hg^0.

II. EXPERIMENTAL

A brief description of the field setting and laboratory activities will be provided in this chapter. Background information on our prior lacustrine-related atmospheric Hg research, studies of volatile Hg and particulars concerning the field programs, geographic location, experimental techniques, collection methods, and analysis can be found in the following works: Fitzgerald and Watras, 1989; Wiener et al., 1990; Fitzgerald et al., 1991; Vandal et al., 1991; and Mason et al., 1992.

A. STUDY REGION

This Hg program has focused on representative clear-water seepage lakes in the sparsely populated temperate rural Northern Highland Lake District of Wisconsin, near Trout Lake in Vilas County (46°00'N, 89°40'W). The geology is dominated by outwash sands and deep glacial tills, and anthropogenic influences are limited. The lakes chosen for study are mod-erately productive (oligotrophic-mesotrophic) and provide a range of pH (4.7 to 7.2), ALK (-5 to 128 μeq L^{-1}), and dissolved organic C (1.5 and 5.6 mg L^{-1}). These variables have

Table 1   **Limnological features of the study lakes in northern Wisconsin.**

Lake	Maximum depth (m)	Area (ha)	pH	ALK (μeq L^{-1})	DOC (mg L^{-1})
Little Rock Reference (LRLR)	6.5	8.1	6.1	26.4	3.8
Little Rock Treatment (LRLT)	10.3	9.8	4.7[a]	−7	2.6
Max	6.1	9.2	5.2	−5	2.3
Vandercook (VDC)	7.2	43.6	6.1	21	3.2
Crystal (CRL)	20.0	37	6.3	9	1.7
Pallette (PAL)	18.2	70	7.25	128	5.06[b]

[a] pH = 4.7 in August 1990.
[b] TOC.

been correlated with the bioaccumulation of Hg in natural waters (Scheider et al., 1979; Hakanson, 1980; Sloan and Schofield, 1983; Lindqvist et al., 1991; Wiener, 1987). The principal limnological characteristics for each lake are presented in Table 1. Additional information on the region and many of its lakes can be found in Eilers et al., 1983, Magnuson et al., 1984, and Watras and Frost, 1989.

Little Rock Lake has been acidified experimentally and received the most detailed study. In 1985, it was divided with a sea curtain into two basins (Table 1), one of which is untreated (reference pH: 6.1) while the other was experimentally acidified. The acidification stopped in 1990 and the current pH is 5.2. Further information on the Little Rock Lake Acidification Project is available in Magnuson et al., 1984; Brezonik et al., 1986; and Watras and Frost, 1989.

B. SAMPLING PERIODS: ELEMENTAL Hg

The lakes have been studied under a variety of limnological conditions. Vandal et al., (1991) reported results from the following seasonal periods: summer (peak stratification), fall (following turnover), and late winter (under ice). This paper focuses on the late summer peak stratification conditions and compares results from August 10 to 24, 1989 with the August 22 to 29, 1990 experiments on the six study lakes (LRLR, LRLT, Max, VDC, CRL, and PAL).

C. SAMPLING PERIODS: ATMOSPHERIC Hg

Atmospheric measurements of gaseous and particulate Hg were made in support of the lake studies. Precipitation was collected on an event basis throughout the year.

III. ANALYTICAL

A. COLLECTION SYSTEMS: AIR

The sampling apparatus and protocols are presented in Fitzgerald et al., 1991. Briefly, air collections for Hg were made using an atmospheric sampling system similar to the one we developed and employed in oceanic studies [Fitzgerald and Gill, 1979; Fitzgerald, 1986, 1989 (review paper)]. The apparatus was installed on a dock at Max Lake. Previous measurements were made at the reference basin of Little Rock Lake (LRLR). The atmospheric Hg sampling stacks consist of the following trapping materials.

- Au: yields a total gaseous Hg determination (TGM).
- Carbotrap®/Au: provides a collection that can be used for direct Hg speciation determinations by GC separation and atomic fluorescence detection (CTS).
- Quartz wool plug is for particulate Hg determinations (PM).

Additional details associated with these various trapping materials can be found in Fitzgerald and Gill, 1979, Kim and Fitzgerald, 1986, and Bloom and Fitzgerald, 1988. Most samples for TGM, PM, and speciation of the gaseous phase (CTS) were collected at Max Lake and analyzed at the Trout Lake Laboratory. Some samples were collected by personnel from the Wisconsin Department of Natural Resources (WDNR) and shipped to the University of Connecticut (UCT) for analysis. Sample volumes ranged from 0.5 to 2.0 m^3 and were collected over a 24-h period.

B. AIR: ANALYSIS

The TGM analyses are conducted by the two-stage gold amalgamation technique (Fitzgerald and Gill, 1979) with detection by atomic fluorescence spectroscopy (AFS). Currently, the overall atmospheric TGM methodology yields a detection limit of about 0.15 ng m^{-3} and a precision of about 15% at 1.5 ng m^{-3}. Particulate Hg is determined following pyrolysis of the quartz wool plug and trapping on gold. Chemical speciation of the gaseous phase is achieved through analysis of the graphitized substrate (Carbotrap®) collections using gas chromatographic separation and AFS detection (Bloom and Fitzgerald, 1988). The detection limit for the determination of monomethyl-Hg and dimethyl-Hg is about 5 pg m^{-3}.

C. COLLECTION SYSTEMS: DEPOSITION—RAIN

All rainfall was collected on an event basis by WDNR personnel employing ultraclean techniques. The rain collection apparatus was operated from a dock either at the reference basin of Little Rock Lake (1989) or at Max Lake (1990). During 1989, custom-made Pyrex® glass and solid Teflon® collectors were used (details in Fitzgerald et al., 1991). A new, lighter funnel, constructed from a molded Teflon® sheet, has been used since 1990. Further particulars are reported in Mason et al., 1992. We do emphasize, however, that the funnel is contained in an acrylic housing, with a removable acrylic lid. It is designed so that rain entering the funnel contacts only Teflon® parts which were rigorously acid-cleaned prior to use.

Teflon® bottles used in rain collections were boiled in aqua regia, followed by a number of additional cleaning steps (Gill and Fitzgerald, 1985). After cleaning, they were filled with 0.5% low-Hg HCl solution, hermetically sealed, and double-bagged in acid-cleaned polyethylene bags. The Teflon® funnel was soaked in a heated bath of 10% HCl for at least 4 weeks prior to use. The funnel was always rinsed before use with distilled deionized water (<3 pM Hg) from the University of Connecticut. The water (Q water) was first distilled, then deionized using a Barnstead NANOpure IIR cartridge deionization system prior to a final subboiling distillation using a quartz still. The sample water used to rinse the funnel (funnel blank) was collected and analyzed along with a sample of the rinse water. Thus, contamination from the funnel itself could be determined.

In between rain events the acrylic housing was covered by large clean polyethylene bags. Prior to sampling, the bags and the acrylic lid were removed, carefully stored, and the funnel rinsed with Q water. The collecting personnel wore rainsuits, used long polyethylene gloves, and were always downwind of the rain collector. The 2-L bottle containing the rinse water was used as the collection bottle in most instances. The time between removal of the lid and collection of the rain was minimized to prevent particle contamination. The funnel is rinsed with about 1 L of the rinse water and the second 1-L aliquot of water is used to take a funnel blank. Half of the aliquot is rinsed through the funnel and collected and the other half is retained without passing through the funnel. As noted, these samples are used to test and monitor the funnels for artifacts and potential contamination. In general, there is no discernable difference between the concentration of Hg in the rinse water and the funnel blank. The funnel is exposed to the atmosphere for the duration of the event. Following the sampling, the rain collectors are carefully covered. An aliquot of rain for ethylation/speciation determinations is decanted into a 100-mL bottle prior to sample acidification. The 100-mL aliquot is frozen and shipped separately to the UCT for analysis.

D. SNOW

Snow samples were taken by scooping snow directly into 1-L Teflon® jars as soon as possible after a significant snowfall (at least 3 in fresh snow). Great care is taken to maintain sample integrity during collection and shipment. During snow collection operations, the sampling personnel wear particle-free nylon suits (reserved for snow collection) and use arm-length plastic gloves. Snow collecting consists of carefully scooping snow directly into the clean Teflon® jar. Snow was collected by a person wearing long polyethylene gloves while moving into the wind and away from possible contamination. The jars were wrenched tight, stored, and transported frozen to the University of Connecticut for analysis.

The snow samples for reactive Hg and total Hg determinations are acidified with low-Hg reagent grade HCl in our class 100 clean laboratory, resealed, and allowed to thaw at room temperature. The sample for chemical speciation is allowed to thaw without acidification. The samples are analyzed immediately after thawing.

E. PRECIPITATION: ANALYTICAL

Three different procedures are used in the analysis of rain and snow samples and these measurements provide information on the forms of Hg in precipitation. All procedures rely on the production of volatile Hg species in solution, which are purged from solution with inert Hg-free gas and trapped on adsorbing substrate. The determination of "reactive" or "acid-labile" Hg (= Hg_R) involves sample acidification to pH = 1, reduction of ionic Hg and labile Hg to Hg^0 with $SnCl_2$, aeration of the solution, and collection on Au (Gill and Fitzgerald, 1987a). The determination of "total" Hg (Hg_T) is similar to the reactive Hg procedure after sample pretreatment with a strong oxidant, BrCl, followed by reduction of the BrCl with NH_2OH-HCL prior to $SnCl_2$ reduction, sparging, and collection on Au (Bloom and Crecelius, 1983). Oxidation of the solution with BrCl will destroy many strong organometal associations and decompose MMHg, rendering bound Hg available for $SnCl_2$ reduction. The procedural blank for the reactive Hg determination is 0.025 ± 0.010 ng and the sample size is generally 250 mL, resulting in a detection limit (defined as three times the standard deviation of the blank) of 0.5 pM. The blank associated with the oxidation technique is 0.1 ± 0.03 ng for a 250 mL sample with a corresponding detection limit of 2 pM.

Quantification and identification of Hg species in precipitation involves the ethylation of dissolved Hg in solution using sodium tetraethylborate. The volatile ethyl-Hg derivatives, as well as other volatile Hg species in solution (i.e., Hg^0 and dimethylmercury), are purged from solution, concentrated on Carbotrap®, separated by cyrogenic GC, and detected by atomic fluorescence. The details of the procedure are outlined in Bloom, 1989. This method allows for the identification and measurement of MMHg as methylethylmercury, labile Hg(II) as diethylmercury, as well as dimethylmercury and elemental Hg^0 without ethylation. The detection limits are 0.05 pM for methylmercury, and 0.1 pM for Hg(II).

F. LAKE WATER COLLECTIONS: DISSOLVED
   GASEOUS Hg

Details associated with our investigations of the cycling of volatile Hg will be found in Vandal et al., 1991. Briefly, the lakes were sampled in intensive experiments conducted in August of 1989 and 1990, in summer stratification. Water was collected by hand with a 8-L Teflon®-coated Go-Flo® sampling bottle using Kevlar® line and a Teflon® weight and

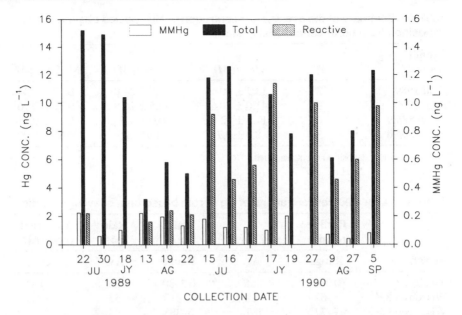

Figure 1 Total, reactive, and methylmercury in rain collected at Little Rock Lake Reference Basin in 1989 and Max Lake in 1990, adapted from Fitzgerald et al., 1991, and Mason et al., 1991.

messenger. The bottle integrates about a 1-m depth range. Sampling was conducted from a fiberglass boat. Usually, one lake was sampled on a given day and the water analyzed immediately (circa 1 to 3 h) in the clean laboratory at the University of Wisconsin Limnological Center at Trout Lake. The samples were maintained near their ambient temperature until analysis.

G. DISSOLVED GASEOUS Hg: ANALYSIS

Dissolved gaseous mercury (DGM) measurements are made by sparging the volatile species from solution using Hg-free argon and trapping on either Au or Carbotrap® (CTS) without pretreatment of the sample. The collection of Hg on Au allows for the measurement of total DGM and the CTS collection allows for separation and identification of the volatile dissolved species by cryogenic GC with AFS detection. The samples were purged in a 2-L Pyrex® bubbler, and a total sample volume of 4 L was used for each determination. The detection limit for total DGM is 5 fM (femtomolar), and 3 fM for dimethylmercury.

IV. RESULTS AND DISCUSSION

A. ATMOSPHERIC CYCLING OF Hg: PRECIPITATION

Since atmospheric deposition is the predominant flux of Hg to lacustrine systems and the open ocean, a knowledge of the speciation of Hg in precipitation is essential to our understanding of the biogeochemical cycling of Hg at the earth's surface. Measurements of Hg_T, Hg_R, and MMHg in rain from the mid-continental rural temperate lacustrine environs of north-central Wisconsin are summarized in Figure 1. A 2-year summary (1989 and 1990) of Hg speciation results in wet deposition (snow and rain) is presented in Table 2.

B. PARTICULATE Hg SCAVENGING

There are several general features evident in the broad geochemical view provided by the average speciation results in Table 2. Firstly, the average Hg_T was similar during both rain

Table 2   **Summary of the average concentration of Hg species observed in wet deposition from north-central Wisconsin.**

Sample period	n^a	Hg_R (pM)	Total (pM)	MMHg (pM)
Rain 1989	12	13.7 ± 10.6	52.5 ± 24.0	0.78 ± 0.34
Rain 1990	9	41.0 ± 20.6	49.3 ± 20.8	0.37 ± 0.16
Snow 88/89	6	17.5 ± 12	30.0 ± 4.5	0.24 ± 0.11
Snow 89/90	3	8.0 ± 0.75	14.9 ± 3.9	0.52 ± 0.20

[a] n = number of deposition events sampled.
Adapted from Mason et al., 1992.

Table 3   **Calculated concentrations of Hg in rain based on a scavenging ratio.**

Region	Particle (pg/m³) Range	Av.	Calc. conc. Range	Rain (pM) Av.	Measured conc. (pM) Av.
Pacific	0.2–6	2.6	0.2–30	6	14
Wisconsin 89[a]	5–62	22	4–310	55	51
Wisconsin 90[a]	7–77	37	6–385	93	49

[a] The average particulate concentration and range (all months) found in Wisconsin during 1989 and 1990.

Note: W = 600 (range 200–1200) and W = [C_{rain} (pg/kg) $\times$ 1.2 kg/m³] $\div$ C_{air} (pg/m³). The average values were calculated using an average scavenging ratio (W) of 600 while the variability was estimated using a range for W of 200 to 1200 and the actual particulate concentration extremes found at these sites. The values for W were taken from the data for Pb reported by Maring et al. (1989).

seasons (52.5 ± 24.0 and 49.3 ± 20.8 pM) while the average Hg_R was higher during 1990 (41.0 ± 20.6 pM) than during 1989 (13.7 ± 10.6 pM). As we have noted previously (Fitzgerald et al., 1991), the difference between Hg_T and Hg_R is not due to MMHg, which is present in very small quantities (≤ 1 pM; see Figure 1). Rather, we suggest that it is indicative of Hg associated with atmospheric particulates, some of which may be strongly bound to organic substances. The influence of atmospheric particulate Hg on the composition in rain (i.e., particulate Hg scavenging hypothesis) was demonstrated using the scavenging ratio approach by Fitzgerald et al., 1991, Iverfeldt, 1991, and Brosset and Lord, 1991 showed that the filterable or coloidal species dominate the Hg_T in rain. Predicted Hg_T concentrations in rain based solely on the scavenging of atmospheric particulate Hg are summarized in Table 3. Recent data from the equatorial Pacific are also included to illustrate and emphasize the integral role particulate Hg plays in determining the composition of Hg in rain.

Further, and as discussed in Mason et al., 1992, the differences in the Hg content between the equatorial Pacific and mid-continental rains correspond to the atmospheric particulate Hg distribution between these regions. For example, the average particulate Hg concentration was 2.6 pg/m³ during the 1990 cruise in the equatorial Pacific and 22 and 37 pg/m³ for Wisconsin in 1989 and 1990, respectively. Moreover, estimates of expected rain concentrations due solely to particulate Hg scavenging are comparable to the observations at both locations (Table 3).

These predictions suggest that the variation in Hg_R and Hg_T between regions and the temporal differences at Wisconsin are a function of the differences in particulate Hg composition and burden. As we have suggested previously (Fogg and Fitzgerald, 1979; Fitzgerald et al., 1983), this depositional behavior indicates that while most of the Hg in the atmosphere

Table 4   **Annual Hg depositional fluxes (μg m^{-2} year^{-1}) in north-central Wisconsin between October 1988 and October 1990.**

Mercury	Wet deposition		Dry deposition	
	1988/89	1989/90	1988/89	1989/90
Total	6.8 ± 2.0	8.7 ± 3.7	3.5 ± 3.0	3.9 ± 3.8[a]
React.	2.5 ± 1.3	7.1 ± 3.5		
Methyl	0.09 ± 0.03	0.07 ± 0.03		

[a] Estimated from a yearly average of 25 ± 23 pg m^{-3} for particulate Hg obtained from the measurements made in the winter and spring of 1989 and the summer data for 1990. A depositional velocity of 0.5 cm s^{-1} is used (Fitzgerald et al., 1991).

is Hg^0 (>97%) it is not oxidized and solubilized in processes leading directly to the formation of precipitation. A more general gas-to-particle atmospheric oxidation mechanism is implicated. Atmospheric Hg deposition is thus analogous to other trace metals, such as Pb, which exist as particles in the atmosphere (Maring et al., 1989; Buat-Menard, 1985).

C. Hg SPECIATION IN PRECIPITATION: GEOCHEMICAL SIGNIFICANCE

Although the washout calculations indicate that oxidation of Hg^0 during raindrop formation and subsequent deposition is not significant, the work of Iverfeldt and Lindqvist, 1986, suggests that oxidation of Hg^0 in clouds could be an important mechanism contributing to Hg in rain. The authors predict an oxidation rate of 0.01 h^{-1} (88 year^{-1}) for conversion of Hg^0 to Hg(II) in clouds, assuming 1 g m^{-3} of liquid water and 23 ppb ozone (a value similar to observed background concentrations). In the absence of ozone, the reaction rate is three orders of magnitude slower. However, a residence time of Hg in the atmosphere of approximately 1 year (Fitzgerald, 1989) based on the global cycle yields an overall conversion rate of approximately 1 year^{-1}, assuming all the Hg in rain is derived from oxidation of Hg^0. This calculation represents a maximum conversion rate. Therefore, the reaction investigated by Iverfeldt and Lindqvist, 1986, which has a substantially larger rate constant, should not be a predominant mechanism for the oxidation of Hg^0 in the atmosphere if the reaction occurs at rates comparable to those found in the laboratory. Munthe and Lindqvist, 1989, and Munthe, 1992, modified this model by suggesting that rapid sulfite (SO$_3^=$) complexation of Hg^{2+} in cloud water would yield [Hg(SO$_3$)$_2^=$] with subsequent reduction of Hg^{2+} to Hg, thereby serving as a potential reverse mediation reaction to limit the net amount of Hg^0 solubilized. However, insufficient amounts of atmospheric S species are available in the atmosphere over most of the earth's surface (i.e., oceans). Thus, other gas-to-particle conversion processes must be providing pathways for formation of the Hg$_R$ compounds found in rain.

As summarized in Table 2, there was a substantial difference between 1988/1989 and 1989/1990 in the Hg$_R$ composition in wet deposition and the associated yearly supply of Hg$_R$ to the lakes. A comparison of the estimated inputs is shown in Table 4. Approximately 3 times more active substrate was introduced in 1989/1990 relative to 1988/1989, yet the Hg$_T$ inputs between years were similar. Such a change in the delivery of readily available Hg(II) substrate to the lakes should have profound short-term implications for the overall biogeochemical cycling of Hg, particularly to the in-lake synthesis of MMHg and Hg^0. Indeed, a direct linkage between deposition of Hg$_R$ and in-lake values of Hg^0 can be demonstrated from the experiments conducted during August of 1989 and 1990.

D. ELEMENTAL Hg CYCLING

As shown by Vandal et al., 1991, the in-lake cycling of Hg^0 is quite important and the evasional fluxes are geochemically significant. In general, the dissolved gaseous Hg (DGM)

Table 5   **Summary: degree of saturation for elemental Hg (Hg^0) in north-central Wisconsin lakes.**

Lake	Date	Hg^0 (fM) range	S (%) range	S (%) Mean and std. dev.
Little Rock—Treat.	August, 1989	83–107	305–345	325 ± 28
Little Rock—Ref.		135–200	500–740	620 ± 170
Pallette		60–355	140–1180	605 ± 530
Vand.		85–163	315–600	458 ± 202
Max		283–297	990–1100	1045 ± 78
Little Rock—Treat.	August, 1990	181–490	920–2790	1703 ± 971
Little Rock—Ref.		214–358	1220–2040	1536 ± 441
Pallette		92–640	340–3650	1476 ± 1370
Vand.		281–570	1630–3250	2440 ± 810
Max		182–546	1040–3110	2075 ± 1035
Crystal		90–785	350–4460	1583 ± 1953
Russett		179–1035	660–4060	2033 ± 1484

fraction in lake waters consists principally of Hg^0 under all sampling conditions, with no significant contribution from volatile organic Hg species, i.e., dimethyl-Hg [detection limit of 3 fM]. Experimentally, the air-water partitioning of Hg^0 is determined from simultaneous measurements of Hg^0 in the atmosphere and in the lakes. The degree of saturation (%S) for Hg^0 in lake water relative to the appropriate temperature-corrected equilibrium with the atmosphere (Sanemasa, 1975) is determined using the following relationship:

$$\%S = [(C_{water} \times H)/C_{air}] \times 100$$
$$\%S > 100 = \text{supersaturation in water}$$
$$H = \text{Henry's Law Constant for } Hg^0$$
C_{water} and C_{air} = the concentration of Hg^0 in water and air, respectively

A summary of the 1989 and 1990 Hg^0 data from the August studies in the Wisconsin seepage lakes is given in Table 5. It is evident and geochemically significant that, in 1990, the lakes were highly supersaturated, with values near the surface ranging from 10.4 (S = 1040%) in Max to 44.6 (S = 4460%) times the equilibrium level in Crystal Lake. In August 1989 supersaturation ranges were from about 1.4 to 12 times the saturation concentration. Thus, these large %S values will translate into higher lake-to-atmosphere fluxes of Hg^0 than reported previously (Vandal et al., 1991). The water-air transfer of Hg^0 is estimated from the thin film gas exchange model using the following relationship:

$$F = K (C_{air} \times H^{-1} - C_{water})$$

where F = gaseous Hg flux into (+) or out of (−) the lake
   C_{air} = air concentration of Hg^0
C_{water} = water concentration of Hg^0
    H = Henry's Law Constant
    K = transfer velocity, 1.5 cm h^{-1} (0.36 m d^{-1}) for August 1989 and 1990
        (21 to 24° C).
A detailed explanation of the lake-air Hg^0 exchange calculation is given in Vandal et al., 1991.

Table 6   **Estimated average evasional fluxes for August 1989 and August 1990 for various north-central Wisconsin lakes.**

Lake	Evasional flux in August 1989 (pmol/m^2 day)	Evasional flux in August 1990 (pmol/m^2 day)
Little Rock — Treat.	25	167
Little Rock — Ref.	50	120
Pallette	85	221
Vandercook	50	92
Max	98	57
Russett	—	143
Crystal	—	274
Average	62 ± 30	153 ± 75
Depositional flux (Hg$_T$)	85	260
Depositional flux (Hg$_T$)	304	301

Note: Fluxes are in pmol m^{-2} day^{-1}, calculated using a transfer velocity of 1.5 cm h^{-1} (0.36 m day^{-1}).

E.   ELEMENTAL Hg PRODUCTION

The dynamic nature and biogeochemical importance of Hg^0 production is demonstrated in Table 6 by comparing evasional fluxes from the north-central Wisconsin study lakes in August 1989 with August 1990 results. It is evident that with the exception of Max lake the effluxes of Hg^0 are about two to six times larger in 1990 and interlake differences are significant. Further, and as summarized in Table 6, the atmospheric deposition of Hg$_R$ was approximately three times greater in August 1990 relative to August 1989. The Hg$_R$ input during August 1990 (260 pmol m^{-2} d^{-1}) was much higher than that of August 1989 (85 pmol m^{-2} d^{-1}). However, as indicated in Table 6, the total Hg (Hg$_T$) input was similar for both periods, 301 pmol m^{-2} d^{-1} vs. 305 pmol m^{-2} d^{-1}. A relationship between Hg^0 evasion and the input of Hg$_R$ is apparent, as the increases in Hg^0 supersaturation appear to respond to the supply of Hg$_R$ and not to Hg$_T$.

The atmospheric depositional fluxes of Hg$_R$ and Hg$_T$ were obtained in the following manner. The average daily wet depositional input to the lakes during August 1989 and 1990 was calculated using the actual rainfall amounts measured in August and the corresponding average reactive (Hg$_R$) and total Hg (Hg$_T$) concentrations for the respective seasons (Fitzgerald et al., 1991; Mason et al., 1992). The dry depositional flux of Hg$_R$ was estimated from the overall yearly Hg$_T$ input, assuming a relative Hg$_R$ concentration similar to that of rain, and using a representative deposition velocity of 0.5 cm s^{-1} (Fitzgerald et al., 1991).

F.   ELEMENTAL Hg PRODUCTION: MECHANISMS

The principal source(s) of the high levels of supersaturation in the surface waters is not yet known. However, photocatalytic reduction of ionic Hg, bacterial demethylation reactions, organic matter, and/or photosynthetic processes may all play a role. We have considered the two most probable sources of Hg^0 (demethylation and direct reduction) in detail, and the results are illustrated in Figures 2 and 3 where we have shown (1) the estimated depositional Hg$_R$ input and evasional fluxes of Hg^0 for Little Rock Lake Treatment Basin and Pallette Lake, respectively, in August 1989 and 1990, and (2) the probable sources of Hg^0—demethylation of Hg^0 and direct reduction of Hg(II). The amounts of MMHg formed in the mixed layer were estimated using the measured epilimnetic Hg$_R$ concentrations in August 1989 (Bloom, N.S., personal communication) and our measurements in 1990, along with water column methylation rates of 0.01 to 0.3% d^{-1}, determined from laboratory spike experiments (Xun et al., 1987; Korthals and Winfrey, 1987; Gilmour and Henry, 1990). The rate of Hg^0

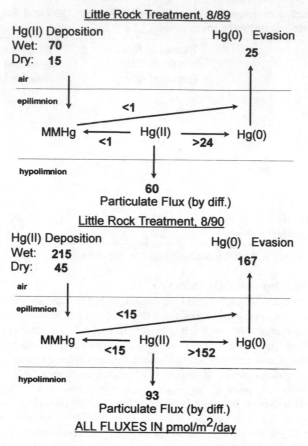

Little Rock Treatment, 8/89

Hg(II) Deposition
Wet: 70
Dry: 15

Hg(0) Evasion
25

Little Rock Treatment, 8/90

Hg(II) Deposition
Wet: 215
Dry: 45

Hg(0) Evasion
167

ALL FLUXES IN pmol/m^2/day

Figure 2 Modeling the potential pathways for the production and evasion of Hg^0 in epilimnion of the treatment basin of Little Rock Lake, Wisconsin. The amounts of Hg^0 produced by demethylation and direct reduction of Hg(II) are estimated and compared to the input of Hg$_R$ by atmospheric deposition for the August 1989 and August 1990 experiments.

formation by demethylation of MMHg must be less than the MMHg formation rate. Thus, the calculations show that demethylation is a minor source of Hg^0 in the epilimnion of the study lakes. Direct reduction of Hg(II) must be the primary source of Hg^0 in the epilimnion of these lakes; a similar situation was found for the mixed layer of the equatorial pacific (Mason and Fitzgerald, 1993).

The observed evasion in Little Rock Lake Treatment Basin could be maintained by Hg^0 formation rates of 8×10^{-7} s^{-1} (7% d^{-1}) in 1989 and 3.3×10^{-7} s^{-1} (2.8% d^{-1}) for 1990. At Pallette Lake, the observed evasion would require Hg^0 formation rates of 3.3×10^{-6} s^{-1} (28% d^{-1}) in 1989 and 1.4×10^{-7} s^{-1} (1.2% d^{-1}) for 1990. Abiotic production of Hg^0 (25 $\times 10^{-7}$ s^{-1}; 22% d^{-1}) in the presence of humic acids has been demonstrated in laboratory studies (Alberts et al., 1974). Moreover, the reaction rates indicate that the system response to Hg$_R$ input is rapid: a pulse input of Hg(II) would be converted to Hg^0, in the absence of other reactions, in about 70 d at a conversion rate of 5% d^{-1} (t$^{1/2}$ = 14 d).

Although the Hg^0 data is limited to one set of measurements at each lake in each season, these results support the postulate that the production and subsequent evasion of Hg^0 is directly linked to the rate of supply of Hg$_R$. The reactive Hg concentration, therefore, is a

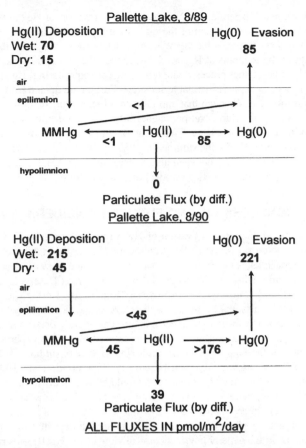

Figure 3 Modeling the potential pathways for the production and evasion of Hg^0 in epilimnion of Pallette Lake, Wisconsin. The amounts of Hg^0 produced by demethylation and direct reduction of Hg(II) are estimated compared to the input of Hg_R by atmospheric deposition for the August 1989 and August 1990 experiments.

measure of the readily available substrate. In addition, the rates of conversion are of a similar order to those estimated for the equatorial Pacific (Mason and Fitzgerald, 1993), suggesting that analogous processes are involved in these two systems.

The similarity between the Hg_T depositional inputs for both seasons suggests further that the unreactive, strongly bound Hg fraction is not directly available for conversion into Hg^0. Prior to any remobilization into a reactive form that can be converted into other Hg species, it is likely that the strongly bound fraction is transported into the anoxic regions of the lakes (the hypolimnion) or the sediment either vertically or horizontally.

In the Wisconsin lakes, there was no significant increase in the Hg^0 concentration during winter ice over (Vandal, G.M., personal communication). Iverfeldt, 1988, attributed differences in the concentrations of dissolved gaseous Hg, reactive, and total Hg in Framvaren fjord between samples collected in September 1983 and February 1985 to the lack of atmospheric input during ice cover. While the lack of Hg^0 formation under the ice could be interpreted as evidence of Hg^0 production being associated with a process requiring light (i.e., primary production or photoreduction), it is also suggestive of substrate limitation in winter through the lack of atmospheric input. The results from these studies, Iverfeldt's work

(1988), and Mason and Fitzgerald (1992) indicate that formation of Hg^0 in the mixed layer of natural systems is a direct function of the rate of supply of available Hg substrate.

In Table 6, the mass balances for the Wisconsin lakes show that, with the exception of Crystal Lake, atmospheric inputs of Hg_R were larger than the evasional losses of Hg^0 during the August 1990 studies. Other processes are removing (or sequestering) Hg_R from the mixed layer and these processes limit the available substrate. Particulate uptake from the epilimnion is an important removal mechanism that can deliver substrate for methylation to the hypolimnion or the sediment interface. Net particulate fluxes required to balance the increase in Hg in the hypolimnion of Little Rock Treatment Basin during summer stratification were estimated at 55 pmol m^{-2} d^{-1} for the summer of 1989 by Hurley et al., 1991. This flux most likely accounts for the difference between atmospheric input and Hg^0 evasion and suggests that this process is a primary competing removal mechanism for epilimnetic Hg(II) substrate.

V. CONCLUSIONS AND FUTURE CONSIDERATIONS

It is evident that the production and evasion of Hg^0 in natural waters is a major feature of the aquatic biogeochemical cycling of Hg. Significant effluxes of Hg^0 have been observed in this study of seepage lakes in Wisconsin as well as in a range of systems such as the open ocean equatorial Pacific (Kim and Fitzgerald, 1986), Davis Creek Reservoir, California (Gill and Bruland, 1992), and drainage lakes in Sweden (Xaio et al., 1991). Mason et al., 1992, have shown that Hg^0 emissions to the atmosphere are proportional to the availability and supply of Hg_R (the Hg(II) substrate) whether it is atmospherically derived as in this work or supplied principally through upwelling as in the equatorial Pacific. It is particularly striking that a large fraction of the Hg_R input to the north-central Wisconsin lakes is returned to the atmosphere. Indeed, and as we suggested previously (Fitzgerald et al., 1991), lakes with limnological conditions favoring Hg^0 production would be less likely to have elevated levels of Hg in fish. Moreover, there should be an inverse relationship between Hg^0 evasion and the accumulation of Hg in sediments for a particular lake (Rada et al., 1992).

Interlake variations in Hg^0 production and evasion are both expected and observed (Table 6). This is consistent with a general physicochemical view of the Hg(II) substrate hypothesis. For example, methylating and reducing processes compete to utilize Hg(II) species, and this active substrate can be rendered inactive by strong sequestering with organic ligands, or inorganic interactions (e.g., with sulfitic ligands) which could reduce the activity of the substrate. Thus, inorganic and organic components, suspended matter, pH, or biological productivity within a lake can serve to alter the availability of the substrate. Vandal et al., 1991, presented evidence suggesting that (1) increasing acidity may reduce the in-lake production of Hg^0, and (2) photosynthetic activity may enhance Hg^0 production. Extensive studies are needed since very few details of the processes affecting production and destruction of Hg^0 are known. There are many questions concerning short time-scale spatial and temporal variability as well as the importance of photoreduction reactions and redox boundaries (i.e., oxic/anoxic transition zones) in the production of Hg^0. In addition, the relationships among phytoplankton productivity and microbial populations (e.g., bacterial reduction) on the activity of Hg^0 should be evaluated. Broadly based Hg^0 investigations are required, particularly those including ancillary biological studies and concurrent methylation investigations. Seasonal information and spatial data for evasion of Hg^0 are limited. This points toward a need to refine efflux estimates from lake waters to the atmosphere and to assess, quantitatively, their influences on the overall cycling of Hg in lake systems.

We suggest that Hg_R found in precipitation and atmospheric particulate matter is derived from the atmospheric oxidation products of Hg^0 in the atmosphere. This form of Hg is labile and highly reactive in aqueous systems and readily available, for example, to participate in competitive reactions associated with methylation, reduction to Hg^0, uptake by biota, and sequestering with humics. The other fraction of the Hg_T in deposition is the operationally defined strongly bound Hg portion ("unreactive" Hg) which gives rise to intricate biogeochemical questions as well. For example, are these particles environmentally active? Available evidence suggests that this fraction is associated with soot and may be strongly bound or sequestered in some type of sulfur-carbon association (Brosset and Lord, 1991). Perhaps this unreactive Hg can be solubilized under anoxic and/or sulfitic conditions to yield a species such as $Hg(HS)_2^0$ which can be bacterially methylated. This process could take place at the sediment water interface in the epilimnetic zone as well as in the low oxygen waters of the hypolimnion. This would be an insidious process where an apparently unreactive component under oxic conditions would yield MMHg in low oxygen zones of the lakes.

The strongly bound Hg components in the Wisconsin atmosphere are likely to have different geographic origins than the Hg_R species. Soot-associated Hg particles, for example, will probably have an anthropogenic source and a local/regional origin. There may be a spectrum of different components in this fraction and the atmospheric residence times should range from days to weeks. At present, we know little about this part of the atmospheric Hg cycle. The Hg_R fraction, as suggested, is most probably derived from the oxidation of Hg^0. However, there is evidence for the presence of water-soluble forms of Hg in stack emissions from coal-fired power plants (Brosset, 1987; Lindqvist et al., 1991). As a consequence, Hg_R will be coupled to the global Hg system with both anthropogenic and natural sources. Also, and depending on location, local/regional anthropogenic contributions of Hg_R may be significant.

These observations illustrate the value of the chemical speciation approach to our developing understanding of the cycling of Hg in nature. Indeed, they force us to ask and address the following general question: How do such speciation changes in the depositional fluxes of Hg affect the cycling of Hg in aquatic systems, and what causes the variation in the Hg_T and Hg_R composition found in deposition? At present, there are no unequivocal answers to questions concerning the sources and variability of the atmospheric Hg species.

ACKNOWLEDGMENTS

We thank our colleagues in the Mercury in Temperate Lakes program for their advice and assistance during this work. We are particularly grateful to Steve Claas, Kent Hatch, and Bill Fitzgerald for air and precipitation collections that were often obtained in adverse weather conditions ranging from polar temperatures to deafening electrical storms. The use of the facilities at the Trout Lake Limnological Station, Center for Limnology, University of Wisconsin-Madison is gratefully noted. Technical assistance and reactor time provided by the Laboratoire Pierre Süe, Centre d'Etudes Nucléaires de Saclay, as well as the use of office and laboratory facilities at the Centre des Faibles Radioactivitiés, CNRS/CEA, Gif sur Yvette, are particularly appreciated. This continuing investigation was supported by the Wisconsin Department of Natural Resources and the Electric Power Research Institute (RP2020–10). Additional support has been provided by a NATO International Collaborative Research Grant (0160/88 and 90). This is contribution No. 250 from the Marine Sciences Institute, The University of Connecticut.

REFERENCES

Alberts, J.J., Schildler, J.E., Miller, R.W., and Nutter, D.E. (1974), Elemental mercury evolution mediated by humic acid, *Science,* 184:895–897.

Bloom, N.S. (1989), Determination of picogram levels of methylmercury by aqueous phase ethylation, followed by cryogenic gas chromatography with atomic fluorescence detection, *Can. J. Fish. Aquat. Sci.,* 46:1131–1140.

Bloom, N.S. and Crecelius, E.A. (1983), Determination of mercury in seawater at subnanogram per liter levels, *Mar. Chem.,* 14:49–59.

Bloom, N.S., Watras, C.J., and Hurley, J.P. (1991), Impact of acidification on the methyl-mercury cycle of remote seepage lakes, *Water Air Soil Pollut.,* 56:1714–1720.

Bloom, N.S. and Fitzgerald, W.F. (1988), Determination of volatile mercury species at the picogram level by low temperature gas chromatography with cold-vapor atomic fluores-cence detection, *Anal. Chim. Acta,* 208:151–161.

Brezonik, P.L., Baker, L.A., Eaton, J.R., Frost, T.M., Garrison, P., Kratz, T.K., Magunson, J.J., Rose, W.J., Shepard, B.K., Swenson, W.A., Watras, C.J., and Webster, K.E. (1986), Experimental acidification of Little Rock Lake, Wisconsin, *Water Air Soil Pollut.,* 31:115–121.

Brosset, C. (1987), The behavior of mercury in the physical environment, *Water Air Soil Pollut.,* 34:145–166.

Brosset, C. and Lord, E. (1991), Mercury in precipitation and ambient air, a new scenario, *Water Air Soil Pollut.,* 56:493–506.

Buat-Menard, P. (1985), Air to sea transfer of anthropogenic trace metals, in: *The Role of Air-Sea Exchange in Geochemical Cycling,* Buat-Menard, P. (Ed.), D. Reidel Publishers, Dordrecht, pp. 477–496.

Crusius, J. and Wanninkhof, R.H. (1990), Refining the gas exchange-wind speed relationship at low wind speeds on Lake 302N with SF_6, *EOS Trans. AGU,* 71:1234.

Eilers, J.M., Glass, G.E., Webster, K.E., and Rogalla, J.A. (1983), Hydrologic control of lake susceptibility to acidification, *Can. J. Fish. Aquat. Sci.,* 40:1896–1940.

Emerson, S. (1975), Gas exchange rates in small Canadian Shield lakes, *Limnol. Oceanogr.,* 20:754–761.

Fagerström, T. and Jernelöv, A. (1972), Some aspects of the quantitative ecology of mercury, *Water Res.,* 6:1193–1202.

Fitzgerald, W.F. (1986), Cycling of mercury between the atmosphere and oceans, in: *The Role of Air-Sea Exchange in Geochemical Cycling,* NATO Advanced Science Institutes Series, P. Buat-Menard (Ed.), D. Reidel Publishers, Dordrecht, pp. 363–408.

Fitzgerald, W.F. (1989), Atmospheric and oceanic cycling of mercury, in: *Chemical Ocean-ography,* Vol. 10, Riley, J.P. and Chester, R. (Eds.), Academic Press, London, pp. 151–186.

Fitzgerald, W.F. and Gill, G.A. (1979), Subnanogram determinations of mercury by two-stage gold amalgamation and gas phase detection applied to atmospheric analysis, *Anal. Chem.,* 51:1714–1720.

Fitzgerald, W.F., Gill, G.A., and Kim, J.P. (1984), An equatorial Pacific Ocean source of atmospheric mercury, *Science,* 224:597–599.

Fitzgerald, W.F., Vandal, G.V., and Mason, R.P. (1991), Atmospheric cycling and air-water exchange of mercury over mid-continental lacustrine regions, *Water Air Soil Pollut.,* 56:745–767.

Fitzgerald, W.F. and Clarkson, T.W. (1991), Mercury and monomethylmercury. Present and future concerns, *Environ. Health Perspect.,* 96:159–166.

Fitzgerald, W.F. and Watras, C.J. (1989), Mercury in surficial waters of rural Wisconsin lakes, *Sci. Total Environ.,* 87/88:223–232.

Fitzgerald, W.F., Gill, G.A., and Hewitt, A. (1983), Air-sea exchange of mercury, in: *Trace Metals in Seawater,* Wong, C.S. et al. (Eds.), NATO Conference Series, IV, Marine Science, Vol. 9, Plenum Press, New York, pp. 297–316.

Fogg, T.R. and Fitzgerald, W.F. (1979), Mercury in southern New England coastal rains, *J. Geophys. Res.,* 84:6987–6988.

Gill, G.A. and Bruland, K.W. (1992), Mercury speciation and cycling in a seasonally anoxic freshwater system: Davis Creek Reservoir. Final Report to the Electric Power Research Institute, Palo Alto, CA.

Gill, G.A. and Fitzgerald, W.F. (1985), Mercury sampling of open ocean waters at the picomolar level, *Deep-Sea Res.,* 32:287–297.

Gill, G.A. and Fitzgerald, W.F. (1987), Mercury in the surface waters of the open ocean, *Global Biogeochem. Cycles,* 3:199–212.

Gill, G.A. and Fitzgerald, W.F. (1987a), Picomolar mercury measurements in seawater and other materials using stannous chloride reduction and two-stage gold amalgamation with gas phase detection, *Mar. Chem.,* 20:227–243.

Gilmour, C.G. and Henry, E.A. (1991), Mercury methylation in aquatic systems affected by acid deposition, *Environ. Pollut.,* 71:31–169.

Håkanson, L. (1980), The quantitative input of pH bioproduction and Hg contamination on the Hg content of fish (pike), *Environ. Pollut. (Ser. B),* 1:285–304.

Hurley, J.P., Watras, C.J., and Bloom, N.S. (1991), Mercury cycling in northern Wisconsin seepage lakes: the role of particulate matter in vertical transport, *Water Air Soil Pollut.,* 56:543–551.

Iverfeldt, Å. (1988), Mercury in the Norwegian fjord Framvaren, *Mar. Chem.,* 23:441–456.

Iverfeldt, Å. (1991). Occurrence and turnover of atmospheric mercury over the Nordic countries, *Water Air Soil Pollut.,* 56:251–265.

Iverfeldt, Å. and Lindqvist, O. (1986), Atmospheric oxidation of elemental mercury by ozone in the aqueous phase, *Atmos. Environ.,* 20:1567–1573.

Kim, J.P. and Fitzgerald, W.F. (1986), Sea-air partitioning of mercury in the equatorial Pacific Ocean, *Science,* 231:1131–1133.

Kim, J.P. and Fitzgerald, W.F. (1988), Gaseous mercury profiles in the tropical Pacific Ocean, *Geophys. Res. Lett.,* 15:40–43.

Korthals, E.T. and Winfrey, M.R. (1987), Seasonal and spatial variations in mercury methylation and demethylation in an oligotrophic lake, *Appl. Environ. Microbiol.,* 53:2397–2404.

Lindqvist, O., Johansson, K., Aastrup, M., Andersson, A., Bringmark, L., Hovsenius, G., Håkanson, L., Iverfeldt, Å., Meili, M., and Timm, B. (1991), Mercury in the Swedish environment—recent research on causes, consequences and corrective methods, Spec. Rep, *Water Soil Air Pollut.,* 55.

Liss, P.S. (1983), Gas transfer: experiments and geochemical implications, in: *Air-Sea Exchange of Gases and Particles,* NATO Advanced Science Institutes Series, P.S. Liss and W.G. Slinn (Eds.), D. Reidel Publishers, Dordrecht, pp. 241–298.

Magnuson, J.J., Bowser, C.J., and Kratz, T.K. (1984), Long-term ecological research on north temperate lakes of the United States, *Verh. Int. Verein. Limnol.,* 22:533–535.

Maring, H., Patterson, C., and Settle, D. (1989), Atmospheric input fluxes of industrial and natural lead from the Westerlies to the mid-north Pacific, in: *Chemical Oceanography,* Vol. 10, Riley, J.P. and Chester, R. (Eds.), Academic Press, London, pp. 84–106.

Mason, R.P., Fitzgerald, W.F., and Vandal, G.V. (1991), The sources and composition of mercury in oceanic precipitation, *J. Atmos. Chem.,* 14:489–500.

Mason, R.P. and Fitzgerald, W.F. (1991), Mercury speciation of open ocean waters, *Water Air Soil Pollut.,* 56:779–789.

Mason, R.P. and Fitzgerald, W.F. (1990), Alkylmercury species in the equatorial Pacific, *Nature,* 347:457–459.

Mason, R.P., Fitzgerald, W.F., and Vandal, G.M. (1992), The sources and composition of mercury in Pacific Ocean rain, *J. Atmos. Chem.,* 14:489–500.

Mason, R.P. and Fitzgerald, W.F. (1993), The distribution and biogeochemical cycling of mercury in the equatorial Pacific Ocean, *Deep-Sea Res.,* 40:1897–1924.

Mierle, G. (1990), Aqueous inputs of mercury to precambian shield lakes in Ontario, *Environ. Toxicol. Chem.,* 9:843–851.

Munthe, J. (1992), The aqueous oxidation of elemental mercury by ozone, *Atmos. Environ.,* 26A:1461–1468.

Munthe, J. and Lindqvist, O. (1989), The aqueous atmospheric chemistry of mercury, Proc. Nordic Symp. Atmospheric Chemistry, Stockholm/Helsinki, December, 1989.

Rada, R.G., Winfrey, M.R., Wiener, J.G., and Powell, D.E. (1987), A comparison of mercury distribution in sediment cores and mercury volatilization from surface waters of selected northern Wisconsin lakes. Final report, Wisconsin DNR, Bureau of Water Resources Management, Madison, WI.

Rada, R.G., Powell, D.E., and Wiener, J.G. (1993), Whole-lake burdens and spatial distribution of mercury in surficial sediments in Wisconsin seepage lakes, *Can. J. Fish. Aquat. Sci.,* 50:865–873.

Sanemasa, I. (1975), The solubility of elemental mercury vapor in water, *Bull. Chem. Soc. Jpn.,* 48:1795–1798.

Scheider, W.A., Jeffries, D.S., and Dillon, P.J. (1979), Effects of acidic precipitation on Precambrian freshwaters in southern Ontario, *J. Great Lakes Res.,* 5:45–51.

Sloan, R. and Schofield, C.L. (1983), Mercury levels in Brook Trout (*Salvelinus fontinalis*) from selected acid and limed Adirondack lakes, *Northeast. Environ. Sci.,* 2:165–170.

Upstill-Goddard, R.C., Watson, A.J., Liss, P.S., and Liddicoat, M.I. (1990), Gas transfer velocities in lakes measured with SF_6, *Tellus,* B42:364–377.

Vandal, G.M., Fitzgerald, W.F., and Mason, R.P. (1991), Cycling of volatile mercury in temperate lakes, *Water Air Soil Pollut.,* 56:791–803.

Wanninkhof, R., Ledwell, J.R., and Broecker, W.S. (1985), Gas exchange-wind speed relation measured with sulfur hexafluoride on a lake, *Science,* 227:1224–1226.

Watras, C.J., Bloom, N.S., Fitzgerald, W.F., Wiener, J.G., Rada, R., Hudson, R.J.M., and Porcella, D.G. (1994), Sources and fates of mercury and methylmercury in remote temperate lakes, in: *Mercury Pollution: Integration and Synthesis,* Watras, C.J. and Huckabee, J.W. (Eds.), Lewis Publishers, Chelsea, MI, chap. I.12.

Watras, C.J. and Frost, T.M. (1989), Little Rock Lake: perspectives on an experimental ecosystem approach to seepage lake acidification, *Arch. Environ. Contam. Toxicol.,* 18:157–165.

Wiener, J.G. (1987), Metal contamination of fish in low-pH lakes and potential implications for piscivorous wildlife, *Trans. N. Am. Wildl. Nat. Resour. Conf.,* 52:645–657.

Wiener, J.G., Fitzgerald, W.F., Watras, C.J., and Rada, R.G. (1990), Partitioning and bioavailability of mercury in an experimentally acidified Wisconsin lake, *Environ. Toxicol. Chem.,* 9:909–918.

Winfrey, M.R., and Rudd, J.W.M. (1990), Environmental factors affecting the formation of methylmercury in low pH lakes, *Environ. Toxicol. Chem.,* 9:853–869.

Xiao, Z.F., Munthe, J., Schroeder, W.H., and Lindqvist, O. (1991), Vertical fluxes of volatile mercury over forest soil and lake surfaces in Sweden, *Tellus,* 41B:267–279.

Xun, L., Campbell, N.E.R., and Rudd, J.W.M. (1987), Measurement of specific rates of net methylmercury production in the water column and surface sediments of acidified and circumneutral lakes, *Can. J. Fish. Aquat. Sci.,* 44:750–757.

Atmospheric Bulk Deposition of Mercury to the Southern Baltic Sea Area

Arne Jensen and Åke Iverfeldt

CONTENTS

I. INTRODUCTION

Although mercury (Hg) for many years has been considered all over the world as a "black list" element with restrictions on its use and on emissions to the environment, it is still a problematic element. Recently, this was demonstrated by the many papers represented at the International Conference: "Mercury as a Global Pollutant" held in Sweden in 1990,[1] as well as by another publication: "Mercury in the Swedish Environment".[2]

Generally, it is recognized that mercury is transported by air masses over long distances.[3,4] It has been shown that the mercury concentration in air in Sweden is related to the long-range transport of mercury from source areas south of Sweden.[2] A clearly decreasing south-north gradient exists in the Nordic countries for mercury in precipitation. This gradient is based on about 20 yearly samples from several localities collected at precipitation events.[4] Similar studies have been performed in the Upper Great Lakes Region.[5] However, to our knowledge, very few studies with regular monthly measurements of mercury in precipitation have been reported that estimate the yearly bulk deposition of mercury.

It is the objective of this study to test the performance of new bulk sampling equipment by setting up duplicate samplers at the same locality. Furthermore, it is the intention to estimate the atmospheric bulk deposition of mercury to the southern part of Sweden and to Denmark, and to show whether there is any difference between samplers placed at an island, coastal areas, and inland sites.

II. MATERIALS AND METHODS

A. SAMPLING EQUIPMENT AND SAMPLING STATIONS

The samplers were constructed by the Swedish Environmental Research Institute (IVL). The monthly samples of precipitation were collected by a specially designed bulk collector of black polyethylene for unbiased Hg sampling, see Figure 1.

All parts of the collector in contact with the samples were made of borosilicate glass. The diameter of the borosilicate funnel was 8.2 cm, which gives an area of 52.8 cm^2. The length

1–56670–066–3/94/$0.00+$.50
© 1994 Lewis Publishers

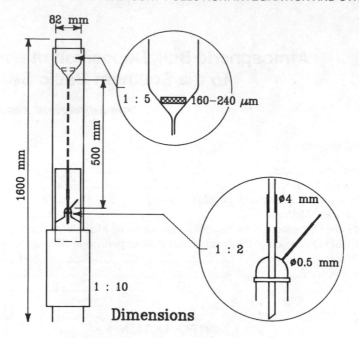

Figure 1   Monthly precipitation and throughfall water bulk collector.

of the protective capillary was 0.5 m. An acid-leached glass filter (porous size 160 to 250 μm) was loosely positioned in the bottom of the funnel of each collector, not to filter the water, but to protect the sample from contamination by litter or insects. Tests showed that the water collected by the funnel passed between the edge of the glass disc and the funnel wall, where the spacing was much larger than the porous size of the glass disc. It is therefore reasonable to assume that all of the particle-associated Hg of atmospheric origin in precipitation and throughfall water will reach the collecting bottle.

The samplers were supplied with electrical heating elements to maintain a temperature inside the sampler above 2 to 3° C. All glassware, including sampling bottles, were extensively cleaned with nitric acid as described in DS 2214[6] or according to standard procedures applied by IVL.[4]

Table 1 summarizes the number of samplers at the different localities and Figure 2 shows a map with the sampling localities indicated. At Ulfborg-Husby and Bornholm, both with forest and coastal stations, the samplers were placed at a distance of 10 to 15 km to determine possible differences between a coastal and an inland forest station. At Aspvreten-Nynäs the distance between the two locations is 1 km. Draget is situated 20 km from Rörvik, with Nidingen island about 10 km from Draget. Sampling was performed monthly at all stations. To all 500 mL sampling bottles 2.5 mL of concentrated hydrochloric acid was pre-added for the preservation of mercury during the sampling period.

B. ANALYTICAL METHODS

At four Danish localities with duplicate samplers (Table 1) samples were analyzed by both laboratories. One sample was analyzed by both laboratories to detect any differences between the analytical techniques. The other sample was analyzed by IVL to detect any difference between duplicate samplers. Samples from the other stations in Denmark were analyzed by FORCE and in Sweden by IVL.

At FORCE, all the samples were analyzed for total mercury content by radiochemical neutron activation analysis (RNAA).[7] About 3 g of the shaken sample was used for the

Table 1   **Sampling localities and number of samplers.**

Locality	Type	Number of samplers
Denmark		
Frederiksborg	Inland, forest	1[a]
Keldsnor	Island, coastal	1[a]
Anholt	Island, coastal	2[a]
Ulfborg	Inland, forest	2[b]
Husby	Coastal	2[b]
Bornholm	Island, coastal	2[b]
Bornholm	Island, forest	2[b]
Sweden[c]		
Aspvreten	Coastal	2
Nynäs	Coastal	2
Hoburg, Gotland	Island, coastal	2
Ven	Island, coastal	2
Rörvik	Coastal	2
Draget	Coastal	2
Nidingen	Island, coastal	2
Gårdsjön	Inland, forest	2 × 3

[a] Analyzed by FORCE.
[b] Analyzed by FORCE and IVL.
[c] Analyzed by IVL.

analysis. Sometimes particles and insects have been found in the sample and these are avoided as far as possible. The detection limit is about 3 ng/L and is estimated for each analytical series. The analytical precision is better than 10% except for results near the detection limit. With every analytical batch an internal quality control material was also analyzed. NBS 1642 B (National Bureau of Standards) certified seawater (1.49 µg/L) was used and an average of 1.523 ± 0.013 µg/L was found.

At IVL the samples were analyzed until November 1990 by plasma emission spectrometry (AES)[8] and after that time by atomic fluorescence spectrometry (AFS).[9] A 50 g sample was used for AES and 10 to 20 g for AFS. In both methods, the total mercury content was determined after destruction by addition of 5 ml BrCl per liter of sample. In both techniques mercury is trapped onto gold traps as a means of preconcentration and interference removal. The detection limit for total mercury for AES is about 0.2 ng/L and for AFS about 0.05 ng/L. The analytical precision is usually between 5 and 10%.

Both IVL and FORCE have participated in the latest intercalibration of mercury in rain water in 1991 initiated by the Paris Commission. The results are not yet published. FORCE and IVL have participated in the ICES 6th round intercalibration for trace metals in estuarine water.[10] In this exercise no bias in the methods was detected.

III. RESULTS AND DISCUSSION OF THE ANALYTICAL AND SAMPLING METHODS

Since both IVL and FORCE have analyzed samples No. 1 at Husby-Ulfborg and at both sites at Bornholm, a paired-sampled t-test[11] was performed. The results of the test show that a significant difference exists between the analytical results from FORCE and IVL when comparing the RNAA and AES methods at 95%, but not at 99% probability level

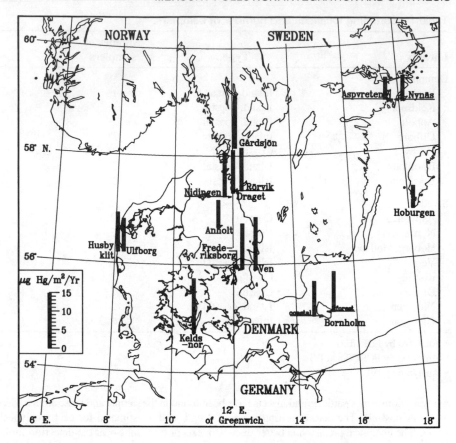

Figure 2 Sampling stations for bulk deposition of mercury in Denmark and southern Sweden and yearly bulk deposition of mercury (μg/m²/year).

Table 2 **Comparison of the different analytical techniques used at FORCE and IVL.**

	RNAA-AES	RNAA-AFS
Number of samples	25[a]	24[a]
Mean difference ng/L	−1.96	−1.72
Standard deviation ng/L	5.00	5.29
Calculated t value	1.96	1.59
t 95% significance	1.71	1.71
t 99% significance	2.48	

[a] One outlier rejected by Grubbs test for outlying observations.

(Table 2). A similar t-test for the RNAA and AFS methods resulted in no significant differences at the 95% probability level between the two methods.

The reason for this difference between RNAA and AES may be that FORCE uses about 3 g per sample whereas IVL uses 50 g. In several samples there were small particles, such as insects and soot. By using a larger sample there is a greater chance of getting a subsample which contains particles such as insects. Of course it is possible to filter the sample before the analysis. However, mercury in precipitation is mostly bound to particulate matter, and

Table 3   **Comparison of duplicate samplers analyzed by IVL.**

	mm Precipitation	μg Hg/m²
Number of samples	42[a]	41[b]
Mean difference	−0.63	−0.01
Standard deviation	3.6	0.31
Calculated t value	1.13	0.20
t 95% significance	1.68	1.68

[a] Two outliers rejected by Grubbs test for outlying observations.
[b] One outlier rejected by Grubbs test for outlying observations.

very low concentrations will be present in the filtrate which increases the risk of contamination of the filtrate during filtration and analysis. For these reasons it was decided to perform the study without filtration.

As shown in Table 1, duplicate samplers were placed at five locations in Denmark. A paired-sampled t-test (Table 3) between the amount of precipitation (in mm) collected by the duplicate samplers shows no significant difference at the 95% probability level. Furthermore, the results in Table 3 for the paired-sampled t-test between the duplicate samplers, where IVL has analyzed both samples, gave no significant statistical difference at 95% probability between the mercury concentrations (ng/L) as well as the bulk precipitation of mercury (μg/m²).

These results indicate that the reproducibility of the samplers for collection of precipitation and bulk deposition of mercury is excellent.

IV. RESULTS AND DISCUSSION OF PRECIPITATION AND BULK DEPOSITION OF MERCURY

Figure 3 shows—as an example representative for all stations—that the bulk deposition of mercury at the two coastal stations, Rörvik and Draget, and the station at the island of Nidingen varies over the course of the year. It further shows that the bulk deposition of mercury at the two coastal stations and at the island station follows nearly the same trends.

Iverfeldt[4] and Glass et al.[5] showed that mercury concentrations in precipitation are negatively correlated with precipitation volumes measured during precipitation events. In the present study, with monthly bulk collection of precipitation, we have found a similar significant correlation with the mercury concentration, as shown in Figure 4. This indicates that the mercury is washed out of the air masses during heavy rainfalls, which has also been shown in several studies[1] during rainstorms. However, it has not previously been demonstrated for monthly sampling.

Table 4 shows the accumulated precipitation and bulk deposition of mercury for the sampling stations for the 12-month period May 1990 to April 1991. Generally, the amount of precipitation is less at the islands and at the coastal stations than at the nearby inland station. However, this is not reflected in the yearly bulk deposition of mercury, as there is no significant difference between the nearby-situated station when performing a paired-sampled t-test. This is in disagreement with the conclusions drawn above concerning the negative correlation between mercury concentration and precipitation. However, the amount of data is too small at each station to detect any significant difference and a possible correlation may be smoothed out by accumulation of the monthly data.

Furthermore, the results in Table 4 and Figure 2 show that the highest bulk deposition of mercury is measured at Keldsnor, Frederiksborg, Ven, and Gårdsjön. The high deposition at Keldsnor station is caused by one high value. If this high value is substituted by the bulk deposition in April 1990, the deposition of Hg becomes 9.9 μg/m². The high values at

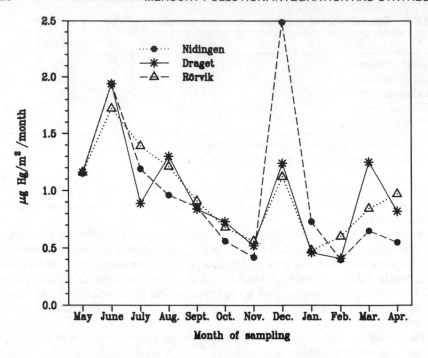

Figure 3 Monthly mercury deposition (μg/m²) at the coastal sampling stations Rörvik and Draget and at the island of Nidingen.

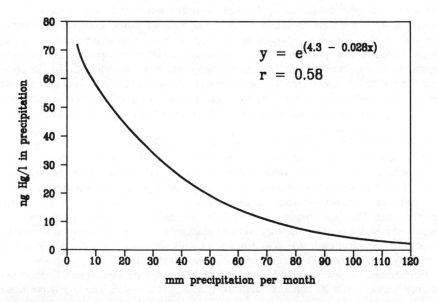

Figure 4 Exponential relation between monthly mercury deposition (ng/l) and precipitation volumes (mm) for all sampling stations.

Table 4   **Bulk deposition of mercury in µg/m²/year in this study and a comparison with other studies in the same area.**

Locality	Area	Year	Precipitation (mm)	µg Hg/m²/year
This study: Denmark				
Ulfborg	Forest	1990–91	875	10.6
Husby	Coastal	1990–91	762	8.6
Anholt	Island	1990–91	527	7.4
Bornholm	Coastal	1990–91	571	9.5
Bornholm	Forest	1990–91	704	10.9
Frederiksborg	Forest	1990–91	602	12.3
Keldsnor	Island	1990–91	446	14.8[a]
This study: Sweden				
Aspvreten	Coastal	1990–91	491	6.1
Nynäs	Coastal	1990–91	479	6.7
Hoburg	Island	1990–91	351	5.9
Ven	Island	1990–91	481	14.2
Rörvik	Coastal	1990–91	719	11.6
Draget	Coastal	1990–91	688	11.6
Nidingen	Island	1990–91	619	11.9
Gårdsjön	Forest	1990–91	880	14.2
Other studies				
Keldsnor[4]	Island	1987–89	430	16.8
Aspvreten[4]	Coastal	1987–89	520	10.1
Rörvik[4]	Coastal	1987–89	770	27.0
Birkenes[4]	Coastal	1987–88	1730	35.1
Copenhagen[14]	City	1976–78	560	65
Zealand, south[14]	Inland	1976–78	600	42
Zealand, south[13]	Inland	1976–78	560	44
Great Britain[13]	Coastal	1989[b]	565	30
Great Britain[13]	Coastal	1989[c]	515	42
Great Britain[13]	Coastal	1989[d]	547	18

[a] 9.9 µg Hg/m² if one high value at 5.72 µg Hg/m² is substituted by the deposition at 0.99 µg/m² measured in April 1990.
[b] Northern part of Scotland, Aberdeen, Station No. GB3.
[c] Northern England, Scarborough, Station No. GB2.
[d] Middle England, Norwich, Station No. GB1.

Frederiksborg and Ven could be explained by the short distance to the heavily populated area around the Sound. Bulk Hg deposition of 12 µg/m²/year has previously been measured at Gårdsjön.[12] Normally, a higher rainfall is found at Gårdsjön, about 1.000 to 1.100 mm.

The mercury bulk deposition at the two isolated islands of Anholt and Gotland (Hoburg) and the coastal stations Aspvreten-Nynäs are lower than at the other stations.

A comparison of the results from this study with other studies in the area is shown in Table 4. During the period 1987–1989 bulk deposition of mercury at several locations in Scandinavia was measured based on a smaller number of precipitation events.[4] From these measurements, an average concentration of mercury in the precipitation was multiplied with the measured yearly precipitation. The deposition calculated by this method is higher than found in the present study for stations Keldsnor, Aspvreten, and Rörvik with about the same amount of precipitation. If the same calculation was used for Aspvreten in the present study the Hg deposition would be 8.2 $\mu g/m^2$ instead of 6.1 $\mu g/m^2$.

The station at Birkenes shows a high deposition of mercury which could be correlated with the high precipitation amount. At the three stations in Great Britain,[13] higher deposition rates have been found than in the present study.

During the period 1976 through 1978 bulk deposition of mercury was measured at three locations in Denmark by FORCE Institutes.[14] As shown in Table 4 the bulk deposition is a factor of 3 to 4 higher than the recent results. Whether this decrease in deposition is caused by a reduced atmospheric input or is partly caused by a different sampling technique is impossible to say. The sampler was a PE funnel placed on the top of a 5-L PE bottle which was sheltered against sunshine by placing it in a gray PVC cylinder. The funnel was placed 2 m above the ground; 100 to 200 mL of a solution containing 64 mL concentrated nitric acid and 2 μg gold tetrachloride per 100 mL was added to the sampling bottle. The analytical methods were the same RNAA as used in the present study. An effect on the Hg level by the material used in the collector cannot be excluded.

In a Danish report about mercury,[15] an average deposition of mercury of 65 $\mu g/m^2$/year in Denmark was used for setting a mercury input budget. A value 4 to 8 times lower is found in the present study. It is further cited in that report that the deposition before 1850, estimated by analysis and dating of peat bogs,[16] was about 15 $\mu g/m^2$/year. This is the same rate as found in the present study. In a recent study of historical deposition rates of mercury in Scandinavian peat bogs[17] the mercury deposition rate before 1850 was estimated to be about 8 $\mu g/m^2$/year at Rörvik and Birkenes, which is a factor of 3 to 4 lower than the recent deposition at Rörvik and Birkenes measured in 1987 through 1988 (Table 4) but close to the present bulk deposition at Rörvik.

The present measurements show that the yearly bulk deposition of mercury to Denmark and the southern part of Sweden with the surrounding sea area varies between 6 to 14 $\mu g/m^2$/year, with the lowest values measured at easternmost stations, including isolated islands which also have the lowest precipitation.

V. CONCLUSION

In 1990 through 1991 a mercury deposition monitoring network with 15 stations was set up in Sweden and Denmark around the southern part of the Baltic Sea and part of the North Sea. Stations on islands, on the coast, and inland were established. Sampling was performed monthly at all stations. At four stations the samples from one of the samplers were analyzed by both Swedish and the Danish laboratories using different analytical techniques. At the other stations the samples were analyzed in the country of collection. The results showed no significant difference between the results of radiochemical neutron activation analysis (RNAA) and plasma emission spectrometry (AES) at the 99% probability level. No significant difference was found for the RNAA and atomic fluorescence spectrometry (AFS) methods.

The reproducibility of the new samplers for collection of precipitation and measurement of bulk deposition of mercury was excellent with no significant difference between the two parameters for the duplicate samplers. In addition, pairs of samplers were separated by a few kilometers at some locations to study the influence of the local geography. Here too, the effect of these distances was insignificant.

The present measurements show that the yearly bulk deposition of mercury to Denmark and the southern part of Sweden with the surrounding sea areas varies between 6 to 14 $\mu g/m^2$/year, with the lowest values measured at the easternmost stations, including isolated islands which also have the lowest precipitation. Further, mercury deposition is shown to be negatively correlated with precipitation volumes.

A comparison was done with other studies in the area. Much lower values were found in the present study than in a Danish study from 1976 through 1978 where 40 to

65 μg/m²/year were found. It is not possible to decide whether this difference is caused by a reduction in inputs and/or different sampling techniques.

ACKNOWLEDGMENTS

This project has been financed by the Nordic Council of Ministers and by the Danish National Agency of Environmental Protection. We would like to thank the staff at IVL and FORCE for analyzing the samples and all those who have been responsible for the sampling stations. Special thanks to Mads Hovmand at National Environmental Research Institute in Denmark who has had the responsibility for setting up all the Danish stations.

REFERENCES

1. Lindqvist, O., Ed., Mercury as an Environmental Pollutant. Refereed papers from the International Conference held in Gävle, Sweden, June 11–13, 1990, *Water Air Soil Pollut.*, 56, 1991.
2. Lindqvist, O., Ed., Mercury in the Swedish environment. Recent research on causes, consequences and corrective methods, *Water Air Soil Pollut.*, 55, 1991.
3. Semb, A. and Pacyna, J.M., Toxic Trace Metals and Chlorinated Hydrocarbons: Sources, Atmospheric Transport and Deposition. Nordic Council of Ministers, NORD, 74, 1988.
4. Iverfeldt, Å., Occurrence and turnover of atmospheric mercury over the Nordic countries, *Water Air Soil Pollut.*, 56, 251–265, 1991.
5. Glass, G.E., Sorensen, J.A., Schmidt, K.W., Rapp, G.R., Yap, D., and Fraser, D., Mercury deposition and sources for the Upper Great Lakes Region, *Water Air Soil Pollut.*, 56, 235–249, 1991.
6. DS 2214, Danish Standard: Water Analysis—Sampling of Natural Water for Analysis of Trace Metals, 1990.
7. Jensen, K. and Carlsen, V., Low level mercury analysis by neutron activation analysis, *J. Radioanal. Chem.*, 47, 121–134, 1978.
8. Iverfeldt, Å., Mercury in the Norwegian Fjord Framvaren, *Mar. Chem.*, 23, 441–456, 1988.
9. Bloom, N.S. and Fitzgerald, W.F., Determination of volatile mercury species at the picogram level by low-temperature gas chromatography with cold-vapor atomic fluorescence detection, *Anal. Chim. Acta* 15, 151–161, 1988.
10. Berman, S.S. and Boyko, V.J., ICES Sixth Round Intercalibration for Trace Metals in Estuarine Water (JMB 6/TM/SW), *Coop. Res. Rep.*, 152, 1987.
11. Duncan, A.J., Quality Control and Industrial Statistics, 5th ed., R.D. Irwin, Burr Ridge, IL, 1986.
12. Iverfeldt, Å., Mercury in forest canopy throughfall water and its relation to atmospheric deposition, *Water Air Soil Pollut.*, 56, 553–564, 1991.
13. Paris Commission, Measurements and Calculations of Atmospheric Input to the North Sea in 1989. Eighth Meeting of the Working Group on the Atmospheric Input of Pollutants to Convention Waters, Bilthoven 7–9 November 1990, Summary Record, Annex 6, 1990.
14. Isotopcentalen, Mercury in Precipitation. Report to Statens Naturvidenskabelige Forsningsråd, 19 pp, 1979 (in Danish).
15. Miljøstyrelsen, Kviksølvredegorelse (Report on Mercury). Redegørelse fra Miljøstyrelsen, 5, 104 pp., 1987 (in Danish).
16. Pheiffer Madsen, P., Peat bog records of atmospheric mercury deposition, *Nature*, 293, 127–130, 1981.
17. Jensen, A. and Jensen, A., Historical deposition rates of mercury in Scandinavia estimated by dating and measurement of mercury in cores of peat bogs, *Water Air Soil Pollut.*, 56, 769–777, 1991.

Measurements of Atmospheric Mercury in the Great Lakes Basin

Gerald J. Keeler, Marion E. Hoyer, and Carl H. Lamborg

CONTENTS

ABSTRACT: To investigate the levels and atmospheric transport of mercury, and to investigate the emission and transformation of atmospheric mercury in source regions, the University of Michigan Air Quality Laboratory (UMAQL) has initiated several studies in the Great Lakes region. Reported here are results from sampling campaigns in two urban areas, Detroit, MI and Chicago, IL, several rural locations in Michigan, as well as over-water measurements on Lake Michigan aboard the research vessel *Laurentian*. Vapor- and particulate-phase Hg samples were collected, along with atmospheric acids and other trace elements, to investigate the urban/industrial sources of atmospheric mercury. We describe here our initial results from the ambient sampling and present our early results from precipitation Hg sampling across the state of Michigan.

This paper will focus on the results of our multi-site atmospheric mercury measurements in rural and urban areas. We will highlight the finding that Hg levels in urban areas are significantly elevated over those measured concurrently in rural areas. Vapor phase Hg levels were on the average 4 times higher (8.7 ng/m^3 vs. 2.0 ng/m^3) in Chicago than in South Haven, MI. Furthermore, a diurnal pattern was observed in the vapor-phase Hg levels measured at the Chicago site. The average concentration (ng/m^3) for AM (8 am to 2 pm) samples was 3.3 times greater than the NIGHT (8 pm to 8 am) samples and the average concentration for PM (2 pm to 8 pm) samples was 2.1 greater than the NIGHT samples (average of NIGHT

1–56670–066–3/94/$0.00+$.50
© 1994 Lewis Publishers

samples, 3.7 ng/m^3). Particle-phase Hg concentrations were also higher, on average, at the Chicago site (98 pg/m^3 vs. 19 pg/m^3) in South Haven.

Precipitation Hg levels measured in Michigan varied from 1.5 to 46.5 ng/L during the March through May period discussed here. Current studies at the UMAQL are investigating the methods utilized, developing new techniques for atmospheric Hg collection and analysis, and speciation of Hg in all phases found in the atmosphere (vapor, particle, aqueous).

I. INTRODUCTION

The importance of mercury as an environmental contaminant stems from its ubiquitous nature which is due largely to the multitude of sources (both anthropogenic and natural), its volatility, mobility, and persistence in the environment. While mercury was once thought to be a threat to human and environmental health only in the locality of large industrial facilities, recent research has found that even remote and pristine waters can have elevated levels of mercury (predominantly methylmercury) in fish. Inherent in our ability to understand the toxicity, bioaccumulation, chemistry, and transport of this ubiquitous contaminant is the requirement that an interdisciplinary and regional approach be applied.

Great progress has been made in recent years in understanding the cycling of mercury in aquatic and terrestrial ecosystems, such that the importance of the atmospheric pathway to these cycles is now well recognized.[1-4] It is well known that Hg is found in the environment in various chemical forms and complexes, some of which are very toxic.[5,6] While the concentration of mercury in the atmosphere in remote locations is typically quite low (ppt), mercury can bioaccumulate in animal tissue, such that even in the presence of extremely low concentrations of mercury in the water column, concentrations of mercury in fish tissue can reach levels that pose a significant human and wildlife health risk.[7] In fact, many states now issue fish consumption advisories because of the high levels of Hg in fish tissue. In Michigan alone, 40 of the 107 lakes studied by the Michigan Department of Natural Resources (MDNR) from 1987 through 1990 were found to contain fish with levels of mercury greater than the public health fish consumption advisory level of 0.5 mg Hg/Kg.[7] Because the relative amounts of atmospheric Hg species may be altered in response to changes in regional tropospheric chemistry,[5,8] the distribution of atmospheric Hg species and the importance of Hg deposition need to be better understood.

To investigate the levels and atmospheric transport of mercury in the Great Lakes basin and to investigate the emission and transformation of atmospheric mercury in urban/source regions, the University of Michigan Air Quality Laboratory (UMAQL) has initiated several studies in the Great Lakes region. Reported here are results from sampling campaigns in two urban areas: Detroit, MI and Chicago, IL and several rural locations: Ann Arbor, South Haven, and Pellston MI, and over Lake Michigan on the Research Vessel *Laurentian*. Samples were collected at these sites as part of three distinct projects: (1) vapor and particulate phase mercury samples were collected at three sites (in Chicago, IL, over Lake Michigan, and downwind on the shore of Lake Michigan) as part of a comprehensive 30-day sampling campaign, the Lake Michigan Urban Air Toxics Study (LMUATS), a cooperative project between the USEPA and UMAQL conducted from July 10 to August 9, 1991; (2) an ongoing 2-year study initiated in March 1991 at three rural sites in Michigan where precipitation is collected on an event basis and subsequently analyzed for total Hg, operationally defined Hg species, and major ions; and (3) two sites in Detroit and one site in Ann Arbor were operated simultaneously on a daily basis during April 1992, during which time vapor- and particulate-phase Hg samples were collected along with atmospheric acids and trace elements, to investigate the urban/industrial sources of atmospheric mercury. The overall objective of our Hg program is to provide an assessment of the atmospheric levels of Hg in the Great Lakes basin and to utilize various meteorological and hybrid modeling techniques to diagnose the major source(s) of this International Joint Commission (IJC)-defined critical pollutant in the Great Lakes region.

II. STUDY SITES

Sampling sites utilized for the intensive 1-month LMUAT study included a site at the Illinois Institute of Technology (IIT) in Chicago, IL, aboard the Research Vessel *Laurentian* (LAU), and a farm near South Haven, MI (SHA). Vapor and particulate mercury measurements were taken as part of the LMUATS to provide accurate mercury measurements for the Great Lakes Region using state-of-the-art clean sampling and analysis techniques, to investigate spatial and temporal variations in vapor and particulate mercury, to investigate the transport and deposition of mercury, and to begin to investigate the potential sources and source regions for the observed mercury.

The objective of the 2-year multi-site study of mercury in precipitation in Michigan is to assess the magnitude, seasonal variation, transport, and sources of atmospheric mercury in this region of the Great Lakes. Samples are collected for each precipitation event at three sites in Michigan: (1) the University of Michigan Biological Station near Pellston, MI (also a National Atmospheric Deposition Program site), (2) the National Dry Deposition Network site northwest of Ann Arbor (Ann Arbor/Dexter), and (3) at the LMUATS site in South Haven, MI. In year two of this project, vapor and particulate mercury sampling will begin at all three sites. This unique database will be analyzed using several source apportionment techniques including principal component analysis, trajectory analysis, and Quantitative Transport Bias Analysis.[9]

Lastly, results of a preliminary investigation of vapor and particulate Hg levels in downtown Detroit, MI and Ann Arbor, MI are presented. These urban/source area measurement programs are critical to our ability to understand the atmospheric chemistry and source-receptor relationships for mercury in the global environment.

III. METHODS

Ultra-trace-metal clean sampling and analytical techniques are required when attempting to obtain accurate, reliable data on concentrations of mercury and other trace metals in environmental samples. All sampling supplies and storage vessels with which the sample comes into contract must be rigorously cleaned and carefully dried before use. We currently employ a 5-step, 11-day cleaning procedure[10] which begins with an acetone rinse, a wash in hot water with 1% Alconox®, careful rinsing in ultrapure water, a 6-h 80° C bath in ultrapure 3 M HCl, followed by thorough rinsing again with ultrapure water. Lastly, two separate periods (3 days, rinsed with ultrapure water, then 7 days) of room temperature soaking in 3.5% ultrapure HNO_3 are performed. The last soaking cycle is completed in a Class 100 clean room, so that after a final rinse with Milli-Q® water (containing <0.3 ng Hg/L) all supplies and sample bottles are allowed to dry before being triple-bagged in ziplocked polyethylene bags for transport to the field.

During all procedures involving sample setup and collection, particle-free gloves were worn and operators were positioned downwind of samples at all times. Because outdoor concentrations of Hg in all forms are lower than those of indoor areas, most of the sample handling was done outdoors.

A. COLLECTION AND ANALYSIS OF VAPOR-PHASE MERCURY

Vapor-phase mercury was captured on gold-coated sand traps (Brooks Rand, Ltd.) with glass fiber pre-filters in acid-cleaned all-Teflon® filter packs. The flow rate was maintained at 0.3 L/min, using Tylan mass flow controllers, which corresponds to sample volumes of about

0.2 and 0.4 m^3 for 12- and 24-h samples, respectively. The Au-coated sand traps efficiently collect most of the mercury species in the atmosphere. However, upon thermal desorbtion at 400° C, all forms of mercury (specifically, mercuric chloride and organomercury forms) are not quantitatively removed from the gold surface as free elemental mercury.[11] Since cold vapor atomic fluorescence spectrometry (CVAFS) only quantifies elemental Hg, these forms pass through the detector unquantified. In this study, the mercury quantified by collection onto Au-coated sand traps will be referred to as vapor-phase mercury. However, since vapor-phase Hg is dominated by Hg^0 in most locations, the concentration measured is probably very close to total gaseous mercury.[8,11] A minimum of one field blank was taken for every six samples collected. During the 30 days of the LMUAT study, vapor-phase samples at SHA were collected for a duration of 12 h (8 am to 8 pm, CDT), and at IIT 12-h daytime vapor-phase samples were collected when the R/V *Laurentian* was in port, and two 6-h daytime (8 am to 2 pm, 2 pm to 8 pm) and one 12-h nighttime sample were collected when the R/V *Laurentian* was at station. Vapor-phase samples on the R/V *Laurentian* were also collected for two 6-h periods during the day (8 am to 2 pm and 2 pm to 8 pm) and for 12 h during the night.

After collection, samples were stored, with Teflon® end-plugs Teflon® taped, inside individual polyethylene tubes, and then triple-bagged in polyethylene ziplocked bags. During LMUATS, samples were stored outdoors in waterproof containers and shipped twice per week to Brooks Rand, Ltd. for analysis. For all subsequent studies, samples were analyzed at the UMAQL. Vapor-phase mercury levels were determined by thermal desorbtion (at 400° C) using the dual amalgamation technique described by Bloom and Fitzgerald[12] followed by CVAFS. In the course of our preliminary studies the method detection limit, determined as 3 times the standard deviation of the field blank, was 45 pg total Hg or 94 pg/m^3. The field blanks were handled exactly like all other samples including field handling, shipping, and analysis. The precision in the vapor-phase Hg determinations was better than 15%, as determined from duplicate analysis of Au-coated sand traps containing 1 ng of Hg, in the LMUATS and early Ann Arbor sampling campaigns. The precision of the vapor Hg analysis was better than 10% for the more recent results given in this chapter.

B. COLLECTION AND ANALYSIS OF PARTICULATE-PHASE MERCURY

All sample collection supplies, including filter packs and cyclones, were constructed of Teflon® or were Teflon® coated. Filter packs, forceps, Teflon® sample storage vials, petri dishes, and other field sampling equipment were rigorously acid-cleaned in the 5-step, 11-day process described above. Glass and quartz fiber filters which were acid-extracted were pre-fired at 500° C for a minimum of 1 h prior to use in sampling. Teflon® filters used for collection of fine-fraction particulate material for subsequent analysis by neutron activation analysis (NAA) did not require pretreatment. Particulate samples were collected using acid-cleaned open-faced Teflon® filter packs at a nominal flow rate of 30 L/min. Exposed filters were placed in 25 mL acid-cleaned Teflon® vials or acid-cleaned petri dishes which were capped tightly, sealed with Teflon® tape, triple-bagged in polyethylene, and frozen until analysis. Field blanks were routinely taken at all sampling sites, with blank values ranging from 100 to 250 pg total Hg. Further information on the particulate-Hg collection and analysis can be found in chapter 7 of this section by Lamborg et al. For completeness, a brief description of the methods are given here.

Extraction and analysis of glass and quartz fiber filters were performed in a class 100 clean room and, in most instances, reagents were further purified to obtain lower blank values. Samples were extracted for 30 min by sonication in a 2 *N* nitric acid/sulfuric acid mixture in Teflon® vials. After extraction, samples were oxidized with BrCl for 1 h. The BrCl oxidation step is needed to break down the organic-Hg forms which are then available for reduction, yielding total acid-extractable Hg. Sample aliquots were analyzed after reduction with NH_2OH and then $SnCl_2$. Mercury was purged from the solution with Hg-free N_2 and was subsequently captured onto a Au-coated sand trap. The total Hg concentration was determined using the dual-amalgamation cold vapor atomic fluorescence technique described above. A calibration curve was generated by spiking vials containing blank filters with varying amounts of a 2 ng/mL standard (in nitric acid and 1% BrCl). The method detection limit for total particulate mercury concentrations performed for the studies reported here is 9 pg/m³. However, our current procedural detection limits are about a factor of two lower due to lower blanks from cleaner reagents utilized for extraction and analysis. The importance of checking the Hg levels in all reagents and acids on a daily, or at least weekly, basis cannot be overstated, especially as Hg content in acids can change over time.

Neutron activation analysis was performed on Teflon® filters collected during the LMUATS study by Dr. Ilhan Olmez at the Massachusetts Institute of Technology. Unlike most other techniques, NAA does not require sample extraction, addition of reagents, or other sample preparation. The only sample handling in the laboratory involves transferring the sample material, in this case, 47 mm Teflon® filters, into acid-cleaned containers. This process takes only a few seconds and the sample material is not exposed to the laboratory environment for any significant time. Samples were irradiated at the MIT Nuclear Reactor Laboratory for 12 h at a thermal flux of 10^{13} n/sec cm². Total Hg determinations were performed using the Hg-196 isotope as it is much more sensitive for Hg determinations due to its higher cross section and shorter half life. Using the Hg-196 isotope also has the advantage of not having any significant spectral interferences as is the case if Hg-203 is used (large interference with Ta and Se).

C. COLLECTION AND ANALYSIS OF MERCURY IN PRECIPITATION

An accurate assessment of mercury in precipitation samples depends on careful consideration of possible inter-conversion processes, contamination sources, sample preservation, handling, and storage. Automatic precipitation collectors with acid-cleaned Teflon®-coated funnels (MIC-B, Thornhill, Ontario) are being used to collect precipitation samples on an event basis.[13] Custom Teflon® adapters, through which the sample passes from the funnel into a rigorously acid-cleaned 10-L borosilicate glass vessel, were made at the University of Michigan. After sample collection, the 10-L bottle is carefully swirled and a sample is poured from this collection vessel into an acid-cleaned 1-L borosilicate glass bottle. The sample bottle is capped, Teflon® taped, triple-bagged in ziplocked, acid-cleaned polyethylene bags, and shipped by overnight mail to the UMAQL for analysis. The Teflon® funnel is rinsed biweekly with ultrapure Milli-Q® water which contains less than 0.3 ng Hg/L. Several liters of ultrapure water are provided for the funnel rinse, of which 500 mL is saved for analysis as the control. The remainder is collected through the funnel into the glass vessel to be analyzed as the field blank. The 10-L glass collection vessel is exchanged with a freshly cleaned bottle on a biweekly schedule as well. Contamination levels from both the collection vessel and the Teflon® funnel can then be determined and tracked. In general, funnel and collection vessel rinses are indistinguishable from the ultrapure water used to rinse them. The 1-L sample bottles are also checked routinely and are consistently free of Hg.

On the day of receipt at UMAQL, a 100-mL aliquot of each sample is poured off for analysis of major ions and pH. The remaining sample volume is then oxidized to 1% bromine monochloride. Analysis of the precipitation for total Hg is performed within 48 h by reduction with NH_2OH and $SnCl_2$, and purging the reduced Hg onto a gold-coated sand trap for subsequent analysis by CVAFS as described above. Experiments are being conducted at UMAQL to determine if Hg is lost from precipitation samples between the time precipitation stops and when they are received at UMAQL. In addition, we are investigating loss of mercury from unoxidized samples after receipt at UMAQL. Stability experiments in our laboratory have indicated that precipitation samples oxidized as described above are stable

Table 1   **Vapor-phase Hg measurements during the LMUATS: Chicago (IIT),
aboard the R/V** *Laurentian* **(LAU) and South Haven (SHA) (ng/m³).**

Site	n	Median	Mean	Min	Max
IIT	58	4.5	8.7	1.8	62.7
LAU[a]	25	2.2	2.3	1.3	4.9
SHA	38	1.8	2.0	1.8	4.3

[a] Sampling Dates: 7/11 to 7/12, 7/25 to 7/27, 8/5 to 8/8.

for a minimum of 72 h after oxidation. In the initial phases of this project our detection limit, defined as the average of the total reagent blanks, (n = 39) was 0.17 ng Hg/100 mL aliquot (three times the standard deviation of the total reagent blank = 0.15 ng Hg/100 mL aliquot). The Hg is believed to have been from the acid being used to make bromine monochloride. Currently, the detection limit for this analysis is 50 pg Hg/100 mL sample aliquot. We have also begun performing analysis for dissolved Hg (purging nonfiltered, nonoxidized samples), and reactive Hg (which involves acidifying nonoxidized samples to pH = 1 with ultrapure HCl before reduction).

IV.   RESULTS

A.   THE LAKE MICHIGAN URBAN AIR TOXICS STUDY

1.   Vapor-Phase Mercury Measurements

The variation in atmospheric Hg concentrations was investigated during a 1-month study in the southern Lake Michigan basin. Vapor-phase mercury concentrations measured at IIT ranged from 1.8 to 62.7 ng/m³ with an average of 8.7 ng/m³ (see Table 1). Over-water measurements performed on the R/V *Laurentian* included 25 vapor-phase mercury samples which were collected during three separate cruises. The vapor-phase mercury concentrations measured on the R/V *Laurentian* were typically quite low and ranged from 1.3 to 4.9 ng/m³ with an average of 2.3 ng/m³.

Concentrations observed in South Haven, MI also were generally low with a range of 1.8 to 4.3 ng/m³ and an average of 2.0 ng/m³. The range in concentrations observed over Lake Michigan and in South Haven were quite similar even though the R/V *Laurentian* was not on the lake for the entire month of study and the two were separated by more than 75 km at most times.

2.   Diurnal Variations in Vapor-Phase Mercury

It has long been suspected that the levels of many hazardous pollutants are higher and more temporally variable in the urban areas than in rural locations. To investigate this hypothesis we sampled more frequently at the Chicago (IIT) site during the LMUATS. At IIT, 18 samples were collected between 8 am to 2 pm (designated as **AM**), 17 samples were collected between 2 pm to 8 pm (**PM**), 11 daytime 12-h samples were collected between 8 am to 8 pm (**DAY**), and 12 nighttime samples from 8 pm to 8 am were collected (**NIGHT**) in order to investigate potential diurnal behavior of vapor-phase mercury. The average concentration (ng/m³) for AM samples was 3.3 times greater than the NIGHT samples and the average concentration for **PM** samples was 2.1 times larger than NIGHT samples (average of NIGHT samples, 3.7 ng/m³). Furthermore, the average of the concentrations of the AM and PM samples together resulted in a mean of 10.1 ng/m³, which closely resembled the 9.9 ng/m³ average vapor-phase concentration for DAY samples.

3.   Particulate Mercury Measurements

Total particulate mercury was measured at the three sites for periods when the R/V *Laurentian* was at station. At the IIT site 16 24-h particulate Hg samples were collected on glass fiber

Table 2   **Particle-phase total Hg measurements during the LMUATS: Chicago (IIT), aboard the R/V *Laurentian* (LAU) and South Haven (SHA) (pg/m³).**

Site	n	Median	Mean	Min	Max
IIt	16	60	98	22	518
LAU[a]	9	24	28	9	54
SHA	18	19	19	9	29

[a] Sampling Dates: 7/23 to 7/27, 8/5 to 8/7.

filters and the concentrations varied from 22 to 518 pg/m³ (Table 2). The average concentration of particle-phase mercury for the period discussed was 98 pg/m³. On the R/V *Laurentian*, nine samples were collected giving an average particulate-phase mercury concentration of 28 pg/m³, with a range of 9 to 54 pg/m³. In South Haven, 18 glass fiber filters were collected with an average particulate-phase mercury concentration of 19 pg/m³ and a range of 9 to 29 pg/m³.

The particle-phase mercury consistently represented a small percentage of the total atmospheric Hg (vapor-phase + particulate Hg) measured during the Lake Michigan Study. Only 1.7% of the total Hg was in the particulate form at IIT, 1.2% at South Haven, and 1.3% on the R/V *Laurentian*. The range in the percentage of mercury found in the particle phase was largest at IIT (0.07 to 7.3%). Particle-phase mercury at South Haven and on the R/V *Laurentian* varied from 0.6 to 1.9% and 0.6 to 2.3% of the total atmospheric Hg observed.

B. ATMOSPHERIC MERCURY DEPOSITED IN MICHIGAN

Mercury in precipitation is thought to be the dominant delivery mechanism for inputs of Hg to the Great Lakes and their basin. The first 3 months of precipitation Hg data available from our network are shown in Figure 1. During the period March through May 1992, 15 samples were collected at Pellston, 14 at South Haven, and 24 at the Ann Arbor/Dexter site. The total Hg concentration in precipitation in Pellston ranged from 1.5 to 46.5 ng/L, in South Haven the concentration ranged from 4.0 to 34.2 ng/L and at the Ann Arbor/Dexter site the mercury concentration ranged from 1.7 to 44.2 ng/L (Table 3). Interestingly, Pellston, the most northern site, recorded the highest and lowest total Hg concentrations observed at the three sites spread across the state. While the average concentrations for this period are not statistically different, South Haven was slightly higher on average than the other two sites during the spring period.

Samples are also being analyzed for major ions to provide additional information needed for understanding the variations of Hg observed across Michigan. Results of the major ion content of these event-based samples will be reported elsewhere. Mixed-layer trajectories are currently calculated in real-time and plotted for each precipitation event in order to diagnose the source regions contributing to Hg deposition in Michigan. Additional meteorological information including wind speed, wind direction, and temperature are also collected on an hourly basis at each of the sites.

C. DETROIT AND ANN ARBOR INVESTIGATIONS

During the month of April 1992 a 10-day intensive pilot study was carried out to investigate the levels of atmospheric acids and Hg at two sites in the city of Detroit, and one site in Ann Arbor. The levels of particulate and vapor-phase Hg were determined using the same methodologies utilized in the LMUATS. The only difference in the protocols utilized in the Detroit Pilot Study was that we ran extra Au-coated sand traps in series to ensure that we could quantify breakthrough of the vapor phase Hg in case of direct plume impact from one of the large sources in the area. Measurements of fine particles (less than 2.5 μm) were also carried out for subsequent analysis by X-ray fluorescence at the EPA.[14] The measurements in Ann

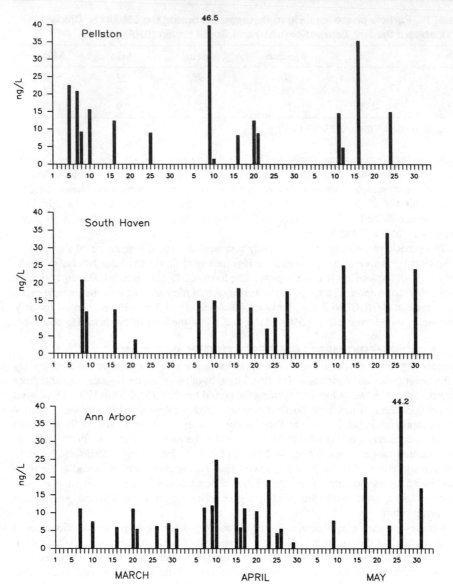

Figure 1 Total mercury concentrations in precipitation at three sites in Michigan in 1992.

Table 3 **Total Hg in precipitation collected at three sites in Michigan (ng/L).**

Site	n	Median	Mean	Min	Max
Pellston	15	12.4	15.8	1.5	46.5
S. Haven	14	15.0	16.4	4.0	34.2
Ann Arbor	24	9.1	11.7	1.7	44.2

Table 4   **Vapor- and particulate-Hg collected at three sites in Michigan during April 1992. Mean and maximum values.**

Site	Vapor (ng/m^3)	Particle (pg/m^3)
Ann Arbor	2.0 (4.4)	100 (207)
Detroit—Site A	>40.8 (>74)	341 (1086)
Detroit—Site B	3.7 (8.5)	297 (1230)

Arbor were performed daily during this intensive study to investigate the potential gradient from the rural upwind site to the two urban/industrial sites in downtown Detroit.

The concentrations of vapor and particulate Hg measured concurrently at the three sites are given in Table 4. Measurements of the vapor-phase Hg at the Detroit Site A were complicated by the fact that the first three samples resulted in off-scale readings (>6 ng Hg/trap) for both the front and back Au-coated sand traps. This was the first time that this phenomenon had been observed in the field and we hypothesized that the capacity of the Au-coated trap had been exceeded. The remaining seven samples were run at one third the flow rate (0.1 L/min) and a third Au-coated sand trap was added in series to the sampling train. The same phenomenon was again observed on two subsequent sampling days, with the front traps being off-scale and the second and third traps having considerable Hg on them. To determine the concentration of vapor-phase Hg sampled we used the reported capacity of Au-coated sand traps of 20 ng total Hg. For the days when Hg collected on both the front and back-up traps exceeded the range quantifiable by the CVAFS analyzer, we used the 20 ng saturation concentration for the front trap. The amount on the back-up trap was calculated as the maximum measured by the analyzer, since we do not have proof that these traps were also saturated. While this probably underestimates the "true" amount of Hg found on the second trap we felt that this approach was defensible and truly represented the minimum vapor-phase Hg concentration. After seeing the same phenomenon at the lower flow rates, we performed experiments in the laboratory to determine the capacity of the Au-coated sand traps used in this study. The Au-coated sand traps were found to have a minimum capacity of approximately 200 ng Hg. This suggests that the actual Hg concentrations observed on those days with break-through was probably much higher than what we estimated here. The levels reported here should be viewed as a lower estimate on the vapor-phase Hg concentrations measured.

It should be noted that the levels of Hg that we are suggesting here are much higher than previously reported anywhere in North America. The downtown Detroit area has several large point sources of Hg that could have contributed to the levels observed. A large municipal waste incinerator is located only 9 km southwest of the Detroit Site A and a large sludge incinerator is located only 5 km from Detroit Site B. Curiously, the break-through phenomenon was only observed at Site A and the highest particulate Hg concentration was observed at Site B. Further trace element data and individual particle analysis will be employed to further define the potential source of the elevated Hg measured in Detroit.

V. CONCLUSIONS

Our recent observations of atmospheric Hg concentrations in the Great Lakes Basin suggest that our understanding of the behavior of Hg in the atmosphere is incomplete. The vapor and particulate mercury concentrations measured during the 1-month study decreased from Chicago (IIT) to downwind sites on the R/V *Laurentian* and in South Haven, MI. Diurnal variation in vapor-phase mercury observed at IIT indicated that samples collected between 8 am and 2 pm may be influenced by local sources impacting the sampling site during typical daytime flow patterns, while predominant nighttime wind patterns (from Lake Michigan) may not result in local point source impacts at IIT.

Vapor phase concentrations measured in South Haven were similar to those measured at other rural and remote locations in the Great Lakes region.[15] Vapor-phase mercury levels measured in South Haven did not demonstrate episodic behavior with flow from the southwest urban source region as did other pollutants measured. However, fine fraction (<2.5 µm) particulate Hg concentrations as determined by NAA did reveal a peak during the main episode with SW transport. These data will be merged with measurements taken for organic and elemental carbon, volatile organic carbon, polyaromatic hydrocarbons, fine and coarse trace elements, and acidic aerosol and gaseous species. Receptor modeling techniques will be applied to the combined data sets to determine sources and source strengths of the observed atmospheric mercury.

While ambient mercury levels at South Haven were uniformly low, these low concentrations are also present in remote environments where the atmosphere is implicated as the dominant source of mercury to waterbodies. In sharp contrast to vapor-phase mercury measurements made in South Haven during LMUATS, ongoing investigations of vapor-phase mercury collected in Ann Arbor indicate that, on occasion, levels measured are an order of magnitude higher than typical rural values.

Particulate mercury concentrations varied widely at IIT, possibly due to local source influence. However, the processes that control formation of particulate mercury are not well understood. Volatilization of mercury from the particle phase during sampling probably represents a small loss of particulate mercury during the 12- to 24-h duration samples at the flow rates used in this study. Experiments being conducted in our lab using Au-coated annular denuders will help define the actual rate of loss (or gain) of Hg, if any, under actual field sampling conditions. Potential artifacts in particulate Hg determinations are suspected to be greatest in source areas or near large point sources where the atmospheric chemistry and Hg speciation is thought to change most dramatically.

Furthermore, it was pointed out earlier that Au-coated sand traps efficiently trap most forms of vapor phase Hg but the CVAFS analysis only determines elemental Hg vapor. Thus, it is very possible that the concentrations of vapor phase Hg measured in the Detroit area were even higher than what we measured, as the dominant form of Hg emitted from the closest large point sources may be mercuric chloride. The Hg (II) vapor would not have been determined but could have been substantial in this airshed. Future studies are being planned in the Detroit area to investigate the atmospheric conversion of Hg downwind of large sources. A failure to observe any Hg (II) in the vicinity of these large sources would suggest a rapid transformation to elemental Hg in the air.

The levels of total Hg in precipitation at the three-site network in Michigan do not reveal a concentration gradient from the northern rural location (Pellston) to southern rural areas of South Haven and Ann Arbor for this period of the study. The precipitation volume appears to be the controlling factor for the differences in deposition observed between these sites (not discussed here). As the study progresses into warmer summer months, the observed similarity in total Hg concentration may change. In addition, analysis of operationally defined reactive and dissolved Hg will aid in understanding the transformation of Hg in precipitation and the consequences for Hg deposition to bodies of water.

ACKNOWLEDGMENTS

We would like to thank our colleagues in the Lake Michigan Urban Air Toxics Study for their assistance with the planning and field operations, including: Tom Holsen, Nasrin Khalili, and Ken Noll (IIT), Gary Evans, Alan Hoffman, Bob Stevens, and Mack Wilkins (EPA). We would also like to thank Dr. Ilhan Olmez of the M.I.T. Nuclear Reactor Laboratory for particulate mercury analyses, as well as the many UMAQL people who helped with site operations for the Ann Arbor and Detroit studies including Tim Dvonch, Ganda Glinsorn, Jennifer Falk, as well as those from Wayne County including Peter Warner, and his staff.

We are also grateful to our dedicated field operators, Mary Barden, Katherine Beverstock and Robert Vande Kopple. Funding for the Michigan Mercury Precipitation Project was provided by The Michigan Great Lakes Protection Fund. We would also like to thank the reviewers of this manuscript for their useful comments and suggestions.

REFERENCES

1. Lindberg, S. E., Turner, R. T., Meyers, T. P., Taylor, G. E., and Schroeder, W. H., Atmospheric concentrations and deposition of Hg to a deciduous forest at Walker Branch watershed, TN, USA, *W.A.S.P,* 56, 577, 1991.

2. Schroeder, W. H., Sampling and analysis of mercury and its compounds in the atmosphere, *Environ. Sci. Technol.,* 16, 394A, 1982.

3. Lindqvist, O. and Rodhe, H., Atmospheric mercury—a review, *Tellus,* 37B, 136, 1985.

4. Glass, G. E., Sorensen, J. A., Schmidt, K. W., and Rapp, G. R., New source identification of mercury contamination in the Great Lakes, *Environ. Sci. Technol.,* 24, 1059, 1990.

5. Lindberg, S. E., Stokes, P. M., Goldberg, E., and Wren, C., Rapporteur's report on mercury, in *Lead, Cadmium, and Mercury in the Environment,* Hutchinson, T. C. and Meema, K., Eds., United Nations Scientific Committee on Problems in the Environment Series, John Wiley & Sons, New York, 1987, 17.

6. Suzuki, T., Imura, N., and Clarkson, T., *Advances in Mercury Toxicology,* Plenum Press, New York, 1991, 1.

7. Michigan Department of Natural Resources, Michigan Fish Contaminant Monitoring Program, Annual Report, MI/DNR/SWQ-91/273, 1991.

8. Ottar, B., Lindberg, S. E., Voldner, E., Lindqvist, O., Mayer, R., Steinnes, E., and Watt, J., Special topics concerning interactions of heavy metals with the environment, in *Control and Fate of Atmospheric Trace Metals,* Pacyna, J. and Ottar, B., Eds., NATO Advanced Science Institute Series, Kluwer Academic Publishers, Dordrecht, Holland, 1989, 365.

9. Keeler, G. J. and Samson, P. J., On the spatial representativeness of trace element ratios, *Environ. Sci. Technol.,* 23, 1358, 1989.

10. Rossmann, R. and Barres, J., Trace element concentrations in near-surface waters of the Great Lakes and methods of collection, storage, and analysis, *J. Great Lakes Res.,* 14, 188, 1991.

11. Schroeder, W. H., Hamilton, M. C., and Stobart, S R., The use of noble metals as collection media for mercury and its compounds in the atmosphere, *Rev. Anal. Chem.,* 8, 179, 1985.

12. Bloom, N. and Fitzgerald, W. F., Determination of volatile mercury species at the picogram level by low-temperature gas chromatography with cold-vapor atomic fluorescence detection, *Anal. Chim. Acta,* 208, 151, 1988.

13. Mierle, G., Aqueous inputs of mercury to Precambrian Shield lakes in Ontario, *Environ. Toxicol. Chem.,* 9, 843, 1990.

14. Stevens, R., USEPA-AREAL, RTP, N.C., personal communication, 1992.

15. Fitzgerald, W. F., Vandal, G. M., and Mason, R. P., Atmospheric cycling and air-water exchange of mercury over mid-continental lacustrine environments, *W.A.S.P.,* 56, 745, 1991.

Diurnal Variations in Mercury Concentrations in the Ground Layer Atmosphere

Kestutis Kvietkus and Jonas Sakalys

CONTENTS

ABSTRACT: Vertical distributions and diurnal courses of mercury concentration in the ground layer atmosphere in Lithuania and Kirghizia are presented. The correlation coefficients of radon (Rn), ozone (O_3), and mercury (Hg) time variations and diurnal course concentrations are analyzed. Mercury vapor in the ground layer atmosphere was collected using silver-coated denuders. Sampling efficiency of elemental mercury vapor with the diffusive samplers coated by silver is 98%. Analysis was performed by an atomic fluorescence analyzer with an absolute detection limit of 0.2 pg.

I. INTRODUCTION

About 95% of mercury in the ground layer atmosphere is in the gaseous phase and mercury vapor, like any gas, participates in atmospheric processes.[1-9] Mercury vapor is a potentially toxic pollutant and at the same time is important as an indicator of the location of minerals. Mercury is an accompanying element in mineral formation; it may also form separate Hg locations. Owing to its high volatility, excessive quantities of mercury may form anomalies that can serve as an indicator of ore bed locations. Behavior of these dispersive aureoles is determined to a great extent by the diurnal cycle, turbulence, and by meteorological parameters that must be taken into account when gaseous mercury measurements are performed in the atmosphere.

Seasonal variations of mercury concentrations have been thoroughly investigated in Sweden by Brosset.[4] Diurnal variations of mercury concentrations over the Pacific Ocean Have been studied by Seiler et al.[5] and by Slemr et al.[6] Brosset suggested that the total Hg in air may be divided into two fractions: the most important one may be a background, probably the result of reemission of Hg by the ground and by natural water. The other fraction is highly dependent on air mass transport direction in the same way as soot. This fraction is mainly of anthropogenic origin. However, according to McNerney et al.[7] and Ozerova,[8] degasing of mercury from soil and from considerable depths of the earth is continuous. In different geographical location there are many strong local and regional sources of the endogenic origin of gaseous mercury which is degased in the air, too. From day to day the continually increasing interest for base and precious metals has led to vigorous exploration programs to locate new ore bodies. The mining industry continually open more and more new ore bodies, releasing significant amounts of gaseous mercury to the atmosphere. Thus, in the atmosphere there are both anthropogenic and endogenic origins of mercury today.

The gaseous mercury origin identification in the atmosphere is possible by considering ozone and radon as good tracers. The behavior of ozone in the atmosphere seems to be similar to the behavior of mercury of anthropogenic origin. The origin of radon in the ground layer atmosphere is considered by Styro[9] to be endogenic and the behavior of radon seems to be similar to the behavior of mercury of endogenic origin. According to a number of earlier measurements by Kvietkus et al.,[2,10] the diurnal variations of mercury concentrations in the ground layer atmosphere are mainly determined by the dominating origin of the mercury source in that area. Thus, the continuous measurements of Hg, Rn, and O_3 together in the ground layer atmosphere will help us to understand which portion of the mercury is related with anthropogenic or endogenic activities and probably to identify the origin and location of the mercury sources.

II. MATERIALS AND METHODS

During recent years, better understanding of many atmospheric aspects of mercury have became available as new analytical and sampling techniques were improved by a number of authors: Brosset,[4] Fitzgerald et al.,[11] Slemr et al.,[6] Fursov,[12] Kvietkus et al.,[19] Bloom et al.,[14] and Xiao et al. [15] New methods for the identification of mineral location via gaseous mercury aureoles in the ground layer atmosphere are currently being developed in various countries by Saukov,[16] McNerney et al.,[7] Fursov,[12] and Kvietkus et al.[17]

Mercury concentration measurements were performed in the ground layer atmosphere in an area free from endogenic mercury sources at the background monitoring station Aisetas in eastern Lithuania. A second sampling station was in an area of strong endogenic sources of mercury and directly above ore deposits at a location of a mercury-stibium belt at South Fergana (Kirghizia). The sampling period was for 30 days in the summer of 1986 (Station Aisetas) and in the summer of 1987 (Station South Fergana). Each day 24 samples of mercury vapor were sampled using two parallel sampling systems. Radon and ozone were measured continuously using devices developed in the Institute of Physics, Lithuania.[18,19] Sampling height varied from soil surface up to 8 m.

Mercury vapor in the ground layer atmosphere was collected by diffusive samplers—silver-coated dunuders developed by Kvietkus[20] to separate its gaseous constituent from the aerosol constituent. The sampling efficiency of elemental mercury vapor with the diffusive silver-coated tubes is 98%. In field conditions the efficiency may change slightly, therefore an identical tube is joined to the first one in a train for the purposes of efficiency control. The analysis was performed by an atomic fluorescence analyzer,[13] "Fluoran", designed by the Institute of Physics, with an absolute detection limit of 0.2 pg (carrier gas—helium).

III. RESULTS AND DISCUSSION

Figure 1 shows fragments of Hg, Rn, and O_3 measurements results at the background station Aisetas. The figure demonstrates a close correlation between mercury and ozone, and a negative correlation between mercury and radon. The measurements were performed during marginal changes in weather, i.e., when it was cloudy and wet and when it was sunny and dry. Mercury, radon, and ozone concentrations can be a little different during cloudy weather owing to great moisture and rain. The troposphere is considered by Matveev[21] to be the main source of ozone. With a change in turbulence, the flow of ozone from the upper atmosphere layers into the lower ones also changes. Soil is the source of radon, and with a change in

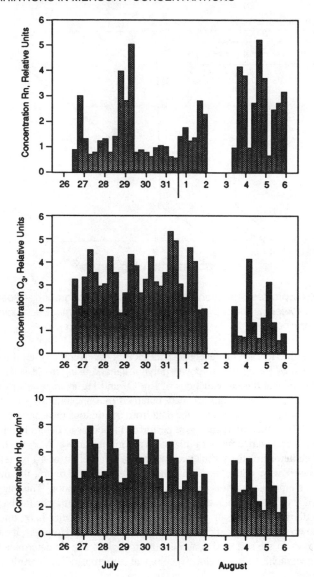

Figure 1 Fragments of Hg, Rn, and O₃ measurements results at background station Aisetas, Lithuania.

atmospheric turbulence radon concentrations in the ground layer atmosphere change also. Thus, there are two different sources of origin. Mercury origin can be identified according to its correlations between radon and ozone.

An additional experiment was made to get more information about the possible origin of mercury at the measurement station Aisetas. During three days (A, B, C), mercury concentrations at various heights were measured (Figure 2A, B, C). Vertical profiles of mercury concentrations presented in Figure 2 testify to the fact that the main source of mercury in the measurement region (Aisetas) is not the soil, but rather the atmospheric transport of air masses saturated with mercury from industrial regions. Curve B indicates possible air purification from mercury vapor as the air masses pass through the forest.

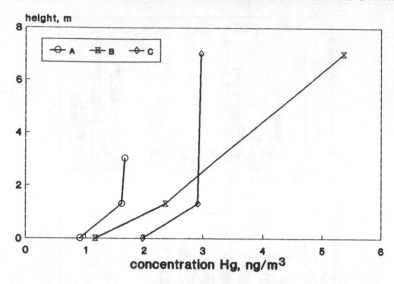

Figure 2 Vertical profiles of mercury concentrations near the ground, Aisetas, Lithuania. Curves A,B,C, represents three different days of measurements. Each point represents at least six determinations.

To reduce the interference of meteorological factors the average of all the experimental data was calculated by means of 5 rd power polynomial and using the least squares method. Figure 3 presents typical diurnal variations of Rn, O_3 and Hg average results of our measurements during a month, including the data obtained by concentrating samples every 2 h with two parallel sampling systems, and the data from continuous measurement of radon and ozone (Aisetas). These measurements were performed in order to determine more precisely the maximum in diurnal variations of the elements mentioned above. Also, the figure presents typical diurnal variations of the turbulence coefficient at a height of 10 m, referred to in a study by Matveev[21] with regard to local time. As seen in the figure, the maximum and minimum of the elements under discussion coincide in the afternoon. This suggests that the turbulent exchange of air masses in the ground layer atmosphere may be the basic factor that determines the diurnal variation of all three elements. On the other hand this process has caused well-known meteorological changes during a 24-h cycle. However, the level of mercury concentrations in the soil, and consequently in the air, can be determined to a certain extent by the geographic position, local geological structure of the earth's crust, and its receptor properties.

During the measurement period at the station Aisetas, the obtained results allow us to conclude that variations of mercury concentration are determined mainly by air mass turbulence at the measurement site and by the geographical-geological position of the measurement site.

The analysis of long-range air mass trajectories did not reveal any noticeable correlation between changes in air masses and mercury concentrations. Nevertheless, the problem of mercury transport with air masses has not been fully investigated as yet. Perhaps we may be able to conclude that today in the Baltic Sea region, the main source of pollution by mercury is the long-range transport of air masses from industrial regions.

According to numerous data in a study by Fursov,[12] the earth's crust contains rather intensive local sources of mercury vapor, for instance, locations of high HgS. In such cases vertical profiles of mercury vapor, in comparison to the profiles presented in Figure 2, may produce reverse variations, i.e., in approaching the surface of the soil the concentration of

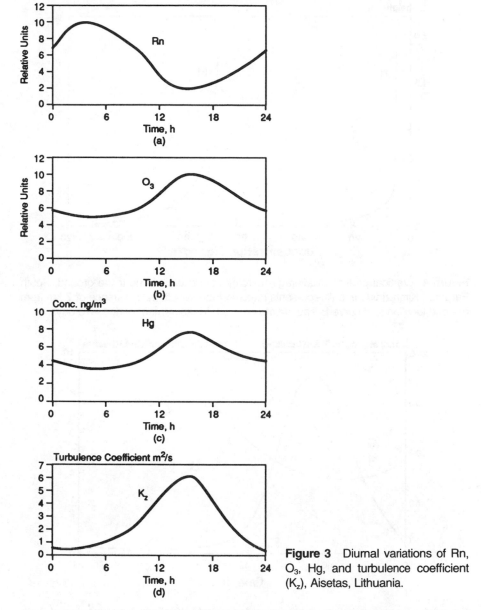

Figure 3 Diurnal variations of Rn, O_3, Hg, and turbulence coefficient (K_z), Aisetas, Lithuania.

mercury vapor increases (see Figure 4). A positive correlation between the concentration of mercury and radon may appear in this case.

Vertical measurements of gaseous mercury concentrations in South Fergana were performed from the ground to a height of 2 m. Diurnal variations were determined in the same way as for Aisetas. During 10 days, 12 measurements per day at two different points were performed and averaged to reduce various interferences. The measurements were performed above a vertically deposited ore bed, with an overlapping loamy soil capacity of 10 m (Figure 4B) and at a distance of 7 km from the source location (Figure 4A). Data are calculated in the same way mentioned above.

The higher mercury concentrations above the ore bed at night (Figure 5A) is caused by the formation of an inversive temperature layer in the atmosphere which prevents a turbulent

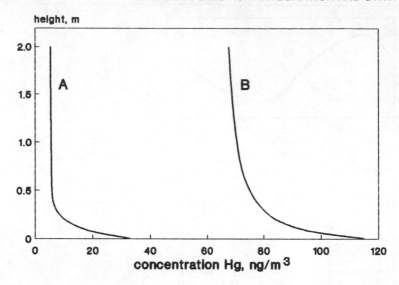

Figure 4 Vertical profiles of average mercury concentrations near the ground, South Fergana, Kirghizia. Curve A represents measurement results at a distance of 7 km from ore bed location and curve B—above the ore bed. Measurement period—10 days.

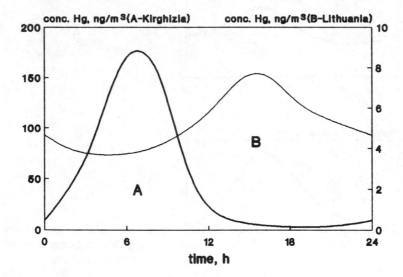

Figure 5 Diurnal variations of Hg concentrations: A—Kirghizia (at a distance of 7 km from ore bed location); B—Lithuania.

exchange of air and mercury transport into the upper layers of the atmosphere, whereas destruction of this layer in the daytime results in a great reduction of mercury concentration in the ground layer atmosphere. Reverse dependence may be determined in the absence of local endogenic mercury sources (Figure 5B). The nocturnal inverse temperature layer prevents the transport of the mercury that was brought (possibly from anthropogenic sources) to the ground layer atmosphere.

The decrease of mercury concentration just above the earth's surface may occur due to the presence of an anthropogenic or any other atmospheric mercury source. Excessively higher mercury concentrations just above the ground testify to the presence of endogenic mercury sources. Having analyzed the diagrams of the vertical distribution of mercury in the ground layer atmosphere and the diurnal variations of mercury concentration, we may draw conclusions about the origin of mercury source in the sampling area. Absolute concentration values and the gradient value (slope of the diagram of concentrations) make it possible to estimate intensity and type of source.

IV. CONCLUSIONS

The position of maximum value in diurnal variations of mercury concentration may be considered a characteristic of the measurement site. Furthermore, when intensive sources of anthropogenic mercury are present in the atmosphere, it is possible to determine the relationship of the strength of the anthropogenic and endogenic sources. Thus, we can provide a method to identify the origin and localization of mercury sources in the atmosphere and on the earth's surface.

The determination of these laws enables to give a more reliable interpretation of the results obtained from a wide region of atmospheric gaseous mercury measurements, to choose the optimal measurement period, and to compare the results obtained during different measurement periods of the diurnal cycle.

ACKNOWLEDGMENTS

The authors are grateful to M. Kulakauskas and to J.A. Afanasov for technical assistance and to J. Krokys for editorial assistance.

REFERENCES

1. Lindqvist, O. and Rhode, H., Atmospheric mercury—a review, *Tellus,* 37B, 136 1985.
2. Kvietkus, K., Investigation of Mercury Concentrations in the Atmosphere, Ph.D. Dissertation, Institute of Physics, Vilnius, 1985.
3. Slemr, F., Schuster, G., and Seiler, W., Distribution, speciation, and budget of atmospheric mercury, *J. Atmos. Chem.,* 3, 407, 1985.
4. Brosset, C., Total airborne mercury and its possible origin, *Water Air Soil Pollut.,* 17, 37, 1982.
5. Seiler, W., Eberling, C., and Slemr, F., Global distribution of gaseous mercury in the troposphere, *Pure Appl. Geophys.,* 118, 964, 1980.
6. Slmer, F., Seiler, W., and Shuster, G., Latitudinal distribution of mercury over the Atlantic Ocean, *J. Geophys. Res.,* 86, 1159, 1981.
7. McNerney, J.J. and Buseck, P.R., Geochemical exploration using mercury vapor, *Econ. Geol.,* 68, 1313, 1973.
8. Ozerova, H.A., *Mercury and Endogenic Ore Bodies Formation,* Nauka, Moscow, 1986, chap. 2 (in Russian).
9. Styro, B.I., *Radioactive Pollutants Removal From the Atmosphere,* Mintis, Vilnius, 1968, chap. 1 (in Russian).
10. Kvietkus, K., Sakalys, J., and Rozenberg, G., Measurement results of the atmospheric mercury concentrations along horizontal and vertical profiles, *Atmos. Phys.,* 10, 69, 1985 (Lithuania).
11. Fitzgerald, W. F. and Gill, G. A., Sub-nanogram determination of mercury by two-stage gold amalgamation and gas phase detection applied to atmospheric analysis, *Anal. Chem.,* 15, 1714, 1979.
12. Fursov, B. Z., *Geochemical Exploration Using Gaseous Mercury Method,* Nauka, Leningrad, 1983 (in Russian).

13. Kvietkus, K., Sakalys, J., Remeikis, V., and Sopauskas, K., The application of the atomic fluorescence method for determining mercury concentrations by a photon counter, *Atmos. Phys.*, 8, 127, 1983.

14. Bloom, N.S. and Fitzgerald, W.F., Determination of volatile mercury species at the picogram level by low temperature gas chromatography with cold-vapor atomic fluorescence detection, *Anal. Chim. Acta*, 117, 391, 1988.

15. Xiao, Z., Munthe, J. and Lindqvist, O., Sampling and determination of gaseous and particulate mercury in the atmosphere using gold coated denuders, in *Proc. Int. Conf. Mercury as an Environ. Pollut.*, Lindqvist, O., Ed., Kluwer Academic Publishers, Dordrecht, 1991, 347.

16. Saukov, A.A. Geochemistry of Mercury, Akad. Nauk. S.S.S.R. Doklady Inst. Geol. Nauk., 73, (Mineralogo-Geokhem. Seriya, 17), 1946.

17. Kvietkus, K., Sakalys, J., Afanasov, J.A., Zemskova, I.I., and Vebra, E., Method for Determination of Origin of Gaseous Mercury Anomalies, U.S.S.R. Patent N4094739, 1986.

18. Girgzdys, A. and Ulevicius, V., A daily course of radon and its short-lived decay products in the ground level air, *Atmos. Phys.*, 10, 22, 1985 (Lithuania).

19. Girgzdiene, R. and Sopauskas, K., An investigation of the spatial-temporal variability of ozone, *Atmos. Phys.*, 10, 33, 1985 (Lithuania).

20. Kvietkus, K., Methods for separation, sampling and analysis of gaseous and aerosol mercury fractions in the air, *Atmos. Phys.*, 11, 145, 1986 (Lithuania).

21. Matveev, L.T., *Basic Meteorology*, Gydrometeoizdat, Leningrad, 1976, chap. 2.

Particulate-Phase Mercury in the Atmosphere: Collection/Analysis Method Development and Applications

Carl H. Lamborg, Marion E. Hoyer, Gerald J. Keeler, Ilhan Olmez, and Xudong Huang

CONTENTS

ABSTRACT: Particulate-phase mercury Hg(p) may constitute a small percentage of the total atmospheric mercury, but is thought to play an important role in deposition of mercury to terrestrial and aquatic ecosystems. This chapter presents techniques that have been utilized to collect and analyze Hg(p) in several recent projects including monitoring in Detroit and Ann Arbor, MI and during the Lake Michigan Urban Air Toxics Study.

Fine (d_a <2.5 μm) and total suspended particle (TSP) samples were collected and analyzed by two methods: (1) acid extraction followed by sparging and dual amalgamation cold vapor atomic fluorescence spectrometry, and (2) neutron activation analysis. The differences in Hg(p) detected in the samples by these two analytical techniques, coupled with the size-segregated data collected, indicate that different forms of particulate mercury may be distinguished, and that differing physical/chemical properties may reflect potentially different sources.

I. INTRODUCTION

In recent years, the behavior of toxic air pollutants has been receiving a great deal of attention from the scientific community. In the U.S. and elsewhere, much of the concern has focused around the deposition of these compounds to terrestrial and aquatic environments where they enter the food chain and can potentially pose a threat to man. Mercury is of special concern, due to its volatility, mobility, and strong tendency to bioaccumulate.[1,2] Most alarming is the

1–56670–066–3/94/$0.00+$.50
© 1994 Lewis Publishers

observation of high levels of mercury in fish from remote inland lakes. Since these lakes do not receive direct discharges of mercury, the atmosphere has been implicated as the major source of this toxic element. In 1991, the state of Michigan identified 74 lakes, out of the 107 lakes surveyed, in which the mercury levels in the fish are considered hazardous enough to advise "reduced consumption".[3] In the Great Lakes basin, the study of the sources, transport, and deposition of atmospheric mercury has gained increasing attention.

Deposition rates and the processes by which mercury enters the water column are still not adequately understood. In particular, the role of the various physical/chemical forms of mercury deposited from the atmosphere has yet to be determined. While vapor phase mercury is thought to constitute the vast majority of the atmospheric mercury burden,[1] particle-phase mercury may actually play a disproportionately large role in the amount of Hg that finds its way into the other environmental compartments.[5] Currently, there is a lack of consensus in the literature as to the importance and magnitude of wet vs. dry deposition of Hg(p), but most researchers have indicated that the particulate form of the metal should not be quickly dismissed.[6]

An important aspect of this research is the ability to accurately collect and effectively analyze Hg(p). While Hg(p) has been quantified in several studies over the last two decades, recent advances in instrumental sensitivity and the timely application of clean techniques have radically altered our knowledge of Hg(p) concentrations and behavior.[6] Although not in widespread application, determination of very low environmental concentrations (pg/m^3) of Hg(p) are currently feasible. Nevertheless, there is a lack of high quality Hg(p) data, and additional Hg(p) sampling and analysis is needed.

The Air Quality Laboratory at the University of Michigan (UMAQL) has developed the capability to reliably collect and analyze Hg(p). These techniques are presently being utilized to gain a wider understanding of the atmospheric mercury cycle. During the process of technique development at UMAQL, investigations into currently used collection and analysis procedures were compared. The two techniques of analysis used in this work are (1) dual-amalgamation, preconcentration, and cold vapor atomic fluorescence spectrometric (CVAFS) detection performed on Hg(p) extracted from glass or quartz fiber filters using a 2 N acid-extraction solution and further treatment (see below), and (2) neutron activation analysis (NAA) performed on Teflon® filters. The two techniques have distinct advantages and both have been used to quantify Hg(p) in past studies. NAA was the technique of choice for many years due to its wide applicability in analyzing for many species simultaneously. Various forms of emission/absorption spectrometry in conjunction with filter extraction (in and out of "clean" or contamination-free environments) have recently gained in popularity. The relative ease, flexibility, sensitivity, and cost-effectiveness of these techniques make them extremely attractive alternatives. Both methods can have comparable detection limits, but as discussed below, may not give comparable results.

Mercury measurements have been recently performed in three studies: (1) a pilot study in metropolitan Detroit, (2) an on-going, long-term study of Hg and atmospheric acidity in the Ann Arbor area, and (3) the Lake Michigan Urban Air Toxics Study performed in the southern Lake Michigan basin. The Hg(p) sampling was typically performed for 24-h periods at a nominal flow rate of 30 liters per minute (LPM) for total suspended particulates through an open-faced 47 mm filter or 10 LPM for fine fraction particulate samples collected through a Teflon®-coated cyclone sampler. However, the technique developed at the UMAQL is sensitive enough that shorter duration samples are possible in most locations.

II. METHODS

Ultra-clean techniques are required when attempting to obtain reliable Hg(p) data. Sampling equipment including the filter packs, forceps, vials, petri dishes, and other field sampling equipment were rigorously acid-cleaned in a 5-step, 11-day process.[7] All sampling equipment, including filter packs and cyclones, were constructed of Teflon® or were Teflon® coated. Glass and quartz-fiber filters were pre-fired at 500° C for 3 to 6 h prior to use in sampling.

During sample collection, particle-free gloves were worn whenever field equipment was handled. Because outdoor concentrations of Hg in all forms are lower than indoor concentrations, much of the handling of filters and filter packs was done outdoors. During monitoring in Ann Arbor, samples were prepared inside a vinyl glove box that contained a tray of activated charcoal to remove gaseous Hg.

Extraction and analysis were performed in a class 100 cleanroom, and in most instances required reagents that were further "purified" to lower blank values and maintain a consistent detection limit for the entire procedure. For acid digestion/CVAFS analysis the samples were extracted in a 10% solution of a 70% nitric acid/30% sulfuric acid mixture (approximately 2 N) in Teflon® vials.[8] Extraction was performed by placing the vials in a sonic bath for 30 min. While this process results in warming, it does not subject the sample to elevated extraction temperatures. Extraction by sonication is the method of choice at UMAQL since it results in the best recovery of spiked standards and a high degree of reproducibility (see Section III). After extraction, the solution was oxidized with BrCl for 1 h, converting all forms of Hg present into the inorganic +2 oxidation state. The sample was reduced with NH_2OH, and $SnCl_2$ was added to convert the Hg^{2+} to Hg^0 which is volatile and liberated from solution by bubbling with Hg-free N_2. The Hg released in this way was collected on Au-coated sand traps, which have been demonstrated to effectively amalgamate vapor-phase Hg.[9,10] This collected Hg was subsequently analyzed using the dual-amalgamation CVAFS method described by Fitzgerald et al.[11,12] A calibration curve was generated by spiking vials containing blank filters with varying amounts of a 2 ng/mL standard (in 1% BrCl). The analytical detection limit for this process (based on 3 times the standard deviation of field blanks) ranges between 100 to 250 pg Hg with a midscale precision of less than 15% variability. The ambient detection limit, for the measurements discussed in this chapter, is approximately 9 pg/m^3 for a 24-h sample taken at a nominal flow rate of 30 LPM.

Neutron activation analysis (NAA) was performed at the Massachusetts Institute of Technology Nuclear Reactor Laboratory. The method of NAA takes advantage of the fact that, when a medium is irradiated with neutrons, some of the nuclei in the target material will be converted into radioactive isotopes. The isotopes so produced will decay at a characteristic rate, in most cases emitting gamma rays of characteristic energies and half-life. In this study, Teflon® filters were irradiated for 12 h at a thermal neutron flux of 10^{13} n/sec cm^2. The irradiation was followed by a "cooling" period of 5 to 6 days and a 12-h counting period with a Ge(Li) detector. Two naturally occurring mercury isotopes are suitable for total Hg determinations, Hg-196 and Hg-202. Although Hg-196 has a lower abundance compared to Hg-202 (0.15 vs. 29.8%) and in spite of the tendency of some analysts to use Hg-202, Hg-196 is much more sensitive for total mercury determinations. This is mainly due to its higher cross section (1600 times higher) and shorter half-life. It is also free of interference. Experiments have been conducted which indicate that encapsulation (in quartz) does not aid in the retention and quantitation of Hg, and so the Teflon® filters were irradiated in polyethylene vials.[13]

Flow rate through the samplers was measured using calibrated rotameters and dummy filter packs (to prevent contamination), and sample volumes were recorded using in-line calibrated dry test meters. Mass flow-controlled and/or valved diaphragm pumping units were used to pull ambient air through the sampling equipment.

III. RESULTS

A. EXTRACTION EXPERIMENTS

Several different extraction techniques for Hg(p) were compared using ambient sample filters collected through open-face filter packs onto glass fiber filters as described above. The results

Table 1 . **Results from three separate extraction experiments.**

Comparison	Treatment	n	Mean pg Recovered	Variability (%)	Comments
I	Acid (20 *N*)	20	1112	27–35	5 Separate
	Acid + water (2 *N*)	20	1326	5–19	trials
II	Heating 80° C for 3 h	4	652	6	Loss during
	Sonication (30 min)	4	740	6	heating?
III	Short BrCl Rxn time	2	177	60	<1 h
	Long BrCl Rxn time	5	277	17	≥1 h

Table 2   **Results from spiked ambient sample recovery tests.**

Treatment	n	Mean percent recovery
Hg standard (spiked extraction solution w/o filter)	5	93
Hg standard (spiked extraction solution with filter)	8	82

of these comparisons are included in Table 1. In each treatment case, equal halves (cut with a new, clean razor blade) of ambient sample filters were handled and analyzed identically so that the effect of the treatment could be examined. For each trial, a 5 mL aliquot of the extraction solution was analyzed using CVAFS. The first trial (I in Table 1) compared extraction (3-h heating in an 80° C water bath) in 2 mL of 70% nitric/30% sulfuric acid mixture (20 *N*) followed by dilution with 18 mL of pure water (this test is designated "Acid" in Table 1). The other half of the filter was extracted in a 20 mL volume of the acid mixture diluted to 10% (2 mL with 18 mL pure water) and this test is designated "Acid + Water". The second comparison (II) was designed to determine if heating the filters was more/less effective than sonication in extracting Hg. One half of a sample filter (for each of 4 filters) was heated in 2 *N* extraction solution in a water bath at 80° C for 3 h while the other half was placed in a sonic bath for 30 min (during which the temperature never rose above 50° C). Finally, the effect of reaction time with the oxidant BrCl was examined by analyzing aliquots from vials that were allowed to oxidize for 5, 30, 60, 90, 120, and 180 min (comparison III, Table 1). The combined results from these trials suggest that the current procedure utilized at UMAQL (acid/water extraction solution, sonication, and 1-h oxidation) is optimal.

Spiked sample recovery tests were also performed to assess the effectiveness of our technique in recovering Hg from filters. The results from these trials are shown in Table 2. The average 82% recovery observed in these tests was comparable to the recoveries of aqueous control standards from blank filters used in numerous calibration curves. Furthermore, the magnitude of the recovery is quite common for acid-extraction techniques used on a variety of matrices.[14]

In order to determine the shelf-life of samples after extraction/oxidation, samples which had remained in an oxidized form in cold (5° C) storage for 6 weeks were reanalyzed. Nineteen samples were reanalyzed and an average increase of 6% was observed. As mentioned above, the analytical precision at mid-range is approximately 15%, and indicates that this is not analytically significant.

B.   RESULTS FROM MONITORING EFFORTS

The particle-phase Hg collection and analysis technique adopted at UMAQL was utilized in three distinct research projects to assess Hg(p) levels in several different environments. Results from these measurements point out some of the important and interesting issues surrounding the behavior and determination of Hg(p).

Table 3 **Summary statistics for Hg(p) from the Detroit pilot study.**

Site	n	Mean	Median	S.D.	Max	Min
Detroit Site A	10	297	212	349	1230	57
Detroit Site B	10	341	69	342	1086	69
Ann Arbor	10	100	49	94	302	20

Note: All concentration values in pg/m³.

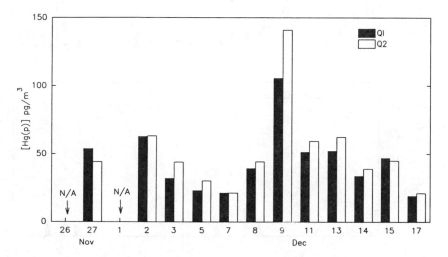

Figure 1 Duplicate particulate-phase mercury measurements performed in Ann Arbor, MI.

1. Detroit Pilot Study

During a 10-day period (April 5 through 14, 1992), 3 sites (2 in Detroit, 1 in Ann Arbor) were operated for the collection of particle and vapor-phase Hg, acidic aerosols and gases, and fine fraction particulate matter (<2.5 μm) for subsequent trace metal analysis. The particulate-phase Hg results are summarized in Table 3. The Detroit sites were located in the downtown area which has a wide variety of potential sources including steel manufacturing, coal-burning power plants, a large municipal waste incinerator, and several sludge incinerators. The concentrations of Hg(p) measured show a wide range of values, with the Detroit values typically being elevated over those observed in Ann Arbor. However, these values are extremely variable, suggesting strong local sources that can be observed when meteorological conditions are conducive to these impacts.

2. On-Going Ann Arbor Monitoring

Currently, samples of Hg(p), as well as total gas-phase Hg and acidic aerosols and gases, are being collected at the University of Michigan School of Public Health building once every sixth day on the national monitoring network schedule. The results of Hg(p) obtained thus far are displayed in Figure 1. On each sample day, two co-located quartz fiber filters in open-faced filter packs (labeled "Q1" and "Q2") were collected.

The concentration values from Ann Arbor during the winter are somewhat surprising since on several occasions concentrations are elevated above the typical background level of 20 to 30 pg/m³. It is suggested that winter time values of gas-phase Hg should be relatively low due to decreased volatility.[6] However, initial results from wintertime monitoring of Hg(p) in Ann Arbor do not follow an analogous trend, suggesting a different seasonal behavior.

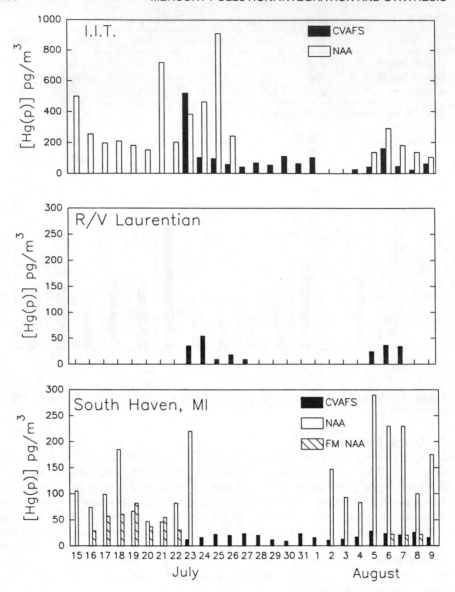

Figure 2 Results of particulate-phase mercury measurements performed during the Lake Michigan Urban Air Toxics Study in July through August, 1991. Analysis shown was performed using cold vapor atomic fluorescence spectrometry (CVAFS) and neutron activation analysis (NAA).

3. Lake Michigan Urban Air Toxics Study

For 1 month during the summer of 1991 (July 9 through August 9) the UMAQL in conjunction with the United States Environmental Protection Agency (USEPA) conducted a 4-site monitoring program on and around southern Lake Michigan. At three of these sites, Hg(p) sampling was conducted using open-faced filter packs with glass-fiber and Teflon® filters. Additionally, Teflon®-coated cyclones were employed upstream of the Teflon® filter packs to collect fine fraction aerosols at the South Haven site. The three sites were located

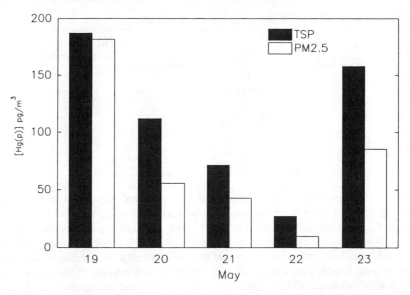

Figure 3 Comparison of particulate-phase mercury in fine fraction (PM) (<2.5 µm) and total suspended particulate (TSP) samples collected in Ann Arbor, MI.

in Chicago (at the Illinois Institute of Technology), in rural western Michigan (South Haven), and on the research vessel, *Laurentian*. Results from these samples are shown in Figure 2. Levels measured are clearly highest in Chicago and lowest in South Haven. However, an interesting distinction becomes apparent when the values for NAA (of TSP and fine particles collected on Teflon® filters) are compared to the results of CVAFS analysis of TSP glass-fiber filters. In South Haven, the NAA results are consistently higher than those from CVAFS, but in Chicago results from NAA and CVAFS are similar. Furthermore, fine mass filters taken in South Haven, analyzed by NAA, are virtually identical to co-located glass-fiber filters analyzed by CVAFS which had no particle size segregation. It seems clear that an operational distinction between "total Hg(p)" (as indicated by NAA) and "acid-extractable Hg(p)" (as indicated by CVAFS) can be made, and that in some cases (South Haven fine mass, and to some extent all sizes of Chicago particulate matter) these two definitions are synonymous.

A better understanding of the observed particulate-phase Hg can be gained by examining the other analyses performed concurrently: (1) the coarse-to-fine mass ratio in South Haven as determined by dichotomous samplers, and (2) TSP and fine mass quartz-fiber filters collected in Ann Arbor. In South Haven, the result of NAA of TSP Teflon® filters correlates significantly with the ratio of coarse-to-fine mass ($r = 0.72, p > 0.001$). This further suggests that the bulk of the Hg(p) measured by NAA in South Haven is located in the coarse fraction and that, because coarse fraction material is most often associated with wind-blown soil material, much of the "total" Hg(p) is of that origin (perhaps being mineral or deposited anthropogenic emissions). However, the coarse/fine separation is not the only characteristic reflecting the behavior of Hg(p). Figure 3 displays the Hg(p) determined by CVAFS for both the TSP and fine fraction particulates as observed in Ann Arbor. Open-faced filter packs and filter packs equipped with Teflon®-coated cyclone inlets, to remove coarse particles, were operated for 24-h periods at 30 and 10 LPM, respectively. The levels observed in both size fractions clearly indicate that "acid-extractable" Hg(p) can be found in both the coarse and fine material, and that in some instances, such as with the bulk of Hg(p) in Ann Arbor, is associated with the fine aerosol size range.

Table 4   **Percent of atmospheric Hg in particle-phase.**

Analysis technique	Site	% Hg in particle-phase (range observed)
CVAFS	IIT	0.1–7.3
CVAFS	South Haven	0.7–11.7
CVAFS	Laurentian	0.6–2.3
CVAFS	Ann Arbor	0.3–20.0
CVAFS	Detroit	0.1–41.0
NAA	IIT	0.3–12.6
NAA	South Haven	2.0–16.2

These results indicate that Hg(p) generated in or near urban/source areas such as Chicago (assuming here that anthropogenic Hg(p) will be acid-extractable) may exist in either the coarse or fine fraction, but that during transport, much of the coarse fraction material which bears Hg(p) may be lost by dry deposition. The end result being that acid-extractable Hg(p) will, in this case, be found predominantly in the fine size range. Furthermore, the evidence of non-acid-extractable Hg(p) in a rural area (South Haven) residing in the coarse fraction and following the coarse-to-fine mass ratios suggests an "acid-extractable"/anthropogenic vs. "non-acid-extractable"/natural Hg(p) distinction.

4. Percent of Atmospheric Hg in Particle-Phase

As mentioned above, the opinions surrounding the importance of particulate-phase Hg have undergone some change over time. Currently, it is felt that Hg(p) contributes a small ($<5\%$) fraction of the total amount of Hg found in the ambient air. The ranges of the percent of atmospheric mercury in the particle-phase (as calculated by comparison to vapor-phase Hg collected concurrently on Au-coated sand traps) measured during these studies are displayed in Table 4. As these results indicate, the range of particulate Hg observed is very large, and suggests that under certain conditions Hg(p) can be a very significant form of atmospheric Hg. Furthermore, the fraction of Hg found in the particle-phase can be extremely variable at any given location. This poses a serious problem when attempting to estimate dry deposition or the fraction of Hg wet deposition related to particle-phase Hg and, subsequently, total Hg deposition.

IV. CONCLUSIONS

As a result of comparison of collection techniques and analysis methods conducted at the UMAQL, and based on application of these techniques to field monitoring for Hg(p), the following conclusions can be made:

• Acid digestion/CVAFS is a sensitive and reliable technique for measuring Hg(p).
• Rural atmospheric levels measured using this technique are comparable to others that have been measured in the Great Lakes basin by other researchers.
• Results from size-segregated collection and different analysis methods (CVAFS vs. NAA) suggest that Hg(p) may be more complex and significant than has been previously suggested.
• Coarse particle (>2.5 μm in aerodynamic diameter) Hg levels may be larger than previously predicted, even in semirural areas such as Ann Arbor, MI.
• Urban levels, as measured by this technique, are significantly elevated above rural values.

ACKNOWLEDGMENTS

We would like to thank our colleagues in the Lake Michigan Urban Air Toxics Study for their assistance with the planning and field operations including: Tom Holsen, Nasrin Khalili, and Ken Noll (IIT), Gary Evans, Alan Hoffman, Bob Stevens, and Mack Wilkins (EPA). We would also like to thank the many UMAQL people who helped with site operations for the Ann Arbor and Detroit studies including Tim Dvonch, Ganda Glinsorn, Jennifer Falk, and Steven Mischler as well as those from Wayne County including Peter Warner and colleagues. We would also like to thank the reviewers of this manuscript for their useful comments and suggestions.

REFERENCES

1. Lindqvist, O. and Rodhe, H., Atmospheric mercury—a review, *Tellus,* 37B, 136, 1985.
2. Clarkson, T. W., Mercury, *J. Am. Coll. Toxicol,* 8, 1291, 1989.
3. Michigan Department of Natural Resources, Michigan fish contaminant monitoring program, 1991 annual report, Report #MI/DNR/SWQ-91/273.
4. Munthe, J. and McElroy, W. J., Some aqueous reactions of potential importance in the atmospheric chemistry of mercury, *Atmos Environ.,* 26A, 553, 1992.
5. Lindberg, S. E., Turner, R. T., Meyers, T. P., Taylor, G. E., and Schroeder, W. H., Atmospheric concentrations and deposition of Hg to a deciduous forest at Walker Branch watershed, TN, USA, *W.A.S.P.,* 56, 557–594, 1991.
6. Schroeder, W. H.,Sampling and analysis of mercury and its compounds in the atmosphere, *ES&T,* 16, 394A, 1982.
7. Rossmann, R. and Barres, J., Trace element concentrations in near-surface waters of the Great Lakes and methods of collection, storage, and analysis, *J. Great Lakes Res.,* 14, 188, 1991.
8. Bloom, N., personal communication, Brooks Rand, Ltd., 1991.
9. Dumarey, R., Dams, R., and Hoste, J., Comparison of the collection and desorption efficiency of activated charcoal, silver and gold for the determination of vapor-phase atmospheric mercury, *Anal. Chem.,* 57, 2638, 1985.
10. Brosset, C. and Iverfeldt, Å., Interaction of solid gold with mercury in ambient air, *W.A.S.P.,* 43, 147, 1989.
11. Fitzgerald, W. F. and Gill, G. A., Subnanogram determination of mercury by two-stage gold amalgamation and gas-phase detection applied to atmospheric analysis, *Anal. Chem.,* 51, 1714, 1979.
12. Bloom, N. and Fitzgerald, W. F., Determination of volatile mercury species at the picogram level by low-temperature gas chromatography with cold-vapor atomic fluorescence detection, *Anal. Chim. Acta,* 208, 151, 1988.
13. Olmez, I., Massachusetts Institute of Technology, personal communication, 1992.
14. Kimbrough, D. E. and Wakakuwa, J., A study of the linear ranges of several acid digestion procedures, *ES&T,* 26, 173, 1992.

Application of Throughfall Methods to Estimate Dry Deposition of Mercury

S. E. Lindberg, J. G. Owens, and W. J. Stratton

CONTENTS

ABSTRACT: Several dry deposition methods for Hg are being developed and tested in our laboratory. These include big-leaf and multilayer resistance models, micrometeorological methods such as Bowen ratio gradient approaches, laboratory controlled plant chambers, and throughfall. We have previously described our initial results using modeling and gradient methods. Throughfall may be used to estimate Hg dry deposition if some simplifying assumptions are met. We describe here the application and initial results of throughfall studies at the Walker Branch Watershed forest, and discuss the influence of certain assumptions on interpretation of the data. Throughfall appears useful in that it can place a lower bound to dry deposition under field conditions. Our preliminary throughfall data indicate net dry deposition rates to a pine canopy which increase significantly from winter to summer (range ~0.1 to 3 $ng/m^{-2}/h^{-1}$), as previously predicted by our resistance model. Atmospheric data suggest that rainfall washoff of fine aerosol dry deposition at this site is not sufficient to account for all of the Hg in net throughfall. Potential additional sources include dry deposited gas-phase compounds (Hg^0 or Hg-II), soil-derived coarse aerosols, and oxidation reactions at the leaf surface.

I. INTRODUCTION

There has been considerable progress in understanding the cycling of Hg in aquatic and terrestrial ecosystems,[1-3] and the importance of atmospheric sources to these cycles is now well recognized. However, very little is known about the processes and rates of dry deposition of Hg to forest/catchment systems. Lab studies suggest that plants accumulate Hg vapor from the air,[4-5] but field data are nearly absent. Mercury in the atmosphere is dominated by vapor forms,[6] and new models have been published to estimate dry deposition velocities (V_d; where dry deposition flux = V_d · air concentration) for several atmospheric gases and vapors to forests.[7-8] At the first conference in this series, we described an application of these models to estimate Hg fluxes in fine aerosol and vapor form to the forest at Walker Branch Watershed in Tennessee.[9] We have since published sensitivity tests of the model, and our initial attempts to apply micrometeorological gradient methods to estimate dry deposition.[10] We discuss here the application of a field-oriented surface analysis method (throughfall), with further reference to a related laboratory method (controlled exposure chambers) to determine the dry deposition of Hg in forest environments. We are currently initiating a study and intercomparison of all these methods, and report below some of our initial results for throughfall.

II. STUDY SITE AND METHODS

The throughfall and air concentration data reported here were collected at Walker Branch Watershed in Oak Ridge, Tennessee during 1990 to 1992. The watershed is a 100 ha decid-uous forest catchment located in hilly terrain, with elevations ranging from 265 to 365 m (Figure 1), and is located at Lat. 35°58' N, Long. 84°17' W.[11] Long-term mean air concen-trations of total Hg vapor and aerosol Hg were determined with iodated activated carbon traps with Teflon® prefilters as described in Lindberg et al.[9] Short-term samples of Hg vapor were collected using the gold-coated sand traps described by Fitzgerald and Gill,[12] through which air was sampled at a constant rate of 250 mL min^{-1} (as determined by mass flow controllers). Both types of absorption tubes collect total vapor phase Hg, but this phase is dominated by elemental Hg vapor.[13] We will refer to the Hg collected with these methods as Hg^0. We collected precipitation on an event basis in a forest clearing and throughfall beneath adjacent pine and deciduous canopies (near site RG 2, Figure 1) using HCl-washed and pre-baked glass bottles and funnels placed in cleaned automatic wet-only collectors.[10] Samples were preserved and treated in the laboratory and analyzed for total Hg by cold vapor atomic fluorescence spectroscopy using the methods described in Bloom and Fitzgerald.[14]

III. FIELD THROUGHFALL STUDIES

A. BACKGROUND

Environmental surfaces, although difficult to simulate either mathematically or physically, have the natural ability to integrate the net result of dry deposition of various airborne com-pounds. Surface analysis methods, such as throughfall, can be used to quantify the amount of deposited material (e.g., Hg) residing on such surfaces for constituents that are not irre-versibly sorbed. Analysis of throughfall in forests (the water which falls to the ground beneath plant canopies during precipitation) has become an increasingly attractive tool for studies of atmospheric fluxes.[15] This method holds some potential advantages over other techniques including the ability to

1. Determine the representativeness of modeled fluxes
2. Scale from point measurements to the landscape
3. Provide long-term mean fluxes
4. Estimate fluxes in highly complex terrain
5. Provide independent field data for validation of experimental micrometeorological approaches

The method has been applied extensively in studies of sulfur deposition,[16-19] and in a limited number of studies of trace metal dry deposition to forests.[20,21]

In its simplest application, long-term mean throughfall fluxes are suggested to be direct estimates of total (wet + dry) atmospheric deposition. This assumes: (1) there is quantitative removal of the deposited material by precipitation falling through the canopy, and (2) foliar leaching by rain of internal plant sources of the element of interest is small. The term "foliar leaching" refers to only that material which enters the plant by root uptake, is incorporated in foliage, and "leaches" into throughfall during a rain event. If this is significant, it will

263

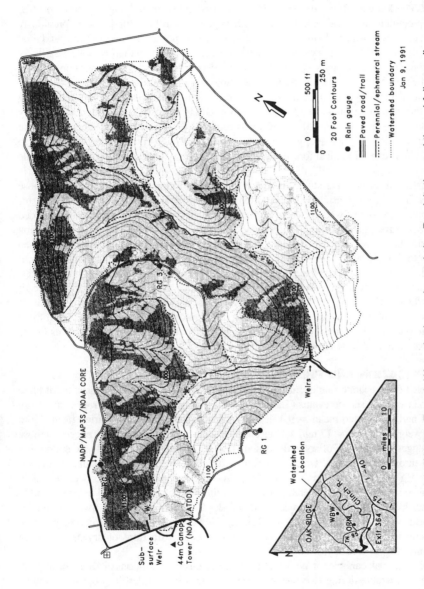

Figure 1. Location and contour map of the Walker Branch Watershed experimental site. Precipitation and throughfall sampling were performed at site RG 2, while atmospheric concentrations of particulate and vapor phase Hg were sampled at RG 2 and at the 44 m Canopy Tower, ~300 m southwest of RG 2.

clearly bias throughfall results. The assumption of minor foliar leaching has been shown to be valid for sulfur,[22] but no such data exist for Hg. However, uptake of Hg by terrestrial plants, even from contaminated soils, seems to be very limited,[5,23] suggesting that internal cycling is small, and that foliar leaching may be unimportant.

Irreversible sorption of Hg by canopy surfaces could occur for both wet and dry deposited materials. The known affinity of dissolved Hg for particulate matter suggests that any free Hg^{2+} in precipitation would bind quickly with vegetation surfaces or particles previously deposited on these surfaces. However, most studies of Hg speciation in rain have found total Hg to be dominated by forms associated with particulate matter, which may be less likely to react directly with canopy sorption sites. Thus, depending on the concentrations of free Hg^{2+} and competing ions, and on the residence time of rain on these surfaces, surface interactions may deplete Hg from incoming rain to some extent. This would result in reduced Hg levels in throughfall, leading to underestimates of total Hg fluxes. Some evidence of canopy absorption of Hg from rain ($\sim$20 to 36%) was reported for a background site at El Yunke Mt. in Puerto Rico.[24] Rain/canopy interactions and water interception losses are known to be highly significant in tropical forests, and this behavior may be less important in temperate forests.

We are aware of no other studies of Hg in throughfall in temperate forests such as those at Walker Branch, but Iverfeldt[25] has reported several months of bulk throughfall data for Hg in boreal forests in Sweden, where concentrations in throughfall consistently exceeded those in bulk precipitation during the growing season. During winter, when dry deposition was presumably lower, Iverfeldt found some instances where Hg concentrations decreased from rain to throughfall (by $\sim$30%). Given the uncertainty in Hg determinations in rain, it is difficult to conclude whether Hg is actually sorbed from incoming rain by canopy surfaces. Nevertheless, we cannot rule out this possibility, and hence, that throughfall may underestimate Hg fluxes. The capability of rain to remove dry deposited Hg from canopy surfaces is unknown, and will be one focus of our controlled laboratory chamber studies.[26] Until these factors are quantified, we must assume that throughfall yields conservative estimates of Hg dry deposition and of total Hg fluxes to forests.

B. APPLICATION AND PRELIMINARY DATA

We sampled five separate rain events during August 1991 to April 1992 consisting of three warm- and two cold-season events (Table 1). Concentrations of total Hg in rain varied from $\sim$8 to 17 ng/L (weighted mean = 9.4 ng/L), while those in throughfall varied from $\sim$9 to 44 ng/L (weighted mean = 17 ng/L). Concentrations in throughfall generally exceeded those in rain, suggesting washoff of some deposited Hg from the canopy. The only exception was the April throughfall sample, which was collected below the leafless oak canopy, and for which the Hg concentration was $\sim$90% of that in the incoming rain. The concentrations in these limited samples are considerably lower than those reported by Iverfeldt[25] for monthly bulk samples collected below a spruce stand at Lake Gardsjön in southern Sweden (precipitation mean = 31 ng/L, throughfall = 48 ng/L). However, it is difficult to compare these data directly since long-term bulk samples contain unknown fractions of dry deposited Hg in addition to Hg in wet deposition.

The pine and oak canopies at Walker Branch produced concentrations of Hg in throughfall that were comparable during two of three events, suggesting similar dry deposition source terms during the antecedent periods for these events. Although our samples were collected as wetfall-only on an event basis, several contained considerable coarse organic particulate matter such as leaf and bark debris. Removal of this material from replicate samples by decanting immediately after collection and prior to acidification and oxidation resulted in somewhat lower concentrations of total Hg (by $\sim$3 to 30%). While this could suggest that some fraction of the Hg in throughfall represents plant-derived material and not dry deposition, it is also possible that this association results from adsorption reactions in the sample

Table 1 Concentrations and fluxes of total Hg in rain and throughfall events collected at Walker Branch Watershed from 8/91 to 4/92.

Event date	Sample type[a]	Hydrologic flux (cm)	Rainfall		Analysis[b]	Hg concentrations (ng/L)	Hg fluxes (ng/m²)
			Duration (h)	Intensity (cm/h)			
8/9/91	Rain	1.1	6	0.18	Total Hg	10.42	115
	Oak TF	0.82			Total Hg	39.22	322
	Pine TF	0.64			Total Hg	44.07	282
	Pine TF	0.64			POM removed	42.58	273
9/7/91	Rain	1.0	0.5	2.00	Total Hg	11.87	119
	Oak TF	1.10			Total Hg	23.53	259
	Pine TF	1.26			Total Hg	25.27	318
	Pine TF	1.26			POM removed	17.03	215
9/24–25/91	Rain	3.6	11	0.33	Total Hg	8.03	291
	Oak TF	3.31			Total Hg	8.85	293
	Pine TF	4.09			Total Hg	15.50	634
	Pine TF	4.09			POM removed	11.44	467
2/14–15/92	Rain	3.21	21	0.15	Total Hg	7.57	243
	Pine TF	2.6			Total Hg	11.55	300
4/10/92	Rain	0.9	1	0.90	Total Hg	17.40	157
	Oak TF	0.87			Total Hg	16.10	140

[a] TF = throughfall.

[b] POM = coarse particulate organic matter which was removed from selected samples by decanting; the pine throughfall samples for 8/9 to 9/24 are field replicates.

collection bottles. Iverfeldt[25] found that ~40% of the total Hg in spruce throughfall was present as Hg-II which readily sorbs to suspended particulate matter. Further data are needed to assess the role of this material.

The total Hg fluxes in individual throughfall events (Table 2) range from ~140 to 630 ng/m^2, and are moderately higher below pine than oak (means of ~380 and 250 ng/m^2, respectively). There are no published event, wet-only data with which to compare these fluxes directly. Iverfeldt[25] reported fluxes in monthly bulk throughfall at Lake Gårdsjön ranging from ~640 to 4300 ng/m^2 (mean 1500), but did not include rainfall amounts for comparison. Using the rainfall weighted mean Hg concentrations in throughfall below the Walker Branch pine canopy with our complete hydrologic record for 1991 to 1992, we can estimate monthly throughfall fluxes for a more direct comparison with the Swedish data. Our estimates are as follows: 2.7 µg/m^2 for the 2-month period August through September 1991 (total throughfall = 13.2 cm), and 13 µg/m^2 for the 7-month period August 1991 through February 1992 (72.0 cm). These rough estimates are comparable to the fluxes reported by Iverfeldt[25] for similar periods during 1989 to 1990: ~5 to 6 µg/m^2 for August through September, and ~11 to 13 µg/m^2 for August through February. If our limited data are representative, they yield a rough estimate of the annual Hg flux in throughfall (~20 µg/m^2, 118.1 cm throughfall) which is close to that for Lake Gårdsjön (17 to 19 µg/m^2), and suggests that there is a significant quantity of Hg reaching the forest floor at Walker Branch in addition to that deposited in wet deposition (~14 µg/m^2/year).[10]

C. INTERPRETATION AND IMPLICATIONS FOR DRY DEPOSITION

Since ions in throughfall have many sources (the incoming rain, dry deposition washoff, and internal foliar leaching), net throughfall (NTF) is commonly used to quantify that portion originating in the canopy:

NTF = (the flux of Hg in throughfall) − (the flux of Hg in incident precipitation)

Assuming that dry deposition is quantitatively washed from the canopy and that wet deposition passes through the canopy without sorption, knowledge of the duration of the antecedent dry period prior to the sampled rain events can be used to compute the mean net dry deposition rate:

Mean dry deposition rate = NTF/antecedent dry period

Two factors complicate these assumptions: (1) as discussed above, previous studies under conditions of low dry deposition rates (in winter and at background sites) suggest moderate canopy uptake of Hg from rain,[24,25] and (2) Hg-II deposited on leaf surfaces may be subsequently reduced to Hg^0 and degassed prior to the next rain event. If these phenomena are important at Walker Branch, then throughfall will underestimate both dry and total deposition of Hg. Since the assumptions for using throughfall data are tenuous for Hg, our preliminary results must be considered as rough approximations of Hg dry deposition. However, they are still useful in that they may provide a lower bound to dry deposition rates under field conditions.

The total Hg fluxes in NTF beneath the pine forest canopy vary from ~60 to 340 ng/m^2, and are generally comparable with those for the adjacent oak canopy (Table 2). We have no explanation for the low NTF flux below the oak canopy for the 9/24–25 sample, other than

Table 2 **Net throughfall (NTF) fluxes of Hg, rain event characteristics, and computed estimates of Hg dry deposition rates based on rain and throughfall collected at Walker Branch Watershed from 8/91 to 4/92.**

Event date	Sample type	Antecedent dry period (h)	Analysis	Net throughfall Hg flux (ng/m²)	Estimated mean dry deposition rate (ng/m²/h)
8/9/91	Rain	318	Total Hg	0	—
	Oak TF		Total Hg	207	0.65
	Pine TF		Total Hg	167	0.53
9/7/91	Rain	189	Total Hg	0	—
	Oak TF		Total Hg	140	0.74
	Pine TF		Total Hg	200	1.06
9/24–25/91	Rain	120	Total Hg	0	—
	Oak TF		Total Hg	2	0.02
	Pine TF		Total Hg	343	2.86
2/14–15/92	Rain	514	Total Hg	0	—
	Pine TF		Total Hg	57.2	0.11
4/10/92	Rain	78	Total Hg	0.00	—
	Oak TF		Total Hg	−16.5	ND

Note: TF = throughfall; NTF = flux in TF − flux in rain; ND = not detectable.

that leaf senescence had begun. Although our data are limited, they suggest a lower NTF flux below both canopies during the winter than summer, similar to the seasonal trend reported by Iverfeldt.[25]

Our estimates of Hg dry deposition rates to the pine canopy derived from the NTF data also suggest a seasonal difference (Table 3). The summer rates are in the low ng/m²/h range, exceeding the rate derived from the February sample by an order of magnitude. The negative NTF flux for the oak canopy during the leafless April period precluded estimating a dry deposition rate, but suggests a very low value as well. We previously published a model of Hg dry deposition to this forest which also indicated much higher input during the summer compared to winter.[9]

We can use our model of dry deposition velocities to estimate the potential contribution of Hg^0 and particulate Hg to the measured NTF fluxes. Ranges of air concentrations representative of the antecedent periods for each NTF sample were determined from nearby monitoring sites[27] and from our earlier data at this site.[9] Dry deposition velocities for aerosol and vapor forms were taken from the model results using meteorological data for these same periods as described in Lindberg et al.[10] Table 3 summarizes the computed range of dry deposition estimates for comparison with those based on NTF. Despite the uncertainty associated with these results, it appears that the NTF-derived fluxes fall between the ranges of contribution of aerosol and vapor forms. If our deposition velocities are accurate, this suggests that dry deposition of aerosol Hg is not sufficient to account for the majority of Hg in NTF, contributing ~10% on average (4 to 23%). In order for aerosol deposition to account for all of the Hg in NTF would require daily mean dry deposition velocities in the range of 0.5 to 3 cm/s. Several field, wind tunnel, and modeling studies suggest that these values are unreasonably large for the fine aerosol fraction in which airborne Hg is expected to exist.[28,29]

Table 3 Mean air concentrations, modeled dry deposition velocities (Vd), and dry deposition rates for Hg, and dry deposition estimates from throughfall beneath a pine canopy at Walker Branch Watershed.

Event date	Representative mean air concentrations[a]		Modeled mean Vd (cm/s)		Modeled dry deposition rates (ng/m²/h)		Estimated dry deposition rates from throughfall (ng/m²/h)
	Hg vapor (ng/m³)	Particulate Hg	Hg vapor	Hg-p	Hg vapor	Hg-p	
8/9/91	5–6	0.02–0.03	0.094	0.11	17–20	0.08–0.12	0.53
9/7/91	4–6	0.02–0.03	0.078	0.11	11–17	0.08–0.12	1.1
9/24–25/91	3–4	0.02–0.03	0.076	0.11	8–11	0.08–0.12	2.9
2/14–15/92	2–3	0.04–0.06	0.009	0.003	0.06–1.0	0.004–0.006	0.11

[a] Measured at or near the throughfall collection sites; Hg vapor includes all gas phases, but consists primarily of Hg^0.

[b] Particulate Hg (Hg-p) consists of total Hg associated with aerosols collected by 1.0 μm pore size Teflon® filter.

There are several other possible sources for the Hg which is washed from the canopy into throughfall:

1. Internal plant Hg
2. A portion of dry deposited Hg^0
3. Coarse aerosol, soil-derived Hg
4. Hg-II compounds dry deposited from the gas phase
5. Hg-II compounds produced on the plant surface by oxidation of Hg^0

We consider the first two sources unlikely to be significant contributors since plant uptake of Hg from soils is limited, and Hg within plant tissues is expected to form strong bonds with sulfur or sulfhydryl groups which are not readily leached by rain. This would apply equally to internal plant Hg derived from soil or stomatal uptake. Published reports that stomatal uptake of airborne Hg^0 is controlled by mesophyl resistance (i.e., reactions between Hg^0 and tissues within the stomatal cavity) support this conclusion.[10,30] However, dry deposited coarse aerosol Hg and Hg-II compounds could be important.

Coarse aerosols exhibit dry deposition velocities larger than those characteristic of fine aerosols.[31] If appreciable airborne Hg occurs in the coarse aerosol fraction, this could account for some of the Hg in NTF. Reliable data on the particle size distribution of airborne Hg has never been published to our knowledge. Hg-II compounds may be the most likely source of the unaccounted for Hg in NTF. Because of their solubility, Hg-II compounds in the gas phase would exhibit sufficiently high deposition velocities to account for the Hg in NTF, as discussed in Lindberg et al.[10] There is doubt that these compounds are important in ambient air, however.[2] On the other hand, the possibility of direct oxidation of Hg^0 by ozone in water films on plant surfaces has been described by Iverfeldt.[25] If the Hg-II produced is stable,[32] then it could provide water soluble Hg compounds for washoff in throughfall.

IV. CONCLUSIONS

Event, wet-only precipitation and throughfall sampling at the Walker Branch forest has provided estimates of Hg dry deposition. Although many of the assumptions in using net throughfall (throughfall flux-rain flux) to estimate dry deposition of Hg have not yet been tested, it appears that net throughfall will be useful for determining a lower bound to dry deposition under field conditions if foliar leaching is not significant. Our preliminary data suggest that rainfall washoff of fine aerosol dry deposition at this site is not sufficient to account for all of the Hg in net throughfall, but that several vapor phase species could be important. Continued sampling of throughfall under a wide range of conditions in combination with solution speciation studies and air concentration measurements is recommended to address the issues raised here. We are pursuing such studies at Walker Branch and other forested sites. In addition, we have initiated detailed studies of the proposed mechanisms of dry deposition using controlled plant chambers with both stable and tracer Hg isotopes, as described in Hanson et al.[26,33]

ACKNOWLEDGMENTS

Research sponsored by the U.S. Department of Energy and the Electric Power Research Institute under contract with ORNL. ORNL is managed by Martin Marietta Energy Systems, Inc., for the U.S. Department of Energy under contract DE-AC05–84OR21400. Publication no. 4053, Environmental Sciences Division, ORNL.

REFERENCES

1. Lindqvist, O., Jernelov, A., Johansson, K., and Rodhe, H., Mercury in the Swedish Environment—Global and Local Sources, SNV Report No. 1816; Swedish Environmental Protection Board, Box 13021, S-17125, Solna, Sweden, 1984.
2. Lindqvist, O., Johansson, K., Aastrup, M., Andersson, A., Bringmark, L., Hovsenius, G., Hakanson, L., Iverfeldt, A., Meili, M., and Timm, B., Mercury in the Swedish environment—recent research on causes, consequences and corrective methods, *Water Air Soil Pollut.*, 55, 1, 1991.
3. Fitzgerald, W. F. and Watras, C. J., Mercury in surficial waters of rural Wisconsin lakes, *Sci. Total Envir.*, 87/88, 1989.
4. Browne, C. L. and Fan, C. S., Uptake of mercury vapor by wheat. An assimilation model, *Plant Physiol.*, 61, 430, 1978.
5. Lindberg, S. E., Jackson, D. R., Huckabee, J. W., Janzen, S. A., Levin, M. J., and Lund, J. R., Atmospheric emission and plant uptake of mercury from agricultural soils near the Almaden mercury mine, *J. Environ. Qual.*, 8, 572, 1979.
6. Brosset, C., Total airborne Hg and its possible origin, *Water Air Soil Pollut.*, 17, 37, 1982.
7. Hicks, B. B., Baldocchi, D. D., Meyers, T. P., Hosker, R. P., Jr., and Matt, D. R., A preliminary multiple resistance routine for deriving deposition velocities from measured quantities, *Water Air Soil Pollut.*, 36, 311, 1987.
8. Meyers, T. P., Huebert, B. J., and Hicks, B. B., HNO_3 deposition to a deciduous forest, *Boundary-Layer Meteorol.*, 49, 395, 1989.
9. Lindberg, S. E., Turner, R. R., Meyers, T. P., Taylor, G. E., and Schroeder, W. H., Atmospheric concentrations and deposition of airborne Hg to Walker Branch Watershed, *Water Air Soil Pollut.*, 56, 577, 1991.
10. Lindberg, S. E., Meyers, T. P., Taylor, G. E., Turner, R. R., and Schroeder, W. H., Atmosphere/surface exchange of mercury in a forest: results of modeling and gradient approaches, *J. Geophys. Res.*, 97, 2519, 1992.
11. Johnson, D. W. and Van Hook, R. I., Eds., *Analysis of Biogeochemical Cycling in Walker Branch Watershed*, Springer-Verlag, Berlin, 1989.
12. Fitzgerald, W. F. and Gill, G. A., Subnanogram determination of mercury by two-stage gold amalgamation and gas phase detection applied to atmospheric analysis, *Anal. Chem.*, 51, 1714, 1979.
13. Schroeder, W. H. and Jackson, R. A., Environmental measurements with an atmospheric mercury monitor having speciation capabilities, *Chemosphere*, 16, 182, 1987.
14. Bloom, N. S. and Fitzgerald, W. F., Determination of volatile mercury species at the picogram level by low-temperature gas chromatography with cold-vapor atomic fluorescence detection, *Anal. Chim. Acta*, 208, 151, 1988.
15. Lindberg, S. E., Cape, J. N., Garten, C. T., Jr., and Ivens, W., Can sulfate fluxes in forest canopy throughfall be used to estimate atmospheric sulfur deposition? A summary of recent results, in Schwartz, S. E. and Slinn, W.G.N., Eds., *Precipitation Scavenging and Air/Surface Exchange Processes*, Hemisphere, Washington, D.C., 1992, 1367.
16. Ivens, W. P. M. F., Kauppi, P., Alcama, J., and Posch, M., Empirical and model estimates of sulfur deposition onto European forests, *Tellus*, 42B, 294, 1990.
17. Bredemeier, M., Forest canopy transformation of atmospheric deposition, *Water Air Soil Pollut.*, 40, 121, 1988.
18. Lindberg, S. E. and Garten, C. T., Jr., Sources of sulfur in forest canopy throughfall, *Nature*, 336, 148, 1988.
19. Cape, J. N., Sheppard, L. J., Fowler, D., Harrison, A. F., Parkinson, J. A., Dao, D., and Paterson, I. S., Contribution of canopy leaching to sulfate deposition in a Scots pine forest, *Environ. Pollut.*, in press.

20. Höfken, K. D., Input of acidifiers and heavy metals to a German forest area due to dry and wet deposition, in *Effects of Accumulation of Air Pollutants in Forest Ecosystems,* Ulrich, B. and Pankrath, J., Eds., D. Reidel, Doredecht, 1983, 57.

21. Lindberg, S. E. and Harriss, R. C., The role of atmospheric deposition in an Eastern U.S. deciduous forest, *Water Air Soil Pollut.,* 15, 13 1981.

22. Garten, C. T., Foliar leaching, translocation, and biogenic emission of ^{35}S in radiolabelled loblolly pines, *Ecology,* 71, 239, 1990.

23. Gilmour, J. T. and Miller, M. S., Fate of a mercuric-mercurous chloride fungicide added to turfgrass, *J. Environ. Qual.,* 2, 145, 1973.

24. Lindberg, S. E. and Harriss, R. C., Mercury in rain and throughfall in a tropical rain forest, *Proceedings Fifth International Conference on Heavy Metals in the Environment,* CEP Consultants Ltd., Edinburgh, U.K., Vol. 1, 1985, 527.

25. Iverfeldt, Å., Mercury in forest canopy throughfall water and its relation to atmospheric deposition, *Water Air and Soil Pollut.,* 56, 553, 1991.

26. Hanson, P. J., Lindberg, S. E., Owens, J. G., and Turner, R. R., Quantifying Hg vapor deposition to forest landscape surfaces: laboratory approaches and initial results. Poster presented at Int. Conf. Mercury as a Global Pollutant, Hyatt Regency, Monterey, California, May 31–June 4, 1992.

27. Bogle, M. A., Oak Ridge National Laboratory, personal communication (May, 1992).

28. Davidson, C. I. and Wu, Y.-L., Dry deposition of particles and vapors, in *Acidic Precipitation,* Vol. 3: Sources, Deposition, and Canopy Interactions, Lindberg S. E., page A., and Norton S., Eds., Springer-Verlag, New York, 1990, 103.

29. Hicks, B. B., Matt, D. R., McMillen, R. T., Womack, J. D., Wesely, M. L., Hart, R. L., Cook, D. R., Lindberg, S. E., de Pena, R. G., and Thomson, D. W., A field investigation of sulfate fluxes to a deciduous forest, *J. Geophys. Res.,* 94, 13003, 1989.

30. Du, S.-H. and Fan, C. S., Uptake of elemental mercury vapor by C_3 and C_4 species, *Environ. Exp. Bot.,* 22, 437, 1982.

31. Davidson, C. I., Lindberg, S. E., Schmidt, J., Cartwright, L., and Landis, L., Dry deposition of sulfate onto surrogate surfaces, *J. Geophys. Res.,* 90, 2121, 1985.

32. Munthe, J., The aqueous oxidation of elemental mercury by ozone, *Atmos. Environ.,* in press.

33. Hanson, P. J., Rott, K., Taylor, G. E., Jr., Gunderson, C. A., Lindberg, S. E., and Ross-Todd, B. M., NO_2 deposition to forest landscape surfaces, *Atmos. Environ.,* 23, 1783, 1989.

The Atmospheric Chemistry of Mercury: Kinetic Studies of Redox Reactions

John Munthe

CONTENTS

I. INTRODUCTION

In the atmosphere, the predominant form of mercury is Hg^0 vapor (95 to 100%).[1-4] This species has a residence time of around one year, which makes long-distance transport feasible. The ultimate fate of atmospheric mercury is wet or dry deposition, of which the former is probably more important[4] although dry deposition is known to be of importance in forests.[24,25] Wet deposition can only occur after the volatile Hg^0 has been oxidized to water soluble forms [e.g., $Hg(II)$].

Several different experimental approaches have been used for investigations of the atmospheric chemistry of mercury. The development of reliable techniques for sampling and analysis of mercury in ambient air was critical for our ability to explore the atmospheric cycling of this toxic element.[5] With analytical tools more readily available, measurements were performed at different locations and times in order to identify the general patterns for occurrence and transport of various forms of mercury.[6]

We have now reached a stage where we need to predict the future rate of deposition of atmospheric mercury and how the deposition will change if measures are taken to reduce the amounts of mercury released to the atmosphere. The effect of changes in concentrations of other pollutants on the deposition of mercury is also of interest. In order to accomplish this, mathematical models are needed that are capable of simulating the chemistry and transport of mercury in the atmosphere with some accuracy. The development of models is dependent on the availability of chemical kinetic data for the relevant processes.

Much interest has been focused on the identification of the major chemical reactions involved in transforming the airborne volatile mercury to more water-soluble forms that can be deposited with precipitation. Different methods have been employed such as correlating measured concentrations of mercury in air and precipitation with concentrations of other pollutants[3,4] or by performing exploratory laboratory experiments.[7] Experiments have also been performed where the chemical and physical properties of the atmospheric mercury is investigated in the field or in the laboratory.[6,8,9]

1–56670–066–3/94/$0.00+$.50
© 1994 Lewis Publishers

This chapter describes a first attempt to use some recently presented kinetic constants and modeling schemes for predictions of the concentration of water-soluble mercury in rain. Although the current knowledge of the atmospheric chemistry of mercury is far from complete, simple steady-state calculations can be performed to estimate the concentration of water-soluble Hg(II) compounds in rainwater. The intention is to provide a starting point for future laboratory investigations which will lead to more comprehensive modeling.

II. IMPORTANT CHEMICAL PROCESSES

A. GAS PHASE CHEMISTRY

Gas phase oxidation of Hg^0 is generally assumed to be slow.[8] The reason for this assumption is the predominance of Hg^0 vapor over oxidized and particulate forms in air. If gaseous Hg^0 was oxidized in air by an oxidant (X) at an appreciable rate, the product HgX or HgX_2 would be found as gaseous molecules in air or adsorbed to particles. Since no significant concentrations of gaseous oxidized or particulate forms of mercury are found[3] the gas phase oxidation process has been assumed to be slow. However, a cyclic process where both oxidation and reduction occurs cannot be ruled out. In this case, the rate of production of water soluble oxidized mercury species would be governed by the rate of reduction and the rates of uptake on solid particles or aqueous droplets. With this type of process, a slow removal of gaseous Hg^0 could take place without the accumulation of significant concentrations of gaseous or particulate forms of divalent mercury. Despite this, the most important removal process for atmospheric mercury is probably heterogeneous oxidation followed by deposition.

Two gaseous processes have been investigated that may be of importance in the atmospheric cycling of mercury; the oxidation of Hg^0 by O_3[10] and by NO_2.[11] The former process may be of some importance in the atmosphere whereas the latter is probably too slow to influence the concentration of Hg^0 in ambient air. Furthermore, the experimental results in these studies may be influenced by heterogeneous processes on the reactor wall surfaces which makes it difficult to extrapolate the results to atmospheric conditions. Yarwood and Niki[12] and Schroeder et al.[13] have discussed other reactions of potential importance in the atmospheric gas-phase chemistry of mercury. From calculations of enthalpies of reactions they conclude that, with gaseous HgO as the end product, the only oxidants that may oxidize gaseous Hg^0 are O_3, $O(^3P)$, $O(^1D)$, H_2O_2, and possibly NO_3. From the discussion given above, it can be concluded that the role of gas-phase processes is somewhat unclear and further research is needed before the environmental importance of these processes can be assessed.

B. AQUEOUS PHASE CHEMISTRY

The role of ozone in the aqueous phase atmospheric chemistry of mercury has been discussed in several papers.[7-9,14] Other oxidants that may be of importance are OH and HO_2,[15] whereas the usually reactive species H_2O_2 does not seem to react at all, or very slowly, with aqueous Hg^0.[15-17] It may, however, serve as a source of radicals in aqueous aerosols. Reduction of the divalent mercury to Hg^0 may occur through reactions with dissolved SO_2.[18] The role of photo-induced reduction of aqueous Hg(II) is not entirely clear, but may prove to be of importance under some conditions. Laboratory studies have shown that $Hg(OH)_2$ can be reduced by simulated sunlight, but the primary processes have not been identified.[17,19,20] The atmosphere also contains some other substances that are capable of reducing Hg(II); for example, CO has been shown to reduce Hg(II) to Hg^0 but at a very slow rate.[21]

III. MODELING THE AQUEOUS ATMOSPHERIC
CHEMISTRY OF MERCURY

A. PRINCIPAL EQUATIONS

The ultimate goal of kinetic studies of reactions occurring in the atmosphere is to produce rate constants that can serve as input to models. If accurate rate constants for the relevant processes are available, models can be designed that are capable of predicting the future rates of depositions. In the case of mercury, a lot of work still remains before reliable models with detailed chemistry can be developed. Large-scale models with simplified chemistry have been used with some success.[22] Although many potentially important chemical processes exist, kinetic constants are only available for a few aqueous redox reactions. These reactions, along with an attempt to model the redox cycling of mercury in clouds and rainwater, i.e., the primary production of water-soluble Hg(II) species, have been described earlier.[14,23] The model is based on the hypothesis that the concentration of divalent mercury in precipitation is governed by reactions occurring in a simple aqueous phase redox cycling of mercury. In this model, aqueous elemental mercury is oxidized by ozone to form divalent mercury (R1).[14]

$$Hg^0 + O_3 \rightarrow HgO + O_2 \qquad K_1 = (4.7 \pm 2.2) \times 10^7 \ M^{-1} \ s^{-1} \qquad (R1)$$

Rate expression:

$$-\frac{d[Hg^0]}{dt} = k_1[Hg^0][O_3]$$

The Hg(II) thus produced can then react with different components of the aqueous phase such as Cl^-, OH^- (which leads to the formation of stable HgX_2 molecules), or dissolved SO_2 (S(IV)) which leads to the formation of $Hg(SO_3)_2^{2-}$ (R2), and the possibility of reduction of the Hg(II) back to Hg^0 (R3).[18]

$$HgSO_3 + SO_3^{2-} \leftrightarrow Hg(SO_3)_2^{2-} \qquad \begin{array}{l} k_{2f} = 1.1 \times 10^8 \ M^{-1} \ s^{-1} \\ k_{2r} = 4.4 \times 10^{-4} \ s^{-1} \end{array} \qquad (R2)$$

$$HgSO_3 \rightarrow \rightarrow Hg^0 \qquad k_3 = 0.6 \ s^{-1} \qquad (R3)$$

Rate expression:

$$-\frac{d[Hg(SO_3)_2^{2-}]}{dt} = \frac{k_{2r}k_3[Hg(SO_3)_2^{2-}]}{k_{2f}[SO_3^{2-}] + k_3}$$

B. STEADY-STATE CALCULATIONS

Here, the aqueous droplet is treated as an aqueous solution where the concentrations of the volatile reactants (Hg^0, O_3, and SO_2) are determined by the gas phase concentrations and the corresponding Henry's Law constants. This enables calculations of steady-state concentrations of divalent mercury in the droplet as a function of the rates of oxidation and reduction and the gas-phase concentrations of Hg^0, O_3, and SO_2. A steady-state concentration of $Hg(SO_3)_2^{2-}$ can be calculated if it is assumed that the HgO produced in reaction (R1) is rapidly converted to $Hg(SO_3)_2^{2-}$ through a ligand addition reaction (R4).

$$HgO + 2HSO_3^- \rightarrow Hg(SO_3)_2^{2-} + 2H^+ \qquad (R4)$$

The rate expressions for reactions (R1), (R2), and (R3) can be combined to give an expression for the steady-state concentration of $Hg(SO_3)_2^{2-}$ (E3).

$$[Hg(SO_3)_2^{2-}]_{ss} = \frac{k_1[Hg^0(aq)][O_3(aq)](k_{2f}[SO_3^{2-}] + k_3)}{k_{2r}k_3}$$

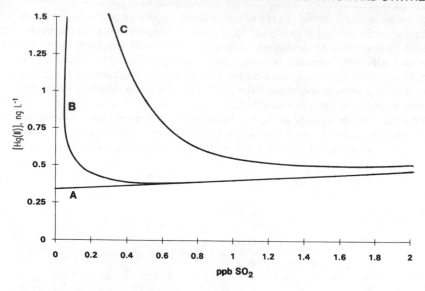

Figure 1 Calculated concentrations of divalent mercury in an aqueous atmospheric aerosols; $A = [Hg(SO_3)_2^{2-}]_{ss}$; $B = A + [Hg(OH)_2]$; $C = B + [HgCl_2] + [HgOHCl]$; pH = 4.5; $[O_3(g)] = 30$ ppb; $[Hg^0(g)] = 3$ ng m^{-3}.

This equation can only be used for predictions of the concentration of $Hg(SO_3)_2^{2-}$. Although the detailed speciation of Hg(II) in cloud- or rainwater is not known, other ligands are present and will influence the concentration of Hg(II). The different ligands will compete to form complexes with Hg(II). Apart from SO_3^{2-}, chloride (Cl$^-$) and hydroxide (OH$^-$) are present in significant concentrations in aqueous atmospheric aerosols. The influence of these ligands on the calculated concentration of $Hg(SO_3)_2^{2-}$ can be estimated using a simple calculation procedure described earlier.[14,23] In short, the calculations are made by calculating a ratio, defined as the total concentration of Hg(II) complexes divided by the concentration of $Hg(SO_3)_2^{2-}$ (E4). The total concentration of Hg(II) complexes is then obtained by multiplying the calculated steady-state concentration of $Hg(SO_3)_2^{2-}$ with this ratio. Some results are shown in Figure 1.

$$\text{Ratio} = \frac{[Hg(II)]_{tot}}{[Hg(SO_3)_2^{2-}]_{ss}} = 1 + \frac{\Sigma[Hg^{2+}][X]^n \beta_{HgX_n}}{[Hg(SO_3)_2^{2-}]_{ss}} = 1 + \frac{\Sigma[X]^n \beta_{HgX_n}}{[SO_3^{2-}]^2 \beta_a}$$

Where $[Hg(II)]_{tot}$, $[Hg^{2+}]$, and $[X]$ are the concentrations of total divalent mercury, the free water-coordinated mercury ion, and the ligand, respectively, and β_{HgX_n} and β_a are the stability constants for the complex HgX_n and $Hg(SO_3)_2^{2-}$, respectively.

The calculated concentrations agree reasonably well with measured concentrations of the operationally defined fraction Hg(IIa) or "reactive mercury",[3] although the levels are too high at low SO$_2$ concentrations. It is also clear that the influence of Cl$^-$ on the concentration of Hg(II) complexes is much greater than that of OH$^-$. Inclusion of Cl$^-$ in the model increases the concentration of Hg(II) by a factor of 2 to 3 at SO$_2$ concentrations around 0.6 ppb. If the same calculation procedure is used to estimate the influence of HS$^-$, CN$^-$, or other ligands that readily form complexes with Hg(II), much higher concentrations are predicted.

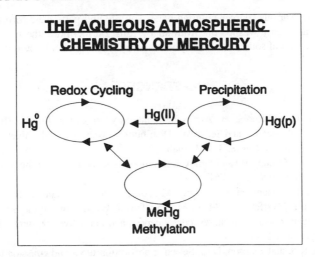

Figure 2 Some potentially important processes in the aqueous atmospheric chemistry of mercury.

IV. SUGGESTIONS FOR FURTHER WORK

The research described in this paper has shown that the concentration of dissolved inorganic mercury in precipitation can be estimated using a simple steady-state model. Further research in this area will lead to improved accuracy, which is needed for predictions of the future rate of deposition of mercury. Some suggestions are given below.

A. REACTION STUDIES

The gas phase process that should be investigated is mainly the reaction of Hg^0 and O_3. The influence on the reaction rate of heterogeneous reactions and humidity, as well as product studies, are of importance. Some other potentially important gas-phase reactions were discussed in Section II.A.

As discussed above, the concentration of dissolved inorganic mercury is heavily influenced by the presence of ligands that are capable of forming stable complexes with Hg(II). This feature of the atmospheric chemistry of mercury should be examined with great care in order to assess the influence of ligands such as CN^-, HS^-, and other compounds containing reduced sulfur. Other redox processes should also be investigated, such as the oxidation of Hg^0 by OH and HO_2 radicals. Processes leading to the formation of particulate phase mercury in aqueous aerosols should also be investigated since a large fraction of the mercury in rain is associated with particles. Reactions of Hg(II) with compounds containing reduced sulfur are probably important in this aspect.

Of great importance, also, is the possibility of formation of methylmercury in the atmosphere. To date, the source of methylmercury in precipitation has not been identified. In Figure 2, some of the most important processes in the aqueous atmospheric chemistry of mercury are presented schematically. If the main reacting species in these processes could be identified and the kinetics of these processes quantified, detailed description of the aqueous atmospheric chemistry of mercury could be used for model calculations and future predictions of the deposition of mercury from the atmosphere.

The chemistry described in this chapter has only taken into account the production of some divalent mercury complexes in aqueous aerosols. A more detailed model would have to include not only a more detailed aqueous- and gas-phase chemistry, but also a quantitative description of the various sources emitting water-soluble and particulate forms of mercury, i.e., forms that are easily washed out from air by rain. Furthermore, the magnitude of anthropogenic and natural sources of atmospheric methylmercury should be investigated.

REFERENCES

1. Lindqvist, O. and Rodhe, H., Atmospheric mercury—a review, *Tellus*, 37B, 136, 1985.
2. Slemr, F., Schuster, G., and Seiler, W., Distribution, speciation, and budget of atmospheric mercury, *J. Atmos. Chem.*, 3, 407, 1985.
3. Iverfeldt, Å., Occurrence and turnover of atmospheric mercury over the Nordic countries, *Water Air Soil Pollut.*, 56, 251, 1991.
4. Lindqvist, O., Johansson, K., Aastrup, M., Andersson, A., Bringmark, L., Hovsenius, G., Håkanson, L., Iverfeldt, Å., Meili, M., and Timm, B., Mercury in the Swedish environment—recent research on causes, consequences and corrective measures, *Water Air Soil Pollut.*, 55, 261, 1991.
5. Braman, R.S. and Johnson, D.L., Selective absorption tubes and emission techniques for determination of ambient forms of mercury in air, *Environ. Sci. Technol.*, 8, 996, 1974.
6. Brosset, C., Total airborne mercury and its possible origin, *Water Air Soil Pollut.*, 34, 37, 1982.
7. Iverfeldt, Å. and Lindqvist, O., Atmospheric oxidation of elemental mercury in the aqueous phase, *Atmos. Environ.*, 20, 1567, 1986.
8. Brosset, C., The behaviour of mercury in the physical environment, *Water Air Soil Pollut.*, 34, 145, 1987.
9. Brosset, C. and Lord, E., Mercury in precipitation and ambient air—a new scenario, *Water Air Soil Pollut.*, 56, 493, 1991.
10. PÝankov, V.A., Kinetics of the reaction between mercury vapor and ozone, *Zhur. Obshchev., Khim. (J. Gen. Chem.)*, 15, 224, 1949.
11. Hall, B., An Experimental Study of Mercury Reactions in Combustion Flue Gases, Thesis, Dept. Inorg. Chem., Chalmers University of Technology and University of Göteborg, S-412 96 Göteborg, Sweden.
12. Yarwood, G. and Niki, H., A Critical Review of Available Information on Transformation Pathways for the Mercury Species in the Atmospheric Environment. Report prepared for Atmospheric Environment Services, Environment Canada, Downsview, Ontario, Canada.
13. Schroeder, W.H., Yarwood, G., and Niki, H., Transformation processes involving mercury species in the atmosphere—results from a literature survey, *Water Air Soil Pollut.*, 56, 653, 1991.
14. Munthe, J., The aqueous oxidation of elemental mercury by ozone, *Atmos. Environ.*, 26A, 1461, 1992.
15. Kobayashi, T., Oxidation of metallic mercury in aqueous solution by hydrogen peroxide and chlorine, *J. Jpn. Air Pollut. Soc.*, 22, 230, 1987.
16. Wigfield, D.C. and Perkins, S.L., Oxidation of elemental mercury by hydroperoxides in aqueous solutions, *Can. J. Chem.*, 63, 275, 1985.
17. Munthe, J. and McElroy, W. J., Some aqueous reactions of potential importance in the atmospheric chemistry of mercury, *Atmos. Environ.*, 26A, 553, 1992.
18. Munthe, J., Xiao, Z.F., and Lindqvist, O., The aqueous reduction of divalent mercury by sulfite, *Water Air Soil Pollut.*, 56, 621, 1991.
19. Xiao, Z.F., Munthe, J., Strömberg, D., and Lindqvist, O., Photochemical behavior of inorganic mercury compounds in aqueous solution, in *Mercury Pollution: Integration and Synthesis*, Watras, C.J. and Huckabee, J.W., Eds., Lewis Publishers, Chelsea, MI, 1994, chap. VI.6.

20. Iverfeldt, Å., Structural, Thermodynamic and Kinetic Studies of Mercury Compounds; Applications Within the Environmental Mercury Cycle, Thesis, Department of Inorganic Chemistry, Chalmers University of Technology and University of Göteborg, S-412 96 Göteborg, Sweden.

21. Golodov, V.A. and Panov, Y.I., *React. Kinet. Catal. Lett.,* 24, 133, 1984.

22. Petersen, G., Iverfeldt, Å., and Munthe, J., Atmospheric mercury species over central and northern Europe. Model calculations and comparison with observations from the Nordic air and precipitation network for 1987 and 1988, *Atmosph. Environ.,* in press.

23. Munthe, J., The Redox Cycling of Mercury in the Atmosphere, Thesis, Department Inorganic Chemistry, Chalmers University of Technology and University of Göteborg, S-412 96 Göteborg, Sweden.

24. Lindberg, S.E., Meyers, T.P., Taylor, G.E., Turner, R.R., and Schroeder, W.H., Atmosphere-surface exchange of mercury in a forest: results of modeling and gradient approaches, *J. Geophys. Res.,* 97, 2519, 1992.

25. Iverfeldt, Å., Mercury in forest canopy throughfall water and its relation to atmospheric deposition, *Water Air Soil Pollut.,* 56, 553, 1991.

Atmospheric Mercury Measurements at a Rural Site in Southern Ontario, Canada

W. H. Schroeder

CONTENTS

I. INTRODUCTION

During the final decade of the 20th century, mercury appears likely to be ranked high among those priority environmental pollutants of continuing concern to environmentalists, legislators, environmental scientists and engineers, politicians, and the citizenry of industrialized nations, as well as developing countries around the globe. The unique physical and chemical attributes of this member of the family of so-called "heavy metals" make it an important component or constituent in a wide variety of commercial/consumer products of which thermometers, barometers, electrical (contact) switches, dental amalgams, and dry cell batteries are just a few examples.

Mercury and its compounds have also been used in enormous quantities in numerous industrial processes, including the production of chemicals utilized in the pulp and paper industry (chlor-alkali installations), as catalysts in the petrochemical industry, in the extraction and purification of deposits of precious metals such as gold and silver, and as additives in paints to prevent the formation of mildew. The myriad beneficial applications of this element have, in the past, not been balanced against the detrimental effects resulting from its release into the environment in large amounts. Indeed, the very success of mercury and its organic as well as inorganic compounds as important industrial commodities virtually guaranteed that they would become notorious environmental pollutants in many jurisdictions.

In addition to the many anthropogenic sources emitting mercury into the biosphere, this metal occurs naturally in the earth's crust and is mobilized by human activities or released from a number of natural sources (e.g., weathering of rocks, crustal erosion, outgassing of soils, volcanic eruptions, geysers and other geothermal activities, and releases from living or decaying vegetation).

Once mercury has been mobilized or released into the environment, the earth's atmosphere plays a very significant role in dispersing this element over large spatial scales—even globally. Consequently, mercury has been measured in air, water, snow, soils, and biota even in the most remote corners of the globe.[1-4] Reliable measurements of air concentrations of

1–56670–066–3/94/$0.00+$.50
© 1994 Lewis Publishers

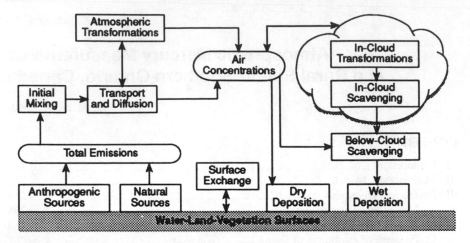

Figure 1 The emission-to-deposition cycle for toxic airborne contaminants.

pollutants over a broad range of spatial and temporal scales is of fundamental importance in understanding the atmospheric pathways and characteristics of toxic airborne contaminants such as mercury. The central role played by air concentration data in a given chemical's emission-to-deposition cycle is illustrated in the conceptual/interpretive framework of Figure 1. This type of information also serves as an important input variable or validation check parameter in mathematical models simulating atmospheric processes such as dispersion, transport, transformation, and removal from the atmosphere.

In 1989, a collaborative scientific project dealing with the modeling of atmospheric processes of mercury was initiated by scientists from the Federal Republic of Germany (GKSS Research Centre, near Hamburg) and from Canada (Ontario Ministry of the Environment, and the Atmospheric Environment Service of Environment Canada).[5-7] Our German colleagues are particularly interested in determining the quantity of mercury being deposited in the North Sea and the Baltic Sea. The Canadian efforts are focused primarily on atmospheric inputs of mercury into the Great Lakes ecosystem. The air concentration data obtained at Egbert, Ontario during March and April 1990 are to be used in evaluating and calibrating the Eulerian model used by the Canadian partners. The location of the field measurement site (at Egbert, Ontario; Lat. 44° 14'N, Long. 79° 47'W; 250 m above sea level), in the setting of a North American geographical context, is given in Figure 2.

II. MATERIALS AND METHODS

Over a period of 46 days (March 15 to April 30), in the late winter and early spring of 1990, total vapor-phase concentrations of mercury were measured in ambient air at the Atmospheric Environment Service (AES) research center located at Egbert, Ontario. This site (see Figure 2) is approximately 65 km north of the metropolitan Toronto area.

For the sake of quality control, method validation, and statistical data analysis, all sampling during this project was conducted in duplicate. For this purpose, two identical custom-made air sampling units were mounted on the roof of the Clean Air Sampling Facility at Egbert. The air intake was thus at a height of approximately 5 m above ground. Figure 3 is a photograph of an air sampling unit (taken at AES headquarters after completion of this study).

Air samples (integrated over a time period of approximately 23 to 24 h) were collected daily and then analyzed immediately after termination of the sample collection step. Sampling commenced at 800 h E.S.T. (12 Z, GMT) each day. With the aid of a Gilian portable air sampling pump (Model HFS 113 A; Gilian Instrument Corporation, Wayne, NJ) ambient air

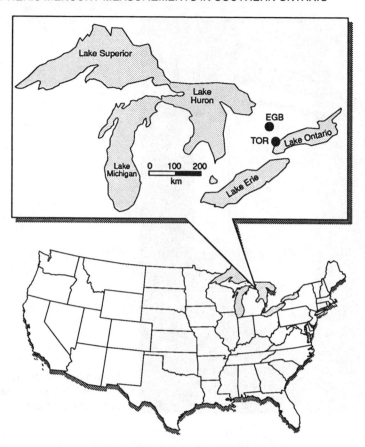

Figure 2   Location of sampling site (Egbert, abbreviated to EGB) in relation to the city of Toronto (abbreviated to TOR).

(at a nominal flowrate of 0.75 L/min) was pulled through an open-face Teflon® filter holder which supported a 47 mm diameter micro-quartz fiber filter (Gelman Sciences Inc., Montreal, Quebec).

This filtration procedure removed any mercury species bound to suspended particulate matter (or aerosols) from the ambient air being sampled. Before use, the filter disks were heated (at 450° C overnight) in a muffle furnace (Model No. F-A1730, Thermolyne, Dubuque, IA) to remove any mercury that may have been adsorbed or otherwise associated with these filter media. Filters were routinely changed at 7-day intervals. Exposed filter media were placed in plastic petri dishes, double-bagged, and stored for later analysis.

After the study was completed, the exposed filter media, along with several unused and untreated filters (as received from the supplier), heat-treated filters, analytical blanks, and field blanks, were sent to a commercial laboratory for chemical analysis. When they were analyzed, it was found that the various types of unexposed filters and associated blanks had relatively high and quite inconsistent mercury levels. As a result, particulate mercury concentrations are not considered to be sufficiently reliable for inclusion in this chapter.

Directly behind (downstream of) the filter holder was a vapor-phase mercury collection unit of the type specifically designed for the Jerome Model 301 gold film mercury detector (Arizona Instrument Corporation, Jerome, AZ) which was employed in this field study. The

Figure 3 Photograph of sampler used for atmospheric mercury measurements.

collection and release efficiencies of the individual vapor-phase mercury collectors were determined before and after this study.

The Jerome gold film detector and each of the vapor-phase mercury collection units were routinely calibrated (each day) via injections of several elemental mercury vapor calibration standards using a gas-tight syringe (Precision Sampling Corporation, Baton Rouge, LA). To eliminate (or at least reduce) calibration curve extrapolation errors, the amount of mercury contained in the calibration standard was matched closely to the amount of mercury detected in the previous sample. A minimum of two successive injections of each calibration standard were made at all times, or as many as were required to achieve a precision (reproducibility) of better than 3% for successive injections. A more detailed account of sampling, sample handling and storage, chemical analysis, and quality assurance/quality control aspects of this study will be published elsewhere, owing to page restrictions in this volume.

The most important sample collection parameters (sampling time, sampling flowrate, sample volume, and collector unit collection and release efficiency) have been summarized in Table 1. Even-numbered collectors (No. 2 and 4) were used on even-numbered calendar days, while odd-numbered collectors (No. 3 and 5) were used on odd-numbered days. The degree of dispersion around the mean value for all data in this table spanned a rather narrow range (0.3 to 3.7%). This parameter was calculated as: % dispersion = (standard deviation/mean) × 100. Accordingly, these sampling parameters contributed only a relatively small amount to the overall measurement uncertainty in our air concentration data set.

Table 1 **Summary of sampling parameters.**

		Sampler A		Sampler B	
		Collector 2	Collector 3	Collector 4	Collector 5
Sampling time	Min.	21.0	17.1	21.0	17.1
(h)	Max.	24.3	24.3	24.3	24.3
	Mean	23.7	23.7	23.6	23.7
	Sdev.	0.3	0.3	0.7	0.3
Flow rate	Min.	0.69	0.73	0.73	0.72
(L/min)	Max.	0.78	0.77	0.78	0.80
	Mean	0.74	0.74	0.76	0.76
	Sdev.	0.01	0.01	0.01	0.01
Sample vol. (L)	Min.	925.4	760.5	934.2	768.7
	Max.	1134.4	1106.6	1119.5	1150.9
	Mean	1054.6	1055.8	1070.2	1080.0
	Sdev.	25.6	21.4	39.8	26.1
Collector	Min.	98.0	91.5	98.0	90.8
efficiency	Max.	99.5	93.0	100.0	92.0
(%)	Mean	98.3	92.3	99.5	91.4
	Sdev.	0.3	0.5	0.5	0.4

Table 2 **Statistical error analysis for co-located atmospheric mercury samplers.**

Statistical parameter under consideration	Corresponding value
Number of paired data	22
Population mean	3.92 (Collector 2)
	4.16 (Collector 4)
Population median	2.90 (Collector 2)
	3.14 (Collector 4)
Mean error term (%)	13.7
Median error term (%)	10.7
Median absolute deviation (M.A.D.)	0.437
Modified M.A.D. (M.M.A.D.)	0.459

III. RESULTS AND DISCUSSION

A. MEASUREMENT UNCERTAINTY

To gain a fuller appreciation of the extent of "error bounds", henceforth called total measurement uncertainty (i.e., the sum of uncertainties associated with each step in the entire measurement procedure, including sampling, sample preparation/handling/storage, and chemical analysis), the results obtained with the two co-located air sampling units used in this study were analyzed statistically. It needs to be pointed out that this analysis (see Table 2) is based on the results obtained with sample collectors No. 2 and 4, since collector No. 3 was found, after the study, to consistently underestimate the actual mercury content in the air that was sampled.

The difference in results obtained with collector No. 3 and its partner (collector No. 5) was large enough to be statistically significant (according to the Student t-test criterion).[8,9] For 22 paired sets of data from collectors No. 2 and No. 4, the value of the t-statistic was 1.34, whereas for the 23 pairs of concentration values derived from collectors No. 3 and

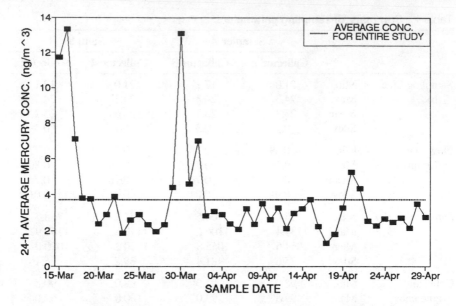

Figure 4 Atmospheric mercury concentrations (total vapor-phase; daily integrated samples) for 6-week study at Egbert, Ontario.

No. 5, the t-statistic was calculated to be 3.76. This latter difference in the values calculated for "t" is highly significant (alpha = 0.001). Hence, the results obtained with collector No. 3 could justifiably be excluded from the data set. As a result, some of the statistical parameters in this table (e.g., mean, median, standard deviation, etc.) may have values that are slightly different than those used later in the context of the entire data set which is based on three sample populations (collectors 2, 4, and 5).

B. ATMOSPHERIC MERCURY CONCENTRATIONS

The daily values (from integrated samples collected over an approximately 24-h time interval) of atmospheric mercury concentrations determined over a 6-week period—March 15 to April 30, 1990—at Egbert, Ontario, are displayed in Figure 4. The values given are for "total vapor-phase mercury" concentrations. To some extent this is an operationally defined quantity since all gaseous mercury species which have passed through (or have been released by) the particulate-phase mercury filter during sampling are expected to be efficiently trapped by the gold surfaces of the vapor-phase mercury collection units and thus will be reported as a component of total vapor-phase mercury. All of the most likely vapor-phase mercury species to be encountered in outdoor air (Hg^0, $HgCl_2$, CH_3HgCl, and CH_3HgCH_3) are efficiently bound (amalgamated) to gold surfaces.[10-12] At locations other than urban areas or industrial sites, particulate-phase mercury generally accounts for only a few percent, at most, of the mass of total airborne mercury (vapor- and particulate-phase constituents).[4,13,14] Furthermore, of the total vapor-phase mercury fraction encountered in the lower atmosphere (troposphere), elemental mercury vapor is now considered to constitute by far the largest fraction.[15,16]

The episodic nature of atmospheric mercury concentrations (at least at this site) from day to day can be clearly seen in Figure 4. Two major episodes for which the air concentrations exceeded 10 ng/m³ occurred at the beginning of the study (mid-March) and in late March to early April. The mercury concentrations measured in ambient air at Egbert spanned a range of 1 decade: from 1.3 to 13.3 ng/m³. The dotted horizontal line, parallel to the x-axis of the graph, indicates the overall average mercury concentration for the 6-week period of

Table 3   **Statistical summary of atmospheric mercury concentration data obtained at Egbert, Ontario during March/April 1990[a].**

Parameter	Corresponding value (ng/m^3)
Arithmetic mean	3.71
Median value	2.90
Geometric Mean	3.22
Range of values	1.3–13.3

[a] Number of consecutive data points = 46.

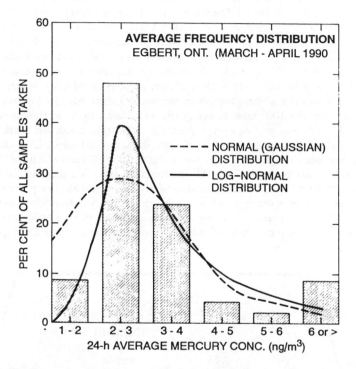

Figure 5   Frequency distribution of air concentration data (compared with Gaussian or normal vs. log-normal distribution).

our study. The complete set of experimental results obtained at Egbert is presented as a statistical summary in Table 3.

C.   FREQUENCY DISTRIBUTION OF DATA

Air quality data are frequently presented in terms of the following statistical parameters: frequency distributions, mean or median values, and standard deviations. It has been found empirically that the majority of air quality data sets conform to a log-normal, rather than a normal (Gaussian) type of frequency distribution. The former type of distribution displays the number of experimental values within selected intervals over the entire range of measurements. The statistics reported most frequently for log-normally distributed air quality data sets are the geometric mean and the geometric standard deviation. The atmospheric mercury concentration data set obtained during this study was examined to determine the type of frequency distribution that has the "best fit" to our experimental data (see Figure 5).

D. AIR PARCEL BACK TRAJECTORIES

Air parcel back trajectories, which trace the routes taken by air pollutants prior to their arrival at a receptor/measurement site, are often a very valuable tool for the identification of specific source locations, or at least a general area or region of a country or continent, giving rise to substantial quantities of a given air pollutant. Such trajectories have been applied widely and frequently in the identification or confirmation of industrial sites or regions now known to be primarily responsible for emissions of acid rain precursor species (especially sulphur and nitrogen oxides) leading to deleterious effects on sensitive terrestrial and aquatic ecosystems.

Five-day (120-h) air parcel back trajectories at three different heights (1000, 925, and 850 millibars), with 6-h time resolution, and with Egbert, Ontario as the end point, were run on the AES supercomputer at Dorval, Quebec using an algorithm developed by Olson et al.[17] Daily trajectories were obtained for 12 Z (1200 h GMT), the time closest to 800 h E.S.T., the beginning of each daily sampling cycle for mercury measurements at Egbert. It is instructive to look at the trajectories corresponding to those days giving rise to the highest (March 16) and the lowest (April 17) atmospheric mercury concentrations measured at Egbert for the entire duration of the field study. Starting with the trajectories of Friday, March 16 (Figure 6A) it is readily seen that all three trajectories arriving at Egbert that morning correspond to southerly air flow. The 1000 mbar trajectory (the one closest to the earth's surface) came from the Gulf of Mexico, passing through the states of Georgia, South and North Carolina, Virginia, Maryland, Pennsylvania, and New York, then travelled across Lake Ontario and over the metropolitan Toronto area on its way to Egbert. The 925 and 850 mbar trajectories travelled virtually on top of one another through Kentucky and Ohio, then crossed Lake Erie and approached Egbert from a south to south-westerly direction. This flow pattern remained generally unchanged over the next 24 h. Considering the highly industrialized areas that all three trajectories passed over on their way to our measurement site, it should not come as a big surprise to the reader that elevated values of mercury (as well as other air pollutants) were observed on that day.

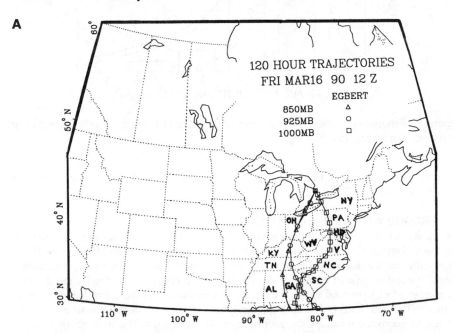

Figure 6 Air parcel back trajectories for the highest (March 16) and lowest (April 17) values of atmospheric mercury concentrations measured at Egbert (March/April 1990).

B

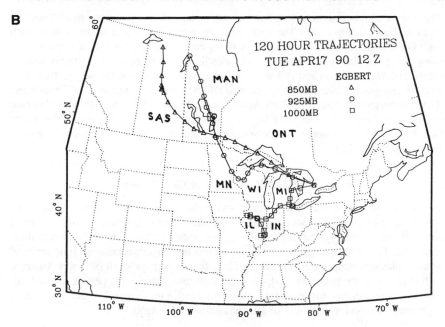

C

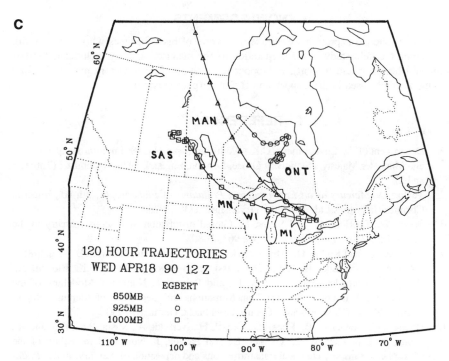

Figure 6 Continued.

The general air circulation pattern that existed, at least during the morning of Tuesday, April 17 is shown in Figure 6B. The 1000 mbar trajectory followed a rather tortuous path through Illinois, Indiana, and Michigan prior to its entry into the Province of Ontario after crossing over Lake Huron. The other 2 trajectories have their origin in the northern parts of Manitoba and Saskatchewan and follow a south-easterly direction on the way to their destination, largely passing over sparsely populated and non-industrialized areas of Canada and the U.S. Furthermore, between 12 Z on April 17 and the same time the following day (see Figure 6C), the three trajectories have swung around in a clockwise direction to the extent that all of them originate either in Saskatchewan or Manitoba, and pass over relatively pristine areas on their way to Egbert.

IV. SUMMARY AND CONCLUSIONS

Over a 6-week time span during late winter and early spring of 1990 (March 15 to April 30), total vapor-phase concentrations of mercury were determined daily in ambient air at Egbert, Ontario, a primarily rural site located approximately 65 km north of the metropolitan Toronto area. To our knowledge, this study represents the longest, continuous time series of daily atmospheric mercury measurements in Canada, and perhaps in all of North America, published so far in the peer-reviewed scientific literature. During this period, ambient air concentrations of mercury varied tenfold, ranging from 1.3 to 13 ng/m^3 of air. The overall average concentration for this study was calculated to be 3.7 ng/m^3.

ACKNOWLEDGMENTS

The author wishes to express his gratitude to several of his colleagues who were helpful, either during the field study or the preparation of the manuscript. They include: J. Markes (technical laboratory/field support), F. Hopper (generation of air trajectory maps), S. Chao (plotting of graphs used in this paper), and B. Kiely (photography).

REFERENCES

1. Department of the Environment (U.K.), Pollution Paper No. 10: Environmental Mercury and Man, Her Majesty's Stationery Office, London, England, 1976, Appendix I (Tables 1 and 2).
2. Taylor, D., *Mercury as an Environmental Pollutant—A Bibliography*, 5th ed., Imperial Chemical Industries Ltd., Brixham, Devonshire, U.K., 1978.
3. Seiler, W., Eberling, C., and Slemr, F., Global distribution of gaseous mercury in the troposphere, *Pure Appl. Geophys.*, 118, 963, 1980.
4. Lindqvist, O. and Rodhe, H., Atmospheric mercury—a review, *Tellus*, 37B, 136, 1985.
5. Petersen, G., Schneider, B., Eppel, D., Grassl, H., Iverfeldt, A., Misra, P.K., Bloxam, R., Wong, S., Schroeder, W.H., Voldner, E., and Pacyna, J., Numerical Modelling of the Atmospheric Transport, Chemical Transformations and Deposition of Mercury, Report GKSS 90/E/24, GKSS Research Centre, Geesthacht, Germany, 1990.
6. Petersen, G., Schneider, B., Eppel, D., Grassl, H., Iverfeldt, A., Misra, P.K., Bloxam, R., Wong, S., Schroeder, W.H., Voldner, E., and Pacyna, J., Numerical modelling of the atmospheric transport, chemical transformations and deposition of mercury, in *Air Pollution Modeling and its Application*, VIII, vanDop, H. and Stern, D.G., Eds., Plenum Press, New York, 1991, 215.
7. Bloxam, R., Wong, S., Misra, P.K., Voldner, E., Schroeder, W.H., and Petersen, G., Modelling the long range transport, transformation and deposition of mercury in a comprehensive Eulerian framework, in *Proc. Int. Symp. Heavy Metals in the Environment*, Farmer, J.G., Ed., CEP Publications, Edinburgh, U.K., Vol. 1, 1991, 326.

8. Laitinen, H.A., *Chemical Analysis—An Advanced Text and Reference,* McGraw-Hill, New York, 1960, ch. 26.
9. Havlicek, L.L. and Crain, R.D., *Practical Statistics for the Physical Sciences,* American Chemical Society, Washington, D.C., 1988, ch. 10.
10. Braman, R.S. and Johnson, D.L., Selective absorption tubes and emission technique for determination of ambient forms of mercury in air, *Environ. Sci. Technol.,* 8, 996, 1974.
11. Schroeder, W.H., Sampling and Analytical Methodology for Mercury and its Compounds in Air—A Selective, Coded Bibliography, Report ARQA-64–79, Atmospheric Environment Service, Downsview, Ontario, 1979.
12. Schroeder, W.H., Hamilton, M.C., and Stobart, S.R., The use of noble metals as collection media for mercury and its compounds in the atmosphere, *Rev. Anal. Chem.,* 8, 179, 1985.
13. Schroeder, W.H., Sampling and analysis of mercury and its compounds in the atmosphere, *Environ. Sci. Technol.,* 16, 394 1982.
14. Fitzgerald, W.F., Mason, R.P., and Vandal, G.M., Atmospheric cycling and air-water exchange of mercury over mid-Continental lacustrine regions, *Water Air Soil Pollut.,* 56, 745, 1991.
15. Schroeder, W.H. and Jackson, R.A., Environmental measurements with an atmospheric mercury monitor having speciation capabilities, *Chemosphere,* 16, 183, 1987.
16. Brosset, C. and Lord, E., Mercury in precipitation and ambient air—a new scenario, *Water Air Soil Pollut.,* 56, 493, 1991.
17. Olson, M.P., Panigas, A., and Oikawa, K.K., A Users Guide to the Interactive Trajectory Computation and Plotting System (ITCAPS), Report AQRB-85–006-T, Atmospheric Environment Service, Downsview, Ontario, 1985.

SECTION III

Terrestrial and Watershed Processes

Mercury in Forest Ecosystems: Risk and Research Needs

Douglas L. Godbold

CONTENTS

I. INTRODUCTION

Although other heavy metals have been shown to be accumulating in forest ecosystems throughout northern and central Europe,[1] mercury as a potential highly toxic pollutant has for the most part been neglected. Outside Scandinavia few measurements of Hg in forest ecosystems have been carried out. Of the few measurements, some have shown relatively high levels of Hg in forests soils.[2,3] In the environment Hg may be present in a number of chemical forms, and ionic Hg can be methylated by microorganisms[4] and by contact with humic substances.[5] All forms of mercury have been shown to accumulate in the humic layer of forest soils. Factors affecting the accumulation and distribution of Hg in forest soils have been the subject of a recent review.[6]

II. PHYSIOLOGICAL EFFECTS OF MERCURY

Studies on the physiological effects of mercury on forest vegetation are few. However, in both algae and higher plants, mercury has been shown to influence a number of primary processes.[7-9] In higher plants, exposure to Hg reduced photosynthesis and transpiration,[9] water uptake,[7] and chlorophyll synthesis.[10] Mercury has been shown to inhibit root growth in several plant species.[11] Intact plasma membranes and functioning transport systems are essential for root growth and nutrient uptake. In plasma membranes isolated from roots of *Zea mays,* Hg affected trans-root and transmembrane potentials and inhibited plasma membrane ATPases.[12,13] Exposure of both Chlorella and roots of *Zea mays* to Hg and organic Hg compounds resulted in changes in membrane integrity and a loss of K and nucleotides from the cells.[14,15]

1–56670–066–3/94/$0.00+$.50
© 1994 Lewis Publishers

Table 1   Total, primary, secondary, and tertiary root length per plant of *Picea abies* seedlings after a 4-week treatment with 10 nM HgCl$_2$ or 1 nM CH$_3$HgCl in nutrient solutions.

Treatment	Root length (cm)			
	Total	Primary	Secondary	Tertiary
Control	58.6 a	15.5 a	39.6 a	6.2 a
1 nM CH$_3$Hg	36.0 b	9.0 b	23.4 b	3.7 b
10 nM Hg	41.5 b	12.2 c	26.3 b	3.1 b

Note: Treatments with no common indices are significantly different ($p > 0.05$) n = 32.

Table 2   Total, primary, secondary, and tertiary root tips per plant of *Picea abies* seedlings after a 4-week treatment with 10 nM HgCl$_2$ or 1 nM CH$_3$HgCl in nutrient solutions.

Treatment	Number of root tips			
	Total	Primary	Secondary	Tertiary
Control	127a	1a	55a	73a
1 nM CH$_3$Hg	81b	1a	43b	38b
10 nM Hg	81b	1a	45b	35b

Note: Treatments with no common indices are significantly different ($p > 0.05$) n = 32.

Much of the work on woody plants has been concerned with the foliar uptake of Hg.[16-19] This work has shown that Hg is taken up from the air by plant leaves and that airborne Hg can be phytotoxic. In natural communities some plant species seem to be tolerant to airborne Hg.[17] Investigations on the effects of Hg on the physiology of forest trees are limited almost entirely to Norway spruce (*Picea abies* L. Karst.).[9,20-22] Norway spruce is economically the most important tree species in northern Europe, and is a dominant species in natural boreal forests in Europe. A model for the physiological basis of mercury toxicity in Norway spruce is described in the following, and discussed in terms of what we presently know about Hg in forest ecosystems.

III.   MERCURY TOXICITY IN NORWAY SPRUCE SEEDLINGS

The model described here is based on a number of experiments.[9,20-22] The methods used have been published in detail elsewhere.[9,20-22] Briefly, the seedlings of *Picea abies* have been exposed to a range of concentrations of HgCl$_2$ or CH$_3$HgCl in nutrients solutions. Mercury in the seedlings was determined by cold vapor atomic absorption spectrophotometry.

A.   ROOT GROWTH

In short-term studies, root elongation of *Picea abies* seedlings was shown to be inhibited by HgCl$_2$ and CH$_3$HgCl.[20] The degree of inhibition increased with an increase in the external concentration. In Tables 1 and 2 the effects of a longer-term exposure (4 weeks) to HgCl$_2$ and CH$_3$HgCl on the root morphology are shown. A 4-week exposure to 10 nM HgCl$_2$ or 1 nM CH$_3$HgCl reduced the length of primary, secondary, and tertiary roots (Table 1). Not only is the growth of roots inhibited by mercury but also their formation (Table 2). Exposure to 10 nM HgCl$_2$ or 1 nM CH$_3$HgCl reduced the number of secondary and tertiary roots formed. The inhibition of formation of tertiary roots was stronger than that of secondary roots.

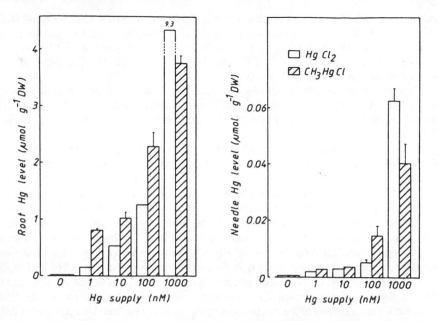

Figure 1 Mercury levels in roots and needles of *Picea abies* seedlings exposed for 7 weeks to nutrient solutions containing $HgCl_2$ or CH_3HgCl, ±S.E. (From Godbold, D.L. and Hüttermann, A., *Physiol. Plantarum*, 74, 270, 1988. With permission.)

Table 3   **Mercury levels in roots and needles of *Picea abies* seedlings after an up to 39 day treatment with 10 nM $HgCl_2$ or 1 nM CH_3HgCl in nutrient solutions, ±S.E.**

	Hg content (nmol g^{-1} Dwt)			
	Roots		**Needles**	
Days	**1 nM CH$_3$Hg**	**10 nM Hg**	**1 nM CH$_3$Hg**	**10 nM Hg**
9	760 ± 120	580 ± 40	1.0 ± 0.2	0.12 ± 0.70
16	820 ± 100	640 ± 40	2.8 ± 1.0	0.33 ± 0.10
26	1010 ± 90	780 ± 60	2.0 ± 0.6	0.33 ± 0.15
39	770 ± 80	660 ± 40	2.7 ± 0.6	0.66 ± 0.10

B.   MERCURY UPTAKE

After a 7-week exposure to mercury in roots at external concentration below 1000 nM Hg, higher levels of Hg were found in roots treated with CH_3HgCl than with $HgCl_2$ (Figure 1). At 1000 nM Hg the reverse was found. In needles, equal Hg levels were found at 1 and 10 nM Hg irrespective of the level and form of Hg supplied. Only at higher external concentrations do needles Hg levels reflect those of the roots. If the accumulation of Hg by roots and needles over time is followed (Table 3), roots exposed to both $HgCl_2$ or CH_3HgCl become rapidly saturated. In needles, a slow accumulation of Hg could be found. Independent of the Hg form supplied and the external concentration, the Hg level in roots exceeded that of the needles by a factor of over 100.

C.   MINERAL NUTRITION

After a 7-week exposure to Hg, the mineral levels in root tips were found to be more strongly affected than those of the whole roots.[20] The effects of exposure to $HgCl_2$ and CH_3HgCl on the mineral levels of 5 mm root tips are shown in Table 4. After exposure to 1000 nM $HgCl_2$ the levels of K, Mg, and Ca decreased. Treatment with 1000 nM $HgCl_2$ increased the levels of Fe in the root tips. In roots exposed to 1 nM CH_3HgCl, the levels of K and Mg in 5 mm root tips decreased, whereas those of Fe increased. At 10 nM CH_3HgCl a statistically significant increase in Fe, Ca, and Zn was found.

Exposure of the roots to 10 nM $HgCl_2$ significantly (Table 4) decreased the contents of Zn and Mn. At 1000 nM $HgCl_2$, the Mg and Ca levels of the needles were also significantly lower. Exposure to 1 and 10 nM CH_3HgCl decreased the levels of all mineral elements determined.

The above data show that the effects of $HgCl_2$ and CH_3HgCl on the mineral nutrition of *Picea abies* are similar. However the levels of $HgCl_2$ which cause such changes in element levels exceed those of CH_3HgCl by about a factor of 100. The loss in roots of primarily cytosolic and vacuolar elements such as K, Mg, and Mn may be due to changes in membrane permeability, as suggested by other authors.[15,23] In short-term experiments [45]Ca uptake has been used to determine membrane integrity. In the presence of $HgCl_2$ in the nutrient solution the uptake of [45]Ca into the root increased dramatically (Table 5). The increase in [45]Ca uptake was strongly dependent upon the level of $HgCl_2$ supplied. In comparison, [45]Ca uptake only slightly increased in the presence of CH_3HgCl, and was independent of the CH_3HgCl supply.

During the duration of the experiment, net Ca uptake into the roots was minimal and the [45]Ca uptake equaled an exchange of isotopic for nonisotopic Ca within the root.[24,25] In *Allium cepa* roots, the Ca content of the cytoplasm was shown to be regulated by a passive Ca influx and an active Ca efflux.[24] The [45]Ca uptake in the presence of $HgCl_2$ suggests that Hg may directly affect the plasma membrane, increasing Ca uptake into the symplast. In comparison to $HgCl_2$, CH_3HgCl does not appear to act directly at the plasma membrane.

D. GAS EXCHANGE

A 7-week exposure to $HgCl_2$ and CH_3HgCl, affected CO_2 uptake more strongly than transpiration.[21] The degree of inhibition of CO_2 uptake for both forms of Hg was dependent upon the external supply. Changes in gas exchange were again induced at CH_3HgCl concentrations about 100 times lower than those of $HgCl_2$. At the lower levels of Hg, Godbold and Hüttermann[21] estimated that inhibition of CO_2 uptake by Hg can be explained by a decrease in chlorophyll levels and stomatal closure. The changes in chlorophyll levels appeared not to be due to the direct effects of Hg in the needles, but rather to a decreased water and nutrient supply.

E. A MODEL OF Hg TOXICITY IN NORWAY SPRUCE

Mercury chloride and CH_3HgCl are extremely toxic compounds which inhibit primary processes at very low concentrations. The greater toxicity and rate of accumulation of CH_3HgCl may be due to its higher lipid solubility,[26] which allows CH_3HgCl to pass through biological membranes.[27] Mercury chloride and CH_3HgCl appear to have a similar mechanism of action, both cause loss of membrane integrity. However, whereas $HgCl_2$ directly affects the plasma membrane, CH_3HgCl may primarily affect the metabolism in the cytoplasm which subsequently affects membrane integrity. In roots of *Zea mays*, Hg affected transmembrane potential and H-ion efflux, and at high concentrations, ATPase activity.[12,13]

The root damage induced by Hg causes a decrease in water and nutrient uptake by the roots and a subsequent lower supply to the needles, which results in lower rates of gas exchange. Davis *et al.*[28] showed that active root tips are a prerequisite for keeping stomata open. At low Hg concentrations the decrease in photosynthesis is mainly due to lower chlorophyll levels as a result of the lower supply of nutrients and water.[21] Thus, at low Hg concentration root damage is the primary mechanism for all other physiological changes.

Table 4   Levels of mineral elements in 5 mm roots tips (n = 3) and needles (n = 8) of *Picea abies* seedlings after a 7-week treatment with various concentrations of $HgCl_2$ or CH_3HgCl in nutrient solutions.

	Root tips					
	K	Mg	Ca	Fe	Zn	Mn
Control	1016 a	91 a	56 a	140 a	2.4 a	7.7 a
10 n*M* $HgCl_2$	997 a	110 b	52 a	133 a	2.4 a	5.0 a
1000 n*M* $HgCl_2$	189 b	34 c	42 b	228 a	2.5 a	2.4 a
Control	1205 a	128 a	86 a	96 a	2.6 a	4.1 a
1 n*M* CH_3HgCl	334 b	49 b	71 a	158 ab	3.2 a	1.2 a
10 n*M* CH_3HgCl	198 c	56 b	148 b	256 b	4.9 b	1.9 a

	Needles					
	K	Mg	Ca	Fe	Zn	Mn
Control	288 a	54 a	96 a	1.5 a	0.8 a	15 a
10 n*M* $HgCl_2$	277 a	53 a	71 a	1.6 a	0.4 b	9 b
1000 n*M* $HgCl_2$	292 a	34 b	54 b	1.7 a	0.4 b	3 c
Control	283 a	53 a	83 a	1.0 a	0.7 a	15 a
1 n*M* CH_3HgCl	240 b	44 b	59 b	0.8 ab	0.4 b	6 ab
10 n*M* CH_3HgCl	194 c	26 c	38 c	0.7 b	0.3 c	4 b

Note: Treatments with no common indices are significantly different ($p > 0.05$).

Table 5   Influence of $HgCl_2$ or CH_3HgCl on ^{45}Ca uptake into roots of *Picea abies* seedlings. Duration of uptake 1 h.

	^{45}Ca μmol g^{-1} Dwt	
Supply μM	$HgCl_2$	CH_3HgCl
0	13.6 d	16.7 a
0.1	16.2 d	19.0 a
0.5	21.3 c	19.4 a
2.0	30.6 b	18.6 a
8.0	46.6 a	20.9 a

Note: Treatments with no common indices in columns are significantly different ($p > 0.05$) (n = 3).

IV.   MERCURY IN FOREST ECOSYSTEMS

A.   MERCURY CONTENT OF HUMUS LAYERS

The humic layer has been shown to be the primary site of Hg accumulation in forest soils, and also provides an important rooting zone for trees. Until recently, few measurements of Hg in forest soils outside Scandinavia were available (Table 6). In Norway and Sweden most Hg levels in the humus layer of forest soils are below 240 ng g^{-1}. In Solling and Hunau, Germany, higher levels of Hg have been reported. Both German sites appear to have unusually high levels of Hg in the humic layer. At the Hunau site 4.3 ng g^{-1} dry weight methyl-Hg has been determined.[29] In the soil solution of the humic layer at Hunau, Hg and methyl-Hg concentrations of 0.2 and 0.02 nM, respectively, were determined. Although much of the Hg and methyl-Hg in solution was bound to fulvic acids,[29] these levels are only 50-fold lower then those shown to severely damage seedlings of *Picea abies*. Both the Solling and Hunau

Table 6   **Levels of Hg in the humus layer of forest soils in Scandinavia and northern Europe.**

	ng g^{-1} Dwt	Ref.
Norway	190	Låg and Steinnes[36]
Sweden	Median 240	Lindqvist[6]
Solling, Germany	880	Lamersdorf[2]
Hunau, Germany	700	Padberg[29]

Table 7   **Levels of Hg in Norway spruce (*Picea abies*) needles in Scandinavia and northern Europe.**

	μg g^{-1} Dwt	Ref.
Huaklampi, Finland	0.04–0.05	Koshi et al.[37]
West coast, Sweden	0.04–0.05	Lindqvist[6]
Solling, Germany	0.06–0.13	Teuwsen[38]
Berchtesgaden, Germany	0.02–0.03	Padberg and May[3a]
Hunau, Germany	0.04–0.11	Padberg and May[3a]
Winterthur, Switzerland	0.06	Wyttenbach and Tobler[39]

[a] Recalculated on a dry wt. basis assuming a 50% needle water content.

site are stocked with Norway spruce and show moderate and severe symptoms of forest decline, respectively.

B.   MERCURY CONTENT OF NORWAY SPRUCE NEEDLES

Table 7 shows the Hg contents of spruce needles at sites in northern Europe and Scandinavia. With the exception of the slightly higher values at Solling and Hunau, Germany, all Hg values determined are in a similar range. In the *Picea abies* seedlings at low Hg supply, the Hg content of the needles was relatively unaffected. This is because little of the Hg taken up by the roots is transported to the needles. Due to the cation exchange capacity of the stem wood in mature trees it is likely that even less Hg is transported to the needles. Only in areas with high atmospheric concentrations of Hg have higher Hg levels in conifer needles been reported.[18] Thus, the Hg content of needles may be a better reflection of atmospheric Hg levels than those found in the forest soil. Lindberg et al.[30] showed that the forest canopy takes up Hg vapor. At the Hunau site only 5% of the total Hg was found to be on the surface of the needles.[29] Padberg[29] calculated that Hg absorbed from the atmosphere by Norway spruce needles and then deposited on the forest floor when the needles fall makes an important contribution to Hg input. This results in a higher Hg deposition to the forest than to surrounding areas, which has also been shown for other heavy metals.[1,2]

C.   MICROORGANISMS

The forest floor supports a large number of microorganisms and invertebrates, most of which play an important role in turnover of nutrients. The microorganisms are in the bulk soils as well as in the rhizospheres of the tree roots. The rhizosphere bacteria and mycorrhizae of tree roots may play an important role for trees in nutrient and water uptake. The effect of heavy metals on soil microorganisms and invertebrates has recently been the subject of an excellent review.[31] It is notable, however that no data are presented for the effects of Hg on soil microorganisms and invertebrates. Based on studies by Stadelmann and Santschi-Fuhrimann[32] and Wilke,[33] Lindqvist[6] calculated that Hg levels found in the mor layer of Swedish forests may be sufficient to reduce microbial respiration. Lindqvist,[6] however, makes the reservation that further research is needed before clear statements can be made.

Table 8 **Levels of Hg in fruiting bodies of mycorrhizal fungi in Finland.**

Fungi	$\mu g\ g^{-1}$ Dwt
Paxillus involutus	0.03
Lactarius rufus	0.08
Russula sp.	0.23
Suillus granulatus	0.82
Lactarius torminosus	0.98
Amanita muscaria	2.6

Data from Kojo, M.-R. and Lodenius, M., *Angew. Botanik*, 63, 279, 1989.

As previously stated, mycorrhizal fungi have been attributed an important role in nutrient and water uptake by trees as well as regulating or protecting trees from the effects of heavy metals. The role of ectomycorrhizae in the metal tolerance of trees has also been the subject of a recent review.[34] The benefits to trees in terms of nutrient uptake and metal tolerance appears to be dependent upon the species of ectomycorrhizae. Fungi vary in their tolerance to heavy metals, and pollution of soils with heavy metals often results in a change in the species composition.[31] Using ectomycorrhizal fungi isolated from Solling, Germany, Pb uptake into *Picea abies* was found to be highly species dependent.[35] As has been determined for other heavy metals, a wide range of concentrations of Hg has been determined in fruiting bodies of various mycorrhizal fungi (Table 8). These data also suggest that the uptake and transport of Hg differs between species.

D. RISKS TO HIGH ELEVATION FORESTS

Investigations of the atmospheric chemistry of Hg has shown that O_3 and H^+ will increase the Hg(II) content of cloud water.[26] High elevation forests are often alternately exposed to acidic cloud water and periods of increased O_3 and UV light. The wet canopy of a forest could thus provide a large surface area for the oxidation of Hg^0 in the atmosphere and substantially contribute to Hg input to forest. If deposition is sufficiently high, Hg-induced membrane damage to needle cells could contribute to leaching of nutrients out of the needles.

V. RISK AND RESEARCH NEEDS

This study shows how little is known about the effects of Hg on vegetation and forest floor microflora and fauna. There is evidence to suggest that in some forest ecosystems in northern Europe Hg levels are sufficiently high to influence trees and microorganisms. However, before the risk of Hg can be properly assessed more must be known about the levels of Hg in, and the input of Hg to, forest ecosystems. This work should be carried out together with an assessment of the effects of Hg on soil and rhizosphere organisms and vegetation.

REFERENCES

1. Mayer, R., Natürliche und anthropogene Komponenten des Schwermetall-Haushalts von Waldöksystemen, *Göttingen Bodenkdl. Ber.,* 70, 1, 1981.
2. Lamersdorf, N., Verteilung und Akkumulation von Spurenstoffen in Waldökosystemen, In: Ulrich, B., (Ed.), *Ber. Forschungszentr. Waldökosyst.,* B36, Göttingen, 1988.
3. Padberg, S. and May, K., Occurrence and behavior of mercury and methylmercury in the aquatic and terrestrial environment, In: Rossbach, M., Schladot, J.D., and Ostapczuk, P. (Eds.), *Specimen Banking,* Springer-Verlag, Berlin, 195, 1992.

4. Jenson, S. and Jernelöv, A., Biolgical methylation of mercury in aquatic organisms, *Nature*, 223, 752, 1969.
5. Nagase, H., Ose, Y., Satao, T., and Ishikawa, T., Methylation of mercury by humic substances in the aquatic environment, *Sci. Total Environ.*, 24, 133, 1982.
6. Lindqvist, O., Mercury in the Swedish environment, *Water Air Soil Pollut.*, 55, 1, 1991.
7. Beauford, W., Barber, J., and Barrington, A.R., Uptake and distribution of mercury within higher plants, *Physiol. Plantarum*, 39, 261, 1977.
8. De Filippis, L.F., Hampp, R., and Zeigler, H., The effects of sublethal concentration of zinc, cadmium and mercury on Euglena: Growth and pigments, *Z. Pflanzenphysiol.*, 101, 37, 1981.
9. Schlegel, H., Godbold, D.L., and Hüttermann, A., Whole plant aspects of heavy metal induced changes in CO_2 uptake and water relations of spruce (*Picea abies*) seedlings, *Physiol. Plantarum*, 69, 265, 1987.
10. Prasad, D.D.K. and Prasad, A.R.K., Altered D-aminolevulinic acid metabolism by lead and mercury in germinating seedlings of bajra (*Pennisetum typhoideum*), *J. Plant Physiol.*, 127, 241, 1987.
11. Mhatre, G.N. and Chapehekar, S.B., Response of young plants to mercury, *Water Air Soil Pollut.*, 21, 1, 1983.
12. Kennedy, C.D. and Gonsalves, F.A.N., The action of divalent zinc, cadmium, mercury, copper and lead on the trans-root potential and H^+ efflux of excised roots, *J. Exp. Bot.*, 38, 800, 1987.
13. Kennedy, C.D. and Gonsalves, F.A.N., The action of divalent Zn, Cd, Hg, Cu and Pb ions on the ATPase activity of a plasma membrane fraction isolated from roots of *Zea mays*, *Plant Soil*, 117, 167, 1989.
14. Lipsy, R.L., Accumulation and physiological effects of methyl mercury hydroxide on maize seedlings, *Environ. Pollut.*, 8, 149, 1975.
15. De Filippis, L.F., The effect of heavy metal compounds on the permeability of chlorella cells, *Z. Pflanzenphysiol.*, 92, 39, 1979.
16. Goran, R. and Siegel, S.M., Mercury induced ethylene formation and abscission in citrus and coleus explants, *Plant Physiol.*, 57, 628, 1976.
17. Siegel, S.M., Siegel, B.Z., Barghigiani, C., Aratini, K., Penny, P., and Penny, D., A contribution to the environmental biology of mercury accumulation in plants, *Water Air Soil Pollut.*, 33, 65, 1987.
18. Bargagli, R. and Maserti, B.E., Mercury in vegetation of the mount Amiata area (Italy), *Chemosphere*, 15(8), 1035, 1986.
19. Tamura, R., Fukuzaki, N., Hirano, Y., and Mizushima, Y., Evaluation of mercury contamination using plant leaves and humus as indicators, *Chemosphere*, 14:11/12, 1687, 1985.
20. Godbold, D.L., Mercury inducued root damage in spruce seedlings, *Water Air Soil Pollut.*, 56, 831, 1991.
21. Godbold, D.L. and Hüttermann, A., Inhibition of photosynthesis and transpiration in relation to mercury-induced root damage in spruce seedlings, *Physiol. Plantarum*, 74, 270, 1988.
22. Godbold, D.L. *Die Wirkung von Aluminium und Schwermetallen auf Picea abies Sämlingen*, Sauerländer's, Frankfurt, 1991.
23. Shieh, Y.J. and Barber, J., Uptake of mercury by chlorella and its effect on potassium regulation, *Planta*, 109, 49, 1973.
24. Macklon, A.E.S., Cortical fluxes and the transport to the stele in excised root segments of *Allium cepa* L., II, Calcium, *Planta*, 122, 131, 1975.
25. Godbold, D.L., Fritz, E., and Hüttermann, A., Aluminium toxicity and forest decline, *Proc. Natl. Acad. Sci. U.S.A.*, 85, 3888, 1988.
26. Carty, A.J. and Malone, S.F., The chemistry of mercury in biological systems, In: *The Biogeochemistry of Mercury in the Environment*, 1, (Ed., J.O. Nriagu), Elsevier North-Holland, 433, 1979.

27. De Filippis, L.F., Localisation of organomercurials in plant cells, *Z. Pflanzenphysiol.*, 88, 133, 1978.

28. Davis, W.J., Metcalfe, J., Lodge, T.A., and da Costa, A.R., Plant growth substances and the regulation of growth under drought, *Aust J. Plant. Physiol.*, 13, 105, 1986.

29. Padberg, S., Quecksilber im terrestrischen Ökosystem. Untersuchungen von Transport- und Umsetzungsmechanismen am Beispiel einer Meßstation im Sauerland, *Berichte des Forschz. Jülich,* 2534, 1991.

30. Lindberg, S.E., Turner, R.R., Meyer, T.P., Taylor, G.E., and Schroeder, W.H., Atmospheric concentrations and deposition of Hg to a deciduous forest at Walker Branch watershed Tennessee, USA, *Water Air Soil Pollut.*, 56, 553, 1991.

31. Tyler, G., Balsberg-Påhlsson, A-M., Bengtsson, G., Bååth, E., and Tranvik, L., Heavy metal ecology of terrestrial plants, microorganisms and invertebrates, *Water Air Soil Pollut.*, 47, 189, 1989.

32. Stadelmann, F.X. and Santschi-Fuhrimann, E., Beitrag zur Absteuerung von Schwerme- tallrichtwerten im boden mit Hilfe von Bodenatmungmessungen. Swiss Federal Research Station for Agriculture and Hygiene of the Environment, Liebfeld, Switzerland, 1987.

33. Wilke, B.-M., Langzeitwirkung potentieller anorganischer Schadstoffe auf die mikrobielle Aktivität einer sandigen Braunerde, *Z. Pflanzenernähr. Bodenk.*, 151, 131, 1988.

34. Wilkins, D.A., The influence of sheating (ecto-) mycorrhizae of trees on the uptake and toxicity of metals, *Agri. Ecol. Environ.*, 35(2–3), 245, 1991.

35. Marschner, P., Einfluss der Mykorrhizierung auf die Aufnahme von Blei bei Fichtenkeim- lingen, Ph.D. Thesis, University of Göttingen, Germany, 1994.

36. Låg, J. and Steinnes, E., Study of mercury and iodine distribution in Norwegian forest soils by neutron activation analysis, In: *Nuclear Techniques in Environmental Pollution,* IAEA, Vienna, 429, 1971.

37. Koski, E., Venäläinen, M., and Nuorteva, P. The influence of forest type, topographical location and season on the levels of Al, Fe, Zn, Cd and Hg in some plants in southern Finland, *Ann. Bot. Fennici,* 25, 365, 1988.

38. Teuwsen, N., Zur Quecksilberkontamination einheimischer Schalenwildarten, Dissertation University of Göttingen, Germany, 1983.

39. Wyttenbach, A. and Tobler, L., The seasonal variation of 20 elements in 1st and 2nd year needles of Norway spruce *Picea abies* (L.) Karts. trees, *Trees,* 2, 52, 1988.

40. Kojo, M.-R. and Lodenius, M., Cadmium and mercury in macrofungi. Mechanims of transport and accumulation, *Angew. Botanik,* 63, 279, 1989.

Spatial Distribution Patterns of Mercury in an East-Central Minnesota Landscape

D. F. Grigal, E. A. Nater, and P. S. Homann

CONTENTS

ABSTRACT: We investigated total mercury (Hg) distribution in relation to forested slope position and to vegetation cover type, ranging from abandoned agricultural field to mature forest, at Cedar Creek Natural History Area in east-central Minnesota. Mean concentration of Hg was 94 ng g^{-1} in the organic layer at the soil surface, 20 ng g^{-1} in the 0 to 10 cm soil layer, and 8 ng g^{-1} in the 10 to 50 cm soil layer. Distribution of Hg across the landscape was significantly related to cover type, with more Hg present in surface soils under forests than in fields. Mercury concentration and mass were related to levels of soil organic matter, and were therefore also related to the time when fields had been abandoned from agriculture and the resulting period of organic matter accumulation. Surface soils in forests are enriched in Hg relative to nitrogen compared to those in fields; processes that affect the variation of Hg with soil organic matter differ in those systems. The Hg concentration in the 10 to 50 cm soil layer was assumed to be a measure of background levels in the uniform soil parent material, and net increase of Hg on the landscape was calculated. Lowest net increase was in fields, 0.3 mg m^{-2}, and highest was in forests, 2.4 mg m^{-2}; net increase in transitional vegetation was also low (0.7 mg m^{-2}). These quantities are comparable to those we have measured in forest soils across the Great Lakes States. Because terrestrial landscapes are the receptors for the majority of Hg deposited from the atmosphere, the accumulation and transport of Hg in those landscapes merits further study.

I.　INTRODUCTION

Mercury (Hg) has been recognized as a potential health threat in relation to consumption of fish in the Great Lakes states and the adjacent Province of Ontario. This recognition has resulted in significant research on both deposition of Hg from the atmosphere[1] and its levels and behavior in aquatic systems.[2-4] Although terrestrial systems occupy a much greater proportion of the landscape in affected regions, levels and behavior of Hg in terrestrial systems have not been intensively investigated. Both accumulation and translocation of Hg in terrestrial systems and movement to aquatic systems are important processes that are poorly understood.

In previous work, we found a gradient in Hg content in soils along a geographic gradient across the Great Lakes states.[5] Increasing Hg levels were associated with an increase in acidic deposition across the gradient. Sampling in that study was restricted to forest stands in nearly level, well-drained sites. In the present study, our interest shifted to a local scale to determine the influence of land use, vegetation structure, and topography on Hg content of soils. We chose an area of very uniform soil parent material so that Hg differences due to intrinsic soil properties would be minimized. Our hypothesis was that levels of Hg in soil would be related to organic matter accumulation and vegetation structure, both of which are influenced by land-use history. Specifically, we hypothesized: (1) more Hg in forests than in fields because of greater organic matter in forest soils and greater interception of atmospherically deposited Hg by forest canopies, and (2) more Hg in fields that were abandoned from agriculture relatively long ago compared to those more recently abandoned because of more time for both Hg deposition and concomitant Hg and soil organic matter accumulation.

II.　METHODS

A.　LOCATION

The study was conducted at Cedar Creek Natural History Area (CCNHA), Minnesota. This 2300-ha area, a National Science Foundation Long Term Ecological Research Site, is located in the Anoka Sand Plain in east central Minnesota ($45° 25'N$, $93° 10'W$). The area is characterized by gently rolling topography, a high groundwater table, organic soils in wetlands, and sandy ($>90\%$ sand) mineral soils on uplands.[6] Upland soils at CCNHA, and at the study sites, are mapped as Alfic Udipsamments (Zimmerman series) and Typic Udipsamments (Sartell series).[6] Annual precipitation averages 66 cm.[6] Climate interacts with the sandy soils to create typically dry sites.

Approximately half of the upland area has been cultivated, but most agricultural fields have been abandoned over the last 60 years, resulting in a series of "old fields" that are undergoing secondary succession.[7] The remaining area is currently closed canopy forests dominated by northern red oak (*Quercus rubra* L.), northern pin oak (*Quercus ellipsoidalis* E.J. Hill), and bur oak (*Quercus macrocarpa* Michx.), but other species are also present, including red pine (*Pinus resinosa* Ait.), white pine (*Pinus strobus* L.), basswood (*Tilia americana* L.), sugar maple (*Acer saccharum* Marsh.), red maple (*Acer rubrum* L.), and paper birch (*Betula papyrifera* Marsh.).

B.　FIELD

Transects, 30 to 50 m in length, were established from the summit to base of three forested slopes. Slopes were approximately 15% and aspect varied among the transects. Five positions were chosen along each transect: summit, shoulder, backslope, footslope, and toeslope (nomenclature follows Ruhe and Walker[8]). Sites were chosen to have relatively constant vegetation along transects. Additional transects were established from seven abandoned agricultural fields and one (never-cultivated) prairie into adjacent mature forests. Transects were placed at fairly level sites to minimize the influence of topography on soil properties.

Along each transect, five positions were chosen, typically two in the field, one on the field-forest transition, and two in the forest. Known dates since abandonment of fields from agriculture ranged from 3 to 63 years. One field had been abandoned more than 65 years before sampling, and the prairie site had never been cultivated.

At each position along the transects, five sampling points were established perpendicular to and within 5 m of the transect. At each sampling point, the organic material (Oi + Oe horizons) on the mineral soil surface was collected from a 15.5×15.5 cm square. No collections of this layer were made from the prairie-to-forest transect because a prescribed burn shortly before sampling had consumed most of that layer. A core (3.75 cm diameter) of mineral soil was taken to a depth of 50 cm at each sampling point. Slopes were sampled in August 1989 and field-to-forest transects in June 1990.

C. LABORATORY

Samples of organic horizons were oven-dried (70° C). Mineral soil samples were air-dried and roots were removed with forceps. For each transect position, samples representing the organic layer, the upper 10 cm of mineral soil, and the 10 to 50 cm layer were composited from samples from the five sampling points. A subsample of each composite was ground and analyzed for Hg. Samples were digested at 145° C with sequential additions of concentrated HNO_3, H_2SO_4, and HCl followed by cold digestion with $KMnO_4$ and $K_2S_2O_8$. Mercury was determined by cold vapor atomic absorption spectrophotometry in conjunction with a gold trap.[9]

A subsample of material was also oven-dried at 105° C and soil organic matter (SOM) was determined by loss-on-ignition at 450° C.[10] Another composite of the upper 10 cm of mineral soil from each sampling point was also ground, oven-dried, and total C and N concentrations were determined with a Carlo Erba CNS analyzer.

D. NUMERICAL

Measured concentrations of Hg were converted to mass per unit area based on the mass of the sampled organic layer, measured mass of the 0 to 10 cm soil layer determined from the core samples, and estimated mass of the 10 to 50 cm layer based on a bulk density-SOM relationship.[11]

Analyses of variance were used to determine significance of differences between variables by transect position, with individual transects treated as blocks.[12] Analyses of variance were also used to determine significance of differences with vegetative cover (field, forest, or transition), irrespective of transect. Separation of means was by Bayes least significant difference (BLSD).[13] Linear regression analyses were by SYSTAT.[14]

III. RESULTS AND DISCUSSION

A. ANALYTICAL PRECISION

Laboratory blanks averaged -0.78 ng g^{-1} (n = 16), Hg recovery in spiked samples averaged 106% (n = 15), and duplicates had a coefficient of variation of 18% (n = 63 pairs). Analyses of National Institute of Standards and Technology (NIST) reference materials were near the upper limits of their confidence intervals (Estuarine Sediment, SRM 1646, mean = 78 ng g^{-1}, n = 9, NIST = 63 ± 12 ng g^{-1}; Citrus Leaves, SRM 1572, mean = 97 ng g^{-1}, n = 6, NIST = 80 ± 20 ng g^{-2}).

Two samples of the 0 to 10 cm horizon, collected at toeslope positions on the forested slopes, were treated as outliers in the analysis. Their Hg concentrations were more than 14 standard deviations from the remainder of the population of samples from that layer. They were excluded from the analyses except in the calculation of net increase of mass of Hg.

Table 1   **Results of analysis of variance of differences in Hg at five stations along three transects from summit to toeslope in forests.**

Layer	Units	Mean	MSE$^{\frac{1}{2}}$[a]	F[b]	d.f.[c]	Prob.
Organic	ng g^{-1}	142.92	28.32	2.14	4,8	0.167
	mg m^{-2}	0.33	0.23	1.08	4,8	0.426
0–10 cm	ng g^{-1}	36.25	6.66	14.71	4,6	0.003
	mg m^{-2}	3.41	0.51	8.69	4,6	0.011
10–50 cm	ng g^{-1}	11.31	4.44	1.24	4,8	0.367
	mg m^{-2}	7.47	2.90	1.24	4,8	0.368

[a] Square root of mean square error.
[b] F-ratio.
[c] Degrees of freedom.

Table 2   **Results of analysis of variance of difference in Hg at five stations along eight transects from abandoned agricultural fields into forests.**

Layer	Units	Mean	MSE$^{\frac{1}{2}}$[a]	F[b]	d.f.[c]	Prob.
Organic	ng g^{-1}	76.09	20.67	19.36	4,24	0.0001
	mg m^{-2}	0.12	0.06	16.05	4,24	0.0001
0–10 cm	ng g^{-1}	14.37	3.45	16.40	4,28	0.0001
	mg m^{-2}	1.74	0.39	12.16	4,28	0.0001
10–50 cm	ng g^{-1}	7.72	3.14	0.78	4,28	0.546
	mg m^{-2}	5.11	2.08	0.78	4,28	0.548

[a] Square root of mean square error.
[b] F-ratio.
[c] Degrees of freedom.

B. FORESTED SLOPE TRANSECTS

Differences in Hg on the forested slopes were significant in the 0 to 10 cm soil layer, but not in either the organic layer or the deeper soil layer (Table 1). In the 0 to 10 cm layer, maximum concentrations (83 ng g^{-1}) and mass (5.4 mg m^{-2}) of Hg were found at toeslope positions, but higher levels were also found on the shoulder of the slope (44 ng g^{-1}; 4.4 mg m^{-2}), with lowest levels at the summit and backslope (about 29 ng g^{-1}; 2.9 mg m^{-2}). Increased Hg in toeslope positions may result from translocation of organic-Hg complexes by interflow or runoff.[15,16] We have no explanation for increased levels in shoulder positions. The lack of a unidirectional change in Hg from summit to toeslope is consistent with irregular patterns of many other soil properties on slopes in this landscape.[17]

C. FIELD-TO-FOREST TRANSECTS

Both concentration and mass of Hg in the organic layer and in the surface mineral soil varied significantly with location along the field-to-forest transects (Table 2). Both measures of Hg were lowest in the field portion of the transects, and increased through the transition into the forest (Figure 1). Neither the concentration nor the mass of Hg in the deeper soil layer (10 to 50 cm) varied significantly along the transect (Table 2). Either the deposition of Hg or its retention appears to differ with land use history and vegetation structure.

D. EXPLANATORY VARIABLES

Soil organic matter is variable in the CCNHA landscape due to land-use history. Forest and savannah soils contain greater amounts of SOM in the surface mineral layer than fields;[18]

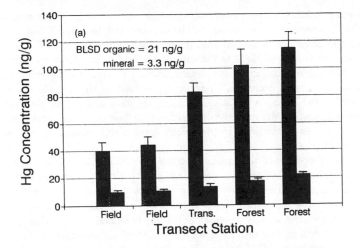

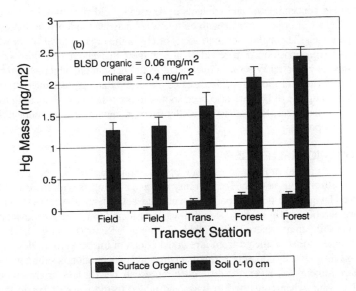

Figure 1 Mercury concentration (a) and mass (b) in soils at five positions along eight transects from abandoned agricultural fields (and a native prairie) into adjacent oak and pine forests at Cedar Creek Natural History Area, Minnesota. Standard error indicated. Mercury concentration of the 10 to 50 cm soil layer did not differ significantly among stations (mean = 7.7 ng g^{-1}, pooled SE = 1.1 ng g^{-1}).

consistent with the reduction in SOM upon cultivation.[19] Detailed sampling of a range of fields of differing periods since abandonment indicates that many soil chemical and microbial properties, including organic matter, increase exponentially, approaching an asymptote about 70 years after abandonment.[18] Concentration of Hg and SOM are closely related in the surface mineral soil layer (r^2 = 0.84, n = 53, prob. = 0.000). In fact, when all mineral soil samples from fields, forests, and transition zones are included in a regression, a strong relationship between Hg (ng g^{-1}) and SOM (%) persists,

$$Hg = 0.74 + 6.6 \text{ SOM}, (r^2 = 0.82, \text{Sy.x} = 4.84, \text{n} = 108, \text{prob.} = 0.000) \quad [1]$$

The relationship is much poorer in the organic layer ($r^2 = 0.02$). These results suggest regulation of Hg retention in the mineral soil by decomposed, humified SOM.

Mercury concentrations decreased with depth (Table 1 and Table 2) for two reasons; (1) atmospheric deposition of Hg is to the soil surface, and hence Hg concentrations at the surface should be higher than in the subsurface, and (2) SOM decreases with depth so that closely associated Hg also decreases. Conversely, the mass of Hg per unit area apparently increases with depth (Table 1 and Table 2) because the mass of the three soil horizons that we sampled increases from the surface downward as both their thickness and density increase. The lesser concentration, multiplied by the greater mass of soil, yields a greater mass of Hg per unit area with depth.

Mercury concentrations of the organic and the mineral surface soil layers of the field samples from the field-to-forest transects are weakly but positively correlated with time since agricultural abandonment (when known) ($r^2 = 0.67$ for the organic layer and 0.38 for the surface soil layer, n = 12). Such trends are lacking in the deep soil layer ($r^2 = 0.05$). Agricultural manipulation of soil decreases SOM; likely decreasing Hg content. Increasing time since abandonment may provide a longer period of accumulation of both Hg from atmospheric deposition and SOM from plant growth.

Increased Hg concentrations and contents in forest vs. field sites may be due to several processes. The loss of organic matter with cultivation[19] may lead to leaching losses of Hg in these clay-poor soils. The higher concentrations of Hg in the organic layer of forests may be due to greater Hg deposition. Fitzgerald et al.[2] suggest that about one third of the total deposition to sites in central Wisconsin may occur as dry deposition, which would be highest on forest sites because of greater surface area of the canopy and greater interaction with air masses. Stomatal uptake of Hg^0 has also been shown to occur in forests,[20] leading to eventual incorporation in leaf litter. Harvest of agricultural crops, including hay, reduces Hg accumulation by this process.

E. ANTHROPIC INFLUENCES

The data can be used to provide an estimate of the net amount of Hg that has been accumulated or lost from soils at CCNHA. Mercury in the soil samples with lowest SOM, those from the deep layer of the field to forest transects, did not differ with transect station (Table 2). The concentration of Hg in those samples was considered to be the background concentration for the soil parent material at CCNHA (7.7 ng g^{-1}, SE = 0.5 ng g^{-1}). This value is exactly the same as that predicted from a regional equation for background levels of Hg in soil based on silt-plus-clay content.[5] If that background concentration is subtracted from that of the surface mineral soil layer, the difference represents net Hg loss or enrichment in the surface. The result, when converted to mass and added to the mass of Hg in the organic layer, can be compared over the landscape.

If the samples are considered to simply represent three classes of vegetative cover—field, transition, and forest—then fields have the lowest net increase in Hg mass (0.30 mg m^{-2}) compared to the transition cover (0.74 mg m^{-2}) and the forest (2.38 mg m^{-2}; F = 21.35; d.f. = 2,52; prob. = 0.000; MSE$^{1/2}$ = 1.08 mg m^{-2}; BLSD = 0.73 mg m^{-2}). These net increases are the same magnitude as those we have reported for forested sites in a transect across the Great Lakes states, ranging from <0.5 mg m^{-2} in forests in northwestern Minnesota to 2.5 mg m^{-2} in eastern Michigan.[5]

Although this difference in Hg increase can be strictly attributed to differences in SOM accumulation among the three cover types (field = 2.71 kg m^{-2} to 10 cm, transition = 3.80 kg m^{-2}, forest = 8.44 kg m^{-2}), that is not a complete explanation. The stoichiometry of the soil organic matter in relation to Hg differs among the three cover types. The molar ratio of Hg to N, both constituents of SOM, is significantly higher in the surface mineral soil layer of the forest (1.49 μmole mole^{-1}) than in fields (0.95 μmole mole^{-1}), or in transition areas (1.07 μmole mole^{-1}) (F = 6.50; d.f. = 2,52; prob. = 0.003, MSE$^{1/2}$ = 0.51 μmole mole^{-1};

BLSD = 0.37 μmole mole^{-1}). These differences imply that different processes are influencing the retention of Hg in those three systems.

IV. CONCLUSIONS

Historical land use has differentiated the upland landscape at CCNHA into fields and closed canopy forests. Although the upland soils have very uniform parent material, the fields contain lower amounts of SOM because of prior cultivation. As we hypothesized, these factors have influenced the distribution of Hg in soils across the landscape. The higher Hg concentrations in the organic layer and surface mineral soil in the forests, and the higher Hg/N ratios in that surface, may be due to higher Hg deposition related to the structure of the forest canopy. Higher Hg concentrations in the surface mineral layer of forests parallel higher levels of SOM, and suggest that Hg retention is controlled by interaction with decomposed, humified organic matter. Leaching losses of Hg associated with loss of SOM due to cultivation of these dry, sandy soils before abandonment is also likely. Controls on Hg deposition and retention in terrestrial ecosystems merit further investigation.

ACKNOWLEDGMENTS

Our thanks to Linda Kernik, Judy Liddell, and Sandy Brovold for technical assistance. Research partially supported by NSF grant BSR 881184 for Long-Term Ecological Research at Cedar Creek Natural History Area, and by projects 25–032 and 25–054 of the University of Minnesota Agricultural Experiment Station. Published as Journal Paper no. 19,911 of that station.

REFERENCES

1. Glass, G. E., Sorenson, J. A., Schmidt, K. W., Rapp, G. R., Jr., Yap, D., and Fraser, D., Mercury deposition and sources for the Upper Great Lakes Region, *Water Air Soil Pollut.*, 56:235–249, 1991.
2. Fitzgerald, W. F., Mason, R. P., and Vandal, G. M., Atmospheric cycling and air-water exchange of mercury over mid-continental lacustrine regions, *Water Air Soil Pollut.*, 56:745–767, 1991.
3. Sorenson, J. A., Glass, G. E., Schmidt, K. W., Huber, J. K., and Rapp, G. R., Jr., Airborne mercury deposition and watershed characteristics in relation to mercury concentrations in water, sediments, plankton, and fish of eighty northern Minnesota lakes, *Environ. Sci. Technol.*, 24:1716–1727, 1990.
4. Swain, E. B., Engstrom, D. R., Brigham, M. E., Henning, T. A., and Brezonik, P. L., Increasing rates of atmospheric mercury deposition in midcontinental North America, *Science*, 257:784–787, 1992.
5. Nater, E. A. and Grigal, D. F., Regional trends in mercury distribution across the Great Lakes states, north central USA, *Nature*, 358:139–141, 1992.
6. Grigal, D. F., Chamberlain, L. M., Finney, H. R., Wroblewski, D. V., and Gross, E. R., Soils of the Cedar Creek Natural History Area. University of Minnesota Agricultural Experiment Station Miscellaneous Report 123–1974, 47p, 1974.
7. Inouye, R. S., Huntly, N. J., Tilman, D., Tester, J. R., Stillwell, M., and Zinnel, K. C., Old-field succession on a Minnesota sand plain, *Ecology*, 68:12–26, 1987.
8. Ruhe, R. V. and Walker, P. H., Hillslope models and soil formation. I. Open systems, *Trans. 9th Int. Congr. Soil Sci.*, 4:551–660, 1968.
9. Gill, G. A. and Fitzgerald, W. F., Picomolar mercury measurements in seawater and other materials using stannous chloride reduction and two-stage gold amalgamation with gas phase detection, *Mar. Chem.*, 20:227–243, 1987.

10. Goldin, A., Reassessing the use of loss-on-ignition for estimating organic matter content in noncalcareous soils, *Commun. Soil Sci. Plant Anal.*, 18(9):1111–1116, 1987.

11. Grigal, D. F., Brovold, S. L., Nord, W. S., and Ohmann, L. F., Bulk density of surface soils and peat in the North Central United States, *Can. J. Soil Sci.*, 69:895–900, 1989.

12. Snedecor, G. W. and Cochran, W. G., *Statistical Methods*, 6th ed., Iowa State University Press, Ames, IA, 1967.

13. Smith, C. W., Bayes least significant difference: a review and comparison, *Agron. J.*, 70:123–127, 1978.

14. Wilkinson, L., SYSTAT: The System for Statistics, SYSTAT Inc., Evanston, IL, 1986.

15. Johansson, K., Aastrup, M., Andersson, A., Bringmark, L., and Iverfeldt, A., Mercury in Swedish forest soils and waters—assessment of critical load, *Water Air Soil Pollut.*, 56:267–281, 1991.

16. Meili, M., The coupling of mercury and organic matter in the biogeochemical cycle—towards a mechanistic model for the boreal forest zone, *Water Air Soil Pollut.*, 56:333–347, 1991.

17. Hairston, A. B. and Grigal, D. F., Topographic influences on soils and trees within single mapping units on a sandy outwash landscape, *Forest Ecol. Manage.*, 43:35–45, 1991.

18. Zak, D. R., Grigal, D. F., Gleeson, S., and Tilman, D., Carbon and nitrogen cycling during old-field succession: constraints on plant and microbial biomass, *Biogeochemistry*, 11:111–129, 1990.

19. Mann, L. K., Changes in soil carbon storage after cultivation, *Soil Sci.*, 142:279–288, 1986.

20. Lindberg, S. E., Turner, R. R., Meyers, T. P., Taylor, G. E., Jr., and Schroeder, W. H., Atmospheric concentrations and deposition of Hg to a deciduous forest at Walker Branch Watershed, Tennessee, USA, *Water Air Soil Pollut.*, 56:577–594, 1991.

Methylmercury Input/Output and Accumulation in Forested Catchments and Critical Loads for Lakes in Southwestern Sweden

Hans Hultberg, Åke Iverfeldt, and Ying-Hua Lee

CONTENTS

I. INTRODUCTION

Forest soils contain large pools of mercury species in the organic layers, including methylmercury (MeHg).[1,2] The proportion of MeHg to other mercury species varies between sites due to differences in atmospheric input, as well as differences in methylation and demethylation rates if these processes are of significant importance in forest soils. In Sweden, previous work on the accumulation and output of mercury from forested catchments has been focused on total mercury and to some extent on operationally defined fractions of the total mercury.[1] Very few data on the atmospheric input of MeHg to small forested catchments are available. Recently it has been shown that the source of MeHg in remote lakes in Sweden was linked to terrestrial input via runoff to the lakes, and atmospheric input via wet deposition.[3] However, in this study, the MeHg data consist of measurements of a few event samples of bulk precipitation and throughfall water collected in the Lake Gårdsjön area, southwestern Sweden.[4] The input of MeHg to the forest floor via litterfall was not included in the study. In a later phase of the ongoing research program on mercury in the Lake Gårdsjön area, the flux of mercury to the forest floor via litterfall was considered. However, only data on the total mercury level in litterfall, collected in December 1989, were reported.[5] This limited data set has been used previously, together with data on the total mercury level in contemporaneous samples of bulk precipitation and throughfall, to estimate the relative importance of atmospheric dry deposition processes for the total mercury input to forested ecosystems.[5] Furthermore, when comparing MeHg in rain and throughfall water at four simultaneous collection sites in the Lake Gårdsjön area, there was no evidence of dry deposition of MeHg.[4] Also the existence of MeHg production on the needles could not be verified. The need for additional studies was clearly documented in this study.

1–56670–066–3/94/$0.00+$.50
© 1994 Lewis Publishers

Besides previously reported total mercury outputs from various catchments and some MeHg concentration measurements at various locations in Sweden,[4] only one study on the transport of methylmercury by surface runoff water has been performed.[3] During 1986 to 1987, the yearly output of methylmercury from a reference catchment (F1) in the Lake Gårdsjön area was determined by monthly sample collections.[3]

Here, we report the first results from an extensive study focused on input fluxes of atmospheric MeHg, and possible accumulation and output from catchments. The data presented are mainly from the 1990 to 1992 period, but older data from the F1 catchment are also included in the evaluation. In a parallel paper, we report related data on MeHg in soil, focusing on the soil pool and mobilization from the pool, and include an attempt to model the fluxes within and from the soil.[6] The information generated may form an important basis for assessment of the best abatement strategy for reduction of mercury emissions to achieve a persistent decrease in the mercury level in fish for human consumption.

II. METHODS

A. SITE DESCRIPTION

The study area is situated in the Lake Gårdsjön area on the west coast of Sweden (58°04'N, 12°01'E).[7-8] Of the three catchments studied, F1 (3.7 ha) is used as a reference, and G2 (0.5 ha) and catchment G1 (0.45 ha), which was covered by a roof in April 1991, are used to study effects of decreased loading of air pollutants like mercury, sulfur, and nitrogen on the biogeochemical turnover and output of MeHg.[9]

The catchments have mature coniferous forests dominated by Norway spruce and lesser Scots pine. The catchments are underlined by gneissic granodiorite and have a thin till cover generally less than 0.5 m thick. Outcrops are characteristic on the sideslopes and topographic divides. The soils are typical iron podsols with an average soil depth less than 50 cm. Lower portions of the catchments are outflow areas with *Sphagnum* moss along a small acid brook (pH about 4).[7,8]

B. SAMPLING AND ANALYSIS

Bulk precipitation and throughfall samples were collected on a monthly basis as integrated samples for the time period. Previous estimates of the total Hg load to the forest, as well as to nearby open fields, were also based on monthly precipitation and canopy throughfall samples.[5] In the present investigation, two lines of bulk collectors were operated in catchment F1 during the entire study, while one line of collectors was used in catchment G2. The positioning of the collectors was in accordance with the collection strategy previously applied (i.e., fixed distances between the collectors). MeHg deposition rates were determined from composite throughfall water samples from the respective lines. In addition, one open field site (T2) with three separate bulk collectors was used in the present study. A detailed description of the bulk collector was presented elsewhere.[5] Total Hg samples were preserved by the addition of 2.5 mL HCl (Merck, Suprapur®) to the collection bottle (0.5 L) before the start of the sampling period. The accuracy of the bulk sampling equipment has been further supported by a more extensive technical evaluation in the study of atmospheric bulk deposition of mercury to the southern Baltic Sea area.[10] The collection of litterfall for MeHg and Hg analysis was performed using previously reported methods.[5] In short, nets were located in parallel with the monthly throughfall collectors. Samples were frozen immediately after collection and were stored frozen in triple plastic bags until analysis.

Surface runoff water samples were collected at a minimum frequency of one sample per month. In addition to MeHg, total mercury levels were always determined for later comparison. Sampling details and the extensive cleaning procedure are described elsewhere.[4,5,11–13] Concentrations of MeHg in water samples and litterfall were determined by GC-CVAFS with aqueous phase ethylation, using extraction/distillation as a preseparation step.[14,15]

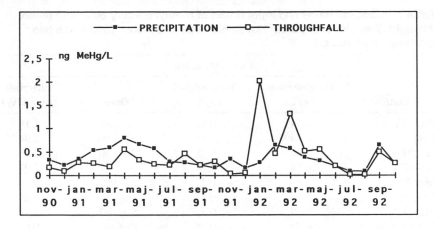

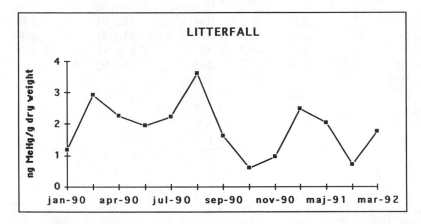

Figure 1 Methylmercury concentrations in precipitation, throughfall, and litterfall during November 1990 to October 1992 at forested catchments F1 and G2 in the Gårdsjön watershed in southwest Sweden.

III. RESULTS AND DISCUSSION

A. EMISSIONS AND ATMOSPHERIC DEPOSITION

Monthly sampling (Figure 1) from October 1990 to November 1991 provided a wet deposition rate estimate of 0.42 g MeHg/km²/year (Table 1). Assuming the difference to be a natural fluctuation, the average annual wet deposition for the period October 1990 to September 1992 was estimated to be 0.41 g MeHg/km²/year.

It is interesting to note the increase in both the concentration and deposition of MeHg during the spring both in 1991 and 1992 (Figure 1 and Table 1). This increase is yet unexplained, but is clearly an important subject for future research.

The previously reported data on MeHg in precipitation in the Lake Gårdsjön area, as well as in canopy throughfall water, were event samples.[4] The few event samples collected in 1988 provide a wet depositional flux estimate of 0.23 g MeHg/km²/year.[16] Calculations of the yearly input of MeHg based on average concentrations and precipitation amounts from a few randomly sampled events are probably not as accurate as the sum of the true monthly deposition rates. For total mercury this is certainly the case,[10] since the level is negatively correlated with the precipitation volume.[11,13] On the other hand, MeHg is probably present

Table 1   **Concentrations and input fluxes of methylmercury by precipitation, throughfall and litterfall during November 1990 to October 1992 in the two catchments F1 and G1.**

	Concentrations			
Date	Precipitation (ng/L)	Throughfall (ng/L)	Date	Litterfall (ng/g DW)
Nov. 90	0.34	0.18	Jan. 90	1.18
Dec. 90	0.24	0.11	Mar. 90	2.94
Jan. 91	0.36	0.28	Apr. 90	2.26
Feb. 91	0.54	0.27	Jun. 90	1.97
Mar. 91	0.6	0.2	Jul. 90	2.24
Apr. 91	0.8	0.57	Aug. 90	3.63
May 91	0.68	0.35	Sep. 90	1.62
Jun. 91	0.59	0.25	Oct. 90	0.6
Jul. 91	0.3	0.24	Nov. 90	0.96
Aug. 91	0.29	0.48	Mar. 91	2.48
Sep. 91	0.24	0.24	May 91	2.03
Oct. 91	0.17	0.3	Sep. 91	0.71
Nov. 91	0.36	0.05	Mar. 92	1.77
Dec. 91	0.17	0.06		
Jan. 92	0.29	2.03		
Feb. 92	0.66	0.48		
Mar. 92	0.58	1.32		
Apr. 92	0.39	1.52		
May 92	0.32	0.56		
Jun. 92	0.22	0.22		
Jul. 92	0.11	0.03		
Aug. 92	0.1	0.03		
Sep. 92	0.66	0.5		
Oct. 92	0.25	0.27		
Average conc.	0.41	0.40		1.87

	Fluxes			
Date	Precipitation g/km^2/mo	Throughfall g/km^2/mo	Date	Litterfall g/km^2
Nov. 90	0.015	0.003	Jan. 90	0.16
Dec. 90	0.034	0.011	Mar. 90	0.29
Jan. 91	0.037	0.022	Apr. 90	0.05
Feb. 91	0.040	0.015	Jun. 90	0.05
Mar. 91	0.034	0.009	Jul. 90	0.04
Apr. 91	0.022	0.009	Aug. 90	0.01
May 91	0.029	0.007	Sep. 90	0.01
Jun. 91	0.122	0.024	Oct. 90	0.03
Jul. 91	0.026	0.013	Nov. 90	0.00
Aug. 91	0.018	0.018	Mar. 91	0.16
Sep. 91	0.030	0.015	May 91	0.14
Oct. 91	0.018	0.020	Sep. 91	0.04
Nov. 91	0.057	0.005	Mar. 92	0.39
Dec. 91	0.026	0.006		
Jan. 92	0.005	0.016		

Table 1 **Continued.**

		Fluxes		
Date	Precipitation g/km²/mo	Throughfall g/km²/mo	Date	Litterfall g/km²
Feb. 92	0.053	0.024		
Mar. 92	0.036	0.050		
Apr. 92	0.047	0.046		
May 92	0.018	0.016		
Jun. 92	0.004	0.002		
Jul. 92	0.006	0.001		
Aug. 92	0.018	0.004		
Sep. 92	0.000	0.017		
Oct. 92	0.000	0.000		
Deposition (g/km²/year)				
Nov. 90/Oct. 91	0.426	0.166		
Nov. 91/Oct. 92	0.333	0.203		
Avg. dep./mo.	0.034	0.015		
Avg. dep./year	0.414	0.184		0.61

as dissolved complexes in precipitation.[4] The difference in behavior is also demonstrated by the fact that the MeHg levels seems to be constant during a rainstorm, while the other mercury forms are subjected to rain-out processes.[17]

The origin of the MeHg found in the precipitation samples collected at the study area is not resolved. MeHg may be emitted from various anthropogenic processes in European source areas, such as waste incineration, but natural emissions of MeHg from the soil may be important. An atmospheric production involving other man-made or natural reactants may also be present. Also, deposition of total Hg by precipitation is not well understood. Other air pollutants such as ozone, sulfur dioxide, and soot are probably of as equal importance in regulating the Hg deposition as direct anthropogenic emissions of Hg.[1,13,18,19]

B. METHYLMERCURY IN THROUGHFALL AND LITTERFALL

The throughfall concentration of MeHg shows a seasonal pattern with the higher concentrations during winter/spring months, such as April 1991 and January to May 1992 (Figure 1 and Table 1). This effect is probably due to the higher concentrations in air and rain. The yearly throughfall deposition of MeHg is calculated to be 0.18 g MeHg/km²/year based on an average monthly deposition of 0.015 g MeHg/km²/year for the period November 1990 to October 1992. The average annual throughfall deposition of about 0.2 g MeHg/km²/year for the whole study period is the same for each of the studied years (Table 1).

Based on sampling from January 1990 to March 1992 by litter traps and litter falling on the roof in catchment G1, the litterfall to the forest floor is estimated to about 300 tons of litter/km²/year. This mass of litter gives a MeHg deposition rate of 0.6 MeHg/km²/year, which gives a total input by throughfall plus litter of 0.8 g MeHg/km²/year. This flux is two times higher than the measured wet depositional flux (0.4 g MeHg/km²/year). This means that dry deposition may be of importance and that dry deposition of MeHg is equal to the wet deposition in the study area.

Part of the MeHg of atmospheric origin will probably be retained in the wax layer of the spruce needles, which is probably not the case for the particle-associated mercury species[13] present in the precipitation. Dissolved MeHg species on the other hand, are hydrophobic, and partitioning is to the organic wax layer rather than the water phase.

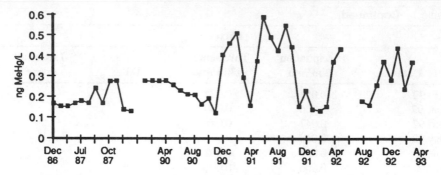

Figure 2 Methylmercury concentrations in runoff from catchment F1 in the Gårdsjön watershed.

Table 2 **Yearly output fluxes of methylmercury and total mercury in runoff from the reference catchment F1.**

	Unit (g/km²/year)	
Year	**Methylmercury**	**Total mercury**
1986/1987	0.14	
1990	0.16	3.5
1991	0.19	3.4
1992	0.15	2.3

In the previous study at Lake Gårdsjön, deposition via litterfall was found to be an important pathway for total Hg to reach the forest floor. The sum of mercury in litterfall plus the additional amount of mercury in throughfall water was higher than input by open field precipitation during the growing season. The higher deposition in the forest is probably due to dry deposition of inorganic mercury to the tree canopies.[5,16]

C. Methylmercury in Catchment Runoff

The concentration of MeHg in runoff was studied from December 1986 to December 1992 in reference catchment F1. Calculations of the yearly flux in runoff were based on water samples and continuous stream flow measurements at the catchment outlets. Figure 2 shows that the MeHg concentration varied during seasons and varied between years, with the highest levels present in autumn after periods of low water flow. The yearly fluxes of both MeHg and total Hg are shown in Table 2. The data from the catchment are from 1986 through 1987 and the three years 1990, 1991, and 1992. The average output flux from the F1 catchment gave an average MeHg export of 0.16 g/km²/year or about 0.2 g/km²/year.

D. INPUT/OUTPUT BALANCE OF CATCHMENTS AND LAKES

Figure 3 shows a mass balance for the average yearly MeHg fluxes for the catchment F1. The input/output fluxes show that a net MeHg accumulation occurs in the catchment. A net retention in the soil was earlier shown for total Hg.[16] The catchment output of MeHg via runoff was about 0.2 g/km²/year, which is 50% of the wet deposition and 25% of deposition via throughfall plus litterfall.

Throughfall plus litterfall input of MeHg (0.8 g/km²/year) is twice the deposition (0.4 g/km²/year), which indicates that dry deposition may be very important. The input/output budget also shows that atmospheric input via throughfall may be the main factor regulating the catchment output of MeHg, and that release from the soil pool (about 100 g/km²) may

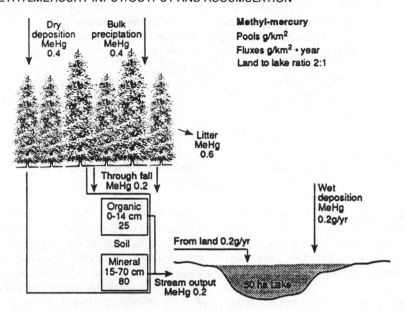

Figure 3 Input/output balance of methylmercury in a coniferous forested catchment and the methylmercury mass balance of a 50 ha lake in southwest Sweden.

be of less importance for the output flux of MeHg. The input/output balance of MeHg shows that soil retention of MeHg was about 75%, which is smaller than for other mercury species where input/output fluxes show that about 95% of the input is retained in the upper organic parts of the soil profile. Atmospheric wet plus dry deposition of inorganic mercury is also high (about 50 g/km^2/year) in the Lake Gårdsjön area[16] and some part of this mercury may be methylated in the soil and show up as MeHg in runoff. Our results suggest, however, that methylation of inorganic mercury in the soil is not a necessary process to explain the observed MeHg output from terrestrial ecosystems. Atmospheric deposition of MeHg alone may be the direct cause of catchment output of MeHg, assuming that the soils act as a less important sink of MeHg than for other Hg species.

E. CRITICAL LOAD OF METHYLMERCURY

Figure 3 gives an example of the MeHg load of a hypothetical lake of 50 ha, with a land-to-lake ratio of 2:1. The figure shows that the direct wet deposition of MeHg to the lake surface and input to the lake via runoff from land are about equal. Current loading rates of about 0.4 g/km^2/year would yield 400 mg/year to a 50 ha lake, or 8 mg/ha/year. Therefore, assuming that all MeHg ends up in fish, a production of 8 kg fish muscle/ha/lake/year would have a MeHg concentration at or exceeding the health standard of 1 mg/kg in fish used for human consumption in Sweden. A majority of Swedish lakes are nutrient poor and have a fish production between 1 to 10 kg/ha/year. This means that if the results of the MeHg balance derived from the Lake Gårdsjön area are applicable to the rest of Sweden, and if most of the MeHg is accumulated in the fish biomass, a large proportion of the Swedish lakes would contain fish exceeding the 1 mg/kg health standard. This may be compared to recent estimates[1] indicating that some 10,000 lakes have part of the fish population exceeding 1 mg/kg and about 40,000 lakes (48%) have fish exceeding 0.5 mg/kg. This would thus partly be a direct result of the atmospheric MeHg deposition, which reaches surface waters directly, and indirectly via leaching from the terrestrial ecosystem.

A simple set of calculations on critical MeHg loading is shown in Table 3. The calculations are based on the assumption that 50% of the total input of MeHg to the aquatic ecosystem

Table 3 **Estimated critical input of methylmercury by catchment runoff in g/km²/year which will cause fish concentrations exceeding 1 mg/kg.**

Fish production kg/ha/year	Land to lake ratio		
	1:1	5:1	10:1
2	0	0	0
5	0.6	0.12	0.06
10	1.6	0.3	0.16

Note: Wet deposition is set to 0.4 g/km²/year to the lake surface and 50% of the input will accumulate in fish.

from the catchment and of the direct wet deposition to the lake surface, will eventually accumulate in fish biomass. The remaining 50% is assumed to be incorporated in the sediments and/or lost at the lake outlet. Table 3 shows the critical MeHg input by catchment runoff in relation to the yearly fish production and the land-to-lake ratio. The estimates are also related to the human health standard—to avoid the MeHg concentration in fish from exceeding the limit of 1 mg/kg. The estimates show that lakes with large drainage areas (high land-to-lake ratio) are very sensitive to inputs of MeHg from the catchment. A lake with a land-to-lake ratio of 10:1 and a fish production of 10 kg/ha/year will be in danger of fish concentrations exceeding 1 mg/kg with an input of only 0.16 g/km/year from the surrounding land, and a direct wet deposition to the lake surface of 0.4 g/km²/year. When the terrestrial area is small—land-to-lake ratio 1:1—and the fish production is small (2 kg/ha/year), the MeHg concentration in fish will still exceed 1 mg/kg if the input by wet deposition is 0.4 g/km²/year and the catchment runoff is 0 g/km²/year. Since the examples shown in Table 3 cover a wide range of lake systems typical of oligotrophic forest lake ecosystems, our data show that the atmospheric deposition of MeHg alone may explain the MeHg content in fish in these lakes. Our calculations show that the atmospheric input to the terrestrial system and directly onto water surfaces of nutrient poor lake ecosystems in northern Europe and North America should not exceed 0.05 to 0.1 g/km²/year in order not to reach concentration of MeHg in fish harmful for human and wildlife consumption. This means that the atmospheric deposition of MeHg should be decreased by 50 to 90%. These rough estimates, however, are based on the limited data that exist on quantitative measurements of MeHg fluxes in natural ecosystems on the Swedish west coast and should therefore be used with caution.

IV. CONCLUSIONS

Natural and anthropogenic emissions and/or atmospheric reactions which result in the formation of MeHg are followed by atmospheric transport and wet/dry deposition to forest ecosystems, and finally be a substantial export by runoff to lakes and streams. This, together with direct wet deposition to water surfaces, is the most probable explanation for elevated MeHg concentrations in fish in remote lakes and streams. Critical load estimates based on emissions and deposition of Hg species other than MeHg are therefore of limited value. These estimates should be reevaluated as new results are produced which are based on real and quantitative measurements on the behavior of MeHg in the environment. The critical load estimates based on MeHg indicate that deposition needs to be reduced by 50 to 90% in order to reach the level of 0.05 to 0.1 g/km²/year, which would result in concentrations in fish which are not harmful for human or wildlife consumption.

ACKNOWLEDGMENTS

The study was conducted with financial support from the Swedish Environmental Research Institute, the CEC STEP programme, the Swedish State Power Board, National Power (U.K.), and the Electrical Power Research Institute (U.S). Thanks are also due to Elsmari Lord, Ingvar B. Andersson, and John Munthe for the analytical work and technical discussions.

REFERENCES

1. Lindqvist, O., Johansson, K., Aastrup, M., Andersson, A., Bringmark, L., Hovsenius, G., Håkansson, L., Iverfeldt, Å., Meili, M., and Timm, B., Mercury in the Swedish environment—recent research on causes, consequences and corrective methods, *Water Air Soil Pollut.*, 55, 261, 1991.
2. Padberg, S. and Stoeppler, M., Studies of transport and turnover of mercury and methylmercury, In: *Metal Compounds in Environment and Life*, Vol. 4, Merian, E. and Haerdi, W., Eds., Science Reviewers, Inc., Wilmington, 1992, 329.
3. Lee, Y.-H. and Hultberg, H., Methylmercury in some Swedish surface waters, *Environ. Toxicol. Chem.*, 9, 833, 1990.
4. Lee, Y.-H. and Iverfeldt, Å., Measurement of methylmercury and mercury in run-off, lake and rain waters, *Water Air Soil Pollut.*, 56, 309, 1991.
5. Iverfeldt, Å., Mercury in forest canopy throughfall water and its relation to atmospheric deposition, *Water Air Soil Pollut.*, 56, 553, 1991.
6. Lee, Y.-H., Borg, G., Iverfeldt, Å., and Hultberg, H., Fluxes and turnover of methylmercury: mercury pools in forest soils, International Conference on Mercury as a Global Pollutant, May 31-June 4, 1992, Monterey.
7. Olsson, B., Hallbäcken, L., Johansson, S., Melkerud, P.-A., Nilsson, S.I., and Nilsson, T., The Lake Gårdsjön area—physiological and biological features, *Ecol. Bull.*, 37, 10, 1985.
8. Hultberg, H., Budgets of base cations, chloride, nitrogen and sulphur in the acid Lake Gårdsjön catchment, SW Sweden, *Ecol. Bull.*, 37, 133, 1985.
9. Hultber, H., Andersson, B.I., and Moldan, F., The covered catchment. An experimental approach to reversal of acidification in a forest ecosystem, Int. Symp. Experim. Manipulat. Biota Biogeochem. Cycl. Ecosyst.—Appr. Meth. Find., 18–20 May, 1992.
10. Jensen, A. and Iverfeldt, Å., Atmospheric bulk deposition of mercury to the southern Baltic sea area, International Conference on Mercury as a Global Pollutant, May 31–June 4, 1992, Monterey.
11. Bloom, N.S. and Crecelius, E.A., Determination of mercury in seawater at subnanogram per litre levels, *Mar. Chem.*, 14, 49, 1983.
12. Iverfeldt, Å., Mercury in the Norwegian fjord Framvaren, *Mar. Chem.*, 23, 441, 1988.
13. Iverfeldt, Å., Occurrence and turnover of atmospheric mercury over the Nordic countries, *Water Air Soil Pollut.*, 56, 251, 1991.
14. Bloom, N.S., Determination of picogram levels of methylmercury by aqueous phase ethylation, followed by cryogenic gas chromatography with cold vapour atomic fluorescence detection, *Can. J. Fish. Aquat. Sci.*, 46, 1131, 1989.
15. Horvat, M., May, K., Stoeppler, M., and Byrne, A.R., Comparative studies of methylmercury determination in biological and environmental samples, *Appl. Organomet. Chem.*, 2, 515, 1988.
16. Driscoll, C.T., Otton, J.K., and Iverfeldt, Å., Trace metals speciation and cycling, In: *Biogeochemistry of Small Catchments*, Moldan, B. and Cerny, J., Eds., John Wiley & Sons, New York, 1994, 299.

17. Bloom, N.S. and Watras, C.J., Observations of methylmercury in precipitation, *Sci. Total Environ.,* 87/88, 199, 1989.
18. Iverfeldt, Å. and Lindqvist, O., Atmospheric oxidation of elemental mercury by ozone in the aqueous phase, *Atmos. Environ.,* 20, 1567, 1986.
19. Munthe, J., The aqueous oxidation of elemental mercury by ozone, *Atmos. Environ.,* 26A, 1461, 1992.

The Relation Between Mercury Content in Soil and the Transport of Mercury from Small Catchments in Sweden

Kjell Johansson and Åke Iverfeldt

CONTENTS

I. INTRODUCTION

Increasing atmospheric deposition rates of mercury during the last century has resulted in its elevated accumulation in forest soils. In the southern part of Sweden and along the Bothnian coast, concentrations in the mor (surficial humus) layer of the soils are markedly elevated. Here, approximately 80% of the accumulated mercury content is of anthropogenic origin.[1]

Generally, the mercury deposited from the atmosphere onto terrestrial surfaces is quickly bound to the organic matter in the upper layer of the soil (i.e., the humic layer). In principle, this layer may be regarded as a filter for mercury in infiltrating precipitation. Thus, the quantities presently found in the humic layer reflect accumulated mercury deposition, mainly from the last century. Despite the fact that emissions have been drastically reduced in the last decades, the current atmospheric deposition rate still promotes an accumulation of mercury in the soils. The ongoing accumulation of mercury in the mor layer in forest soils may, in the future, or perhaps even today, have an effect on the microorganisms and on biological processes in the soils. A reduction of about 80% in the deposition rate has to be achieved to reach a balanced input/output budget for the soils in Scandinavia.[1] Also, in the north-central U.S. an increased content of mercury in forest soils has been indicated, which probably is due to deposition of anthropogenic mercury.[2]

Mercury is transported from soils to watercourses and lakes, resulting in a fivefold increased mercury level in fish compared to background levels in southern and central Sweden. In about 10,000 of about 83,000 lakes in Sweden, the mercury concentrations in predatory fish exceed the Swedish limit value of 1 mg/kg.[3] Many studies have shown that dissolved humic matter acts as a carrier of mercury from soils to inland waters.[4-6] In Sweden, a synoptic study from watersheds in three geographically separate areas has previously demonstrated that the main factors controlling the run-off amounts of mercury are the content of humic matter in the waters and the mercury concentrations in the humic substances. Other factors, such as acidification of soils and waters are of secondary importance.[7]

In this chapter we focus on the factors influencing the long-term run-off of mercury from small catchments. The objective is to assess the impact of increased mercury content in forest soils on the mercury load on inland waters. This knowledge is important to understand the processes controlling mercury transport from soils to waters, and to predict the future loading

1–56670–066–3/94/$0.00+$.50
© 1994 Lewis Publishers

of mercury on lakes in relation to various atmospheric mercury deposition scenarios. The evaluation is based on data from a 3-year (1985 to 1987) regional study of the mercury content in run-off waters from small catchments located in different parts of Sweden. The results are also used to discuss the influence of land use operations in forest areas on the loadings of mercury to lakes.

II. METHODS

Monthly sampling was performed in stream waters from 15 small catchments at 11 locations in Sweden, during a 3-year period (Figure 1). The size of the catchment areas range from about 20 to 100 ha. The catchments areas are underlain by poorly weathered pre-Cambrian (granite-gneiss) bedrock, which is covered by till of varying thickness (roughly <2 m).[8]

The samples were taken directly in borosilicate glass bottles from the central section of the streams. Plastic gloves were used during the sampling to avoid contamination. All the cleaning, sampling, and handling were performed according to established ultratrace metal-free procedures. Immediately after collection, the samples were sent to the laboratory and analyzed for Hg(II) which is defined analytically below. After this analytical step the remaining part of the various samples, preserved by HCl and BrCl additions, was stored in darkness at 5° C until the subsequent analysis of total mercury.

The nomenclature used in the present work and the analytical procedure applied are based on the reducibility of the various mercury compounds.[9–11] The fraction named Hg(II) is made up of mercury compounds reducible to elemental mercury (Hg^0) with the strong reducing agent $NaBH_4$. A solution of 10% freshly made $NaBH_4$ was added directly to the sample aliquot without any preceding HCl addition/pretreatment. Total mercury was determined after HCl pretreatment (5 mL L^{-1} HCl; Merck, Suprapur®), followed by addition of the very strong oxidizing agent BrCl (5 ml L^{-1}), prereduction with $NH_2OH \cdot HCl$ and a final reduction step using 10% $SnCl_2$ in 3% H_2SO_4.[10,12]

After the various chemical treatments, the water sample was purged with Hg-free He gas (about 0.5 L min^{-1} for 30 min) and the volatized Hg^0 was preconcentrated on a gold trap. The sorbed mercury was then detected by double amalgamation He-direct current plasma atomic emission spectrometry.[9–11,13–15] In general, the data on mercury levels in water discussed here will be represented by the Hg(II) fraction, unless otherwise stated.

The water flow and chemical variables were taken from other research programs which were running in parallel in these areas.[16–18] TOC was determined in samples from 9 catchments, while data on absorbance at 420 nm (a Swedish standard method to determine organic matter in water) had to be converted to TOC in samples from the remaining 6 sites. A TOC value of 10 mg L^{-1} corresponds to 0.21 measured as absorbance, according to previously reported data.[7]

In 7 of the catchments, the mercury and organic matter content in the mor layer have been determined in samples from 6 or 12 locations in each catchment.[19] These data were used in the present study.

III. RESULTS AND DISCUSSION

In areas where the stream samples for analysis of mercury and other water-chemical parameters were taken on the same day, there is a good correlation between the Hg(II) concentration and the content of organic matter in the water. The linear correlation coefficient varied between 0.6 and 0.9 (Figure 1). The positive covariation between these two parameters suggests an important role for humic matter as a controlling factor for the transportation of mercury from soils through watersheds. This is also in accordance with the results from a previously reported synoptic study.[7] The transport of humic matter from these catchments is mainly governed by hydrologic factors. High transport of humic matter and mercury is found

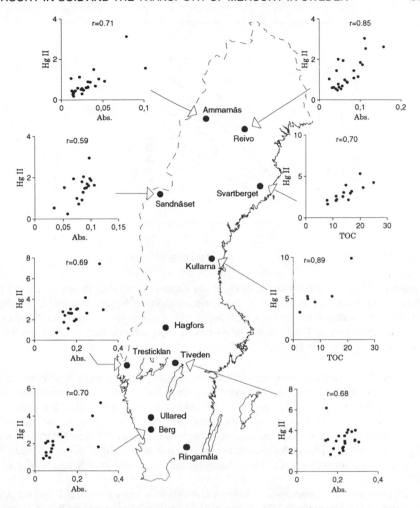

Figure 1   Concentrations of Hg (ng/L) and organic matter (absorbance and TOC, mg/L) in run-off waters from small catchments in different parts of Sweden.

during periods of greater recharge rates, which also give rise to higher ground water table conditions.

However, in some samples the content of mercury per humic matter diverged from the average values. In Tiveden, Tresticklan, and Ammarnäs (see Figure 1), the highest mercury concentrations were observed during periods of rapidly increasing water flow after a long, dry period. These periods are also characterized by a higher than normal proportion of coarser humic material in the waters. If these few data are accurate, they imply that these short periods may be of great importance for the fluxes of mercury from small catchments. In a monitoring program based on monthly samplings, these short events are probably not fully represented. Hence, the calculated annual run-off amounts of mercury may be underestimated.

Figure 1 also shows a variable, site-specific content of mercury in streamwater in the different areas. This could be due to varying levels of mercury in the soils between the catchments. In order to evaluate a possible coupling, the concentration of mercury in organic matter in the soils of seven of the catchments[19] and in the corresponding streamwaters were

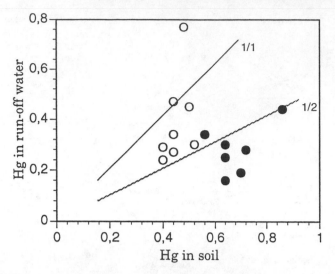

Figure 2   Median concentrations of Hg II per organic C (mg/kg) in run-off waters from small catchment areas in relation to the concentrations of Hg per organic C (mg/kg) in the soils (mor layer) of the areas. Catchments with major lakes or larger peatland extensions are indicated by filled circles, while catchments without these characteristics are represented by open circles.

compared. No correlation was observed between the content of mercury in soils and in run-off waters in these catchments. The eight remaining catchment areas were added to this evaluation. For these areas, the content of mercury in the mor layer was estimated from a nationwide mapping of the mercury content in forest soils.[20] The complete data set did not show a better coupling between the content of mercury in soils and the level in run-off waters (Figure 2). Obviously, the mercury content in humic matter is modified during transport from the soils to the run-off waters. This change is related to site-specific properties of the catchments.

In Figure 2 some characteristics of the catchments have been indicated. From the data in this figure, it can be concluded that the presence of major lake basins and larger extensions of peatland in the catchments have a significant effect on the content of mercury in the humic matter found in the run-off waters. Apparently, in these catchments the concentration of mercury in humic matter is lower than in catchments where the soilwater is transported from the till soils directly to the running waters.

The general pattern of the mercury content in organic matter in soils, waters, and lake sediments in small catchments in southern and central Sweden is shown in Figure 3. In this figure, the concentration of total mercury in streamwaters has been calculated by multiplying the Hg(II) content by a factor of 1.3 (an average ratio between total mercury and Hg II concentrations found in these streamwaters).[20] Overall, the content of mercury in humic matter is about the same in the soils of the catchments and the run-off waters from the soils. This indicates that in these catchment types, the transport of mercury is mainly controlled by the mercury content in the soil and the humic matter content of streamwaters. However, if the stream flow passes through a lake basin where the water retention time is more than about 6 months, the content of mercury in the humic matter is lowered. A possible explanation is the occurrence of a humic matter separation process in lake basins, where coarser particles settle to the sediments of lakes. Presumably these particles may be richer in mercury per

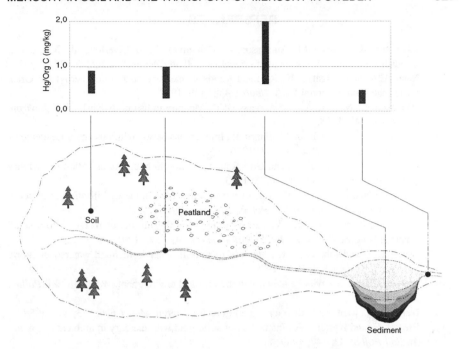

Figure 3 Concentrations of Hg per organic C in soils (mor layer), run-off waters, and sediments[20] in small catchments in southern and central Sweden.

organic matter than the smaller ones.[20] This hypothesis is supported by the fact that the concentrations of mercury in the sediment are significantly higher than in the humic matter in the outlet water from lakes (Figure 3). The lower content of mercury in run-off water from catchments, where larger areas of peatland are present, may be caused by the efficient retention of airborne mercury in the top layer of the peatland and a very slow transport of the humic matter through the bogs to the streamwaters.

In view of the results presented here, some implications for land use management operations can be discussed, to reduce the transport of organic matter (and mercury) from the terrestrial compartment to streamwaters and lakes. In general, the management operations should be performed in such a way that the high flow events and drainage of run-off waters directly from till soils to lakes are avoided. For example, ditching of forest soils should not be carried out all the way into lakes. In clear-cutted areas, new vegetation should be introduced in the area as quickly as possible, in order to shorten the period of high water flow.

IV. CONCLUSION

The main conclusions from this paper can be summarized as follows:

1. There is a good correlation between the content of mercury and humic matter in run-off waters from small catchments in Sweden. The transport of mercury from soils is controlled by the fluxes of humic matter, but no simple correlation with the mercury content in soil can be found.
2. The content of mercury in organic matter in run-off waters is site-specific. The level is lower in streamwater below major lake basins or larger extensions of peatland.

REFERENCES

1. Johansson, K., Aastrup, M., Andersson, A., Bringmark, L., and Iverfeldt, Å., Mercury in Swedish forest soils. Assessment of critical load, *Water Air Soil Pollut.*, 56, 267, 1991.
2. Nater, E.A. and Grigal, D.F., Regional trends in mercury distribution across the Great Lakes states, north central USA, *Nature*, 358, 139, 1992.
3. Håkanson, L., Nilsson, Å., and Andersson, T., Mercury in fish in Swedish lakes, *Environ. Pollut.*, 49, 145, 1988.
4. Benes, P., Gjessing, E.T., and Steinnes, E., Interactions between humus and trace elements in fresh water, *Water Res.*, 10, 711, 1976.
5. Lodenius, M., Seppänen, A., and Uusi-Rauva, A., Sorption and mobilization of mercury in peat soil, *Chemosphere*, 12, 1575, 1983.
6. Mierle, G. and Ingram, R., The role of humic substances in the mobilization of mercury from watersheds, *Water Air Soil Pollut.*, 56, 349, 1991.
7. Johansson, K. and Iverfeldt, Å., Factors influencing the run off of mercury from small watersheds in Sweden, *Verh. Int. Verein. Limnol.*, 24, 2200, 1991.
8. Iverfeldt, Å. and Johansson, K., Mercury in run-off water from small watersheds, *Verh. Int. Verein. Limnol.*, 23, 1626, 1988.
9. Brosset, C., The behavior of mercury in the physical environment, *Water Air Soil Pollut.*, 34, 145, 1987.
10. Iverfeldt, Å., Mercury in the Norwegian fjord Framvaren, *Mar. Chem.*, 23, 441, 1988.
11. Brosset, C. and Iverfeldt, Å., Interaction of solid gold with mercury in ambient air, *Water Air Soil Pollut.*, 43, 147, 1989.
12. Bloom, N.S. and Crecelius, E.A., Determination of mercury in seawater at sub-nanogram per liter level, *Mar. Chem.*, 14, 49, 1983.
13. Brosset, C., Total airborne mercury and its possible origin, *Water Air Soil Pollut.*, 17, 37, 1982.
14. Iverfeldt, Å. and Lindqvist, O., Distribution equilibrium of methyl mercury chloride between water and air, *Atmos. Environ.*, 16, 2917, 1982.
15. Iverfeldt, Å., Structural, Thermodynamic and Kinetic Studies of Mercury Compounds; Applications Within the Environmental Mercury Cycle, Ph.D. Thesis, Chalmers University of Technology and University of Göteborg, Sweden, 1984.
16. Nihlgård, B., Liming the soil in small catchments areas, background information 1984, Report, University of Lund, 1985.
17. Rosén, K., Supply, loss and distribution of nutrient in three coniferous watersheds in central Sweden, Swedish University of Agricultural Science, Rep. 41, 70 pp, 1982.
18. Bernes, C., Giege, B., Johansson, K., and Larsson, J.E., Design of an monitoring programme in Sweden, *Environ. Monit. Assess.*, 6, 113, 1986.
19. Bringmark, L., Monitoring of soil chemistry and soil biology in reference areas of the Swedish Monitoring Programme, Swedish Environmental Protection Agency, Rep. 3802, (In Swedish, English summary), 21 pp., 1990.
20. Lindqvist, O., Johansson, K., Aastrup, M., Andersson, A., Bringmark, L., Hovsenius, G., Håkanson, L., Iverfeldt, Å., Meili, M., and Timm, B., Mercury in the Swedish environment—recent research on causes, consequences and corrective methods, *Water Air Soil Pollut.*, 55, 261, 1991.

Fluxes and Turnover of Methylmercury:
Mercury Pools in Forest Soils

Ying-Hua Lee, Gunnar Ch. Borg, Åke Iverfeldt, and Hans Hultberg

CONTENTS

I. INTRODUCTION

It is now recognized that the mercury (Hg) problem in Scandinavia and North America is of a regional and chronic character, due to the diffuse deposition of Hg from the atmosphere. The increase of Hg in remote lakes is mainly a result of the increased load of Hg to those lakes, originating from atmospheric deposition and aggravated by acidification.[1] Most of the Hg deposited on the soil is immobilized by organic material and accumulates in the mor layer and shallow soil horizons. Aastrup et al.[2] reported that only a small percentage (<0.1%) of the Hg in soil is transported annually in runoff from soil to surface water. Nevertheless, this terrestrial input can make up 25 to 75% of the Hg reaching lakes in southern and central Sweden.[3]

Previous studies of Hg transport processes did not focus specifically on methylmercury (MeHg), which is the most toxic form of ambient Hg and is most prone to accumulation in the aquatic food chain. Techniques for analysis of MeHg have recently been refined,[4-5] which makes studies of MeHg transport processes possible as well as allowing us to separate the dynamic behavior of MeHg from other Hg forms. The present investigation is part of an ongoing study concerning the mercury cycle in a coniferous forest ecosystem. In this chapter we present the results from investigations carried out over a 9-month period from 1991 to 1992. As part of the study, the water flow in the soil was simulated using the SOIL model.[6] The outcome of the study provides a basic understanding of the dynamics of terrestrial MeHg, and both the fluxes within the watershed and transport from the soil to the aquatic ecosystem.

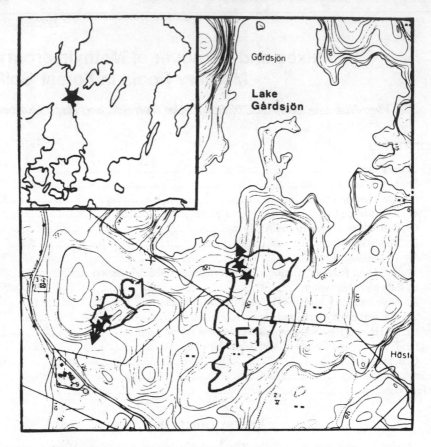

Figure 1   Site location of Lake Gårdsjön and catchments F1 and G1. Triangle = runoff stations and star = lysimeter plots.

The main objectives of the present investigation are

- To quantify the MeHg (and Hg-tot) pool in the soil of two small catchments
- To study the mobilization of MeHg (and Hg-tot) in the forest soil pool
- To estimate the flux of MeHg (and Hg-tot) within and from soil to surface water

II.   STUDY SITE AND METHODS

A.   SITE DESCRIPTION

Two small micro-catchments, F1 (3.7 ha) and G1 (0.45 ha), in the Lake Gårdsjön area were chosen for the investigation (Figure 1). Lake Gårdsjön is located 50 km north of Göteborg and approximately 10 km from the Swedish west coast (58°4'N, 12°3'E), and has a prolonged acidification history. The G1 catchment has been covered by a roof since April 1991 as part of the catchment manipulation experiments performed in the Lake Gårdsjön area. The catchments are forested with mature Norway spruce and some Scots pine, except in the upper part of F1 which is occupied by dense stands of young spruce and pine. The catchments are located on gneissic granodiorite with a thin till cover on fairly steep hillslopes. The thin till cover is frequently interrupted by outcrops of bedrock. The mean soil depth of catchment G1 is 43 cm, with a depth in some areas exceeding 200 cm. The mean soil depth is greater in catchment F1 than in G1, but in the fringes and in the upper part of the catchment the soil layer is thin with areas of bare bedrock. The soils are typical orthic podsols developed on the thin deposit of sandy-silty till with approximately 5% clay content.

B. ESTIMATION OF THE MeHg AND Hg-tot STORES

The soil samples were collected from seven to eight different areas in each catchment. Different vegetation types were considered in the choice of the topsoil sample locations and samples were taken at about 10 cm depth. At one of the sampling locations in each catchment, a vertical soil profile was sampled from the litter layer to the BC-horizon. MeHg and Hg-tot concentrations, water content, as well as loss on ignition were determined in these soil samples. In estimating the MeHg and Hg-tot stores in the F1 and G1 catchments, the MeHg and Hg-tot concentrations found at different horizons were multiplied by the horizon thickness and the bulk densities. The bulk densities of the topsoils were roughly measured. The densities in the other soil horizons were estimated according to literature.[7] The mean soil depths of 53 and 43 cm for catchments F1 and G1, respectively, were used in the estimation of the soil stores of MeHg and Hg-tot. The estimations were rather rough and based on a few samples from the AB to BC horizons. An extensive sampling program is presently being undertaken, and the results from this study will be reported in a forthcoming paper.

C. SOIL WATER SAMPLING AND ANALYSES

Soil water was sampled with ''Prenart'' PTFE suction lysimeters, which have both an inert character and good hydrologic conductivity. The pore size of the lysimeters is 5 μm. These lysimeters were installed in the main soil horizons, from 10 to 70 cm in depth in the podsol profiles, at several sites along the slope. MeHg and total Hg (Hg-tot) were analyzed in the soilwater. Sampling was performed once a month at two plots in the recharge and discharge area of F1 as well as in the discharge area and outflow area of G1 (Figure 1). Sampling of soil water was carried out during a 7-month period, starting in August 1991.

The extensive cleaning procedure, which is part of the water sample collection and analytical method for Hg-tot used in the present study, has been described previously.[8–9] The concentrations of MeHg in soil and soil water were determined using a GC-CVAFS technique with aqueous phase ethylation and using an extraction/distillation preseparation step.[5,10]

D. THE SOIL MODEL

The SOIL model was originally described in detail by Jansson and Halldin[6] and has recently been improved.[11] The model is based on fundamental physical principles for water and heat flow. The basic structure of the model is a depth profile of the soil. The soil profile can be divided into a number of layers. For each layer and each boundary between layers, these basic principles are considered. Two coupled partial differential equations, derived from Darcy's and Fourier's laws for water and heat flow, can be solved with an explicit numerical method. The model uses standard daily meteorological input data as driving variables. Model parameters are soil and plant properties. The soil properties include soil water tension (pF curves), hydraulic conductivity, thermal conductivity, and heat capacity. The plant properties are those controlling water uptake and evapotranspiration.

1. Measurements

Daily meteorological data (temperature, relative humidity, windspeed, precipitation, and global radiation) were provided from the measurement station at the Gårdsjön area. The drainage of catchment F1 was measured at a weir at the outlet of the catchment. The ground water level was recorded in several tubes. The physical properties of the soil in catchments F1 and G1 in the Gårdsjön area were determined, and the water retention properties (pF-curves) were measured layerwise for different depths, as well as for the saturated hydraulic conductivity (the k-value).[12]

2. Model Parameterization and Performance of the Simulation

The seasonal variation of variables governing evapotranspiration and water uptake presupposes that the growing season lasts from mid-April until mid-October, following the definitions of Odin et al.[13] In winter the leaf area index of the vegetation was assumed to be 5.5, and in the height of the summer, 7. The surface resistance of the vegetation in the model parameterization was set to 300 s/m in the winters and to 90 s/m during the summers.

The seasonal development of the roots regarding depth and distribution was based on general knowledge. This means that the values used in the present study were based on published root distributions.[14,15] A fitting technique was applied to improve the agreement between the measured and the simulated drainage.

In the parameterization of the SOIL model prior to the simulations, the soil profile of catchment F1 was divided into different layers. No bedrock layer was considered because the simulations were performed in such a way that the ground water level comprised the lower boundary condition, which was within the above-mentioned soil depth.

In the simulation process two subperiods were more closely examined, namely autumn and winter 1991. We assume that the differences in the water flows in the recharge and discharge areas were small. According to the measurements of the F1 runoff water flow, the total water flow during the studied period constitutes 82% of the flow during the whole hydrology year. The simulated ground water level (g.w.l.) rises during late summer/early autumn up to a magnitude of 20 to 25 cm. This simulated high level is in agreement with the normal seasonal variability of the recorded level in the g.w.l. tubes, even though the absolute value varies somewhat between different parts of the catchment. During summer, when the low g.w.l. causes greater difference between recharge and discharge areas in the catchments, the water flows are very low, as can be seen in the cumulative diagram (Figure 6). The MeHg and Hg-tot levels in the soil water were not recorded during this period. In this case, we therefore assume that it is possible to use the water flow simulated with the SOIL model to calculate both the MeHg and Hg-tot fluxes for the recharge and discharge areas. Because of the validation possibilities (g.w.l. and total runoff) and the measured soil properties (water retention properties and hydraulic conductivity), it was possible to estimate the distribution of the water flows between soil layers.

III. RESULTS AND DISCUSSION

A. DISTRIBUTION OF MeHg AND Hg IN THE SOIL PROFILE

For both catchments, the distribution of MeHg and Hg-tot (ng/g d.w. as Hg) in the soil profile shows a clear accumulation in the mor layer, and a strong indication of MeHg and Hg mobilization to the lower horizons, especially in the catchment G1 (Figure 2). For catchment G1, the MeHg and Hg-tot contents in upper and lower B-horizons are large—nearly at the same level as in the mor layer. This may be explained by the high permeability of the soil, hence, the soil water percolates easily through the B-horizons in the unsaturated zone. The loss on ignition determination (as a relative measure of organic carbon) for soil samples reveals that the variations of the loss on ignition with soil depth have a pattern similar to the distribution curves of MeHg and Hg-tot. This indicates that the strong association of MeHg and Hg with the soil organic material causes retention of MeHg and Hg. The variation of the MeHg/Hg-tot ratio (from 0.16 to 2) in relation to the soil depth (Figure 3) indicates that the mobilization of MeHg occurs more readily than other forms of Hg. The reason may be that other forms of Hg(II) are more strongly bonded to soil organic material and thus are more "fixed" in the soil than MeHg. The changes of the MeHg/Hg-tot ratio with the soil depth could also indicate a loss of Hg(II) from deeper horizons or a production of MeHg at depth.

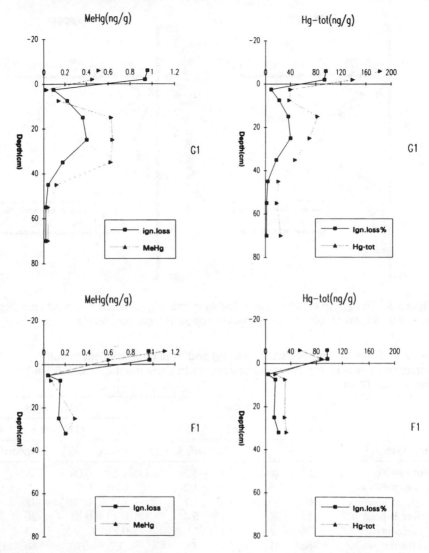

Figure 2 The distribution of methylmercury (MeHg) and total mercury (Hg) in soil profiles from catchments F1 and G1. The part of the vertical axis between –20 and 0 corresponds to the layer of fresh and old litter (ng/g d.w. as Hg).

B. SOIL STORES OF MeHg AND Hg

The estimated soil stores of MeHg and Hg are given in Table 1. The total Hg store in the mor layer of catchments F1 and G1 is 4.2 and 3.5 kg/km², respectively, which is very close to the value of 3.6 kg/km² previously found in Tiveden.[2] However, the total Hg store in the soil, taking the whole soil profile into account, is larger in G1 (18 kg/km²) than in Tiveden (8.8 kg/km²) and F1 (13 kg/km²). The total MeHg store in the F1 mor layer and in the whole soil profile of F1 is 32 and 86 g/km², respectively, compared to 18 and 123 g/km², respectively, for G1. According to a recent investigation by Hultberg et al.,[16] the yearly output of MeHg from catchments by runoff is about 25% of the atmospheric input. The output is

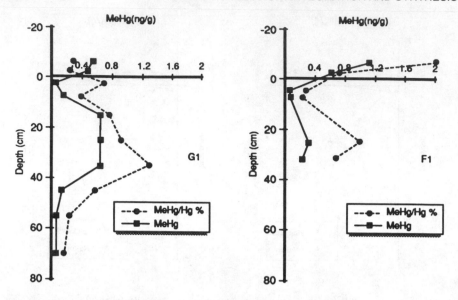

Figure 3   The variation of the ratio of MeHg to total Hg (dimensionless) with soil depth, as well as the distribution of the MeHg concentration (ng/g d.w. as Hg).

Table 1   **Store of methylmercury (MeHg) and total mercury (Hg) in soil horizons as well as for the whole soil profile in catchment G1 (0.0045 km²) and F1 (0.037 km²).**

	Total Hg stores				MeHg stores			
	F1	G1	F1	G1	F1	G1	F1	G1
Soil layer	(g)		(kg/km²)		(g)		(g/km²)	
Mor (n=8)	153	16	4.2	3.5	1.2	0.08	32	18
E-horizon[a]	15	6	0.4	1.2	0.04	0.01	1	1
Upper B-hor.[a]	167	35	4.5	7.7	1.0	0.20	28	45
Lower B-hor.[a]	115	24	3.1	5.4	0.73	0.27	20	59
C-horizon[a]	36		1		0.2		5	
Total store	486	81	13	18	3.2	0.6	86	123

[a] Single or duplicate samples were analyzed.

0.2 g/km²/year for F1 and 0.14 g/km²/year for G1, which is about 0.2% and 0.1%, respectively, of the total MeHg store in these catchments. The percentage of Hg-tot from the soil transported annually in runoff is much less for both catchments, i.e., about 0.02% for F1 and 0.01% from G1. The lower yearly output of MeHg and Hg-tot by runoff from G1 compared to F1 is consistent with the greater total stores of MeHg and Hg-tot in catchment G1.

Examining the MeHg and Hg-tot stores in the individual soil horizons in G1, most of the total soil store is found in the B-horizon of this catchment (about 85 and 73%, respectively). In F1 the portion of MeHg and Hg-tot in the B-horizon is about 56 and 59%, respectively, of the total soil store. Considering that G1 and F1 are located very close to each other (Figure 1), the atmospheric Hg input to the catchments should be of similar magnitude. The difference in the total MeHg and Hg-tot soil profile stores between G1 and F1 could be due to different hydrological flow pathways in the two catchments (or too few data).

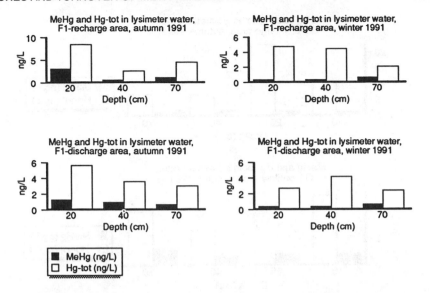

Figure 4   The mean concentrations of MeHg and total Hg in lysimeter water collected at different soil depths in catchment F1 during the autumn and winter periods, 1991.

C.   MeHg AND Hg-tot CONCENTRATION IN SOIL WATER

The data on MeHg and Hg-tot concentrations in soil water have been divided into two groups (autumn and winter, 1991). Because of the shortage of lysimeter water at some sampling occasions, the calculations of mean values of the MeHg and Hg-tot concentrations in soil water are based on lysimeter data from two to four occasions. The lysimeter data are presented in Figures 4 and 5. For both catchments, the mean values of the MeHg concentration in soil water are much greater in autumn than in winter. For catchment F1 the MeHg concentration ranged from 0.4 to 2.6 ng/L in autumn. For G1 the MeHg level varied between 0.5 and 1.4 ng/L in autumn. Below 10 cm soil depth in both catchments, high concentrations of MeHg are found in soil water with low organic content. Further, a high and comparable MeHg concentration in soil water is found in F1 and G1 during autumn, despite the fact that no direct atmospheric wet deposition of MeHg and Hg is possible to the roof-covered catchment G1. This supports a previously stated hypothesis[17] which suggests that: (1) the terrestrial MeHg and Hg-tot immobilized in the mor layer can be remobilized from soil to soil water as a result of the mineralization of organic substances in the soil, and transported by the water flow, and (2) a major part of the MeHg in soil water with a low organic content is probably not bound to humic or fulvic substances. Generally, a warm summer promotes the activity of microorganisms and hence increases mineralization of the organic substances in the topsoil, especially in the late summer and autumn when the soil humidity is high.

The Hg-tot concentrations ranged from 2.2 to 8.6 ng/L and 1.8 to 23 ng/L for F1 and G1, respectively, and the difference in concentration between the two periods was not as pronounced as for MeHg. The variation in the MeHg/Hg-tot ratio was greater in soil water than in soil, varying from 0.02 to 0.08 ng/L and from 0.01 to 0.3 ng/L for F1 and G1, respectively.

D.   TRANSPORT AND FLUXES OF MeHg AND Hg

A time period between March 1, 1990 and February 29, 1992 was used in the model simulation. In order to achieve reliable simulated hydrologic variables at the start of the

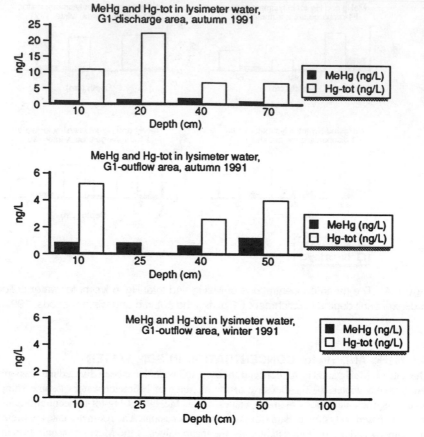

Figure 5   The mean concentrations of MeHg and Hg-tot in lysimeter water collected at different soil depths in catchment G1 during the autumn and winter, 1991.

investigation period, the simulation period started prior to the measurements of Hg in soil water and in the soil. A comparison of the simulated and the measured total runoff (denoted as mm = L/m^2) is graphically presented in Figure 6.

A selection of the simulated water flows are presented graphically in Figure 7. The flows are calculated on a daily basis, but the figures given represent the cumulative flow during the whole simulation period. The precipitation amount (corrected to account for losses due to aerodynamic effects) during the simulation period is 2090 mm (= L/m^2), of which 1490 mm penetrated into the soil and the rest is intercepted in the tree canopy (590 mm) or evaporated from the soil surface (30 mm). The total plant water uptake is 640 mm. The water uptake by plants from each soil layer is graphically presented in Figure 7. The total evapo-transpiration is 1260 mm. These simulated values may be regarded as estimations, but the simulation work allows some validation.

In Figure 7, the changes in the storage (moisture) in each soil layer and in the water level are not shown and therefore the different sums do not end up evenly. However, these variables are also calculated on a daily basis by the SOIL model.

As a basis for calculation of Hg fluxes, the two subperiods, autumn and winter (1991), were chosen. The specific transport of MeHg (and Hg, ng/m^2 or mg/km^2) was calculated by multiplying the water flow (mm or L/m^2) with the mean MeHg (and Hg-tot) concentration (ng/L) of the lysimeter water. Both the horizontal and the vertical specific transports of MeHg

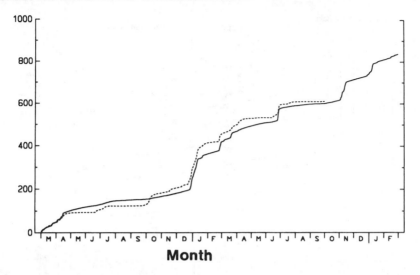

Figure 6 Cumulative measured (broken line) and simulated (solid line) drainage of the catchment F1, Gårdsjön. The unity is mm = L/m².

and Hg-tot are illustrated in Figure 8. Retention of MeHg in the soil occurs when MeHg is transported by a vertical water flow from the upper soil layer down to deeper soil layers. During autumn 1991 the total MeHg accumulation in soil was 58% of the specific transport of MeHg calculated for a 20-cm soil depth in the recharge area and 52% in the discharge area, respectively. The MeHg retention becomes smaller during the winter, i.e., 16% in the recharge area and 21% in the discharge area. The specific total horizontal transport of MeHg is 0.24 g/km² in the recharge area and 0.09 g/km² in the discharge area, respectively, during autumn 1991. The respective winter values are 0.012 g/km² and 0.016 g/km², which are much smaller than the values found during autumn.

The average specific transport of MeHg, calculated over a 7-month period, for these two areas is 0.179 g/km². The total water flow during this period constitutes 82% of the total flow during the whole hydrology year (April 1, 1991 and March 31, 1992). Using 0.179 g/km² in the calculation of the yearly total MeHg flux, a flux of 0.22 g/km²/year is found, which is very close to the measured output flux in runoff from the catchment (0.2 g/km²/year).[16-18] This strongly supports the assumption that the transport of terrestrial MeHg in soil by water is responsible for the major part of the MeHg output by runoff water in these two areas.

The same calculation for total Hg results in a specific transport of 1.12 g/km² during the 7-month study period. This corresponds to a flux of 1.36 g/km²/year for one hydrology year. For Hg-tot the calculated yearly transport from the areas contributes to about 45% of the measured output in the runoff.[16] The reason for this difference needs to be further investigated.

The retention of MeHg and Hg-tot within the soil takes place in the upper 40 cm of the soil profile, which explains the vertical profiles of the MeHg and Hg-tot stores, i.e., more than 50% of the MeHg and the Hg-tot store are found in the B-horizon.

IV. CONCLUSIONS

A strong indication of a MeHg and Hg mobilization to lower horizons is supported both by the observed MeHg and Hg stores in soil profile and by the results from the simulated mobilization of MeHg and Hg-tot in soil by the water flow.

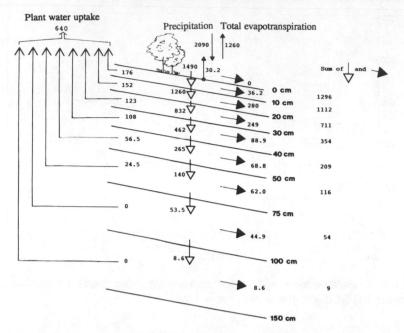

Figure 7   The simulated (SOIL model) water flows from March 1990 to February 1992. The unity is mm = L/m². Because of changes in water storage of the individual soil layers and the varying ground water level, the numbers of the water fluxes do not add up (see text). The total evapotranspiration is 1270 mm (canopy interception = 590 mm, plant water uptake = 640 mm, evaporation from soil surface = 30 mm). The upward-directed arrows at the left represents the plant uptake of water via the roots from the respective soil layer. Water flows between the layers are presented as a net sum (sometimes a capillary rise occurred). The horizontal water flows (filled arrow heads) from each layer will contribute to the ground water and, in this case, end up as stream water at the weir of the catchment.

The total MeHg store in the whole soil profile is larger in catchment G1 (123 g/km²) than in F1 (86 g/km²). More than 50% of the total soil store is found in the B-horizon in both catchments. Similar findings are also observed for total Hg. The total Hg stores are 18 kg/km² in G1 and 13 kg/km² in F1, respectively.

The soil water is generally low in organic carbon below 10 cm depth in both catchments. The MeHg concentration in the soil water from catchment G1 was at the same high level as in catchment F1 during the autumn period, despite no direct atmospheric wet deposition of MeHg and Hg-tot occurred in the roof-covered catchment G1. This supports the hypothesis that (1) terrestrial MeHg and Hg immobilized in the mor layer can be remobilized from soil to soil water, as a result of mineralization of the organic substances in the soil, and transported by the water flow, and (2) a major part of the MeHg in soil water of low organic content is probably not bound to high molecular weight organic substances (i.e., humic or fulvic substances).

The accumulation of MeHg in the soil occurs when MeHg is transported by a vertical water flow from upper soil layers down to a deeper soil layer. During autumn 1991, it was calculated that more than 50% of the MeHg flux at 20 cm depth soil in the recharge and discharge areas accumulated in the soil. The MeHg accumulation becomes smaller during winter.

The calculated total mean horizontal MeHg transport is found to be 0.22 g/km²/year, which is very close to the measured output flux by runoff from these catchments (0.2 g/km²/year).[16,17] This means that the transport of terrestrial MeHg in soil by the water

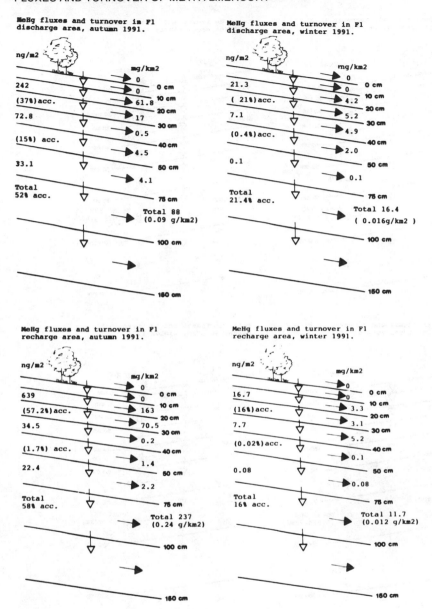

Figure 8 Horizontal and vertical MeHg and total Hg fluxes in the recharge and discharge areas of catchment F1 during autumn and winter periods, 1991; the open arrow and closed arrow denote the specific vertical and horizontal transports of MeHg (or Hg-tot), respectively.

flux contributes the major part of MeHg output by the runoff water. For Hg-tot the calculated yearly horizontal transport by water flow from these areas is 1.36 g/km^2/year, which is only about 45% of the measured output by the runoff water.

The accumulation of retention of MeHg and Hg-tot within the soil are predominantly taking place in the upper 40 cm, which explains the distribution of the MeHg and Hg stores in the soil profile, i.e., more than 50% of the MeHg and Hg stores are found in the B-horizon.

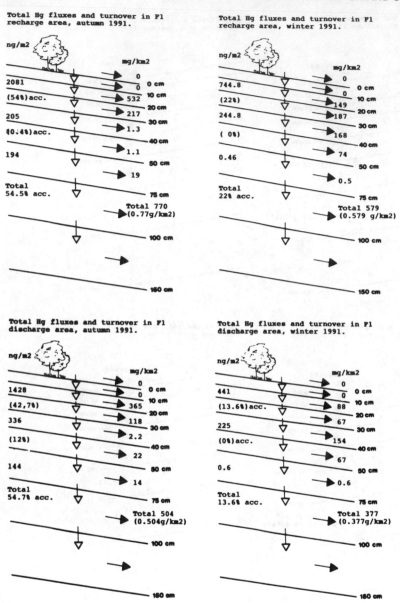

Figure 8   Continued.

ACKNOWLEDGMENTS

The study was performed as part of a project within the CEC STEP program and was also supported by the Swedish State Power Board (VATTENFALL). Thanks are due to Elsemarie Lord for technical assistance in the analytical work and to Jan Tobisson and Ulf Carlsson for their assistance in the field, and to Ulf Nyström and Ingvar Andersson for valuable discussions.

REFERENCES

1. Lindqvist, O., Johansson, K., Aastrup, M., Andersson, A., Bringmark, L., Hovsenius, G., Håkansson, L., Iverfeldt, Å., Meili, M., and Timm, B., Mercury in the Swedish environment-recent research on causes, consequences and corrective methods, *Water Air Soil Pollut.*, 56, 26, 1991.

2. Aastrup, M., Johnson, J., Bringmark, E., Bringmark, L., and Iverfeldt, Å., Occurrence and transport of mercury within a small catchment, *Water Air Soil Pollut.*, 56, 155, 1991.

3. Iverfeldt, Å. and Johansson, K., Mercury in runoff from small watersheds, *Verh. Int. Verein. Limnol.*, 23, 1626, 1988.

4. Lee, Y. H. and Mowrer, J., Determination of methylmercury in natural waters at the sub-nanograms per litre level by capillary gas chromatography after adsorbent preconcentration, *Anal. Chim. Acta*, 221, 259, 1989.

5. Bloom, N. S., Determination of picogram levels of methylmercury by aqueous phase ethylation, followed by cryogenic gas chromatography with cold vapour atomic fluorescence detection, *Can. J. Fish, Aquat. Sci.*, 46, 1131, 1989.

6. Jansson, P.-E. and Halldin, S., Model for annual water and energy flow in layered soil, In: *Comparison of Forest and Energy Exchange Models*, Halldin, S., Ed., Society for Ecological Modelling, Copenhagen, 1979, pp. 145–163.

7. Espeby, B., Water Flow in a Forested Till Slope–Field Studies and Physically Based Modelling, Rep. Trita-Kut No. 1032, Royal Institute of Technology, Dept. of Land and Water Resources, Stockholm, 1989. 33 pp.

8. Bloom, N. S. and Crecelius, E. A., Determination of mercury in seawater at subnanogram per litre levels, *Mar. Chem.*, 14, 49, 1983.

9. Iverfeldt, Å., Occurrence and turnover of atmospheric mercury over the Nordic countries, *Water Air Soil Pollut.*, 56, 251, 1991.

10. Horvat, M., May, K., Stoeppler, M., and Byrne, A. R., Comparative studies of methyl-mercury determination in biological and environmental samples, *Appl. Organomet. Chem.*, 2, 515, 1988.

11. Jansson, P.-E., Simulation Model for Soil Water and Heat Conditions, Report 165, Division of Agricultural Hydrotechnics, Swedish University of Agricultural Sciences, Uppsala, 1991, 72 pp.

12. Nyström, U., Transit time distributions of water in two small forested catchments, *Ecol. Bull.*, 37, 97, 1985.

13. Odin, H., Eriksson, B., and Perttu, K., Temperature climate maps for Swedish forestry, Report in Forest Ecology and Forest Soil, 1983, 57 pp.

14. Espeby, B., Photogrammetric method for studying the physical environment of coarse heterogeneous soils, Water in the Unsaturated Zone, NHP-Seminar, January 29–30, 1986, Ås, Norway, NHP-Report no. 15, 1986, 65–74.

15. Johansson, P.-O., Estimation of groundwater recharge in sandy till with two different methods using groundwater level fluctuations, *J. Hydrol.*, 90, 183, 1987.

16. Hultberg, H., Iverfeldt, Å., and Lee, Y. H., Accumulation and output of methylmercury from forested catchments in southwestern Sweden, In: *Mercury Pollution: Integration and Synthesis*, Watras, C.J. and Huckabee, J.W., Eds., Lewis Publishers, Chelsea, MI, 1994, chap. III.3.

17. Lee, Y. H. and Iverfeldt, Å., Measurement of methylmercury and mercury in runoff, lake and rain water, *Water Air Soil Pollut.*, 56, 309, 1991.

18. Lee, Y. H. and Hultberg, H., Methylmercury in some Swedish surface waters, *Environ. Toxicol. Chem.*, 9, 833, 1990.

Mercury in Terrestrial Ecosystems: A Review

Martin Lodenius

CONTENTS

ABSTRACT: Mercury contamination in aquatic ecosystems is well known and fish is in most cases the main dietary source of this metal. Our knowledge of the occurrence and accumulation mechanisms of mercury in terrestrial ecosystems is much more sparse.

In the soil mercury is strongly bound to organic matter and particles. The mobilization of metals from the soil is affected by a great number of factors, some of which are interacting. The importance of humic matter for the sorption of mercury is well documented. This binding is almost unaffected by fluctuations in pH and Cl concentration. The humic matter is obviously the main carrier in the transport of mercury from terrestrial to aquatic ecosystems and it is of principal importance for the bioaccumulation in aquatic food chains. A strong uptake of mercury has been found in earthworms and some other soil organisms, but they seem to have an effective excretion mechanism which prevents any further biological accumulation.

The uptake of metals from the soil into plants is influenced by several soil quality factors including soil type, pH, organic matter, and ion exchange capacity. Acid atmospheric deposition has been assumed to increase the solubility of some metals and also increase the uptake of these elements into plants. This seems not to be the case for mercury, which is strongly bound to organic matter even at low pH values. Mercury in soil solution may be phytotoxic even in small concentrations. Toxic effects on tree roots are mainly determined by the amount of soluble metal

1–56670–066–3/94/$0.00+$.50

ions. In this sense the total amount of metal is less important than the amount of soluble metal.

Terrestrial plants normally contain only small amounts of mercury. The uptake of this metal from the soils is poor and only small amounts of organic mercury compounds have been detected. Mercury may be readily absorbed through the leaves directly from the air, but there is also a reemission, which reduces the concentrations in leaves.

I. INTRODUCTION

Mercury belongs to the group of compounds considered most hazardous for the environment. As an element, it is persistent and it may form fat soluble, readily bioaccumulating forms. Most of the mercury in human diet comes from aquatic food chains, mainly fish. However, as mercury in the aquatic food chains normally originates from terrestrial sources, it is important to know the behavior, especially the mobilization processes, of this metal in terrestrial ecosystems.

Mercury has been widely used in agriculture and for many industrial purposes as well. Consequently mercury may be found in agricultural land, municipal sludge, and municipal and industrial wastes. Mercury emitted from industrial processes or by burning of wastes or coal is exceptionally well spread via atmospheric transport and may travel hundreds of kilometers through the atmosphere. This anthropogenic mercury may be transported via air and water and bioaccumulated in the environment.

II. SOIL

The sorption and mobilization of mercury in soil is affected by many factors; physical, chemical, and biological. It is well known that soil properties affect the biota, but it is important to remember that the biota also may affect the physical environment, like soil pH, soil texture, etc.

Mercurial fungicides have been widely used in agriculture. Most of the mercury is obviously retained at the soil surface, but in some cases significant amounts of mercury may leach out. Considerable amounts of municipal and industrial wastes containing mercury have been put into landfills.

A. THE ACIDIFICATION PROBLEM

Acid atmospheric deposition has been assumed to increase the solubility of several metals and thereby increase the uptake of these elements into plants. Simulated acidification has been reported to cause significantly increased concentrations and fluxes of several metals.[1] The soluble metals may either leach to deeper soil layers, or to the ground or surface water, or move into vascular plants through the roots. Consequently, mobilization of metals in the soil does not necessarily result in increased metal concentrations of the vegetation. Acidification of the soil may even cause a decrease in metal content of vascular plants.

Acidification of precipitation affects both terrestrial and aquatic ecosystems. It is known that fish in acid lakes usually contain more mercury than fish from more neutral lakes. This is obviously not caused by increased leaching of mercury from the soil, but is a result of some other chemical and biological processes. Elevated mercury concentrations have also been recorded from fish in man-made lakes and lakes, where the water level is regulated for hydroelectric and flood control purposes. Also fish from lakes receiving drainage water from ditched or tilled bogs and forests may contain elevated concentrations. In these cases mercury has obviously been mobilized from the inundated soil, attached to humic compounds, methylated, and bioaccumulated in aquatic food chains.[2]

Data concerning effects of pH on the sorption of mercury in soils are partly contradictory. The maximum sorption of mercury for several soil types is in the pH range 4.75 to 6.50. At low pH-values the pH-related changes in sorption are insignificant. In many soils the leaching of mercury decreases with decreasing pH.[3]

The chloride ion reduces the sorption of mercury significantly, and the influence of chloride ions on the adsorption of mercury may be more important than that of hydrogen ions. Addition of chloride also shifts the region of maximum sorption to higher pH values.[4]

B. THE ROLE OF HUMIC MATTER

The humus layer is very important for the forest ecosystem. Most of the plant roots are found here and most of the biological activity occurs in this layer. Many metals are strongly but reversibly bound to the humic matter of the topsoil. More attention should be paid to investigations concerning sorption, mobilization, transformation, and transport mechanisms in the humus layer. Especially, the interface between soil and roots and the mycorrhizal system are interesting when considering the mechanisms of metal uptake.

The importance of humic matter for the sorption of mercury is well known. Even if mercury may also be bound to pure quartz sand, the mercury is easily removed by acid rain. The humic matter is obviously the principal carrier in the transport of mercury from terrestrial to aquatic ecosystems and the bioaccumulation in aquatic food chains.[5,6] Mercury may be translocated in soil horizons and even lost from the surface soil through evaporation to the atmosphere. This seems to be at least partly a result of microbial activity. Mercury loss increases with increasing temperature.[7,8]

Many mercury compounds are strongly toxic to fungi and other microorganisms and have, therefore, been used as fungicides and bactericides. Mercury compounds may inhibit CO_2- evolution, dehydrogenase activity, and nitrification. Inorganic mercury compounds (Hg^0, HgO, $HgCl_2$, Hg_2Cl_2) have a strong inhibitory effect on *Azotobacter* and nitrifying microorganisms. Ag(I) and Hg(II) are the most effective in inhibition of nitrogen mineralization.[9] However, mercury compounds may even stimulate the growth of certain microorganisms like *Actinomycetes*.[10]

III. PLANTS

A. VASCULAR PLANTS

The uptake of mercury from soil into vascular plants is usually poor and the background concentrations in plant biomass are low. The mercury concentrations are usually lower in fruits, grain, and leaves than in other parts of plants. The background concentrations of mercury in plant foodstuffs usually vary within the range 0.003 to 0.1 mg/g.[11] The uptake is most efficient in young plants and mercury forms, which may be absorbed include Hg^0, Hg^{2+}, CH_3Hg^+ and $C_2H_5Hg^+$.

The uptake of metals from the soil into plants is determined by several soil quality properties including physicochemical factors like soil type, pH, organic matter, and ion exchange capacity, and biological factors like microbial activity and plant species. Physical factors like ploughing or tilling may mobilize several elements, including mercury, and make them available to plants[12] (Figure 1). A decrease in soil pH increases the solubility of most metals.

Vascular plants may absorb Hg^0 directly from the air, and therefore can be used for monitoring airborne mercury pollution. Different plant species have different capabilities of absorbing mercury from the air. Most of this mercury is absorbed through the leaves, but a small part is translocated to the roots. Vascular plants may also release mercury and this ability is highly dependent on the plant species.[13] The uptake of inorganic mercury seems to be more effective than that of organic mercurials. Catalase appears to play an important role in mercury vapor uptake and this is especially for gramineous C_3 species while this mechanism is of lesser importance for C_4 species.[14]

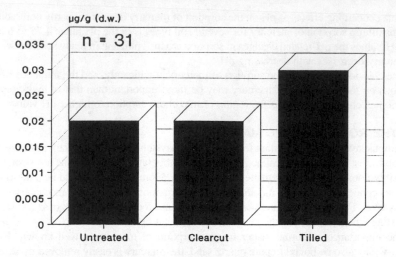

Figure 1 Mercury concentrations in leaves of *Vaccinium myrtillus* after three different forest treatments. The mean mercury concentration for tilled plots is significantly higher (ANOVA: $p < 0.001$) than for the other plots.[12]

An inhibition of photosynthesis (CO_2 uptake) was observed after exposure of spruce seedlings to mercury concentrations of 1 nM (0.2 µg/L) CH_3Hg or 10 nM (2.7 µg/L) $HgCl_2$ in the soil solution. The inhibition was associated with a reduced chlorophyll content and transpiration in the needles and, for CH_3Hg, a increased shoot/root ratio.[15] Also, the root elongation is inhibited at 100 nM Hg.

B. BIOLOGICAL MONITORING USING MOSSES AND LICHENS

The occurrence and spreading of mercury in the ecosystems is rather different from that of most other metals. Due to the high volatility of elemental mercury this metal is spread very effectively, and significant reemission from receiving surfaces occurs. Monitoring of airborne metal pollution may be carried out by analyzing the deposition in a gauge or by studying the accumulation in natural or transplanted vegetation.

Biological monitoring of mercury and other heavy metals has been used for monitoring industrial and urban emissions and for estimating the regional distribution pattern. In many countries the spread of airborne mercury has been monitored more or less regularly by using different biological material such as mosses, moss bags (moss transplants), and epiphytic lichens, which absorb great amounts of mercury from the air. The accumulation of mercury in three indicator materials (moss bags, mosses, and lichens) around a Finnish chlor-alkali plant reveal the same spreading pattern. The negative correlation between mercury accumulation and distance from the pollution source is very strong for all materials.[16]

Moss and lichen tissue absorb atmospheric (elemental) mercury effectively. However, naturally growing mosses (e.g., *Pleurozium schreberi* or *Hylocomium splendens*) and lichens (e.g., *Hypogymnia physodes*) are often lacking from industrial and urban areas. The relatively weaker uptake of mercury in *Hypogymnia* growing in industrial areas, compared with those from areas without sulfur emissions, may be due to the stunted growth of the lichen in industrial areas. In such cases, it may be better to use the moss bag method (*Sphagnum* spp.), which gives accurate and repeatable data relative to the accumulation of mercury by vegetation. By using this technique it is also possible to estimate the accumulation during a certain period (e.g., 2 months) and in severely polluted areas. The method is simple and inexpensive.

Table 1   Deposition of mercury around a Finnish chlor-alkali plant (Äetsä) calculated on basis of moss bag data.[16]

	Distance (km)			
	0–1	1–5	5–20	20–100
Deposition $\mu g/m^2 \cdot$ year	1200	260	29	16
Total deposition kg/year[a]	3.8	19	25	230
% Emission	1	5	6	58

[a] Background deposition (8 $\mu g/m^2 \cdot$ year) subtracted.

The net deposition of mercury from a chlor-alkali plant may be estimated on the basis of moss bag values. In the vicinity of a Finnish chlor-alkali plant the total annual mercury deposition is estimated to more than 1 mg/m². Most of the mercury emitted is nevertheless deposited far away from the plants (Table 1). Rough estimates determined by using this method are in agreement with budget calculations for emission, reemission, and deposition.[16]

C. HISTORICAL MONITORING USING TREE RINGS

Tree rings have been used for historical monitoring of metal emissions. Horizontal movement of some elements in some tree species makes this method uncertain. Mercury emissions have been successfully monitored by using the Japanese cedar tree.[17] In birch (*Betula verrucosa*) there is obviously a horizontal transport (between rings formed during different years) of mercury, which means that this species cannot be used for this purpose[16] (Figure 2).

IV. FUNGI

A. MERCURY UPTAKE IN MACROFUNGI

The uptake of metals in fungi is in many respects different from that of plants. Most macrofungi contain significantly more zinc and copper than green plants, and the strong accumulation of mercury and cadmium in certain species is another example of these differences. There are pronounced differences between different ecological and systematic groups of fungi. The highest concentrations of heavy metals have been recorded in species growing on lawns. These concentrations are also found in uncontaminated environments. The factors governing the accumulation of metals in fungi are poorly known. It has been assumed that the accumulation of mercury could depend on the sulfhydryl, disulfide, and methionine groups of the proteins.

Low concentrations of mercury are found in most fungi: usually less than 5 $\mu g/g$ (d.w.). The mean concentration of lawn decomposer species is clearly higher than that of mycorrhizal fungi or wood decomposer fungi (Figure 3). High concentrations have been recorded in species such as *Agaricus* and in *Boletaceae*. Moreover there is a very broad variation of mercury concentrations within the genus *Agaricus*. The mean mercury concentration of *Agaricus edulis* is less than 5 mg/kg while that of *A. comtulus* is almost 100 mg/kg. Only small concentrations of methylmercury may be detected from the four *Agaricus* samples analyzed. The binding of mercury in *Langermannia gigantea* seems to concentrate in high molecular weight compounds without any peak for low molecular weight proteins.[18]

Mercury is quite unevenly distributed within the fruiting bodies. In *A. campestris* the lamellae contain significantly more mercury than the rest of the cap and the stalk in both young and fully developed specimens.

B. CATALASE ACTIVITY AND MERCURY UPTAKE

The decomposition of hydrogen peroxide by catalase is obviously related to the decomposing capacity of fungi. There are significant differences in the catalase concentration of different

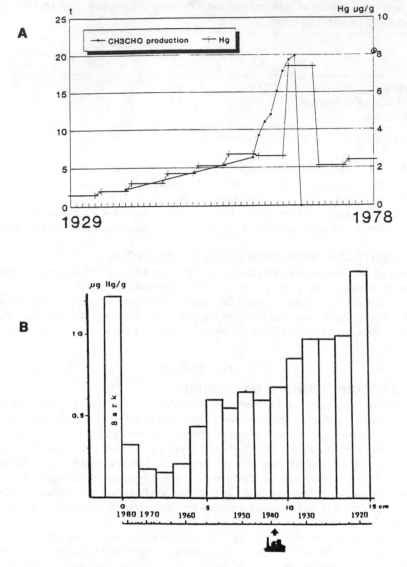

Figure 2 Historical monitoring of airborne mercury using tree rings. (A) Japanese cedar tree near an acetaldehyde factory, production stopped in 1967. Mercury in tree rings reflects emissions from the factory.[17] (B) Birch (*Betula verrucosa*) near a chlor-alkali factory, production began in 1939. Mercury has moved horizontally in the wood.[16]

species. The fruiting bodies growing on lawns contain significantly more catalase than those growing in forests. The fruiting bodies growing in forests (mainly mycorrhizal fungi) have, with few exceptions, low catalase activity and low mercury concentration.[19] Hence, the positive correlation found between the mercury concentration and the catalase activity is not surprising. A positive correlation has also been demonstrated between the uptake of mercury from air and the catalase activity.

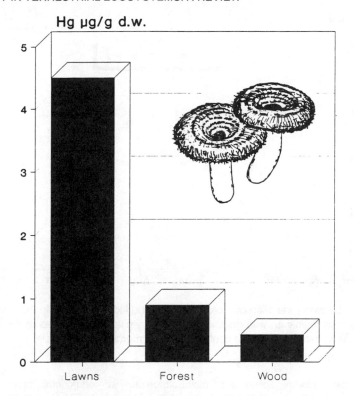

Figure 3   Mean mercury concentrations of fruiting bodies of macrofungi according to the growth site.[18]

C. BINDING OF MERCURY TO SULFHYDRYL GROUPS

In vascular plants, often more than 95% of the SH-groups are in glutathione. In plants the glutathione-cysteine plays an essential role in the oxidation-reduction processes. Also, in macrofungi the sulfhydryl concentration may play a role in the uptake of mercury and cadmium. On average, about 70% of the sulfhydryl groups are protein bound. The intraspecific differences are small while there are considerable differences between different species and groups of species. The amount is highest in the lawn decomposing species. In the mycorrhizal fungi the values are considerably lower and the proportion of non protein-bound SH-groups is higher. Surprisingly, within the genus *Agaricus* the highest concentrations of mercury are found in specimens with very low concentrations of sulfhydryl groups. In mycorrhizal species there is a strong correlation between the mercury concentration and the SH-concentration. This correlation is slightly stronger for free SH-groups than for the protein-bound groups. There are no significant correlations between the SH-concentration and any other metal.[18]

Mercury is accumulated more effectively by fungi when the concentrations of these metals in the substrate are low. This would indicate an uptake and transport of this metal together with a carrier compound—possibly the same as for essential metals. The correlations between metals in different species indicate a possible uptake of mercury with copper and/or manganese. Within the fungus cap the metals might be bound and the carrier be liberated for new transport.[20] In the species *Agrocybe aegerita*, zinc competes with cadmium but not with mercury in the uptake of metals into the fruiting bodies.[21]

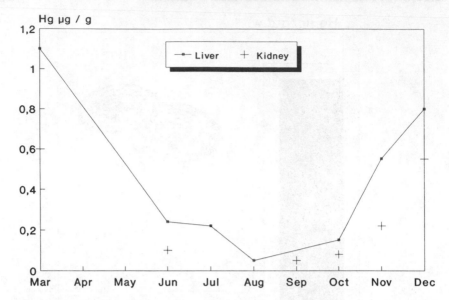

Figure 4 Seasonal variation of mercury in liver and kidney of hare. In the summertime the hares feed mainly on wild grasses and herbs while at other times of the year they feed to a great extent on cereals contaminated by mercury fungicides.[26]

If the metal binding agents were metallothioneine-like compounds, three SH-groups would be required for the binding of one metal ion.[22] Even if the SH-concentration of fungi usually seems to be enough for binding considerable amounts of heavy metals, it cannot be capable for binding all of the heavy metals. In some cases the sulfhydryl concentration might be a limiting factor.

V. ANIMALS

A. MAMMALS AND BIRDS
The toxicity of methylmercury was demonstrated dramatically in the mid-1950s when many wild terrestrial mammals and birds were found dead or suffering from severe mercury poisoning in Sweden due to its use as a seed dressing. Factors like species, sex, age, diet, and season affect the uptake and excretion of mercury. Normally, the diet seems to be more important for the mercury concentration in mammals than the species. The heavy metal concentrations in terrestrial mammals are typically low. The mercury level is normally higher in predatory mammals than in herbivorous species. Background mercury values are usually below 0.2 µg/g (fresh wt.) in soft tissues.[23] For example, the mercury concentrations of Finnish reindeer are very low: approximately 0.01 µg/g (fresh wt.) in muscle and fat, 0.09 µg/g in the liver, and 0.16 µg/g in the kidney.[24] Elevated concentrations in terrestrial mammals due to the use of mercurial fungicides have been reported in Poland and Austria[25,26] (Figure 4). The elevated mercury concentrations found near chlor-alkali plants affect the developmental stability (measured as genetic aberrations and asymmetry) of shrews.[27]

Although the biomagnification is much less pronounced than in aquatic food chains, the concentrations are normally higher in carnivores than in herbivores. In mammalian species obtaining part or all of the food supply from aquatic sources (mink, otter, seals, etc.) the concentrations are often much higher than in nonaquatic, terrestrial carnivore species. Few investigations have dealt with methylation processes in terrestrial ecosystems. In piscivorous

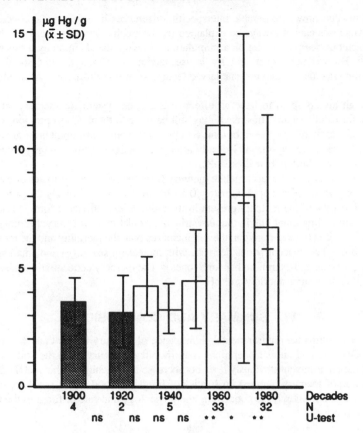

Figure 5 Mercury concentrations in museum samples of feathers of Finnish sparrow hawks (*Accipiter nisus*) from the beginning of this century. Data from the earliest decades (cross-hatched bars) have been reduced by excluding samples that probably were externally contaminated.[30]

mammals (otter and mink), methylmercury represents 30 to 50% of the total mercury concentration[23] while only 0 to 8.5% was found in small terrestrial mammals.[28] Typically, the total concentrations (calculated on a fresh weight basis) follow the order: hair > liver > kidney > muscle = brain.

The use of methylmercury in agriculture caused an intoxication of terrestrial food chains in several countries in the 1950s and 1960s until the use of this compound was banned. In many countries substantial amounts of methoxyethyl mercury have been used until recently, and in some cases are still used, for seed dressing. The agricultural use of mercury cannot be seen in the mercury concentrations of feathers of the kestrel (*Falco tinnunculus*).[29] On the contrary, there is a clear peak in feathers of the sparrow hawk (*Accipiter nisus*), with the highest concentrations being in feathers collected in the 1960s. It is interesting to note that these concentrations (Figure 5) seem to follow the industrial use of mercury much better than the agricultural use.[30]

B. INVERTEBRATES

It has been suggested that heavy metals could have an adverse effect on the natural regulation of insect pest in forests and diminish the natural pest resistance of trees. Thus, the metals would be a key factor in the multistress disease. Studies concerning the metal concentrations in forest insects show considerable interspecific differences in heavy metal concentrations. These differences cannot always be explained by the trophic levels in the food chains. Effective excretion seem to inhibit bioaccumulation of heavy metals in many terrestrial invertebrates.[31] For example, gnat and fly larvae feeding on fungi do not show a higher accumulation rate than 1, and for gnat larvae feeding on mycorrhizal fungi it usually is much lower.[32]

Although insects seem to have an effective excretion system for mercury, at least for inorganic forms of mercury, this metal may still be toxic to them. Concentrations of 1 to 10 mg/kg significantly reduce the embryonic and postembryonic development of grasshoppers. Exposure to dietary methylmercury clearly affects the activity of meal bugs (*Tenebrio molitor*) but does not shorten their lifespan.[33]

Earthworms may readily accumulate mercury from the soil. The estimates of earthworm-soil ratio vary widely: from approximately 0.4 (dry:wet) to approximately 20 (wet:wet) with between 8 to 13% of the mercury present in the animals in methylated form.[28] Earthworms obviously are an important link in the terrestrial accumulation of mercury. Mercury concentrations of between 0.5 to 5 mg/kg in the soil increases both the mortality and the reproduction of earthworms. The concentrations decrease with increasing size (age) and small specimens show faster uptake and excretion than large ones.[34] The mercury concentrations also increase at low soil temperatures and low soil pH.[35]

VI. SUMMARY AND CONCLUSIONS

In forest ecosystems the environmental significance of mercury is less than in aquatic ecosystems. Ditching and similar operations may mobilize mercury from the soil, but acidification of the environment obviously does not increase the leaching of this metal. The strong accumulation of mercury into certain species of fungi is a food safety problem, but it also reminds us that our knowledge concerning the physiology and role of fungi in the terrestrial ecosystem is poor.

REFERENCES

1. Bergqvist, B., Leaching of metals from a spruce forest soil as influenced by experimental acidification, *Water Air Soil Pollut.*, 31, 901, 1987.
2. Verta, M., Rekolainen, S., and Kinnunen, K., Causes of increased fish mercury levels in Finnish reservoirs, *Publ. Water Res. Inst., Nat. Bd. Waters, Finland*, 65:44, 1986.
3. Zvonarev, B.A. and Zyrin, N.G., Zakonomernosti sorbzii rtuti pochvami. I. Vlijanie pH na sorbziju rtuti pochvami (Relationships governing the sorption of mercury in soils I. Effect of pH on sorption of mercury in soils; in Russian), *Vestnik Moskovskovo Universiteta, Pochvovedenie*, 4, 43, 1982.
4. Lodenius, M., Sorption of mercury in soils, In: *Encyclopedia of Environmental Control Technology 4: Hazardous Waste Containment and Treatment*, Cheremisinoff, P., Ed., Gulf Publishing Ltd, Houston, 1990, 339.
5. Lodenius, M., Seppänen, A., and Autio, S., Sorption of mercury in soils with different humus content, *Bull. Environ. Contam. Toxicol.*, 39, 593, 1987.
6. Lodenius, M. and Seppänen, A., Kvicksilvrets binding till humus i vatten (Abstract: Binding of mercury to aquatic humus), *Vatten*, 40, 302, 1984.
7. Dudas, M. and Pawluk, S., The nature of mercury in chernozemnic and luvisolic soils in Alberta, *Can. J. Soil Sci.*, 56, 413, 1976.
8. Landa, E.R., Soilwater content temperature as factors in the volatile loss of applied mercury(II) from soils, *Soil Sci.*, 126, 44, 1978.
9. Liang, C.N. and Tabatai, M.A., Effects of trace element on nitrogen mineralization in soils, *Environ. Pollut.*, 12, 141, 1977.

10. Maliszewska, W., Dec, S., Wierzbicka, H., and Woznikowska, K., The influence of various heavy metal compounds on the development and activity of soil micro-organisms, *Environ. Pollut.*, A37, 195, 1985.

11. Barudi, W. and Bielig, H.J., Gehalt an Schwermetallen (Arsen, Blei Cadmium, Quecksilber) in oberirdish wachsenden Gemüse und Obstarten, *Z. Lebensm. Unters. Forsch.*, 170, 254, 1980.

12. Leinonen, K., The influence of soil preparation on the levels of aluminum, manganese, iron, copper, zinc, cadmium and mercury in *Vaccinium myrtillus*, *Chemosphere*, 18, 1581, 1989.

13. Siegel, S.M., Puerner, N., and Speitel, T., Release of volatile mercury from vascular plants, *Physiol. Plant.*, 32, 174, 1974.

14. Du, S.-H. and Fang, S.C., Catalase activity of C_3 and C_4 species and its relationship to mercury vapor uptake, *Environ. Exp. Bot.*, 23, 347, 1983.

15. Godbold, D.L. and Hüttermann, A., Whole plant aspects of methyl mercury, mercury and lead toxicity in spruce seedlings, *Proc. Int. Conf. Heavy Metals in the Environ.*, New Orleans, Sept., 1987, Vol. 2, 253, 1987.

16. Lodenius, M., Biological monitoring of airborne mercury, Brasser, L.J. and Mulder, W.C., Eds., In: *Man and His Ecosystem—Proc. 8th World Clean Air Congress*, The Hague, Sept. 11–15, 1989, Vol. 3, Elsevier, Amsterdam, 159, 1989.

17. Arai, M., Kawabe, H., and Hagino, N., Historical evidence of mercury pollution using tree ring of Japanese Cedar (*Cryptomeria japonica*) (in Japanese), *Ann. Rep. Study Group Niigata Biol. Educ.*, 15, 49, 1980.

18. Kojo, M.-R. and Lodenius, M., Cadmium and mercury in macrofungi—mechanisms of transport and accumulation, *Angew. Botanik*, 63, 279, 1989.

19. Lamaison, J.L., Pourrat, A., and Pourrat, H., Activité catalasique des macromycefes. Obtention d'une catalase purifiée, *Ann. Pharm. Fr.*, 33, 441, 1975.

20. Brunnert, H. and Zadrazil., F., Translocation of cadmium and mercury into the fruiting bodies of *Agrocybe aegerita* in a model system using agar platelets as substrate, *Eur. J. Appl. Microbiol. Biotechnol.*, 12, 179, 1981.

21. Brunnert, H. and Zadrazil, F., The influence of zinc on the translocation of cadmium and mercury in the fungus *Agrocybe aegerita* (a model system), *Angew. Botanik*, 59, 469, 1985.

22. Cherian, M.G. and Goyer, R.A., Metallothioneins and their role in the metabolism and toxicity of metals, *Life Sci.*, 23,1, 1986.

23. Wren, C., A review of metal accumulation and toxicity in wild mammals, *Environ. Res.*, 40, 210, 1986.

24. Lodenius, M., Poroilla alhaiset elohopeapitoisuudet (Summary: Low mercury concentrations in Finnish reindeer), *Ympäristö ja Terveys*, 15, 162, 1984.

25. Krynski, A., Kaluzinski, J., Wlazeiko, M., and Adamowski, A., Contamination of roe deer by mercury compounds, *Acta Theriol.*, 27, 499, 1982.

26. Tataruch, F., Heavy metal residues in wildlife—indicators for environmental pollution, *Proc. IVth Int. Conf. Bioindicatores Deteriorisationis Regionis, 1982*, 323, 1986.

27. Pankakoski, E., Hyvärinen, H., and Koivisto, I., Äetsän kloloritehtaan vaikutusympëristön pikkkunisäkkäiden elohopeapitoisuuteen. [Influence of a chlor-alkali plant in Aetsä on the mercury concentrations of small mammals], *Ympäristö ja Terveys*, 23, 458, 1992.

28. Bull, K., Roberts, R., Inskip, M., and Goodman, G., Mercury concentration in soil, grass, earthworm and small mammals near an industrial emission source, *Environ. Pollut.*, 12, 135, 1977.

29. Lodenius, M. and Kuusela, S., Mercury content in feathers of the Kestrel (*Falco tinnunculus L.*) in Finland, *Ornis Fennica*, 62, 158, 1985.

30. Solonen, T. and Lodenius, M., Mercury in Finnish Sparrow hawks *Accipiter nisus*, *Ornis Fennica*, 61, 58, 1984.

31. Nuorteva, P., Nuorteva, S.-L., and Suckcharoen, S., Bioaccumulation of mercury in blow-flies collected near the mercury mine of Idrija, Yugoslavia, *Bull. Environ. Contam. Toxicol.*, 24, 515, 1980.

32. Lodenius, M., Mercury contents of dipterous larvae feeding on macrofungi, *Ann. Ent. Fenn.*, 47, 63, 1981.

33. Schütt, S. and Nuorteva, P., Metylkvicksilvrets inverkan på aktivitet hos Tenebrio molitor L. (Col., Tenebrionidae). (Decrease in activity caused by methylmercury in adult Tenebrio molitor beetles), *Acta Entomol. Fenn.*, 42, 78, 1983.

34. Honda, K., Nasu, T., and Tatsukawa, R., Metal distribution in the earthworm, *Pheretima hilgendorfi*, and their variations with growth, *Arch. Environ. Contam. Toxicol.*, 13, 427, 1984.

35. Braunschweiler, H., The effects of acidification and season on the metal contents of the earthworm *Deandrobaena octaedra* in Finnish forest soils. University of Jyväskylä, Congress Publications 3. XI International Colloqium on Soil Zoology, Jyväskylä, Finland, 10–14 August 1992, p. 215, 1992.

Mercury Release and Transformation from Flooded Vegetation and Soils: Experimental Evaluation and Simulation Modeling

Kenneth A. Morrison and Normand Thérien

CONTENTS

I. INTRODUCTION

In reservoirs created within the last two decades, increased mercury levels have been found in fish.[1-5] In general, these reservoirs have not been subjected to industrial effluents containing mercury. Because mercury concentrations in fish from adjacent undisturbed lakes have not increased to any significant extent, a change in atmospheric loading is not the cause. Therefore, this mercury must have been present in the terrestrial environment prior to flooding, and has been released into the aquatic environment entirely as a result of reservoir flooding, i.e., decomposition of flooded vegetation and soils, dissolving of surficial mercury deposited from the atmosphere and, possibly, shoreline erosion. Addition of organic matter via decomposition may have an additional effect—the stimulation of mercury methylation.

Reservoir creation results in the flooding and eventual decomposition of the soils and vegetation within the reservoir limits. This releases organic material and nutrients into the reservoir waters,[6-10] modifying water quality and possibly primary productivity. It is known that increases in these materials will increase the rate of Hg methylation.[3,11-17] The flooding of bogs and acidic soils can reduce water pH, also found by some researchers to influence methylation, although mercury may be bound by the peat or by humic acids.[18-20] While many

1–56670–066–3/94/$0.00+$.50

studies have found relationships between lower pH and fish mercury levels, some contradictory results indicate that such effects must be interpreted in the context of other limnological characteristics.[21,22]

The soils themselves are important mercury sources in reservoirs.[1-3] It has not yet been established if leaching of mercury from flooded vegetation is important, but the stimulation of fish mercury levels by immersed vegetation has been demonstrated.[14,15] While reservoir creation is known to be associated with increased mercury accumulation in fish,[4,16,21,23] it is not clear whether the effects are due simply to increased dissolved mercury or are predominantly due to increased methylation.

Despite the agreement on possible sources of the problem, it has until recently been impossible to quantify mercury release and transformation after reservoir creation due to limitations in analytical techniques. Thus, most experimental studies along these lines have been indirect, and for the most part, qualitative. As a result, abilities to predict impacts of proposed reservoirs from the point of view of mercury are currently severely limited.

The first objective of this study was the experimental investigation and quantification of the release of mercury from immersed samples of soils and vegetation under controlled laboratory conditions. The second objective of the study was the parallel measurements with time of organic carbon, nitrogen, and phosphorus fractions released by the decomposing materials under these conditions. Finally, the third objective was the integration of the results on mercury release to the water column into an existing model of decomposition of flooded vegetation and soils in reservoirs.

II. EXPERIMENTAL WORK

There were two experimental series to the project, but examination here is limited to the second series. The essential approach for both series was to simulate in microcosms the decomposition of flooded vegetation, soil cover, and soil after flooding due to reservoir inundation. Measurements were made on mercury concentrations as well as various water quality parameters to provide information on the evolution of these parameters after flooding—a necessary initial step to understanding the impact of flooding on mercury in the aquatic systems. More detail is available elsewhere.[24,25]

A. MATERIALS AND METHODS

1. Field Site and Sample Collection

Samples of typical vegetation, soils and soil cover, as well as river water were collected from unperturbed areas in the vicinity of the LG-4 development on La Grande Rivière in Québec, Canada (54° N 73° W). Species were selected on the basis of dominance, and are thus representative of the majority of ground cover in the region. Tree samples consisted of small branches with attached foliage, and the species collected were black spruce (*Picea mariana*) and alder (*Alnus* sp.). Soil cover was Sphagnum moss (*Sphagnum* sp.) and lichen (*Cladonia* sp.), and these samples included roots and the thin litter layer. The soil samples were surface layers with organic soil cover layers removed, and were thus essentially mineral in content.

2. Experimental Setup

Sample material was placed in Teflon® TFE baskets in 26-L borosilicate glass tanks, the tanks were then sealed with glass covers. Through a hole in each cover was a Teflon® TFE fitting with Teflon® TFE tubes for water addition-extraction, and Teflon® FEP tubes for air entry and exit as well as a very fine permeable Teflon® FEP tube for oxygenation where desired. Into the tanks were added 22 L of water from the study area. Purified air was passed through the headspace in each container. The setup is shown in Figure 1.

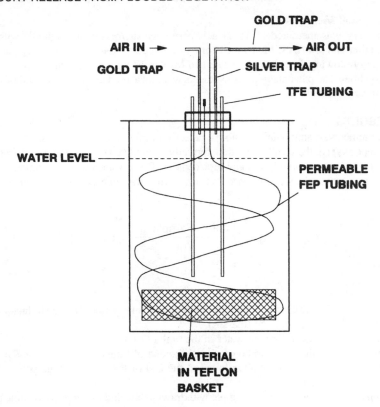

Figure 1 Layout of an experimental tank.

On each sampling date, 12-L water samples were withdrawn from each tank and filtered on a 5 μm glass-fiber filter for chemical analyses, and were replaced with an equal volume of distilled deionized water. Filtration was used to limit examination to dissolved species. All wetted parts of the pumping and filtering system were made of Teflon®. Also, on approximately a monthly basis, measurements of temperature, pH, and dissolved oxygen concentrations were made by closed-circuit recirculation pumping of the recipient water through a specially constructed cell made of Teflon® and housing the different sensors.

All vegetative samples weighed 201.5 g (±1) and soil was 500 g. Duplicate samples of all substrates were maintained at ~18.5° C with oxygenation, and duplicate samples of spruce and lichen (dominant vegetation types) were treated to ascertain the effects of certain reservoir characteristics on rates of release and/or transformation. At the beginning of this series a breakdown of the Teflon® pump occurred, and initial sampling was delayed. The treatments applied were as follows:

Dissolved oxygen. In controls, dissolved oxygen was maintained in the containers via immersed Teflon® tubing pressurized with oxygen. The contrast in this case was to allow the containers to go anoxic by not providing any external oxygen supply.

pH. In the first experimental series, it was noted that the water samples went slightly acidic, and therefore, during the second series, control of pH involved the addition of a base to increase pH to the 6 to 6.5 range. Lithium hydroxide was used due to its low mercury content.

Temperature. The sealed containers were placed in an insulated water bath, the temperature of the bath being maintained in the 4 to 6° C temperature range to simulate conditions found on reservoir bottoms.

3. Analytical Methods

Total mercury was measured using an acid/permanganate/persulfate method[26] with gold amalgamation and CVAAS. Methylated forms of mercury were determined by gas chromatography following the approach of Lee and Mowrer,[27] with preconcentration on sulfhydryl cotton fiber. The other chemical constituents were measured using routine methods, most from Standard Methods.[26]

B. RESULTS

No measurements of ammonia were above detection limits. For all other dissolved constituents except oxygen, the results were transformed to calculated concentrations corrected for dilutions due to sample withdrawal and replacement. These values were determined by calculating incremental changes in concentrations during the time period between sampling dates as:

$$Q_i = Conc_i - [\frac{(Vol - Sample_{i-1})}{Vol} \times Conc_{i-1}] \qquad [1]$$

where:

Q_i	=	the change in concentration during time period "i" due to factors other than dilution (mass/L)
Vol	=	the volume of water in the tank (22 L)
$Conc_i$	=	the concentration found at the end of time period "i" (mass/L)
$Conc_{i-1}$	=	the concentration found at the end of the previous time period (mass/L)
$Sample_{i-1}$	=	the volume of sample withdrawn at the end of the previous time period and replaced with pure water (L).

The corrected concentration for a time period is then the sum of the incremental changes up to and including that time period. While these values do not account for any losses via volatilization or sorption, they do correct for reductions in concentrations due to sample withdrawal and replacement with pure water, and are thus somewhat easier to interpret. Results for Hg in the control tanks are shown in Figure 2. More detailed results are presented elsewhere.[24]

C. INTERPRETATION OF RESULTS

The two tree species had highest release of carbon, phosphorus, and apparent color as well as the maximum amount of dissolved total mercury and methylmercury. Alder produced the highest methylmercury levels. The other three substrates had similar release of total mercury, but methylmercury was very low for soils, and was fairly low for moss compared to the others. These two substrates also released low amounts of Kjeldahl nitrogen as compared to the others, which were all similar. For both spruce and lichen, anoxia increased carbon and color and decreased nitrates. Phosphates were slightly increased for lichen, but were hardly affected for spruce. Anoxia increased total mercury release for both substrates. Accumulated methylmercury was only slightly affected in either substrate, but the temporal patterns were modified with earlier production in lichen and slightly later for spruce. Increased pH did not affect final methylmercury concentrations for either substrate, but did modify patterns of release of total mercury for both. For carbon, pH control increased amounts released from spruce, while for Kjeldahl nitrogen, amounts were reduced for lichen. Otherwise, there were no other pH effects on the two substrates. Temperature reduction slightly accelerated the

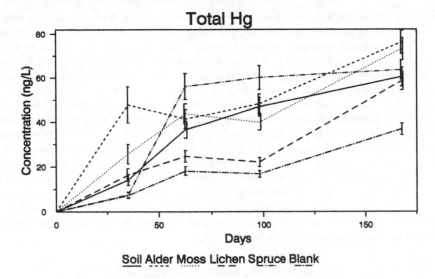

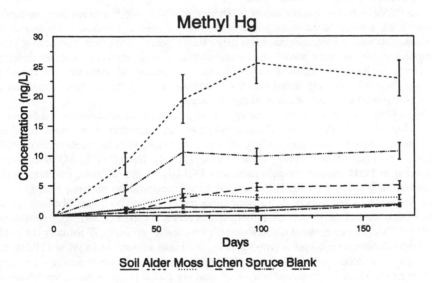

Figure 2 Mean total and methyl-Hg concentrations in the control tanks, corrected for dilutions.

initial release of total mercury for lichen but not the final concentration, but reduced release for spruce. Methylmercury production was reduced for both substrates at the colder temperature, and both phosphates and Kjeldahl nitrogen were reduced for lichen, while carbon was increased for spruce.

D. DISCUSSION OF EXPERIMENTAL RESULTS

All substrates released important amounts of mercury. Results from these experiments did not demonstrate a difference in methylation rates due either to dissolved oxygen or pH. With the exception of the soil samples, there was a significant drop in oxygen and pH in all tanks containing sample material, and methylation was found. However, there was no clear effect on methylation of allowing anoxia or of maintaining pH at higher levels. Also, in all tanks except soil there was important release of organic materials into the water. Due to CO_2 production, bacterial activity will have the effect of reducing dissolved oxygen and pH,[5] and therefore bacterial activity, oxygen, and pH reductions are concomitant effects in newly flooded reservoirs. However, decomposition of lignocellulose is very slow under anoxic conditions,[8,28] and low pH slows decomposition of materials other than peat.[29-31] It would thus seem quite possible that where effects of lowered dissolved oxygen or pH on increased fish mercury have been reported, the relationships have been correlative and not due to direct effects of either dissolved oxygen or pH on methylation. The reductions in methylation with colder temperatures are entirely consistent with reduced microbial activity and decomposition at lower temperatures.[8,28,29] The methylation decreases at the lower temperature found in this study for spruce were consistent with changes in release from sediment,[32] while the reductions for lichen were even greater. The reductions for lichen were more in proportion with decreases in phosphates and Kjeldahl nitrogen, and thus nutrients may have been more limiting in the lichen tanks than in the spruce tanks.

III. MODELING MERCURY RELEASE

A. CONCEPTUAL DESCRIPTION

The model described here is the model of Thérien and Morrison,[10] adapted from an earlier model.[33] It is a model based on the concepts of system dynamics, kinetics, and conservation of mass, thus using differential equations rather than algebraic ones. While appearing more complicated than algebraic models, it is more adaptable to situations occurring in different reservoirs, particularly in terms of varying hydraulic regimes and scenarios. The model is entirely defined by the differential equations, although a computer program is required to solve (integrate) them. Details are available elsewhere.[24]

For a given type of substrate (soil or type of vegetation) a known quantity is submerged due to inundation (TOMF). From this material, there are three immediate losses: some act immediately as particulate matter (POM), and some are immediately leachable as either labile organic matter (DOM) or refractory organic matter (HUM). The rest of the TOMF is slowly converted to POM by various organisms, and POM is further attacked, resulting in the liberation of DOM and HUM. DOM and HUM are degraded by bacteria, with O_2 consumption and CO_2 production. At the same time, the PO_4^{-3} and NH_4^+ are liberated. These components are absorbed by other microorganisms and/or particulates, and NH_4^+ is oxidized to NO_3^-. This latter component is also absorbed. Dissolved "nonmethyl" mercury (HGNM) from the substrate is initialized as proportional to the initial amounts of DOM and HUM that are released at flooding. Methylmercury (HGME) is initialized as not present—i.e., zero concentration. There is subsequent release of mercury that is proportional to the release of DOM and HUM. This nonmethyl mercury is subsequently methylated by bacterial action, and this methylation is considered as a Monod-type function of DOM. Both mercury compartments have losses due to volatilization or sorption to particulates. Demethylation returns some methylmercury to the nonmethyl form. Figure 3 shows the transfers of this conceptual model. The initial quantity of material as well as the particulates are stationary and are actual masses, while the other components are concentrations in solution. Obviously, the two stationary components can only be decomposed if flooded, and thus are influenced by water levels. The dissolved components are affected by water levels via hydraulic effects and the inputs from the stationary components (i.e., whether or not they are flooded).

There are thus 11 simultaneous differential equations in the model, CO_2 not being calculated explicitly. Numerical integration of these equations over time produces the different

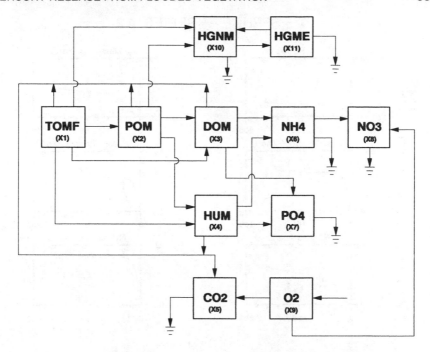

Figure 3   Flows in the conceptual model (see text for explanation of terms).

concentrations for the case of no hydraulic effects. In the case where hydraulic effects must be considered, these effects are included in the appropriate differential equations.

B. MODEL CALIBRATION
Calibration was carried out in two separate stages. The first stage was the calibration of the decomposition model to the nutrient data, ignoring mercury. Once this stage was completed, the model was then calibrated to mercury observations using the nutrient constant values found in the first stage for the decomposition aspects of the model. Because temperature is an integral part of the model and data on two temperatures (18.5 and 5° C) were available for lichen and spruce, simultaneous calibration was carried out for these two substrates at the two temperatures.

C. MODEL RESULTS
The model fit reasonably well for carbon, phosphates, Kjeldahl nitrogen, and both forms of mercury for the substrates other than lichen. For nitrates the model gave less satisfactory results, but this is due to the weighting during the optimization. The model results were compared to the observations for spruce at 18.5 and 5° C (Figures 4 and 5).

D. APPLICATION OF THE MODEL TO RESERVOIRS
1. General Considerations
The model as described so far addresses only the simple kinetics, but does not explicitly include hydraulic processes. In reservoirs, these processes are important, since the volume of water is active in terms of inflows and outflows, and consequently there are varying surface areas and volumes. It is therefore necessary to include these effects in a model for it to be useful in making predictions.

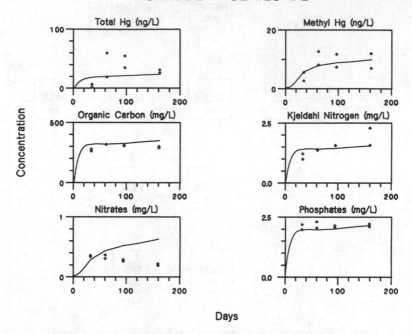

Figure 4   Comparison of observations and predictions for spruce at 18.5° C.

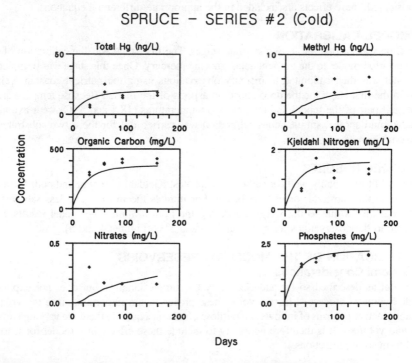

Figure 5   Comparison of observations and predictions for spruce at 5° C.

2. Data Requirements

Data requirements vary with the degree of precision required for the simulation. In the most rigorous applications, the requirements for each volumetric zone under consideration are

- The types and quantities of vegetation and soil to be flooded, and this information divided by vertical strata
- Daily flowrates for all inflows and outflows
- The concentrations of all constituents in all inflows on a daily basis
- The initial concentrations of all constituents in the volume prior to flooding
- The annual pattern of temperature in the volume after flooding

As precision requirements decrease, simplifications can be made. Inflowing concentrations can be averaged over different periods, as can the flow rates themselves. Decomposable material can be distributed to strata on a quadratic basis, or stratification can be neglected completely. A mean volume can be used instead of the additional differential equation. Initial concentrations can be approximated from other lakes or rivers, as can temperature patterns. Other simplifications can also be made.

IV. SUMMARY

The experimental results of this study demonstrated the release of mercury by all substrates and demonstrated clear differences in methylmercury among substrates. The use of these results to calibrate the decomposition model and to add a mercury submodel means that a tool now exists that can be used to estimate mercury release in proposed reservoirs, as well as to compare relative impacts of different construction and operation scenarios.

REFERENCES

1. Cox, J.A., Carnahan, J., Dinunzio, J., McCoy, J., and Meister, J., Source of mercury in new impoundments, *Bull. Environ. Contam. Toxicol.,* 23, 779, 1979.
2. Meister, J.F., Dinunzio, J., and Cox, J.A., Source and level of mercury in a new impoundment, *J. Am. Water Works Assoc.,* 71, 574, 1979.
3. Bodaly, R.A., Hecky, R.E., and Fudge, R.J.P., Increases in fish mercury levels in lakes flooded by the Churchill River diversion, northern Manitoba, *Can. J. Fish. Aquat. Sci.,* 41, 682, 1984.
4. Boucher, R., Schetagne, R., and Magnin, E., Teneur en mercure des poissons des réservoirs La Grande 2 et Opinaca (Québec, Canada) avant et après la mise en eau, *Rev. Franç. Sci. Eau,* 4, 193, 1985.
5. Phillips, G.R., Medvick, P.A., Skaar, D.R., and Knight, D.E., Factors affecting the mobilization, transport, and bioavailability of mercury in reservoirs of the Upper Missouri River Basin, Fish and Wildlife Technical Report #10, United States Fish and Wildlife Service, 64 p, 1987.
6. Baxter, R.M., Environmental effects of reservoirs, In: *Microbial Processes in Reservoirs,* Gunnison, D., Ed., Dr. W. Junk Publishers, Dordrecht, Holland, 1985, chap. 1.
7. Brannon, J.M., Chen, R.L., and Gunnison, D., Sediment-water interactions and mineral cycling in reservoirs, In: *Microbial Processes in Reservoirs,* Gunnison, D., Ed., Dr. W. Junk Publishers, Dordrecht, Holland, 1985, chap. 7.
8. Godshalk, G.L. and Barko, J.W., Vegetative succession and decomposition in reservoirs, In: *Microbial Processes in Reservoirs,* Gunnison, D., Ed., Dr. W. Junk Publishers, Dordrecht, Holland, 1985, chap. 4.

9. Gunnison, D., Engler, R.M., and Patrick, W.H., Chemistry and microbiology of newly flooded soils: relationship to reservoir-water quality, In: *Microbial Processes in Reservoirs,* Gunnison, D., Ed., Dr. W. Junk Publishers, Dordrecht, Holland, 1985, chap. 3.

10. Thérien, N. and Morrison, K., Modèle prévisionnel de la qualitè des eaux du réservoir hydro-electrique LG-2: considération des zones hydrauliques actives et stagnantes, *Rev. Int. Sci. Eau,* 1, 11, 1986.

11. Bisogni, J.J., Jr., Kinetics of methylmercury formation and decomposition in aquatic environments, In: *The Biogeochemistry of Mercury in the Environment,* Nriagu, J.O., Ed., Elsevier/North-Holland, Amsterdam, 1979, chap. 10.

12. D'Itri, F., *The Environmental Mercury Problem,* CRC Press, Boca Raton, FL, 124, 1972.

13. Jackson, T.A., Methyl mercury levels in a polluted prairie river-lake system: seasonal and site-specific variations and the dominant influence of trophic conditions, *Can. J. Fish. Aquat. Sci.,* 43, 1873, 1986.

14. Hecky, R.E., Bodaly, R.A., Ramsey, D.J., and Strange, N.E., Enhancement of mercury bioaccumulation in fish by flooded terrestrial materials in experimental ecosystems, Canada-Manitoba Agreement on the Study and Monitoring of Mercury in the Churchill River Diversion, Appendix, 35 p., 1986.

15. Hecky, R.E., Bodaly, R.A., Strange, N.E., Ramsey, D.J., Anema, C., and Fudge, R.J.P., Mercury bioaccumulation in yellow perch in limnocorrals simulating the effects of reservoir creation, *Can. Data Rep. Fish. Aquat. Sci.,* 628, 158, 1987.

16. Rudd, J.W.M. and Turner, M.A., The English-Wabigoon River system. V. Mercury and selenium bioaccumulation as a function of aquatic primary productivity, *Can. J. Fish Aquat. Sci.,* 40, 2251, 1983.

17. Compeau, G.C. and Bartha, R., Sulfate-reducing bacteria: principal methylators of mercury in anoxic estuarine sediments, *Appl. Environ. Microbiol.,* 50, 498, 1985.

18. Lodenius, M. and Seppanen, A., Kvicksilvrets bindning till humus i vatten (Binding of mercury to aquatic humus), *Vatten,* 40, 302, 1984.

19. Lodenius, M., Seppanen, A., and Autio, S., Sorption of mercury in soils with different humus content, *Bull. Environ. Contam. Toxicol.,* 39, 593, 1987.

20. Lodenius, M., Seppanen, A., and Autio, S., Leaching of mercury from peat soil, *Chemosphere,* 16, 1215, 1987.

21. Rodgers, D.W., Watson, T.A., Langan, J.S., and Wheaton, T.J., Effects of pH and feeding regime on methylmercury accumulation within aquatic microcosms, *Environ. Pollut.,* 45, 261, 1987.

22. Richman, L.A., Wren, C.D., and Stokes, P.M., Facts and fallacies concerning mercury uptake by fish in acid stressed lakes, *Water Air Soil Pollut.,* 37, 465, 1988.

23. Wren, C.D. and MacCrimmon, H.R., Mercury levels in the sunfish, *Lepomis gibbosus,* relative to pH and other environmental variables of PreCambrian Shield lakes, *Can. J. Fish. Aquat. Sci.,* 40, 1737, 1983.

24. Morrison, K.A., A Study of Factors Affecting the Release and Transformation of Mercury in Hydroelectric Reservoirs, Ph.D. Thesis, Department of Chemical Engineering, University of Sherbrooke, Canada, 1991, 190 p.

25. Morrison, K.A. and Thérien, N., Experimental evaluation of mercury release from flooded vegetation and soils, *Water Air Soil Pollut.,* 56, 607, 1991.

26. American Public Health Association—American Water Works Association—Water Pollution Control Federation, *Standard Methods for the Examination of Water and Wastewater, 16th ed.,* Amer. Pub. Health Assoc., Washington. 1985. 1289p.

27. Lee, Y.H. and Mowrer, J., Determination of methylmercury in natural waters at the subnanogramms per litre level by capillary gas chromatography after adsorbent preconcentration, *Anal. Chim. Acta,* 221, 259, 1989.

28. Brinson, M.M., Lugo, A.E., and Brown, S., Primary productivity, decomposition and consumer activity in freshwater wetlands, *Annu. Rev. Ecol. Syst.,* 12, 123, 1981.

29. Webster, J.R. and Benfield, E.F., Vascular plant breakdown in freshwater ecosystems, *Annu. Rev. Ecol. Syst.,* 17, 567, 1986.

30. Rochefort, L., Vitt, D.H., and Bayley, S.E., Growth, production, and decomposition dynamics of *Sphagnum* under natural and experimentally acidified conditions, *Ecology,* 71, 1986, 1990.

31. Schoenberg, S.A., Benner, R., Armstrong, A., Sobecky, P., and Hodson, R.E., Effects of acid stress on aerobic decomposition of algal and aquatic detritus: direct comparison in a radiocarbon assay, *Appl. Environ. Microbiol.,* 56, 237, 1990.

32. Wright, D. and Hamilton, R., Release of methyl mercury from sediments: effects of mercury concentration, low temperature, and nutrient addition, *Can. J. Fish. Aquat. Sci.,* 39, 1459, 1982.

33. Thérien, N. and Spiller, G., A mathematical model of the decomposition of flooded vegetation in reservoirs, in *Simulating the Environmental Impact of a Large Hydroelectric Project,* Thérien, N., Ed., Simulation Proceedings Series, Vol. 9, No. 2, Society for Computer Simulation, La Jolla (Calif.), 1981, chap. 9.

SECTION IV
Bioaccumulation

Amplification of Mercury Concentrations in Lake Whitefish (*Coregonus clupeaformis*) Downstream from the La Grande 2 Reservoir, James Bay, Québec

Denis Brouard, Jean-François Doyon, and Roger Schetagne

CONTENTS

ABSTRACT: In the James Bay region, and throughout the area of the Canadian Shield, methylmercury occurs naturally and is bioaccumulated by fish to varying degrees in different species. The creation of the La Grande hydroelectric complex reservoirs has led to significant increases in mercury levels in fish. Downstream from the reservoirs, and especially in the fluvial section of the La Grande River immediately below the LG 2 reservoir, mercury concentrations in 400-mm lake whitefish (*Coregonus clupeaformis*) were 2.5 times higher than those in fish caught upstream of the LG 2 power house. These differences in mercury concentrations were attributed to a change in feeding habits of the lake whitefish towards a more piscivorous diet, induced by the continuous inputs of food from the LG 2 reservoir into the La Grande River. In order to verify the hypothesis of change in feeding habits of lake whitefish, field work was conducted in the LG 2 reservoir and in the fluvial section of the La Grande River in 1990.

A sampling survey of organisms drifting in the river and the analysis of stomach contents of lake whitefish provided evidence of large numbers of small cisco (*Coregonus artedii*) being flushed into the La Grande River through the LG 2 power house. Lake whitefish >450 mm captured immediately below the LG 2 power house exhibited significantly higher percentages of fish in their diet, and mercury levels in their stomachs and flesh were five times higher than those in fish caught upstream in the LG 2 reservoir. At river sites located 42 km and 70 km below the reservoir, the proportion of fish in the diet and mercury levels in flesh of lake whitefish decreased compared to those in specimens caught immediately below the reservoir. In addition, new species of forage fish were found in the stomachs of specimens captured at the stations further downstream. These results suggest that the input of fish from the reservoir into the river is limited to a relatively short reach below the LG 2 power house. Increased bioaccumulation of mercury in fish reported below other impoundments within the La Grande complex and in Labrador will be discussed.

I. INTRODUCTION

In reservoirs and water diversion areas, processes leading to the methylation of mercury by microorganisms and its subsequent bioaccumulation in the food chain are similar to those observed in natural environments. However, rates of methylation are increased by the sudden flooding of terrestrial organic matter leading to intense decomposition during the first years of the impoundment.[1-4] The phenomenon of biomagnification causes methylmercury increases at each higher trophic level in the food chain.[5] Certain fishes, occupying niches relatively high in the food web, are good indicators of the degree of change in the availability of this metal in the aquatic environment. For example, piscivorous pike (*Esox lucius*) exhibit higher total mercury concentrations than similar-sized benthivorous and planktonivorous species such as lake whitefish (*Coregonus clupeaformis*).[6]

The creation of the La Grande hydroelectric complex (Figure 1) led to a significant increase in mercury levels in fishes living in the reservoirs. Depending on the species and reservoir studied, concentrations of up to five times the natural level were found.[6-8] Immediately below the control structures and power stations, mercury bioaccumulation was higher than in the upstream reservoirs, particularly in nonpiscivorous species including lake whitefish. In the fluvial section of the La Grande River, the mercury levels of lake whitefish were generally 2.5 times higher than those of specimens from the La Grande 2 (LG 2) reservoir.[7] Similar differences have also been observed in other downstream environments: namely, below the Caniapiscau reservoir, the Fontanges basin, the Opinaca reservoir, and also below the Smallwood reservoir in Labrador.[6] This phenomenon has not yet been explained in the literature on this subject.

Lake whitefish usually feed on benthic invertebrates.[9] However, this species can sometimes act as an opportunistic feeder and eat whatever is available, including small fish.[10] Observed differences in mercury concentrations between lake whitefish from the reservoir and those from the river could be due to a change in feeding habits. It was hypothesized that individuals downstream from the reservoir may take advantage of a continuous outflow of food, particularly small fish, from impoundments by becoming piscivorous.[11] These inputs from the reservoir would also supply the numerous seagulls found immediately below the LG 2 power house during the summer.[12]

In order to verify this hypothesis, we identified organisms drifting from the LG 2 reservoir into the river, and described and compared the diet of lake whitefish captured above (reservoir) and below (river) the LG 2 power house.

II. STUDY AREA

The La Grande hydroelectric complex development is located in the watershed of the La Grande River, on the eastern side of James Bay (Figure 1). The whole region lies on the Canadian Shield, one of the oldest known geological formations, which is composed of igneous and metamorphic rock. Prior to hydroelectric development, the territory was free of any industrial activity and was sparsely occupied by aboriginal people following traditional lifestyles in a terrain of scattered coniferous forests with numerous peat bogs. Surface water

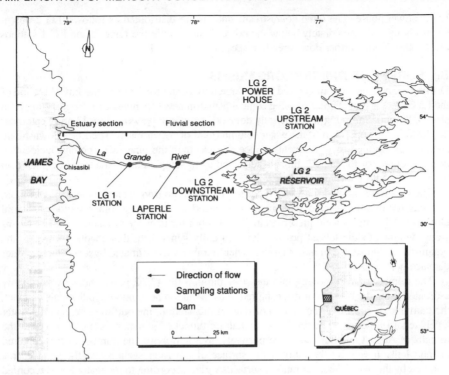

Figure 1 La Grande hydroelectric complex and location of sampling stations.

in the region is naturally colored by humic and fulvic organic acids, well oxygenated, slightly acidic (pH 5.9 to 6.9), with low mineral content (conductivity 8 to 30/μS cm). It is also relatively rich in organic matter and poor in nutrients (0.004 to 0.01 mg/L total P). The most common nonpiscivorous fish species in the region include the longnose sucker (*Catostomus catostomus*), the white sucker (*Catostomus commersoni*), the lake whitefish (*Coregonus clupeaformis*) and the cisco (*Coregonus artedii*). Piscivorous species include the northern pike (*Esox lucius*), the walleye (*Stizostedion vitreum*), and the lake trout (*Salvelinus namaycush*).

The La Grande project resulted in the creation of five large reservoirs, ranging in surface area from 765 to 4275 km^2. The LG 2 reservoir, flooded in 1978, was the first to be created. The reservoir submerged a large land tract (92% of the total reservoir area of 2835 km^2) and has both a short water residence time (6.9 months) and a short filling period (13 months). Before reaching James Bay, the water flowing through the LG 2 power house has to travel 80 km along the fluvial section of the La Grande River and 37 km along the river's estuary section (Figure 1). Since the LG 2 power house was put into service, its mean annual flow has doubled to approximately 3400 m^3/s as a result of the diversion of the Opinaca and Eastmain rivers from the south, and of the Caniapiscau river from the east. Our study area included the fluvial part of the river and the area of the LG 2 reservoir located immediately above the power house (Figure 1).

III. MATERIAL AND METHODS

A. SAMPLING STATIONS

The study was conducted at four sampling stations situated above and below the LG 2 power house. The LG 2 upstream station, located within the reservoir, is immediately above the

LG 2 power house. The three downstream stations are distributed as follows: LG 2 down-stream is located immediately below the power house, while the Laperle and LG 1 stations are 42 and 70 km further downstream, respectively (Figure 1).

B. SURVEY OF DRIFTING ORGANISMS

Drifting organisms were monitored at the three downstream stations in the fluvial section of the La Grande River (Figure 1) using conical plankton nets (opening: 1 m, length: 4 m, mesh size: 0.95 mm). Sampling was generally done once a month between July 13 and September 20, 1990 in the upper 2 m of the water column (total of 39 samples). The LG 2 downstream station was sampled more frequently (once a week) than the other two stations; its location near the outlets of the tailrace tunnels permitted us to sample and characterize outputs from the reservoir into the river.

The nets were moored for 5 to 7 h per set at the station immediately below the reservoir, whereas the duration of the sampling was reduced to 2 h at the other stations to avoid clogging. All samples were preserved in 70% alcohol for later identification. The organisms were identified to the lowest possible level (usually genera). In this chapter, however, the organisms are presented in major groups. Their number was estimated by a volumetric water displacement method.

The number of fish moving into the river through the LG 2 power house was roughly estimated using the number of fish captured in the plankton nets moored in the main channel, the bathymetric profile of the section of river at this station, the outflow rate, and the water levels recorded hourly at the outlet of the tailrace tunnels. Estimates were calculated for the months of August and September, respectively, by multiplying the number of fish captured in the plankton nets by the ratio of the surface of the river section and the surface area sampled by the nets. The latter ratio of surfaces varied according to the water levels recorded at the tailrace tunnels.

C. LAKE WHITEFISH DIET AND MERCURY CONTENT

Experimental fishing was conducted at each of the four stations once a month from June to September, 1990. Additional sampling was also carried out immediately below the LG 2 power house during October and November. Fish were captured with experimental gill nets (length: 46 m, height: 2 m, mesh: 76 or 102 mm) set for 6 to 18 h. All whitefish captured were measured (total length in mm) and weighted (±10 g). A minimum of 30 lake whitefish stomachs with sufficient contents (>25% of total volume of the stomach), were collected during each sampling at both the station located immediately above and the one immediately below the LG 2 power house. Of these 30 stomachs, 25 were used for identification of organisms and 5 to determine the mercury concentrations of the stomach content. Further downstream the number of stomachs collected (for identification only) was lower due to the poor catch of lake whitefish per unit of effort. The stomachs were preserved in alcohol until their contents could be identified, or frozen for subsequent analysis of the total mercury content. In addition to stomachs, at least 10 samples of flesh were also taken from lake whitefish at every station for analysis of total mercury. The methodology used to identify prey items was the same as that used in the survey of drifting organisms. In this paper, the occurrence is defined as the percentage of all samples containing at least one of the specified food items (micro algae, plant debris, zooplankton, benthos, and fish). Total mercury con-centrations (wet weight) in stomach contents and flesh of whitefish were determined by a certified laboratory. The procedure used, developed by Environment Canada (Naquadat No. 80601–2), involves acid digestion and atomic absorption spectrophotometry to determine dosage.[13]

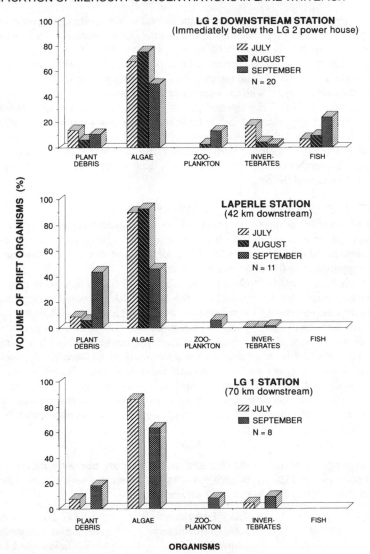

Figure 2   Spatiotemporal changes in the proportions of various drifting organisms in the La Grande River.

IV.   RESULTS

A.   DRIFTING ORGANISMS

The predominant groups of organisms flowing from the reservoir into the fluvial section of the La Grande River included micro algae, plant debris (bryophytes and vascular plants), benthic invertebrates (mainly insects), zooplankton, and fish (Figure 2). At each station, algae were present during every sampling period. However, the proportional volume of these organisms decreased during September at all three sampling sites.

At the LG 2 downstream station, fish were the second most important group of organisms in the samples of August (9% of volume) and September (23% of volume). Their occurrence in samples rose from 20% in July to 60% in August and 100% in September. Plant debris was the second most important component at the Laperle and LG 1 stations. Aquatic insects (especially Diptera) and zooplankton were the third or fourth most important groups drifting in the river: their rank varying with location and season (Figure 2).

Fish collected at the LG 2 downstream station were generally whole and consisted essentially of underyearling cisco (20 to 50 mm in total length) or yearling cisco (90 to 130 mm in total length). The average total number of cisco moving downstream was roughly estimated as 700/h (min.: 208, max.: 1696 in August (five samplings) and 1500/h (min.: 644, max.: 1716) in September (three samplings). This is likely an underestimate since a certain number of fish can avoid the nets.

B. LAKE WHITEFISH DIET AND MERCURY CONTENT

The major groups of organisms found in stomachs of lake whitefish captured in the reservoir were in order of importance: benthic invertebrates, plants debris, zooplankton, and fish (Figure 3). Fish accounted for less than 2% of the total volume of the stomach contents and occurred in approximately 6% of all stomachs examined.

Immediately below the LG 2 power house, fish were second to benthos in proportional volume in the stomach contents (Figure 3). At this station, fish accounted for approximately one third of the total volume of food ingested and were found in almost 40% of the stomachs examined. The occurrence and proportional volume of prey fish (mainly cisco) in the stomachs of lake whitefish of all sizes were significantly higher ($p < 0.01$, Chi-square using contingency analysis table) below the power house (respectively, 42.0 and 32.6%) than above (respectively, 7.1 and 1.0%) (Table 1). These differences were also reflected in the mercury levels of stomach contents, which were nearly five times higher below (0.87 mg Hg/kg) compared to within the reservoir station (0.19 mg Hg/kg). Mercury concentrations in whitefish muscle also was significantly higher ($p < 0.01$, t-test) below the reservoir station than within it (1.86 vs. 0.40 mg Hg/kg).

Table 1 **Importance of fish in the diet and mean mercury concentrations (stomach contents and flesh) for lake whitefish of all sizes caught immediately above and immediately below the LG 2 power house.**

	N[a]	LG 2 Upstream station (immediately above the LG 2 power house)	N	LG 2 Downstream station (immediately below the LG 2 power house)
Occurrence of fish in stomachs (%)	113	7.1	460	42.0[b]
Volume of fish in stomach contents (%)	113	1.0	460	32.6[b]
Total mercury in stomachs (in mg/kg)	15	0.19 ± 0.01	13	0.87 ± 0.50[c]
Total mercury in flesh (in mg/kg)	29	0.40 ± 0.20	19	1.86 ± 1.08[c]

[a] N = Number of samples.
[b] Significant differences between stations (Chi square test using contingency table; $p < 0.01$).
[c] Significant differences between stations (Mann-Whitney U test, $p < 0.01$).

These differences in feeding habits and mercury levels are especially evident in large specimens of whitefish (Table 2, Figure 4). In whitefish of less than 450 mm total length captured within the reservoir, forage fish composed less than 1% of the stomach volume and were found in less than 10% of all stomachs analyzed. In contrast, forage fish made up 22% of stomach volume and occurred in 32% of the stomachs of similar-sized whitefish captured below the power house. Even at sizes <450 mm, the differences in feeding habits led to a substantial increase in mercury levels found in stomach contents (0.12 to 0.71 mg Hg/kg) and flesh (0.29 to 1.07 mg Hg/kg), comparing specimens taken from the LG 2 upstream station to those taken from the LG 2 downstream station.

For lake whitefish longer than 450 mm, the change in diet and the effects on bioaccumulation became particularly obvious. In the reservoir, fish composed 1% of the volumetric percentages and occurred in 4% of stomachs examined. These percentages were not significantly different from whitefish below 450 mm (Table 2). This contrasts with values of approximately 85% for both contents and occurrence in stomachs of similar-sized lake whitefish collected at the LG 2 downstream station (Table 2). Consequently, the mercury levels

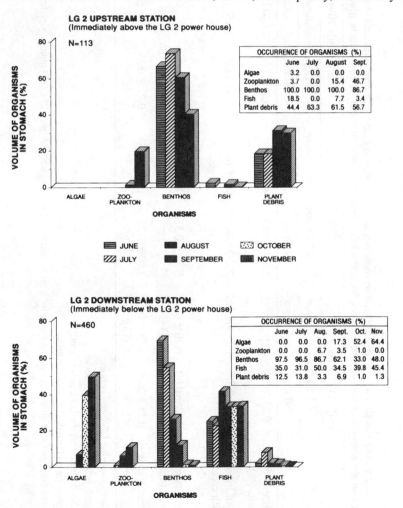

Figure 3   Food items found in lake whitefish caught immediately above and immediately below the LG 2 power house.

Table 2 Importance of fish in the diet and mean mercury concentrations (stomach contents and flesh) for lake whitefish of two length classes caught above (reservoir) and below (river) the LG 2 power house (number of samples is given in parentheses).

	LG 2 reservoir				La Grande River			
	LG 2 Upstream station[a]		LG 2 Downstream station[b]		Laperle station[c]		LG 1 Station[d]	
	<450 mm	>450 mm	<450 mm	>450 mm	<450 mm	>450 mm	<450 mm	>450 mm
Occurrence of fish in stomachs (%)	9.23 (65)	4.16 (48)NS	32.36 (380)	87.50 (80)*	6.50 (15)	29.40 (34)NS	0.00 (6)	27.30 (11)NS
Volume of fish in stomach contents (%)	0.84 (65)	1.27 (48)NS	21.77 (380)	84.15 (80)**	4.75 (15)	22.60 (34)NS	0.00 (6)	13.70 (11)NS
Total mercury in stomachs (mg/kg)	0.12 ± 0.00 (1)	0.20 ± 0.05 (14)	0.71 ± 0.31 (11)	1.75 ± 0.33 (2)**	—	—	—	—
Total mercury in flesh (mg/kg)	0.29 ± 0.08 (16)	0.54 ± 0.21 (13)**	1.07 ± 0.23 (11)	2.95 ± 0.76 (8)**	0.77 ± 0.26 (12)	1.84 ± 1.02 (10)**	0.85 ± 0.39 (6)	1.87 ± 0.64 (5)**

[a] Immediately above the LG 2 power house.
[b] Immediately below the LG 2 power house.
[c] 42 km downstream from the LG 2 power house.
[d] 70 km downstream from the LG 2 power house.

Note: (*) indicate significant differences between size for a given station using Chi square test (p<0.05); (**) indicate significant differences between size using Mann-Whitney U test (p<0.05); NS = no significant difference.

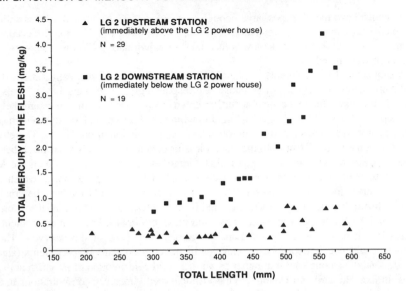

Figure 4 Relationships between length and mercury concentrations in the flesh of lake whitefish captured immediately above and immediately below the LG 2 power house.

found in the stomachs and flesh of these fish, respectively, were nine (0.20 to 1.75 mg Hg/kg) and five (0.54 to 2.95 mg Hg/kg) times higher at the LG 2 downstream station than in the reservoir. At the LG 2 downstream station all concentrations were significantly higher in whitefish longer than 450 mm than those below this size, while no such differences were observed between sizes in the reservoir.

Further downstream, at the Laperle and LG 1 stations, fish were of less importance in the diet of lake whitefish compared to the LG 2 downstream station. Their occurrence and volumetric percentages in the stomachs of whitefish larger than 450 mm were significantly lower ($p < 0.01$, Chi-square) than the values obtained at the LG 2 downstream station (Table 2). In addition, these variables showed no significant differences between whitefish less than 450 mm and those greater than that length for both stations. Furthermore, the species preyed upon by whitefish at Laperle and LG 1 stations, respectively, located 42 and 70 km below the power house, were different (stickleback and sculpin) from those identified in the stomach contents of specimens at the LG 2 downstream station (cisco).

These results suggest that whitefish from sites located further downstream in the fluvial section of the river are feeding upon locally produced fish rather than on those drifting from the reservoir into the river. This results in a lower mercury burden for whitefish at the Laperle and LG 1 stations (1.84 and 1.87 mg/kg compared to 2.95 mg/kg at LG 2 downstream station, for fish >450 mm).

V. DISCUSSION

Our data show that large numbers of small cisco were flushed through the turbines of the LG 2 power house. Generally, the cisco were found to remain unharmed or only stunned despite their passage through the turbines. As a direct consequence of this continuous input of small fish from the reservoir into the river system, the diet of the lake whitefish found below the dam has changed markedly, toward a more piscivorous orientation. The change in diet was most obvious for individuals longer than 450 mm. Whitefish have small subterminal mouths well adapted to feeding on insect larvae and small molluscs. Most likely, only large whitefish have mouth gapes large enough to swallow fish. As a result of this shift, a noticeable increase in the bioaccumulation of mercury in the tissue of lake whitefish occurred. Concentrations up to five times the levels measured in individuals >450 mm caught within the reservoir were noted.

The stomachs of lake whitefish contained less fish at further downstream stations than immediately below the LG 2 power house. Correspondingly, there were lower levels of mercury in the lake whitefish caught at further downstream sites than in fish caught below the LG 2 power house. The fish species found in whitefish stomach contents differed among sites, suggesting that consumption is controlled by local production of prey items. Therefore, it appears that the input of fish from the reservoir is limited to a relatively short reach below the LG 2 power house. We also suggest that forage fish produced downstream are less vulnerable to predation than are the small cisco flushed through the turbines, which are probably stunned for a certain period of time. The preceding would explain the smaller amount of fish in the diet of specimens captured further downstream in the fluvial section of the river, and the lower mercury concentrations in these specimens. The mercury content of whitefish is still higher at LG 1 and Laperle stations than those of the reservoir. These differences could not be explained by changes in mercury contents of prey items (algae, plankton, benthos) among sites, but rather by an increased consumption of prey fish at these stations. In fact, Brouard and Doyon[12] found identical methylmercury concentrations in the longnose sucker (young of the year) at LG 1 and Laperle stations compared to the reservoir sampling site. This indicates that the prey items of these young specimens contain similar mercury levels throughout the stations.

The apparent abundance of small vulnerable cisco is confirmed by the seagull feeding site immediately below the power house. An analysis of 40 seagull stomachs taken immediately below the LG 2 power house showed that 36 contained fish.[12] The 12 stomachs containing identifiable fish showed an average of 5 whole cisco (mean total length of 95 mm). Two of these stomachs contained as many as a dozen cisco each.

Similar increases of mercury levels through bioaccumulation have also been reported below the Smallwood reservoir in Labrador.[6] In fact, 6 years after the Churchill Falls power house was brought into service, measured mercury levels were three times higher for a standardized-length of 400 mm in whitefish caught in the Churchill river than for those caught in the reservoir (1.01 vs. 0.32 mg Hg/kg in flesh).[6]

In the La Grande complex, Brouard et al.[6] and Verdon et al.[7] also found higher mercury levels in lake whitefish captured immediately downstream from the Opinaca reservoir (Boyd-Sakami diversion), the Caniapiscau reservoir, and the Fontanges basin than in whitefish caught within these impoundments. Below these water masses, where there are control structures rather than power houses, the increase in mercury levels was not as great as that reported below the LG 2 and Churchill Falls power houses. This can probably be attributed to the different hydraulic characteristics of control structures, where there are no turbines and the hydraulic head is generally lower than for power houses. Fish passing through water level control structures would be less vulnerable to predation downstream than those flowing through turbines in power houses. This would explain the lower concentrations of mercury observed in whitefish located below these control structures. Moreover, in the Caniapiscau and Fontanges impoundments, the absence of cisco could result in a reduction in the rate of fish flowing through the control structures.

ACKNOWLEDGMENTS

Special thanks are due to D. Dussault and J. Mercier for field assistance and M. Bilodeau for identification of stomach contents. We also wish to acknowledge Dr. P.G.C. Campbell, M. Laperle, D. Messier, R. Verdon, and Dr. F.G. Whoriskey for helpful comments on the manuscript. This study was supported by Hydro-Québec.

REFERENCES

1. Verta, M., Rekolainen, S., Mannio, T., and Surma-Aho, K., The origin and level of mercury in Finnish forest lakes, Publications of the Water Research Institute, National Board of Waters, Finland, 1986, No. 65, 21.

2. Verta, M., Rekolainen, S., and Kinnunen, K., Causes of increased fish mercury levels in Finnish reservoirs, Publication of the Water Research Institute, National Board of Waters, Finland, 1986, No. 65, 44.

3. Ramlal, P.S., Anema, C., Furutani, A., Hecky, R.E., and Rudd, J.W.M., Mercury methylation studies at Southern Indian Lake, Manitoba: 1981–1983, Canadian Technical Report Fisheries Aquatic Sciences, 1987, No. 1490.

4. Bodaly, R.A., Hecky, R.E., and Fudge, R.J.P., Increase in fish mercury levels in lakes flooded by the Churchill River diversion, northern Manitoba, *Can. J. Fish. Aquat. Sci.*, 45, 1131, 1984.

5. Surma-Aho, K., Paasivirta, J., Rekolainen, S., and Verta, M., Organic and inorganic mercury in the food chain of some lakes and reservoirs in Finland, Publication of the Water Research Institute, National Board of Waters, Finland, 1986, No. 65, 59.

6. Brouard, D., Demers, C., Lalumière, R., Schetagne, R., and Verdon, R., Summary report. Evolution of mercury levels in fish of the La Grande hydroelectric complex, Québec (1978–1989), Joint report, Vice-Présidence Environnement, Hydro-Québec and Groupe Environnement Shooner inc., Montréal, Canada, 1990.

7. Verdon, R., Brouard, D., Demers, C., Lalumière, R., Laperle, M., and Schetagne, R., Mercury evolution (1978–1988) in fishes of the La Grande hydroelectric complex, Canada, *Water Air Soil Pollut.*, 56, 405, 1991.

8. Boucher, R., Schetagne, R., and Magnin, E., Teneur en mercure des poissons des réservoirs La Grande 2 et Opinaca (Québec, Canada) avant et après la mise en eur, *Rev. Fr. Sci. Eau*, 4,193, 1985.

9. Scott, W.B. and Crossman, E.J., Freshwater fishes of Canada, *Fish. Res. Board Can. Bull.*, 184, 966, 1973.

10. Machniak, K., The effects of hydroelectric development on the biology of northern fishes (reproduction and population dynamics). I. Lake whitefish (*Coregonus clupeaformis* (Mitchill). A literature review and bibliography, Fisheries and Marine Services Technical Report, Freshwater Institute, Winnipeg, Canada, 1975, No. 527.

11. Messier, D. and Roy, D., Concentrations en mercure chez les poissons au complexe hydroélectric de La Grande rivière (Québec), *Naturaliste Can. (Rev. Ecol. Syst.)*, 114, 857, 1987.

12. Brouard, D. and Doyon, J.-F., Recherches exploratoires sur le mercure au complexe La Grande (1990), report presented by Groupe Environnement Shooner inc. to Vice-présidence Environnement, Hydro-Québec, Montréal, Canada, 1991.

13. Agemian, H. and Cheam, V., Simultaneous extraction of mercury and arsenic from fish tissues, and an automated determination of arsenic by atomic absorption spectrometry, *Anal. Chim. Acta*, 101,193, 1978.

Chapter IV.2

Earthworm Bioaccumulation of Mercury from Contaminated Flood Plain Soils

Dean Cocking, Mary Lou King, Lisa Ritchie, and Robert Hayes

CONTENTS

ABSTRACT: Earthworms, *Lumbricus* spp., bioaccumulate Hg under field and laboratory conditions in amounts which are dependent on soil concentration and duration of exposure. Maximum total tissue Hg concentrations in the laboratory cultures are generally only one fifth the 10.0 to 14.8 μg Hg g^{-1}dw observed in individuals collected from contaminated soils ($\sim$21 μg Hg g^{-1}) on the South River flood plain at Waynesboro, VA. Bioconcentration occurred under field conditions in "uncontaminated" control soils ($\sim$0.2 μg Hg g^{-1}); however, total tissue Hg concentrations (0.4 to 0.8 μg Hg g^{-1}dw) were only 1 to 5% of those for earthworms collected from contaminated soils. High individual variability limited demonstration of soil pH effects on uptake under field conditions; however, uptake was significantly enhanced from slightly acidic soils (pH 5.9 to 6) in laboratory culture.

I. INTRODUCTION

In many moist temperate terrestrial ecosystems more than three fourths of the soil animal biomass is composed of earthworms. They benefit soil structure by increasing aggregate formation, aeration, moisture-holding capacity, organic matter breakdown, and soil turnover. Their biology and ecology has received considerable attention.[1-4] Earthworms have been used as bioindicators for contamination of soils by heavy metals and organic and inorganic chemicals by numerous investigators including Ma,[5] Ireland,[6] and Roberts and Dorough.[7] International guidelines have been published by OECD[8] and EEC[9] for the use of earthworms to test chemicals for acute toxicity and van Gestel et al.[10] have recently proposed a standardized method for using the effects of toxic substances on growth and sexual development of *Eisenia andrei* as a measure of impact. Research by van Rhee[11] and others has shown that earthworms are able to tolerate high environmental concentrations of heavy metals, either by not absorbing them, accumulating them in a nontoxic form, or excreting them efficiently. This biodiscrimination may explain why different species at the same site can contain different concentrations of metals, and why there are different maximum tolerance levels to a specific metal. For example, Ireland and Richards[12] found that Pb content in *Dendrobaena rubida* was significantly higher than in *Lumbricus rubellus* and there was also a species difference in Zn tolerance. Thus there must be some type of species selectivity. One possible mechanism to explain this would be differential absorption of the two metals by the intestine while another would be differences in feeding patterns.

Earthworms pass a mixture of both organic and inorganic materials through their guts when feeding. Earthworms tend to be in one of two groups: deep-working species or shallow-working subsurface species. The members of the deep-working group move through the full depth of available soil and subsoil and even down to the parent rock. The shallow-working or subsurface feeders are seldom found very deep in the soil and their activities are confined to the top 15 cm. Surface-feeding worms, such as *L. terrestris* which lives in deep burrows, consume large amounts of organic matter by drawing leaves and other materials into the mouth of their burrows in addition to ingesting soil. Smaller earthworms, such as *L. castaneus* and *Eisenia foetida,* which feed on the woodland litter, produce numerous casts that are almost entirely made up of fragmented litter, whereas larger species which consume a large proportion of soil have less organic matter in their casts.[3] Other examples are *Lumbricus rubellus,* which is predominately a litter feeder, and *Eisenia veneta hibernica* f. *typica,* a raw humus feeder.[13] Some differences in metal content could therefore also be accounted for by differences in feeding habits if, as is generally the case, heavy metal distribution is heterogeneous.

Earthworm influence on the redistribution of Cd, Zn, and Cs within the soil has been reported by Van Hook,[14] and many studies have found that earthworms not only bioaccumulate, but also bioconcentrate heavy metals in their tissues.[15-18] In some cases, earthworms appear to have some sort of mechanism for regulating the heavy metals because the concentrations within their tissues are not proportional to the concentration in the soil. One example of this is the apparent ability of *Lumbricus rubellus* to regulate Cu, Zn, and Mn in its tissues.[13] The amount of metal within an earthworm also varies with age and over time. For example, Hartenstein et al.[18] observed a significant increase in the uptake of Cd, Cu, and Zn over time. Thus the amount of heavy metal within an earthworm may be small initially; however, as time passes and the earthworm remains in the contaminated soil, uptake will continue and the concentration of the metal within the earthworm may increase.

These invertebrates may exert a significant influence on the distribution of trace elements in soils and in food chains by altering metal concentrations. Predators of earthworms, notably birds, mammals, and amphibia, may accumulate heavy metals such as Cd, Zn, Cu, and Pb by consuming contaminated earthworms. Casts produced by earthworms living in contaminated environments may contain metals which, by passing through the alimentary tract, may become more available to plants. The decay of dead earthworms may also release heavy metals which, in turn, become available to vegetation.[6] While the relationship between terrestrial earthworms and heavy metals has been studied for many years, Hg uptake has not been extensively considered and is frequently not included in terrestrial studies which survey heavy metals.

A. MERCURY IN TERRESTRIAL ECOSYSTEMS

Many aspects of Hg contamination in terrestrial ecosystems are fundamentally different in comparison with the aquatic Hg contamination problem due to the obvious contrasts in structure and community physiognomy. In a soil profile, Hg tends to accumulate in surface horizons.[19] This accumulation could be due to the strong, but reversible, adsorption of Hg to the organic matter. Even a low humus content is enough to adsorb significant amounts of Hg.[20] Rogers[21] has shown that humus promotes the leachability of Hg and causes nonbiological methylation in the soil.

When a Hg compound is introduced into soil, the initial adsorption process is the likely determinant of the element's fate. Possible mechanisms of inorganic Hg immobilization in soil systems include: formation of insoluble forms, adsorption by soil colloids, and complexing by organic ligands. Mercury adsorption is dependent on several factors; among them the chemical form of Hg applied, the amount and chemical nature of inorganic and organic soil colloids, the type of cations on the exchange complex, the redox potential, and pH.[22-24] Mercury that is not adsorbed is eventually volatilized, precipitated, leached, or taken up by plants.[19]

The chemical form of Hg in soil is also dependent on a number of factors including salt content, pH, and composition of the soil solution. Andersson[25] relates that the chemical form of the element influences its retention and mobility in the soil and its transference to neighboring reservoirs and ecosystems. Chemical transformations occur with changing redox conditions and may be mediated by microbial activity. Lodenius et al.[26] found that a high pH value will promote the solubility of Hg, and a low pH causes a stronger binding to the organic matter, thus promoting the bioaccumulation of Hg as the organic matter is ingested by detritivores. However, they state that the possibility of mobilization of Hg from mineral soil from lowered pH still exists. Since soil pH is an important factor in the cycle of mercury in the terrestrial environment, soil acidification is of interest. The reaction of soil to acid additions is complex and dependent on numerous soil parameters such as type of clay present, base saturation, the presence of easily weatherable minerals, the ionic composition of the precipitation, and texture, kind, and amount of organic matter.[27]

B. MERCURY IN FLOOD PLAIN SOILS AT WAYNESBORO, VA

The importance of heavy metals in the metabolism of earthworms, the relative scarcity of information on Hg in terrestrial ecosystems, and the recognition of the presence of mercury in a specific local terrestrial environment stimulated our interest in the characteristics of mercury bioaccumulation by these organisms. The Hg-contaminated soil which was used in this study was obtained from the South River flood plain in Waynesboro, VA. Mercury was released into this environment by the E. I. du Pont de Nemours and Company (Du Pont) at Waynesboro, VA, while using mercuric sulfate as a catalyst in the production of acetate fibers from 1929 to 1950. Soils on the Du Pont property and sediment and water levels of the South River have been found to contain elevated levels of Hg. Initial samples of river bottom sediment taken after the discovery of mercury contamination in 1976 showed mercury levels up to 240 μg Hg g^{-1} downstream from Du Pont, in contrast to the <1 μg Hg g^{-1} concentrations upstream.[28] The present mercury contamination of the flood plain has been postulated to be the result of flooding in the aftermath of large storms, the latest of which was in November 1985. Average 100-year flood plain soil mercury concentrations for the first 25 miles downstream from the Du Pont plant were calculated to be 10 μg Hg g^{-1}.[29] High mean values of 34.5 μg Hg g^{-1} were measured at one site by Bolgiano[29] and 22.0 μg Hg g^{-1} at another by Cocking et al.[30]

We selected three contaminated sites in forest, shrubland, and an old field for study and sampled soil, plant, and animal ecosystem compartments for the presence of Hg. It was found to be extensively distributed throughout the ecosystems in highly variable amounts and some of the highest concentrations in lyophilized animal tissues were reported for earthworms (*Lumbricus* spp.). The mean of 20 samples collected from the forested Waynesboro, VA site during 1983 to 1984 was 28.1 ± 6.7 μg Hg g^{-1}dw, and samples collected from the old field site during this same period and subsequently in 1987 and 1988 ranged from means of 10.0 to 14.8 μg Hg g^{-1}dw.[30]

Soil analyses obviously can indicate whether or not a site is contaminated with a heavy metal. However, since animals and plants can differentially accumulate metal and pass it through the food chain, soil analyses alone seldom accurately reflect the potential biohazards posed by given metals. Valuable information on the extent of penetration into biotic compartments of the ecosystem could be supplied through bioaccumulation studies with animals and these data can supplement those obtained from abiotic samples. Earthworms have been identified as suitable indicator species in the decomposer food webs in the terrestrial ecosystem because they possess convenient features of large size and ubiquitous distribution. The current study looks more closely at old field data and several experiments carried out using earthworm cultures maintained in growth chambers under laboratory conditions which were designed to characterize uptake of Hg under natural and more controlled assay conditions.

II. METHODS

The primary purpose of this study was to examine the accumulation and/or concentration of Hg in the earthworms *L. terrestris* (L.) and *L. rubellus* (Hoff.), two species inhabiting different soil niches which are native to eastern North America, which have been collected from the flood plain of the South River.

Total soil Hg concentration was the major variable being examined; however, Ma[31] indicates that field-plot studies have suggested that pH, organic matter content, and cation exchange capacity of the soil also play a significant role in governing the uptake by lumbricid earthworms of trace metals other than Hg. Analysis of effects of organic matter content and cation exchange capacity are beyond the scope of this study, but a preliminary investigation of the impact of changes in pH was attempted. Bengtsson et al.[32] determined the effects of Cu, Cd, and Pb pollution in acidified soils of Sweden on the earthworm *Dendrobaena rubida* (Sav.), and reported that highest tissue metal concentrations were not always associated with the lowest soil pH, but varied with the metal and its concentration in the soil. The uptake of Hg by earthworms is also likely to be affected by soil pH, as Hg adsorption to organic and mineral fractions in the soil and solubility of metals are pH dependent.

Pooled samples of both *Lumbricus* species were found to accumulate Hg in the preliminary 1983 to 1984 study.[30] The total Hg concentration was shown to be dependent on the extent of Hg contamination in the soil. The concentration ratio, determined by dividing the mean Hg concentration in *Lumbricus* by the mean Hg concentration in collected soil, averaged 0.6 (60%) and was always less than 1 in areas of high Hg levels, indicating bioaccumulation but not bioconcentration. Earthworms from low Hg level sites had elevated tissue Hg concentration ratios, ranging from 5 (500%) to 8 (800%), indicating that bioconcentration had occurred.

A. FIELD EXPERIMENT SITE SELECTION AND TREATMENT

Two natural old field sites in use by Cocking et al.[30] were selected for the field experiment. Both are located on the 100-year flood plain of the South River, VA with deposited sediment composing a portion of the soil. A 30 × 80 m grid had been previously established at each site. The uncontaminated, or control site is located ~9 km upstream from the downtown Waynesboro, VA, Du Pont property, and is Hg pollution free, with the mean soil concentration being 0.2 ± 0.2 µg Hg g^{-1}. In April 1985, a randomized block method was utilized to assign treatments for the alteration of pH among the quadrats. At the control site, 8 of the 24 plots were limed to increase pH to achieve a value near neutrality, 8 received an addition of sulfur to decrease pH, and 8 remained untreated and comprised the mid-pH quadrats.[33]

Approximately 4 km downstream from the Du Pont property is an experimental site where Hg contamination has occurred.[30] A similar grid was established and soil Hg content was measured in the quadrats. The deposition of Hg-laden sediment within these quadrats was highly variable, ranging from background concentrations to as high as 48 µg Hg g^{-1}. The site mean soil concentration was 21 ± 11 µg Hg g^{-1}. These quadrats were treated similarly to those at the control site for alteration of pH.

No significant differences in earthworm Hg concentration related to the pH alterations were observed at either field site and therefore the data were pooled to examine the effects of soil Hg concentration as a main effect without a pH interaction. Failure to demonstrate significant differences due to soil pH alteration in the field does not mean that they were not present.

B. FIELD COLLECTION

Earthworm collection took place in May 1987 and March 1988, the earliest possible times in the spring when ground moisture was optimum for earthworm collection. Earthworms were collected near the center of each 100 m^2 quadrat. Earthworms of both *Lumbricus* species in the area were collected and composited for analysis; however, the majority of earthworms collected were *L. rubellus*. The number of worms contributing to each tissue sample varied from a few to many, depending on abundance in the quadrat. Composite samples of three soil cores in the immediate area of earthworm collection were obtained and physically mixed for concurrent analysis of mercury content and soil solution pH.

The earthworms were returned to the laboratory where they were washed with cold tap water, placed into glass finger bowls lined with Whatman #1 filter paper, and allowed to naturally void gut contents. The filter paper was changed every 12 h over a 48-h period and then the earthworms were rinsed in deionized water and frozen in 25 mL polystyrene sample vials until lyophilization. This was considered to be a sufficient time for voiding their guts since a preliminary study of earthworms collected from Hg-contaminated soil which contained initial concentrations of 7 to 9 µg Hg g^{-1}dw for uncleared worms declined to ~4 µg Hg g^{-1}dw within 24 h and did not continue to decrease in the succeeding 24-h period. In order to complete the analysis of the earthworms, the frozen worms were thawed at room temperature, pureed with 10 to 15 mL of 1.5% HNO_3, and freeze-dried using a VirTis® 10–104 LD lyophilizer for approximately 4 h. The earthworm powder was transferred to rubber-stoppered glass vials and stored in a freezer at 0° C.

C. SOIL ANALYSIS

Soil samples for Hg analysis were air-dried and pulverized. Field soils which had not received additions of lime or sulfur were sieved using a No. 100 USA Standard soil sieve to remove all stones and debris from the sample. Since non-soil material in the laboratory soils had been removed before the addition of vermiculite, these soils were not sieved. A 1:5 suspension with deionized water was prepared, vigorously shaken for 30 s, and allowed to settle for 20 min prior to pH determination.

D. MERCURY ANALYSIS

Mercury bound in the soils and in lyophilized earthworms was released into solution using a modified Knechtel and Fraser[34] wet digestion procedure, as detailed by Cocking et al.[30]

Boiling chips, soil or earthworm samples of 250 mg or 500 mg, respectively, and a vanadium pentoxide catalyst were placed in 250-mL flasks. Following the addition of 10 mL concentrated HNO_3, the flasks were heated for 5 min at 150° C. After a period of cooling, 15 mL of concentrated H_2SO_4 was added and the flasks reheated to 150° C for an additional 15 min. The solution was then diluted to 25 mL with deionized water, stored in acid-washed sample vials, and placed in a 0° C freezer until time of analysis.

A Perkin-Elmer® 3030 atomic absorption spectrophotometer (The Perkin-Elmer Corporation, Norwalk, CT) equipped with a MHS-10 mercury hydride system was used to determine the total Hg concentration of each sample. The accuracy of these determinations was verified with NBS orchard leaves, river sediment, and comparison with analyses carried out by CMI Environmental Laboratory, Mt. Sidney, VA as described in Cocking et al.[30]

E. LABORATORY EXPERIMENTS

The earthworm *L. terrestris* was used exclusively for laboratory models due to its ability to survive in artificial culture. Earthworms purchased commercially from Wilcox Bait and Tackle, Norfolk, VA were maintained in Hg-free peat humus (<1 µg Hg g^{-1}dw). Quality control analysis of tissues of individuals from these shipments indicated negligible amounts of Hg (<0.4 µg Hg g^{-1}dw) which was comparable to the quantities of Hg accumulated by earthworms at the field control site.

Plastic containers 17 cm high and 15 cm in diameter and approximately 75% filled were used to contain cultures for the laboratory experiments. The composition of the soil, soil pH, soil Hg concentration, duration of the experiment, and culture conditions were altered to address various questions. The details of the culture conditions will be described along with the results obtained from the individual studies.

III. RESULTS

A. EARTHWORM TISSUE Hg CONCENTRATIONS IN THE FIELD

As indicated in Section II, there were no significant differences in earthworm tissue Hg concentration due to soil pH in the field study, Experiment I. There was, however, a highly significant difference ($p > 0.0001$, $F_{(1,36)} = 119.98$) between earthworm tissue concentrations in composite samples of organisms collected from individual quadrats at the high Hg site in comparison with those at the low level upstream control location. The direct relationship between earthworm tissue Hg and the soil Hg concentration from which the organisms were collected is illustrated in Figure 1a. The difference between bioconcentration in very low Hg content soils and bioaccumulation in the more contaminated soils is evident in Figure 1b.

B. EARTHWORM TISSUE Hg CONCENTRATIONS IN
CULTURES OF DIFFERING SOIL Hg CONTENT

Soil obtained from the Hg-contaminated flood plain field site on the South River near Waynesboro, VA was mixed and sifted through a screen to obtain a fine, homogeneous sample. Different proportions of this soil were combined with vermiculite-rich potting soil to obtain dry soil mixtures with a gradient of concentrations. For Experiment II, each concentration was replicated in two culture containers. Soil samples were taken from each replicate for mercury analysis, the soil was wetted with 300 mL of distilled water, and 20 earthworms (*Lumbricus* sp.) were placed into each container. Forty worms were randomly selected and divided into ten groups for analysis of baseline mercury levels and served as controls. Each container was placed inside a Warren Sherer® Model CEL 36–10 growth chamber set at 10° C with continuous light.

The earthworms were harvested after 30 days, and then 3 weeks later the experiment was replicated with a new group of earthworms placed in the previously used soils. Pre-experiment control earthworm tissues for the first group of animals initially contained

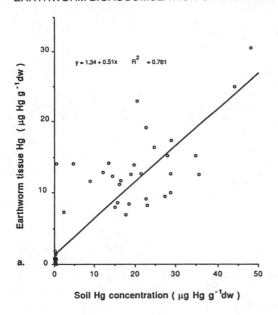

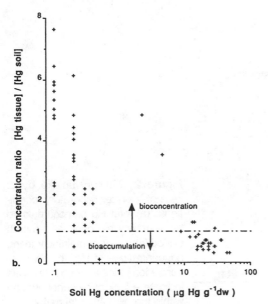

Figure 1 Mercury uptake by earthworms under field conditions: (a) total Hg in pooled *L. rubellus* and *L. terrestris* tissue following 48-h gut purging as a function of the total Hg concentration in the soil from which they were collected; (b) the ratio of tissue Hg to soil Hg indicating bioconcentration from low ambient Hg environments.

traces of Hg (<0.2 µg Hg g^{-1}dw) and those for the second had only slightly more (<0.4 µg Hg g^{-1}dw). The soil Hg concentrations in each culture at the beginning of the experiments differed from that at the end of the 3-month period. A linear model was used to estimate the mean soil concentration for each culture at the midpoint of time during each run, and those values are used as the soil concentrations for Experiment II in Figure 2a. The ''fresh soil only'' curve is based on the first run of the experiment and the ''all cultures'' curve utilizes all of the data from both experiments. Figure 2b presents the concentration ratios obtained from each culture, which were virtually all below 1, and indicates that only bioaccumulation was occurring.

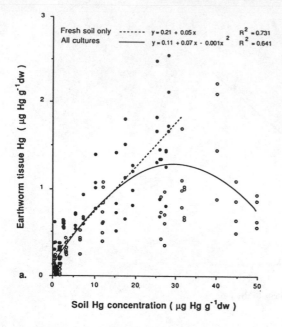

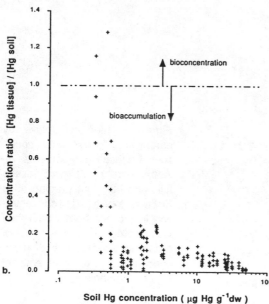

Figure 2 Mercury uptake by *L. terrestris* under laboratory conditions: (a) total Hg bioaccumulated by earthworms over a 1-month period from soils containing varying concentrations of Hg; (b) low concentration ratios at all soil Hg levels indicate a lack of bioconcentration within the experimental period.

C. EARTHWORM Hg UPTAKE OVER TIME IN CULTURES WITH HIGH Hg SOIL

Sifted soil collected from the old field Hg-contaminated site ($\sim$21 μg Hg g^{-1}) was used for three sets of cultures, Experiment III, that were sampled at different times after the introduction of Hg-free earthworms ($<$0.2 μg Hg g^{-1}dw). No supplementary potting soil was added. In each case, a large group of earthworms (30 to 40 *L. terrestris*) were placed in the culture soils on day 0 and then samples of 5 individuals were removed on specified days. The cultures were monitored during the experiments and kept moist while held at 10 to 12° C in a 12-h photoperiod. All dead individuals were removed and discarded.

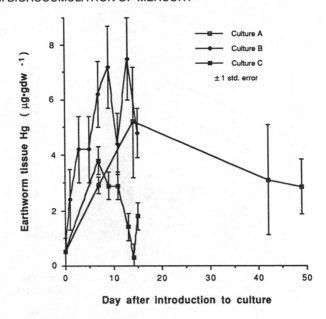

Figure 3 Uptake of mercury by previously uncontaminated adult *L. terrestris* during a period of 1 to 2 weeks following introduction to laboratory culture soils containing 21 µg Hg g⁻¹.

    The results of these experiments are summarized in Figure 3. Culture A included nine containers. Three were acidified and three were neutralized in the same manner as the original field study. Again, no differences significant at the $\geq 0.05\%$ level in Hg uptake by earthworms living in different pH levels were observed and the data were pooled. In Culture B, earthworms were placed into ten replicates of the Hg-contaminated soil (pH of 7), and Culture C was a repeat introduction into the same soils used in Culture B. Both Cultures A and B reflected an increase in earthworm tissue Hg content over the first 2 weeks of the experiments, which was dependent on the length of time the organisms were exposed to Hg-contaminated soil. The longer-term Culture A reflected little additional uptake after this period and Culture B also appeared to no longer be accumulating additional Hg after this initial period. Culture C reflected uptake during the first week and then showed a decline in earthworm tissue Hg content thereafter.

D. THE EFFECT OF SOIL pH ON EARTHWORM Hg UPTAKE IN CULTURE

A more complex laboratory study, Experiment IV, was formulated which would include the factors of pH and soil mercury content over a longer period of time. Contaminated soils from the high mercury and control old field sites were collected, homogenized, and mixed in a ratio of two parts vermiculite to three parts soil to increase friability. The original intent was to mix the soils in order to obtain an intermediate soil Hg level. However, the concentrations attained for intermediate and high Hg (28.6 and 33.1 µg Hg g⁻¹) were so close that the data are pooled from both sets of containers as a high Hg soil condition for comparison with the low Hg soils (0.3 µg Hg g⁻¹). High and low Hg cultures were placed at 10 to 12° C in complete darkness for the duration of the experiment. The pH of the soils was adjusted in the same manner as the field study to attain slightly acidic (pH 5.9 to 6) and neutral (pH 6.5 to 7) cultures.

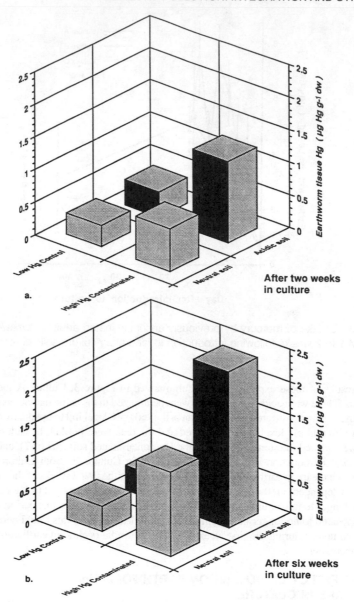

Figure 4 Enhanced mercury uptake by previously uncontaminated adult *L. terrestris* from pH 5.9 to 6 laboratory culture soils in comparison with that from cultures with more neutral (pH 6.5 to 7) conditions after a period of (a) 2 weeks and (b) 6 weeks following introduction into Hg-contaminated soils.

The initial earthworm tissue Hg concentrations were <0.4 µg Hg g^{-1}dw and 44 individuals were introduced per container on Day 0. Sampling of three earthworms per container took place over a 6-week period. The individual daily samples were highly variable and the sample size was not sufficient for statistically significant time course analysis. However, when samples collected between Days 11 and 19 and Days 38 and 45, roughly 2 and 6 weeks, respectively, were pooled, a picture of a statistically significant general uptake pattern emerged, as shown in Figure 4. An ANOVA of these pooled data indicates significant main effects for soil Hg concentration ($p > 0.0001$, $F_{(1,120)} = 41.04$), soil acidity ($p > 0.007$, $F_{(1,120)} = 7.55$), and time in culture ($p > 0.005$, $F_{(1,120)} = 8.21$) as well as significant interactions of soil Hg concentration with both pH ($p > 0.01$, $F_{(1,120)} = 6.52$) and time ($p > 0.008$, $F_{(1,120)} = 7.34$). In this study, not only did the uptake by earthworms appear to be dependent on soil Hg concentration, but the individuals living in the more acidic conditions were accumulating greater amounts of Hg than those in neutral soils.

E. EARTHWORM MORTALITY

While this study was not directed toward evaluating Hg toxicity, survival and appearance of the worms remaining in cultures were observed throughout the laboratory experiments. In general, prolonged maintenance of cultures led to the degeneration of conditions and earthworms died. This is not unexpected since soils were not replaced during the experiments and additional organic matter for food was not added. No significant patterns emerged indicating that Hg toxicity was a primary factor. For example, survival expressed as a percent of zero mortality on Day 38 of Experiment IV in soils with near neutral pH was 41% in the low Hg cultures and 41.4% in the high Hg cultures. Corresponding values in pH 5.9 to 6 cultures were 35 and 31.9%, indicating that acidity, or the elemental sulfur addition, was more detrimental than Hg. Thirty-day survival rates in Experiment II also showed no correlation with Hg concentration and averaged 83.8% when fresh soils were used and 57% when they were reused for the second replication in time. Quantitative population data were not obtained in the field Experiment I, but densities may have been lower in the sulfur-supplemented low pH soils.

IV. DISCUSSION

This study confirms previous observations, such as those of Helmke et al.[35] and Siegel et al.,[36] that earthworms bioaccumulate Hg from contaminated environments. This is consistent with research on other heavy metals in which relatively large amounts are accumulated by various species of earthworms.[11–14,18,31,32,37–39] As shown with other heavy metals, soil and earthworm tissue Hg content were positively correlated under both field and laboratory culture conditions. However, in the laboratory, the relationship is not overriding and may be disrupted by interactions such as the reuse of culture soils, which may have altered organic content, build-up of metabolic byproducts, or other alterations to conditions. There was considerable year to year and within-site variation in the tissue content of earthworms living under field conditions. While not examined systematically, it is likely that seasonal variations are also present. These factors all complicate the potential use of field collected earthworms as a quantitative index of contamination in situ.

Quantities of Hg accumulated by the earthworms were considerably less in laboratory studies than those observed in the field. One of the primary factors for this may have been the shorter period of exposure to the Hg under culture conditions. Another potential difference in the field is that all developmental stages were exposed to Hg rather than just the mature individuals. Furthermore, genetic selection is likely to have occurred in the field populations and those individuals are likely to be more tolerant of Hg and have possibly acquired physiological mechanisms for sequestering accumulated Hg.

The use of field collected soils to prepare laboratory cultures as a means of assessment of Hg contamination is also potentially affected by sources of variability such as organic matter content, presence of interacting ions, differences in the forms of Hg present, accumulation of other toxins, and environmental conditions during culture such as photoperiod, soil pH, temperature, and moisture availability. Some of these factors, particularly pH and organic content, are difficult to reproduce and maintain over the several weeks necessary to obtain Hg uptake. This makes the development of a satisfactory repeatable laboratory assay very elusive. It is much easier to determine how much Hg of a particular form is toxic to earthworms, than to use earthworms to assay the environment. For example, Mac et al.[40] attempted to develop a 10-d earthworm bioaccumulation bioassay for Hg, Zn, and PCBs in freshwater sediments. It was successful for PCBs, but no Hg or Zn uptake from contaminated materials was observed. In contrast, van Gestel et al.[41] used 28-d and 45-d earthworm bioassays to evaluate bioavailability of Hg in contaminated soils and observed uptake. However, information about statistical variability and details about processing earthworm tissues were not included in the report.

Concentration ratios indicate whether bioconcentration above ambient concentrations is occurring through the process of biomagnification. Generally, heavy metal concentration factors tend to decline with increasing levels of soil contamination.[31] This has also been observed in this study, and for the most part, only bioaccumulation was observed. However, the fact that earthworm bioconcentration can apparently occur from soils with very low Hg concentrations has a potential impact on ecosystems if global ambient atmospheric deposition rates increase. This would be a potential entry pathway into terrestrial food webs. For example, Ma and Broekhuizen[42] studied badgers (*Meles meles*) in the Meuse river watershed of the Netherlands. They attribute the high Hg (up to 3.6 µg Hg g^{-1}dw) and other heavy metal levels in kidney cortex of individuals, in part, to the high frequency of earthworms in their diet. This could also be the source of the high 17.8 to 29.8 µg Hg g^{-1}dw concentrations in short-tailed shrew (*Blarina brevicauda*) lyophilized kidney of individuals reported by Cocking et al.[30] from the Waynesboro, VA sites.

Fleming and Richards[43] distinguished between absorbed metals and those which have been "internalized" or incorporated within the tissues. They suggest that to obtain accurate heavy metal absorption rates this distinction must be made, particularly in the case of animals of small size where the surface area:volume ratio is high and the practicalities of isolating surface-bound extracellular compounds are difficult to overcome. They concluded that the supracuticular mucoid coat of *Eisenia foetida*, derived from epidermal mucous cells, contains traces of many naturally occurring heavy metals. They also showed that 68% of total body wall cadmium in *L. rubellus* incubated under high cadmium conditions for 26 d was surface-bound. This indicates that the earthworm's mucoid coat obstructs heavy metal internalization. Many questions, such as where Hg is distributed, remain to be answered more fully. This internalized and adsorbed partitioning of Hg, for example, may have an impact on the use of concentration ratios for comparisons of bioavailability. However, to the predator of the earthworm, this distinction is of no consequence, since all mercury associated with the earthworm is taken into the body and potentially assimilated. The mole, *Talpa europea*, favors the earthworm for its food source and stores them after biting off several anterior segments, preventing escape.[3] Ma,[44] using earthworms and moles as indicators of metal bioavailability in terrestrial environments, has found that the accumulation of cadmium, lead, and zinc in moles reflects the bioavailability of the metals in the earthworms, and does not necessarily reflect the metal concentration in the soil. Critical levels of cadmium toxicity in moles were exceeded in soils which had relatively low concentrations of the metal and low soil pH.

Another area of physiological research which bears on this study is the determination of reasons that earthworms are apparently able to live under conditions with high soil Hg concentrations. The chemical form of the Hg certainly plays a role and this was not investigated in our study. Beyer et al.[45] found that additions of MeHg of 25 ppm (µg Hg g^{-1}) were toxic to earthworms and that 85 ppm was bioconcentrated in tissues when soil concentrations were on the order of 5 ppm. Methylation of inorganic Hg in soils increases toxicity and alters the bioavailability. This would subsequently have major impact on transfers through the food chains of the ecosystem.

Ma[31] has indicated that soil pH is a prime factor for predicting the uptake and accumulation of Cd and Zn as well as Pb. It was concluded that lowering the pH leads to increased desorption of metal cations due to competition with H$^+$ ions, and that soil pH is the most important single factor determining the solubility of Zn in the soil solution. In another study investigating the influence of soil pH and organic matter on the uptake of Cd, Zn, Pb, and Cu by earthworms, Ma[38] found that the lowering of soil pH strongly promotes the uptake of the metals Cd, Zn, and Pb, at a mean soil pH of 4.6, with soil pH in the study ranging from 3.5 to 6.1. Multiple regression analysis was used to determine the relative importance of soil pH and organic matter (mean of 5.0%, range of 2.2 to 8.6%) as affecting metal uptake by earthworms. This analysis showed the best fit equations for Cd and Zn were described by a log transformation with soil pH as the only factor. Similarly, the organic matter content of the soil played a significant role only in the earthworm uptake of Pb. Preliminary results of this study indicate that reductions in soil pH may also increase Hg bioaccumulation by earthworms. However, complex impacts of increased soil acidity on the health of earthworms which would interact with effects on availability leave this topic as one requiring extensive study before any generalizations can be made. If, however, earthworm tissue Hg concentrations do increase with modest acidification, then interactions of the Hg cycle with acid rain and the S cycle would be involved, and this may prove to be one of the important points of focus for the study of Hg in terrestrial environments.

ACKNOWLEDGMENTS

The authors wish to thank Peter Nielsen, who helped in the set-up of the analytical protocol, and Nicholas Mastrota, Ronald Thomas, Jane Walker, and Deanna Ward, who have also contributed to this study. We are likewise appreciative of the support of James Madison University, which was the sole source of physical and financial resources for this project.

REFERENCES

1. Satchell, J. E., Some aspects of earthworm ecology, In: *Soil Zoology,* Kevan, D. K. McE., Ed., Butterworths, London, 1955, 180.
2. Satchell, J. E., Lumbricidae, In: *Soil Biology,* Burgess, A. and Raw, F., Academic Press, London, 1967, 259.
3. Edwards, C. A. and Lofty, J. R., *Biology of Earthworms,* Chapman and Hall, London, 1972.
4. Lee, K. E., *Earthworms. Their Ecology and Relationships with Soils and Land Use,* Academic Press, Orlando, FL, 1985.
5. Ma, W.-C., *Regenwormen als bioindicators van bodemverontreiniging,* Bodembeschermingsreeks nr. 15,1, Staatsuitgeverij, The Hague, 1983.
6. Ireland, M. P., Heavy metal uptake and tissue distribution in earthworms, In: Satchell, J. E., Ed., *Earthworm Ecology: From Darwin to Vermiculture,* Chapman and Hall, New York, 1983, 247.
7. Roberts, B. L. and Dorough, H. W., Relative toxicities of chemicals to the earthworm *Eisenia foetida, Environ. Toxicol. Chem.,* 3, 67, 1984.
8. OCED, Guideline for Testing of Chemicals, no. 207, Earthworm, Acute Toxicity Test. Adopted 4 April, 1984.
9. EEC, EEC Directive 79/831, Annex V, Part C. Methods for the Determination of Ecotoxicity, Level I, C(II)4: Toxicity for Earthworms. Artificial soil test. DG XI/128/82 Final, 1985.
10. van Gestel, C. A. M., van Dis, W. A., Dirven-van Breemen, E. M., Sparenburg, P. M., and Baerselman, R., Influence of cadmium, copper, and pentachlorophenol on growth and sexual development of *Eisenia andrei* (Oligochaeta; Annelida); *Biol. Fertil. Soils,* 12, 117, 1991.
11. van Rhee, J. A., Effects of soil pollution on earthworms, *Pedobiologia,* 17, 201, 1977.

12. Ireland, M. P. and Richards, K. S., The occurrence and localization of heavy metals and glycogen in the earthworms *Lubricus rubellus* and *Dendrobaena rubida* from a heavy metal site, *Histochemistry*, 51, 153, 1977.

13. Ireland, M. P., Metal accumulation by the earthworms *Lubricus rubellus, Dendrobaena veneta* and *Eiseniella tetraedra* living in heavy metal polluted sites, *Environ. Pollut.*, 19, 201, 1979.

14. van Hook, R. I., Cadmium, lead, and zinc distributions between earthworms and soils: potentials for biological accumulation, *Bull. Environ. Contam. Toxicol.*, 12, 509, 1974.

15. Gish, C. D. and Christensen, R. E., Cadmium, nickel, lead, and zinc in earthworms from roadside soil, *Environ. Sci. Technol.*, 7, 1060, 1973.

16. Ireland, M. P., Metal content of *Dendrobena rubida* (Oligochaeta) in a base metal mining area, *Oikos*, 26, 74, 1975.

17. Ireland, M. P. and Richards, K. S., Glycogen-lead relationship in the earthworm *Dendrobaena rubida* from a heavy metal site, *Histochemistry*, 56, 55, 1978.

18. Hartenstein, R., Neuhauser, E. F., and Collier, J., Accumulation of heavy metals in the earthworm, *Eisenia foetida, J. Environ. Qual.*, 9(1), 23, 1980.

19. Adriano, D. C., *Trace Elements in the Terrestrial Environment*, Springer-Verlag, New York, 1986, 298.

20. Lodenius, M., Seppanen, A., and Autio, S., 1987. Sorption of mercury in soils with different humus content, *Environ. Contam. Toxicol.*, 39, 593, 1987.

21. Rogers, R. D., Methylation of mercury in agricultural soils, *J. Environ. Qual.* 5(4), 454, 1976.

22. Hogg, T. J., Stewart, J. W. B., and Bettany, J. R., Influence of chemical form of mercury on its adsorption and ability to leach through soils, *J. Environ. Qual.*, 7, 440, 1978.

23. Aomine, S. and Inoue, K., Retention of mercury by soil. II. Adsorption of phenylmercuric acetate by soil colloids, *Soil Sci. Plant Nutr.*, 13, 195, 1967.

24. Inoue, K. and Aomine, S., Retention of mercury by soil. III. Adsorption of mercury in dilute phenylmercuric acetate solutions, *Soil Sci. Plant Nutr.*, 15, 86, 1969.

25. Andersson, A., Mercury in soils, in Nriagu, J. O., Ed., *The Biogeochemistry of Mercury in the Environment*, Elsevier/North Holland, Amsterdam, 1979, 79.

26. Lodenius, M., Seppanen, A., and Uusi-Rauva, A., Sorption and mobilization of mercury in peat soil, *Chemosphere*, 12, 1575, 1983.

27. Wiklander, L., The acidification of soil by acid precipitation, *Grundforbattring*, 26, 155, 1973.

28. Bolgiano, R. W., Mercury Contamination of the South, South Fork Shenandoah, and Shenandoah Rivers, Commonwealth of Virginia State Water Control Board, Richmond, VA. Basic Data Bulletin No. 47, 1980.

29. Bolgiano, R. W., Mercury Contamination of the Flood Plains of the South and South Fork Shenandoah Rivers, Commonwealth of Virginia State Water Control Board, Richmond, VA. Basic Data Bulletin No. 48, 1981.

30. Cocking, D., Hayes, R., King, M. L., Rohrer, M. J., Thomas, R., and Ward, D., Compartmentalization of mercury in biotic components of terrestrial flood plain ecosystems adjacent to the South River at Waynesboro, VA, *Water Air Soil Pollut.*, 57–58, 159, 1991.

31. Ma, W.-C., The influence of soil properties and worm-related factors on the concentration of heavy metals in earthworms, *Pedobiologia*, 24, 109, 1982.

32. Bengtsson, G., Gunnarsson, T., and Rundgren, S., Effects of metal pollution on the earthworm *Dendrobaena rubida* (Sav.) in acidified soils, *Water Air Soil Pollut.*, 28, 361, 1986.

33. Thomas, R. L., Effects of pH Alteration on the Accumulation of Mercury By Vegetation in South River, VA, flood plain soil. MS Thesis (unpublished), James Madison University, Harrisonburg, VA, 1987.

34. Knechtel, J. R. and Fraser, J. L., Wet digestion method for the determination of mercury in biological and environmental samples, *Anal. Chem.*, 51(2), 315, 1979.

35. Helmke, P. A., Robarg, W. P., Korotev, R. L., and Schomberg, P. J., Effects of soil-applied sewage sludge on concentrations of elements in earthworms, *J. Environ. Qual.,* 8, 322, 1979.

36. Siegel, S. M., Siegel, B. Z., Puerner, N., and Speitel, T., Water and soil biotic relations in mercury distribution, *Water Air Soil Pollut.,* 4, 9, 1975.

37. Beyer, W. N., Hensler, G., and Moore, J., Relation of pH and other soil variables to concentrations of Pb, Cu, Zn, Cd, and Se in earthworms, *Pedobiologia,* 30, 167, 1987.

38. Ma, W.-C., Edelman, T., van Beersum, I., and Jans, T., Uptake of cadmium, zinc, lead, and copper by earthworms near a zinc-smelting complex. Influence of soil pH and organic matter, *Bull. Environ. Contam. Toxicol.,* 30, 424, 1983.

39. Morgan, J. E. and Morgan, A. J., Differences in the accumulated metal concentrations in two epigeic earthworm species (*Lumbricus rubellus* and *Dendrodrilus rubidus*) living in contaminated soils, *Bull. Environ. Contam. Toxicol.,* 47, 296, 1991.

40. Mac, M. J., Nocughi, G. E., Hesselberg, R. J., Edsall, C. C., Shoesmith, J. A., and Bowker, J. D., A bioaccumulation bioassay for freshwater sediments, *Environ. Toxicol. Chem.,* 9, 1405, 1990.

41. van Gestel, C. A. M., Adema, D. M. M., de Boer, J. L. M., and de Jong, P., The influence of soil clean-up on the bioavailability of metals, In: Wolf, K., van den Brink, W. J., and Colon, F. J., Eds., *Contaminated Soil '88,* Kluwer Academic Publishers, Norwell, MA, 1988, 63.

42. Ma, W.-C. and Broekhuizen, S., Belasting van dassen *Meles meles* met zware metalen: invloed van de verontreinigde massuiterwaarden? (Possible influence of the polluted fore-lands of the river Meuse on the heavy-metal burden of badgers *Meles meles* in the Netherlands), *Lutra,* 32, 139, 1989.

43. Fleming, T. P. and Richards, K. S., Localization of adsorbed heavy metals on the earthworm body surface and their retrieval by chelation, *Pedobiologia,* 23, 415, 1982.

44. Ma, W.-C., Heavy metal accumulation in the mole, *Talpa europea,* and earthworms as an indicator of metal bioavailability in terrestrial environments, *Bull. Environ. Contam. Toxicol.,* 39, 933, 1987.

45. Beyer, W. N., Cromartie, E., and Moment, G. B., Accumulation of methylmercury in the earthworm *Eisenia foetida,* and its effect on regeneration, *Bull. Environ. Contam. Toxicol.,* 35, 157, 1985.

Mercury Concentration in Perch (*Perca fluviatilis*) as Influenced by Lacustrine Physical and Chemical Factors in Two Regions of Russia

Terry A. Haines, Victor T. Komov, and Charles H. Jagoe

CONTENTS

ABSTRACT: We conducted field studies of lakes in Darwin National Reserve in 1989, and in Karelia in 1991. Darwin National Reserve is 300 km north of Moscow at the west end of Rybinsk Reservoir. The terrain is generally flat and consists of thick sandy till covered with hardwood forest. Karelia is north of St. Petersburg and consists of hilly terrain underlain with Precambrian bedrock. In each case, remote lakes were visited and sampled for water and fish, primarily perch *Perca fluviatilis*. Water samples were analyzed for pH, color, specific conductance, and major cations and anions. Fish were weighed and measured, and dorsal muscle tissue was collected and analyzed for mercury. In both locations acidic lakes (acid-neutralizing capacity <0) were common. Acidic lakes were both clear and colored and the dominant anion in both types was sulfate, indicating that the lakes were acidic because of atmospheric deposition of strong acids and not because of organic acids. Mercury content of fish ranged from 0.06 to 3.04 µg/g wet weight and was highest in acidic and in colored lakes. Fish from high-pH lakes were low in mercury regardless of lake color. Mercury content in fish from low-pH lakes varied widely and was highest in fish from colored, seepage lakes. Regressions including lake pH and color, separated by drainage type, explained about 70% of the variance in fish mercury content. Mercury seems to enter the lakes by atmospheric deposition, as there are no local sources. Organic acids are believed to increase mercury bioavailability in lakes by transporting mercury from terrestrial regions and possibly contributing to methylation. Also, mercury methylation is probably enhanced in acidic lakes, increasing bioavailability.

I. INTRODUCTION

Mercury contamination of fish in waters of low acid-neutralizing capacity is a widespread problem in northern Europe and North America.[1,2] Concentrations in fish are apparently increasing,[1,3] as are atmospheric concentrations.[4] Further, some evidence indicates that lake acidification increases mercury concentration in fish. This evidence includes correlative studies,[5–7] laboratory studies,[8] and lake manipulation studies.[9] The mechanisms by which acidification affects mercury accumulation by fish, however, remains controversial. For example, in Sweden mercury accumulation in fish is believed to be controlled primarily by lake productivity and humic content, and the correlation between fish mercury content and lake pH may result from the covariation of pH with other environmental variables.[1,10]

Relatively little is known concerning environmental contamination in the former Soviet Union, but available information indicates serious environmental degradation.[11] Atmospheric deposition of sulfur compounds in Europe has been estimated with a model developed by the Cooperative Programme for Monitoring and Evaluation of the Long Range Transmission of Air Pollutants in Europe (EMEP). This model[12] estimates an annual deposition rate of 800 to 1,640 eq S/ha for northern Russia, which is comparable to or exceeds that for regions in Scandinavia where lake acidification and contamination of fish with mercury have occurred.

In undertaking this study, our first objective was to confirm the previously reported presence of clear- and brown-water acidic lakes in two areas of Russia, one containing seepage lakes located in till, and one containing primarily drainage lakes on bedrock. Our second objective was to determine whether fish inhabiting these lakes contained elevated levels of mercury. We attempted to sample lakes with a wide range of physical and chemical conditions (acidic to neutral pH, clear to brown color, seepage and drainage hydrology) to ascertain how these conditions affected mercury burdens in fish.

II. METHODS

Sampling was conducted in two phases. In May and June 1989, lakes in Darwin National Reserve, Yaroslavl Region, were sampled, and in June 1991 lakes in the Karelia Autonomous Republic were sampled (Figure 1). Darwin National Reserve is a protected natural area that was created in 1943. It is located at the western end of the Rybinsk Reservoir and contains thick tills and sandy soils.[13] In Karelia, we visited lakes in the vicinity of Suoyarvi, which were primarily highly colored lakes, and Kondopoga, which were primarily clear-water lakes. Both areas were underlain by granitic bedrock covered with thin soil.[13] Most lakes were in forested catchments without dwellings or roads. An exception was Suoyarvi lake, which has a settlement at the southern end.

At each lake, we collected a water sample by immersion of cleaned polyethylene bottles just below the surface near the point of maximum depth. Bottles were tightly capped, held in the dark, and within a few hours analyzed for pH (Orion Model SA210 meter equipped with a Ross combination electrode), acid-neutralizing capacity (ANC, measured by inflection point titration), and apparent color (visual comparison with platinum-cobalt standards). The remaining water sample from each lake was refrigerated and later analyzed for major cations and anions. Calcium, magnesium, and sodium were determined by flame atomic absorption spectrophotometry; chloride, fluoride, sulfate, and nitrate by ion chromatography; and total aluminum by graphite furnace atomic absorption spectrophotometry.

Perch (*Perca fluviatilis*) were collected by angling from various points around the shoreline of each lake, except for Suoyarvi lake, where fish were captured with a variable-mesh gill net. Fish were placed in plastic bags and frozen ($-5°$ C) within a few hours; they were kept frozen until analyzed for mercury. Fish were transported to the Institute for Biology of Inland Waters, Borok, in 1989, and the Institute for Research on Northern Fishes, Petrozavodsk, in 1991. The fish were thawed, weighed to the nearest gram, and measured (total length) to the nearest millimeter. Scales were removed for age determination but these later

Figure 1 Map of Europe showing location of the sampling sites.

proved to be unreadable. Each fish was placed on an acid-washed Plexiglass® plate, and a 2- to 3-g sample of skeletal muscle was dissected from the left side, adjacent to the dorsal fin and extending from the rib cage to the center of the dorsal surface. All dissecting instruments were stainless steel, and all instruments and glassware were cleaned with 10% nitric acid and rinsed with distilled water. Moisture content was determined by placing a portion of the tissue in a dried, tared glass dish and weighing before and after drying at 105° C. The remainder, typically 1 to 2 g, was placed in a tared glass beaker, weighed, and digested in a 1:1 mixture of nitric acid and hydrogen peroxide.[14]

Fish were processed in batches of 5 to 10. Two fish from each batch were randomly selected for use in estimation of the accuracy and precision of mercury determinations. For these fish, the tissue was divided into two nearly equal portions. One portion of tissue was spiked with a known amount of inorganic mercury, and all portions were digested and analyzed. A sample of National Institute of Science and Technology (NIST) tuna reference material and a reagent blank were included with each batch of samples. After digestion, solutions were cooled and diluted to 25 mL with distilled water, and 1- to 2-mL portions were analyzed for total mercury with a gold film analyzer (Arizona Instrument Company, Jerome, Arizona, U.S.). Each digestate was analyzed in duplicate. Mercury in all samples exceeded our calculated detection limit of 5 ng. The recovery of mercury from spiked samples averaged 95.4% (range 90 to 102%). The percent difference between replicate samples averaged 2.9% (range 1.7 to 5.2%). The measured mercury content of NIST tuna averaged 0.93 μg/g (range 0.88 to 0.96 μg/g), and all analyses were within the certified range for this material.

Statistical analyses were performed with SAS/STAT computer programs (SAS Institute, Cary, North Carolina, U.S.). Mercury data were transformed according to the function $\log_{10}(Hg + 1)$ to approximate a normal distribution. Other variables were transformed with an appropriate power function if the distribution was non-normal. Preliminary Pearson product-moment correlations indicated that fish mercury concentration was significantly correlated with fish length and especially weight ($r = 0.32$, $p = 0.0001$), and that lake pH was significantly correlated ($p < 0.05$) with all lacustrine variables except color and sulfate. Fish mercury data were subsequently analyzed with a one-way analysis of covariance with weight as a covariate. Least-square means of fish mercury concentration by lake, which

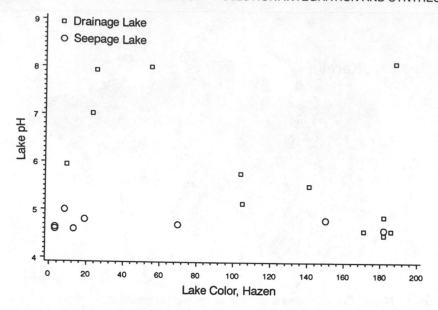

Figure 2   Distribution of the study lakes by pH, color, and drainage type.

represented mercury concentration normalized by fish weight, were then analyzed by multiple regression against lake pH, color, and sulfate. All statistical calculations were performed with least-square mean mercury concentrations.

III. RESULTS

The lakes surveyed varied in pH from 4.6 to 8.1 and in color from 3 to 188 Hazen (Figure 2, Table 1). The combined samples of drainage and seepage lakes were relatively uniformly distributed across the range of color values, whereas all seepage lakes surveyed were acidic (pH $\leq$ 5). Of the chemical and physical factors measured, drainage and seepage lakes differed significantly (Mann-Whitney U-test, $p < 0.05$) in pH, ANC, calcium, magnesium, and sodium (Table 2). Lakes in the Yaroslavl region were predominately seepage lakes, whereas the lakes in Karelia were predominantly drainage lakes.

Fish collected ranged in weight from 8 to 515 g and in mercury content from 0.06 to 3.04 μg/g wet weight (0.76 to 17.9 μg/g dry weight; Table 3). If two unusually large fish were omitted from the Tyomnoye lake sample, the maximum weight was 196 g and the maximum mercury content was 1.03 μg/g wet weight (6.87 μg/g dry weight).

Fish from seepage lakes had significantly higher mercury content than fish from drainage lakes (t-test, $p = 0.01$), and mercury concentration in fish was inversely related to lake pH (Figure 3). Mercury concentrations were low in all high-pH lakes (pH >5.5); however, there was considerable variation in mercury concentration in fish among low-pH lakes, especially lakes with pH <5 (Figure 3). Inspection of the relation of fish mercury content to lake pH and apparent color indicated that lake color may have been an important additional factor, and that the relation to pH differed between seepage and drainage lakes (Figure 4). Accordingly, stepwise regression analysis was performed with the nonintercorrelated variables pH, color, and sulfate, with the data stratified by lake drainage type. Color was more important than pH in the regression for seepage lakes (Table 4). Moreover, the relation between fish mercury and lake pH was not significant for drainage lakes. The inclusion of lake sulfate did not significantly improve the regression for seepage lakes.

Table 1   **Physical and chemical characteristics of the study lakes.**

Lakes	Area (ha)	Depth (m)	Drainage type	Color (Hazen)	pH	ANC (μeq/L)	SO₄ (μeq/L)
Darwin Lakes							
Dorojiv	200	3	Seepage	13	4.6	−41	71
Dubrovskoye	20	2	Seepage	182	4.6	−50	64
Hotavets	160	3	Drainage	188	8.1	227	32
Motykino	2	4	Seepage	19	4.8	−38	49
Rybinsk	395,000	30	Impoundment	55	8.0	1,369	420
Tyomnoye	20	2	Seepage	70	4.7	−53	52
Uteshkovo	5	3	Seepage	150	4.8	−33	29
Karelia Lakes							
Blue Lamba	307	4	Seepage	3	4.6	−17	67
Chuchyarvi	112	5	Seepage	8	5.0	−9	48
Grushna Lamba	3	5	Seepage	3	4.6	−19	48
Ilyakalkenyarvi	104	6	Drainage	171	4.6	−25	54
Kabozero	210	3	Drainage	141	5.5	13	60
Lamba Vegarous	7	4	Drainage	182	4.5	−39	75
Leukunyarvi	—	—	Drainage	182	4.9	−9	68
Sargozero	200	4	Drainage	25	7.9	252	29
Suoyarvi	6,070	5	Drainage	104	5.8	23	83
Uros	426	3	Drainage	9	5.9	16	72
Vegarousyarvi	1,880	5	Drainage	105	5.1	−2	49
Venderskoye	998	6	Drainage	23	7.0	167	44
Vuontelenyarvi	394	3	Drainage	186	4.6	−24	68
Mean values							
Drainage Lakes[a]	1,045	4.2	Drainage	120	5.8	56	58
Seepage Lakes	84	3.5	Seepage	56	4.7	−33	53

[a] Rybinsk Reservoir omitted.

IV. DISCUSSION

The observation of elevated mercury content in fish inhabiting waters with low pH has been widely reported.[1,2] Our results confirm these findings and extend them to two regions of Russia not previously investigated. However, the phenomenon is much more complex than a simple lake acidity-fish mercury relation. An extensive study of mercury accumulation by fish in Swedish forest lakes led to the conclusion that concentrations were determined by the bioavailability of mercury to lower trophic levels, and that lake humicity, productivity, and acidity controlled the bioavailability of the metal.[1] Our results indicate that acidity and humicity are important factors in affecting fish mercury content in Russian lakes (we did not investigate productivity). Concentrations of mercury in fish from high-pH (pH >5.5) lakes were low regardless of color, indicating that acidity was necessary—but not sufficient—to enhance mercury accumulation by fish in these lakes. In low-pH lakes, mercury content in fish was higher in the more colored lakes, but the regressions differed between seepage and drainage lakes indicating that hydrologic character is an important variable.

The mercury content of perch from remote lakes in Russia was comparable to that reported for this species from similar lakes in Sweden and Finland, disregarding the two large fish from Tyomnoye Lake that had unusually high mercury concentrations. In Swedish forest lakes, mercury concentrations were 0.2 to 2.0 μg/g dry weight in "small" perch and 0.6 to

Table 2   **Mean, median, and range of lake physical and chemical variables.**

Variable	All lakes (N = 20)			Drainage lakes (N = 12)			Seepage lakes (N = 8)		
	Mean	Median	Range	Mean	Median	Range	Mean	Median	Range
Area[a] (ha)	618	180	2–6070	1045	302	7–6070	84	20	2–307
Depth (m)	5.3	4.0	2–30	6.5	4.0	2.8–30	3.5	3.5	2–5
Color (Hazen)	91	87	3–188	114	123	9–188	56	16	3–182
pH[b]	5.5	4.8	4.5–8.1	6.0	5.6	4.5–8.1	4.7	4.7	4.6–5.0
ANC[b] (µeq/L)	86	−9	−53–1369	166	14	−33–1369	−33	−36	−53−−7
Ca[b] (µeq/L)	134	56	18–1300	192	62	48–1300	46	43	18–80
Mg[b] (µeq/L)	79	30	11–820	114	38	24–820	26	25	11–49
Na[b] (µeq/L)	42	33	14–158	55	42	29–158	22	22	14–33
SO$_4$ (µeq/L)	74	53	29–420	88	64	29–420	53	50	36–71
Cl (µeq/L)	42	21	10–240	47	20	12–240	35	24	10–112

[a] Rybinsk Reservoir (area = 395,000) omitted.
[b] Significant difference between drainage and seepage lakes (Mann-Whitney U-test, $p < 0.05$).

6.0 µg/g dry weight in "large" perch.[1] The mercury content of perch muscle tissue in 11 Swedish lakes with pH 5.2 to 6.2 ranged from 0.04 to 0.29 µg/g wet weight.[15] In Finland, mercury concentrations ranged 0.03 to 0.53 µg/g in perch from circumneutral lakes and 0.15 to 0.63 in fish from acidic lakes.[16] The larger fish from the most acidic lakes in our study had mercury concentrations considered potentially harmful to fish consumers, both human and wildlife.[17,18]

The surveyed areas of Russia have received considerable atmospheric loadings of sulfur. Karelia annually receives 420 to 800 eq/ha, comparable to deposition rates in southern Finland, Sweden, and Norway. The Yaroslavl region receives 800 to 1640 eq/ha per year, comparable to northern Germany.[12] The critical loadings of acidity have been estimated at 200 eq/ha per year or less for Karelia and 200 to 500 eq/ha per year for the Yaroslavl region; this critical loading may be exceeded by 500 to 1000 eq/ha per year in Karelia and by 200 eq/ha per year in the Yaroslavl region.[19] Indeed, we found no acidic drainage lakes in Darwin Reserve; the seepage lakes were precipitation dominated as reflected in their low ionic content, and were seemingly not influenced by the chemistry of the soils or surficial material.

Other authors have reported differences between seepage and drainage lakes with respect to factors affecting fish mercury content. In Michigan and Wisconsin, U.S. fish mercury content was negatively related to dissolved organic carbon in seepage lakes but not in drainage lakes.[5] A negative relation between lake color and fish mercury content was also found in Wisconsin seepage lakes;[20] however, these results are the opposite of the positive relation usually reported for fish mercury content and aqueous organic carbon or color,[1,6,21] and as found in our study. Inasmuch as seepage lakes have little or no terrestrial watershed, differences from drainage lakes may be related to the transport of mercury to lakes from terrestrial sources.

Humic matter controlled the solubility and watershed export of mercury deposited in precipitation in Sweden and Canada.[22,23] Further, water concentrations of mercury were highly correlated with water color,[24] and most of the mercury in clearwater lakes was deposited in the sediment whereas most was retained in the water column in humic lakes.[10]

Table 3 **Mean and range of fish weight, moisture content, and mercury concentration in fish, and least-square mean mercury concentration in fish from the study lakes.**

Lake	N	Weight (g) mean (range)	Moisture (%) mean (range)	Mercury, µg/g wet weight	
				Arithmetic mean (range)	Least square mean
Blue Lamba	10	22 (18–28)	85.6 (81.1–89.7)	0.34 (0.26–0.44)	0.41
Chuchyarvi	10	59 (18–133)	90.6 (83.2–93.9)	0.10 (0.08–0.13)	0.08
Dorojiv	10	62 (31–94)	80.9 (78.0–86.2)	0.50 (0.36–0.71)	0.47
Dubrovskoye	10	27 (23–30)	83.6 (77.8–88.1)	0.64 (0.56–0.68)	0.71
Grushba Lamba	7	38 (28–48)	81.5 (76.9–86.0)	0.30 (0.20–0.43)	0.33
Hotavets	10	62 (37–150)	88.6 (80.4–93.8)	0.11 (0.08–0.16)	0.09
Ilyakalkenyarvi	10	35 (20–78)	82.4 (72.0–94.2)	0.28 (0.19–0.36)	0.31
Kabozero	10	35 (24–47)	81.5 (61.7–87.2)	0.31 (0.22–0.39)	0.34
Lamba Vegarous	8	55 (28–80)	80.5 (75.9–85.1)	0.40 (0.30–0.52)	0.39
Leukunyarvi	10	30 (8–54)	88.2 (82.8–96.3)	0.21 (0.17–0.24)	0.25
Motykino	10	60 (43–75)	81.4 (79.3–87.9)	0.57 (0.43–0.97)	0.54
Rybinsk	9	86 (59–163)	80.8 (78.1–82.6)	0.19 (0.08–0.45)	0.11
Sargozero	10	18 (8–27)	85.2 (75.0–89.6)	0.12 (0.09–0.18)	0.19
Suoyarvi	10	98 (53–131)	81.5 (79.6–83.6)	0.29 (0.19–0.38)	0.18
Tyomnoye	8	155 (50–515)	79.6 (74.6–83.4)	1.06 (0.45–3.04)	0.56
Uros	10	19 (14–21)	83.7 (76.1–90.3)	0.12 (0.06–0.18)	0.19
Uteshkovo	5	45 (18–98)	80.4 (75.7–85.8)	0.78 (0.54–0.97)	0.80
Vegarousyarvi	10	26 (17–45)	79.5 (76.9–82.8)	0.34 (0.20–0.44)	0.41
Venderskoye	9	41 (12–107)	79.8 (72.9–81.0)	0.15 (0.10–0.24)	0.17
Vuontelenyarvi	9	43 (16–196)	84.5 (76.9–93.0)	0.53 (0.32–1.03)	0.54

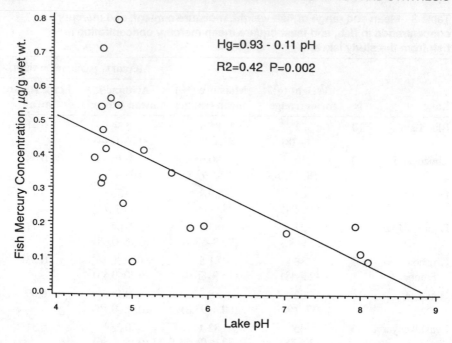

Figure 3   Relation between lake pH and least square mean fish mercury content.

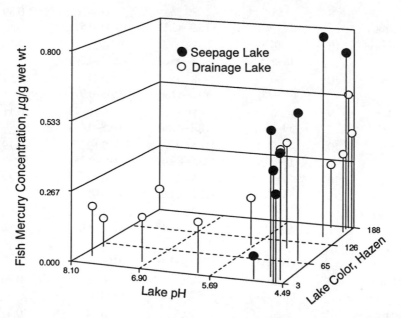

Figure 4   Relation between lake pH, color, and least square mean fish mercury content stratified by lake drainage type.

Table 4   **Statistically significant ($p \leq 0.05$) multiple regression equations, correlation coefficients, and probabilities for stepwise regression of lake physical and chemical variables on fish mercury content.**

Lake type	Number of dependent variables	Equation	R^2	P
All	1	Hg = −0.0347 pH + 0.317	0.46	0.001
	2	Hg = −0.0336 pH + 0.000144 color + 0.298	0.49	0.0035
	3	Hg = −0.0331 pH + 0.000143 color − 0.000019 sulfate + 0.297	0.50	0.0119
Seepage	1	Hg = 0.000711 color + 0.128	0.56	0.0332
	2	Hg = 0.000671 color − 0.173 pH + 0.946	0.69	0.05
Drainage	2	Hg = −0.0260 pH + 0.000030 color + 0.252	0.68	0.0058
	3	Hg = −0.0256 pH + 0.0000284 color − 0.0000167 sulfate + 0.251	0.68	0.0216

Note: Analyses are presented for all lakes, and stratified by lake drainage type. Units of measurement are Hg = $\log_{10}$ (μg/g wet wt. + 1), color = Hazen, sulfate = μeq/L.

Thus humic matter may influence fish mercury content by affecting the mercury loading to a lake and by retaining mercury in the water column, where it is available for uptake in biota.

V. SUMMARY AND CONCLUSIONS

We surveyed 20 lakes in two regions of Russia to determine the mercury content of fish and its relation to lake characteristics. The lakes surveyed ranged in pH from 4.5 to 8.1 and in color from 3 to 188 Hazen. The mercury content of fish ranged from 0.06 to 3.04 μg/g wet weight, which was comparable to that reported for this species from forest lakes in Sweden and Finland. Inasmuch as these were all remote lakes with no local sources of pollution, atmospheric deposition was presumed to be the source of acidity and mercury. Lake acidity and color, or humic content, were the major lake characteristics related to fish mercury content. Fish from high-pH lakes were low in mercury regardless of other lake characteristics. Fish from low-pH lakes varied widely in mercury content; fish from colored and seepage lakes tended to be higher in mercury than those from clear or drainage lakes. Regressions including lake pH and color, separated by drainage type, explained about 70% of the variance in fish mercury content. Acidity and color may affect fish mercury content by regulating loading and the bioavailability of mercury to lower trophic levels in these lakes. Seepage lakes may differ from drainage lakes because of differences in watershed mercury input.

ACKNOWLEDGMENTS

We appreciate financial support provided by the U.S. Fish & Wildlife Service, the U.S. Environmental Protection Agency, the U.S. Department of Energy (through contract DE-AC09–76SR00–819 to the University of Georgia's Savannah River Ecology Laboratory), and the Institute for Biology of Inland Waters, Russian Academy of Sciences. Research facilities were provided by the Institute for Biology of Inland Waters, Borok, and the Institute for Research on Northern Fishes, Petrozavodsk, Russia.

REFERENCES

1. Lindqvist, O., Johansson, K., Aastrup, M., Anderson, A., Bringmark, L., Hovsenius, G., Håkanson, L., Iverfeldt, Å., Meili, M., and Timm, B., Mercury in the Swedish environment—recent research on causes, consequences and corrective methods, *Water Air Soil Pollut.*, 55, 1, 1991.

2. Spry, D.J. and Wiener, J.G., Metal bioavailability and toxicity to fish in low-alkalinity lakes: a critical review, *Environ. Pollut.*, 71, 243, 1991.

3. Swain, E. and Helwig, D., Mercury in fish from northeastern Minnesota lakes: historical trends, environmental correlates, and potential sources, *J. Minn. Acad. Sci.*, 55, 103, 1989.

4. Slemr, F. and Langer, E., Increase in global atmospheric concentrations of mercury inferred from measurements over the Atlantic Ocean, *Nature*, 355, 434, 1992.

5. Grieb, T.M., Driscoll, C.T., Gloss, S.P., Schofield, C.L., Bowie, G.L., and Porcella, D.B., Factors affecting mercury accumulation in fish in the Upper Michigan Peninsula, *Environ. Toxicol. Chem.*, 9, 919, 1990.

6. McMurty, M.J., Wales, D.L., Scheider, W.A., Beggs, G.L., and Dimond, P.E., Relationship of mercury concentrations in lake trout (*Salvelinus namaycush*) and smallmouth bass (*Micropterus dolomieui*) to the physical and chemical characteristics of Ontario lakes, *Can. J. Fish, Aquat. Sci.*, 46, 426, 1989.

7. Häkanson, L., Andersson, T., and Nilsson, Å., Mercury in fish in Swedish lakes—linkages to domestic and European sources of emission, *Water Air Soil Pollut.*, 50, 171, 1990.

8. Ponce, R.A. and Bloom, N.S., Effects of pH on the bioaccumulation of low level, dissolved methylmercury in rainbow trout (*Oncorhynchus mykiss*), *Water Air Soil Pollut.*, 56, 631, 1991.

9. Wiener, J.G., Fitzgerald, W.F., Watras, C.J., and Rada, R.G., Partitioning and bioavailability of mercury in an experimentally acidified Wisconsin lake, *Environ. Toxicol. Chem.*, 9, 909, 1990.

10. Meili, M., Fluxes, pools, and turnover of mercury in Swedish forest lakes, *Water Air Soil Pollut.*, 56, 719, 1991.

11. Khabibullov, M., Crisis in environmental management of the Soviet Union, *Environ. Manag.*, 15, 749, 1991.

12. Chadwick, M.J. and Kuylenstierna, J.C.I., The relative sensitivity of ecosystems in Europe to acidic depositions. A preliminary assessment of the sensitivity of aquatic and terrestrial ecosystems, *Perspect. Energy*, 1, 71, 1991.

13. Ager, D.V., *The Geology of Europe*, McGraw-Hill, New York, 1980, chap. 3.

14. FAO/SIDA, Manual of methods in aquatic environment research. IX. Analyses of metals and organochlorines in fish, *FAO Fish. Tech. Pap.*, 212, 1, 1983.

15. Paulsson, K. and Lundbergh, K., Treatment of mercury contaminated fish by selenium addition, *Water Air Soil Pollut.*, 56, 833, 1991.

16. Verta, M., Mannio, J., Iivonen, P., Hirvi, J.-P., Järvinen, O., and Piepponen, S., Trace metal in Finnish headwater lakes—effects of acidification and airborne load, In: *Acidification in Finland*, Kauppi, P., Kenttamies, K., and Anttila, P., Eds., Springer-Verlag, New York, 1990, 883.

17. Wiener, J.G., Metal contamination in fish in low-pH lakes and potential implications for piscivorous wildlife, Trans. 52nd N. Am. Wildl. & Nat. Resour. Conf., 645, 1987.

18. Grant, L.D., Indirect health effects associated with acidic precipitation, In: *Acidic Deposition: State of Science and Technology, Summary Report of the U.S. National Acid Precip. Assess. Program*, Irving, P.M. Ed., National Acid Precipitation Assessment Program, Washington, D.C., 1991, chap. 23.

19. Hettelingh, J.-P., Howning, R.J., and de Smet, P.A.M., Mapping critical loads for Europe, *Coordination Center for Effects, Tech. Rep.*, 1, 3, 1991.

20. Cope, W.G., Wiener, J.G., and Rada, R.G., Mercury accumulation in yellow perch in Wisconsin seepage lakes: relation to lake characteristics, *Environ. Toxicol. Chem.*, 9, 931, 1990.

21. Verta, M., Mercury in Finnish forest lakes and reservoirs: anthropogenic contribution to the load and accumulation in fish, Water and Environ. Res. Inst., National Board. Waters and the Environ., Finland, 6, 5, 1990.

22. Iverfeldt, Å. and Johansson, K., Mercury in run-off water from small watersheds, *Verh. Int. Verein. Limnol.*, 23, 1626, 1988.

23. Mierle, G. and Ingram, R., The role of humic substances in the mobilization of mercury from watersheds, *Water Air Soil Pollut.*, 56, 349, 1991.

24. Meili, M., Iverfeldt, Å., and Håkanson, L., Mercury in the surface water of Swedish forest lakes—concentrations, speciation and controlling factors, *Water Air Soil Pollut.*, 56, 439, 1991.

Mercury in the Food Chains of a Small Polyhumic Forest Lake in Southern Finland

Martti Rask, Tarja-Riitta Metsälä, and Kalevi Salonen

CONTENTS

ABSTRACT: The food chains of a 0.4-ha pond were sampled during 1982 to 1984 for mercury determinations in order to compare the Hg concentrations at different trophic levels and to detect possible changes after clear cutting and subsequent burning of logging waste in the catchment area.

The concentrations of aquatic mosses were 0.06 to 0.09 μg Hg g^{-1} (dw). In periphyton the concentrations were similar. In crustacean zooplankton the concentrations were higher, between 0.1 and 0.4 μg Hg g^{-1} (dw). The concentrations measured from copepods were generally lower (0.1 to 0.2 μg g^{-1}) than those from cladocerans (0.2 to 0.4 μg g^{-1}). Among macroinvertebrates, the Hg concentrations varied from 0.1 to 0.2 μg g^{-1} (dw) in ephemeropterans and trichopterans to 0.5 to 0.6 and even >1 μg g^{-1} in water boatmen and some coleopterans.

The two fish species of the lake, the European perch, *Perca fluviatilis,* and the northern pike, *Esox lucius,* had mean mercury concentrations of 0.42 μg g^{-1}, ww (2.1 μg g^{-1}, dw). Both species showed clear age and size dependency in mercury enrichment: large and old individuals had the highest concentrations. Generally, the mercury concentrations of the examined biota varied seasonally quite widely and no clear differences from year to year could be detected.

I. INTRODUCTION

In boreal forest areas, high Hg concentrations in fish have been recorded in lakes unaffected by direct discharge of mercury into their catchment.[1-3] In Sweden several thousand lakes have been estimated to have fish mercury concentrations of >1 μg g^{-1}.[4] In Finland, the corresponding number of lakes is about 3000.[5] These lakes are typically small, with low ionic strength, low pH, low productivity, and high amounts of dissolved humic substances in the water.[1-6]

Although there are many reports on the enrichment of mercury in food chains, there is less detailed knowledge of biota other than fish. If other trophic levels have been included in studies, mercury concentrations have usually been determined from composite samples of zooplankton or zoobenthos.[7] Recently, more thorough analyses have been published on the mercury dynamics in lower trophic levels of aquatic ecosystems.[5,8]

In 1982 to 1984, we took samples, as a part of a forest lake research project,[9-11] from different trophic levels of a small forest pond for mercury analyses. The aim of the study was to obtain a general view of the distribution of mercury in this lake and to examine whether the clear cutting of forest in the catchment of the lake would affect the mercury concentrations of biota. This chapter summarizes the mercury records of the study.

II. MATERIAL AND METHODS

A. THE LAKE AND ITS CATCHMENT AREA

The study lake, Lake Nimetön, is an 11-m deep, 0.4-ha forest pond with a catchment area of 0.34 km² located in the Evo State Forest, 100 km north from Helsinki. The catchment is characterized by morainic soils overlying granodiorite and gneiss bedrock.[9] Spruce and Scots pine forest is the dominating vegetation. The water of the lake has low ion content, slight acidity, and high concentrations of humic substances (Table 1). No complete spring or autumn mixing of water mass usually takes place. The lake is therefore steeply stratified with an oxygenated epilimnion of 2 to 4 m in the growing season and an anoxic hypolimnion almost all the time. It was calculated that 65 to 90% of outflowing water originated from the two inlets of the lake. The theoretical residence time was 2 to 6 months, but in practice the uppermost epilimnion can be flushed by throughflow in a few days whereas the residence time for the hypolimnion may be years.

The forest surrounding the lake was clear cut, except just around the lake, in winter 1981 to 1982. The logging waste was then burned in May 1983. Almost half (46%) of the

Table 1 **Some properties of Lake Nimetön. Values for water quality are from the uppermost 1 m water layer in May through September 1982 to 1984.**

	Mean	**Range**
Surface area, km²	0.004	
Catchment area, km²	0.34	
Volume, m³	3.6 × 10³	
Maximum depth, m	11	
Mean depth, m	8	
Theoretical residence, months	2–6	
pH	6.0	5.8–6.2
Alkalinity, mmol L^{-1}	0.08	0.07–0.09
Conductivity, mS m^{-1} 20° C	3.5	3.3–3.7
Color, mg Pt L^{-1}	255	225–300
Total P, µg L^{-1}	47	39–51
Total N, µg L^{-1}	620	560–670
Chlorophyll a, mg m^{-3}	18	12–25
Pp, mg, C m^{-3} d^{-1}	240	225–260
R, mg, C, m^{-3} d^{-1}	185	160–225

Note: Values for chlorophyll a primary production of phytoplankton (Pp), and respiration of plankton (R) are from Rask et al.[12]

catchment area was treated like this.[9] Clear cutting removed the wind protection of large trees, resulting in complete autumn turnover of the water mass. The forestry treatments caused a slight eutrophication of the lake including the decrease of oxygenated epilimnion to 1.5 m in summer 1983, and some increase in pH, alkalinity, and nutrients.[12] The primary production of phytoplankton, chlorophyll *a* concentration, and respiration of plankton were also higher during 1982 to 1984 (Table 1) than in the years before the catchment manipulations.[12]

B. THE SAMPLES

Samples for the determination of mercury were collected from different trophic levels in 1982 to 1984. For primary producers, three species of aquatic mosses (*Fontinalis antipyretica, Sphagnum* sp., and *Drepanocladus* sp.) were sampled. Samples were also taken from the periphyton community growing on glass plates that were tied to a wooden frame and kept in the surface water (0.5 m) of the lake.

Zooplankton samples were hauled from depths of 0 to 2 m with plankton nets of 50 and 300 μm mesh size and taken to the laboratory. Cladocerans and copepods were separated to species-specific samples of thousands of individuals by generating small bubbles which accumulated under the carapace of the cladocerans and lifted them to the water surface, where they were collected with a spoon. The bubbles were generated by introducing carbonated water from a siphon bottle. The cladocerans (*Holopedium gibberum*) were separated from other cladocerans by filtering the samples through a 1000-μm net. *Bosmina longispina* and *Daphnia cristata* could not be separated from each other while being in the same sample, but due to their different life cycles samples from both species could be obtained. Calanoid copepods, mostly *Eudiaptomus gracilis,* anesthetized by carbonated water, were collected with a siphoning tube. Cyclopoid copepods, mostly *Cyclops scutifer,* were less sensitive to carbon dioxide and remained swimming in the water.

Macroinvertebrate samples were taken with a dip net of mesh size 0.5 mm from or below the floating *Sphagnum* mat that surrounds the lake. The sampling depth was 0 to 1 m. Animals were picked and identified in the laboratory. Fish (European perch and northern pike) were caught with wire traps of 1-cm-square mesh. Individuals of the age group 0+ of both species were dip-netted.

C. MERCURY DETERMINATIONS

The Hg concentrations were determined as total Hg from HNO_3-H_2SO_4 (1:4) digestion using cold vapor atomic absorption spectrophotometry[13] and a Perkin-Elmer Coleman MAS-50 analyzer (The Perkin-Elmer Corporation, Norwalk, CT). Before determinations samples of primary producers, zooplankton, and macroinvertebrates were dried at 40° C. The minimum sample weight was 0.2 g (dw). Fish samples, i.e., dorsal axial muscle tissue without skin, were deep-frozen until analyzed. The total number of samples was 14 for primary producers, 40 for zooplankton, and 57 for macroinvertebrates. A total of 39 perch and 22 pike with length ranges of 3.1 to 25.5 cm and 6.2 to 50.5 cm, respectively, were examined.

III. RESULTS AND DISCUSSION

A. MERCURY CONCENTRATIONS IN BIOTA OF LAKE NIMETÖN

The mercury concentrations of *Fontinalis antipyretica* were 0.06 to 0.12 μg g^{-1} (dw). In *Drepanocladus* sp. and *Sphagnum* sp. the concentrations were 0.09 and 0.11 μg g^{-1}, respectively. In the periphyton community, the Hg concentration was 0.05 to 0.11 μg g^{-1}. These values are low and comparable to those recorded in so-called clean areas.[14]

The mercury concentrations in crustacean zooplankton varied from 0.02 μg g^{-1} (dw) in *Eudiaptomus gracilis* in May 1982, to 0.60 μg g^{-1} (dw) in *Holopedium gibberum* in July 1983 (Figure 1). The concentrations of cladocerans exceeded those recorded in copepods; the difference was significant in 1982 and 1983 (*t*-test, $p < 0.01$). There were also wide

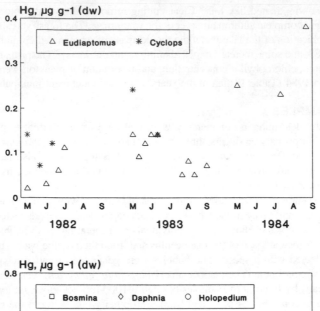

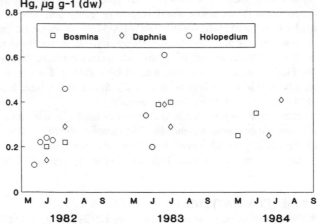

Figure 1 Mercury concentrations of the copepods *Eudiaptomus gracilis* and *Cyclops* sp. (top) and the cladocerans *Bosmina longispina, Daphnia cristata,* and *Holopedium gibberum* (bottom) in 1982 to 1984.

seasonal intraspecific variations. For example, in 1982 the mercury concentration of *Holopedium gibberum* ranged from 0.12 to 0.46 µg g^{-1} (dw). This high variation may be due to the short life cycles of crustacean zooplankton species: samples taken at different times of the growing season represented different generations. Also seasonal changes in abiotic factors such as hydrology have been shown to affect the bioavailability of mercury.[2] Generally, the mercury concentrations of zooplankton recorded in this study were quite similar to observations from other lake ecosystems in Fennoscandia.[5,8]

Macroinvertebrates contained 0.08 to 1.4 µg g^{-1} (dw) mercury (Table 2). In Lake Nimetön, where the profundal detrivores are absent due to anoxic conditions in the hypolimnion, no observations were made on unexpectedly high mercury concentrations in detrivores as observed elsewhere.[8]

The mercury concentrations of the European perch in Lake Nimetön varied from 0.1 to 0.15 µg g^{-1} (ww) in 4- to 6-cm-long fish of age group 0+, to 0.5 to 0.6 µg g^{-1} (ww) in

Table 2   **Mean mercury concentrations (μg g^{-1}, dw) of macroinvertebrates in Lake Nimetön in 1983 and 1984.**

	1983			1984		
	Mean	**SD**	**n**	**Mean**	**SD**	**n**
Insecta						
Ephemeroptera						
Leptophlebia vespertina	0.19	0.06	2	0.14		1
Odonata						
Aeschna grandis	0.20	0.08	4	0.11		1
Aeschna juncea	0.20	0.09	4	0.16	0.06	4
Agrion hastulatum	0.31	0.09	2	0.14		1
Cordulia aenea	0.31	0.27	3	0.22	0.11	3
Leucorrhinia sp.	0.35	0.12	3	0.18	0.08	5
Hemiptera						
Notonecta glauca	0.46		1	0.58		1
Trichoptera						
Limnephilus lunatus	0.14	0.06	6	0.08	0.04	3
Coleoptera						
Dytiscus sp.	0.32		1	0.59		1
Graphoderes cireneus	1.40		1			
Arachnida						
Araneae						
Argyroneta aquatica	0.36	0.09	4	0.30	0.09	4
Crustacea						
Malacostraca						
Asellus aquaticus	0.36	0.33	2			

20- to 25-cm-long fish of ages 6 to 8 years. The Hg concentration correlated significantly both with the length ($R^2 = 0.71$) and with the age ($R^2 = 0.74$) of perch.[15] Similar relations were recorded between the length ($R^2 = 0.83$) and age ($R^2 = 0.84$) of northern pike and their mercury concentrations. The length-corrected mean mercury content for a 1-kg pike in Lake Nimetön was 0.75 μg g^{-1} (ww)[16] but the measured concentrations were quite similar to those from perch (Figure 2).

B. PATTERNS IN BIOACCUMULATION OF Hg IN THE FOOD CHAIN

Evidence from fish Hg measurements suggests that the bioavailability of mercury is highest in humic soft water lakes.[5-8] This may be related to the role of allochthonous organic matter as the main source of energy in such lakes[17] and to the importance of humic substances as a carrier of Hg from the catchment area to the lake.[8]

Interspecific differences in the mercury concentrations of zooplankton most probably reflect differences in species-specific feeding strategies. It was shown in a radiotracer study[10] that the copepod *Eudiaptomus gracilis* in Lake Nimetön prefers relatively more autotrophic food (algae) in comparison to cladocerans that feed more on bacteria. This may explain the differences in mercury concentrations of copepods and cladocerans. Bacteria are

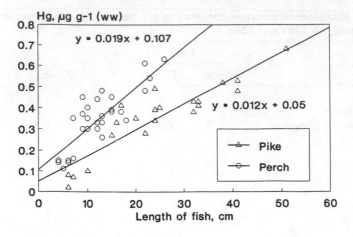

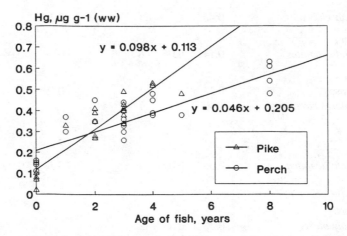

Figure 2 Mercury concentrations in the European perch and the northern pike of Lake Nimetön in relation to the fish length (top) and the age (bottom).

important in transferring Hg coupled with allochthonous organic matter to the food chains of humic lakes.[8]

Among the macroinvertebrates, plant or detritus feeding species like the ephemeropteran *Leptopohlebia vespertina* and the trichopteran *Limnephilus lunatus* has lower Hg concentrations (<0.2 μg g^{-1}, dw) in comparison with predatory dragon fly larvae, water boatmen, and water spiders, which had values of 0.2 to 0.6 μg g^{-1}. The highest mercury concentration from an invertebrate, 1.4 μg g^{-1}, was measured from a predatory water beetle *Graphoderes cireneus*.

The bioaccumulation of mercury in perch took place quite rapidly. The concentrations in 0+ fish (0.55 to 0.80 μg g^{-1}, dw) already exceeded the values recorded from zooplankton and most macroinvertebrates (Figure 3). The concentrations in 1+ perch <10 cm in length were 1.5 to 2.3 μg g^{-1} (dw)—similar to 1- to 2-year-old pike. After the first 2 years of life the rate of mercury accumulation in perch slowed down. The proportion of zooplankton in

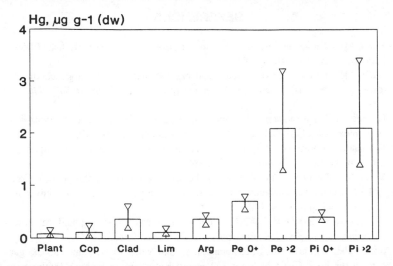

Figure 3 A comparison of mercury concentrations of biota from different trophic levels of the food chains. Mean values and range are given. Plant = primary producers; Cop = copepods; Clad = cladocerans; Lim = caddis larva *Limnephilus lunatus;* Arg = water spider *Argyroneta aquatica.* For perch (Pe) and pike (Pi), mean concentrations are given for young-of-the year (0+) and fishes older than 2 years (>2).

the diet of 0+ and 1+ perch in Lake Nimetön was 75 to 80%.[11] The rapid accumulation in the youngest perch probably occurred through the following pathway: humic substances—(bacteria)—zooplankton—perch.

C. DIFFERENCES BETWEEN THE YEARS—EFFECTS OF CLEAR CUTTING?

In primary producers the Hg concentrations of *Fontinalis antipyretica* and the periphyton community were similar in 1982 and 1983. In zooplankton some increase in mercury concentration took place (Figure 1). For example, in cladocerans the concentrations were significantly higher in 1983 than in 1982 (*t*-test, $p < 0.05$), but not in copepods. This could be due to changes in oxygen conditions that affect the methylation of Hg,[18] but other factors also, like changes in water quality or increased productivity of the plankton community, especially the autotrophic production,[12] could contribute to the changes. In macroinvertebrates the effects of catchment manipulation could not be assessed due to lack of samples in 1982.

The mean Hg concentration of adult perch was 0.30 µg g^{-1} (ww) in 1982, 0.43 in 1983, and 0.37 in 1984. For pike, the length-corrected mean values for a 1-kg fish were 0.75 to 0.84 µg g^{-1} (ww) in 1982 to 1984. Thus, the mercury concentrations of fish were at a quite constant level during the study period and no signs of the effects of catchment manipulation were recorded.

The sampling period, lasting for only three years, is with no doubt too short for a reliable assessment of the potential effects of clear cutting on mercury dynamics in the food chains. The increase in the Hg concentrations of zooplankton was the only change that could be due to the catchment manipulation. The short food chain, humic matter—bacteria—zooplankton, and its short life cycle, could make such a quick response possible, but again, due to the short study period, the natural variation may mask the potential effects.

REFERENCES

1. Verta, M., Changes in fish mercury concentrations in an intensively fished lake, *Can. J. Fish. Aquat. Sci.,* 47, 1888, 1990.
2. Meili, M., The coupling of mercury and organic matter in the biogeochemical cycle—towards a mechanistic model for the boreal forest zone, *Water Air Soil Pollut.,* 56, 333, 1991.
3. Lathrop, R.C., Rasmussen, P.W., and Knauer, D.R., Mercury concentrations in walleyes from Wisconsin (USA) lakes, *Water Air Soil Pollut.,* 56, 295, 1991.
4. Björklund, I., Borg, H., and Johansson, K., Mercury in Swedish lakes—its regional distribution and causes, *Ambio,* 13, 118, 1984.
5. Verta, M., Mercury in Finnish Forest Lakes and Reservoirs: Anthropogenic Contribution to the Load and Accumulation in Fish, Academic dissertation, Helsinki University, Yliopistopaino, Helsinki, 1990, 140.
6. Winfrey, M.R. and Rudd, J.W.M., Environmental factors affecting the formation of methylmercury in low pH lakes, *Environ. Toxicol. Chem.,* 9, 853, 1990.
7. Surma-aho, K., Paasivirta, J., Rekolainen, S., and Verta, M., Organic and Inorganic Mercury in the Food Chain of Some Lakes and Reservoirs in Finland. Publications of the Water Research Institute, National Board of Waters, Finland, 65, 59, 1986.
8. Meili, M., Mercury in Boreal Forest Lake Ecosystems, Doctoral dissertation, Uppsala University, Lindbergs Grafiska HB, Uppsala, Sweden, 1991, 7.
9. Arvola, L., Salonen, K., and Rask, M., Chemical budgets for a small dystrophic lake in southern Finland, *Limnologica,* 20, 243, 1990.
10. Salonen, K. and Arvola, L., A radiotracer study of zooplankton grazing in two small humic lakes, *Verh. Int. Verein Limnol.,* 23, 462, 1988.
11. Rask, M., Differences in growth of perch (*Perca fluviatilis* L.) in two small forest lakes, *Hydrobiologia,* 101, 139, 1983.
12. Rask, M., Arvola, L., and Salonen, K., Effects of catchment deforestation and burning on the limnology of a small forest lake in southern Finland, *Verh. Int. Verein. Limnol.,* 25, 525, 1993.
13. Armstrong, F.A.J. and Uthe, J.F., Semiautomated determination of mercury in animal tissue, *At. Absorpt. Newsl.,* 10, 101, 1971.
14. Lodenius, M., Aquatic plants and littoral sediments as indicators of mercury pollution in some areas in Finland, *Ann. Bot. Fennici,* 17, 336, 1980.
15. Metsälä, T. and Rask, M., Mercury concentrations of perch, *Perca fluviatilis* L., in small Finnish headwater lakes with different pH and water colour, *Aqua Fennica,* 19, 41, 1989.
16. Rask, M. and Metsälä, T.-R., Mercury concentrations in northern pike, *Esox lucius* L., in small lakes of Evo Area, southern Finland, *Water Air Soil Pollut.,* 56, 369, 1991.
17. Salonen, K. and Hammar, T., On the importance of dissolved organic matter in the nutrition of zooplankton in some lake waters, *Oceologia,* 68, 246, 1986.
18. Matilainen, T., Verta, M., Niemi, M., and Uusi-Rauva, A., Specific rates of net methylmercury production in lake sediments, *Water Air Soil Pollut.,* 56, 595, 1991.

Mercury in Vegetation of the Precambrian Shield

Pat E. Rasmussen

CONTENTS

I. INTRODUCTION

Plants readily accumulate Hg from surrounding soil, water, and air to levels of concentration that depend on the species of the plant, the ambient Hg concentration, and the chemical form of the Hg.[1,2] The collection of data describing Hg in different types of vegetation is fundamental to ecological and health studies for at least three reasons, the most notable being the ease with which Hg enters the food chain via plants.[3] Hg concentrations in vegetation must also be known to quantify biogenic sources of Hg to the atmosphere[4] and to natural and artificial aqueous systems.[5] Nriagu[6] found that biogenic sources of Hg and other metals tend to be ignored or badly underestimated in inventories of natural sources. Finally, certain species serve as ideal biomonitors of ambient Hg levels in the environment.[2,7] Vegetation sampling has been used successfully to monitor source strength and/or proximity to unknown sources in both natural and anthropogenic contamination problems.[2,3,7]

In the present study, more than two dozen plant species, vascular and nonvascular, were investigated for their potential usefulness as Hg indicators. Representatives were included from all four divisions of the Plant Kingdom—thallophytes, bryophytes, pteridophytes, and spermatophytes. Approximately 1,100 vegetation samples were collected over a 2-year period from a study area of about 150 km² described in Rasmussen et al.[8] Results from the first

1–56670–066–3/94/$0.00+$.50
© 1994 Lewis Publishers

year suggested that certain nonvascular plants were better Hg indicators than trees, in terms of their ability to concentrate Hg and their relative abundance in the study area.[8] This chapter summarizes the results of the second year's vegetation study, in which particular attention was paid to nonvascular species.

II. METHODOLOGY

A. LOCATION

The study area is located west of the town of Huntsville, Ontario, Canada about 250 km north of Toronto. Huntsville (long. 79°12', lat. 45°20') lies near the southern margin of Canada's Shield region, within the Grenville Structural Province. The study area encompasses a chain of lakes characterized by anomalously high levels of Hg in the fish.[9] There is no known industrial point source of Hg in the Huntsville area.

B. SAMPLE COLLECTION

Most species were collected from well-drained upland sites, and a few were collected from wetland sites and streams. Particular care was taken to identify and minimize unwanted sources of variation and contamination that might be introduced during sampling, processing, and analysis. The vegetation samples were collected with unpowdered vinyl gloves and only stainless steel clippers and plastic gardening tools were used. Samples were triple-bagged in the field using ziplocked polyethylene bags. Samples were stored frozen until time of preparation and analysis.

C. PROCESSING AND ANALYSIS

All samples were rinsed with distilled deionized water to ease the removal of foreign particles. The samples were then oven-dried for 24 h at 60° C. The analytical scheme, detailed previously,[8] consisted of a 6-h hot digestion using HNO_3 and H_2SO_4 in a 1:4 ratio, followed by dilution, $SnCl_2$ reduction, and cold-vapor AAS detection. The analytical work was performed in a clean laboratory facility outfitted with gold air filters to adsorb airborne contaminant Hg.

D. QUALITY CONTROL

Three procedural blanks and replicate NBS standards were included in each batch of digests and were subjected to exactly the same treatment and conditions as the samples. Hg levels in the blanks ranged from 0.002 to 0.020 ng mL^{-1}. The detection limit, defined as three times the standard deviation of the blank, was 1.3 ng g^{-1} calculated for a dry sample weight of 0.100 g. NBS#1575 Pine Standard (certified value 0.15 ± .05 µg g^{-1}) and NBS#1572 Citrus Standard (certified value 0.08 ± .02 µg g^{-1}) were chosen for their similarity to the natural samples in terms of Hg content and matrix composition. The Hg concentrations consistently fell within the certified range for both pine (observed value 0.115 ± .005 µg g^{-1}; N = 5) and citrus (observed value 0.089 ± .010 µg g^{-1}; N = 22) reference material.

E. SOURCES OF VARIATION

It is very important when sampling vegetation for biomonitoring purposes to compare tissue of the same age and the same organ from the same species.[2] Absolute consistency in tissue collection, preparation, and analysis is necessary to minimize unwanted sources of variation, the equivalent of "baseline noise." In this study, different sources of variation were quantified separately in order to determine their relative significance.

1. Analytical Variation

Unwanted variation may be introduced in two aspects of the analytical process: (1) chemical interferences caused by inadequate sample digestion, and (2) sample heterogeneity caused by inconsistent sample preparation. Variation between AAS peaks for separate aliquots of

Table 1 **Hg in foliage of vascular plants of the Canadian Shield.**

Vegetation type	Number of collection sites	Hg concentration (ng/g)	
		Range	Median
Trees			
Abies balsamea	55	20.0–65.5	28.7
Acer saccharum	18	16.5–40.8	24.2
Wildflowers			
Maianthemum canadense Desf.	53	17.2–72.9	30.1
Aquatic macrophytes			
Pontedaria cordata f. taenia	23	17.6–82.9	44.0
Potamogeton sp.	3	54.1–81.8	63.0
Spargangium sp.	2	16.8–33.9	—
Sagittaria sp.	1	44.3	—
Gramineae family	2	20.5–52.5	—
Utricularia vulgaris L.?	1	29.7	—
Clubmosses			
Lycopodium dendroideum	20	32.0–201.3	80.6
Lycopodium annotinum	10	46.3–134.4	90.0
Lycopodium lucidulum	5	63.7–187.2	103.3
Lycopodium clavatum	1	32.2	—

one digest averaged 2.2% COV and ranged from 0.0 to 10.9% COV (N = 480 digests). Larger variation in the AAS response often signified interference from residual NO_x. Interferences also occurred in digests containing high levels of Ca (more than 4% dry wt.), indicated by the formation of a white $CaSO_4$ precipitate in the cooling digest.[10] Problems with sample heterogeneity were indicated by significant differences (>10%) in Hg concentrations of replicate digest of one sample, most often caused by accidental mixing of tissue ages and different organs. Through careful sample processing, this source of variation was kept almost as low as the instrumental variation, averaging 3.7% COV and ranging from 0.1 to 11% COV (N = 61 samples).

2. Within-Site Variation

Natural within-site variation was quantified for each species by comparing Hg contents of two or more plants growing in the same habitat within one site (defined as an area of 10 m^2), using tissue of the same age and the same organ from each plant. Under these strict sampling controls, within-site variation averaged 18% COV and ranged from 2.6 to 52.9% COV (N = 23 sets). These results suggest that a minimum of two or three separate plants of the same species should be sampled at each site in order to obtain a representative Hg concentration for that site.

3. Between-Site Variation

Variation in Hg content between sites is reported in Tables 1 and 2 for each vegetation type. None of these data sets had a normal frequency distribution. Frequency distribution curves were either polymodal or strongly positively skewed, suggesting the presence of two or more Hg concentration populations for each species. Sources of anomalous Hg levels in the Huntsville watershed were of particular concern,[10] and as part of the forthcoming spatial analysis of these data threshold values will be calculated using the statistical methods of Sinclair[11]

Table 2   **Hg in nonvascular plants of the Canadian Shield.**

Vegetation type	Number of collection sites	Hg concentration (ng/g)	
		Range	Median
Moss			
Sphagnum sp.	21	21.4–92.2	41.2
Polytrichum commune Hedw.	61	12.4–224.6	53.0
Dicranum polysetum Sw.	6	39.1–126.4	79.1
"Other" acrocarpous moss	11	32.0–139.7	60.4
Pleurozium schreberi (epiphytic)	57	46.0–270.8	139.9
Pleurozium schreberi (ground)	56	35.1–286.9	95.3
Brachythecium plumosum (Hedw.) B.S.G.	8	81.6–155.4	133.5
"Other" pleurocarpous moss	22	72.9–224.5	135.0
Lichen			
Cladina mitis (Sandst.) Hustich	4	30.7–43.4	33.0
Cladina rangiferina (L.) Nyl.	5	17.2–51.6	24.3
Cladonia phyllophora Ehrh. ex Hoffm.	3	33.0–56.4	46.4
Cladonia cristatella Tuck.	3	40.3–66.9	58.4
Cladonia cervicornis (Ach.) Flot. ssp. verticillata (Hoffm.) Aht	3	28.0–49.4	43.7
Mushrooms			
Various species	6	18.2–222.2	104.0
Aquatic algae			
Lemania fucina Bory (sedentary)	1	56.9	—
Green algae (planktonic)	1	150.0	—
Aquatic bacteria			
Leptothrix sp.	1	63.0	—

and Miesch[12] to distinguish anomalous concentrations lying outside the range of background variation.[10] Data from at least 50 sampling sites are recommended to statistically define the background and anomalous concentration populations using these methods.

III.   RESULTS AND DISCUSSION

A.   Hg CONTENT OF VASCULAR PLANTS
1.   Spermatophytes

In the Huntsville study area the most common tree species are balsam fir (*Abies balsamea*) and sugar maple (*Acer saccharum*). Both species have been previously reported to be useful indicators for Au and other metals.[13–15] Hg concentrations (Table 1) were determined in 2-year-old balsam needles and in maple leaf material only. Twig material was excluded based on an earlier observation[8] that foliage contained higher Hg levels than twigs from the same branch. The range in Hg concentrations (Table 1) was narrow compared to the range in other species, suggesting that these species are perhaps not the best choice for Hg bioindicators. Kovalevsky[2] found that in certain Siberian conifers the external and middle bark layers contained higher Hg concentrations than other parts of the plant. It is possible that an alternate tissue type, such as bark, might be a better choice than foliage for surveying, but such alternatives were not tested in this study.

Wild lily-of-the-valley (*Maianthemum canadense*) was included because of its abundance in the study area, but no previous information was available regarding its usefulness as a Hg indicator. Three sterile shoots, each comprised of a solitary leaf, were collected at 53 sites. The dried leaves, which weighed about 30 mg apiece, were combined into one sample to represent each site. Stem material was excluded. Hg concentrations ranged from 17 to 73 ng g^{-1} (Table 1) and displayed a bimodal frequency distribution, suggesting that wild lily-of-the-valley may be a useful Hg indicator plant.

Pickerelweed (*Pontedaria cordata f. taenia*) was found in 23 streams in the study area and was the most common aquatic macrophyte. This plant consists of submerged sterile rosettes of narrow ribbon-like leaves rooted in the soft sediment. In the remaining streams other species of identical growth form and habitat were collected (Table 1). At all sites, three separate plants of one species were collected and the leaf material was combined into one sample for analysis. The Hg levels were comparable to those of Siegel et al.[7] who report Hg contents ranging between 15 (±2) ng g^{-1} and 132 (±18) ng g^{-1} in the aquatic macrophyte *Potamogeton,* sampled in a mercuriferous fault zone in British Columbia.

2. Pteridophytes

Of the Pteridophyte division, the only representatives collected were the clubmosses, genus *Lycopodium.* The clubmosses displayed a wide range of Hg concentrations (Table 1). Tree clubmoss (*Lycopodium dendroideum*) showed promise as a bioindicator as it was fairly abundant, had a wide range of Hg concentrations, and a distinctly bimodal frequency distribution. Hg contents (Table 1) were measured in leaves only.

B. Hg CONTENT ON NONVASCULAR PLANTS
1. Bryophytes

Erdman[16] described bryophytes as the ideal sampling medium because they are non-barrier to most metals, they have a high surface-to-volume ratio and a perennial growth habit.[16] Hg enters bryophytes by direct absorption through the cell walls, and in two aquatic liverworts it was discovered that Hg accumulates as fine metacinnabar (HgS) crystals.[17]

Preliminary work in the first year indicated that red-stemmed feathermoss (*Pleurozium schreberi*) was the most common terrestrial pleurocarpous species encountered in the study area. This species was previously listed as a useful Hg biomonitor in the Precambrian Shield.[18] In August of the second year, 113 samples were collected: 57 from living deciduous tree trunks (epiphytic habitat) and 56 from soil or rocks of the forest floor (Table 2). Hg concentrations for all samples of *Pleurozium schreberi* ranged from 35.1 to 286.9 ng g^{-1} (Table 2).

It was observed that a sample of epiphytic *P. schreberi* collected from a deciduous tree trunk generally contained more Hg that a sample collected from the ground near the base of the same tree. Further investigation confirmed that Hg contents tended to be higher in epiphytic feathermoss than in ground feathermoss from the same site by a factor of 1.4, averaged for 23 sites. Based on this observation it was concluded that the two subpopulations should be treated separately. The median Hg content of the subpopulation collected from soil or rock ("ground feathermoss") was 95.3 ng g^{-1} while the median Hg content of the epiphytic subpopulation was 139.9 ng g^{-1} (Table 2).

Other species of terrestrial pleurocarpous moss, which included *Brachythecium plumosum,* showed a similarly wide range of Hg concentrations (Table 2). Shacklette[19] and Mouvet et al.[20] found that differences in metal accumulation capacity between moss species are so slight that they can be ignored for biomonitoring purposes. It would be advantageous to be able to ignore differences between species and pool the Hg data in Table 2 for all the pleurocarpous mosses. However, there were insufficient data comparing Hg contents of different species within the same sites to support this practice.

Of the terrestrial acrocarpous mosses, haircap moss (*Polytrichum commune*) was the most abundant, displayed a wide range of Hg concentrations (Table 2), and a strongly positively skewed frequency distribution. It was frequently found growing within a mat of pleurocarpous moss. Haircap moss has been previously cited as a useful indicator species for mineral prospecting.[2] *Dicranum polysetum* was the second most common acrocarpous species. The range of Hg concentrations of the acrocarpous mosses was nearly as wide as that of the pleurocarpous mosses, but the medians were much lower (Table 2). Enough within-site comparative data was collected to conclude that the Hg concentrations of the two moss groups, acrocarpous and pleurocarpous, clearly belong to different populations and should not be pooled.

Peat moss (*Sphagnum* sp.) was the only plant collected from low, wet land. Only living plants were sampled, and only the newest growth, i.e., the green tufts, were snipped off for analysis. Although Hg concentrations in *Sphagnum* were among the lowest of the mosses studied (Table 2), the polymodal frequency distribution of the between-site data suggests this genus may be an excellent indicator of differences in Hg levels between different wetlands.

2. Thallophytes

In general, the usefulness of the thallophytes as Hg bioindicators was limited by their sporadic occurrence in the forest environment. Because of the importance of the thallophytes as components of the food chain, however, several samples from different groups (summarized in Table 2) were collected for comparison with values reported in the literature.

The lichens displayed a narrow range of Hg concentrations (Table 2), comparable to the range of 4 to 40 ppb reported for samples of genus *Cladonia* collected in Alaska, Iceland, and Massachusetts.[21] *Cladonia* has been previously listed as a useful Hg indicator for mineral prospecting.[2] Bargagli et al.[22] found a close relationship (r = 0.807; N = 47) between Hg concentrations in soil and in the epiphytic lichen *Parmelia sulcata* near an abandoned cinnabar minesite on Mt. Amiata, Italy. They concluded that the lichen absorbed Hg from the atmosphere, the dominant source of which was local soil degassing.[22]

Hg levels in the mushrooms varied over a wide range of concentrations (Table 2), an observation previously made by others.[23,24] Kuusi et al.[23] determined that mean Hg contents of various mushroom species ranged from 0.03 to 4.1 μg g^{-1} in rural areas of Finland and 0.09 to 14.1 μg g^{-1} in the Helsinki area. Compared to these concentrations,[23,24] Hg levels in the mushrooms sampled in the Huntsville area were low.

Two forms of aquatic algae collected from streams (Table 2) contained Hg concentrations in the same range as previously reported for several lakes in this vicinity.[25] Attached filamentous algae have been proposed as a monitor of the bioavailability of Hg[25] and other metals.[26] No literature data were available for comparison with the Hg levels determined in the aquatic siderophilic bacteria (Table 2).

IV. ROLE OF VEGETATION IN INTRODUCING Hg INTO THE FOOD CHAIN

A review of the literature indicates that methylmercury, the most important metabolite, occurs in all types of environmental samples, but is more dominant in plants than in soil and sediments. The proportion of total Hg that is methylmercury in uncontaminated soil and sediment is generally reported to be less than 1 or 2%,[27,28] whereas in aquatic algae methylmercury comprises up to 100% of total Hg.[25] Much attention is focused on biomethylation of Hg in the water column, but equally important are reports that Hg can be methylated by multicellular plants.[24,29,30] Common agricultural plants can change inorganic Hg into organic forms of

Hg.[29] Cappon[31] determined that in garden vegetables, methylmercury comprises about 13% of total Hg.

Algae are highly effective absorbents of inorganic Hg, to the extent that algae have been proposed by Barkley[32] as a "biological exchange resin" to remove Hg from industrially contaminated groundwater. Barkley's study[32] indicated that the absorption of inorganic Hg by algae was effective at varied levels of pH, alkalinity, and TDS concentration, unlike conventional ion exchange resins.[32] Neither bivalent nor monovalent cations at high concentrations (10,000 mg L^{-1}) interfered with the metal binding capacity of the algae.[32]

Since algae have the capacity to effectively absorb inorganic Hg in diverse geochemical conditions,[32] it is probable that algae play a key role in Hg biomagnification in widely different settings—from the hardwater, neutral to alkaline lakes of the Florida Everglades to the softwater, acidic lakes of the Precambrian Shield. Models aimed at predicting Hg levels in fish and mammals need to include the important function of vegetation in introducing Hg into the aquatic and terrestrial food chain.

V. CONCLUSIONS AND RECOMMENDATIONS

1.   An examination of sources of variation indicated that Hg contents vary by an average of 18% COV between plants of the same species within one site (defined as an area of 10 m^2) using strict sampling and analytical controls. Sampling at least three separate plants of one species is recommended to obtain a representative Hg value for one site.

2.   All species which were sampled at 10 or more sites displayed greater between-site variation than within-site variation. Frequency distribution curves of between-site Hg concentrations were polymodal or strongly positively skewed. Data from a minimum of 20 separate sites were required to gain a sense of the variability in Hg content that could be expected in a species. Sampling from a minimum of 50 sites in a study area of 150 km^2 is recommended to statistically define background and anomalous concentration in one species.

3.   Species which displayed the widest variation in Hg contents between sites, such as certain mosses, had the greatest potential as Hg bioindicators while those whose between-site variation was only marginally greater than within-site variation, such as certain trees, had the least potential. The best bioindicators display a wide range of Hg concentrations with a bimodal or strongly skewed frequency distribution, reflecting well-separated background and anomalous concentration populations.

4.   Considering both factors of abundance in the study area and Hg accumulation behavior, the following species were identified as potentially useful Hg bioindicators: red-stemmed feathermoss, haircap moss, tree clubmoss, and wild lily-of-the-valley (for high, dry ground); living peat moss (for low, wet land); and pickerelweed (for streams).

5.   The usefulness of thallophytes as bioindicators was limited by their sporadic occurrence in the terrestrial environment. In terms of Hg bioaccumulation, however, thallophytes probably play a key role in introducing Hg into the food chain in a wide range of geochemical conditions.

ACKNOWLEDGMENTS

Sincere thanks go to Andrew Devaney for his capable assistance in the field, to Greg Mierle and Ron Ingram (Ontario Ministry of the Environment) for the use of their clean lab facilities, to Jon Van Loon (University of Toronto) and Dave Gardner (Ontario Ministry of Natural Resources) for critically reading an early version of the manuscript, and to Pak Yau Wong, Robert Ireland and A. Dugal (Botany Department, National Museum of Natural Sciences, Ottawa) for species identification. This study is part of the author's Ph.D. Thesis (under the supervision of Jerome Nriagu and Sherry Schiff, Earth Sciences Department, University of

Waterloo) and was jointly funded by the Ontario Ministry of the Environment, an NSERC Strategic Grant (P.I. Pamela Welbourn) and Ontario Graduate Scholarship.

REFERENCES

1. Jonasson, I.R. and Boyle, R.W., The biogeochemistry of mercury, In: *Effects of Mercury in the Canadian Environment,* N.R.C.C. Publication No. 16739, Ottawa, Canada, 1979, chap. 2.

2. Kovalevsky, A.L., Mercury-biogeochemical exploration for mineral deposits, *Biogeochemistry,* 2, 211, 1986.

3. Jonasson, I.R. and Boyle, R.W., Geochemistry of mercury and origins of natural contamination of the environment, *Bull. Can. Inst. Min. Metal.,* 65, 32, 1972.

4. Siegel, S.M. and Siegel, B.Z., Temperature determinants of plant-soil-air mercury relationships, *Water Air Soil Pollut.,* 40, 443, 1988.

5. Hecky, R.E., Bodaly, R.A., Strange, N.E., Ramsay, D.J., Anema, C., and Fudge, R.J.P., Mercury bioaccumulation in yellow perch in limnocorrals simulating the effects of reservoir formation, *Can. Data Rep. Fish. Aquat. Sci.,* 628, 1987, (v + 158 p).

6. Nriagu, J.O., A global assessment of natural sources of atmospheric trace metals, *Nature,* 338, 47, 1989.

7. Siegel, S.M., Siegel, B.Z., Lipp, C., Kruckberg, A., Towers, G.H.N., and Warren, H., Indicator plant-soil mercury patterns in a mercury-rich mining area of British Columbia, *Water Air Soil Pollut.,* 25, 73, 1985.

8. Rasmussen, P.E., Mierle, G., and Nriagu, J.O.N., The analysis of vegetation for total mercury, *Water Air Soil Pollut.,* 56, 379, 1991.

9. Ontario Ministry of the Environment and Ontario Ministry of Natural Resources, Guide to Eating Ontario Sport Fish, 14th ed., Queen's Printer for Ontario, Toronto, 1990, 161 pp.

10. Rasmussen, P.E., The Environmental Significance of Geological Sources of Mercury: A Precambrian Shield Watershed Study. Ph.D. thesis, Earth Sciences Department, University of Waterloo, Ontario, Canada, 1993, 379 p.

11. Sinclair, A.J., Selection of threshold values in geochemical data using probability graphs, *J. Geochem. Explor.,* 3, 129, 1974.

12. Miesch, A.T., Estimation of the geochemical threshold and its statistical significance, *J. Geochem. Explor.,* 16, 49, 1981.

13. Dunn, C.E., Biogeochemistry as an aid to exploration for gold, platinum and palladium in the northern forests of Saskatchewan Canada, *J. Geochem. Explor.,* 25, 21, 1986.

14. Cohen, D.R., Hoffman, E.L., and Nichol, I., Biogeochemistry: a geochemical method for gold exploration in the Canadian Shield, *J. Geochem. Explor.,* 29, 49, 1987.

15. Belanger, J.R., Prospecting in glaciated terrain: an approach based on geobotany, biogeochemistry, and remote sensing, *Geol. Surv. Can. Bull.,* 387, 38, 1988.

16. Erdman, J.A., Biogeochemistry in mineral exploration, *J. Geochem. Explor.,* 21, 123, 1984.

17. Satake, K., Shibata, K., and Bando, Y., Mercury sulphide (HgS) crystals in the cell walls of the aquatic bryophytes, *Jungermannia vulcanicola* Steph. and *Scapania undulata* (L.) Dum., *Aquat. Bot,* 36, 325, 1990.

18. Hutchinson, T.C., Workshop Summary, In: *Lead, Mercury, Cadmium and Arsenic in the Environment,* (SCOPE 31), Hutchinson,T.C. and Meema, K.M., Eds., John Wiley & Sons, New York, 1987.

19. Shacklette, H.T., The use of aquatic bryophytes in prospecting. *J. Geochem. Explor.,* 21, 89, 1984.

20. Mouvet, C., Andre, B., and Lascombe, C., Aquatic mosses for the monitoring of heavy metals in running freshwaters: comparison with sediments, In: *Heavy Metals in the Environment,* Lindberg, S.E. and Hutchinson, T.C., Eds., CEP Consultants, Edinburgh, U.K., 2, 1987, 424.

21. Siegel, S.M., Siegel, B.Z., Puerner, N., Speitel, T., and Thorarinsson, F., Water and soil biotic relations in mercury distribution, *Water Air Soil Pollut.,* 4, 9, 1975.

22. Bargagli, R., Barghigiani, C., Siegel, B.Z., and Siegel, S.M., Accumulation of mercury and other metals by the lichen *Parmelia sulcata* at an Italian minesite and a volcanic area, *Water Air Soil Pollut.,* 45, 315, 1989.

23. Kuusi, T., Laaksovirta, K., Liukkonen-Lilja, H., Lodenius, M., and Piepponen, S., Lead, cadmium and mercury contents of fungi in the Helsinki area and in unpolluted control areas, *Z. Lebensm. Unters. Forsch.,* 173, 261, 1981.

24. Kojo, M.-R. and Lodenius, M., Cadmium and mercury in macrofungi—mechanisms of transport and accumulation, *Angew. Botanik,* 63, 279, 1989.

25. Stokes, P.M., Dreier, S.I., Farkas, M.O., and McLean, A.N., Mercury accumulation by filamentous algae: a promising biological monitoring system for methyl mercury in acid-stressed lakes, *Environ. Pollut. (Ser. B),* 5, 1983, 255.

26. Bailey, R.C. and Stokes, P.M., Evaluation of filamentous algae as biomonitors of metal accumulation in softwater lakes: a multivariate approach, In: *Aquatic Toxicology and Hazard Assessment: Seventh Symposium,,* ASTM STP 854, Cardwell, R.D., Purdy, R., and Bahner, R.C., Eds., American Society for Testing and Materials, Philadelphia, 1985, 5.

27. Di Giulio, R.T. and Ryan, E.A., Mercury in soils, sediments and clams from a North Carolina peatland, *Water Air Soil Pollut.,* 33, 205, 1989.

28. Revis, M.W., Osborne, T.R., Holdsworth, G., and Hadden, C., Distribution of mercury species in soil from a mercury-contaminated site, *Water Air Soil Pollut.,* 45, 105, 1989.

29. Rogers, R.D. and Gay, D.D., Methylation of Mercury in a Terrestrial Environment, National Technical Information Service, Springfield, VA, Report No. PB 248221/AS, 1975.

30. Bressa, G., Cima, L., and Costa, P., Methylation of mercury by the mushroom *Pleurotus ostreatus,* In: *Heavy Metals in the Environment,* Lindberg, S.E. and Hutchinson, T.C., Eds., CEP Consultants, Edinburgh, U.K., 2, 1987, 258.

31. Cappon, C.J., Uptake and speciation of mercury and selenium in vegetable crops grown on compost-treated soil, *Water Air Soil Pollut.,* 34, 353, 1987.

32. Barkley, N.P., Extraction of mercury from groundwater using immobilized algae, *J. Air Waste Manage. Assoc.,* 41, 1387, 1991.

You Are What You Eat and a Little Bit More: Bioenergetics-Based Models of Methylmercury Accumulation in Fish Revisited

D. W. Rodgers

CONTENTS

ABSTRACT: Bioenergetics-based equations of contaminant accumulation were coupled to an existing fish bioenergetics model to describe methylmercury (MeHg) accumulation of yellow perch and lake trout under a variety of environmental conditions. Diet accounted for the large majority of MeHg uptake in simulations incorporating the much lower concentrations of waterborne MeHg which are consistent with recent measurements. Consequently, MeHg accumulation was particularly sensitive to factors affecting uptake of dietary MeHg. These include not only the MeHg concentration of the diet and the efficiency of MeHg assimilation, but also factors affecting total food consumption, such as the caloric content of the diet. Predicted MeHg accumulation by fish was also significantly affected by the clearance term, with a temperature dependent elimination term providing the best fit to the data. Bioenergetics-based models thus provide a robust process for describing MeHg accumulation in fish which can utilize any available information with respect to fish biology (water temperature, diet, and growth rates) and ambient MeHg concentration. These models may be especially useful in identifying critical processes and pathways contributing to elevated MeHg concentrations in fish and in assessing the effects of environmental perturbation.

I. INTRODUCTION

Regardless of its source, the principal concern with mercury in aquatic ecosystems is the accumulation of mercury as methylmercury (MeHg) by fish. Methylmercury accumulation by fish is of particular concern because consumption of MeHg-contaminated fish comprises the major route for transfer of mercury from the aquatic environment to fish-eating birds and mammals, including man. Methylmercury may be formed from inorganic mercury and other organic mercury compounds by a variety of biological and abiological mechanisms. In freshwater environments, the majority of MeHg appears to be generated by bacteria in the superficial sediments and water column.[1,2] Once formed, methylmercury is efficiently accumulated and transferred by biota, including fish. Fish accumulate MeHg as a result of their efficient uptake from water and diet,[3-7] coupled to their very slow clearance of assimilated MeHg.[7-11]

Because of their critical position in the transfer of MeHg to man, measurements of mercury in fish are routinely available for many waters. Measurements of mercury in fish are generally reliable, because the mercury concentrations in fish are sufficient (>0.01 µg/g) that the analytical procedures are relatively straightforward, and are not subject to the extreme problems associated with measurement of mercury and MeHg in ambient water. Further, supplementary data of fish growth rates, diet, and water temperature are frequently available from fisheries organizations or can be inferred from the species life history.

Consistent with the central role of fish in the dynamics of environmental mercury, a variety of models have been applied to describe MeHg accumulation by fish.[12-15] Bioenergetics-based models[12,13] have been particularly useful in describing MeHg accumulation by fish as common units of energy are used to describe uptake of MeHg from food and water, and elimination of assimilated MeHg. These models have been employed to examine a variety of aspects of MeHg accumulation in fish, ranging from species differences in MeHg accumulation of pike and walleye in Lake Simcoe, Ontario[16] to the potential effects of acidification of MeHg accumulation in walleye.[17] A bioenergetics-based approach has also comprised a critical component in general models of contaminant accumulation in aquatic food chains.[18,19] However, bioenergetics-based models have not been incorporated into many system models of the environmental behavior of mercury (i.e., Blaylock[14]). even though the uncertainty concerning MeHg accumulation by fish constitutes a significant component of the total uncertainty of these models. Harris' model of mercury transfer in aquatic systems,[15] which uses the bioenergetics-based model of Norstrom et al.[13] to describe MeHg uptake by fish, is a notable exception in this regard.

Unfortunately, current models of MeHg accumulation in fish have largely failed to provide a robust mechanism for comparative analysis of MeHg accumulation in fish using the relatively limited data which are normally available from field surveys. In this chapter, I develop a flexible framework for coupling expressions describing the uptake and loss of MeHg by fish to estimates of fish metabolism and growth derived from an existing fish bioenergetics model. The framework is then applied to describe MeHg accumulation of yellow perch and lake trout, using field growth rates and temperature regimes.

II. MODELS AND EQUATIONS

A. BIOENERGETICS

To estimate bioenergetic parameters, I employed the PC-based *A Generalized Bioenergetics Model of Fish Growth*© [20] developed by Hewett and Johnson at the University of Wisconsin-Madison. The model summarizes bioenergetic data for 12 species of freshwater fish, including the major North American salmonids, percids, and esocids.* The model, or its components,

---

*As this manuscript was in final review, *Fish Bioenergetics₂ Model,* an upgrade of the *Generalized Bioenergetics Model of Fish Growth for Microcomputers* was issued. The upgrade provides additional species and is much more user friendly than the original model.

have been applied to a range of fisheries investigations, with a variety of fish species including sea lamprey,[21] alewife,[22] walleye and yellow perch,[23] and lake trout.[24] Additional species can be added if information on their bioenergetics is available or can be inferred from the existing data base. The model can be run in a variety of ways, depending on the questions asked and the information available. In my applications, I used the model to estimate the food consumption (**R,** g/week) and metabolism of fish (**Q,** kcal/week) under specified field growth and temperature regimes. To be consistent with the earlier simulations of Norstrom et al.,[13] all computations were iterated and saved weekly, with the resulting output providing the basis for subsequent calculations of MeHg accumulation.

B. MeHg ACCUMULATION

Following the model of Norstrom et al.,[13] the change in MeHg (dP/dt) within a fish was equated to the difference between uptake from diet and water and clearance of assimilated MeHg, with all terms described in common units of energy.

$$dP/dt = \text{Uptake (diet)} + \text{Uptake (water)} - \text{Clearance} \qquad (1)$$

MeHg uptake from diet was calculated as the product of the efficiency of uptake of MeHg from the diet (e_{pf}—reference value 0.8), the concentration of MeHg in the diet (C_{pf} g/g), and the total food consumption (R), calculated in the bioenergetics program.

$$\text{Uptake (diet)} = e_{pf} * C_{pf} * R \qquad (2)$$

Uptake of MeHg from water was calculated as the product of the efficiency of uptake of MeHg from water (e_{pw} reference value 0.12), the concentration of MeHg in the water (C_{pw} g/g) and the quantity of water passing over the gills of the fish V (g).

$$\text{Uptake (water)} = e_{pw} * C_{pw} * V \qquad (3)$$

The quantity of water passing over the gills of the fish was determined from the metabolic rate of the fish Q (kcal), which was calculated in the bioenergetics program, divided by efficiency of oxygen uptake from water (e_{ox} reference value 0.75), with this quantity multiplied by the caloric equivalent of oxygen (q_{ox} reference value 3.42 kcal/g) and the concentration of oxygen in the water (C_{ox}).

$$V = (Q/e_{ox}) * C_{ox} * q_{ox} \qquad (3a)$$

It was assumed that the water was saturated with oxygen at all times, with C_{ox} (mg O_2/L) calculated as a function of water temperature (T, $^\circ$C), for $0 \leq T \geq 25$.

$$C_{ox} = 14.45 - (0.413 * T) + (0.00556 * T^2) \qquad (3b)$$

Alternately, in Model 2, MeHg uptake from water was calculated from the relationship between MeHg uptake and the product of oxygen consumption (O_2con) and MeHg concentration observed by Rodgers and Beamish[6] for rainbow trout. Following Equation 6 of Rodgers and Beamish:[6]

$$\text{Uptake (water)} = \{10^{1.445} * [(O_2\text{con} * C_{pw} * 10^9)^{1.134}]\} * (168/10^9) \qquad (4)$$

The factors of 10^9 and $168/10^9$ converted the units used by Rodgers and Beamish[6] to the g/week employed by Norstrom et al.[13] O_2con (mg O_2/h) was calculated by dividing Q by $(168 * q_{ox})$.

Following Norstrom et al.,[13] clearance of incorporated MeHg was calculated as the product of W^ζ, the weight of the fish (g) raised to the power ζ (reference value 0.58), the body burden of pollutant in the fish (Pg), and the clearance coefficient (k_{cl} reference value 0.202 $g^{-\zeta}$/wk).

$$\text{Clearance} = W^\zeta * P * k_{cl} \tag{5}$$

In alternate equations, the clearance coefficient k_{cl} was made dependent on temperature either by setting clearance to zero at $T < 5°$ C,

$$k_{cl} = 0 \text{ at } T < 5° C \tag{5a}$$

or by making k_{cl} dependent on temperature with a Q_{10} of 2, using a base temperature (Tb) of $15°$ C for yellow perch and $10°$ C for lake trout:

$$k_{cl}(T) = k_{cl} * 2^{((T-Tb)/10)} \tag{5b}$$

III. SIMULATIONS

A. OTTAWA RIVER PERCH

As described by Norstrom et al.,[13] an Ottawa River yellow perch was assumed to hatch in the 18th week of the year, with growth defined by a von Bertalanffy equation

$$L = a * (1-e^{-b * (t-c)}) \tag{6}$$

where L is length (mm), t is age in weeks, and the parameters a, b, and c were 343 mm, 0.00313/week, and -26 weeks, respectively. Growth in length was converted to weight by the equation

$$V = \lambda * L^\eta \tag{6a}$$

where V is the predicted weight of the fish, and the constants λ and η were calculated as $17.4 * 10^{-6}$ g/mm^η and 3.02, respectively. The 1972 Ottawa River temperature regime of Norstrom et al.[13] was employed in all perch simulations, as was their value for MeHg concentration in the diet ($C_{pf} = 33 * 10^{-9}$ g/g) and caloric content of diet (1 kcal/g).

The estimated food consumption and water flow across the gills calculated for yellow perch from age 0 to 5+, were approximately 10% greater using the bioenergetics model of Hewett and Johnson[20] than the values estimated by Norstrom et al.[13] (Table 1). In comparing specific cohorts, the model of Hewett and Johnson[20] gave substantially lower estimates of annual food consumption and water flow for age 0+ yellow perch, but higher estimates of food consumption and water flow for yellow perch from age 3+ on, with the differences increasing with age. The differences between the two models seem to lie within the limits expected in bioenergetic modeling, given the known difficulties in estimating food consumption of wild fish, particularly young fish.[25] Further, Norstrom et al.[13] noted that for the first growing season of the perch, their predicted metabolic rate substantially exceeded their estimates of active metabolism (the theoretical limit of metabolism).

When the bioenergetic output from the Hewett and Johnson model was incorporated into the MeHg accumulation model, the predicted MeHg concentration of a 150-g yellow perch

Table 1　**Annual food consumption and water flow across gills as estimated from bioenergetic models of Norstrom et al.[13] and Hewett and Johnson.[20]**

Age	Estimated annual food consumption (Kcal/year)		Estimated annual water flow across gills (G * 10⁶/year)	
	Norstrom	Hewett	Norstrom	Hewett
0+	33.5	24.3	0.82	0.43
1+	106	104	2.69	2.45
2+	204	210	5.35	5.34
3+	307	333	8.25	8.87
4+	410	461	11.2	12.7
5+	498	584	13.9	16.6
0–5+	1559	1716	42.2	46.4

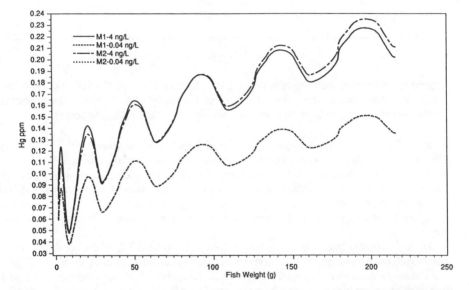

Figure 1　Estimated mercury concentrations of Ottawa River yellow perch using uptake Models 1 and 2 at waterborne MeHg concentrations of 4 and 0.04 ng/L. Note that at waterborne MeHg concentrations of 0.04 ng/L the two models are virtually superimposed.

was ≈0.2 μg/g (Figure 1) at a waterborne MeHg concentration of 4 ng/L (the reference value of the model of Norstrom et al.[13]). This concentration (0.2 μg/g) was roughly one third greater than the concentration of ≈0.15 μg/g reported by Norstrom et al.[13] At a waterborne MeHg concentration of 0.04 ng/L, however, the predicted MeHg concentration of a 150-g yellow perch was 0.13 to 0.14 μg (Figure 1), which was ≈10% lower than the observed values. Measured concentrations of waterborne mercury have decreased by more than two orders of magnitude in the past decade, largely as a result of "clean" sampling and laboratory methods.[1] Consequently, waterborne MeHg concentrations in the range of 0.04 ng/L are more likely within the expected range of natural waters than values of 4.0 ng/L.[1,15] The model of Norstrom et al.[13] may thus have achieved a reasonable fit to observed perch MeHg concentrations through a serendipitous combination of underestimation of bioenergetic parameters and overestimation of waterborne MeHg concentration.

The most significant effect of decreasing the concentration of waterborne MeHg was the accompanying shift in the relative proportion of MeHg uptake from water and diet, with

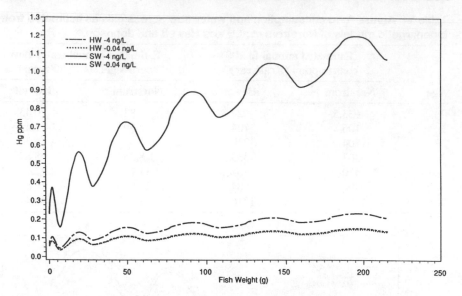

Figure 2 Estimated mercury concentrations of Ottawa River yellow perch in hard and soft water at waterborne MeHg concentrations of 4 and 0.04 ng/L. Note that at waterborne MeHg concentrations of 0.04 ng/L the two models are virtually superimposed.

dietary uptake becoming paramount. At a C_{pw} of 4 ng/L, diet accounted for 70 to 75% of MeHg uptake, with the relative significance of diet decreasing slightly with age; but at a C_{pw} of 0.04 ng/L, diet accounted for greater than 95% of MeHg uptake at all ages. Although consistent with the earlier simulations of Norstrom et al.,[13] the present results clearly indicate that at these low concentrations of waterborne MeHg, accumulation will be most sensitive to factors affecting uptake from diet but relatively insensitive to factors affecting uptake from water.

Uptake of MeHg from water was marginally greater in Model 2, where uptake of waterborne MeHg was described by Equation 4 rather than Equation 3 of Norstrom et al.[13] (Figure 1). The slight differences were only apparent at the higher MeHg concentration (4 ng/L), however, and even then only for the larger and older fish (Figure 1). Accordingly, Model 2 was used in subsequent simulations, as it was based on the observed relationship between MeHg uptake and the product of oxygen consumption and MeHg concentration,[6] and was less complicated to compute.

Another advantage of Model 2 was that it was easily modified to incorporate changes in the relative efficiency of uptake of waterborne MeHg. Thus, the threefold increase in the efficiency of uptake of MeHg relative to oxygen observed in soft water[26] was incorporated into Equation 4 as

$$\text{Uptake (water)} = (10^{1.445 + \log(3)}) * [(O_2\text{con} * C_{pw} * 10^9)^{1.134}] * (168/10^9) \quad (4a)$$

When the threefold increase in the efficiency of uptake of waterborne MeHg observed in soft water was incorporated into the Ottawa River perch simulations, the estimated MeHg concentration of a 150-g perch increased approximately fivefold to ≈ 1 µg/g at a C_{pw} of 4 ng/L (Figure 2). However, the predicted MeHg concentration of a 150-g perch increased by less than 5% at a C_{pw} of 0.04 ng/L; again reflecting the relative lack of importance of MeHg uptake from water at this low concentration of waterborne MeHg.

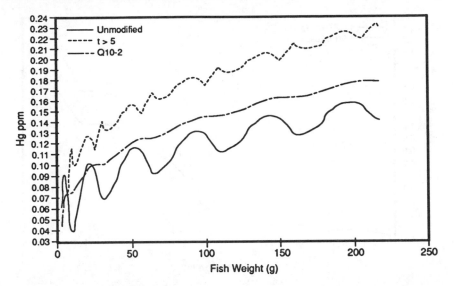

Figure 3 Estimated mercury concentrations of Ottawa River yellow perch using three clearance functions at waterborne MeHg concentrations of 0.04 ng/L.

As applied in these simulations, the model of Norstrom et al.[13] produced dramatic annual oscillations in MeHg concentration (Figures 1 and 2) with significant decrease of MeHg in the overwintering period. The overwinter decrease of MeHg resulted from the reduction of the temperature-dependent uptake functions to very low values in the overwinter period, while loss of MeHg continued at much the same rate as clearance was independent of temperature and metabolic rate. Because clearance was inversely proportional to weight ($W^{-0.58}$), the cycles were most apparent for small fish, with the amplitude of the oscillations decreasing as the fish grew. However, existing field studies indicate that both organic and total mercury concentrations are relatively stable in overwintering fish, including yearling Ottawa River yellow perch.[27] Consequently, the clearance term was modified to incorporate a temperature-dependent term either by eliminating clearance at T $<5°$ C (Equation 5a) or making K_{cl} dependent on temperature by introducing a Q_{10} factor (Equation 5b).

When the modified elimination terms were incorporated into the Ottawa River perch simulations, the amplitude of the annual oscillations was markedly damped (Figure 3). At a C_{pw} of 0.04 ng/L, the predicted MeHg concentration of a 150-g perch increased to ≈0.2 µg/g if there was no elimination at T $<5°$ C. This is roughly equivalent to the MeHg concentration the perch attained at a C_{pw} of 4.0 ng/L using the unmodified model. With the Q_{10} model, the predicted MeHg concentration of a 150-g perch was ≈0.155 µg/g, at a C_{pw} of 0.04 ng/L, which corresponds well with the observed values. Accordingly, the use of the Q_{10} temperature-dependent elimination term would appear to provide the best fit to the Ottawa River perch data.

B. TADENAC LAKE TROUT

These simulations are based on a study by MacCrimmon et al.[28] of mercury accumulation of lake trout in Tadenac Lake, a Precambrian Shield lake in central Ontario, without apparent input of excess mercury from either natural or anthropogenic sources. In brief, 2-year-old lake trout (47 g, 0.04 µg/g Hg) were stocked and then recaptured from 2 to 6 years after stocking; larger and older native lake trout were also caught over the same time period. The lake trout consumed a diet comprised almost entirely of invertebrates from age 2 to 4; between ages 4 to 5 the trout shifted to a mixed diet of invertebrates and fish (largely rainbow smelt);

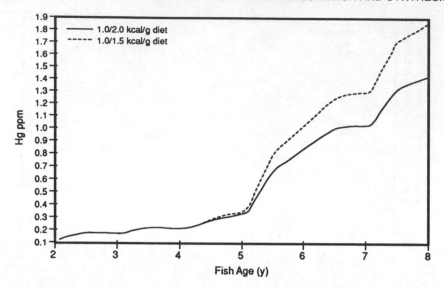

Figure 4 Estimated mercury concentrations of Tadenac Lake lake trout as affected by the caloric content of the diet.

from age 6 on, the trout consumed only fish. Coincident with the change in diet, both growth and MeHg accumulation of the lake trout increased dramatically between 4 to 6 years of age, so that at age 8, the lake trout have a mean weight of ≈1660 g and a mercury concentration of ≈1.0 µg/g.[28] In these simulations, I used a value of 0.04 ng/L for the concentration of waterborne MeHg, and dietary MeHg concentrations of 0.033 µg/g for invertebrates[13] and 0.26 for fish.[28] Also, these simulations used the unmodified elimination function of Norstrom et al.[13] because of the relatively large size of these fish, and it was assumed that the fish spawn annually at age 6 and beyond, losing 6.8% of their mass on spawning.[20]

At a waterborne MeHg concentration of 0.04 ng/L, more than 98% of estimated MeHg uptake of the lake trout was through diet. Consequently, MeHg accumulation was particularly sensitive to changes in variables affecting food consumption. Figure 4 illustrates the effect of the caloric content of the prey fish (smelt) on MeHg accumulation of the lake trout. Because the estimated food consumption decreases with an increase in the caloric content of the diet, a 33% increase in the caloric value of smelt, from 1500 to 2000 cal/g, resulted in an equivalent decrease (≈31%) in the predicted MeHg concentration of an 8-year-old lake trout. Under either dietary regime, the predicted MeHg concentrations of ≈1.9 µg/g at a smelt caloric value of 1500 cal/g and ≈1.45 cal/g at a smelt caloric value of 2000 cal/g are substantially greater than the observed values of ≈1.0 µg/g. This overestimate is exacerbated by the fact that measured values of the caloric value of smelt appear to lie in the range of 1500 cal/g[29] rather than the 2000 cal/g which is representative of adult alewife.[20] However, the predicted MeHg concentrations of the lake trout could be reduced somewhat by using the average MeHg concentration of a 15-cm smelt (0.22 µg/g) for the dietary MeHg concentration rather than the average MeHg concentration of 0.26 µg/g.

The value of the assimilation efficiency of dietary MeHg (e_{pf}) also significantly affected predicted MeHg accumulation, as seen in Figure 5, which demonstrates the effect of changes in the value of e_{pf} from 0.8 to 0.2. The value of 0.8 used by Norstrom et al.[13] lies at the higher end of range observed in controlled experimental conditions;[7,13] 0.65 is at the lower end of this range, and a value of 0.2 has been suggested by Phillips and Gregory[30] for pike consuming a "naturally contaminated" diet. In the present simulations, a value of 0.65

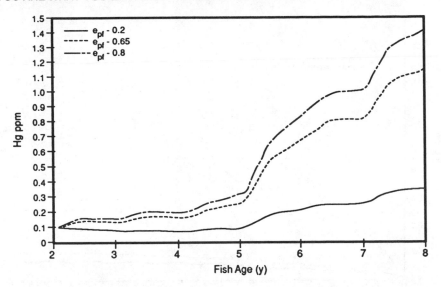

Figure 5 Estimated mercury concentrations of Tadenac Lake lake trout as affected by efficiency of assimilation (e_{pf}) of dietary MeHg.

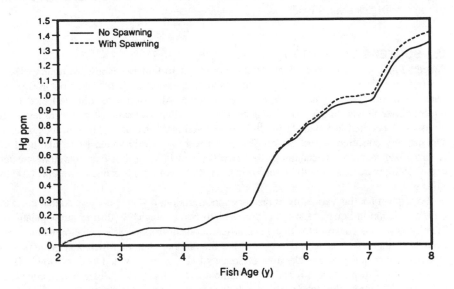

Figure 6 Effects of spawning on estimated mercury concentrations of Tadenac Lake lake trout.

appeared to simulate most closely the observed MeHg concentrations, while a value of 0.2 was insufficient to account for the observed MeHg accumulation.

The importance of food consumption in determining MeHg accumulation of the lake trout is further emphasized in Figure 6, which compares predicted MeHg accumulation of the fish with and without spawning. When the weight loss associated with spawning was included, the total food consumption required to meet observed growth targets increased, with a corresponding increase in predicted MeHg accumulation.

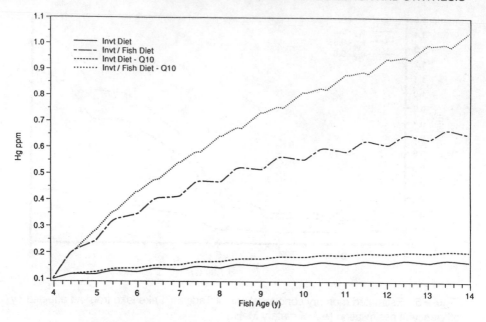

Figure 7 Estimated mercury concentrations of Squeers Lake lake trout as affected by diet and MeHg clearance functions.

C. SQUEERS LAKE TROUT

Squeers Lake, a Precambrian Shield lake in northwestern Ontario, is also without apparent input of excess mercury from either natural or anthropogenic sources, but lake trout in Squeers Lake grow more slowly than those of Tadenac Lake. Although lake trout collected from Squeers Lake ranged in age from less than 4 to more than 18 years, all fish were less than 50 cm and weighed less than 1150g.[31] The observed relations among length, weight, age, and mercury concentration of Squeers Lake fish were relatively weak. Most lake trout of all sizes had mercury concentrations between 0.1 and 0.2 μ/g, but concentrations greater than 0.25 μg were occasionally observed in a few fish over a wide range of weights (200 to 1100 g).[31]

To account for this variability in mercury concentration, it was hypothesized that the lake trout population in Squeers Lake was divided into two ecologically distinct subpopulations. This hypothesis is consistent with general observations of trophic differentiation within populations of the genus *Salvelinus*.[32] Fish in the first subpopulation, which was presumed to represent the majority of the lake trout in Squeers Lake, were assumed to grow slowly (250 g at age 4 to 450 g at age 14) and to consume a low caloric, low MeHg invertebrate diet (759 cal/g; 12.5 ng/g) throughout their life. In the second subpopulation, which was presumed to be relatively uncommon, fish were assumed to grow more rapidly (250 g at age 4 to 625 g at age 14) and to consume a mixed diet of invertebrates and fish, between ages 4 to 6, but then consume only fish (1500 cal/g; 100 ng/g) from age 6 on. In these simulations, a value of 0.04 ng/L was used for the concentration of waterborne MeHg, and it was assumed that the fish spawn annually at age 8 and beyond, losing 6.8% of their mass on spawning.

The predicted MeHg concentrations of the two subpopulations are illustrated in Figure 7. Lake trout consuming the low caloric, low MeHg invertebrate diet maintain a relatively low MeHg concentration (<0.2 μg/g) throughout the simulation, while lake trout switching to a fish-based diet continue to accumulate much higher concentrations of MeHg. As with Ottawa River perch, the use of a Q_{10} temperature-dependent elimination term substantially damped annual oscillations in MeHg concentration but also resulted in increased MeHg concentrations.

A sample of lake trout from Squeers Lake would thus largely consist of fish from the slow-growing population with low MeHg concentrations. A few fish from the second subpopulation, with much higher MeHg concentrations, would also be caught, however, with their frequency in the sample contingent on the prevalence of the second subpopulation. The next step of this modeling process will likely include the use of probabilistic models, in which population responses are estimated from a series of individuals exposed to statistical distributions of key parameters.[33]

IV. SUMMARY AND CONCLUSIONS

Bioenergetics-based models provide a robust mechanism to predict MeHg concentrations in fish by applying relatively uncomplicated field data such as observed growth, diet, and temperature regime in combination with available information concerning ambient MeHg concentrations. These models may be especially useful in identifying critical processes and pathways contributing to elevated MeHg concentrations in fish and in predicting the effects of environmental perturbation.

Diet was responsible for the large majority of MeHg uptake in simulations employing the much lower waterborne MeHg concentrations which are consistent with recent measurements. Accordingly, simulations of MeHg accumulation were particularly sensitive to factors affecting uptake of dietary MeHg and relatively insensitive to factors affecting uptake of waterborne MeHg. Factors affecting uptake of MeHg from the diet included not only the obvious factors such as the concentration of MeHg in the diet and the efficiency of MeHg assimilation, but also factors affecting total food consumption such as the caloric value of the diet. Given the critical importance of diet to MeHg accumulation by fish in the above simulations, detailed experimental studies of the factors affecting uptake of MeHg from diet are clearly required.

The simulations of MeHg accumulation by fish were also significantly affected by the clearance term. Use of the clearance term from an earlier model of mercury accumulation in Ottawa River perch,[13] in which clearance was independent of temperature and metabolic rate, produced dramatic annual oscillations in predicted MeHg concentrations, with a significant decrease of MeHg in the overwintering period. The overwinter decrease of MeHg resulted from the reduction of the temperature-dependent uptake functions to very low values during the overwinter period, while loss of MeHg continued unabated. When modified temperature-dependent elimination terms were incorporated into the Ottawa River perch simulations, the amplitude of the annual oscillations was markedly damped, with a Q_{10} temperature-dependent elimination term providing the best fit to the Ottawa River perch data. Information on the factors affecting of MeHg clearance is limited, in part because of the slow rates of clearance, particularly in large fish. Again, detailed studies are required to resolve the effects of key metabolic parameters on MeHg clearance in different species of fish. These studies should be integrated with the studies of dietary uptake recommended above.

The above simulations tracked MeHg accumulation of an average fish exposed to "average" environmental conditions. In the "real" world, however, individual fish are unlikely to encounter "average" environmental conditions. Rather, an individual fish may be expected to encounter a range of MeHg concentrations, in both diet and water. Thus, factors such as the variability of ambient MeHg concentrations or dietary selection may be critical to understanding the accumulation of MeHg in individual fish. Probabilistic models, in which population responses are estimated from a collection of individuals exposed to statistical distributions of key parameters, may be the next step of this modeling process.

ACKNOWLEDGMENTS

I thank Tony deFreitas and Reed Harris for numerous long and tortuous discussions in which various aspects of this paper were fleshed out. I also thank Chris Wren and John Parks for providing data on lake trout in Tadenac and Squeers Lakes, respectively.

REFERENCES

1. EPRI Technical Brief. Mercury in the Environment, TB.ENV.50.8.90., 1990.
2. Beijer, K. and Jernelov, A., Methylation of mercury in aquatic environments, In: *The Biogeochemistry of Mercury in the Environment,* Nriagu, J.O., Ed., Elvsevier/North Holland Biomedical Press, New York, 1979, 203.
3. Olsen, K.R., Bergman, H.L., and Fromm, P.O., Uptake of methylmercury chloride and mercuric chloride by trout: a study of uptake pathways into the whole animal and by erythrocytes in vitro, *J. Fish. Res. Board Can.,* 30, 1293, 1973.
4. deFreitas, A.S.W., Gidney, M.J., McKinnon, A.E., and Norstrom, R.E., Factors affecting whole body retention of methylmercury in fish, in Proc. Hanford Life Sciences Symposium on Biological Implications of Metals in the Environment, Sept. 29–Oct. 1, Richland Wash., Energy Res. Dev. Adm. Symp. Ser. 15, 1975, 441.
5. Phillips, G.R. and Buhler, D.R., The relative contributions of methylmercury from food or water to rainbow trout (*Salmo gairdneri*) in a controlled laboratory environment, *Trans. Am. Fish. Soc.,* 107, 853, 1978.
6. Rodgers, D.W. and Beamish, F.W.H., Uptake of waterborne methylmercury by rainbow trout (*Salmo gairdneri*) in relation to oxygen consumption and methylmercury concentration. *Can. J. Fish. Aquat. Sci.,* 38, 1309, 1981.
7. Rodgers, D.W. and Beamish, F.W.H., Dynamics of dietary methylmercury in rainbow trout, *Salmo gairdneri, Aquat. Toxicol.,* 2, 271, 1982.
8. Jarvenpaa, T., Tillander, M., and Miettinen, J.K., Methylmercury: half-time of elimination in flounder, pike and eel, *Suom. Kem.,* B43, 439, 1970.
9. Ruohulta, M. and Miettinen, J.K., Retention and excretion of ^{203}Hg-labeled methylmercury in rainbow trout, *Oikos,* 26, 385, 1975.
10. Matida, Y., Kumuda, H., Kimura, S., Saiga, Y., Nose, T., Yokote, M., and Kawatsu, H., Toxicity of mercury to aquatic organisms and accumulation of the compounds by the organisms, *Bull. Freshwater Res. Lab. Tokyo,* 21, 197, 1971.
11. Sharpe, M.A., deFreitas, A.S.W., and McKinnon, A.E., The effect of body size on methylmercury clearance by goldfish (*Carassius auratus*), *Environ. Biol. Fish,* 2, 177, 1977.
12. Fagerstöm, R., Åsell, B., and Jernelöv, A., Model for accumulation of methyl mercury in northern pike, *Esox lucius, Oikos,* 25, 14, 1974.
13. Norstrom, R.J., McKinnon, A.E., and deFreitas, A.S.W., A bioenergetics-based model for pollutant accumulation by fish. Simulation of PCB and methylmercury residue levels in Ottawa River yellow perch (*Perca flavescens*), *J. Fish. Res. Board Can.,* 33, 248, 1976.
14. Blaylock, G., Ed., *Mercury in Aquatic Ecosystems,* Biomovs (BIOspheric MOdel Validation Study) Technical Report 7, Swedish National Institute of Radiation Protection, Stockholm, Sweden, 1990.
15. Harris, R.C., A Mechanistic Model to Examine Mercury in Aquatic Systems, M.Sc. Thesis, McMaster University, Hamilton, Ontario, Canada, 1991.
16. Mathers, R.A. and Johansen P.H., The effects of feeding ecology on mercury accumulation in walleye (*Stizostedion vitreum*) and pike (*Esox lucius*) in Lake Simcoe, *Can. J. Zool.,* 63, 2006, 1985.
17. Jensen, A.L., Modelling the effect of acidity on mercury uptake by walleye in acidic and circumneutral lakes, *Environ. Pollut.,* 50, 285, 1988.

18. Thomann, R.V., Equilibrium model of fate of microcontaminants in diverse aquatic food chains, *Can. J. Fish. Aquat. Sci.*, 38, 280, 1981.

19. Thomann, R.V., Bioaccumulation model of fate of organic chemical distribution in aquatic food chains, *Environ. Sci. Technol.*, 23, 699, 1989.

20. Hewett, S.H. and Johnson, B.L., *A Generalized Bioenergetic Model of Fish Growth for Microcomputers*, WIS-SG-87–245, University of Wisconsin-Madison, Wisconsin, U.S., 1987.

21. Kitchell, J.F. and Breck, J.E., Bioenergetics model and foraging hypothesis for sea lamprey (*Petromyson marinus*), *Can. J. Fish. Aquat. Sci.*, 37, 2159, 1980.

22. Stewart, D.J. and Binkowski, F.P., Dynamics of consumption and food conversion by Lake Michigan alewives: an energetics modelling synthesis, *Trans. Am. Fish. Soc.*, 115, 643, 1986.

23. Kitchell, J. F., Stewart, D.J., and Weininger, D., Applications of a bioenergetics model to yellow perch (*Perca flavescens*) and walleye (*Stizostedion vitreum vitreum*), *J. Fish. Res. Board Can.*, 34, 1922, 1977.

24. Stewart, D.J., Weininger, D., Rottiers, D.V., and Edsall, T.A., An energetics model for lake trout, *Salvelinus namaycush*. Application to the Lake Michigan population, *Can. J. Fish. Aquat. Sci.*, 40, 681, 1983.

25. Brett, J.R. and Groves, T.D.D., Physiological energetics, In: *Fish Physiology, Volume VIII, Bioenergetic and Growth*, Hoar, W.S., Randall, D.J., and Brett, J.R., Eds, Academic Press, New York, 1979, chapter 6.

26. Rodgers, D.W. and Beamish, F.W.H., Water quality modifies uptake of waterborne methylmercury by rainbow trout (*Salmo gairdneri*), *Can. J. Fish. Aquat. Sci.*, 40, 824, 1983.

27. Rodgers, D.W. and Qadri, S.U., Growth and mercury accumulation in yearling yellow perch, *Perca flavescens*, in the Ottawa River, Ontario, Environ. Biol. Fish., 7, 377, 1982.

28. MacCrimmon, H.R., Wren, C.D., and Gots, B.L., Mercury uptake by lake trout, *Salvelinus namaycush*, relative to age, growth, and diet in Tadenac Lake with comparative data from other PreCambrian Shield lakes, *Can. J. Fish. Aquat. Sci.*, 40, 114, 1983.

29. Foltz, F.F. and Norden, C.R., Seasonal changes in food consumption and energy content of smelt (*Osmerus mordax*) in Lake Michigan, *Trans. Am. Fish. Soc.*, 106, 230, 1977.

30. Phillips, G.R. and Gregory, R.W., Assimilation efficiency of dietary methylmercury by northern pike (*Esox lucius*), *J. Fish. Res. Board Can.*, 36, 1516, 1979.

31. Parks, J., Unpublished data, 1991.

32. Balon, E.K., Ed., *Charrs, Salmonid Fishes of the Genus* Salvelinus, Dr. W. Junk Publisher, The Hague, The Netherlands, 1980.

33. DeAngelis, D.L., Godbout, L., and Shuter, B., An individual-based approach to predicting density-dependent dynamics in smallmouth bass populations, *Ecol. Modelling*, 57, 91, 1991.

Methylmercury Levels in Fish Tissue from Three Reservoir Systems in Insular Newfoundland, Canada

D. A. Scruton, E. L. Petticrew, L. J. LeDrew,
M. R. Anderson, U. P. Williams, B. A. Bennett, and E. L. Hill

CONTENTS

ABSTRACT: The concentration of methylmercury in fish tissue was monitored by Newfoundland and Labrador Hydro and the Canadian Department of Fisheries and Oceans in three hydroelectric reservoir systems (Bay d'Espoir-Upper Salmon, Hinds Lake, and Cat Arm) in Newfoundland, Canada between 1982 and 1990. Three salmonid species were sampled from the reservoirs including brook trout (*Salvelinus fontinalis*), ouananiche or landlocked Atlantic salmon (*Salmo salar*), and Arctic charr (*Salvelinus alpinus*). The data were subdivided into combinations of reservoir, species, and year to evaluate spatial and temporal changes. The relationships between fish fork length and mercury concentration in the data sets were evaluated using least squares linear regression. Standardized lengths of fish were used to predict the mercury concentrations in each subset, allowing an evaluation of trends in mercury concentrations following reservoir flooding.

Flooding began in the reservoirs of the Bay d'Espoir system during the late 1960s with additional reservoirs being created in 1982 to 1983. A rise and fall in mercury concentrations in standard length ouananiche over the 8-year study period was observed in both recently flooded systems (Great Burnt Lake and Cold Spring Pond), while the third reservoir (Long Pond), located downstream and flooded in 1967, exhibited only a slight rise and fall during this same period. Trends in the data indicated that the ouananiche mercury concentrations in Bay d'Espoir could return to background levels by 1993 to 1995, 10 to 12 years after flooding. In the two reservoirs where brook trout exhibited significant change (Hinds Lake and Cold Spring Pond), mercury concentration in standard length fish returned to background levels in approximately 7 years following flooding. The Cat Arm Reservoir, an acidic highly colored system, demonstrated an increasing trend in

mercury accumulation in Arctic charr throughout the postflooding period (1985 to
1990). An age class or cohort of fish, followed through the 8-year sampling period,
showed a marked increase in fish mercury concentrations in these reservoirs. The
slope of the mercury increase (ppm Hg/year), or uptake rate, was highest in the
Bay d'Espoir ouananiche populations.

I. INTRODUCTION

Increases in the mercury concentration of fish tissue have been observed following reservoir
development in areas remote from any point source of pollution.[1-4] Elevated mercury levels
in postimpoundment fish have been shown to be associated with an increase in bacterial
degradation of submerged material which results in methylation of inorganic mercury.[3] Several earlier studies speculated that elevated mercury levels in fish were transitory and that
these concentrations would drop 3 to 5 years after flooding.[1,2] Recent studies, based on longer
term and larger data sets, for the La Grande system (Quebec, Canada)[5] and Finnish reservoirs,[6] indicated that fish tissue mercury levels could return to preimpoundment levels over
periods approximating 20 to 30 years.

A study of five Newfoundland reservoirs was conducted to monitor changes in mercury
concentrations in fish tissue. Newfoundland and Labrador Hydro, in cooperation with the
Canadian Department of Fisheries and Oceans, implemented research initiatives and effects
monitoring programs as commitments arising from the environmental assessment of each of
the reservoir developments. The objective of these monitoring studies was to document the

Table 1   **Characteristics of reservoirs and control lakes sampled.**

Reservoir	Surface area (ha)[a]	Area of flooded land (ha)[a]	Water residence time (days)	Mean depth (m)	Impoundment period
Bay d'Espoir reservoirs					
Great Burnt Lake	5,500	2,500 (45%)	3	10	Nov. 1982 to May 1983
Cold Spring Pond	1,700	700 (41%)	2	8	Nov. 1982 to May 1983
Long Pond	21,000	13,000 (62%)	45	16	Dec. 1966 to May 1967
Rocky Pond (Control)	790	N/A	Unknown	Unknown	N/A
Hinds Lake reservoir					
Hinds Lake	4,654	2,141 (46%)	144	Unknown	Nov. 1979 to May 1980
Eclipse Pond (Control)	390	N/A	Unknown	Unknown	N/A
Cat Arm reservoir					
Cat Arm	5,240	4,300 (78%)	199	18	May 1984 to June 1985

[a] Surface and flooded areas are at full supply level (FSP). Measurements are based on predevelopment
design estimates from 1:50,000 scale topographic maps.

evolution of mercury levels in fishes in the reservoirs as well as monitor possible human health concerns arising from recreational consumption of reservoir fish. The Canadian guideline for human consumption of fish and fish products recommends a maximum allowable level of mercury in fish tissue of 0.5 ppm. This limit has been demarcated in several of the following figures to identify fish species and reservoirs of concern. Results from 8 years of monitoring were available for analysis, allowing an investigation of both temporal and spatial trends in mercury concentrations in fish tissue.

II. RESERVOIR CHARACTERISTICS

The reservoirs of three major hydroelectric developments on the Island of Newfoundland were sampled and analyzed for mercury concentrations in fish tissue through the period of 1982 to 1990. Fish populations in control lakes were sampled in two of the systems (Hinds Lake and Bay d'Espoir), while the third (Cat Arm) had two years of preflooding data collected to allow comparisons. The reservoir areas, flooding dates, and morphometric characteristics of the impoundments and control lakes sampled are given in Table 1. Figure 1 shows the locations of the reservoirs.

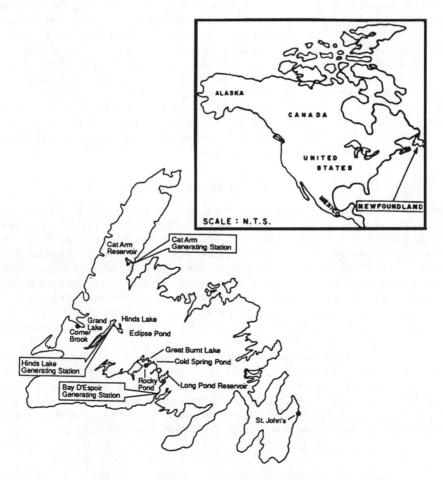

Figure 1 Map of Newfoundland showing the location of the reservoirs studied.

The Hinds Lake Reservoir was created in 1979–1980 and has a total surface area of 4,654 ha, of which 2,141 ha were flooded. This reservoir is located on the Buchans Plateau in west-central Newfoundland and contains both brook trout (*Salvelinus fontinalis*) and landlocked salmon or ouananiche (*Salmo salar*). A smaller undisturbed water body, Eclipse Pond, was used as a control to evaluate the significance of mercury changes in Hinds Lake fish.

Cat Arm Reservoir is located on a plateau of the Long Range Mountains on the Great Northern Peninsula—4,300 ha of the Cat Arm watershed were flooded in 1984 to create a reservoir of 5,240 ha. This system contains populations of brook trout and Arctic charr (*Salvelinus alpinus*). No control pond was sampled in this system; however, lakes in the impoundment area were sampled for 2 years prior to complete flooding.

The Bay d'Espoir hydroelectric development is comprised of five large reservoirs which direct flows from a drainage area of 5,141 km^2 in south-central Newfoundland (Figure 1). A schematic diagram of this complex reservoir system is provided in Figure 2. This development was constructed in two phases, with Long Pond Reservoir being created by flooding 13,000 ha to result in a 21,000-ha impoundment in 1966 to 1967. Four other reservoirs were created upstream of Great Burnt Lake between 1966 and 1969. The Upper Salmon hydroelectric development, a second project located within the Bay d'Espoir drainage system, resulted in the flooding of three additional lakes in 1982 to 1983, including Great Burnt Lake (2,500 ha flooded to result in a 5,500 ha impoundment) and Cold Spring Pond (700 ha flooded resulting in a 1,700 ha reservoir. Water from the original upstream reservoirs continued to flow through Great Burnt Lake, then through Cold Spring Pond, and eventually into Long Pond. Consequently, the Bay d'Espoir system consists of a series of reservoirs which, over 25 years, were flooded at two different times, or received inflows from upstream reservoirs flooded at different times. This study evaluates bioaccumulation of mercury in three of the reservoirs in this extensive system: Long Pond, Cold Spring Pond and Great Burnt Lake. Rocky Pond, which drains into Cold Spring Pond but which receives no water from impounded areas, was used as a control. The Bay of d'Espoir system contains all three species studied; however, only data for brook trout and ouananiche were examined as low sample sizes of Arctic charr precluded statistical analysis.

III. METHODS

Fish were collected using experimental gillnets supplemented with angling. The sampling was conducted in late August of each year. All fish sampled were weighed (g), sized (fork length, mm), sexed, and aged (scales and/or otoliths). A fillet was taken and frozen for subsequent mercury analysis. Analysis for mercury was carried out using an adaptation of standard methods[7,8] which included acid digestion of the sample, oxidation by permanganate, titration of excess permanganate, reduction to elemental mercury vapor, and measurement by atomic absorption spectrophotometry. All samples were analyzed in the Department of Fisheries and Oceans Inspection Laboratory, which has a detection limit of 0.01 ppm. All samples were analyzed in duplicate, with the criterion for acceptance or rejection being a 10% error in the mean of the duplicates.

The data were subdivided into groups representing each species, by reservoir and year. There were no noted differences in size at age by sex for any of these species, therefore sex was not differentiated in the data analysis. The data were logarithmically transformed ($\log_{10}$) to meet the requirements of least squares linear regression. The relationship between log fork length and log mercury concentration was evaluated for all data subsets. Only those subsets that exhibited significant linear regressions ($p < 0.05$) were used in the subsequent analysis.

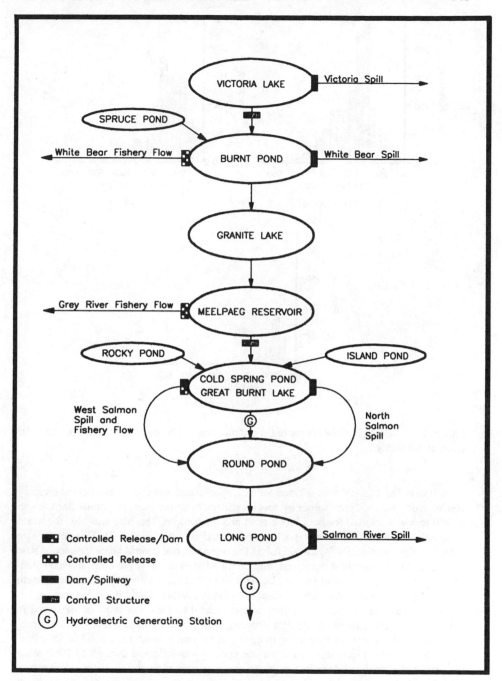

Figure 2 A schematic diagram of reservoirs associated with the Bay d'Espoir-Upper Salmon hydroelectric developments.

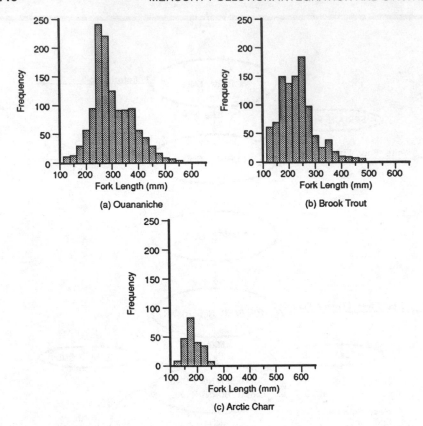

Figure 3 Fork length distributions of (a) ouananiche, (b) brook trout, and (c) Arctic charr used in this study.

A standardized length[5] was selected for each species and was used to predict the mercury concentration for each data subset by species per reservoir per year. A standardized length of 250 mm was selected for both brook trout and ouananiche. This size was also the mode of samples of both species among all reservoirs and controls (Figure 3). This modal size represented predominantly 3-year-old fish in the reservoirs and control lakes sampled. Arctic charr at 175 mm were most numerous; however, owing to the slow growing, stunted nature of this population, 3-year-old fish in Cat Arm were best represented by a 150-mm length class and this size was consequently adopted as the standardized length for this species.

These data were plotted against time and compared to the concentrations measured in control populations. Length at age data were regressed for each species and reservoir combination in order to predict length for an age class moving through the period of sampling. The 1982 2-year-old age class of the various species was followed through to 1990 using the predicted annual fork length to estimate mercury concentrations from the relationships for each reservoir.

IV. RESULTS

Patterns in brook trout mercury levels were observed in three of the five reservoirs (nonsignificant linear regressions between log fork length and log mercury concentration restricted the use of data from Great Burnt Lake and Cat Arm Reservoir). Hinds Lake exhibited a peak

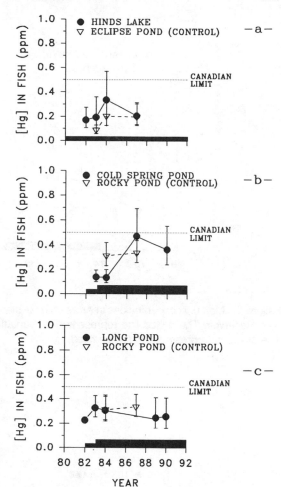

Figure 4 Mercury concentrations in brook trout of standardized length (250 mm) for (a) Hinds Lake, (b) Cold Spring Pond, and (c) Long Pond. The dotted line represents the Canadian limit for human consumption. The black bar represents time since flooding.

in mercury concentrations in 1984, 4 years following flooding, with a return to levels similar to the control pond by 1987 (Figure 4a). In the Bay d'Espoir system, brook trout in Cold Spring Pond showed a significant rise in mercury concentration by 1987 (Figure 4b), 4 years following the completion of flooding (1983). Tissue mercury levels for this population showed a return to concentrations similar to the Rocky Pond control by 1990. Long Pond, downstream of Cold Spring Pond and Great Burnt Lake, exhibited only a slight rise and fall in mercury concentrations in brook trout over the 8-year sampling period (Figure 4c).

Arctic charr in the Cat Arm system showed a continuous increase in mercury levels over the 6 years following the 1984 flooding (Figure 5). Standard length (150 mm) charr had reached the Canadian guideline concentration of 0.5 ppm Hg in 1989. This trend of increasing bioaccumulation of mercury continued through 1990.

Hinds Lake ouananiche demonstrated peak values in 1983, 3 years following flooding (Figure 6). The increase of mercury concentrations in fish in this lake was moderate relative to tissue levels measured in the Eclipse Pond control. Mercury levels in ouananiche from the three reservoirs in the Bay d'Espoir system, along with the values for the Rocky Pond control, are shown in Figure 7. Fish mercury concentrations fluctuated around control values during the period of flooding, from 1982 to 1983. Great Burnt Lake, Cold Spring Pond, and Long Pond demonstrated elevated values above the control levels by 1984. These elevated levels

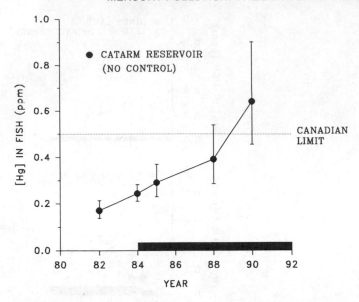

Figure 5   Mercury concentrations in Arctic charr of standardized length (150 mm) for Cat Arm Reservoir. The dotted line represents the Canadian limit for human consumption. The black bar represents time since flooding.

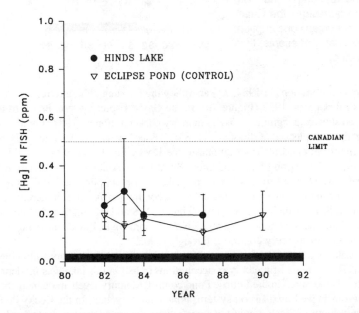

Figure 6   Mercury concentrations in ouananiche of standardized length (250 mm) for Hinds Lake. The dotted line represents the Canadian limit for human consumption. The black bar represents time since flooding.

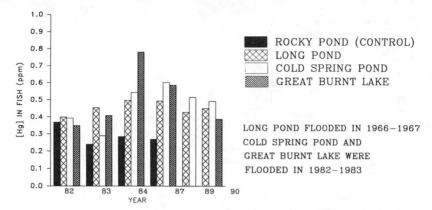

Figure 7 Mercury concentrations in ouananiche of standardized length (250 mm) in the Bay d'Espoir system.

increased or were maintained in 1987, while a general decrease, still in excess of control values, was observed in 1989 and 1990.

Mercury concentrations in the 2-year-old age class (cohort) were followed from 1982 through to 1990 in each of the reservoirs. Concentrations in ouananiche from the Bay d'Espoir reservoirs, as compared to those from the Rocky Pond control, are shown in Figure 8. Ouananiche in the control never exceeded the 0.5 ppm Canadian guideline level; however, ouananiche at age 4 (the first year following flooding) surpassed this guideline level in all three reservoirs. Figure 9 illustrates mercury concentrations for both ouananiche and brook trout in the Hinds Lake reservoir and Eclipse Pond control populations. The elevation of mercury is less pronounced in this reservoir, as compared with the Bay d'Espoir system, with the levels in brook trout just exceeding the 0.5 ppm guideline at age seven, 7 years after flooding. Figure 10 represents the brook trout cohort as it ages in Cold Spring Pond and Long Pond of the Bay d'Espoir system. Brook trout in Cold Spring Pond demonstrate a rapid increase in mercury above control values, exceeding the 0.5 ppm guideline by age 4, while the rate of mercury increase in Long Pond brook trout is lesser, exceeding guideline levels by age 7.

The slopes derived from linear regression of the date (Figures 8 through 10) represent the uptake rate at which each species is accumulating mercury (ppm/year) in each reservoir. Table 2 presents these calculated rates of mercury uptake for all species and reservoirs. The Bay d'Espoir ouananiche exhibit the highest rates of mercury uptake. The Arctic charr in Cat Arm have an uptake rate that is lower than the Bay d'Espoir ouananiche but exceeds both Hinds lake species (ouananiche and brook trout). Brook trout uptake rates exhibit a smaller range (0.085 to 0.199 ppm Hg/year) than Bay d'Espoir ouananiche (0.181 to 0.260 ppm Hg/year). Great Burnt Lake and Cat Arm brook trout demonstrated no significant relationship between fish size and mercury concentration and therefore were excluded from the cohort analysis.

V. DISCUSSION

The response of the five reservoirs to flooding was observed by two methods: (1) following the mercury concentrations in a standardized fish over time, and (2) tracking the changes in mercury concentration in a cohort as it ages. These two approaches provided different perspectives on the response of mercury levels in fish to reservoir flooding. Following the mercury concentrations in fish of a standardized length over the period following flooding

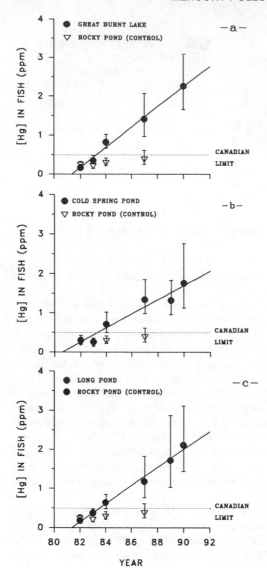

Figure 8 Mercury concentrations estimated for a cohort of ouananiche that were 2 years old in 1982, in (a) Great Burnt Pond, (b) Cold Spring Pond, and (c) Long Pond. The dotted line represents the Canadian limit for human consumption. The solid line represents the regression relating mercury fish concentration to time.

allowed changes in the quality of the reservoir habitat (increased exposure to methylmercury) to be assessed. The standard length selected (3-year-old age class) best reflected the reservoir conditions fish have been exposed to. These changes are compared to mercury concentrations in standardized fish from a control water body to evaluate the return period of the reservoirs. The return period is defined here as the number of years required for the mercury concentrations of the standardized length fish to return to background levels (as determined from the control).

Tracking the mercury concentration changes in a cohort as it aged allowed an examination of the bioaccumulation rates in a species and comparison among reservoirs. The cohort of fish that were 2 years old in 1982 were followed. As flooding occurred in the various reservoirs at different periods (Hinds Lake, 1980; Bay d'Espoir-Upper Salmon development, 1982 to 1983; Cat Arm 1984), this cohort reflected the initial response to impoundment in Bay d'Espoir, the response during and after flooding in Cat Arm, and the observed response

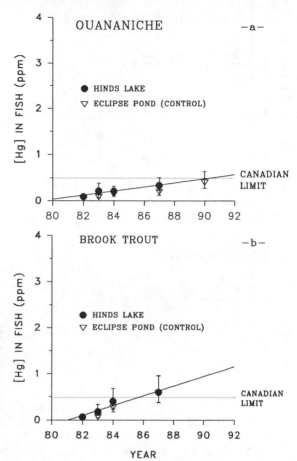

Figure 9 Mercury concentrations estimated for a cohort of fish from Hinds Lake, that were 2 years old in 1982, for (a) ouananiche and (b) brook trout. The dotted line represents the Canadian limit for human consumption. The solid line represents the regression relating mercury fish concentration to time.

in Hinds Lake fish after the initiation of monitoring (i.e., the initial response was missed). Several authors have noted that the first several years following flooding exhibit the most extreme changes in mercury concentrations in biota.[1,2,4,5]

The selected standardized lengths represented relatively small (150 and 250 mm) and young fish (predominantly 3- to 4-year-olds). These length classes were selected such that fish mercury concentrations would best represent exposure to increased methylmercury levels resulting from impoundment and would not be affected by increasing bioaccumulation arising from other influences, such as a shift to piscivorous feeding behavior at larger size classes. The range in ages of smaller fish tended to be narrower, which also implied less error in the exposure period of fish to increased levels of methylmercury. For example, the length at age relationship determined for each species and reservoir indicated a 500 mm brook trout showed an age range, between reservoirs of 5 to 8 years, whereas at 250 mm the age ranged between 3 and 4 years. The brook trout and ouananiche in these reservoir systems are predominantly recruited from the associated stream/river habitats. Samples from streams in the Bay d'Espoir system showed 96% of ouananiche captured were 1 year or less and 88% of brook trout were 1 year or less, indicating considerable recruitment to reservoirs by the age of 2 years;[9] 3-year-old fish would therefore best reflect changes in the reservoir environment.

A. RESERVOIR RESPONSE TIME

The mercury concentrations of standardized length fish over time reflects changes in the level and rate of mercury accumulation in fish of a constant size. Other systems studied in this

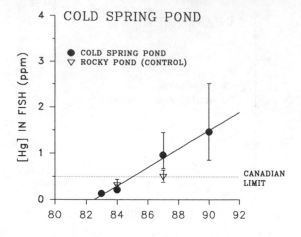

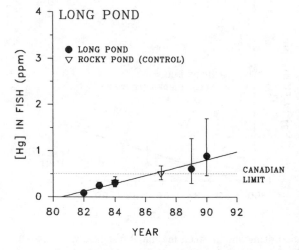

Figure 10 Mercury concentrations estimated for a cohort of brook trout that were 2 years old in 1982 in (a) Cold Spring Pond and (b) Long Pond. The dotted line represents the Canadian limit for human consumption. The solid line represents the regression relating mercury fish concentration to time.

manner have demonstrated return periods in the order of 20 to 30 years,[5,6] however, it appeared that brook trout in two of the systems in this study rebounded in a shorter time period, approximately 7 years (Figure 3a, b). The single Arctic charr population (Cat Arm) allowed no estimation of a return period, as after 6 years concentrations were still increasing (Figure 4). The Cat Arm reservoir represented the largest recently flooded area (almost twice as large as Great Burnt Lake and Hinds Lake, over six times the flooded area of Cold Spring Pond, Table 1) and for this reason may require a longer period to return to control levels. Ouananiche in Hinds Lake showed signs of stabilizing mercury concentrations at or near background levels 4 to 7 years following flooding. In the Bay d'Espoir system, mercury concentrations in ouananiche, in each of the three reservoirs, were still above background levels during the last year of sampling. The peak of the mercury concentration occurred 1 to 4 years following the 1982 to 1983 flooding, and concentrations were still descending towards background levels in 1990 (Figure 6). The trend indicated that mercury concentrations in ouananiche in the Bay d'Espoir system could return to background levels by 1993 to 1995, or 10 to 12 years after flooding. These return times are shorter than those suggested from studies of the La Grande complex in Quebec, Canada.[5] The differences may reflect the smaller scale of the Newfoundland developments or may be a function of using a smaller standardized length fish. The younger, smaller standard size fish in this study may be more representative of changes in reservoir chemistry than the 5-year-old fish used in the Hydro Quebec study.

Table 2 **Mercury uptake rates for the 1982 2-year-old cohort for salmonid populations in three reservoir systems in insular Newfoundland (1982–1990).**

Species and reservoir	Uptake rate ppm Hg/year (S.E.)		r^2	n	P
Ouananiche					
Bay d'Espoir					
Great Burnt Lake	0.260	(0.15)	0.99	5	0.0004
Cold Spring Pond	0.181	(0.23)	0.94	6	0.0015
Long Pond	0.231	(0.12)	0.99	6	0.0001
Hinds Lake	0.045	(0.12)	0.88	4	0.0635
Arctic Charr					
Cat Arm	0.125	(0.18)	0.94	5	0.0063
Brook Trout					
Hinds Lake	0.150	(0.20)	0.94	4	0.0332
Bay d'Espoir					
Great Burnt Lake	0.199	(0.15)	0.99	4	0.0053
Long Pond	0.085	(0.12)	0.95	5	0.0052

Note: r^2 is percentage variation explained by the regression, **n** is the number of years included in the regression, and *P* is the probability that such a relationship is occurring by chance.

The limited magnitude of response evident in the Long Pond reservoir fish (Figure 4c and 7) is of interest as the introduction of water from newly flooded upstream areas represented the second perturbation to this reservoir. The initial flooding of Long Pond occurred in 1966 to 1967, 15 years prior to the upstream flooding of Cold Spring Pond and Great Burnt Lake, but represented the largest flooded area of all reservoirs included in this study. No data regarding the initial response were obtained; however, monitoring was initiated in 1982 when it began to receive water from the two newly flooded systems upstream. The receiving water from flooded areas has been shown to have a diminishing downstream effect on riverine fish[5] and, in this case, the dampened but similar response to the upstream impoundments was noted for both ouananiche and brook trout in Long Pond.

B. MERCURY UPTAKE RATES

Tracking the mercury concentration over time in the 2-year-old year class, in the first year of sampling (1982), allowed determination of mercury uptake rates for each species and reservoir from the slope of the age-mercury concentration relationship. The uptake rates presented in Table 2 are similar to those found in the literature, which are calculated from age-mercury concentration relationships of a perturbed population (i.e., in the literature mercury uptake rates were determined from the mercury content by age for the sample population in a 1-year period). Sampling of Lake Tyrifjorden (Norway) between 1978 and 1982, following a 1970 ban on mercury inputs from industry and agriculture, indicated accumulation rates of 0.52, 0.35, and 0.05 ppm Hg/year for brown trout (*Salmo trutta*), northern pike (*Esox lucius*), and Arctic charr, respectively.[12] The three species were piscivorous, as are ouananiche in this study after reaching approximately age four.[10] Bay d'Espoir ouananiche exhibited the highest uptake rates (Table 2), exceeding that of Hinds Lake ouananiche. This pattern among reservoirs may reflect the scale of the perturbations (flooding), however, it is difficult to quantify as the Bay d'Espoir system experienced cumulative downstream effects from multiple flooding periods.

Hinds Lake reservoir was clear cut before flooding and the removal of organic material from the flood zone has been suggested as a means of reducing the impact of mercury release to the aquatic system.[13] The lower bioaccumulation rates of Hinds Lake species may reflect this reduction of available organic matter. These rates may also reflect the fact that monitoring in Hinds Lake was not initiated until 2 years after flooding and the calculated rates do not reflect the initial postimpoundment response.

The higher uptake rates in ouananiche reflected, in part, faster growth rates. The Bay d'Espoir ouananiche exhibited the steepest growth slopes (length at age), which may imply either a reservoir effect following flooding or differing species and/or reservoir feeding behavior. Ouananiche are noted to begin piscivory slightly earlier and to a much greater degree than the brook trout in lakes in the Bay d'Espoir system.[10] This piscivorous feeding behavior in ouananiche could lead to more rapid bioaccumulation of mercury through the food chain. It may be of interest to ascertain if the relatively rapid return periods, noted in the standardized length analysis, is reflected in a new cohort of fish at a future date. A cohort analysis of 2-year-olds from 1990 could be followed and comparison of bioaccumulation rates between the two cohorts (the 1982 and the 1990) would elaborate on initial findings regarding reservoir return periods.

VI. CONCLUSIONS

Relatively short (7 to 12 years) return periods were noted for mercury content in 3-year-old fish in insular Newfoundland reservoirs. This may reflect the small flooded areas (1,000 to 2,500 ha) associated with recent developments; however, the effect on the return period of the larger Cat Arm impoundment (4,300 ha) is uncertain at this stage. Continued monitoring in the Cat Arm reservoir would allow the response of the Arctic charr to be evaluated in the context of the large flooded area. Monitoring of the Bay d'Espoir system should be continued in order to test the prediction that background levels of fish mercury could be reached by 1993. Continued monitoring would also allow a new cohort of fish to be tracked through time to provide refinement of the conclusions presented here with respect to reservoir return periods after flooding.

It is important to note that the "return periods" referred to here are defined as the time period for mercury in a standard length fish to return to preimpoundment levels. This does reflect the fact that larger, older fish in the population will be exhibiting elevated mercury levels at the same time that standard length fish have normalized to preimpoundment levels. Consequently, the evolution of mercury levels in fish tissue in reservoirs should consider both the bioavailability of mercury, as addressed by the standard length and return period analysis, as well as the biomagnification of mercury in fish as they age, as addressed by the cohort analysis.

ACKNOWLEDGMENTS

The authors would like to acknowledge the Inspection Branch of the Canadian Department of Fisheries and Oceans (DFO), specifically K. Kennedy and R. Benson, for valuable assistance in laboratory analyses of fish tissue samples for mercury content. R. McCubbin and M. Barnes, also of DFO, were instrumental in designing and implementing several of the effects monitoring programs contained in this study. M. Barnes and D. Stansbury (DFO) have also previously analyzed data collected in the early years of the monitoring program. LeDrew, Fudge and Associates Limited, L.G.L. Limited, and G.F. Pope and Associates were contracted by Newfoundland and Labrador Hydro to complete some of the fish collections associated with the monitoring programs.

REFERENCES

1. Abernathy, A. R. and Cumbie, P. M., Mercury accumulation by largemouth bass (*Micropterus salmoides*) in recently impounded reservoirs, *Bull. Environ. Contam. Toxicol.*, 17:595, 1977.

2. Cox, J.A., Carnahan, J., DiNunzio, J., McCoy, J., and Meister, J., Source of mercury in new fish impoundments, *Bull. Environ. Contam. Toxicol.*, 23:779, 1979.

3. Jackson, T.A., The mercury problem in recently formed reservoirs of Northern Manitoba (Canada): effects of impoundment and other factors on the production of methyl mercury by microorganisms in sediments, *Can. J. Fish. Aquat. Sci.*, 45:97, 1988.

4. Bodaly, R.A., Hecky, R.E., and Fudge, R.J.P., Increases in fish mercury levels in lakes flooded by the Churchill River diversion, Northern Manitoba, *Can. J. Fish. Aquat. Sci.*, 41:682, 1982.

5. Verdon, R., Brouard, D., Demers, C., Lalumiere, R., Laperle, M., and Schetagne, R., Mercury evolution (1978–1988) in fishes of the La Grande hydroelectric complex, Quebec, Canada, *Water Air Soil Pollut.*, 56:405, 1991.

6. Verta, M., Rekolainen, S., and Kinnunen, K., Causes of Increased Fish Mercury Levels in Finnish Reservoirs, Publications of the Water Research Institute, National Board of Waters, Finland, No. 65, 1986.

7. Armstrong, F.A.J. and Uthe, F.J., Semi-automated determination of mercury in animal tissue, *Atomic Absorption News*, 10:101, 1971.

8. Uthe, F.J., Armstrong, F.A.J., and Stainton, M. P., Mercury determination in fish by wet digestion and flameless atomic absorption spectrophotometry, *J. Fish. Res. Board Can.*, 27:805, 1970.

9. Newfoundland and Labrador Hydro, Environmental Policy Department, Upper Salmon Hydroelectric Development: Environmental Impact Statement, 1980.

10. LeDrew, Fudge, and Associates, Fisheries Component Study, Island Pond Hydroelectric Development: Environmental Impact Study, 1991.

11. Lee, Y.H. and Iverfeldt, A., Measurement of methyl mercury and mercury in run-off and rain waters, *Water Air Soil Pollut.*, 56:309, 1991.

12. Skurdala, J., Qvenild, T., and Skogheim, O.K., Mercury accumulation in five species of freshwater fish in Lake Tyrifjorden, south-east Norway, with emphasis on their suitability as test organisms, *Environ. Biol. Fishes*, 14:233, 1985.

13. Canadian Electrical Association, Mercury Release in Hydroelectric Reservoirs, Canadian Electrical Association, Report No. 185 G 399, 1986.

Mercury in Yellow Perch from Adirondack Drainage Lakes (New York, U.S.)

Howard A. Simonin, Steven P. Gloss,
Charles T. Driscoll, Carl L. Schofield, Walter A. Kretser,
Ralph W. Karcher, and John Symula

Contents

ABSTRACT: Yellow perch (*Perca flavescens*) were collected from 12 drainage lakes located in the Adirondack Park, New York, U.S. in order to better document mercury concentrations in fish in this region. The fish were aged and tissue samples analyzed for mercury concentration by the New York State Department of Environmental Conservation laboratory at Hale Creek and by Syracuse University. Water chemistry data collected by the Adirondack Lakes Survey Corporation were used to relate fish mercury concentration to lake surface water chemistry. Mercury levels were found to exceed the New York State guidelines of 1 ppm in several large perch from three of the lakes. Using age 4+ yellow perch, among-lake comparisons showed that pH, acid neutralizing capacity, dissolved inorganic carbon, calcium, conductivity, and magnesium were the water quality variables directly correlated with mercury concentrations. Air-equilibrated pH was the best single predictor of mercury concentrations in age 4+ yellow perch ($r^2 = 0.61$) and in perch 200 to 300 mm in total length ($r^2 = 0.49$). Within an individual lake, fish age, length, and weight were directly related to the mercury level in the muscle tissue of the fish. Air-equilibrated pH of the lake surface water and total length of the fish were used to create a model predicting mercury concentrations in yellow perch from Adirondack drainage lakes.

I. INTRODUCTION

The bioaccumulation of mercury by fish in remote lakes with low acid neutralizing capacity (ANC) has been documented by numerous studies,[1-4] and Spry and Wiener[5] recently published a critical review of the subject. In certain waters, mercury concentrations in fish have prompted concern with regard to fish consumption by humans and wildlife. Human health advisories have been issued by a number of states and provinces, including New York State, recommending against consumption of fish from such waters.

Because of the abundance of low-ANC surface waters in the Adirondack region of New York State and the impacts of acidic deposition,[6,7] researchers have suggested that mercury concentrations be monitored more intensively in fish from this region.[8,9] Bloomfield et al.[3] conducted a detailed study of factors regulating mercury concentrations in smallmouth bass (*Micropterus dolomieui*) in Cranberry Lake. Sloan and Schofield[2] collected brook trout (*Salvelinus fontinalis*) from limed (base treated) and untreated ponds and found no significant differences in mercury concentrations in the fish. Heit and Klusek[10] measured the concentrations of selected trace elements in white suckers (*Catostomus commersoni*) and bullheads (*Ictalurus nebulosus*) from two Adirondack lakes and found that mercury was the only element that appeared to bioaccumulate in these fish. The New York State Department of Environmental Conservation annually conducts a statewide toxic substances monitoring program.[9] Through this program, mercury concentrations have been measured in several fish species from Adirondack lakes. However, more detailed data from additional Adirondack lakes are needed to gain a better understanding of the factors regulating mercury concentrations in fish tissue in this region.

Rather than conduct a comprehensive survey, we studied physical and chemical factors related to mercury concentrations in fish from a number of Adirondack waters. These data were used to make comparisons with similar studies from other regions. This study was designed to coincide with the final year of the Adirondack Lakes Survey Project,[11] in which water chemistry and fisheries data were collected from over 1400 Adirondack waters over a 4-year period. Approximately 85% of the lakes in the Adirondacks are drainage lakes, including those examined here. Yellow perch (*Perca flavescens*) was selected as a study species because yellow perch are common in Adirondack waters, and comparative data on perch from other regions are available.[12,13]

II. METHODS

Potential study lakes were selected based on existing fisheries survey data. Yellow perch were collected during the fall of 1987 (September 2 to October 21) from 12 lakes and ponds in the upper Hudson and Mohawk-Hudson watersheds of the Adirondack Park (Figure 1)[11] Collections were made during the fourth year of fisheries surveys conducted by the Adirondack Lakes Survey Corporation. Fish were collected overnight using gill nets. Yellow perch from the study lakes were obtained from the Adirondack Lakes Survey Corporation field staff, placed on ice, and transported to the Department of Environmental Conservation Rome Field Station, Rome, NY. Each fish was individually weighed, measured (total length), and assigned an identification number. Scales for aging were taken from just below the anterior edge of the dorsal fin. Individual yellow perch were then placed in plastic bags and frozen. Staff at the Cornell University Shackleton Point Field Station aged the fish from acetate impressions of the scales.

After age estimation, individual fish were selected for analysis to provide mercury data on a range of age groups from each lake and to ensure an adequate sample of 4-year-old (the most common age) fish for inter-lake comparisons. These fish were sent to the Department of Environmental Conservation Hale Creek Field Station in Gloversville, NY, for analysis. The remaining fish were analyzed at Syracuse University.

At the Hale Creek Field Station perch were individually filleted, and fillets (including the skin and ribs) were ground and homogenized. Tissue samples were then digested with sulfuric acid and analyzed for mercury with a procedure similar to that described by Grieb et al.[14] Mercury samples were analyzed with a Perkin-Elmer Mercury Analyzer System, model 50A (The Perkin-Elmer Corporation, Norwalk, CT), and data were reported as micrograms of total mercury per gram of wet tissue sample. Duplicate samples, sample blanks, and National Institute of Standards and Technology (NIST) RM-50 albacore tuna reference samples were analyzed as part of the quality assurance program.

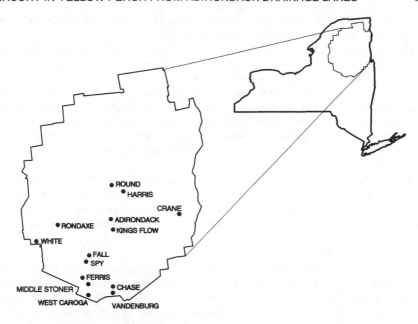

Figure 1 Map of the Adirondack Park in New York State (U.S.) indicating location of the study lakes.

The yellow perch samples analyzed at Syracuse University were prepared in a different manner. Muscle tissue was obtained from the right side of the fish, anterior of the dorsal fin, posterior to the gill, and vertically from the lateral line to the top of the fish. Skin and bone were removed from the muscle tissue samples. The analytical methods used at Syracuse University were comparable to those used by the Department of Environmental Conservation.[14] The quality assurance program used at Syracuse University included analysis of NIST reference samples, spiked samples, and a number of split fish samples analyzed at both Syracuse University and Battelle Pacific Northwest Laboratory in Seattle, Washington.[14]

Lake water samples were collected by Adirondack Lakes Survey Corporation staff from August 4 to 20, 1987, before the fish sampling. One-liter samples were collected at the 1.5-m depth near the middle of each lake. These samples were stored on ice and transported to the Adirondack Lakes Survey Corporation analytical laboratory in Ray Brook, NY, where they were analyzed for 24 physical and chemical parameters. Analytical instrumentation, methods, quality assurance, and quality control methods are described in detail by Kretser et al.[11] The parameters measured include laboratory pH, air-equilibrated pH, acid neutralizing capacity (ANC), aluminum, ammonium, calcium, chloride, dissolved inorganic carbon (DIC), dissolved organic carbon (DOC), fluoride, iron, lead, magnesium, manganese, nitrate, potassium, silica, sodium, specific conductance, sulfate, phosphorus, true color, and zinc. Prior to measurement of the metals the water samples were filtered through a 0.4 μm filter.

Detailed bathymetric maps of each study lake were prepared by Adirondack Lakes Survey Corporation staff and lake volume, mean depth, and other morphometric parameters determined. Lake areas and watershed areas were determined from U.S. Geological Survey 7.5 min maps. Flushing rate was estimated as the product of watershed area and mean annual runoff divided by total lake volume. The impoundment status (whether a dam was present on the outlet) of each water body was identified by on-site inspection.

The summer surface water chemistry data collected by the Adirondack Lakes Survey Corporation were the most comprehensive data for these study lakes. Air-equilibrated pH ranged from 5.72 to 7.85, and ANC ranged from 9.5 to 634 μeq/L in the 12 lakes sampled

(Table 1). The lakes with DOC concentrations exceeding 5.0 mg/L had ANC values greater than 85 μeq/L and air-equilibrated pH values greater than 6.8.

The lake classification scheme of Driscoll et al.[7] was used to classify the study lakes. Chase and Ferris Lakes, the two most acidic waters of the 12, were classified as "thin till, clear water" lakes. Fall Lake and Round Pond were classified as "thick till, colored water" lakes. Six of the lakes (Adirondack, Harris, Middle Stoner, Spy, Vandenburg, and West Carogo) were classified as salt-impacted waters ($Cl^- > 20$ μeq/L). This pattern was most likely due to the close proximity of these lakes to roads where salt was used as a de-icer during winter. These six waters were also clear-water systems. Kings Flow and Crane Pond were carbonate-influenced waters, and two of the salt-impacted waters (Adirondack and West Caroga) were also carbonate-influenced lakes.

A total of 372 yellow perch were obtained from the 12 lakes. The number of perch collected from each lake ranged from 7 to 53. After the scales from these fish were aged, 134 perch were selected and analyzed at the Department of Environmental Conservation laboratory, and the remainder were provided to Cornell University for sample processing and analysis by the laboratory at Syracuse University. A total of 131 yellow perch were analyzed at Syracuse University.

The ages of the yellow perch ranged from 2+ to 11+ years, age 4+ being the most abundant age group; 58 of these age 4+ fish were analyzed for mercury to make among-lake comparisons. Ferris Lake and Round Pond contained most age groups of yellow perch, with several large 10+ or 11+ age fish. Other lakes, such as Spy, West Caroga, and Middle Stoner, had yellow perch of age 6+ or less. Crane Pond had the largest age 4+ yellow perch (mean total length = 231 mm) and Adirondack Lake had the heaviest (mean weight = 149 g). Age 4+ yellow perch from Chase Lake were the smallest (mean total length = 146 mm and mean weight = 33 g). Lake depth and conductivity were positively correlated with total length of the age 4+ yellow perch.

The original study was designed to utilize only the analytical data from the Department of Environmental Conservation laboratory, but it was later decided that it would be possible to double the number of data observations and therefore strengthen the study findings by combining these data with those from Syracuse University. Data sets of equal size were randomly selected from the Department of Environmental Conservation data and the Syracuse University data to compare fish of equal age and from the same lake. This comparison revealed that the mean concentration in the Syracuse University data was 0.11 μg/g lower than the mean for the Department of Environmental Conservation data ($p < 0.001$). Further evaluations suggested that the discrepancy in mercury concentrations between the two laboratories appeared to be consistent across the range of concentrations analyzed.

Measurements of the NIST RM-50 albacore tuna also showed a significant difference between the two laboratories ($p < 0.001$). The mean of 10 NIST tuna samples at the Department of Environmental Conservation laboratory was 1.012 μg/g, and the mean of 39 samples analyzed by the Syracuse University laboratory was 0.88 μg/g. The NIST reports a value of 0.95 ± 0.1 μg/g as the value which encompasses the means of analyses of this material by a number of different laboratories. Analytical data obtained from both the Department of Environmental Conservation and Syracuse University laboratories were within the acceptable range. Analysis of four procedure blanks by the Department of Environmental Conservation laboratory also indicated a mean value equivalent to 0.05 μg/g.

Because of the systematic discrepancy in the analyses by the two laboratories, the full data set was adjusted by adding 0.07 μg/g to each mercury value generated by the Syracuse University laboratory and by subtracting 0.06 μg/g from each mercury value of the Department of Environmental Conservation laboratory. These adjustments, which were based on the NIST sample analyses, allowed us to use the full data set to test a number of hypotheses. To evaluate differences in mercury concentrations among the 12 lakes, the data from the age

Table 1 Selected surface water chemistry and physical parameters of 12 Adirondack study lakes (data from Adirondack Lakes Survey Corporation[1]).

			Estimated flushing rate (times/ year)			Summer surface water chemistry 8/87							
Lake area (ha)	Watershed area (ha)	Mean depth (m)		Field pH	Air Eq. pH	ANC (µeq/L)	DOC (mg/L)	Conductance (µmho/cm)	SO_4 (mg/L)	Total P (µg/L)	Ca (mg/L)	Al (µg/L)	Cl (mg/L)
Ferris Lake 48	578	3.6	3.0	5.81	5.72	10	4.0	18	5.36	7	1.52	70	0.28
Chase Lake 26	446	4.4	3.0	5.49	5.77	10	4.7	17	4.66	7	1.41	63	0.41
Spy Lake 151	904	5.1	1.0	6.29	6.75	73	3.4	32	4.58	23	2.45	32	3.06
Middle Stoner Lake 33	988	3.5	5.4	6.27	6.82	69	3.5	34	5.17	3	2.55	17	3.32
Vandenberg Pond 53	1367	1.2	14	6.29	6.87	89	6.0	27	5.00	25	2.71	14	1.11
Round Pond 90	1497	2.5	4.2	5.89	7.03	157	8.3	34	6.04	4	4.18	31	0.34
Fall Lake 10	4241	2.2	171	6.30	7.04	199	12.3	34	4.22	48	3.63	113	0.29
Kings Flow 79	5275	0.6	81	6.36	7.12	128	7.1	25	4.27	13	3.46	102	0.25
Crane Pond 68	2068	11.6	1.3	6.38	7.17	141	3.6	31	6.27	5	3.80	<2	0.43
Harris Lake 116	9133	3.5	17	6.55	7.32	194	6.7	42	4.80	6	4.78	51	2.29
West Caroga Lake 129	1413	8.8	0.9	6.95	7.36	182	4.3	55	5.12	3	4.51	46	7.25
Adirondack Lake 80	378	2.2	1.9	7.12	7.85	634	5.4	78	4.11	23	11.95	8	2.83

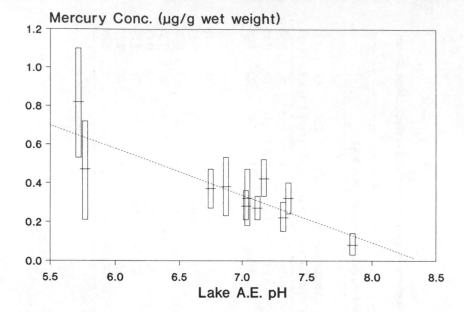

Figure 2 Relation between mercury concentrations in age 4+ yellow perch and lake summer surface air-equilibrated pH in 11 Adirondack drainage lakes. The horizontal lines indicate the mean, and the bar indicates ±1 standard deviation ($r^2 = 0.75$).

4+ yellow perch were used along with the individual lake water chemistry and physical characteristics.

All statistical analysis of the data was conducted with the Statistical Analysis System (SAS)[15] primarily using regression procedures. Among-lake comparisons were made with the data on total mercury concentrations in age 4+ yellow perch, and with the lake physical and chemical variables. Data from two additional Adirondack Lakes Survey lakes were included in several analyses because mercury data were available. Within-lake comparisons were made with the variables that separated individual fish (age, length, and weight). To create a predictive model of mercury concentrations in perch, the most important variable (highest r^2) identified in among-lake comparisons was used along with one of the significant variables identified during within-lake statistical tests.

III. RESULTS

Air-equilibrated pH was the best predictor of mercury concentrations in the age 4+ yellow perch ($r^2 = 0.61$). When the mean mercury concentrations of the age 4+ yellow perch from each lake were regressed against air-equilibrated pH the r^2 term increased to 0.75 (Figure 2). Variability in mercury concentrations in age 4+ fish from the same lake was considerable, as demonstrated by the standard deviation bars in Figure 2. This variability may have been caused by errors in aging, analytical variability, or biological variation in the diet and metabolism of individual fish. Ferris and Chase Lakes, the two most acidic lakes in the study and the only two classified as "thin till, clear water" lakes, had perch with the highest mean mercury concentrations. Adirondack Lake, which had the highest pH, had perch with the lowest mean mercury concentration. Other variables that were also highly correlated ($p < 0.0001$) with the mercury concentration of age 4+ yellow perch included ANC, DIC, calcium, conductivity, magnesium, and field pH (Table 2). Variables that were not as strongly correlated ($p < 0.05$) with mercury concentrations in the fish included DOC, sodium, SO_4,

Table 2   **Lake variables significantly ($p < 0.0001$) correlated with mercury concentrations in age 4+ yellow perch.**

	r^2
Air Eq. pH	0.61
Log_{10} (ANC)	0.61
Log_{10} (DIC)	0.60
Log_{10} (Ca)	0.54
Log_{10} (Conductivity)	0.42
Log_{10} (Mg)	0.41
Field pH	0.31

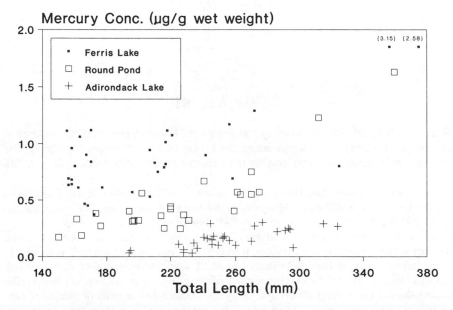

Figure 3   Relation between mercury concentration and total length of individual yellow perch from three Adirondack lakes.

lake area, and watershed area. Variables that were not related to mercury concentrations in the age 4+ yellow perch included true color, total phosphorus, total aluminum, chloride, mean depth, flushing rate, the ratio of watershed area to lake area, the ratio of watershed area to lake volume, fish total length, and fish weight.

The variability in mercury concentrations observed in the age 4+ fish was evident in all sizes of yellow perch (Figure 3). The individual fish data for Ferris Lake, Round Pond, and Adirondack Lake also show that for these waters there are distinct differences in mercury concentrations among these three samples.

The fact that a number of the lakes were "salt-impacted" did not appear to influence our findings. When these lakes were excluded from the data set air-equilibrated pH remained the key variable, and other relationships remained unchanged.

Additional mercury data were available for yellow perch from two other Adirondack lakes, White Lake (pH = 7.29) and Lake Rondaxe (pH = 6.26), which were studied by Gloss et al.[12] These fish were not aged; therefore, we analyzed our data plus the data from these two lakes by using the mean mercury concentrations in yellow perch 200 to 300 mm

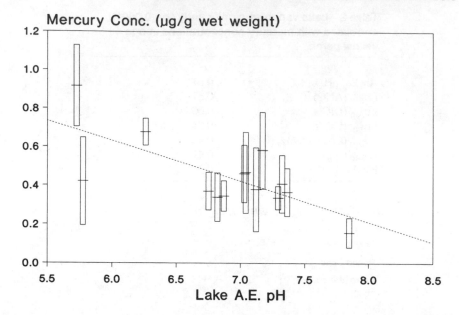

Figure 4   Relation between mercury concentrations in yellow perch 200 to 300 mm in total length and summer surface air-equilibrated pH in 14 Adirondack drainage lakes. The dashed line indicates the mean, and the bar indicates ±1 standard deviation ($r^2 = 0.49$).

in total length from each lake. The ages of the yellow perch in this group of fish ranged from age 3+ to 9+. Air-equilibrated pH was again the best predictor of mercury concentrations in the yellow perch in this size group ($r^2 = 0.49$, Figure 4).

For a given lake, fish age most frequently showed the strongest correlation with tissue mercury concentrations. In several lakes, the oldest fish collected were 6 or 7 years old. In Ferris Lake and Round Pond individuals 10 and 11 years old were collected. Adirondack Lake had low mean mercury concentrations in all age groups of yellow perch (Table 3). The mean values of the mercury concentrations by age showed a clear trend of increasing concentration with increasing age. When the full data set was used to evaluate the relationships between mercury concentration and the total length, weight, or age of the fish, all three variables were significantly correlated. Fish weight had the highest coefficient of determination ($r^2 = 0.21$). This coefficient was not greater because of the variability in mercury concentrations among lakes.

Total length of the fish and air-equilibrated pH of lake surface water from the 12 Adirondack study lakes were used to create an empirical model to predict mercury concentrations in yellow perch from Adirondack drainage lakes. Fish total length was used so that the model could be easily applied. The multiple regression obtained was:

$$\text{Hg concentration } (\mu g/g) = 1.754 - 0.315 \text{ (pH)} + \qquad\qquad (1)$$
$$0.004 \text{ (total length in mm)} \qquad (R^2 = 0.51)$$

The standard errors of the slopes were 0.022 for pH and 0.0003 for total length, and both slopes were highly significant ($p < 0.001$).

The predicted mercury concentrations of 0.5 µg/g and 1.0 µg/g are shown in Figure 5 for yellow perch from Adirondack drainage lakes with fish total length and lake pH as predictors. Mean yellow perch data from White Lake (pH = 7.29, total length = 239 mm,

Table 3 Mean mercury concentrations (μg/g wet weight) in Adirondack yellow perch at different ages.

	Air Eq. pH	Age (years)									
		2+	3+	4+	5+	6+	7+	8+	9+	10+	11+
Ferris Lake	5.72	0.63(2)	0.74(8)	0.82(8)	0.72(4)	1.02(3)	0.91(3)	1.03(2)	—	1.98(2)	2.58(1)
Chase Lake	5.77	—	0.32(5)	0.47(2)	0.40(3)	0.33(8)	0.42(8)	—	—	—	—
Spy Lake	6.75	0.08(1)	0.13(3)	0.37(2)	0.42(1)	—	—	—	—	—	—
Middle Stoner Lake	6.82	0.26(2)	0.28(8)	—	0.28(4)	0.41(7)	—	—	—	—	—
Vandenburg Pond	6.87	0.15(3)	0.30(7)	0.38(3)	0.34(5)	0.31(3)	0.40(5)	0.33(1)	—	—	—
Round Pond	7.03	—	0.17(1)	0.28(8)	0.39(5)	0.30(4)	0.62(3)	0.54(3)	0.57(2)	1.23(1)	1.63(1)
Fall Lake	7.04	0.11(2)	0.27(1)	0.32(6)	—	0.49(3)	0.67(3)	0.25(1)	0.65(1)	—	—
Kings Flow	7.12	0.18(1)	0.28(5)	0.27(8)	0.30(5)	0.25(5)	0.53(2)	0.32(1)	0.70(1)	—	—
Crane Pond	7.17	—	0.55(1)	0.42(4)	0.67(1)	0.66(1)	0.82(2)	—	—	—	—
Harris Lake	7.32	—	0.31(5)	0.22(5)	—	0.24(2)	0.38(1)	0.49(2)	—	—	—
West Caroga Lake	7.36	0.23(5)	0.36(8)	0.32(6)	0.71(1)	0.36(1)	—	—	—	—	0.83(1)
Adirondack Lake	7.85	—	—	0.08(6)	0.07(4)	0.15(5)	0.19(7)	0.27(4)	0.20(4)	—	—
Mean		0.23	0.34	0.36	0.43	0.41	0.55	0.46	0.53	1.61	1.68

Note: Lakes are arranged in order of increasing pH, and (n) is the number of individual fish analyzed.

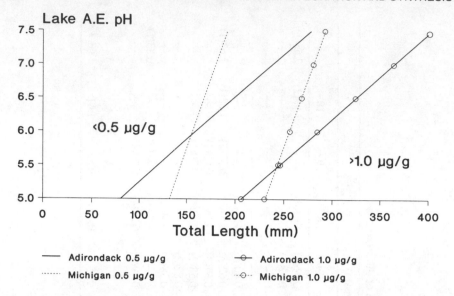

Figure 5 Plots of predictive equations indicating lake pH and fish length combinations associated with mercury concentrations <0.5 or >1.0 µg/g fish consumption advisory guidelines. These equations were developed with data from the Adirondacks (this study) and the Upper Peninsula of Michigan (Gloss et al.[12]).

Hg concentration = 0.335 µg/g) and Lake Rondaxe (pH = 6.26, total length = 270 mm, Hg concentration = 0.675 µg/g) appear to fit this predictive model quite well.

For comparison, data from drainage lakes in the Upper Peninsula of Michigan[12] were also evaluated by multiple regression. Using air-equilibrated pH for these lakes[16] and the yellow perch length and mercury concentration data, the predictive equation obtained was:

$$\text{Hg concentration (µg/g)} = 0.466 - 0.124\,(\text{pH}) + \qquad\qquad (2)$$
$$0.005\,(\text{total length in mm}) \qquad (R^2 = 0.67)$$

The standard errors of the slopes were 0.012 for pH and 0.0003 for total length; both were highly significant ($p < 0.001$).

Comparing the predicted mercury concentration for the Upper Peninsula of Michigan with the predicted concentrations for the Adirondacks (Figure 5) it appears that lake pH may be a more important variable regulating mercury concentrations in yellow perch in the Adirondacks than in Michigan lakes.

IV. DISCUSSION

In the present study we examined empirical relationships between lake characteristics and mercury concentrations in yellow perch. Our results from Adirondack waters show that lake pH strongly influences mercury concentrations in yellow perch. Also, fish age, length, and weight are key variables correlated with mercury concentration.

The yellow perch has been widely used in assessments of mercury contamination in fresh waters. Suns and Hitchin[17] analyzed yearling yellow perch from 16 Ontario lakes to assess the relationship between whole body mercury concentrations and water quality. Cope et al.[13] conducted a similar study with age 2 yellow perch from 10 Wisconsin seepage lakes. Yellow perch were the most abundant fish collected by Grieb et al,[14] in a study of mercury concentrations in fish tissue for the Upper Peninsula of Michigan. In experimentally acidified Little Rock Lake, Wisconsin, total mercury concentrations in whole age 1 yellow perch were compared to fish from the unacidified section of the lake.[18] Each of these studies yielded results consistent with our findings in Adirondack lakes—lake pH was negatively correlated with the mercury concentrations in fish. Our findings contradict those reported by Heit et al.,[19] who studied three Adirondack lakes of varying pH and found higher mercury levels in the yellow perch from a circumneutral lake.

Gutermann and Lisk[20] reported analytical differences in mercury concentrations in brown trout (*Salmo trutta*) fillets after skinning and fat removal. We adjusted our laboratory data based on the analyses of the NIST albacore tuna samples by both the Syracuse University and Department of Environmental Conservation laboratories. After these adjustments, we were unable to detect any significant differences in the two data sets. Yellow perch have less fat than brown trout,[9] and skinning the fillets may not alter the mercury concentrations.

Fish mercury concentrations have previously been found to be directly related to the age, length, and weight of the fish.[2,3,14,21] Bloomfield et al.[3] reported that the increase in mercury concentration with length is a curvilinear relationship, with mercury increasing in larger increments as the fish grow. We observed a similar relationship for several of our study lakes. This may be due to a diet shift in the larger fish to become more piscivorous.

Our results for 12 Adirondack drainage lakes showed that mercury concentrations in perch were weakly negatively correlated with DOC concentrations in lake water. However, our study did not include any low-ANC, high-DOC waters. In Michigan drainage lakes, mercury concentrations in yellow perch were not significantly correlated with DOC, whereas in seepage lakes a significant negative correlation was observed.[14] Positive correlations with DOC were reported by McMurtry et al.[22] for mercury concentrations in lake trout (*Salvelinus namaycush*) and smallmouth bass from Ontario lakes.

Impoundment status may affect the bioavailability of mercury in lakes;[23] but a clear relationship between impoundment status and mercury in yellow perch was not evident in our study. A larger data set would be needed to adequately test the hypothesis that impoundment results in increased mercury concentrations in fish. We similarly did not find a significant correlation between mercury concentrations and the watershed area/lake volume ratio, as reported by Suns and Hitchin.[17]

An equation predicting mercury concentration based on fish length and lake pH may be a useful tool for identifying lakes with potentially high mercury concentrations in yellow perch. These lakes may also have additional piscivorous species of fish with high mercury concentrations.[4,13,17] Wiener et al.[4] developed a multiple regression equation for walleyes which was very similar to the one we developed for yellow perch. In their study, lake pH and total length of the fish accounted for 69% of the variation in mercury concentration.

Based on data from the Adirondack Lakes Survey Corporation,[11] 346 of the 1469 Adirondack waters surveyed contained yellow perch. Most of these waters were drainage lakes, 67 had a summer surface pH less than 6.0 and 96 had a summer surface pH less than 6.5. Given these observations and using the predictive equation for Adirondack lakes, we estimate that more than 100 waters in the Adirondacks contain large perch exceeding the 1 μg/g mercury guideline. Additional waters which may not contain perch may have piscivorous fish with high mercury concentrations, and others may be influenced by flooding of formerly forested areas. Meacham Lake is an Adirondack lake with a summer surface water pH of 7.16 which was recently found to contain yellow perch with mercury concentrations exceeding 1 μg/g.[24]

A larger Adirondack data set should be gathered to increase our understanding and ability to predict mercury concentrations in fish tissue. The role of lake physicochemical factors in determining mercury concentrations could be better understood by studying a larger number of waters. Only two of our lakes had air-equilibrated pH values less than 6.0 and ANC less than 50 µeq/L. Fish from these lakes had the highest mercury concentrations and strongly influenced our evaluation of the data. Clearly, data from additional low pH waters and low ANC lakes with a range of DOC concentrations would help refine the empirical relationships developed in this study. Determinations of mechanisms and pathways regarding mercury bioaccumulation in fish would require a more comprehensive study of individual lake systems to answer remaining questions regarding mercury biogeochemistry.

ACKNOWLEDGMENTS

We thank the Adirondack Lakes Survey Corporation staff for making the fish collections and providing the water chemistry data. The Empire State Electric Energy Research Corporation funded their work. Paul VanValkenburg assisted in the initial data collection, and Tony Vandervalk of Cornell University aged the yellow perch scales. James Gallagher prepared Figure 1 and assisted in obtaining the necessary data. We are grateful to Sharon Kohler for typing and revising the manuscript.

REFERENCES

1. Wren, C.D. and MacCrimmon, H.R., Mercury levels in the sunfish, *Lepomis gibbosus*, relative to pH and other environmental variables of precambrian shield lakes, *Can. J. Fish. Aquat. Sci.*, 40, 1737, 1983.
2. Sloan, R. and Schofield, C.L., Mercury levels in brook trout (*Salvelinus fontinalis*) from selected acid and limed Adirondack lakes, *Northeastern Environ. Sci.*, 2, 165, 1983.
3. Bloomfield, J.A., Quinn, S.O., Scrudato, R.J., Long, D., Richards, A., and Ryan, F., Atmospheric and watershed inputs of mercury to Cranberry Lake; St. Lawrence County, New York, In: *Polluted Rain*, Toribara, T.Y., Miller, M.W., and Morrow, P.E., Eds., Plenum Press, New York, 1980, 175.
4. Wiener, J.G., Martini, R.E., Sheffy, T.B., and Glass, G.E., Factors influencing mercury concentrations in walleyes in northern Wisconsin lakes, *Trans. Am. Fish Soc.*, 119, 862, 1990.
5. Spry, D.J. and Wiener, J.G., Metal bioavailability and toxicity to fish in low-alkalinity lakes: a critical review, *Environ. Pollut.*, 71, 243, 1991.
6. Sullivan, T.J., Historical changes in surface water acid-base chemistry in response to acidic deposition. Report 11, NAPAP State of Science and Technology, National Acid Precipitation Assessment Program, Washington, DC, 1990, 11–151.
7. Driscoll, C.T., Newton, R.M., Gubala, C.P., Baker, J.P., and Christensen, S.W., Adirondack Mountains, In: *Acidic Deposition and Aquatic Ecosystems, Regional Case Studies*, Charles, D.W., Ed., Springer-Verlag, New York, 1991, chap. 6.
8. Quinn, S.O. and Bloomfield, N., Acidic Deposition, Trace Contaminants and Their Indirect Human Health Effects: Research Needs, New York State Dept. of Environmental Conservation, Albany, NY, 1985, 54.
9. Sloan, R., Toxic Substances in Fish and Wildlife: Analyses Since May 1, 1982, Volume 6, New York State Dept. of Environmental Conservation, Albany, NY, 1987, 10.
10. Heit, M. and Klusek, C.S., Trace element concentrations in the dorsal muscle of white suckers and brown bullheads from two acidic Adirondack lakes, *Water Air Soil Pollut.*, 25, 87, 1985.
11. Kretser, W., Gallagher, J., and Nicolette, J., Adirondack Lakes Study 1984–1987, An Evaluation of Fish Communities and Water Chemistry, Adirondack Lakes Survey Corporation, Ray Brook, NY, 1989.

12. Gloss, S.P., Grieb, T.M., Driscoll, C.T., Schofield, C.L., Baker, J.P., and Landers, D., Mercury Levels in Fish From the Upper Peninsula of Michigan (ELS Subregion 2B) in Relation to Lake Acidity, EPA/600/3–90/068, U.S. Environmental Protection Agency, Washington, D.C., 1990, 102.

13. Cope, W.G., Wiener, J.G., and Rada, R.G., Mercury accumulations in yellow perch in Wisconsin seepage lakes: relation to lake characteristics, *Environ. Toxicol. Chem.*, 9, 931, 1990.

14. Grieb, T.M., Driscoll, C.T., Gloss, S.P., Schofield, C.L., Bowie, G.L., and Porcella, D.B., Factors affecting mercury accumulation in fish in the upper Michigan peninsula, *Environ. Toxicol. Chem.*, 9, 919, 1990.

15. SAS Institute, SAS/STAT Guide for Personal Computers, version 6.03, Cary, NC, 1988.

16. Kanciruk, P., Eilers, J.M., McCord, R.A., Landers, D.H., Brakke, D.F., and Linthurst, R.A., Characteristics of Lakes in the Eastern United States. Volume III: Data Compendium of Site Characteristics and Chemical Variables, EPA/600/4–86/007c, U.S. Environmental Protection Agency, Washington, D.C., 1986, 439.

17. Suns, K. and Hitchin, G., Interrelationships between mercury levels in yearling yellow perch, fish condition and water quality, *Water Air Soil Pollut.*, 50, 255, 1990.

18. Wiener, J.G., Fitzgerald, W.F., Watras, C.J., and Rada, R.G., Partitioning and bioavailability of mercury in an experimentally acidified Wisconsin lake, *Environ. Toxicol, Chem.*, 9, 909, 1990.

19. Heit, H., Schofield, C., Driscoll, C.T., and Hodgkiss, S.S., Trace element concentrations in fish from three Adirondack lakes with different pH values, *Water Air Soil Pollut.*, 44, 9, 1989.

20. Gutenmann, W.H. and Lisk, D.J., Higher average mercury concentration in fish fillets after skinning and fat removal, *J. Food Safety,* 11, 99, 1991.

21. Scott, D.P. and Armstrong, F.A.J., Mercury concentration in relation to size in several species of freshwater fishes from Manitoba and northwestern Ontario, *J. Fish. Res. Board,* 29, 1685, 1982.

22. McMurtry, M.J., Wales, D.L., Scheider, W.A., Beggs, G.L., and Dimond, P.E., Relationship of mercury concentrations in lake trout (*Salvelinus namaycush*) and smallmouth bass (*Micropterus dolomieui*) to the physical and chemical characteristics of Ontario lakes, *Can. J. Fish. Aquat. Sci.*, 46, 426, 1989.

23. Jackson, T.A., Biological and environmental control of mercury accumulation by fish in lakes and reservoirs of northern Manitoba, Canada, *Can. J. Fish. Aquat. Sci.*, 48, 2449, 1991.

24. Skinner, L.C. FY 92–93 Health Advisory Memorandum, New York State Dept. of Environmental Conservation, April 7, 1992, 11.

SECTION V

Modeling of Aquatic Ecosystems

Modeling the Biogeochemical Cycle of Mercury in Lakes: The Mercury Cycling Model (MCM) and Its Application to the MTL Study Lakes

Robert J. M. Hudson, Steven A. Gherini,
Carl J. Watras, and Donald B. Porcella

CONTENTS

ABSTRACT: As a part of the Mercury in Temperate Lakes (MTL) study in Wisconsin, U.S., we have developed a mechanistic model of the biogeochemical cycle of mercury in lakes. The Mercury Cycling Model (MCM) is a deterministic simulation model that incorporates the major processes that transport mercury across lake boundaries—atmospheric deposition, gas exchange, inflow and outflow of water, and burial in sediments; chemically transform it—reduction, methylation, and demethylation; and lead to its accumulation in aquatic biota—uptake, depuration, and trophic level transfer. In this chapter, we discuss the theory of mercury biogeochemical cycling, apply a simplified, steady-state version of the MCM to analyze field data from the MTL study, and examine mechanistic issues using the dynamic MCM model.

Theories of mercury biogeochemistry are based on knowledge of the aqueous speciation of mercury and the mechanisms of its reactions—considerable gaps in

1–56670–066–3/94/$0.00+$.50

our knowledge of both remain, however. Our approach, therefore, is to use what is known about mercury biogeochemistry to formulate hypothetical rate and equilibrium expressions. Using a simplified, steady-state version of the MCM model, we then examine the ability of these expressions to describe the field data of the MTL study. For example, although it is known that Hg^{2+} and CH_3Hg^+ ions are complexed by the hydroxide, chloride, sulfide, and humic acids present in lakewater, uncertainties in the strength of organic complexation and the concentrations of competing metals make equilibrium calculations speculative. Applying our model of seston-water partitioning for Hg^{II} and CH_3Hg^{II} to the MTL data, we examine potential mechanisms of mercury uptake by phytoplankton and the strength of organic complexation in these lakes. Our analysis suggests greater than 70% organic complexation of both Hg^{II} and CH_3Hg^{II} and significant accumulation of CH_3Hg^{II} by plankton. To model the in-lake cycling of mercury, we derive theoretical rate laws, corresponding to a variety of hypothesized mechanisms for the reactions of the lacustrine mercury cycle and test their ability to predict the observed concentrations of Hg^0, Hg^{II}, and CH_3Hg^{II} in lakewater. Only a limited number of mechanisms were consistent with the data. In modeling bioconcentration factors (BCF_{FISH}) for CH_3Hg^{II} in fish, we found that the binding of CH_3Hg^{II} by DOC explains the observed dependence of BCF_{FISH} on DOC, and that calcium inhibits bioaccumulation, likely at trophic levels above phytoplankton. Our results suggest observed correlations between lake pH and fish mercury content arise from generally higher CH_3Hg^{II} concentrations at low pH and lower bioconcentration factors at high pH.

To investigate issues relating to food chain accumulation and the generation of CH_3Hg^{II} in anoxic waters in greater detail, we employed the dynamic MCM model. We hypothesize that passive uptake of the neutral $Hg(SH)_2^0$ species by methylating bacteria is an explanation for the apparently high methylation rate of Hg^{II} in sub- and anoxic waters.

I. INTRODUCTION

Reports of elevated levels of mercury in fish from remotely located, low pH lakes have generated public concern about the scope of environmental mercury contamination.[1-4] Because of mercury's toxicity to humans and wildlife, these findings may have significant public policy implications if the elevated levels of mercury in fish are caused by lake acidification[5] or recent increases in atmospheric mercury deposition.[6,7] Consequently, the quantity and sources of mercury in atmospheric deposition and the environmental factors that govern its biogeochemical cycling and bioaccumulation in lake ecosystems are the subjects of much scientific inquiry.

These questions have proved difficult to answer in the past. The primary reason is that methods of sampling and analysis that are sufficiently sensitive and contamination-free to obtain accurate measurements of environmental mercury levels have been developed only recently. Using these new methods, limnologists and oceanographers have discovered that mercury exists at extremely low concentrations in surface waters—as low as 1 ng/L total with 5 to 20% methylmercury—and in precipitation—10 ng/L total with about 1% methylmercury.[8-13] These concentrations are 2 to 3 orders of magnitude lower than the levels of mercury typically employed in experimental studies, raising doubts about the relevance of such studies for understanding mercury biogeochemistry.

A knowledge of mercury chemistry and its microbial transformations has, however, enabled scientists to predict some features of the mercury cycle in natural waters that are consistent with recent observations.[14-16] Along with the basic reactions of the mercury cycle, the chemical principles which govern the behavior of mercury in the environment—the control of trace element biogeochemistry by chemical speciation—are generally known.[17] Several conceptual models[14,18] and a limited number of mechanistic simulation models[19] have been formulated on the basis of these principles. In addition, correlation analyses of fish mercury survey data have identified several key limnological parameters that influence levels of fish mercury, most notably pH and DOC,[4] and have been used to derive empirical models of fish mercury content.[1] However, the mechanisms linking lake properties, the chemical speciation of mercury, and its accumulation in fish have not yet been identified or quantified.

Partly for this reason, the Mercury in Temperate Lakes (MTL) study in northern Wisconsin was conducted to quantify the fluxes of mercury in lake ecosystems and to investigate the influence of key limnological parameters on the cycle of mercury in lakes.[20] Seven seepage lakes, spanning a range of pH and DOC concentrations, were selected for the study. The MTL study has provided one of the most extensive data sets of mercury cycling in lakes available to date; several reports of its findings appear elsewhere in this volume. As a part of this study, the Mercury Cycling Model (MCM) was developed. The MCM mechanistically simulates the transport and transformations of mercury in natural waters based on the concept that aqueous chemical speciation governs the rates of these processes (Figure 1). The goal in developing the model has been to aid in the analysis of the MTL results and to provide a basis for assessing the impact of mercury in other seepage lake systems.

In this chapter, we apply the MCM to analyze the dependence of the lacustrine mercury cycle on limnological properties. In doing so, we will consider several hypotheses proposed in the literature to explain these dependencies (Table 1). Our analysis involves three steps. In the first we consider the chemical principles governing the environmental cycling of mercury. Second, we apply a simplified, steady-state version of the model to data from the MTL study to test the explanatory power of competing hypotheses for the mechanisms of the reactions governing the mercury cycle. Generally, different mechanisms imply different dependencies on mercury speciation and limnological parameters. In the third step, we use the dynamic model to investigate issues that cannot be addressed using the steady-state model.

Table 1 **Significant hypotheses concerning mercury biogeochemical cycling and bioconcentration examined in this chapter.**

Hypotheses	References
DOC compounds strongly bind Hg^{II} and CH_3Hg^{II}, reducing CH_3Hg^{II} bioavailability	Mantoura et al. (1978),[25] Lee and Hultberg (1990),[66] and Grieb et al. (1990)[4]
Hg^{II} reduction is more rapid in high pH lakes, limiting the substrate available for methylation	Brosset (1987),[21] Winfrey and Rudd (1990),[5] and Fitzgerald et al. (1991)[13]
Methylation/demethylation ratio is higher in low pH lakes, increasing the net production of CH_3Hg^{II}	Xun et al. (1987)[92]
Microbial methylation in lakes is stimulated by sulfate	Winfrey and Rudd (1990),[5] and Gilmour and Henry (1991)[80]
External loads of CH_3Hg^{II} to lakes are sufficient to account for observed outputs	Hultberg et al. (this volume)[93]
Calcium inhibits CH_3Hg^{II} accumulation in fish	Wren and MacCrimmon (1983)[2] and Rodgers and Beamish (1983)[94]
Chloride stimulates CH_3Hg^{II} bioconcentration by passive uptake/diffusion processes	Gutknecht (1981),[43] Boudou et al. (1983),[95] and Mason and Morel (1992)[73]

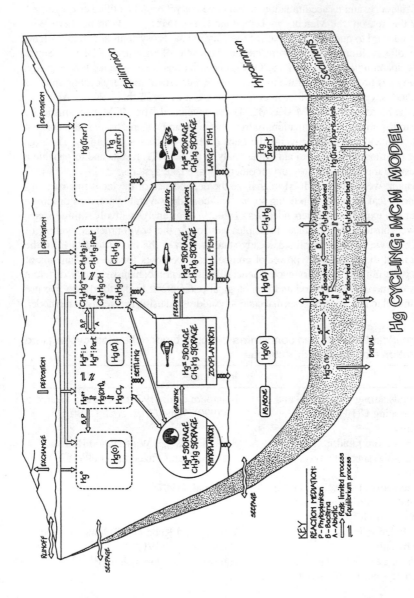

Figure 1 Mercury cycling as represented in the MCM model. (Reactions of hypolimnion components are similar to those in the epilimnion with allowances for differences in chemistry). The MCM model utilizes 14,000 lines of computer code to represent the illustrated pools and processes.

This methodology yields testable predictions of the rate laws for the key processes of the mercury cycle. Verification of these rate laws or identification of more appropriate ones by further experimental and field research will yield a model of the lacustrine mercury cycle accurate enough to predict concentrations in fish from mercury loadings and limnological properties.

II. THEORY OF MERCURY BIOGEOCHEMICAL CYCLING

Chemical speciation serves as the conceptual foundation for modeling the environmental biogeochemical cycle of mercury and most other metals. The term "speciation" is commonly used to include the important physical and chemical distinctions in the forms in which mercury or other solutes may be found. The most common fractionation of water samples involves the separation of particulate and dissolved forms of a substance. A second type of speciation involves identifying the chemical nature of solutes by distinguishing species engaged in rapid, reversible equilibrium reactions, i.e., inorganic and organic complexes of metal ions. Finally, speciation can involve differentiating forms of trace elements that are not rapidly interconvertible by equilibrium reactions, e.g., alkylated metals and different redox states of metals. All three kinds of speciation are relevant to the environmental cycling of mercury.

The importance of speciation lies in the fact that different species behave differently, e.g., particulate species settle and those associated with dissolved or colloidal organic matter are generally unavailable to biota.[17] Consequently, the rates of the transport and transformation processes making up the environmental cycle of mercury are determined by the availability of various mercury species to the governing biogeochemical processes (Table 2). To represent this fact, the model rate law for a non-equilibrium process p that transforms or transports mercury between two chemical forms or physical compartments:

$$Hg^A \rightarrow Hg^B \tag{1}$$

includes the fraction of Hg^A present as the rate-controlling species (α_P^A), to account for the direct dependence on mercury speciation, and an apparent first-order rate constant (k_P^*),

$$\text{Rate of process } p = k_P^* \cdot \alpha_P^A \cdot [Hg^A] \tag{2}$$

to account for all other dependencies on lake physical (f_P^{phys}), biological (f_P^{bio}), and chemical (f_P^{chem}) properties

$$k_P^* = k_P^o \cdot f_P^{phys} \cdot f_P^{chem} \cdot f_P^{bio} \tag{3}$$

We are able to separate the dependence on mercury speciation from other factors because mercury is a subtoxic, trace constituent in most lakes.

Despite recent progress in elucidating the lacustrine mercury cycle, neither the mechanisms for several key processes nor the equilibrium speciation of divalent and methyl-mercury have been established. For this reason, we begin with an overview of the principles applicable in formulating models of each.

A. MERCURY SPECIATION IN LAKES

Current methods of speciating mercury have identified four important chemical forms of mercury in freshwater systems: elemental mercury, monomethylmercury, and the "reactive" and "nonreactive" fractions of mercury(II).[21] (Dimethylmercury, which occurs in marine

Table 2 Speciation dependence of hypothetical mechanisms of mercury transport and transformation processes.

Process	Description	Rate controlling step	Rate determining species			Comments
			Hg^0	Hg^{II}	CH_3Hg^{II}	
Biotic						
Facilitated transport (or transformation)	Metal binds to cell surface sites followed by transmembrane transport (or transformation)	Transmembrane transport (or transformation)	n.a.	Hg^{2+}	CH_3Hg^+	Transport rates modulated by availability of essential trace metals; Competition from H^+ for binding to sites
		Complexation by cell surface sites	n.a.	$Hg^{II}_{INORGANIC}$	$CH_3Hg^{II}_{INORGANIC}$	Same as above; free ion and complexes with OH^- and Cl^- assumed to be labile
		Solute diffusion to cell surface	n.a.	$Hg^{II}_{INORGANIC}$	$CH_3Hg^{II}_{INORGANIC}$	Sensitive to organism size
Passive diffusion	Permeable, neutral species diffuse through membrane	Transmembrane diffusion	Hg^0	$Hg(Cl)^0_2$	CH_3HgCl^0 CH_3HgOH^0	
		Solute diffusion to cell surface	Hg^0	$Hg^{II}_{INORGANIC}$	$CH_3Hg^{II}_{INORGANIC}$	Sensitive to organism size

Abiotic

Process	Process description	Mechanism	Species A	Species B	Species C	Remarks
Air-water exchange	Volatile species diffuse between water and air	Liquid phase diffusion	Hg^0	n.a.	n.a.	
		Gas phase diffusion	n.a.	$HgCl_2$	CH_3HgCl^0 CH_3HgOH^0 CH_3Hg^+	
Scavenging by particles	Equilibrium partitioning of metal between dissolved and settling particulate species	Particle sedimentation and burial	n.a.	Hg^{2+}		Effects of other metals uncertain; DOC reduces sorption by particles
Reduction	Reduction of Hg^{II}	Photoreduction of Hg^{II}-humate complex	n.a.	Hg^{II}_{HUMATE}	n.a.	
		Photoreduction of $Hg(OH)_2^0$	n.a.	$Hg(OH)_2^0$	n.a.	
		DOC-mediated thermal reduction	n.a.	Hg^{2+}	n.a.	
Methylation	Methylation of Hg^{II}	Humus-mediated methylation	n.a.	Hg^{II}_{HUMATE} or Hg^{2+}	n.a.	

waters,[22] is below detection limits in the freshwaters examined to date[13,23]). Of these forms of mercury, only elemental and monomethylmercury are analytically well-defined; the reactive/nonreactive fractionation of mercury(II) is operational. Since even strong inorganic complexes of Hg^{2+} are "reactive" in the presence of $SnCl_2$ under acidic conditions,[21] the "nonreactive" fraction likely corresponds to strong organic complexes of Hg^{2+} [12] or to uncharacterized, stable Hg compounds produced during the combustion of coal.[21] On the basis of this analytical speciation, we have defined four aqueous chemical *components* of mercury in the MCM: Hg^0, CH_3Hg^{II}, Hg^{II}, and Hg^{Inert} (Figure 1). While the model distinguishes labile Hg^{2+} complexes (the component Hg^{II}) from kinetically inert species of mercury(II) (the component Hg^{Inert}) for the purposes of this analysis we will consider that mercury(II) exists only as labile species within the lake water column. We make this assumption out of necessity, since it is not known whether or at what rates the nonreactive and reactive forms of Hg^{II} from atmospheric deposition may interconvert within lakes and since analytical chemists do not agree about the reliability and interpretation of "nonreactive" Hg measurements in lakewater.[24]

Each *component* is comprised of the chemical *species* that are in thermodynamic (psuedo) equilibrium with one another, but not with the species of other components. For example, in lake epilimnia the Hg^{II} component includes the species in equilibrium with the Hg^{2+} ion, e.g., $Hg(OH)_2^0$, $HgCl_2^0$, $HgOHCl^0$, humate-bound Hg^{II}, abiotic particulate species of Hg^{II}, etc. Some aspects of the equilibrium speciation of mercury in freshwaters, such as complexation of Hg^{2+} and CH_3Hg^+ by inorganic ligands, are well known (Figure 2). However, reliable measurements of the binding strength of natural DOC compounds are not yet available. The only reported complexation constants for Hg^{II} by humic acid[25] were obtained at pH 8 using undoubtedly high mercury concentrations relative to natural systems. Nevertheless, the strength of the Hg^{II}-humic complexes they report are comparable to Hg^{II}-thiols.[26] Humic acids do contain thiols[27] as well as other functional groups that are also capable of binding CH_3Hg^{II} strongly.[26,28] Dissolved elemental mercury does not engage in coordination reactions or adsorb strongly to natural organic matter, as its weak partitioning into organic solvents indicates.[29] As a consequence, essentially all dissolved Hg^0 should be able to freely exchange with the atmosphere or penetrate cell membranes.

A further complication in determining the aqueous speciation of mercury results directly from its picomolar ambient concentrations—its equilibrium speciation may be influenced by strong ligands present at concentrations below present detection limits and by other trace metals not measured in most limnological studies of mercury. Low concentrations of strong ligands could completely overwhelm the effects of more abundant, weaker ligands. For example, with Cu^{2+} complexation by humic acids an approximately inverse-squared relationship between binding site strength and abundance has been observed.[30] Were such a relationship to hold for mercury, the difference in Hg^{2+} ion concentrations which would be observed under equilibrium with strong or weak humic acid binding sites could easily be two of orders of magnitude (Figure 2A, B). Hydrogen sulfide is another strong (inorganic) ligand present in undefined quantities in lake epilimnia. A sulfide concentration of 1 nM, comparable to that observed in surface seawater,[31] would dominate the speciation of Hg^{II} and CH_3Hg^{II} (Figure 2C). Even if ligand concentrations were known, the presence of other trace metals, such as aluminum,[32] copper, or iron, that compete with mercury for binding to organic ligands and sulfide, would increase the uncertainty of speciation calculations.

Finally, slow kinetics may prevent mercury from attaining complexation equilibrium. For example, Hg^{II} deposited in the lake may rapidly bind to more numerous weak humic sites and then react at slower rates to form the strongest complexes. Kinetics of complexation and ligand exchange are also influenced by the presence of other metals. For example, despite strongly favorable equilibria, Hg^{2+} at high concentrations displaces Cu^{2+} bound by humic acid very slowly,[33] and Cu^{2+} complexation by EDTA in seawater, where it is principally CaEDTA, is dramatically slowed relative to complexation by uncomplexed or protonated

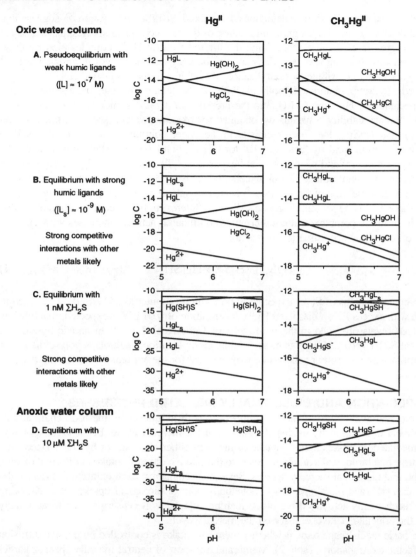

Figure 2 Calculated equilibrium aqueous speciation of Hg^{II} and CH$_3$Hg^{II}. Stability constants for association with ligands (L) in DOC derived in Section III.A. Strong organic ligands (L$_s$) assumed to be 10^4-times stronger but 10^2-times less abundant than L. Stability constants for inorganic complexes are for 11° C.[65] With H$_2$S present, HgS(s) precipitation is presumed kinetically slow. Calculations assume 3 mg L^{-1} DOC and 7 μM Cl and 5 pM Hg.

EDTA species.[34] Equilibration with trace ligands could be slowed by any metal present in quantities sufficient to bind a majority of the ligands, e.g., Al, Fe, and perhaps Cu in freshwaters. Until environmental chemists characterize these effects for mercury, we must cautiously interpret chemical speciation calculations.

In anoxic waters, complexation by sulfide is generally assumed to control speciation for both Hg^{II} and CH$_3$Hg^{II}.[18,26,35] Measurements of reduced sulfur speciation in marine porewaters[36] and equilibrium calculations for mixed $S(0)/S(-II)$ systems[37] suggest that polysulfides may comprise a significant fraction of the reduced sulfur species, particularly at neutral or greater pH. Calculations using the limited data on Hg^{2+} complexation by polysulfides suggests that Hg^{II}-polysulfides may also be a significant species in anoxic waters.[38] Thiol ligands may contribute to solubilization of Hg^{II} in sulfidic waters as well.[26]

In order to compare anoxic hypolimnia with oxic waters, we have calculated the equilibrium speciation of Hg^{II} and CH_3Hg^{II} in the presence of 10 μM sulfide (Figure 2D). Using the Hg^{II}-humate stability constants we obtained for oxic lakewaters (see Section III.A) and assuming that HgS(s) does not precipitate in the water column, we calculate that sulfide outcompetes humic acid by factors of 10^4 for CH_3Hg^{II} and 10^{18} for Hg^{II}. In sediments, the chemistry of mercury and of its potential ligands is poorly understood. Hg^{II} in sediments is correlated with organic matter,[39] but since sulfides also covary with organic matter a controlling solid phase cannot be inferred.[40] Mercuric sulfide is a highly stable mineral that should theoretically be present in anoxic sediments. Interestingly, the concentration of soluble mercuric bisulfide complexes at equilibrium with cinnabar actually increases with the concentration of sulfide according to the reaction:

$$HgS(s) + H_2S \leftrightarrow Hg(SH)_2^0 \qquad (4)$$

For the concentrations of $S(-II)$ expected in porewaters, about 100 nM based on $FeS_{(s)}$ equilibrium with excess Fe^{II}, a dissolved Hg^{II} concentration of 0.01 pM is expected, much below the 50 pM levels observed in the porewaters of Little Rock Lake.[41] In anoxic hypolimnia with total $S(-II) > 10$ μM, Hg^{II} is more soluble. Complexation by thiols,[35] polysulfides,[38] or equilibrium with a different solid phase may account for the apparently supersaturated concentrations in sediments.

B. SPECIATION AND BIOLOGICALLY MEDIATED PROCESSES

The biological availability of mercury influences both its uptake by the food chain and its involvement in biologically mediated reactions of the mercury cycle. Because most biologically mediated processes require uptake of mercury either into cells or by cell surface sites, we assume that biological availability refers to the relative rates of uptake of different methyl, divalent, and elemental mercury species. Several mechanisms, each exhibiting different dependencies on trace metal speciation, potentially control biological uptake rates. Knowing which mechanisms actually drive biotic mercury uptake is necessary for us to accurately model the effects of lake chemistry on the mercury cycle.

The basic mechanistic issue is whether mercury uptake is controlled by passive diffusion or by facilitated transport (Table 2). Membrane transport of neutral mercury species, particularly Hg^0 and CH_3HgCl^0 and $HgCl_2^0$, by passive means is rapid in lipid bilayer membranes, [42,43] while charged complexes with inorganic ligands and hydrophilic organic chelators are generally not membrane permeable. The passive uptake of neutral mercury species, for example $HgCl_2^0$, may be written as:

$$HgCl_2^0 \rightarrow Hg_{CELL}^{II} + 2Cl^- \qquad (5)$$

where Hg_{CELL}^{II} denotes intracellular Hg^{II}. The rate of passive uptake by this reaction is then:

$$HgCl_2^0 \text{ uptake rate} = P_{HgCl_2} \cdot [HgCl_2^0] \cdot A_{CELL} \qquad (6)$$

where P_{HgCl_2} is the permeability of the cell membranes to $HgCl_2^0$ (cm s^{-1}) and A_{CELL} is the specific surface area of the cells (cm^2 kg-cell^{-1}). Permeability coefficients depend on the properties of both solutes and membranes. Studies using synthetic bilayer membranes have found values of P_{HgCl_2} between 2.6×10^{-3} and 1.3×10^{-2} cm s^{-1}.[42,43] On the basis of octanol-water partition coefficients, P_{CH_3HgCl} and P_{CH_3HgOH} should, respectively, be about twice and one tenth P_{HgCl_2} in a given membrane.[42,44] Passive diffusion provides a baseline uptake rate that facilitated transport may supplement.

Facilitated transport of mercury occurs via specific mercury transport proteins, such as may be induced in mercury-resistant bacteria,[45] and possibly via nonspecific uptake through transport systems for nutrient metals in nonmercury-stressed aquatic organisms. For example, Hg^{II} could enter cells through the uptake systems whose normal physiological function is to acquire zinc, iron, or other essential metals. Facilitated uptake of toxic metals is a consequence of imperfect selectivity in nutrient uptake systems and has been previously observed in marine phytoplankton with copper and manganese[46] and with cadmium and iron.[47] Essential trace metal transport generally involves complexation of the metals by specific extracellular or surface-associated ligands ($X_{SURFACE}$). Although all aqueous species may react with the transport ligand at some rate, inorganic complexes generally react the most readily. For example, the major inorganic complexes of Hg^{II} all react with $X_{SURFACE}$:

$$Hg^{2+} + X_{SURFACE}^{-} \rightleftharpoons HgX_{SURFACE}^{+} \tag{7a}$$

$$HgOH_2 + X_{SURFACE}^{-} \rightleftharpoons HgX_{SURFACE}^{+} + 2OH^{-} \tag{7b}$$

$$HgCl_2 + X_{SURFACE}^{-} \rightleftharpoons HgX_{SURFACE}^{+} + 2Cl^{-} \tag{7c}$$

$$HgX_{SURFACE}^{+} \xrightarrow{\;k_{in}\;} Hg_{CELL}^{II} + X_{SURFACE}^{-} \tag{7d}$$

As long as the $HgX_{SURFACE}^{+}$ complex is stronger than the inorganic complexes, Hg^{II} complexes with hydroxide and chloride should react at rates roughly similar to Hg^{2+}.[48] Similarly, all the major CH_3Hg^{II} inorganic species should react with free surface sites at comparable high rates, as has been found for other strong ligands.[49]

When k_{in}, the rate constant for intracellular absorption of Hg^{II} from $HgX_{SURFACE}^{+}$, is slow relative to the formation and dissociation reactions (Equations 7a–c), the transport ligand and solution are effectively in equilibrium with respect to Hg^{2+} complexation and the concentration of $HgX_{SURFACE}^{+}$ is proportional to "free metal ion" concentration or activity:

$$[HgX_{SURFACE}^{+}] = B_{HgX} \cdot [X_{SURFACE}^{-}] \cdot [Hg^{2+}] \tag{8}$$

For this thermodynamically controlled case, the uptake rate is proportional to the Hg^{2+} ion concentration:

$$Hg(II) \text{ uptake rate} = \frac{k_{in} \cdot X_T \cdot [Hg^{2+}]}{1 + \beta_{HgX} \cdot [Hg^{2+}]} \tag{9}$$

where X_T is the total concentration of the transport ligands. Although this is the best known dependence of metal transport rates on aqueous speciation, equilibrium between the site and the solution may not always be attained. In such kinetically controlled cases, the rate of uptake reflects the complexation rates of all labile species, not just that of the free metal ion.[48]

Efficient uptake kinetics mandate that the transport ligands generally be undersaturated with respect to the nutrient metal being transported[48] and we assume that concentrations of mercury species are not saturating in the absence of gross contamination by mercury. There is evidence, however, that protonation of transport sites affects uptake rates of metals such as Al,[50] Zn, and Cd.[51] Thus, we define a protonation reaction for the transport ligand:

$$H^{+} + X_{SURFACE}^{-} \xrightarrow{\;\beta_{HX}\;} HX_{SURFACE} \tag{10}$$

where β_{HX} is the proton association constant. Since we do not know the values of k_{in} and X_T, we reformulate Equation 9 as:

$$Hg^{II}\ uptake\ rate = k_X \cdot \frac{[Hg^{2+}]}{1 + \beta_{HX} \cdot [H^+]} \tag{11}$$

where k_X is the mass specific mercury uptake rate constant (L kg-cell^{-1} d^{-1}). A similar rate law proportional to the concentration of inorganic Hg^{II} species may be written for kinetically controlled transport. Rates of facilitated transport are also influenced by feedback between the number and type of transport sites and the trace element nutritional status of the organism.[52] Cellular feedback would enhance relative rates of mercury uptake when essential metals have low availability.

C. DIFFUSION LIMITATION OF BIOTIC MERCURY UPTAKE

At high cellular uptake rates, slow diffusion through aqueous media can cause the transported species to become depleted at the cell surface, thereby limiting the maximum uptake rate of any solute. Diffusion-limited uptake of nutrients, such as P, Fe, and Zn, has been observed in both freshwater and marine phytoplankton.[48,53,54] Since diffusion limitation affects large cells to a greater extent than small cells, it is likely to exert a greater influence on bioaccumulation by phytoplankton than on bacterial mercury transformation processes.

When diffusion-limited rates are attained, cellular uptake may exhibit a speciation dependence that differs from that of the actual membrane transport step (Table 2). Consider uptake of Hg^{II} via facilitated transport as in Equation 6. Since most inorganic complexes of Hg^{2+} react readily with strong organic ligands,[48] such as likely constitute the transport sites, all the inorganic species, not only Hg^{2+}, will be depleted at the cell surface. The diffusion-limited uptake rate then depends on the total concentration of species which react in the transport step as well as complexes which dissociate to reacting species within the diffusive boundary layer.[55] Likewise, passive uptake of Hg^{II} and CH$_3$Hg^{II} is limited by the concentrations of species that interconvert with the permeable forms within the diffusive boundary layer.

Jackson and Morgan[55] have shown that the necessary condition for interconversion of an unavailable metal complex (ML) and an available species (M) with and within the diffusive boundary layer is that the characteristic time constant for complex formation ($\tau_{COMPLEXATION}$) be less than the time constant for diffusion through the boundary layer ($\tau_{DIFFUSION}$). The complexation time constant for a second-order complexation reaction with forward rate constant k_1 is

$$\tau_{COMPLEXATION} = 1/(k_1 \cdot [L']) \tag{12}$$

at an available ligand concentration [L']. The diffusion time constant is given by:

$$\tau_{DIFFUSION} = R^2_{CELL}/D \tag{13}$$

where R_{CELL} is the cell radius and D is the diffusivity of the metal species. For a 2.5-μm-radius cell, $\tau_{DIFFUSION}$ is about 6×10^{-3} s. The lability of water bound in the inner coordination sphere of Hg^{2+} ions and complexes with chloride and hydroxide[56] makes the interconversion of Hg^{2+} and its complexes with OH$^-$ and Cl$^-$ rapid compared to diffusion. Interconversion of CH$_3$Hg$^+$ inorganic complexes is similarly rapid.[49]

Diffusion-limited uptake of inorganic Hg^{II} and CH$_3$Hg^{II} constrains the plankton-water partition coefficients ($K^*_{PLANKTON}$) attainable in lakes. Because phytoplankton cells continually divide, the partitioning into live cells reflects a balance of rates—uptake vs. depuration

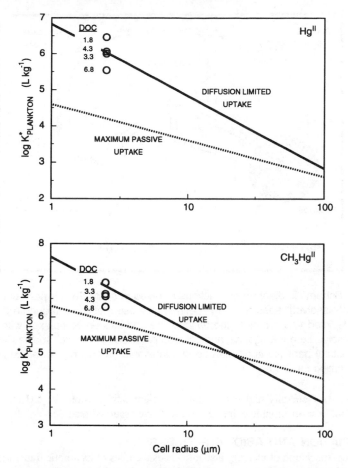

Figure 3 Size-dependence of plankton-water partition coefficients ($K^*_{PLANKTON}$) for Hg^{II} and CH_3Hg^{II}. The partition coefficients at diffusion-limited uptake rates (solid lines) and for passive uptake in the absence of depuration (dashed lines) are calculated for pH 6 and 3 mg L^{-1} DOC with organic complexation calculated as in Section III.A. The points indicated on each graph (open circles) are the values of $K^*_{PLANKTON}$ calculated from fitting the MTL data for the pH 6 ± 0.2 lakes. The actual DOC concentrations are indicated for each lake.

and growth dilution (Appendix 1)—that is fundamentally different from adsorption onto detrital or inorganic particles. Maximum mercury uptake rates can be calculated a priori from cell geometry and measured physical constants. Such calculations are particularly straight-forward for small (diameter <100 μm) phytoplankton, since the geometric and fluid flow considerations are simple.[57] As illustrated for the example of a pH 6 lake containing 3 mg L^{-1} DOC (Figure 3), the diffusion-limited uptake line constitutes the absolute upper limit for plankton-water partition coefficients. Because the diffusion-limited flux to a spherical cell is proportional to its radius, the mass partition coefficient (effectively normalized by cell volume) is proportional to R^{-2}_{CELL}. Also depicted is the membrane permeability-limited uptake line, which represents the maximum partition coefficient that could be observed in the absence of facilitated transport. In this case, uptake rates are proportional to membrane area and the partition coefficient is proportional to R^{-1}_{CELL}. Thus, irrespective of the operative uptake mechanism, small cells are able to accumulate Hg^{II} and CH_3Hg^{II} to much higher levels than large

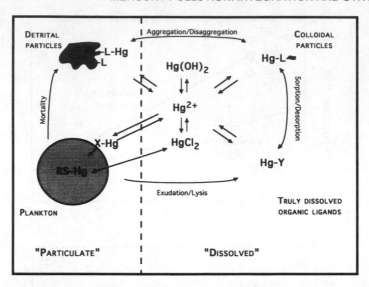

Figure 4   Schematic diagram of reactions involved in seston-water partitioning of Hg^{II}. Apparently "dissolved" species of Hg^{II}, defined as those that pass through a 0.2–0.4 µm filter, are depicted to the right of the dashed line and include complexes with inorganic ligands, dissolved organic ligands, and colloidal particles. "Sestonic" species include Hg^{II} bound to detrital particles and taken up by planktonic organisms. An analogous set of reactions applied for CH_3Hg^{II}.

cells (Figure 3). Generally higher partition coefficients are possible for CH_3Hg^{II} than Hg^{II} because of the weaker organic complexation we have assumed (see Section III.A).

D. SPECIATION AND ABIOTIC PROCESSES

The chemical speciation of mercury also influences the rates of its abiotic transformation and physical transport in aquatic systems (Table 2). Scavenging by detrital sorbents is governed by equilibrium between mercury adsorbed on particles and in solution (Figure 4). Only neutral species of mercury volatilize significantly, leading to the control of gas exchange by the concentrations of species such as Hg^0, CH_3HgCl^0, and $Hg(OH)_2^0$.[58] The reduction of Hg^{II} to Hg^0 may be mediated by humic substances[59] or photochemical reactions involving $Hg(OH)_2^0$.[21,60] Humic acids have also been implicated in the methylation of Hg^{II}.[61,62] As with biologically mediated reactions, significant uncertainties remain about the mechanisms, rates, and rate laws of abiotic reactions of mercury at ambient concentrations in lakes.

III. MODELING THE MERCURY IN TEMPERATE LAKES (MTL) STUDY DATA

Watras et al.[63] have presented an overview and analysis of the field data from the MTL study. In this section we analyze the MTL data in greater detail by developing steady-state models from the principles described in Section II and examining their ability to explain the observations. Since the aqueous speciation of Hg^{II} and CH_3Hg^{II} has not been quantified in lakewaters, we begin by estimating the strength of mercury-DOC associations through modeling particle-water partitioning. This is possible because of the direct relationship between partitioning and aqueous speciation. Using this description of speciation, we then model the key processes which generate the observed water-column concentrations of Hg^{II}, CH_3Hg^{II}, and Hg^0. Finally, we analyze the process by which CH_3Hg^{II} accumulates in fish.

At each stage of model development, we combine what is known of the biogeochemical processes with representations of a variety of hypothetical rate-controlling mechanisms. Using the mechanistic models that best fit the data, we address questions concerning which factors control the lacustrine mercury cycle, e.g., how do pH and DOC affect mercury speciation and bioavailability vs. their influence on rates of limnological processes? Since in many cases we infer information that requires verification through more detailed field measurements and experiments, this analysis should be regarded as an investigation of the explanatory power of different mechanistic hypotheses.

The data used in this analysis are the averages of all measurements obtained as a part of the MTL study and are therefore biased towards ice-free periods of the year. We consider three aqueous mercury components: Hg^{II}, CH_3Hg^{II}, and Hg^0. "Observed" Hg^{II} measurements are calculated from measured total mercury (Hg_T) minus CH_3Hg^{II} and Hg^0. We also use average values of limnological parameters. Significant positive correlations exist between pH and lake depth and between DOC and chlorophyll a in the seven lakes studied. In the following discussion we separate out the effects of the autocorrelations among limnological parameters using appropriate models. Except as noted, all data used in this analysis are reported in Watras et al.[63]

For each hypothetical rate or equilibrium expression being tested, we fit a model to the data by varying one or more parameters to minimize the root-mean-squared relative error (ε) between observed data and model output for the seven MTL lakes. We employed a multi-dimensional, downhill simplex method[64] to identify minima. Repeating the optimization procedure from a large number (≥ 100) of randomly generated starting points identified global minima. We also calculated constants for linear regressions between model results and observed values as additional measures of the goodness of fit of each model. Good fits should have a high Pearson correlation coefficient (r) and a slope (m) close to unity.

A. SESTON-WATER PARTITIONING AND AQUEOUS SPECIATION OF MERCURY

The partitioning of Hg^{II} and CH_3Hg^{II} between "dissolved" and "particulate" states results from the competition for mercury among ligands in solution, on colloids and abiotic particles, and in cells (Figure 4). Because of this competition, partition coefficients are directly related to aqueous speciation, e.g., they vary inversely with the abundance and affinity of "dissolved" ligands relative to "particulate" ligands and are directly proportional to the bioavailability of mercury to plankton. Thus, partition coefficients should be systematically related to lake chemistry. As expected if DOC compounds strongly complex Hg^{II}, an inverse relationship between the apparent seston-water partition coefficient K^*_{SESTON} and DOC does exist for Hg^{II} in the MTL study lakes (Figure 5). Why the K^*_{SESTON}-DOC relationship is so weak for CH_3Hg^{II} (Figure 5), however, requires further examination. In this section we model the data for particle-water partitioning of Hg^{II} and CH_3Hg^{II} in order to infer their aqueous speciation, particularly their organic complexation equilibria, and to identify the species that govern their rates of uptake by plankton.

The only well-characterized complexation reactions of Hg^{2+} and CH_3Hg^+ are with inorganic ligands.[65] Some studies also point to significant binding of Hg^{II}[25] and CH_3Hg^{II}[66] by DOC compounds, but the degree of organic complexation in lakes is poorly quantified. In developing our speciation model, we assume that the concentration of mercury-binding organic ligands is proportional to the concentration of DOC, i.e., the DOC has reasonably similar chemical properties in all of the lakes. All the MTL lakes, except Russet Lake (color ≈ 20 to 40 PCU, DOC = 6.8 mg L^{-1}), are clearwater lakes with primarily autochthonous DOC. Thus, concentrations of chlorophyll a and DOC are positively correlated ($r^2 = 0.74$). Most of the lakes also contain some humic acids derived from shoreline sphagnum—Russett has the most, while Crystal Lake (DOC = 1.8 mg L^{-1}) contains almost none.

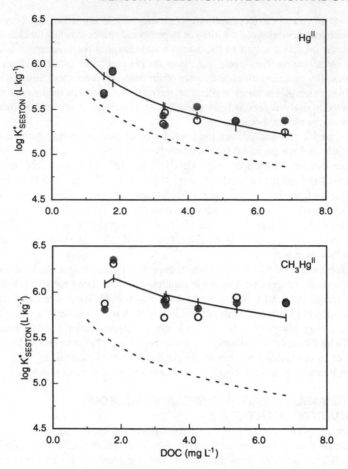

Figure 5 DOC-dependence of apparent seston-water partition coefficients (K^*_{SESTON}) for Hg^{II} and CH_3Hg^{II}. Average observed value for each MTL lake (shaded circles) is shown with calculated values for plankton/particle model (open circles) and for the abiotic particle model with $n_L^{DOC} = n_L^{POC}$ (dashed lines) and with optimized n_L^{POC}/n_L^{DOC} (solid lines).

The bulk of suspended particles in the MTL lakes consists of viable cells (10 to 40% by weight) and detrital organic matter (most of the remainder).[63] (Viable cell mass was estimated by assuming a 100:1 ratio of organic matter to chlorophyll *a*.) Both are potentially important classes of mercury-accumulating particles. Phytoplankton have been shown to accumulate Hg^{II} both extra- and intracellularly.[51,67] A significant role of detrital POC as a sorbent is suggested by correlations between the organic matter and mercury contents of sediments.[39] The concentrations of potential inorganic sorbents have not been quantified in these lakes, but the total levels of Al, Mn, and Fe are low enough that in most cases only a small fraction of the particulate matter could be inorganic. For example, the mixed layers of Little Rock Lake's two basins typically have 20 µg L^{-1} Al, 10 to 100 µg L^{-1} Fe, and 10 to 80 µg L^{-1} Mn[68] out of about 1 mg L^{-1} seston. Although we do not consider inorganic solids in our model (see also below), in systems where metal concentrations are substantially higher they could be important sorbents, as has been suggested for manganese oxides in a seasonally anoxic reservoir.[69]

Here, we explore the ability of a model based on two simplifying hypotheses to explain the magnitudes and DOC dependence of the seston-water partition coefficients (see Appendix

1 and Figure 4). The first hypothesis is that plankton and detrital organic particles are the primary types of mercury-accumulating particles. They differ, however, in that dividing plankton cells have a potential for effectively irreversible uptake, while mercury sorbed to detrital particles should be at equilibrium with the aqueous phase. Thus, K^*_{SESTON} is the sum of apparent partition coefficients into plankton ($K^*_{PLANKTON}$) and detrital particles ($K^*_{PARTICLES}$):

$$K^*_{SESTON} = X_{PLANKTON} \cdot K^*_{PLANKTON} + (1 - x_{PLANKTON}) \cdot K^*_{PARTICLES} \qquad (14)$$

where $X_{PLANKTON}$ is the fraction of the particulate material consisting of viable cells. The second hypothesis, that the mercury-binding sites in DOC and POC are similar in strength and abundance, presumes DOC and detrital POC share a common origin or interconvert to a significant degree. This approximation should be particularly good when DOC is primarily colloidal and aggregation-disaggregation kinetics between colloidal DOC and filterable POC are rapid.[70]

In modeling K^*_{SESTON}, we calculated the equilibrium speciation in the different lakes using fitted stability constants (β_{ML}) for Hg^{2+} and CH_3Hg^+ complexation by ligands (L) in DOC and detrital POC (see Appendix 1),

$$Hg^{2+} + L^- \xleftrightarrow{\ \beta_{HgL}\ } HgL^+ \qquad (15)$$

which also undergo a protonation reaction

$$H^+ + L^- \xleftrightarrow{\ \beta_{HL}\ } HL \qquad (16)$$

Since the ligands are uncharacterized at present, we assume they are chemically similar to the strongest Cu^{2+} binding sites in humic acid, for which complexation data are consistent with a $\beta_{HL} \gg 10^8 \ M^{-1}$.[71] Using $\beta_{HL} = 10^{10}$ for the strong metal binding sites yields consistent stability constants from the Cu^2-humate association data of Cabaniss and Schumann[71] ($\beta_{CuL} = 10^{11.5}$ for the strongest sites) and Mantoura et al.[25] ($\beta_{CuL} = 10^{11.4}$ after adjusting for β_{HL}). In addition, the fit to the Hg^{II} partitioning data improved for β_{HL} much greater than the highest lake pH observed.

We first consider whether equilibrium binding to DOC and detrital POC alone, i.e., an abiotic model with $K^*_{PLANKTON} = 0$, can explain the observed seston-water partitioning data. Morel and Gschwend[70] (see Appendix 1) have shown that the particle-water partition coefficient for a metal equals the inverse of the concentration of dissolved organic matter, within a factor of 2 to 3, under certain conditions. These conditions are that (1) colloidal organic matter constitutes a major fraction of the DOC, (2) most of the apparently "dissolved" metal is bound to colloidal DOC, and (3) colloidal and larger particulate matter are of similar makeup, such as when reversible coagulation and peptization interconverts larger particles and colloidal matter. As shown in Figure 5, with equal ligand densities in DOC and POC this abiotic partitioning model represents the DOC-dependence of K^*_{SESTON} for Hg^{II} well; the factor of 2 underestimate in magnitudes is within the model's expected range of uncertainty. For CH_3Hg^{II}, this model overestimates the DOC dependence of K^*_{SESTON} and underestimates the magnitudes by a factor of 12. To explain the enrichment in Hg^{II} and CH_3Hg^{II} of POC relative to DOC (on a unit carbon basis) using this model requires that either ligands are degraded as cellular debris is broken down to colloidal size or the ligands in colloidal DOC are diluted with ligand-poor DOC.

By increasing the metal-complex stability constants (β_{ML}) and density of ligands in detrital POC relative to DOC ($n_L{}^{POC}/n_L{}^{DOC}$), we obtained reasonable relative errors in modeled

K^*_{SESTON} values for both Hg^{II} ($\varepsilon = 0.37$) and CH_3Hg^{II} ($\varepsilon = 0.30$) using the abiotic, particles-only model (Table 3). The equilibria calculated using the optimal β_{ML} suggests that organic complexes make up more than 99% of the dissolved Hg^{II} and 17 to 70% of dissolved CH_3Hg^{II} in the MTL lakes. Fitting the Hg^{II} data only constrains the minimum value of β_{HgL}, since any value causing organic complexation of Hg^{II} to dominate its aqueous speciation fit equally well. Weaker complexation of CH_3Hg^{II} is necessary with this model to explain the unsystematic DOC-dependence of CH_3Hg^{II} partitioning (Figure 5). The modest correlations between modeled and observed K^*_{SESTON} for Hg^{II} ($r^2 = 0.46$; $m = 0.55$) and for CH_3Hg^{II} ($r^2 = 0.53$; $m = 0.30$) indicate this model does not well represent some factors that affect partitioning.

Alternatively, the mercury enrichment of seston relative to DOC may result from the accumulation of Hg^{II} and CH_3Hg^{II} in viable plankton cells (Figure 4). Plankton potentially accumulate more mercury than detritus due to their passive and facilitated transport mechanisms and intracellular ligands. We assumed the plankton grew with a specific rate of 1 d^{-1}, a rate consistent with field measurements of chlorophyll a and primary production. (Note also that maximum phytoplankton growth rates at $11°C$ are about 1.2 d^{-1} according to Eppley.[72]) For a mean cell radius of 2.5 µm, we calculated $K^*_{PLANKTON}$ for a variety of plausible mechanisms of Hg^{II} and CH_3Hg^{II} uptake. We obtained the best fit to the Hg^{II} data ($\varepsilon = 0.23$; $r^2 = 0.93$; $m = 0.99$) with facilitated transport by sites with $\beta_{HX} = 10^4$ in equilibrium with Hg^{2+} (Equation 10). Calculated uptake rates were influenced by diffusion limitation in some lakes. Note that the β_{ML} obtained by applying this model are sensitive to parameter values that influence cellular uptake rates, but other reasonable values—$\beta_{HX} = 10^7$, a lower specific growth rate, and a range of cell radii (1 to 10 µm)—yielded comparable improvements in fits relative to the abiotic model (Table 3). Thus, our inference of significant planktonic Hg^{II} uptake is reasonably independent of the specific parameter values used.

We also tested other hypothetical mechanisms of planktonic Hg^{II} uptake. Rates controlled by concentrations of $Hg^{II}_{INORGANIC}$ (kinetically controlled facilitated uptake) and by $Hg(Cl)^0_2$ (passive uptake) yielded much poorer fits than Hg^{2+} (thermodynamically controlled facilitated uptake) (Table 3). We note that the inverse relationship of K^*_{SESTON} and DOC for Hg^{II} (Figure 5) mandates that only species whose concentrations in these lakes decrease systematically with DOC despite pH variations can exert significant control over the seston-water partitioning of Hg^{II}. Total concentrations of inorganic Hg^{II} species do not, due to the strong pH dependence of the complexation equilibria of $Hg(OH)^0_2$, the inorganic species in greater abundance. Both Hg^{2+} and $HgCl^0_2$ concentrations do vary inversely with DOC, but passive $HgCl^0_2$ uptake rates are too slow compared to cellular growth to permit high values of $K^*_{PLANKTON}$ to be attained (see Figure 3).

Methylmercury accumulated in seston to a greater extent than mercury(II) (Figure 5), making it more likely that viable cells account for the poor fit of the abiotic model (Table 3). The best fit to the K^*_{SESTON} data for CH_3Hg^{II} ($\varepsilon = 0.18$; $r^2 = 0.92$; $m = 0.85$) was obtained assuming kinetically controlled facilitated transport, i.e., uptake rates proportional to $CH_3Hg^{II}_{INORGANIC}$, via a site with $\beta_{HX} = 10^{10}$ and is much improved over the abiotic model. As for Hg^{II} uptake, the strength of organic complexation of CH_3Hg^{II} is such that diffusion limitation influences uptake by phytoplankton (Figure 3) and the estimated β_{ML} is sensitive to assumptions about the size and growth rate of phytoplankton. Note that obtaining the best fits with this model required us to assume either that the passive permeability of CH_3HgCl^0 is actually at the low end of the range of expected value, i.e., 20% of the highest estimate, or that depuration is rapid enough that the passive uptake becomes irrelevant.

Because passive absorption of CH_3Hg^{II} is well known and may be the predominant mechanism of uptake in estuarine phytoplankton,[73] we examined what conditions permit uptake via this mechanism to explain the observed partitioning data best. No good fit was possible for $\beta_{HL} = 10$. Assuming that $\beta_{HL} = 4$ and increasing P_{CH_3HgCl} by 10% allowed us to obtain a reasonably good fit using solely passive uptake ($\varepsilon = 0.28$; $r^2 = 0.79$; $m = 0.62$). The

Table 3 **Model fits to MTL seston-water partitioning data.**

Parameter modeled _Model formulation_ Mechanism—rate controlling species	Best fit $\log \beta_{ML}$ (L kg-C^{-1})	RMS[a] relative error ε	Calculated vs. observed regression constants[b] r^2	 Slope
(A) K^*_{SESTON} Hg^{II}				
(i) Particles only[c] ($n_L^{POC} = 2.2 \cdot n_L^{DOC}$)	>21.5	0.373	0.458	0.541
(ii) Plankton uptake[d] ($n_L^{POC} = n_L^{DOC}$)				
Facilitated transport—[Hg^{2+}] control				
$\log \beta_{HX} = 4$; $\log k_X = 14.9$				
$\log \beta_{HX} = 7$	17.0	0.225	0.931	0.975
$\mu = 0.5$; $\log \beta_{HX} = 4$	17.0	0.230	0.896	0.841
$R_{cell} = 1.0 \, \mu m$; $\log \beta_{HX} = 4$	17.3	0.225	0.931	0.982
$R_{cell} = 10 \, \mu m$; $\log \beta_{HX} = 4$	17.8	0.226	0.931	0.987
	15.6	0.227	0.913	0.700
Facilitated transport—[Hg$^{II}_{INORGANIC}$] control	15.9	0.404	0.266	0.400
$\log \beta_{HX} = 4$				
Passive transport only—[HgCl$^0_2$] control	15.9	0.581	0.182	0.327
(B) K^*_{SESTON} CH$_3$Hg^{II}				
(i) Particles only[c] ($n_L^{POC} = 12.4 \cdot n_L^{DOC}$)	10.9	0.295	0.526	0.301
(ii) Plankton uptake[d] ($n_L^{POC} = n_L^{DOC}$)				
Facilitated transport— [CH$_3$Hg$^{II}_{INORGANIC}$] control				
$\log \beta_{HX} \geq 7$; $\log k_d \geq 2$; $\log k_X = 14.3$	12.0	0.184	0.923	0.853
$\log \beta_{HX} \geq 7$; $\log k_d = -2$	12.2	0.425	0.557	0.525
$\log \beta_{HX} \geq 7$; $P_o = 0.0026 \, cm \, s^{-1}$	12.0	0.221	0.891	0.813
$R_{cell} = 1.0 \, \mu m$; $\log \beta_{HX} = 10$; $\log k_d \geq 2$	12.8	0.215	0.907	0.895
$R_{cell} = 10 \, \mu m$; $\log \beta_{HX} = 10$; $\log k_d \geq 2$	10.6	0.292	0.393	0.243
Facilitated transport—[CH$_3$Hg$^+$] control				
$\log \beta_{HX} = 10$; $\log k_d \geq 3$	11.3	0.205	0.840	0.529
Passive transport only—[CH$_3$HgCl^0] control				
$\log \beta_{HL} = 10$	12.2	0.748	0.001	0.017
$\log \beta_{HL} = 4$; $P_{CH_3HgCl} = 2.5 \times P_o$	7.3	0.277	0.787	0.620

[a] Definition of RMS relative error (ε) in parameter x: $\varepsilon = \sqrt{\dfrac{\sum\limits_{i=1}^{n_{LAKES}} \left[\left(x_i^{calc} - x_i^{obs} \right)/x_i^{obs} \right]^2}{n_{LAKES}}}$

[b] Correlation constants for linear regressions with calculated values as the dependent variable. r is the Pearson correlation coefficient.

[c] $K^*_{PLANKTON} = 0$.

[d] For plankton, $\mu = 1 \, d^{-1}$, depuration rate constant $(k_d) = 10^{-2} \, d^{-1}$, and $R_{CELL} = 2.5 \, \mu m$ ($K_{DIFFUSION} = 10^8 \, L \, kg^{-1}$) unless otherwise specified. M_{CELL} calculated assuming 0.2 kg-dry wt L-cell^{-1}[96]. Passive uptake of HgCl$^0_2$, CH$_3$HgCl^0, and CH$_3$HgOH^0 assumed to occur in addition to other mechanisms with $P_o = 0.013 \, cm \, s^{-1}$, $P_{HgCl_2} = P_o$, $P_{CH_3HgCl} = 2 \times P_o$, $P_{CH_3HgOH} = 0.1 \times P_o$.

measures of the fit are intermediate between those of the abiotic model and those of the best fit with plankton. While this result suggests passive uptake should not be dismissed as a possible mechanism, we note that the β_{ML} ($\approx 10^{6.3}$ M^{-1} for 10 µmol mg-C^{-1} carboxylic acid sites) found in fitting the data is higher than expected for a site with such a low β_{HL}, e.g., $10^{3.2}$ M^{-1} for acetate, unless electrostatic attraction is significant.[74]

Our modeling of the partitioning data suggest significant complexation of Hg^{II} and CH$_3$Hg^{II} by autochthonous DOC compounds. The equilibrium constant for the Hg^{2+}-DOC association, 10^{17} L kg-C^{-1}, yielded 93.7 to 99.7% organic complexation for the MTL lakes, while that of the CH$_3$Hg$^+$-DOC complex, 10^{12} L kg-C^{-1}, yielded 72 to 97% organic complexation. To compare these values to measured stability constants of organic ligand-mercury complexes, we employ an estimate of 0.3 to 0.6 mol S kg-C^{-1} in humic matter[27] to obtain $\beta_{HgL} = 10^{17.5}$ M^{-1} and $\beta_{CH_3HgL} = 10^{12.5}$ M^{-1}. The Hg^{2+} constant appears significantly lower than the values measured for freshwater humic substances,[25] 10^{20} to 10^{23} M^{-1} (adjusted for $\beta_{HL} = 10^{10}$), and estimated for a generic thiol,[26] $10^{22.1}$ M^{-1}. Perhaps Hg^{2+} is bound by relatively weak organic ligands due to the presence of competing metals or to pseudoequilibrium conditions. Similarly, CH$_3$Hg^{II} appears to be bound by ligands weaker than cysteine, which has a stability constant of $10^{15.7}$ M^{-1}.[49] The strong organic complexation suggested by our modeling demonstrates the importance of obtaining direct measurements of the complexation of mercury by DOC compounds in lakes.

A measure of the uncertainty in these estimates of β_{ML} is given by calculating the upper limits to organic complexation predicted by the partitioning model. These upper limits are defined by the concentrations of uncomplexed mercury that give the values of $K^*_{PLANKTON}$ needed to fit the K^*_{SESTON} data when mercury is taken up at diffusion-limited rates. Since diffusion limitation of uptake is already a factor for Hg^{II} and CH$_3$Hg^{II}, the extent of complexation cannot be more than a few times the value we have estimated without causing $K^*_{PLANKTON}$ to be too low to contribute significantly to K^*_{SESTON} (see Figure 3). If smaller cells controlled partitioning, approximately sixfold higher β_{ML} would be predicted (Table 3). Alternatively, if the abiotic model is more realistic, no upper limit can be proposed. Although inferring trace metal speciation from sestonic metal content is imprecise, the correspondences observed between trace metal speciation and experimentally derived models of trace metal uptake by phytoplankton in marine systems[54] suggest our attempt is warranted.

While the plankton model fits the partitioning data quite well, the fit alone does not prove our assumptions correct. Field data do provide some support for our central result concerning the role of plankton in seston-water partitioning. The coincidence of layers of particulate Hg$_T$ and plankton in these lakes[68] suggests that accumulation in plankton is significant. There is indirect evidence that scavenging by hydrous ferric oxides (HFO) does not control mercury removal from the water column—Hg^{II} and CH$_3$Hg^{II} build up in the hypolimnia of the MTL lakes during the summer, but not in the winter anoxic period, in contrast to iron which is released from sediments during both periods.[75] Since manganese oxides are reduced prior to iron as anoxia develops, manganese oxides likely do not control Hg^{II} scavenging either. In addition, we may compare our estimates of the affinities of detrital POC for mercury and those of inorganic materials. By assuming that all of the iron in the water column is particulate HFO, we calculate from published sorption constants[76] that iron can account for about 0.025% of the binding of Hg^{II} to seston in pH 6 lakes. Similar calculations for methylmercury are less precise since sorption constants must be estimated from a linear free-energy relationship,[76] but they do suggest that sorption on HFO could be significant. Further field research into this issue is needed.

In summary, modeling the major limnological factors influencing K^*_{SESTON} for Hg^{II} requires that strong complexation by DOC be invoked and that the twofold stronger association of Hg^{II} with particles than with DOC on a unit carbon basis be explained. Models which account for the DOC-POC difference as a greater accumulation in plankton relative to detrital POC ($K^*_{PLANKTON}$ is eightfold greater than $K^*_{PARTICLES}$ on average in the best fit model) agree with the observations much more closely than simply assuming that POC has a greater density of mercury-binding sites.

For CH_3Hg^{II}, the absence of a strong K^*_{SESTON}-DOC relationship (Figure 5) is the result of two factors—somewhat weaker organic complexation of CH_3Hg^{II} than Hg^{II} and a positive chlorophyll a-DOC correlation in these lakes. Since in high DOC lakes a larger proportion of viable plankton in the seston partially compensates for increased association with DOC, a stronger CH_3Hg^{II}-DOC association is predicted when plankton are included—13-fold greater β_{CH_3HgL}—than with the abiotic model. Consistent with this explanation, our best fit to the data was obtained with significant organic complexation of CH_3Hg^{II} and with values of $K^*_{PLANKTON}$ 35-fold greater than $K^*_{PARTICLES}$ on average.

B. STEADY-STATE MODEL OF THE MERCURY CYCLE IN SEEPAGE LAKES

Field data from MTL and other studies have identified the key transport and transformation processes of the mercury cycle in seepage lakes (Figure 6). However, the dependencies of the rates of these processes on mercury speciation and limnological properties have not been empirically determined in general. To investigate these dependencies, we simplified the dynamic MCM to a set of steady-state mass conservation and rate equations (Appendix 2). We limited the steady-state MCM to three components—Hg^0, Hg^{II}, and CH_3Hg^{II}—with only the essential reactions—reduction, methylation, and demethylation—and transport processes—deposition, gas exchange, and scavenging by particles. The model includes the additional assumptions that storage of mercury in the biota is at steady state and that no net transformation of mercury between forms occurs in zooplankton and fish. Using this model, we test the ability of different rate laws of the form of Equation 2 to fit the observed concentrations of dissolved Hg^0, Hg^{II}, and CH_3Hg^{II} in the MTL lakes. Because the mean residence times of mercury in these lakes vary from 5 to 10 months for Hg_T to a few days for Hg^0,[63] there is a variability inherent in the natural systems which require a dynamic model to simulate accurately. Nevertheless, we proceed with our ''steady-state'' analysis by seeking to represent

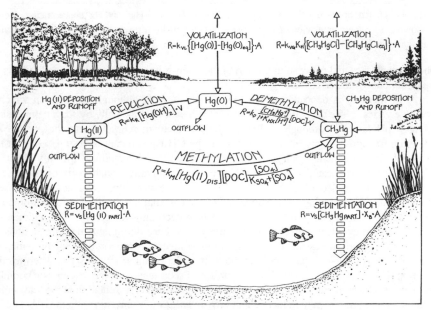

Figure 6 Schematic depiction of the lacustrine mercury cycle in the steady-state MCM. The rate laws for each major process derived from fitting the MTL study data are shown.

Table 4 **Optimum steady-state model fits to MTL dissolved mercury concentration data.**

Mercury component simulated[a] (processes included)	RMS relative error ε	Calculated vs. observed regression constants	
		r^2	Slope
Hg^{II}			
Scavenging only—fit to [Hg^{II}]	0.458	0.542	0.708
Scavenging + reduction—fit to [Hg^{II}] and [Hg^0]	0.368	0.894	1.697
Scavenging + reduction + demethylation	0.354	0.865	1.363
Scavenging + reduction + demethylation + methylation	0.267	0.859	1.073
CH$_3$Hg^{II}			
Scavenging only—fit to [Hg^{II}]	1.138	0.592	0.603
Scavenging + reduction − [Hg^{II}] and [Hg^0]	2.728	0.570	1.099
Scavenging + reduction + demethylation	0.416	0.724	0.218
Scavenging + reduction + demethylation + methylation	0.312	0.792	0.816
Hg^0			
Scavenging only—fit to [Hg^{II}]	0.854	0.393	0.000
Scavenging + reduction—fit to [Hg^{II}] and [Hg^0]	0.405	0.754	1.125
Scavenging + reduction + demethylation	0.286	0.662	0.956
Scavenging + reduction + demethylation + methylation	0.219	0.732	0.876

[a] Except as noted, fits were optimized for sum of errors in Hg^{II}, CH$_3$Hg^{II}, and Hg^0 concentrations. Results shown are for the combination of mechanisms that fit best for a given set of processes, as described in text (see Table 5).

the average rates and concentrations during the open-water season, a choice that reflects the seasonal bias of the MTL data.

Most mercury is deposited in the MTL lakes as Hg^{II}, although the fraction which is reactive varies.[77] The depositional flux of Hg^{II} is balanced principally by scavenging, except in high pH lakes.[63] Although settling particles contain mercury in both the particle and plankton fractions, the progressive decrease in the CH$_3$Hg^{II} content of seston, sediment trap material, and sediments[35] suggests extensive release of intracellular mercury from the plankton following death. A decrease in Hg$_T$ content and particle-water partition coefficients, which primarily reflects Hg^{II} behavior, also occurs between the water column and sediments.[41] Simulations using the dynamic MCM suggest that nearly all of the planktonic Hg^{II} and CH$_3$Hg^{II} component must be released before becoming mixed with the sediments in order to avoid excessive depletion of epilimnetic mercury during summer (simulations not shown). Therefore, we assumed that removal to sediments is proportional solely to mercury concentrations in abiotic particles calculated from the equilibrium constants determined in Section III.A. If detrital particles near the surface of sediments or at the oxic-anoxic boundary are near equilibrium with epilimnetic waters, this simplified representation of the scavenging-sedimentation process may in fact be quite reasonable. With scavenging as the only sink for Hg^{II}, the best model fit to the dissolved Hg^{II} data (Table 4) has a relative error of 46% (ε = 0.46) with a modest correlation between modeled and observed Hg^{II} concentrations (r^2 = 0.54, m = 0.71).

The importance of Hg^0 evasion in the mercury budgets of some of the MTL lakes[13,63] suggests that accounting for reduction should improve the fit of the steady-state model. In agreement with the field studies, the error and correlation coefficient for Hg^{II} concentrations improve markedly over the scavenging only case ($\varepsilon = 0.37$, $r^2 = 0.89$), although the slope of the modeled-observed regression ($m = 1.70$) was greater than is ideal ($m = 1$). Upon including reduction, the regression slope for Hg^0 concentrations also improved from zero to 1.13 ($\varepsilon = 0.41$; $r^2 = 0.75$).

To obtain this model fit, we assumed that the reduction rate was controlled by the $Hg(OH)_2^0$ species (Table 5), because no positive Hg^0 regression slope was obtainable when Hg^{II} species whose concentrations decrease with increasing pH, such as Hg^{2+} and $HgCl_2^0$, were assumed to control reduction rates. This speciation dependence may correspond to photochemical reduction of $Hg(OH)_2^0$ [21,60] being the primary mechanism of Hg^0 production. A reasonable model fit was also obtained for rate expressions proportional to $Hg^{II}_{INORGANIC}$, which primarily consists of $Hg(OH)_2^0$ except at the lowest pH values (Figure 2), and either DOC or chlorophyll a. The DOC dependence was introduced to represent an assumption of organic carbon-limited (DOC) growth in heterotrophic bacteria.[78] Reduction by either bacteria ($f_R^{bio} = [DOC]$) or phytoplankton ($f_R^{bio} = [chl\ a]$) at rates controlled by Hg^{II} complexation kinetics could explain such a dependence. Although humic-mediated reduction could also be consistent with a DOC dependence, experimental investigations suggest abiotic reduction is more rapid at low pH,[79] presumably since Hg^{2+} is more easily reduced than $Hg(OH)_2^0$, and therefore is not consistent with $Hg^{II}_{INORGANIC}$ control of the rate. Direct reduction of Hg^{II}-humate complexes appears to be unlikely due to the low slope ($m = 0.45$) of the calculated-observed correlation for $Hg^{II}L$ as the controlling species.

That Hg^0 production in the MTL lakes appears to be pH-dependent has been noted previously by Fitzgerald et al.[13] and Watras et al.[63] By accounting for both the effects of pH on Hg^{II} speciation and lake depth on Hg^0 loss rates (see Appendix 2), our analysis demonstrates that the pH-Hg^0 relationship is a chemical effect of pH and is not caused by the correlation of pH with depth in the MTL lakes. The pH-dependence of Hg^{II} reduction results primarily from the inverse relationship of pH and organic complexation of Hg^{II}. As pH increases, the association of Hg^{II} with humic acid lessens due to the formation of $Hg(OH)_2^0$:

$$HgL^+ + 2H_2O \leftrightarrow H^+ + Hg(OH)_2^0 + HL \qquad (17)$$

Methylmercury concentrations in lakes reflect a balance between the rates of deposition, gas exchange, scavenging, methylation, and demethylation. The best quantified source of methyl mercury to the lakes is the depositional flux—about 0.1 $\mu g\ m^{-2}\ year^{-1}$.[63] In addition, gas exchange can provide a source to the lake of comparable magnitude. Here we have assumed that the most volatile methylmercury species, CH_3HgCl^0, controls air-water exchange, and that its dissolved concentration at equilibrium reflects its concentration in rain (see Appendix 3). Sediment burial, the best quantified sink of CH_3Hg^{II}, approximately equals the depositional flux.[63] Note that although a higher proportion of CH_3Hg^{II} than Hg^{II} is partitioned into seston (see Section III.A), it comprises only about 1% of the sediment mercury burden.[63] We therefore selected a burial efficiency for CH_3Hg^{II} of 0.1, which yielded net particle scavenging fluxes of Hg^{II} and CH_3Hg^{II} comparable to their proportions in sediments. With this approach, we in effect subtract mineralization in surficial sediments from the gross water column scavenging rate.

To investigate whether atmospheric deposition alone could be the principal source of CH_3Hg^{II} for the MTL lakes, we attempted to fit the CH_3Hg^{II} concentration data without invoking in-lake methylation. We examined all combinations of the five best-fitting reduction rate expressions and plausible demethylation rate laws and found the best fit by assuming demethylation rates reflect CH_3Hg^+ reacting at a site with $\beta_{HX} = 10^{10}$ ($\varepsilon = 0.42$, $r^2 = 0.72$, $m = 0.22$). The low slope of the modeled-observed regression, which corresponds to low

Table 5 Tests of hypothetical process rate expressions.[a]

Process modeled—fit to Hg concentrations *Mechanism and key dependencies* Rate-controlling species	Mechanism label	Mechanisms of other processes	RMS relative error ε	Calculated vs. observed regression constants		Combined RMS relative error[b] Σε
				r^2	Slope	
(A) Reduction—fit to [Hg⁰] and [Hg^{II}] (scavenging and reduction only)			Parameters for Hg⁰			
(i) Abiotic mechanisms						
Hg(OH)$_2^0$ photoreduction	R$_1$		0.405	0.754	1.122	
Hg^{2+} reduction by DOC			0.736	0.168	−0.055	
Hg$^{II}_{HUMATE}$ photo- or thermal reduction	R$_2$		0.311	0.747	0.451	
(ii) Facilitated by bacteria—f$_R^{bio}$ ∝ [DOC]						
Hg^{2+}; log β$_{HX}$ = 10	R$_3$		0.314	0.764	0.448	
Hg^{2+}; log β$_{HX}$ = 7			0.390	0.529	0.288	
Hg$^{II}_{INORGANIC}$; log β$_{HX}$ = 7			0.471	0.681	1.169	
Hg$^{II}_{INORGANIC}$; log β$_{HX}$ = 4	R$_4$		0.283	0.672	0.962	
(iii) Facilitated uptake by phytoplankton—f$_R^{bio}$ ∝ [chl a]						
Hg^{2+}; log β$_{HX}$ = 10; R$_{cell}$ = 2.5 μm	R$_5$		0.406	0.826	0.519	
Hg$^{II}_{INORGANIC}$; log β$_{HX}$ = 4; R$_{cell}$ = 2.5 μm			0.352	0.682	1.016	
(B) Methylation[c]—fit to [Hg⁰], [Hg^{II}] and [CH$_3$Hg^{II}]			Parameters for CH$_3$Hg^{II}			
(i) Abiotic mechanisms						
Atmospheric input only	M$_0$	R$_4$, D$_1$	0.416	0.724	0.218	1.056
Hg$^{II}_{HUMATE}$	M$_1$	R$_1$, D$_1$	0.364	0.688	0.747	0.879

497

(ii) Bacterial methylation—$f_M^{bio} \propto [DOC] \cdot [SO_4]/(K_{SO_4} + [SO_4])$

		Parameters for CH₃Hgᴵᴵ			
Hgᴵᴵ_DISSOLVED; $K_{SO_4} = 100\ \mu M$[d]					
M₂	R₁, D₁	0.312	0.792	0.815	0.798
$K_{SO_4} = 0\ \mu M$					
M₃	R₁, D₁	0.362	0.686	0.737	0.878
Hg²⁺; $K_{SO_4} = 0\ \mu M$; log β_HX = 4					
M₄	R₄, D₁	0.314	0.818	0.496	0.954
HgCl₂; $K_{SO_4} = 100\ \mu M$; log β_HX = 0					
M₅	R₄, D₁	0.370	0.661	0.347	1.009
Hgᴵᴵ_INORGANIC; $K_{SO_4} = 100\ \mu M$; log β_HX = 4					
M₆	R₃, D₂	0.369	0.835	0.527	0.926

(C) Demethylation—fit to [Hg⁰], [Hgᴵᴵ] and [CH₃Hgᴵᴵ]

(i) Bacterial demethylation—$f_D^{bio} \propto [DOC]$

		Parameters for CH₃Hgᴵᴵ			
CH₃Hg⁺; log β_HX = 10					
D₁	R₁, M₂	0.312	0.792	0.815	0.798
CH₃Hgᴵᴵ_INORGANIC; log β_HX = 10					
D₂	R₃, M₆	0.369	0.835	0.527	0.926
CH₃HgCl⁰					
D₃	R₁, M₄	0.353	0.706	0.498	1.036

[a] Biotic processes have $f_P^{bio} \propto 1/(1 + \beta_{HX} \cdot [H^+])$; log β_HX = 0 when no other value is indicated.

[b] Sum of RMS relative errors in Hgᴵᴵ, CH₃Hgᴵᴵ, and Hg⁰.

[c] All results in parts B and C obtained with [DOC] dependence for methylation (M) and demethylation (D). For some combinations of mechanisms, better fits were obtained without the [DOC] terms. The best overall fit (R₁;M₂;D₁) had both M and D dependent on [DOC].

[d] Best fit rate constants, as defined in Equation 3 and Appendix 2, are: $k_R^o = 0.15$ d⁻¹, $k_M^o = 1.4 \times 10^3$ L kg-C⁻¹ d⁻¹, $k_D^o = 1.8 \times 10^{10}$ L kg-C⁻¹ d⁻¹, and $v_s = 0.37$ m d⁻¹.

predicted values in the high CH_3Hg^{II} concentration range, suggests that additional methylation occurs in the lakes, a conclusion supported by other arguments for the MTL lakes.[63]

To examine in-lake methylation and demethylation, we tested combinations of species and lake parameter dependencies corresponding to the hypothesized mechanisms for each process (Table 5B and C). For the five best-fitting reduction rate laws (Table 5A), we examined all combinations of hypothesized methylation and demethylation rate expressions (Table 5B and C). For each case, we also tested methylation rates with and without a sulfate dependence and demethylation and methylation rates with and without a DOC dependence ($f^{bio} = [DOC]$). We obtained reasonable fits to the CH_3Hg^{II} data for a limited number of speciation dependencies. The closest fit was with methylation rates proportional to concentrations of total dissolved Hg^{II} with $f_M^{bio} = [DOC] \cdot [SO_4]/(K_{SO_4}] + [SO_4])$ in Equation 3. The SO_4 dependence suggests that sulfate-reducing bacteria may control the methylation rate.[80] Demethylation was best modeled as a process dependent on facilitated transport of CH_3Hg^+ ion by a site with $\beta_{HX} = 10^{10}$ and with $f_D^{bio} = [DOC]$, consistent with bacterial mediation of the process via a specific demethylation reaction (Table 5). The poor fit obtained assuming passive diffusion of CH_3HgCl^0 argues that passive transport is not the mechanism of uptake by the bacteria effecting demethylation, a finding consistent with the occurrence of specific transport sites in bacteria that demethylate using the mercurial resistance systems.[81]

While the dependence of methylation on DOC and SO_4 are explainable as reflections of the microbial processes involved, the dependence on total dissolved Hg^{II} concentrations does not directly correspond to any known mechanism. It implies that all Hg^{II} species are equally available, which could happen if dissolved Hg^{II} species were transported to an environment where the normal species distinctions vanish. One such locale could be low oxygen environments with sulfide present. Since only trace levels of sulfide are required to completely control Hg^{II} speciation (Figure 2), either micro zones in aquatic particles or the oxic-anoxic boundary region could fit the description. Alternatively, diffusion-limited uptake by large particles or flocs could increase the diffusion time constant sufficiently to allow dissociation of organic Hg^{II} complexes to become significant within the diffusive boundary layer (see Section II.B). Methylation in either type of micro zone could be dispersed throughout the epilimnion, giving the depth dependence assumed in the steady-state model most directly. Other mechanisms, e.g., entrapment in hypolimnia or diffusion towards anoxic zones, could yield dependencies of methylation on lake depth that differ from the one assumed in our analysis.

We also investigated scenarios for methylation of mercury in sediments. If sediments are the predominant site of methylation in these lakes, a dependence on the sediment Hg^{II} accumulation rate may be apparent. Poor correlations with low slopes were observed with or without a SO_4 dependence (results not shown), consistent with the fact that sedimentation fluxes of Hg^{II} [63] and dissolved concentrations of CH_3Hg^{II} are poorly correlated in these lakes ($r^2 = 0.10$). However, observed concentrations of dissolved CH_3Hg^{II} are roughly proportional to dissolved Hg^{II} (Figure 7). Note that the steady-state MCM also predicts an overall correlation between dissolved CH_3Hg^{II} and total dissolved Hg^{II} in the MTL lakes.

While true in situ rates of methylation and demethylation are unknown, a comparison of our modeled rates and values obtained for surface waters in these lakes using radiometric assays is worthwhile.[82] When normalized to total dissolved concentrations of substrate, the rate constants for methylation from the best-fit model—0.07 to 0.2% per day—are about sevenfold higher than the assay values while the demethylation rate constants—0.1 to 0.2% per day—fall within the range of assay values—0.06 to 0.5% per day. The close correspondence for demethylation is surprising considering the high concentrations of substrate used in the assays and may reflect the control of CH_3Hg^{II} speciation by an organic ligand present in great excess of ambient levels of mercury, e.g., relatively numerous metal-binding sites in the DOC.

The model rate expressions also define average fluxes for the ice-free period in the MTL lakes. The best fit results suggest that the rate of in-lake methylation varies from about 2 to

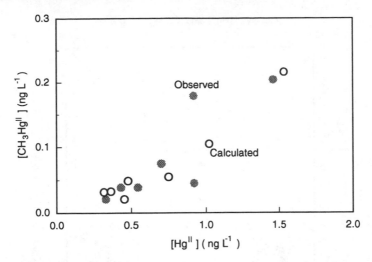

Figure 7 Relationship of dissolved CH_3Hg^{II} and Hg^{II} concentrations in MTL lakes— values observed in MTL study[63] (solid circles) and calculated using steady-state MCM (open circles).

20 times the fluxes of CH_3Hg^{II} deposition plus gas exchange, with the methylation rates being highest in the high DOC lakes, e.g., Russett Lake. Comparison of our Hg^0 evasion fluxes with those of Watras et al.[63] indicates that we correctly predict the three lakes with highest fluxes, but overestimate the evasion rate significantly in Pallette Lake. We calculate that in Russett Lake, Hg^0 is produced primarily from demethylation.

While the degree to which this set of mechanism-based rate laws for the key reactions of the mercury cycle fits the observed data is heartening, we have a limited ability to demonstrate which mechanisms are actually operating in situ because of the small number of lakes modeled and the sensitivity of our results to model assumptions and parameters. For example, weaker DOC-CH_3Hg^{II} associations, $\beta_{CH_3HgL} = 10^{11.3}$, yielded a somewhat better overall fit to the concentration data ($\Sigma\varepsilon = 0.72$), but worsened the fit to the K^*_{SESTON} data. Also, while the best overall fit included a sulfate dependence for methylation, the support for this conclusion is limited. When the assumption that methylation and demethylation rates are proportional to DOC concentrations is removed, i.e., some other factor controls bacterial activity, we found the best fit with different mechanisms of all three reactions and methylation independent of sulfate. This overall fit is only slightly worse than the best fit ($\Sigma\varepsilon = 0.82$ vs. $\Sigma\varepsilon = 0.80$) and the fit to CH_3Hg^{II} concentrations is comparable ($\varepsilon = 0.31$; $r^2 = 0.95$; $m = 0.60$). Although greater certainty awaits results of further field and experimental study, such as in the current Mercury Accumulation Pathways and Processes (MAPP) study, our model describes the field data exceptionally well (Table 4 and Figure 8).

Finally, although we use a single Hg^{II} component in the steady-state model, we do not imply that all Hg^{II} species in the lake are truly at equilibrium. The observation that measurements of "reactive" mercury are pH dependent in the MTL lakes (Figure 9) suggests that despite the analytical uncertainties,[24] there is a systematic variation in the reactivity of labile Hg^{II} species with pH. This would be consistent with our analysis here if the reactive Hg determination measures all inorganic and weak organic complexes of Hg^{II}. Then the weaker organic complexation at higher pH, as we assumed for Hg^{II} above, would cause a relatively larger loss of reactive Hg^{II} via reduction to Hg^0 and subsequent evasion for high pH lakes. This would make the reactive mercury measurements consistent with the hypothesis that deposition of reactive Hg^{II} controls reduction rates.[77]

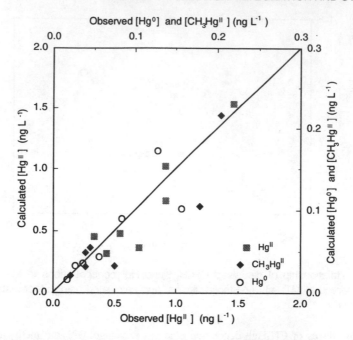

Figure 8 Comparison of average concentrations of dissolved Hg^{II}, CH$_3$Hg^{II}, and Hg^0 observed in MTL study with values calculated using steady-state MCM.

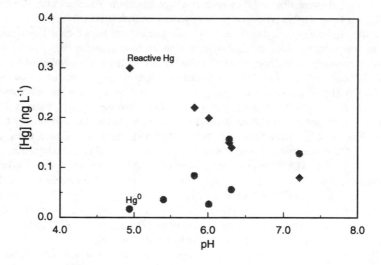

Figure 9 pH dependence of average "reactive" Hg (diamonds) and Hg^0 (circles) concentrations in MTL study lakes.

C. ACCUMULATION OF METHYLMERCURY IN FISH

Because of the variety of processes involved in producing methylmercury and in its accumulation in aquatic food chains, it is necessary to distinguish factors governing the production of CH$_3$Hg^{II} (considered in the Section III.B) from factors which govern its bioaccumulation in order to understand the observed correlations of fish mercury content and limnological properties. For example, bioconcentration factors observed in age-1 yellow perch from the

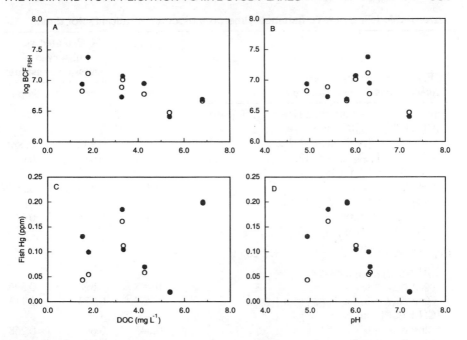

Figure 10 DOC- and pH-dependence of fish mercury concentrations and bioconcentration factors (BCF$_{FISH}$)—average of MTL data for age-1 yellow perch (shaded circles) and calculated values (open circles).

MTL study (Figure 10) are not systematically related to pH,[11] but are inversely related to DOC, while the converse holds for fish mercury content.[63] In this section we examine the role pH, DOC, and other limnological parameters play in CH_3Hg^{II} bioaccumulation by modeling methylmercury bioconcentration factors for age-1 yellow perch observed in the MTL lakes.[63] We also apply these bioconcentration factors in conjunction with the lake mercury cycle model to predict fish mercury levels from mercury inputs and lake properties.

Methylmercury is strongly concentrated in fish relative to water—by at least a factor of $10^{6.5}$ on a dry weight basis.[12,63] The net effect of the processes which concentrate CH_3Hg^{II} in fish is conventionally quantified as a bioconcentration factor (BCF$_{FISH}$) defined by the equation:

$$BCF_{FISH} \equiv \frac{[CH_3Hg]_{FISH}}{[CH_3Hg]_{DISSOLVED}} \qquad (18)$$

In order to simplify comparisons between concentration factors in fish and in plankton, we calculate both on a dry weight basis. However, fish mercury concentrations are expressed in terms of the conventional wet weight basis (dry weight $\approx 0.2 \times$ wet weight).

In our first approach to modeling BCF$_{FISH}$, we postulate, as we did for plankton-water partitioning, that various aqueous CH_3Hg^{II} species govern the bioconcentration process. Values of BCF$_{FISH}$ were calculated according to the equation:

$$BCF_{FISH} = BCF_i \cdot g_{chem} \cdot g_{bio} \cdot \frac{[CH_3Hg_i]}{1 + \beta_{HX} \cdot [H^+] + \beta_{CaX} \cdot [Ca]} \qquad (19)$$

Table 6   **Modeling of MTL fish methylmercury bioconcentration factor data.**

BCF$_{FISH}$ model assumptions *Species controlling uptake rates*	RMS relative error ε	Calculated vs. observed regression constants	
		r^2	Slope
Dependence on solution speciation (Eq. 19)			
[CH$_3$Hg$^{II}_{INORGANIC}$] control			
log β_{HX} = 10; log β_{CaX} = 7.2;	0.234	0.957	0.660
log BCF$_{INORGANIC}$ = 12.7	0.897	0.011	−0.015
log β_{HX} = 10; log β_{CaX} = 0			
[CH$_3$Hg$^+$] control	0.352	0.546	0.390
log β_{HX} = 10; log β_{CaX} = 5.2			
[CH$_3$HgCl^0] control^a	0.321	0.835	0.515
log β_{HL} = 4; log β_{ML} = 7.3; log β_{CaX} = 10			
Dependence on K$^*_{PLANKTON}$^{a} (Eq. 20)			
[CH$_3$Hg$^{II}_{INORGANIC}$] control			
log β_{CaY} = 0	0.488	0.898	0.396
log β_{CaY} = 8; log F$_{transfer}$ = 3.1	0.257	0.953	0.713
[CH$_3$HgCl^0] control^b			
log β_{CaY} = 0	0.545	0.698	0.268
log β_{CaY} = 8	0.322	0.921	0.545

^{a} All parameters used to define K$^*_{PLANKTON}$ are from best fit scenario for each mechanism from Table 3.

^{b} Assumes P$_{CH_3HgOH}$ = 0.05 × P$_{CH_3HgCl}$.

where [CH$_3$Hg$_i$] is the concentration of the species controlling uptake by the food chain and g$_{chem}$ and g$_{bio}$ are means of accounting for speciation-independent influences of limnological parameters. The possible influence of competition from H$^+$ and Ca^{2+} for CH$_3$Hg^{II} transport[83] is included via the β_{HX} and β_{CaX} terms, and the speciation of aqueous CH$_3$Hg^{II} is calculated from the DOC association constants used to fit the seston-water partitioning data.

This model (Equation 19) fit the BCF$_{FISH}$ data best (ε = 0.24; r^2 = 0.96; m = 0.66) with total inorganic CH$_3$Hg^{II} concentrations controlling uptake into the food chain and with β_{HX} = 10^{10} and β_{CaX} = 10$^{7.2}$ (Table 6). The CH$_3$Hg^{II} speciation and pH dependencies are the same as for the mechanism controlling uptake into plankton in Section III.A, while calcium inhibition was not considered there. In contrast to the BCF$_{FISH}$ fit (Table 6), accounting for calcium competition improves the fit to seston-water partitioning data only slightly—from ε = 0.18 to ε = 0.17. Thus, the principal site of Ca-CH$_3$Hg^{II} competition is likely not at the level of uptake into plankton. Since in these lakes CH$_3$Hg^{II} uptake by fish occurs mostly via food intake,[84] the observed dependence on aqueous speciation likely reflects the mechanism of transport at the trophic level with the greatest impact on bioaccumulation. While the similarity of the dependencies on CH$_3$Hg^{II} speciation for plankton and fish suggests that uptake by phytoplankton is the major bioconcentration step, a similar mechanism could be acting in zooplankton as well.

Our second approach to modeling BCF$_{FISH}$ data begins from the hypothesis that incorporation of CH$_3$Hg^{II} into phytoplankton may be the principal route of uptake into the aquatic food chain. If so, a relationship between seston- or plankton-water partition coefficients and BCF$_{FISH}$ should be apparent. A comparison of observed CH$_3$Hg^{II} seston-water partitioning

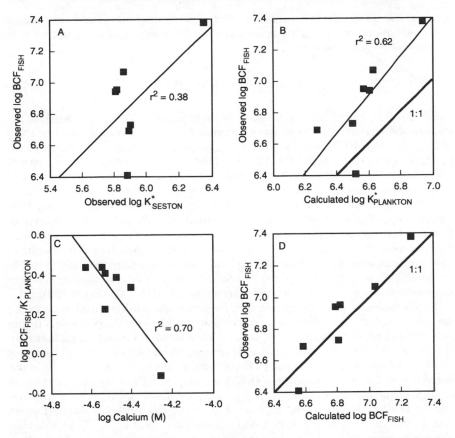

Figure 11 Analysis of fish bioconcentration factors—A: correlation of observed BCF_{FISH} and K^*_{SESTON}; B: correlation of observed BCF_{FISH} and calculated $K^*_{PLANKTON}$; C: dependence of observed $BCF_{FISH}/K^*_{PLANKTON}$ ratio on calcium concentrations; and D: comparison of observed BCF_{FISH} with values calculated using Equation 20 (see Table 6).

data in the MTL lakes (Figure 11A) and bioconcentration factors in perch yields a poor correlation ($r^2 = 0.38$). Except for Crystal Lake, which has both the highest values of K^*_{SESTON} and BCF_{FISH}, BCF_{FISH} appears to be largely independent of K^*_{SESTON}. A much stronger correlation ($r^2 = 0.62$) exists between model estimates of $K^*_{PLANKTON}$ and the observed BCF_{FISH}, however (Figure 11B), consistent with the fact that the same speciation dependence yielded the best fits for CH_3Hg^{II} partitioning into both plankton and fish (Tables 3 and 6). While BCF_{FISH} and $K^*_{PLANKTON}$ clearly are correlated, they do not vary in constant proportion. We therefore postulate a model in which other lake chemical and biological parameters modify the proportionality between BCF_{FISH} and $K^*_{PLANKTON}$:

$$BCF_{FISH} = F_{transfer} \cdot g_{chem} \cdot g_{bio} \cdot K^*_{PLANKTON} \qquad (20)$$

where $F_{transfer}$ is a trophic transfer ratio.

Correlations between the ratio of BCF_{FISH} to $K^*_{PLANKTON}$ (linear and logarithmic), and lakewater concentrations of H^+, chloride, DOC, and chlorophyll a are all weak ($r^2 \leq 0.27$). Lakewater calcium, however, exhibits a strong ($r^2 \geq 0.70$) negative correlation (Figure 11C). The best known effect of calcium on metal accumulation in fish is an inhibition of direct uptake at fish gill surfaces.[83] To explain the effect observed here, we speculate that calcium also inhibits dietary absorption of methylmercury. When modeling the calcium effect as an inhibition of trophic transfer, i.e.,

$$g_{chem} = 1/(1 + \beta_{CaY} \cdot [Ca]) \tag{21}$$

the fit ($\varepsilon = 0.26$; $r^2 = 0.95$; $m = 0.71$), as shown in Figure 11.D, was comparable to that obtained above assuming a solution species dependence (Table 6). We also tested the ability of the passive CH_3HgCl^0 uptake model for $K^*_{PLANKTON}$ to predict fish bioconcentration factors (Table 6). The fit of this model ($\varepsilon = 0.32$; $r^2 = 0.92$; $m = 0.55$), although poorer than with inorganic CH_3Hg^{II} control, was still fairly good when a calcium inhibition term was included.

By accounting for the effect of pH and DOC on CH_3Hg^{II} speciation and the effect of calcium on its absorption, our model yields an inverse relationship of BCF_{FISH} and DOC similar to that observed in the field data (Figure 10A). We note that the greatest deviation from the observed values is for Crystal Lake, which may have DOC of lower humic content than the others. Thus the BCF_{FISH}-DOC relationship is largely explainable as the consequence of a significant association between CH_3Hg^{II} and DOC in lakewaters that reduces its bioavailability. As also observed in the field data, our model BCF_{FISH} predicts no consistent relationship of pH and bioconcentration factors among the MTL lakes (Figure 10B).

Finally, we examine the correspondence of mercury levels in fish predicted by combining our models of lake Hg cycling and bioconcentration in fish with observed values (Figure 10C and D). The mean error ($\varepsilon = 0.32$) and calculated-observed correlation parameters ($r^2 = 0.76$; $m = 0.92$) are reasonably good, especially considering that the predictions are based on calculated, not observed, lakewater CH_3Hg^{II} concentrations. The greatest errors in predicted fish mercury levels occur at low DOC levels (Figure 10C) and reflect a combination of slight under-predictions of BCF_{FISH} (Figure 10A) and dissolved CH_3Hg^{II} concentrations. The combined models also predict the trend of increasing fish mercury at lower pH values, with the exception of the Little Rock Lake treatment basin value at pH ≈ 5. Since this lake also has the lowest DOC, it is likely that the inaccuracy stems from our modeled dependence on DOC.

D. SUMMARY AND SYNTHESIS

Our steady-state analysis has identified a set of mechanisms that define a model of mercury cycling and bioaccumulation capable of describing the data from the MTL seepage lakes. The steady-state MCM includes four limnological parameters that most directly affect fish mercury levels—pH, DOC, calcium, and possibly sulfate. Although the interactions between the processes and lake properties are numerous and potentially complex, the widespread observation of negative correlations between fish mercury and lakewater pH or alkalinity suggests that either a few factors predominate or several act together to generate the overall relationship. Observed relationships between lake DOC and fish mercury are less consistent. How these relationships may arise through the effects of lake chemistry on both the concentrations and bioconcentration of CH_3Hg^{II} in lakes is the subject of this section.

We first consider the influence of pH. Fitzgerald et al.[13] have argued that increased reduction of Hg^{II} in high pH lakewaters may be an important mechanism for decreasing the supply of substrate for in-lake methylation. Our analysis confirms that reduction is a highly pH-dependent process (Equation 17) and that between pH 6 and 7.5 dissolved Hg^{II} decreases approximately threefold with reduction, but increases slightly without reduction (Figure 12A). Since Hg^{II} is the substrate for methylation, the effects of reduction on CH_3Hg^{II} concentrations roughly parallel the Hg^{II} variations. Thus, reduction should be a major contributor to decreases in fish Hg over the pH range 6–8.

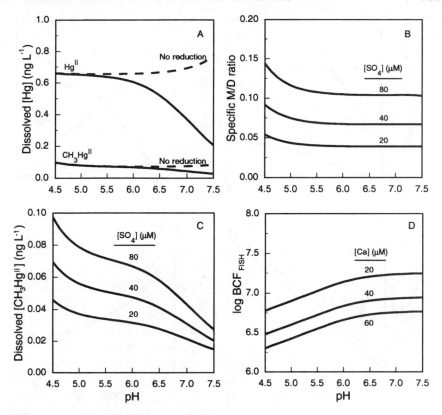

Figure 12 Direct pH-dependence of (A) dissolved Hg^{II} and CH_3Hg^{II} concentrations; (B) specific methylation/demethylation rates; (C) dissolved CH_3Hg^{II} concentrations; and (D) CH_3Hg^{II} bioconcentration factors in fish (from Equation 20). Steady-state MCM calculations assume 3 mg L^{-1} DOC, 7 μM Cl, 0.8 mg L^{-1} detritus, 80 μM SO_4 except as indicated, and a mean lake depth of 5 m.

Increases in the rates of methylation (M) relative to demethylation (D) at low pH also may contribute to increases in fish mercury. Estimates of in situ methylation and demethylation rates obtained with radioisotope techniques suggest that specific rates of methylation and demethylation are pH sensitive.[5] In the steady-state MCM, methylation rates are proportional to total dissolved Hg^{II}, which causes the specific methylation rate to be independent of pH. Since demethylation slows at low pH, the net result is an approximately 40% increase in the M/D ratio at pH 4.5 relative to M/D at pHs above 5.5 (Figure 12B). In the MCM, methylation rates are also a function of sulfate concentrations, which contribute to acidification, providing for an approximately 3.5-fold span of M/D ratios in the MTL lakes. The combined changes in reduction and M/D ratios yield a strong pH and sulfate dependence of CH_3Hg^{II} concentrations (Figure 12C).

Our bioconcentration model predicts a positive pH dependence below pH $\approx$ 6 at constant calcium concentrations (Figure 12D). Field evidence from the MTL study, however, suggests bioconcentration factors in fish are not systematically dependent on pH.[11] Since our model fit the MTL data well (Figure 10B), the independence of BCF_{FISH} and pH in the field data may result from variations in DOC and calcium. The interdependence of calcium and pH in lakewaters suggests that pH-BCF_{FISH} correlations in lakes could differ from the simple pH dependence shown in Figure 12D.

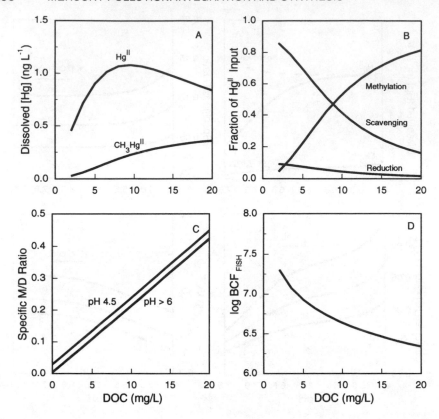

Figure 13   Direct DOC-dependence of (A) dissolved Hg^{II} and CH$_3$Hg^{II} concentrations; (B) fraction of Hg^{II} input flux consumed by indicated reaction; (C) specific methylation/demethylation rates; and (D) CH$_3$Hg^{II} bioconcentration factors in fish (from Equation 20). Steady-state MCM calculations assume pH = 6, 7 μM Cl, 0.8 mg L^{-1} detritus, 40 μM SO$_4$, and a mean lake depth of 5 m.

Our modeling also suggests DOC compounds exert an influence on fish mercury levels by changing aqueous speciation and microbial activity. At constant pH, the MCM predicts marked effects of DOC on Hg^{II} and CH$_3$Hg^{II} concentrations (Figure 13A). Over the range of our calibration, 1.5–7 mg L^{-1} DOC, Hg^{II} and CH$_3$Hg^{II} concentrations increase with DOC, a result consistent with the Hg$_T$ observations from the MTL lakes.[63] Extrapolation of the rate laws indicates a maximum in Hg^{II} should occur near 8 mg L^{-1} DOC. Examination of the rates of scavenging, reduction, and methylation (Figure 13B) demonstrate that the increase between 1 to 8 mg L^{-1} DOC must arise from decreases in scavenging, while the decrease after this point must be due to methylation becoming the dominant sink for Hg^{II}. Methylation/demethylation ratios also increase with DOC according to the model (Figure 13C), since DOC decreases the availability of CH$_3$Hg^{II} but not Hg^{II}. Finally, as noted previously, BCF$_{FISH}$ is inversely proportional to DOC (Figure 13D).

Besides these direct effects of pH, calcium, sulfate, and DOC on CH$_3$Hg^{II} concentrations, we note that indirect effects can also arise through their influence on acid-base chemistry. Calcium and sulfate are the primary determinants of the acid neutralizing capacity (ANC) of lakes, which together with the concentrations of weak acids such as carbonic acid and DOC

A

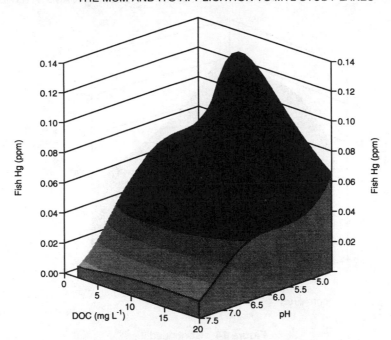

Figure 14 Dependence of (A) fish mercury and (B) BCF$_{FISH}$ on pH and DOC calculated for age-1 yellow perch using dissolved CH$_3$Hg^{II} from steady-state MCM and BCF$_{FISH}$ from Equation 20. For each pH-DOC point, ANC was calculated and calcium determined from Equation 22 (with sulfate held constant at 40 μM) as described in text. Calculations assume 7 μM Cl, 0.8 mg L^{-1} detritus, and a mean lake depth of 5 m.

determines the pH. Using the average ratios of total base cations to calcium, 1.9, and acid anions to sulfate, 1.3, in northeastern U.S. lakes,[85] we can reasonably approximate the ANC as:

$$ANC \approx 1.9 \cdot [Ca^{2+}] - 1.3 \cdot [SO_4^{2-}] \qquad (22)$$

where all concentrations are in eq L^{-1}. The major weak acid in freshwaters is generally carbonic acid, H$_2$CO$_3^*$,[86] whose concentration we approximate here as a threefold excess over equilibrium with atmospheric CO$_2$. DOC compounds can also be important weak acids, with about 15 μeq mg-C^{-1} of sites of varying β_{HL}.[87] Since pH is related positively to ANC and negatively to weak acid concentrations, lakewater pHs increase with increasing calcium and decrease with increasing sulfate and DOC concentrations.

In order to examine the concerted effects of lake chemistry on fish mercury, we applied the steady-state MCM at varying pH and DOC concentrations. Lake depth and the concentrations of chloride, suspended particulate matter, and sulfate were held constant. For each pH-DOC combination, calcium was calculated using Equation 22 with values of ANC defined by the pH, DOC, and partial pressure of carbon dioxide. Calculated mercury levels in age-1 yellow perch exhibit a strong dependence on both pH and DOC (Figure 14A). The increase in fish mercury at low pH results partly from an increase in dissolved CH$_3$Hg^{II} concentration (Figure 12C). However, the apparent pH dependence of BCF$_{FISH}$ when changes

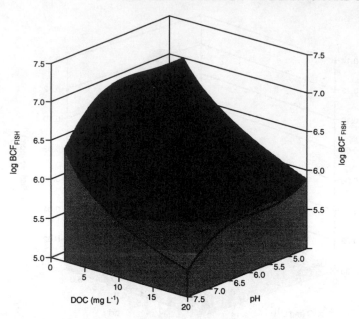

Figure 14   Continued.

in calcium are included (Figure 14B) differs markedly from that predicted for constant calcium (Figure 12D). The calcium-mediated decrease in BCF_{FISH} at pH >6 accentuates the decline in CH_3Hg^{II} concentrations over the same pH range, yielding consistently lower levels of fish mercury despite variations in DOC.

The predicted effect of DOC is more complex. Again, over the 1.5 to 7 mg L^{-1} DOC range of calibration, the model predicts a positive correlation between fish mercury and DOC, particularly at low pH. The decrease in fish mercury above about 8 mg L^{-1} DOC (at low pH) reflects the relatively greater decrease in BCF_{FISH} than increases in CH_3Hg^{II} concentrations at higher DOC (Figure 13). Because it comes from extrapolating beyond the calibration range of the model, the existence of the predicted optimum is uncertain. However, relatively low levels of fish mercury from high DOC seepage lakes have been observed by Grieb et al.,[4] who argued that decreases in the bioavailability of CH_3Hg^{II} caused the effect. While this study also supports the suggestion that DOC decreases the bioavailability of CH_3Hg^{II}, the effects of DOC on lake CH_3Hg^{II} levels may cause its relationship to fish mercury levels to be variable.

Despite the variability introduced by DOC, the steady-state MCM predicts that fish mercury concentrations should increase at low pH (Figure 15), a prediction consistent with the widespread observation of correlations between these two variables. A similarly strong relationship is predicted for fish mercury and calcium (results not shown), which also significantly correlates with fish mercury in some studies.[4] We note that despite a direct dependence of methylation rates on sulfate concentrations in the model, correlation between fish mercury and pH or calcium, but not sulfate, were observed when calcium and sulfate were varied simultaneously at constant DOC to generate a range of pH values from 4 to 7.5 (results not shown). Thus, since the range of calcium concentrations observed in lakes is generally wider than for sulfate,[85] the lack of correlation between sulfate and fish mercury should not be taken as evidence of a lack of an effect of sulfate.

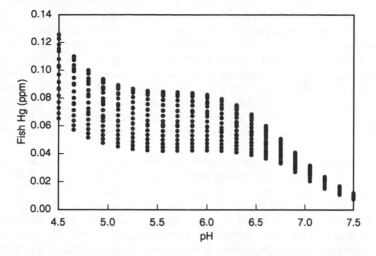

Figure 15 Overall relationship between pH and fish mercury for all points used to generate Figure 14A.

IV. THE MERCURY CYCLING MODEL (MCM)

A. DESCRIPTION

The dynamic MCM employs the chemical speciation principles described and applied throughout this paper, but includes the greater temporal and spatial resolution necessary for dynamic simulations of lake mercury biogeochemistry. We designed the model to have a degree of complexity suited to both the resolution of the available data from the MTL study and the scientific and management questions to be addressed. Because questions remain about which processes control the cycling of mercury in lakes and the rate laws of these processes, the model incorporates considerable flexibility in specifying reactions and the factors which control their rates. To increase its user-friendliness, we developed the model for use on Macintosh personal computers with a graphical user interface (Extend®, Imagine That, San Jose, CA).

The MCM employs a simple compartmental structure to represent the different geochemical environments and trophic levels in lake ecosystems (See Figure 1). The lake is divided into epilimnetic, hypolimnetic, and sedimentary compartments, while the food chain is represented by four trophic levels. Each physical compartment is assumed to be homogeneous with time-varying physical and chemical parameters. Ancillary chemical and physical properties of each compartment are specified by the user rather than modeled dynamically. Ecosystem dynamics are simulated in order to obtain coherent productivity and trophic-transfer fluxes for modeling bioaccumulation and biomagnification. Within the lake and ecosystem, mercury is subdivided into components (see Section II.A) that are either chemically or physically distinct on the time scales of interest, i.e., days to months.

The boundaries of the lake are the atmosphere, lake margins, and permanent sediments. The user must specify the data necessary to calculate input fluxes from the atmosphere (mean atmospheric concentrations of Hg^0, Hg^{II}, CH_3Hg^{II}, and particulate Hg^{II}) and via water transport into the lake (concentrations of each mercury component in groundwater and tributaries). Biogeochemical reactions within the watershed or atmosphere are not considered. Burial in permanent sediments represent a one-way flux out of the surficial sediments.

The physical representation of the lake includes a dynamic water budget to determine total lake volume based on input data for precipitation, stream flow, in-seepage, evaporation, stage-discharge relationships, and hyposometric curves. Rather than dynamically model thermal stratification, a monthly mean epilimnion depth is specified by the user. The model calculates the advective flow between layers needed to account for the specified volume changes and the diffusive transport across the thermocline defined as a velocity. Gas-exchange rates, determined from two-film theory,[88] depend on wind speeds and the air-water partitioning of neutral mercury species. Abiotic particle dynamics are represented by user-defined concentrations of detrital particles, mean settling velocities, and sediment re-suspension rates.

For each component in the water column and sediment compartments, the model computes the equilibrium speciation during each timestep based on the user-defined ancillary chemical data and equilibrium constants. Because the concentrations of mercury present in most aquatic systems are so low, saturation of dissolved, particulate, and biological transport ligands is highly unlikely. Thus, the equilibrium calculations for each component simply requires determining speciation fractions, or side reaction coefficients, and are independent of that component's total concentration. As discussed in Section II, the rates of both abiotic and biotic processes are defined by rates laws with constants particular to each species.

The dynamic MCM also contains a simple ecosystem model. The aquatic biota are represented by a linear food chain comprised of phytoplankton, zooplankton, and planktivorous and piscivorous fish. The model calculates a self-consistent set of carbon fluxes associated with production, respiration, mortality, and predation at each trophic level. The equations which govern the ecosystem behavior are derived from standard ecosystem models which simulate biomass at different trophic levels.[89] The trophic levels do not represent particular species or organism ages, but the aggregate behavior of the organisms at the same trophic level. Each trophic level will have a mean age that corresponds to its biomass turnover time.

The mercury contained in organisms of each trophic level is divided into Hg^{II} and CH_3Hg^{II}. Dietary mercury absorption efficiencies are specific to each form of mercury. Absorption from solution and depuration are the two mechanisms of direct exchange between organism and environment included in the model. Direct absorption rates depend on mercury speciation and rate constants specific to each trophic level. For fish, the rate of direct absorption by the population depends on respiration rate, a reflection of the rate at which water is pumped past the gills. This formulation is derived from a bioenergetics model applied to bioaccumulation in individual fish.[90] Depuration rates are expressed as simple first-order loss terms. Indirect exchanges of mercury between the organisms and environment occur via absorption of dietary mercury and losses due to mortality and predation, which are defined by ecosystem biomass fluxes.

For each component of mercury in each geochemical and ecosystem compartment, a mass conservation equation accounting for all input, loss, transport, and transformation processes is solved every timestep. All of these differential equations are linear with respect to mercury since no higher-order reactions between different forms of mercury and no processes saturable in mercury concentrations are included. The set of differential equations is solved using a fully implicit solution technique and LU decomposition of the resulting matrix.[64] A similar approach is used to solve the ecosystem differential equations. Each second-order expression is reduced to linear form by using an explicit dependence on the concentration of the predator organisms, i.e., their initial concentration for the timestep, and an implicit dependence on the concentration of the prey. The resulting set of equations is stable and conserves mass consistently. A monthly timestep, comparable to the temporal resolution of the available data, is employed throughout the model.

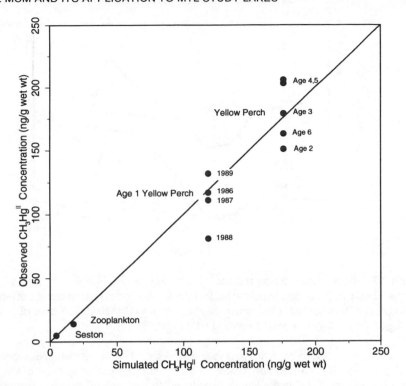

Figure 16   Correspondence of observed CH_3Hg^{II} concentrations in biota from Little Rock Lake reference basin[63,91] and calibration values from the dynamic MCM.

B.   APPLICATION

In the remainder of this chapter, we illustrate how the MCM can be applied to analyze hypotheses about the mechanisms governing the geochemistry and bioaccumulation of mercury in lakes. First, we calibrated the model to yield values of the principal dissolved forms of mercury and the concentrations of CH_3Hg^{II} in biota comparable to those observed in the reference basin of Little Rock Lake. The simulated concentrations in the biota exhibit food chain magnification of mercury, as observed in the natural system (Figure 16). Our simulations suggest that mercury uptake in both herbivorous and piscivorous fish occurs largely (>99%) via the food chain. Uptake of dissolved methylmercury by zooplankton may be significant, however.

We also investigated a possible mechanism to explain the dependence of mercury methylation rates on total dissolved Hg^{II} concentrations (Section III.B). We assumed that mercury uptake by methylating bacteria is a passive process, e.g., $HgCl_2$ controls methylation rates in oxic waters. In oxic waters, this species comprises 1% or less of the total Hg^{II} (Figure 2). In anoxic hypolimnia, however, nearly all of the Hg^{II} is converted to $Hg(SH)_2^0$ (Figure 2) which by virtue of its neutral charge may also be passively permeable in biological membranes. We base this speculative suggestion on the fact the H_2S engages in hydrogen bonding to a lesser degree that H_2O, as evidenced by its greater volatility. Thus, $Hg(SH)_2^0$ may exhibit a passive permeability closer to that of $HgCl_2^0$ than that of $Hg(OH)_2^0$, which is impermeable in membranes. According to this hypothesis, the availability of Hg^{II} for methylation, 100-fold in sulfidic waters when kinetic factors hinder HgS(s) precipitation.

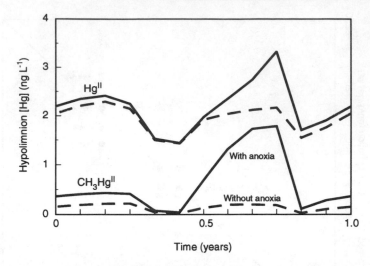

Figure 17 Hypolimnetic concentrations of total Hg^{II} and CH_3Hg^{II} with and without summer anoxia. In the anoxic hypolimnion, $20 \, \mu M$ sulfide and kinetic hindrance of HgS(s) precipitation was assumed. Mechanism causing increased methylation is discussed in text. Simulation of a typical year beginning on January 1.

This effect of sulfide on methylation rates can be observed in our simulation of a stratified lake (Figure 17). Only with sulfide present does a significant summer increase in CH_3Hg^{II} occur. (The buildup of Hg^{II} in the sulfidic hypolimnion is the result of decreased partitioning to and scavenging by particles.) Thus, if methylation primarily occurs in anoxic or suboxic hypolimnia or partially stratified bottom waters, passive absorption of $Hg(SH)_2^0$ complexes could partly account for the observation of methylation by sulfate-reducing bacteria[80] and for the availability of DOC-associated Hg^{II} for methylation observed in our steady-state modeling (Section III.B). The significant accumulation of CH_3Hg^{II} observed in anoxic hypolimnia of the MTL lakes,[11] as well as stratified impoundments,[69] is also consistent with this hypothesis.

V. CONCLUSIONS

We have applied the Mercury Cycling Model (MCM) to analyze the water column mercury speciation and fish mercury data from the Mercury in Temperate Lakes study in Wisconsin, U.S. Our approach has been to fit the field data using models that combine what is known of the aqueous speciation of mercury with the speciation dependencies implied by different hypothetical mechanisms for the major processes of the lacustrine mercury cycle. Our modeling efforts imply the following conclusions:

- Partitioning into plankton, particularly for CH_3Hg^{II}, accounts for the relatively higher affinities of Hg^{II} and CH_3Hg^{II} for seston than for dissolved organic matter in the MTL lakes.
- In-lake methylation is the major source of CH_3Hg^{II} in the MTL lakes, although gas exchange provides a source comparable to atmospheric deposition.
- Reduction of Hg^{II} and subsequent volatilization of Hg^0 is a major sink of mercury in high pH lakes. Inhibition of this mechanism at low pH leads to significant increases in the production of CH_3Hg^{II}, confirming the "substrate limitation hypothesis" of Fitzgerald and co-workers.

- Overall methylation/demethylation ratios increase at low pH. Methylation may also be stimulated by sulfate, a finding consistent with a link to sulfate redox cycling for this process. Both effects may contribute to the increases in methylmercury in low pH lakes.
- Dissolved organic carbon (DOC) compounds bind both Hg^{II} and CH_3Hg^{II} to a significant degree. Although organic complexation of CH_3Hg^{II} reduces its bioavailability to the food chain, DOC may also stimulate bacterial activity and increase the supply of Hg^{II} for methylation, causing the net effect of increases in lake DOC on fish mercury to be variable.
- Fish bioconcentration factors for CH_3Hg^{II} are inversely related to calcium. This competitive effect likely arises at a trophic level higher than phytoplankton. Although our results also suggest the direct effect of low pH reduces CH_3Hg^{II} accumulation in phytoplankton, covariation of pH and calcium in lakes leads to a weak pH dependence of fish bioconcentration factors below pH 6, with some decrease above this value.
- Increases in CH_3Hg^{II} concentrations at low pH combine with decreases in bioconcentration factors at high pH to generate the widely observed negative correlation between lake pH and fish mercury content.

ACKNOWLEDGMENTS

This work was supported by the Electric Power Research Institute, the Wisconsin Department of Natural Resources, and the Empire State Electric Energy Research Corporation. The authors thank all the members of the MTL team for sharing their data and ideas during the course of developing this model. Special thanks are due to R. Mason and G. Gill for numerous helpful discussions and suggestions during the course of model development and to K. W. Bruland for providing an environment conducive to the completion of this work. The comments of the anonymous reviewers are also appreciated.

APPENDIX 1: SESTON-WATER PARTITIONING OF MERCURY

A. DEFINITIONS

$$K^*_{SESTON} = \frac{[Hg^j_{SESTON}]}{[Seston] \cdot [Hg^j_{DISSOLVED}]}$$

$$= x_{PLANKTON} \cdot K^*_{PLANKTON} + (1 - x_{PLANKTON}) \cdot K^*_{PARTICLES}$$

(A.1–1)

where

K^*_{SESTON}	= Seston-water partition coefficient for mercury component j, Hg^j	(L kg^{-1})
$[Hg^j_{SESTON}]$	= Sestonic concentration of Hg^j	(mol L^{-1})
	= $[Hg^j_{PARTICLES}] + [Hg^j_{PLANKTON}]$	
$[Hg^j_{PARTICLES}]$	= Concentration of Hg^j associated with detrital particles	(mol L^{-1})
$[Hg^j_{PLANKTON}]$	= Concentration of Hg^j associated with plankton	(mol L^{-1})
[Seston]	= Concentration of detrital particles and plankton	(kg L^{-1})
	= [Particles] + [Plankton]	
[Particles]	= Concentration of filterable, detrital particles	(kg L^{-1})
[Plankton]	= Mass concentration of viable plankton	(kg-plankton L^{-1})

$[Hg^j_{DISSOLVED}]$ = Concentration of "apparently dissolved" Hg^j $\quad$ (mol L^{-1})

$\qquad\qquad$ = $[Hg^j_{COLLOIDS}] + [Hg^j_{SOLUTES}]$

$[Hg^j_{COLLOIDS}]$ = Concentration of Hg^j in colloids $\qquad\qquad\qquad$ (mol L^{-1})

$[Hg^j_{SOLUTES}]$ = Concentration of "truly dissolved" Hg^j $\qquad$ (mol L^{-1})

$x_{PLANKTON}$ = Fraction of seston as viable plankton

$K^*_{PLANKTON}$ = Plankton-water partition coefficient for Hg^j $\qquad\qquad\qquad\qquad\qquad\qquad\qquad$ (L kg-plankton^{-1})

$K^*_{PARTICLES}$ = Abiotic particle-water partition coefficient for Hg^j $\qquad\qquad\qquad\qquad\qquad\qquad\qquad$ (L kg-particles^{-1})

B. EQUILIBRIUM PARTITIONING TO ABIOTIC PARTICLES AND COLLOIDS

$$K^*_{PARTICLES} \equiv \frac{[Hg^j_{PARTICLES}]}{[Particles] \cdot [Hg^j_{DISSOLVED}]} \quad\quad (A.1-2)$$

$$= \alpha^j_{z+} \cdot \alpha^L_- \cdot n^{PARTICLES}_L \cdot \beta^{PARTICLES}_{ML}$$

where

α^j_{z+} $\quad$ = Fraction of apparently dissolved Hg^j as free metal ion

α^L_- $\quad$ = Fraction of deprotonated mercury-binding sites in particles

$n^{PARTICLES}_L$ = Density of mercury-binding sites in particulate matter $\qquad\qquad\qquad\qquad\qquad$ (mol kg^{-1})

$\beta^{PARTICLES}_{ML}$ = Mercury-particle association constant $\qquad\qquad$ (L mol^{-1})

$$K^*_{COLLOIDS} \equiv \frac{[Hg^j_{COLLOIDAL}]}{[Colloids] \cdot [Hg^j_{DISSOLVED}]} \quad\quad (A.1-3)$$

$$= \alpha^j_{z+} \cdot \alpha^L_- \cdot n^{COLLOIDS}_L \cdot \beta^{COLLOIDS}_{ML}$$

where

$[Colloids]$ = Concentration of colloidal particles $\qquad\qquad$ (kg L^{-1})

α^L_- $\qquad$ = Fraction of deprotonated mercury-binding sites in colloids

$n^{COLLOIDS}_L$ = Density of mercury-binding sites in colloidal particles $\qquad\qquad\qquad\qquad\qquad\qquad$ (mol kg^{-1})

$\beta^{COLLOIDS}_{ML}$ = Mercury-colloid association constant $\qquad\qquad$ (L mol^{-1})

Assumptions for abiotic partitioning model (after Morel and Gschwend[70]):

(i) $\quad$ Colloidal matter comprises major part of DOM:

$$[Colloids] \approx [DOM]$$

(ii) $\quad$ Binding to colloidal DOM dominates "dissolved" species:

$$[Hg^j_{COLLOIDAL}] \approx [Hg^j_{DISSOLVED}]$$

(iii) $\quad$ Detrital particles and colloidal matter bind mercury similarly:

$$n^{PARTICLES}_L \approx n^{COLLOIDS}_L \qquad\qquad \beta^{PARTICLES}_{ML} \approx \beta^{COLLOIDS}_{ML}$$

Then:

$$K^*_{PARTICLES} \approx \frac{[Hg^j_{PARTICLES}]}{[Particles] \cdot [Hg^j_{COLLOIDS}]}$$

$$\approx \frac{\alpha^j_{z+} \cdot \alpha^L_- \cdot n^{PARTICLES}_L \cdot \beta^{PARTICLES}_{ML}}{\alpha^j_{z+} \cdot \alpha^L_- \cdot n^{COLLOIDS}_L \cdot \beta^{COLLOIDS}_{ML} \cdot [DOM]} \approx \frac{1}{[DOM]} \qquad (A.1-4)$$

C. STEADY-STATE PARTITIONING INTO PLANKTON

$$K^*_{PLANKTON} \equiv \frac{[Hg^j_{PLANKTON}]}{[Plankton] \cdot [Hg^j_{DISSOLVED}]} \qquad (A.1-5)$$

1. Partition Coefficient with Respect to All Bioavailable Inorganic Species

$$K^{INORGANIC}_{PLANKTON} = \frac{\sum_{i=1}^{n_{inorganic}} \alpha^j_i \cdot \{k^i_X/(1 + \beta_{HX} \cdot [H^+]) + k^i_P}{\alpha^j_{INORGANIC} \cdot (\mu + k_d) \cdot M_{CELL}} \qquad (A.1-6)$$

where

$n_{inorganic}$	= Total number of inorganic species	
k^i_X	= Rate constant for facilitated uptake of species i	(L cell^{-1} d^{-1})
k^i_P	= Rate constant for passive uptake of species i	(L cell^{-1} d^{-1})
α^i_j	= Fraction of apparently dissolved Hg^j as species i	
$\alpha^j_{INORGANIC}$	= Fraction of apparently dissolved Hg^j as inorganic species	
μ	= Specific growth rate of phytoplankton	(d^{-1})
k_d	= Depuration rate constant	(d^{-1})
M_{CELL}	= Average dry weight of phytoplankton cells of size R$_{CELL}$	(kg cell^{-1})

2. Rate Constants for Passive Uptake

$$k^i_P = P_i \cdot A_{CELL} \qquad (A.1-7)$$

where

P_i	= Membrane permeability coefficient for Hg^j species i	(dm d^{-1})
A_{CELL}	= Average surface area of phytoplankton cells	(dmls2cell^{-1})

3. Diffusion-Limited Partition Coefficient with Respect to Inorganic Species (see also Ref. 48)

$$K_{DIFFUSION} = \frac{4 \cdot \pi \cdot R_{CELL} \cdot D}{\mu \cdot M_{CELL}} \qquad (A.1-8)$$

where

D	= Diffusion coefficient for inorganic Hg^j species	(dm^2 d^{-1})
R_{CELL}	= Average radius of phytoplankton cells	(dm)

4. Apparent Plankton-Water Partition Coefficient (with Diffusion Limitation)

$$K^*_{\text{PLANKTON}} = \frac{\alpha^j_{\text{INORGANIC}} \cdot K^{\text{INORGANIC}}_{\text{PLANKTON}} \cdot K_{\text{DIFFUSION}}}{K^{\text{INORGANIC}}_{\text{PLANKTON}} + K_{\text{DIFFUSION}}} \tag{A.1--9}$$

APPENDIX 2: STEADY-STATE MCM EQUATIONS

A. MASS BALANCE EQUATIONS

Hg^{II}:
$$J_{HgII} - R - M - S_{HgII} - Q_{HgII} = 0 \tag{A.2--1}$$
$$\phantom{J_{HgII}}(1) \quad\; (2) \quad (3) \quad\; (4) \quad\quad (5)$$

CH_3Hg^{II}:
$$J_{CH_3HgII} + M - S_{CH_3HgII} - D - V_{CH_3HgII} - Q_{CH_3HgII} = 0 \tag{A.2--2}$$
$$\phantom{J_{CH_3HgII}}(1) \quad\quad (3) \quad\quad (4) \quad\; (6) \quad\quad (7) \quad\quad\; (5)$$

Hg^0:
$$R + D - V_{Hg0} - Q_{Hg0} = 0 \tag{A.2--3}$$
$$(2) \quad (6) \quad\;\; (7) \quad\quad (5)$$

B. STEADY-STATE CONCENTRATIONS

$$[Hg^{II}_{DISSOLVED}] = J_{HgII}/\{Z \cdot (\alpha^{HgII}_R \cdot k^*_R + \alpha^{HgII}_M \cdot k^*_M + \tau^{-1}) + \tag{A.2--4}$$
$$v_s \cdot K^*_{PARTICLES,HgII} \cdot [\text{Particles}] \cdot x^{HgII}_{BURIED}$$

$$[CH_3Hg^{II}_{DISSOLVED}] = \{J_{CH3HgII} + k_{vg} \cdot K_H \cdot [CH_3HgCl_{eq}] +$$
$$\alpha^{HgII}_M \cdot k^*_M \cdot [Hg^{II}_{DISSOLVED}] \cdot Z\}/\{Z \cdot (\alpha^{CH3HGII}_D \cdot k^*_D + \tau^{-1}) + \tag{A.2--5}$$
$$\alpha^{CH3HgII}_V \cdot k_{vg} \cdot K_H + v_s \cdot K^*_{PARTICLES,CH_3HgII} \cdot [\text{Particles}] \cdot x^{CH3HgII}_{BURIED}\}$$

$$[Hg^0_{DISSOLVED}] = \{k_{vl} \cdot [Hg^0_{eq}] + (\alpha^{HGII}_R \cdot k^*_R \cdot [Hg^{II}_{DISSOLVED}] + \tag{A.2--6}$$
$$\alpha^{CH3HgII}_D \cdot k^*_D \cdot [CH_3Hg^{II}_{DISSOLVED}]) \cdot Z\}/\{Z/\tau + k_{vl}\}$$

where

$$Z = \text{Mean depth of lake} \tag{m}$$

C. PROCESSES AND DEFINITIONS

1. Total Inputs of Mercury Component J To Lake (J_{Hgj})

$$J_{Hgj} = \hat{J}_{Hgj} \cdot A + I_{Hgj} \tag{A.2--7}$$

where

$\hat{J}_{Hgj}$ = Wet and dry deposition flux   ($\mu g\ m^{-2}\ d^{-1}$)

J_{HgII} = 0.039 $\mu g\ m^{-2}\ d^{-1}$ during open water season[13]

J_{CH_3HgII} = 0.00029 $\mu g\ m^{-2}\ d^{-1}$ during open water season[13]

A   = Lake surface area   (m^2)

I_{Hgj} = Inseepage of mercury component j   ($\mu g\ d^{-1}$)

2. Reduction of Hg^{II} to Hg^0 (R)

$$R = k^*_R \cdot \alpha^{HgII}_R \cdot [Hg^{II}_{DISSOLVED}] \cdot V \tag{A.2--8}$$

where

k^*_R = Apparent reduction rate constant (d^{-1})

α^{HgII}_R = Fraction of dissolved Hg^{II} consisting of species controlling reduction rate

V   = Lake volume (m^3)

3. Methylation of Hg^{II} in Water Column and Surficial Sediments (M)

$$M = k_M^* \cdot \alpha_M^{HgII} \cdot [Hg_{DISSOLVED}^{II}] \cdot V \qquad (A.2–9)$$

where

k_M^* = Apparent demethylation rate constant (d^{-1})

α_M^{HgII} = Fraction of dissolved Hg^{II} consisting of species
        controlling methylation rate

4. Scavenging and Sediment Burial (S_{Hgj})

$$S_{Hgj} = x_{BURIED}^j \cdot K_{PARTICLES,j}^* \cdot [Hg_{DISSOLVED}^j] \cdot [Particles] \cdot v_s \cdot A \qquad (A.2–10)$$

where

x_{BURIED}^j = Fraction of sedimented component j that is buried $(L\ kg^{-1})$

$K_{PARTICLES,j}^*$ = Abiotic particle-water partition coefficient of Hg^j $(kg\ L^{-1})$

$[Particles]$ = Concentration of detrital particles (m^{d-1})

v_s = Effective particle settling velocity

5. Out-Seepage of Dissolved Species (Q_{Hgj})

$$Q_{Hgj} = [Hg_{DISSOLVED}^j] \cdot V/\tau \qquad (A.2–11)$$

where

τ = Hydraulic residence time of lake (d)

6. Demethylation in Water Column (D)

$$D = k_D^* \cdot \alpha_D^{CH_3Hg^{II}} \cdot [CH_3Hg_{DISSOLVED}^{II}] \cdot V \qquad (A.2–12)$$

where

k_D^* = Apparent demethylation rate constant (d^{-1})

$\alpha_D^{CH_3HgII}$ = Fraction of dissolved CH_3Hg^{II} as species
          controlling demethylation rate

7. Volatilization Loss of Hg^j from Lake (V_{Hgj})

(a) $V_{Hg0} = k_{vl} \cdot ([Hg^0] - [Hg_{eq}^0]) \cdot A \qquad (A.2–13)$

where:

k_{vl} = Volatilization rate constant for liquid-phase limited solutes
        = 0.2 m d^{-1} (After Watras et al.[63]) $(m\ d^{-1})$

$[Hg^0]_{eq}$ = Concentration of Hg^0 at equilibrium with atmosphere

(b) $V_{CH_3Hg^{II}} = k_{vg} \cdot K_H \cdot \{[CH_3HgCl^0] - [CH_3HgCl_{eq}^0]\} \cdot A \qquad (A.2–14)$

where

k_{vg} = Volatilization rate constant for gas-phase $(m\ d^{-1})$
            limited solutes
          = 310 m d^{-1} based on an average wind speed
            of 2.5 m s^{-1} (after Liss 88)

K_H = Henry's law constant for CH_3HgCl^0 $(\mu g\ m^{-3})$

$[CH_3HgCl_{eq}^0]$ = Concentration of CH_3HgCl^0 at equilibrium with
            atmosphere

APPENDIX 3: GAS EXCHANGE OF Hg^0 AND CH_3Hg^{II}

A. HENRY'S LAW

Dissolved concentrations of Hg^j at equilibrium with atmosphere:

$$K_H = \frac{\{Hg^j_{GASEOUS}\}}{\alpha_V^{Hg\,j} \cdot [Hg^j_{DISSOLVED}]} \tag{A.3–1}$$

where

K_H = Dimensionless Henry's law constant

$\{Hg^j_{GASEOUS}\}$ = Concentration of Hg^j in the atmosphere   ($\mu g\ m^{-3}$)

$\alpha_V^{Hg\,j}$ = Volatile fraction of dissolved Hg^j

B. ELEMENTAL MERCURY

The atmospheric concentration of elemental mercury has been measured directly—1.57 pg L^{-1}.[13] Given the following equilibrium data:

$$K_H = 0.245 \text{ at } 15^\circ C^{97}$$
$$\alpha_V^{Hg0} = 1.00$$

the equilibrium concentration of Hg^0 in lakewater is:

$$[Hg^0_{eq}] = 1.57 \text{ pg } L^{-1}/0.245 = 6.4 \text{ pg } L^{-1}$$

C. METHYLMERCURY

The atmospheric concentration of methylmercury is below current detection limits and must be inferred from its concentration in rain and from its aqueous speciation. The most volatile methyl mercury species is CH_3HgCl^0. For the MTL lakes, $pH_{rain} \approx 4.7$ and $[Cl^-_{rain}] \approx 3\ \mu M$. Assuming organic complexation is negligible in rain, we calculate for $15^\circ C$:

$$\alpha_V^{CH3HgII} = \frac{[CH_3HgCl_{RAIN}]}{[CH_3Hg^{II}_{RAIN}]} = 0.42$$

$$[CH_3Hg^{II}_{RAIN}] = 160 \text{ pg } L^{-1}$$

$$K_H = 1.6 \times 10^{-5} \text{ at } 15^\circ C \text{ for } CH_3HgCl^{58}$$

$$\{CH_3Hg_{GASEOUS}\} = 1.6 \times 10^{-5} \cdot 0.42 \cdot 160 \text{ pg } L^{-1} \cdot 1000 \text{ L m}^{-3} = 1.1 \text{ pg m}^{-3}$$

This value is below the 5 pg m^{-3} detection limit of Fitzgerald et al.[13] Note that because CH_3HgOH^0 is 100-fold less volatile and is less abundant than CH_3HgCl^0 in rain, it is not likely to be the dominant gaseous species (see also Iverfeldt and Lindquist[58]).

The concentration of CH_3HgCl in lakewater at equilibrium with the atmosphere is

$$[CH_3HgCl_{eq}] = [CH_3HgCl_{RAIN}] = \alpha_V^{CH3HgII} \cdot [CH_3Hg^{II}_{RAIN}]$$
$$= 0.42 \cdot 160 \text{ pg } L^{-1} = 67 \text{ pg } L^{-1}$$

REFERENCES

1. Hakanson, L., The quantitative impact of pH, bioproduction and Hg-contamination on the Hg-content of fish (pike), *Environ. Pollut. (Ser. B)*, 1, 285, 1980.
2. Wren, C. D. and MacCrimmon, H. R., Mercury levels in the sunfish, *Lepomis gibbosus*, relative to pH and other environmental variables of Precambrian shield lakes, *Can. J. Fish. Aquat. Sci.*, 40, 1737, 1983.
3. Weiner, J. G., Metal contamination of fish in low-pH lakes and potential implications for piscivorous wildlife, *Trans. N. Am. Wildl. Nat. Resourc. Conf.*, 52, 645, 1987.

4. Grieb, T. M., Driscoll, C. T., Gloss, S. P., Schofield, C. L., Bowie, G. L., and Porcella, D. B., Factors affecting mercury accumulation in fish in the Upper Michigan Peninsula, *Environ. Toxicol. Chem.,* 9, 919, 1990.

5. Winfrey, M. R. and Rudd, J. W. M., Environmental factors affecting the formation of methylmercury in low pH lakes, *Environ. Toxicol. Chem.,* 9, 853, 1990.

6. Swain, E. B., Engstrom, D. R., Brigham, M. E., Henning, T. A., and Brezonik, P. L., Increasing rates of atmospheric mercury deposition in midcontinental North America, *Science,* 257, 784, 1992.

7. Benoit, J. M., Fitzgerald, W. F., and Damman, A. W. H., Historical atmospheric mercury deposition in the mid-continental United States as recorded in an ombrotrophic peat bog, In: *Mercury Pollution: Integration and Synthesis,* Watras, C. J. and Huckabee, J. W., Eds., Lewis Publishers, Chelsea, MI, 1994, chap. II.2.

8. Fitzgerald, W. F., Gill, G. A., and Hewitt, A. D., Air-sea exchange of mercury, In: *Trace Metals in Seawater,* NATO Conference Series IV: Marine Sciences, Vol. 9, Wong, C. S., Boyle, E., Bruland, K W., Burton, J. D., and Goldberg, E. D., Eds., Plenum Press, New York, 1983, 297.

9. Lindquist, O., Jernelov, A., Johansson, K., and Rodhe, H., Mercury in the Swedish Environment, Global and Local Sources, Report No. SNV PM 1816, Swedish Environmental Protection Agency, 1984.

10. Bloom, N. S. and Watras, C. J., Observations of methylmercury in precipitation, *Sci. Total Environ.,* 87/88, 199, 1989.

11. Bloom, N. S., Watras, C. J., and Hurley, J. P., Impact of acidification on the methylmercury cycling of remote seepage lakes, *Water Air Soil Pollut.,* 56, 477, 1991.

12. Gill, G. A. and Bruland, K. W., Mercury speciation in surface freshwater systems in California and other areas, *Environ. Sci. Technol.,* 24, 1392, 1990.

13. Fitzgerald, W. F., Mason, R. P., and Vandal, G. M., Atmospheric cycling and air-water exchange of mercury over mid-continental lacustrine regions, *Water Air Soil Pollut.,* 56, 745, 1991.

14. Fagerstrom, T. and Jernelov, A., Some aspects of the quantitative ecology of mercury, *Water Res.,* 6, 1193, 1972.

15. Wollast, R., Billen, G., and MacKenzie, F. T., Behavior of mercury in natural systems and its global cycle, In: *Ecological Toxicology Research: Effects of Heavy Metal and Organohalogen Compounds,* NATO Conference Series, Environmental Science Research: Vol. 7, McIntyre, A. D. and Mills, C. F., Eds., Plenum Press, New York, 1975, 145.

16. Brosset, C., The mercury cycle, *Water Air Soil Pollut.,* 16, 253, 1981.

17. Morel, F. M. M. and Hering, J. G., *Principles and Applications of Aquatic Chemistry,* Wiley-Interscience, New York, 1993, 405.

18. Bjornberg, A., Hakanson, L., and Lundbergh, K., A theory on the mechanisms regulating the bioavailability of mercury in natural waters, *Environ. Pollut.,* 49, 53, 1988.

19. Fontaine, T. D., III, A non-equilibrium approach to modeling toxic metal speciation in acid, aquatic systems, *Ecol. Modell.,* 22, 85, 1983/1984.

20. Watras, C. J., Bloom, N. S., Fitzgerald, W. F., Hurley, J. P., Krabbenhoft, D. P., Rada, R. G., and Wiener, J. G., Mercury in temperate lakes: a mechanistic field study, *Verh. Int. Verein. Limnol.,* 24, 2199, 1991.

21. Brosset, C., The behavior of mercury in the physical environment, *Water Air Soil Pollut.,* 34, 145, 1987.

22. Mason, R. P. and Fitzgerald, W. F., Alkylmercury species in the equatorial Pacific, *Nature,* 347, 457, 1990.

23. Vandal, G. M., Mason, R. P., and Fitzgerald, W. F., The cycling of volatile mercury in temperate lakes, *Water Air Soil Pollut.,* 56, 791, 1991.

24. Bloom, N. S., Influence of analytical conditions on the observed "reactive mercury" concentrations in natural freshwaters, In: *Mercury Pollution: Integration and Synthesis,* Watras, C. J. and Huckabee, J. W., Eds., Lewis Publishers, Chelsea, MI, 1994, chap. VI.2.

25. Mantoura, R. F. C., Dickson, A., and Riley, J. P., The complexation of metals with humic materials in natural waters, *Est. Coastal Mar. Sci.,* 6, 387, 1978.

26. Dyrssen, D. and Wedborg, M., The sulphur-mercury(II) system in natural waters, *Water Air Soil Pollut.,* 56, 507, 1991.

27. Frimmel, F. H., Immerz, A., and Niedermann, H., Complexation capacities of humic substances isolated from freshwater with respect to copper(II), mercury(II), and iron(II,III), In: *Complexation of Trace Metals in Natural Waters,* Kramer, C. J. M. and Duinker, J. C., Eds., Martinus Nihoff/Dr. W. Junk, The Hague, 1984, 329.

28. Zepp, R. G., Baughman, G. L., Wolfe, N. L., and Cline, D. M., Methylmercuric complexes in aquatic systems, *Environ. Lett.,* 6, 117, 1974.

29. Moser, H. C. and Voigt, A. F., Dismutation of the mercurous dimer in dilute solutions, *J. Am. Chem. Soc.,* 79, 1837, 1957.

30. Sunda, W. G. and Hanson, P. J., Chemical speciation of copper in river water. Effect of total copper, pH, carbonate and dissolved organic matter, in *Chemical Modeling in Aqueous Systems,* ACS Symposium Series, Vol. 93, Jenne, E. A., Ed., American Chemical Society, Washington, D.C., 1979, 147.

31. Cutter, G. A. and Krahforst, C. F., Sulfide in surface waters of the western Atlantic Ocean, *Geophys. Res. Lett.,* 15, 1393, 1988.

32. Bruland, K. W., Personal communication, 1992.

33. Powell, H. K. J. and Florence, T. M., Effect of mercury(II) on metal complex labilities determined by direct-current anodic stripping voltammetry on a thin mercury film electrode, *Anal. Chim. Acta,* 228, 327, 1990.

34. Hering, J. G. and Morel, F. M. M., Slow coordination reactions in seawater, *Geochim. Cosmochim. Acta,* 53, 611, 1989.

35. Hurley, J. P., Watras, C. J., and Bloom, N. S., Distribution and flux of particulate mercury in four stratified seepage lakes, In: *Mercury Pollution: Integration and Synthesis,* Watras, C. J. and Huckabee, J. W., Eds., Lewis Publishers, Chelsea, MI, 1994, chap. I.6.

36. Luther, G. W., III, Giblin, A. E., and Varsolona, R., Polarographic analysis of sulfur species in marine porewaters, *Limnol. Oceanogr.,* 30, 727, 1985.

37. Williamson, M. A. and Rimstidt, J. D., Correlation between structure and thermodynamic properties of aqueous sulfur species, *Geochim. Cosmochim. Acta,* 56, 3867, 1992.

38. Gardner, L. R., Organic versus inorganic trace metal complexes in sulfidic marine waters— some speculative calculations based on available stability constants, *Geochim. Cosmochim. Acta,* 38, 1297, 1974.

39. Weiner, J. G., Fitzgerald, W. F., Watras, C. J., and Rada, R. G., Partitioning and bioavailability of mercury in an experimentally acidified Wisconsin lake, *Environ. Toxicol. Chem.,* 9, 909, 1990.

40. Campbell, P. G. C., Lewis, A. G., Chapman, P. M., Crowder, A. A., Fletcher, W. K., Imber, B., Luoma, S. N., Stokes, P. M., and Winfrey, M., Biologically Available Metals in Sediments, Report to National Research Council of Canada, 1988.

41. Hurley, J. P., Krabbenhoft, D. P., Babiarz, C. L., and Andren, A. W., Cycling of mercury across the sediment/water interface in seepage lakes, In: *Environmental Chemistry of Lakes and Reservoirs,* ACS Symposium Series, Baker, L. A., Ed., American Chemical Society, Washington, D.C., 1994, 425–449.

42. Bienvenue, E., Boudou, A., Desmazes, J. P., Gavach, C., Georgescauld, D., Sandeaux, J., Sandeaux, R., and Seta, P., Transport of mercury compounds across bimolecular lipid membranes: effect of lipid composition, pH and chloride concentration, *Chem.-Biol. Interactions,* 48, 91, 1984.

43. Gutknecht, J., Inorganic mercury (Hg^{2+}) transport though lipid bilayer membranes, *Memb. Biol.,* 61, 61, 1981.

44. Faust, B. C., The octanol/water distribution coefficients of methylmercuric species: the role of aqueous-phase chemical speciation, *Environ. Toxicol. Chem.,* 11, 1373, 1992.

45. Robinson, J. B. and Tuovinen, O. H., Mechanisms of microbial resistance and detoxification of mercury and organomercury compounds: physiological, biochemical, and genetic analyses, *Microb. Rev.*, 48, 95, 1984.

46. Sunda, W. G., Barber, R. T., and Huntsman, S. A., Phytoplankton growth in nutrient rich seawater: importance of copper-manganese cellular interactions, *J. Mar. Res.*, 39, 567, 1981.

47. Harrison, G. J. and Morel, F. M. M., Antagonism between cadmium and iron in the marine diatom *Thalassiosira weisflogii*, *J. Phycol.*, 19, 495, 1983.

48. Hudson, R. J. M. and Morel, F. M. M., Trace metal transport by marine microorganisms: implications of metal coordination kinetics, *Deep-Sea Res.*, 40, 129, 1993.

49. Rabenstein, D. L., The aqueous solution chemistry of methylmercury and its complexes, *Acc. Chem. Res.*, 11, 100, 1978.

50. Gensemer, R. W., The effects of pH and aluminum on the growth of the acidophilic diatom *Asterionella ralfsii* var. *americana*, *Limnol. Oceanogr.*, 36, 123, 1991.

51. Stary, J., Havlik, B., Kratzer, K., Prasilova, J., and Hanusova, J., Cumulation of zinc, cadmium and mercury on the alga *Scenedesmus obliquus*, *Acta Hydrochim. Hydrobiol.*, 11, 401, 1983.

52. Morel, F. M. M., Hudson, R. J. M., and Price, N. M., Limitation of productivity by trace metals in the sea, *Limnol. Oceanogr.*, 36, 1742, 1991.

53. Mierle, G., Kinetics of phosphate transport by *Synechococcus leopoliensis* (cyanophyta): evidence for diffusion limitation of phosphate uptake, *J. Phycol.*, 21, 177, 1985.

54. Sunda, W. G. and Huntsman, S. A., Feedback interactions between zinc and phytoplankton in seawater, *Limnol. Oceanogr.*, 37, 25, 1992.

55. Jackson, G. A. and Morgan, J. J., Trace metal-chelator interactions and phytoplankton growth in seawater media: theoretical analysis and comparison with reported observations, *Limnol. Oceanogr.*, 23, 268, 1978.

56. Margerum, D. W., Cayley, G. R., Weatherburn, D. C., and Pagenkopf, G. K., Kinetics and mechanism of complex formation and ligand exchange, In: *Coordination Chemistry*, ACS Monograph Series, Vol. 174, Martell, A., Ed., American Chemical Society, Washington, D.C., 1978, 1.

57. Lazier, J. R. N. and Mann, K. H., Turbulence and the diffusive layers around small organisms, *Deep-Sea Res.*, 36, 1721, 1989.

58. Iverfeldt, A. and Lindqvist, O., The transfer of mercury at the air/water interface, In: *Gas Transfer at Water Surfaces*, Brutsaert, W. and Jirka, G. H., Eds., D. Reidel, Dordrecht, 1984.

59. Alberts, J. J., Schindler, J. E., Miller, R. W., and Nutter, D. E., Jr., Elemental mercury evolution mediated by humic acid, *Science*, 184, 895, 1974.

60. Lindquist, O. and Iverfeldt, A., Mercury in the Swedish environment: transformation and deposition processes, *Water Air Soil Pollut.*, 55, 49, 1991.

61. Weber, J. H., Reisinger, K., and Stoeppler, M., Methylation of mercury(II) by fulvic acid, *Environ. Technol. Lett.*, 6, 203, 1985.

62. Lee, Y.-H., Hultberg, H., and Andersson, I., Catalytic effect of various metal ions on the methylation of mercury in the presence of humic substances, *Water Air Soil Pollut.*, 25, 391, 1985.

63. Watras, C. J. et al. Sources and fates of mercury and methylmercury in Wisconsin lakes, In: *Mercury Pollution: Integration and Synthesis*, Watras, C. J. and Huckabee, J. W., Eds., Lewis Publishers, Chelsea, MI, 1994, chap. I.12.

64. Press, W. H., Flannery, B. P., Teukolsky, S. A., and Vetterling, W. T., *Numerical Recipes in C: The Art of Scientific Computing*, Cambridge University Press, Cambridge, 1988, 305.

65. Smith, R. M. and Martell, A. E., *Critical Stability Constants*, Vol. 4: Inorganic Ligands, Plenum Press, New York, 1976.

66. Lee, Y.-H. and Hultberg, H., Methylmercury in some Swedish surface waters, *Environ. Toxicol. Chem.*, 9, 833, 1990.
67. Davies, A., An assessment of the basis of the mercury tolerance in *Dunaliella tertiolecta*, *J. Mar. Biol. Ass. U.K.*, 56, 39, 1976.
68. Watras, C. J. and Bloom, N. S., The vertical distribution of mercury species in Wisconsin lakes: accumulation in plankton layers, In: *Mercury Pollution: Integration and Synthesis*, Watras, C. J. and Huckabee, J. W., Eds., Lewis Publishers, Chelsea, MI, 1994, chap. I.11.
69. Gill, G. A. and Bruland, K. W., Mercury Speciation and Cycling in a Seasonally Anoxic Freshwater System: Davis Creek Reservoir, Report, Electric Power Research Institute, Palo Alto, CA, 1991.
70. Morel, F. M. M. and Gschwend, P. M., The role of colloids in the partitioning of solutes in natural waters, In: *Aquatic Surface Chemistry*, Stumm, W., Ed., Wiley-Interscience, New York, 1987, 405.
71. Cabaniss, S. E. and Shuman, M. S., Copper binding by dissolved organic matter. I. Suwanee Rilver fulvic acid equilibria, *Geochim. Cosmochim. Acta*, 52, 185, 1988.
72. Eppley, R. W., Temperature and phytoplankton growth in the sea, *Fish. Bull.*, 70, 1063, 1972.
73. Mason, R. P. and Morel, F. M. M., The accumulation of mercury and methylmercury by microorganisms, *Eos: AGU 1992 Fall Meeting Supplement*, 1992, 303.
74. Bartschat, B. M., Cabaniss, S. E., and Morel, F. M. M., Oligoelectrolyte model for cation binding by humic substances, *Environ. Sci. Technol.*, 26, 284, 1992.
75. Hurley, J. P., Watras, C. J., and Bloom, N. S., Mercury cycling in a northern Wisconsin seepage lake: the role of particulate matter in vertical transport, *Water Air Soil Pollut.*, 56, 543, 1991.
76. Dzombak, D. A. and Morel, F. M. M., *Surface Complexation Modeling: Hydrous Ferric Oxide*, Wiley-Interscience, New York, 1990.
77. Fitzgerald, W. F., Mason, R. P., Vandal, G. M., and Dulac, F., Air-water cycling of mercury in lakes, In: *Mercury Pollution: Integration and Synthesis*, Watras, C. J. and Huckabee, J. W., Eds., Lewis Publishers, Chelsea, MI, 1994, chap. II.3.
78. Travnik, L. J., Availability of dissolved organic carbon for planktonic bacteria in oligotrophic lakes of differing humic content, *Microb. Ecol.*, 16, 1988.
79. Skogerboe, R. K. and Wison, S. A., Reduction of ionic species by fulvic acid, *Anal. Chem.*, 53, 228, 1981.
80. Gilmour, C. G. and Henry, E. A., Mercury methylation in aquatic systems affected by acid deposition, *Environ. Pollut.*, 71, 131, 1991.
81. Belliveau, B. H. and Trevors, J. T., Mercury resistance and detoxification in bacteria, *Appl. Organometal. Chem.*, 3, 283, 1989.
82. Winfrey, M. R. and Rislove, L. J., Mercury methylation and demethylation in Little Rock, Pallatte, Vandercook, Crystal and Russett Lakes, In: *Mercury in Temperate Lakes: 1990 Annual Report*, Electric Power Research Institute, Palo Alto, CA, 1991, chap. 7.
83. Spry, D. J. and Weiner, J. G., Metal bioavailability and toxicity to fish in low-alkalinity lakes: a critical review, *Environ. Pollut.*, 71, 243, 1991.
84. Weiner, J. G., Powell, D. E., and Rada, R. G. Mercury accumulation by fish in Wisconsin seepage lakes: relation to lake chemistry and acidification, In: *Mercury in Temperate Lakes: 1990 Annual Report*, Electric Power Research Institute, 1991, chap. 6.
85. Brakke, D. F., Landers, D. H., and Eilers, J. M., Chemical and physical characteristics of lakes in the northeastern United States, *Environ. Sci. Technol.*, 22, 155, 1988.
86. Stumm, W. and Morgan, J. J., *Aquatic Chemistry: An Introduction Emphasizing Chemical Equilibria in Natural Waters*, Wiley-Interscience, New York, 1981, chap. 4.
87. Munson, R. K. and Gherini, S. A., Influence of organic acids on the pH and ANC of Adirondack lakes, *Water Resources Res.*, 29, 891, 1993.

88. Liss, P. S., Processes of gas exchange across an air-water interface, *Deep-Sea Res.,* 20, 221, 1973.

89. Tetra Tech, Methodology for Evaluation of Multiple Power Plant Cooling System Effects. Vol. II: Technical Basis for Computations, Report to Electric Power Research Institute, Palo Alto, CA, 1979.

90. Norstrom, R. J., McKinnon, A. E., and DeFreitas, A. S. W., A bioenergetics-based model for pollutant bioaccumulation by fish. Simulation of PCB and methylmercury residue levels in Ottawa River yellow perch *(Perca flavescens), J. Fish. Res. Board Can.,* 33, 248, 1976.

91. Watras, C. J. and Bloom, N. S., Mercury and methylmercury in individual zooplankton: implications for bioaccumulation, *Limnol. Oceanogr.,* 37, 1313, 1992.

92. Xun, L., Campbell, N. E. R., and Rudd, J. W. M., Measurements of specific rates of net methylmercury production in the water column and surface sediments of acidified and circumneutral lakes, *Can. J. Fish. Aq. Sci.,* 44, 750, 1987.

93. Hultberg, H., Iverfeldt, A., and Lee, Y.-H., Methylmercury input/output and accumulation in forested catchments and critical loads for lakes in southwestern Sweden, In: *Mercury Pollution: Integration and Synthesis,* Watras, C. J. and Huckabee, J. W., Eds., Lewis Publishers, Chelsea, MI, 1994, chap. III.3.

94. Rodgers, D. W. and Beamish, F. W. H., Water quality modifies uptake of waterborne methylmercury by rainbow trout *(Salmo gairdneri), Can. J. Fish. Aquat. Sci.,* 40, 824, 1983.

95. Boudou, A., Georgescauld, D., and Desmazes, J. P., Ecotoxicological role of the membrane barriers in transport and bioaccumulation of mercury compounds, In: *Aquatic Toxicology,* Nriagu, J. O., Ed., John Wiley & Sons, New York, 1983, 117.

96. Strathman, R. R., Estimating the organic carbon content of phytoplankton from cell volume or plasma volume, *Limnol. Oceanogr.,* 12, 411, 1967.

97. Sanemasa, I., The solubility of elemental mercury vapor in water, *Bull. Chem. Soc. Jpn.,* 48, 1795, 1975.

SECTION VI
Analytical Methodology and Chemistry

Importance of New Specific Analytical Procedures in Determining Organic Mercury Species Produced by Microorganism Cultures

Franco Baldi and Marco Filippelli

CONTENTS

ABSTRACT: The microbial transformation of Hg is examined and Hg alkylation is studied under laboratory conditions by accurate and reliable new methods of mercury speciation in bacterial cultures. The unreliability of conventional methods to detect methylmercury in microbial cultures is discussed and consequently two new methods are developed: (1) the enzymatic determination of methylmercury based on its specific transformation to methane by whole cells of *Pseudomonas putida* strain FB-1, and (2) a method based on determination of the molecular "fingerprint" of volatile methylmercury hydride derivatized from methylmercury chloride by sodium borohydride by purge and trap gas chromatography in line with Fourier transform infrared spectroscopy (PT/GC/FTIR). Methylmercury chloride degradation was observed in axenic cultures of *Desulfovibrio desulfuricans,* which is described as a good Hg-methylator. The main species detected from methylmercury transformation during a 15-day experiment was dimethylmercury together with traces of methane and ionic mercury. Additions of $HgCl_2$ to *D. desulfuricans* cultures led to HgS formation, to Hg(0) volatilization which was determined in the headspace of low-sulfate cultures, and to biosynthesis of methylmercury. The aerobic heterotrophic *Pseudomonas putida* strain KT 2440, which is known to produce large amounts of vitamin B_{12} in PS4 medium, also

1–56670–066–3/94/$0.00+$.50
© 1994 Lewis Publishers

produced high concentrations of methylmercury under laboratory conditions. Moreover, it was observed by PT/GC/FTIR that methylcobalamine itself reacted with $HgCl_2$ in the presence of $NaBH_4$ immediately producing volatile and stable dimethylmercury. No traces of monomethylmercury were ever detected in this reaction. The possible production of monomethylmercury and/or dimethylmercury by bacteria is discussed.

I. INTRODUCTION

Chemical forms of Hg in the environment arise from industrial wastes and natural sources, and are transformed by microbial activity. Microorganisms play an important role in maintaining a Hg species at a steady-state equilibrium. The microbial transformation pathways in the environment and under laboratory conditions are

1. Enzymatic reduction of inorganic mercury to elemental Hg by narrow-spectrum Hg-resistant strains[1,2]
2. Enzymatic cleavage of the C-Hg bond in organomercurials by broad-spectrum Hg-resistant bacteria[1,2]
3. Precipitation of ionic Hg as mercury sulfide by sulfate-reducing bacteria or other H_2S producers[3-6]
4. Methylation of inorganic mercury by methylcobalamine[7,8] or other vitamin B_{12}-like compounds.[9]

Other Hg biotransformations, such as the oxidation of elemental mercury or biological formation of dimethylmercury, have not yet been sufficiently investigated.

Current methods for determination of methylmercury and dimethylmercury in microbial cultures and complex environmental matrices are sometimes not very accurate and/or reliable. Methylmercury analysis in bacterial matrices is carried out by a variant of Westöö's extraction method,[10] with the final determination of methylmercury by gas chromatography and electron capture detector (GC/ECD). Radioisotope techniques based on the indirect determination of toluene-extractable ^{203}Hg as methylmercury are also widely used, especially in in situ experiments.[11] Atomic absorption spectrophotometry has been used to detect total Hg in toluene layers from microbial cultures.[12] More specific and accurate techniques such as gas chromatography in line with mass spectrometry (GC-MS) is also used for dimethylmercury determination. Recently, other techniques have been used to determine methylmercury or other organomercurials in microbial cultures during an intercalibration exercise such as the methylmercury distillation technique[13] and organomercurial ethylation with tetraethylborate followed by cryogenic gas chromatography with cold vapor atomic fluorescence detection.[13,14]

The aim of this chapter is to demonstrate that nonspecific methods can lead to an overstimulation of misinterpretation of MeHg in microbial matrices, and to an incorrect measure of other Hg species formed in microbial cultures. To reduce misinterpretation in Hg speciation, two new methods for specifically detecting organomercurials in bacterial cultures were recently developed: an enzymatic method[15] based on the specific conversion of methylmercury (no other methylmetals are enzymatically converted) to methane by organomercurial lyase. Other organomercurials such as ethylmercury and phenylmercury can also be detected. A second method is based on the determination of the "fingerprint" of volatile methylmercury hydride (CH_3HgH) obtained from the derivatization of CH_3HgCl with $NaBH_4$ in purge and trap gas chromatography in line with Fourier transform infrared spectroscopy (PT/GC/FTIR).[16] This technique can also be used in untreated samples to speciate dimethylmercury from monomethylmercury in headspace analysis during chemical and microbial transformation of Hg(II).

II. MATERIALS AND METHODS

A. CULTURES OF MICROORGANISMS

Heterotrophic aerobic bacteria were isolated from freshwater flowing through cinnabar deposits on Monte Amiata (Southern Tuscany) as previously reported.[17,18] A collection of 150 strains of narrow- and broad-spectrum Hg-resistant Gram positive and Gram negative bacteria were stored as axenic cultures in cryovials with 25% glycerol at $-80°$ C and constantly controlled for viability. These bacteria were refreshed in nutrient broth (Difco) and transferred to Nelson's medium for Hg transforming tests. Nelson's medium (NeM), suitable for Hg toxicity tests because of its low content of sulfydryl groups,[19] contains (per liter): 5 g casamino acid (Difco), 2 g D-glucose, 1 g yeast extract (Difco), and 0.1 g $MgSO_4$ amended with 10 μg/ml $HgCl_2$.

A broad-spectrum Hg-resistant strain of *Pseudomonas putida*, strain FB-1, used for the enzymatic detection of organomercurials, was grown routinely in NeM medium and an overnight culture was centrifuged, rinsed twice with 0.9% NaCl solution, and resuspended in double-strength NeM. Then 1 ml of dense cell suspension > 1 mg cells/mL dry wt.) was mixed with 1 mL of extracted sample in thiosulfate aqueous solution for enzymatic determination of methylmercury.[15]

Pseudomonas putida strain KT 2440, a vitamin B_{12} producer, kindly supplied by Dr. B. Cameron, was used to study inorganic mercury methylation under laboratory conditions. This strain was grown on medium PS4, which is suitable for industrial vitamin B_{12} production,[20] containing per liter: 5 g D-glucose, 5.8 g glutamic acid, 10 g NZ case, 10 g betaine, 0.045 g 5,6-dimethylbenzimidazole, 1.5 g $MgSO_4 \times 7\ H_2O$, 3 g $(NH_4)_2HPO_4$, 0.02 g $MnSO_4$, 0.03 g $ZnSO_4 \times 7H_2O$, 0.03 g $FeSO_4 \times 7H_2O$, 0.005 g $MoO_3Na \times 2H_2O$, 0.12 g $CoCl_2 \times 6H_2O$, and 0.9 g KCl spiked with 2.5 μg of $HgCl_2$. The 3-day-old culture was analyzed for methylmercury.

A strain of *Desulfovibrio desulfuricans* which methylates inorganic mercury by a vitamin B_{12}-like compound was kindly supplied by Prof. R. Bartha. This strain was grown anaerobically in 50 mL serum vials in two different media: Postgate's medium C (MC), having a high concentration of sulfate, and Postgate's medium D (MD) with traces of sulfate, as reported by Berman et al.[9] Two sets of samples (with low and high sulfate contents) were spiked, respectively, with 100 μg/mL of $HgCl_2$ and 100 μg/mL of CH_3HgCl (in ethanol solution). The analysis for Hg species transformation was performed in the liquid culture and in the headspace.

B. METHYLMERCURY DETERMINATION IN CULTURES

Desulfovibrio desulfuricans was grown in 50 mL of MD and MC media amended with 100 μg/mL of $HgCl_2$ and incubated at $28°$ C. After 2 days of incubation, for methylmercury analysis the whole samples were treated with 10 mL of 3 M NaBr in H_2SO_4 (11% plus 2 mL of $CuSO_4$ solution (0.5 M) and then methylmercury was extracted with 25 mL of toluene in a 125 mL separatory funnel. The toluene layer was water-cleaned by repeated additions of anhydrous sodium sulfate. The methylmercury was extracted again with 5 mL of thiosulfate aqueous solution (0.01 N).[11]

The same extraction was performed in 2 mL samples of *D. desulfuricans* cultures and methylmercury residue determined after 100 μg/mL CH_3HgCl spikes in 15-day experiments. For *Pseudomonas putida* strain KT 2440, the procedure for the extraction of methylmercury from 50 mL of culture was the same procedure as for the sulfate-reducing bacterium.

Extraction of methylmercury from 150 Hg-sensitive and Hg-resistant strains spiked with 1 μg of $HgCl_2$ was performed by simple acidification of the samples with 1 mL of conc. HCl and by N_2 purging of the Hg species with $NaBH_4$ solution (20%).[16]

The analytical determination of methylmercury in thiosulfate solution was performed in the following ways:

1. Gas chromatography with electron capture detection after reextraction in toluene and copper sulfate, as reported by Filippelli.[12]
2. Graphite furnace and atomic absorption spectrophotometry (GF-AAS) in thiosulfate solution.[12]
3. The aqueous thiosulfate solution was mixed with a dense culture of a broad-spectrum Hg-resistant *P. putida* strain FB-1 and after 4 h of incubation methylmercury was determined from the derivative methane produced enzymatically in the headspace of 2 mL micro-reaction vessels capped with mininert valves for gas sampling (Supelco). Operating conditions were as described in Baldi and Filippelli.[15]
4. Methylmercury was determined by its derivatization to methylmercury hydride with 0.5 mL added to a 5 mL of sample with purge and trap gas chromatography in line with a Fourier transform infrared spectroscopy (Nicolet model 20 SXB). Operating conditions were as described in Filippelli et al.[16]

C. DIMETHYLMERCURY DETERMINATION

Sets of sulfate-reducing bacteria cultures were grown overnight and 100 μg/mL of CH_3HgCl as Hg in ethanol (95%) was spiked after 24 h into cultures of *D. desulfuricans*. The cultures were incubated from 1 to 15 days at 28° C. Dimethylmercury was determined by PT/GC/FTIR on different days in untreated cultures of *D. desulfuricans* by heating 5 mL of liquid sample to 80° C for 5 min. Operating conditions were the same as described for methylmercury hydride.[16] Dimethylmercury determination was linear up to 200 μg and the coefficient of variation was 5.3%.

Determination of dimethylmercury by the methylcobalamine (Sigma) reaction with $HgCl_2$ was also performed by PT/GC/FTIR—0.05 mL of a 4% solution of the reducing agent $NaBH_4$ and 1 mL of 0.05 M $CuSO_4$ solution were separately added to 5 mL reacting solution to induce dimethylmercury production under N_2 stripping for 5 min.

D. ELEMENTAL MERCURY DETERMINATION

Elemental mercury above aerobic and anaerobic cultures of different strains was determined by sampling 1 mL of the headspace with a 1 mL syringe for gas sampling and injecting it into the cell of an atomic absorption spectrophotometer (Perkin Elmer, model 300S).[17,18] The standard curve was obtained by injecting at room temperature different headspace volumes (from 0.1 to 1 mL) of vapor from metallic mercury sealed in a mininert-capped vial (Supelco). The temperature was always recorded for each set of analyses in order to compute the concentration of elemental mercury in the headspace. The detection limit was 0.2 ng/mL. The coefficient of variation of five replicate analyses was 6.1%.

E. DETERMINATION OF IONIC MERCURY

Ionic Hg was determined in each culture of *D. desulfuricans* spiked with 100 μg/mL of CH_3HgCl at different times. A 1-ml sample was collected with a syringe and mixed with 1 mL of toluene. The sample was vortexed for 1 min and this operation was repeated twice in order to completely extract the methylmercury chloride. The aqueous residue (without bacteria) was mineralized with 1 mL of concentrated HCl and was heated at 60° C for 1 h. The sample was made up to 10 mL with double-distilled water. The concentration of Hg was determined by chemical atomization of ionic mercury with 10% $SnCl_2$ solution with cold vapor atomic absorption spectrophotometry (CV/AAS). Accurate determination was made by serial addition of $HgCl_2$ to the same sample. The coefficient of variation of five replicate analyses was 5.3%.

F. VOLATILE HYDROCARBON DETERMINATION

Volatile hydrocarbons were determined from methylmercury and other organometal microbial degradation as well as enzymatic transformation. Methane was determined in the headspace of *D. desulfuricans* cultures spiked with methylmercury chloride. Hydrocarbons were detected in the headspace by sampling 0.25 mL and injecting it into a gas chromatograph (Hewlett Packard model 5890A) equipped with a flame ionization detector (FID). A 30 m × 0.53 mm GS-Q gas-solid open tubular megabore column (J & W Scientific) was operated isothermally at 70° C with a helium flow rate of 30 mL/min or at programmed temperatures from 70 to 150° C for benzene enzymatically derivatized by phenylmercury degradation. The injection port was held at 200° C and the detector at 250° C. The integrated peaks were recorded by an integrator (Hewlett Packard 3396 A).

III. RESULTS AND DISCUSSION

After 4 years of research in the framework of the European Community Program "Origin and Fate of Methylmercury", mercury speciation in axenic cultures from private bacteria collections and Hg-polluted environments was carried out with both new and conventional methodologies. In the beginning, most of the work was carried out on the enzymatic mercurial transformation: ionic mercury reduction to elemental mercury and cleavage of the C-Hg bond of organomercurials. Afterwards the research was focused on microbial biosynthesis of methylmercury, and consequently new methodologies for organomercurial detection in bacterial cultures were developed because of the lack of reproducibility of conventional methodologies.

A. ENZYMATIC TRANSFORMATION OF MERCURIALS

The two types of Hg biotransformations [enzymatic reduction of Hg(II) to Hg(0) and the cleavage of C-Hg bond in CH_3HgCl to form CH_4 and Hg(0)] gave accurate results based on reproducibility of whole cells and cell-free extract assays vs. gaseous Hg(0) from a metallic mercury droplet under standard conditions of pressure and temperature. The Hg(II) was rapidly reduced to elemental Hg(0) by aerobic heterotrophic narrow-spectrum Hg-resistant bacteria isolated from naturally polluted areas around cinnabar mines, as reported in several investigations.[17,18] The Hg(0) was determined in the headspace by cold vapor atomic absorption spectrophotometry. This methodology was a good alternative to that based on ^{203}Hg radioisotope disappearance from bacterial cultures.[22] Studies of methylmercury or other organomercurial degradation by broad-spectrum Hg-resistant strains can be investigated by determining Hg(0) or the respective hydrocarbons in the headspace of bacterial cultures.[21]

B. UNRELIABILITY OF CERTAIN ORGANOMERCURIAL DETERMINATIONS

The methylmercury synthesis by microorganisms was difficult to investigate. The determination of CH_3Hg^+ in bacterial cultures was not always accurate with the older conventional methods. The main problem was the specificity and reliability of the method rather than sensitivity for the determination of methylmercury in bacterial cultures. The common methods used for detecting methylmercury in our axenic cultures gave conflicting results and internal standards could not be used until an intercalibration exercise for methylmercury determination in axenic cultures of strain ND-132 sulfate-reducing bacteria was performed.[13]

Several screenings for methylmercury formation in pure cultures of heterotrophic aerobic bacteria isolated from rivers and creeks flowing through a cinnabar deposits of Southern Tuscany gave controversial results: (1) in about 15 out of 106 strains, a Hg-soluble species was detected in the benzene layer and in one strain, 2A-4, 2.8% of Hg was soluble in the solvent (Figure 1).[2] A more accurate method for Hg speciation in bacterial cultures was constituted by a purge and trap on gas chromatography on line with a mercurometer (PT/GC/AAS),[15] and it did not show any methylmercury formed after acidification and

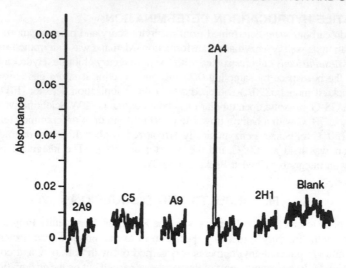

Figure 1 Determination of total Hg by GF/AAS in thiosulfate aqueous solution after solvent extraction in several strains (2A-9, C-5, A-9, 2A-4, and 2H-1) of heterotrophic aerobic bacteria isolated from a cinnabar mine area. Strain 2A-4 produces 2.8% of solvent-soluble mercury. (From Baldi, F., Olson, G.J., and Brinckman, F.E., *Geomicrobiol. J.*, 5, 1, 1985. Copyright 1985, American Chemical Society. With permission.)

$NaBH_4$ addition in 2-day-old aerobic heterotrophic cultures (Figure 2)—only Hg(0) was formed from Hg-resistant strains.

This uncertainty of finding methylmercury questioned if the methodology to detect this organomercurial was suitable for microbial cultures. Whether total Hg in the thiosulfate aqueous solution from the toluene layer means the presence of methylmercury, that it was easy to find it in many cultures. Whether the use of different techniques as GC/ECD, the methylmercury concentration resulted much lower or undetectable. Conventional methods for methylmercury analysis in bacteria matrices produced the following problems:

1. Acid hydrolysis with HCl 1 *N*, used for methylmercury extraction from upper organism tissue, does not work on bacterial cultures, especially for sulfate-reducing bacteria where a large amount of H_2S is developed. Instead, H_2SO_4 11% in 3 *M* NaBr solution should be used rapidly in the extraction mixture because it degrades methylmercury;
2. The retention time of a methylmercury chloride peak detected by GC-ECD at low level was almost indistinguishable among other peaks by other halogenated compounds in solution;
3. The absorbance of total Hg determined after solvent extraction and then determined by GF-AAS always gave an overestimation of methylmercury—in the intercalibration exercise almost 15 times higher.[21]

Additions of light-weight molecules from biological origin such as vitamin B_{12}, Coenzyme M, iodomethane, and methyl mercaptan yielded high concentrations of Hg (2 to 3% of total Hg added) in the toluene layer. A series of sequential extractions of samples in copper sulfate-toluene gave a significant decrease of extractable Hg in the toluene layer. This finding was also confirmed by the radioisotope method with $^{203}Hg(II)$ additions.[22]

The unreliability of methylmercury determination in bacterial cultures with conventional methods finally resulted in development of two new specific methods:

1. An enzymatic method in which methylmercury was specifically converted to methane. This volatile hydrocarbon was formed stoichiometrically from methylmercury by the

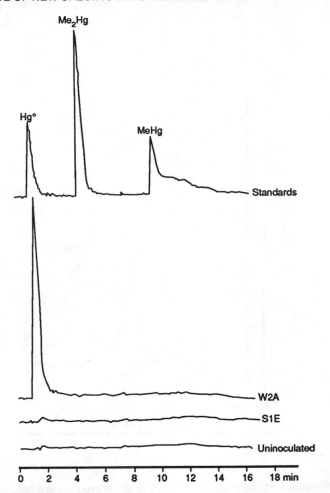

Figure 2   Speciation of Hg by adding NaBH$_4$ and stripping with N$_2$ in axenic cultures of Hg-sensitive (strain S1-E) and Hg-resistant (strain W2-A) heterotrophic aerobic bacteria isolated from a cinnabar mine area by purge and trap gas chromatography in line with a mercurometer (PT/GC/AAS).

    enzyme organomercurial lyase. Other methyl-metals were not transformed enzymatically to methane.[16] Instead, other organomercurials degraded to their respective hydrocarbons such as ethane from ethylmercury and benzene from phenylmercury (Figure 3). Methyl-mercury additions at ng levels onto tuna tissue were recovered and were significantly transformed to methane by a whole cell of *Pseudomonas putida* strain FB 1.[13]

2.   A method based on the detection of a methylmercury "finger-print" by volatilizing methylmercury chloride to methylmercury hydride with NaBH$_4$ (Figure 4), purging the sample with N$_2$, condensing it in a trap ($-120°$ C), then separating it by GC, and determining the concentration by Fourier transform infrared spectroscopy (PT/GC/FTIR).[14]

C.   ORGANOMERCURIAL BIOSYNTHESIS

A study on Hg biosynthesis with these two new methods was carried out in different cultures of aerobic and anaerobic bacteria. A strain of sulfate-reducing bacterium, *Desulvovibrio desulfuricans,* was to demonstrate production of methylmercury[9] through a vitamin B$_{12}$-like

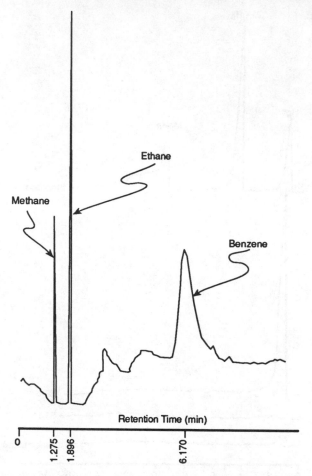

Figure 3 Enzymatic determination of 1 µg/mL methylmercury, ethylmercury, and phenylmercury by their derivatization to the respective hydrocarbons (methane, ethane, and
benzene) with whole cell cultures of *Pseudomonas putida* strain FB-1 in a 2 mL micro-
reaction vessel (headspace 0.7 mL). Hydrocarbons were determined by GC/FID.

pathway, but on the contrary, it was shown to degrade CH_3HgCl spiked in this culture.
Surprisingly, the methylmercury was not toxic at such high concentrations (100 µg/mL
CH_3HgCl) as it was for aerobic bacteria[1] and it was transformed in days by *D. desulfuricans*,
in axenic culture, to dimethylmercury (Figure 5); traces of methane (4 µg/mL), and inorganic
mercury (0.4 µg/ml) were also determined, respectively, in 8.59 ± 0.6 mL headspace and in
1.0 mL of liquid culture. Disappearance of methylmercury (up to 40%) from liquid culture
was also observed in 15 days.

The chemical transformation of methylmercury by H_2S to dimethylmercury and β-HgS precipitation has already been reported by Rowland et al.[24] One year later, Craig and Bartlett[25]
confirmed the chemical volatilization of CH_3HgCl by H_2S in aqueous solution. This conversion occurred through the organic Hg intermediate $(CH_3Hg)_2S$ to produce the stable volatile form, $(CH_3)_2Hg$, but methane and inorganic Hg were not detected. This is the first report
of methylmercury transformation in an axenic culture of *D. desulfuricans*, which was demonstrated to synthesize methylmercury from Hg(II) by a cobalamine pathway,[9] and conversely, to degrade this organomercurial to dimethylmercury, methane, and inorganic

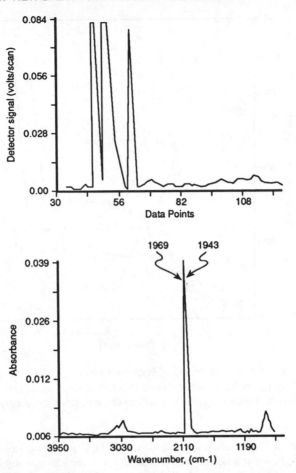

Figure 4 Determination of PT/GC/FTIR of 5 μg of methylmercury chloride derivatized to methylmercury hydride by NaBH₄. Chromatogram of the methylmercury hydride retention time (data points) and its spectrum (below) with two close absorbance peaks at 1968 and 1943 cm⁻¹ of the Hg-H bond.

mercury. No elemental Hg(0) was observed after 15 days of incubation. Instead, addition of HgCl₂ to *D. Desulfuricans* culture in medium without sulfate (medium D) synthesized 6.23 ± 0.85 ng/ml of methylmercury and up to 40 ng/ml of Hg(0) after days incubation at 28° C (Figure 6). The high concentration of Hg° above the saturation level (27.81 ng/mL) was due mostly to the high pressure built up in the headspace, so the volume (1 mL) injected in the AAS cell was at higher pressure than the Hg° above the metallic Hg for standard volumes.

In the presence of a high sulfate content all the Hg(II) was transformed into HgS, which visually precipitated on the bottom of the serum bottle of *D. desulfuricans*. A discrepancy occurs when the methylmercury formed through cobalamine-like pathway was likely transformed to dimethylmercury by H₂S from the dissimilatory reduction of sulfate.

Under aerobic conditions the heterotrophic *Pseudomonas putida* strain KT 2440, a vitamin B₁₂ producer,[18] was able to methylate inorganic mercury aerobically in a 3-day incubation at 28° C in an orbital shaker at 180 r.p.m. Methylmercury was also determined (110 ng/mL) in 3-day-old culture by previous HCl acidification and toluene extraction.

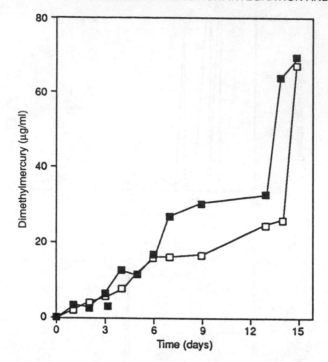

Figure 5 Formation of dimethylmercury from methylmercury transformation (initial concentration 100 µg/mL) in 50 ml liquid culture of *Desulfovibrio desulfuricans* during a 15-day period in Postgate's media with high sulfate (■) and low sulfate concentrations (□).

Experiments on methylcobalamine reaction with $HgCl_2$ were performed with PT/GC/FTIR. Surprisingly, the reaction at once produced 100% dimethymercury in accordance to a ratio of 2:1 if a reducing agent such as $NaBH_4$ was added (Figure 7). The important result was that monomethylmercury was never formed in solution. Without adding any compounds the reaction did not take place for at least 5 h, suggesting that Hg(II) was steady bound by methylcobalamine. Moreover, once dimethylmercury was formed it was stable (e.g., no CH_4 was detected). On the contrary, if HCl (1 *N*) or H_2SO_4 (11%), the solutions most used to extract methylmercury from cultures, the 50 µ*M* of dimethylmercury was transformed completely to monomethylmercury in 4 h. Research is in progress to understand this mechanism of double Hg methylation.

IV. CONCLUSIONS

New methodologies are important for the more specific and direct detection of organomercurial production by microorganisms. The two methods used here, the enzymatic and the PT/GC/FTIR techniques, are very promising for accurate data on mercury speciation. The PT/GC/FTIR method will be especially useful for distinguishing new Hg forms, without chemical pretreatment of samples.

The results achieved in 4 years of research did not solve the question of the origin of the methylmercury produced by microorganisms. On the contrary, they opened new questions which will be investigated in the near future. For example, *D. desulfuricans,* a H_2S producer, transforms most of the methylmercury to dimethylmercury and can also methylate small amounts of inorganic Hg (at ng level) under low sulfate concentrations. It seems possible

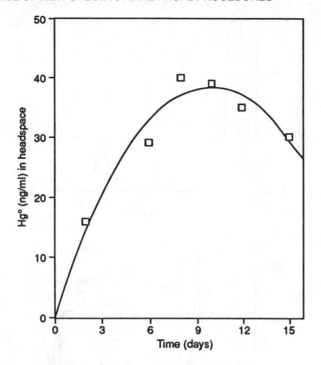

Figure 6 Formation of elemental Hg(0) in the headspace of 50 mL culture of *Desulfovibrio desulfuricans* during a 15-day exposure to 100 µg/mL HgCl$_2$ in Postgate's sulfate medium incubated at 28° C.

that methylation and demethylation can occur in the same microorganisms or that monomethylmercury is just dimethylmercury which is transformed by acid extraction to monomethylmercury.

The finding of dimethylmercury production by an anaerobic culture of *D. desulfuricans* spiked with monomethylmercury, and from the reaction between methylcobalamine and HgCl$_2$ strongly suggests that the measurement of dimethylmercury in the environment will be useful to understand and to study under different points of view several environmental processes such as:

1. Controversial results about abiotic vs. biotic methylation of Hg;
2. Methylmercury increases at low pH in the northern lake ecosystems;
3. Methylmercury concentrations at the floc zone of sediments (probably due to dimethylmercury migration toward the sediment surface);
4. Microbial toxicity of dimethylmercury vs. monomethylmercury;
5. When and how dimethylmercury degrades to monomethylmercury or to other products.

Methods to detect dimethylmercury in environmental matrices will be useful to study the biogeochemical cycle of Hg, where finally dimethylmercury could be permanently included, at least as a biological source of methylmercury, in the environment.

ACKNOWLEDGMENT

This work was financially supported by EEC grant EV4V–136–I and EV4V–0125–I.

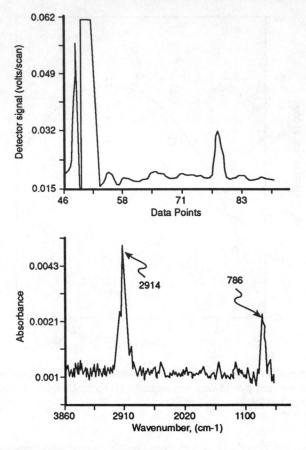

Figure 7 Determination by PT/GC/FTIR of dimethylmercury retention time (upper) and spectrum (below) from 50 μM of methylcobalamine and 25 μM of $HgCl_2$ (ratio 1:2) as Hg after 5 min N_2 stripping at 80° C and addition of 0.5 mL of $NaBH_4$ 4% solution.

REFERENCES

1. Robinson, J.B. and Tuovinen, O.H., Mechanisms of microbial resistance and detoxification of organomercury compounds: physiological, biochemical, and genetic analyses, *Microbiol. Rev.,* 48, 95, 1984.
2. Silver, S. and Misra, T.K., Plasmid-mediated heavy metal resistances, *Ann. Rev. Microbiol.,* 42, 717, 1988.
3. Fagerström, T. and Jernelöv, A., Formation of methylmercury from pure mercury sulphide in aerobic organic sediment, *Water Res.,* 5, 121, 1971.
4. Yamada, M. and Tonomura, K., Microbial methylation of mercury in hydrogen sulphide-evolving environments, *J. Ferm. Technol.,* 50, 901, 1972.
5. Campeau, G. and Bartha, R., Effects of sea salt anions on the formation and stability of methylmercury, *Bull. Environ. Contam. Toxicol.,* 31, 486, 1983.
6. Campeau, G. and Bartha, R., Methylation and demethylation of mercury under controlled redox, pH, and salinity conditions, *Appl. Environ. Microbiol.,* 50, 498, 1985.
7. Wood, J.M., Kennedy, F.S., and Rosen, C.G., Synthesis of methylmercury compounds by extracts of methanogenic bacterium, *Nature,* 220, 173, 1968.
8. Pan-Hu, H.S. and Imura, N., Physiological role of mercury methylation in *Clostridium cochlearium* T-2C, *Bull. Environ. Contam. Toxicol.,* 29, 290, 1982.

9. Berman, M., Chase, T., and Bartha, R., Carbon flow in mercury biomethylation by *Desulfovibrio desulfuricans, Appl. Environ. Microbiol.,* 56, 298, 1990.

10. Westöö, G., Determination of methylmercury compounds in foodstuffs, *Acta Chem. Scand.,* 20, 2131, 1966.

11. Furutani, A. and Rudd, J.W.M., Measurement of mercury methylation in lake water and sediment samples, *Appl. Environ. Microbiol.,* 40, 770, 1980.

12. Filippelli, M., Determination of trace amounts of organic and inorganic mercury in biological materials by graphite furnace atomic absorption spectrophotometry and organic mercury speciation by gas chromatography, *Anal. Chem.,* 59, 116, 1986.

13. Padberg, S., Iverfeldt, A., Lee, Y.-H., Baldi, F., Filippelli, M., May, K., and Stoeppler, M., Determination of low-level methylmercury concentrations in water and complex matrices by different analytical methods—a methodological intercomparison, *Mercury Pollution: Integration and Synthesis,* Watras, C. J. and Huckabee, J. W., Lewis Publishers, Chelsea, MI, 1994, Chap. VI.5.

14. Bloom, N., Determination of picogram levels of methylmercury by aqueous phase ethylation, followed by cryogenic gas chromatography with cold vapour atomic fluorescence detection, *Can. J. Fish. Aquat. Sci.,* 46, 1131, 1989.

15. Baldi, F. and Filippelli, M., New method for detecting methylmercury by its enzymatic conversion to methane, *Environ. Sci. Technol.,* 25, 302, 1991.

16. Filippelli, M., Baldi, F., Brinckman, F.E., and Olson, J.G., Methylmercury determination as volatile methylmercury hydride by purge and trap gas-chromatography in line with Fourier Transform Infrared Spectroscopy (PT/GC/FTIR), *Environ. Sci. Technol.,* 26, 1457, 1992.

17. Baldi, F., Olson, G.J., and Brinckman, F.E., Mercury transformation by heterotrophic bacteria isolated from cinnabar and other metal sulfide deposits in Italy, *Geomicrobiol, J.,* 5, 1, 1985.

18. Baldi, F., Filippelli, M., and Olson, G.J., Biotransformation of mercury by bacteria isolated from a river collecting cinnabar mine waters, *Microbiol. Ecol.,* 17, 263, 1989.

19. Nelson, J.D., Blair, W.R., Brinckman, F.E., Colwell, R.R., and Iverson, W.P., Biodegradation of phenylmercury acetate by mercury-resistant bacteria, *Appl. Microbiol.,* 26, 321, 1973.

20. Cameron, B., Briggs, K., Pridmore, S., Brefort, G., and Crouzet, J., Cloning and analysis of genes involved in coenzyme B_{12} biosynthesis in *Pseudomonas denitrificans, J. Bacteriol.,* 171, 547, 1989.

21. Baldi, F., Cozzani, E., and Filippelli, M., Gas chromatography/Fourier transform infrared spectroscopy for determining traces of methane from biodegradation of methylmercury, *Environ. Sci. Technol.,* 22, 836, 1988.

22. Summers, A.O. and Silver, S., Mercury resistance in a plasmid-bearing strain of *Escherichia coli, J. Bacteriol.,* 112, 1228, 1972.

23. Bernhard, M., Unpublished data, 1992.

24. Rowland, J.R., Davies, M.J., and Grasso, P., Volatilization of methylmercury chloride by hydrogen sulphide, *Nature,* 265, 718, 1977.

25. Craig, P.J. and Bartlett, P.D., The role of hydrogen sulphide in environmental transport of mercury, *Nature,* 275, 635, 1978.

26. Imura, N., Sukegawa, E., Pan, S.K., Nagao, K., Kim, J.Y., Kwan, T., and Ukita, T., Chemical methylation of inorganic mercury with methylcobalamine, a vitamin B_{12} analog, *Science,* 172, 1248, 1972.

27. Jensen, S. and Jernelöv, Å., Biological methylation of mercury in aquatic organisms, *Nature,* 223, 753, 1969.

28. Bertilisson, L. and Neujahr, H.Y., Methylation of mercury compounds by methylcobalamine, *Biochemistry,* 10, 2805, 1971.

29. Andren, A.W. and Harriss, R.C., Methylmercury in estuarine sediments, *Nature,* 245, 256, 1973.

Influence of Analytical Conditions on the Observed "Reactive Mercury" Concentrations in Natural Freshwaters

Nicolas S. Bloom

CONTENTS

ABSTRACT: Researchers estimate the level of "reactive mercury" (Hg_r, variously called $Hg(II^a)$, "easily reducible", "acid-labile", or "ionic"), by $SnCl_2$ reduction of acidified samples. Although the effect of analytical conditions on the recovery of Hg_r has been investigated for open-ocean waters, the method has thus far been applied to freshwater samples without critical examination. In this study, the influence of pH, time of acid storage, and filtering were investigated on natural waters ranging in pH from 4.7 to 8.1, and in DOC from 3 to 27 mg/L^{-1}. Ultraclean sampling and handling techniques were used throughout to allow investigation at the ambient levels of 1 to 3 ng·L^{-1} total Hg. Acidification affected the observed concentration of Hg_r in a manner that varied complexly with length of time after acidification. Generally, acidification to pH 2 resulted in Hg_r concentrations which were immediately and temporarily higher than those observed in fresh, unacidified samples. As the sample was stored at low pH, however, the fraction of Hg_r decreased with time, until after several weeks the concentration was much lower than in the fresh samples. In addition, the observed [Hg_r] vs. time curve was not a smooth one, and varied amongst different water types. The fraction of mercury found as Hg_r, as well as the residual mercury released on subsequent bubblings of the same water sample, increased dramatically with increasing pH above 8. Filtering experiments carried out both before and after acidification indicate that mercury is lost to the particulate phase after acidification—probably due to the coagulation of dissolved humic substances.

541

I. INTRODUCTION

The source of mercury to the aquatic environment appears largely to be in the form of inorganic Hg compounds, either from direct atmospheric deposition or terrestrial runoff.[1-4] Once in the aquatic system, an interactive web of biologically mediated reactions can convert inorganic mercury to methylmercury,[5-6] elemental mercury,[1,7] or particulate-bound Hg.[8] The production of methylmercury is of particular importance, due to its extreme toxicity and biomagnification in the aquatic foodchain.[9] The reactions forming other Hg species are equally important, however, as they may serve as direct competitors to methylation, and so ultimately control the amount of methylmercury produced in an aquatic system.[1] In any attempt to model the ecological pathways resulting in the production of methylmercury in the environment, it is necessary to have an understanding of the chemical nature of the substrate for these reactions. Generally, it is believed that this substrate is represented by the so-called "reactive mercury" (Hg_r, variously called "easily reducible", "acid labile", "ionic", "reactive", or "$Hg(II^a)$").[1,3,10]

Hg_r, determined by Sn(II) reduction of an acidified water sample, has a long history of accurate low-level quantification, especially by chemical oceanographers.[11-14] This is due both to the low blanks attainable by the relatively reagent-free procedure, and the early observation that, for clean seawater, Hg_r measured in stored acidified samples is virtually the same as total Hg.[11,13] Because of its demonstrated accuracy and utility in open-ocean mercury studies, an operative speciation scheme soon became accepted where Hg_r, taken to include all labile ionic Hg forms, was determined by immediate Sn(II) reduction of an acidified sample, while "total Hg" was measured in the same way on a sample stored at <pH 2 for several weeks.[11,13,14]

When the procedure was applied to more organic and particulate-rich waters, researchers observed that no longer did the acid storage method yield results similar to the total mercury (Hg_t, after chemical oxidation).[13] Thus, over the last decade, the operatively defined fractionation has subtly evolved for freshwater and estuarine studies, such that today the acid-storage Sn(II)-reducible Hg is called Hg_r, while Hg_t is determined by strong chemical oxidation followed by Sn(II) reduction.[13,15,19] The effect of pH and acid storage has not been investigated for natural freshwaters, and the supposition that the Hg_r measured after acidification represents the biogeochemically reactive "species" has not been critically evaluated.

Recent studies have indicated that operative definitions in general, and Hg_r in particular, may be geochemically uninterpretable. Early, it was demonstrated that for estuarine samples the acid-storage Hg_r is not similar to either the Hg_r determined by reduction of fresh samples, or to Hg_t.[13] Additionally, studies in Wisconsin over the past two years have indicated that of the species commonly measured (Hg_t, CH_3Hg, and Hg_r), the Hg_r data are much more highly variable (Figure 1) and that the variability appears virtually incoherent—showing no common trends between lakes with seasonality, covariance with other Hg species, etc.

The method also is of questionable theoretical value: while purportedly attempting to estimate a biogeochemically reactive form, it yet subjects the sample to long exposure at pH conditions never experienced in nature. Dramatically lowering the pH has the potential to either overestimate the biologically available Hg by releasing chelated or adsorbed Hg which would otherwise be unavailable,[20] or to underestimate the parameter due to coagulation of and coprecipitation with dissolved humic matter.[21] With these concerns in mind, this detailed study of the effects induced by the analytical procedure on the measured value for Hg_r was undertaken.

II. EXPERIMENTAL

A. SAMPLE COLLECTION

Water samples were collected by hand-dipping into acid-cleaned Teflon® bottles, while employing ultraclean trace-metal-free handling techniques.[15,22] Unpreserved samples were

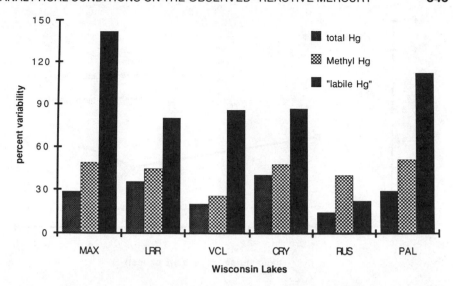

Figure 1 Relative standard deviation of all epilimnetic volume-weighted mean water concentrations for six Wisconsin lakes during 1990 to 1991. Variability represents both field and analytical components, for n = 6 to 21 dates (2 to 6 water samples per date mean) per lake. Lakes not identified in the text are Little Rock Reference Basin (LRR), Vandercook (VCL), Crystal (CRY), and Pallette (PAL).

packed in dark boxes at 0 to 4° C, and shipped by overnight courier to the laboratory in Seattle. Processing and analysis for labile species occurred within 48 h of collection. Storage experiments have shown that little change in speciation occurs over a period of several weeks in unpreserved samples when stored cool, in the dark, and in Teflon® bottles (Figure 2). Many measurements of mercury adsorbed to the walls of Teflon® bottles following the experiments have shown negligible (<5%) losses over the time course of the experiments.

Samples were collected primarily from the study lakes of the Mercury in Temperate Lakes (MTL) project in the Northern Highland Lakes District of Wisconsin. These lakes have been described in detail in many other publications,[8,23,24] and so only the mean salient features are given in Table 1. The lakes, which are remotely located and relatively unperturbed by local anthropogenic activity, were chosen for the MTL study because they span a wide range of in situ pH and DOC. One of the lakes, Little Rock Treatment Basin, is being artificially acidified with H_2SO_4 at a rate of 0.5 pH unit every 2 years as part of an acid precipitation study. In addition to these lakes, some data were collected for Onondaga Lake, a mercury-contaminated urban lake located in the center of Syracuse, NY.[25]

B. ANALYTICAL METHODS

Mercury was generally determined on specifically pretreated water samples by reduction with $SnCl_2$, purging with nitrogen for 20 min, dual stage amalgamation, and cold vapor atomic fluorescence spectrometry (CVAFS).[13,14] The detection limit of the method is approximately 0.02 ng/L as Hg. Total mercury was determined on 100 mL aliquots of water which were first cold oxidized for at least 1 h with 0.002 N BrCl, prereduced with NH_2OH to eliminate interfering free halogens, and then fully reduced with the addition of 0.5 mL of 20% $SnCl_2$ in 20% HCl.

For the determination of acid-labile Hg_r, the samples were first acidified with 10 mL low-mercury HCl per liter of sample (pH 0.93). After storage in acid for the given amount of time, an aliquot of 100 mL was measured into the reduction vessel by weighing, and analyzed

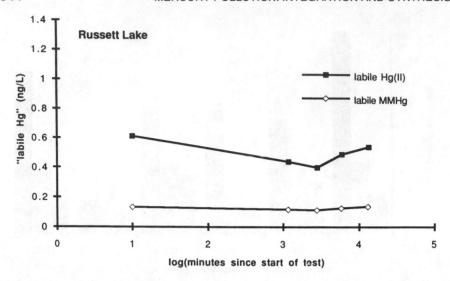

Figure 2 Storage experiment for labile mercury speciation in unpreserved, unfiltered Russett Lake water. The scale spans to approximately 10 days. Samples were kept in the dark, at 0 to 4° C, in sealed Teflon® bottles. Analysis was by aqueous phase ethylation.

Table 1 **General geochemical parameters of the lakes sampled for this study.**

Lake	Location	Description	pH	(mg/L)	Hgt
				**Mean epilimnetic chemistry 1991**	
Little Rock	N. Wisconsin	Acidification study, seepage, dimictic	4.7	3	1.1
Max	N. Wisconsin	Seepage	5.2	3	1.3
Mary	N. Wisconsin	Drainage, meromictic	5.9	24	5
Onondaga	Syracuse, N.Y.	Hg polluted, eutrophic, drainage, calcite saturated	8.1	na	11.2
Russett	N. Wisconsin	Humic rich, seepage	6.1	7	2.1

as above after the addition of 0.5 mL of 20% $SnCl_2$ in 20% HCl. Samples were always thoroughly shaken prior to aliquoting, due to the potential for organic matter precipitation upon acidification. Freshwater samples which were analyzed for Hg_r immediately were aliquoted as above prior to acidification, and 0.5 mL of 20% $SnCl_2$ in 20% HCl as added immediately before purging (final solution pH 2.3).

For ambient pH reduction, a solution of 10% $SnCl_2$ and 30% sodium citrate was prepared. When a 100 mL volume of lake water was reduced with 100 μL of this buffered Sn(II) solution, the pH of the water was never observed to change by more than 0.2 units. For pH dependency experiments, predetermined aliquots of 1% KOH, 1% HCl, and 10% sodium citrate were added to 100 mL aliquots of lake water, and allowed to equilibrate for 2 to 4 h. The samples were then reduced with 100 μL of the Sn(II)-citrate solution, as above. The pH was always determined on the bubbler solution after analysis, to avoid potential contamination.

Table 2   **Typical analytical variability for Hg species reported.**

Species	pH	Age	Reactant	Hg concentration, ng/L			
				Mean	SD	N	RSD (%)
Hg_r (UF)	0.9	2 h	Sn(II)	0.535	0.033	4	6
Hg_r (UF)	0.9	10 d	Sn(II)	0.183	0.012	3	7
Hg_r (UF)	6.1	Fresh	Sn(II)	0.262	0.047	6	18
Hg_r (UF)	5.0	Fresh	NaBEt$_4$	0.610	0.056	8	9
Hg_t (UF)	0.9	2 h	BrCl/Sn(II)	1.99	0.19	5	10
Hg_t (0.2 µm)	0.9	2 h	BrCl/Sn(II)	1.03	0.14	4	14

Note: All samples were from Russett Lake, June and July 1991. UF samples were unfiltered.

Labile Hg using aqueous phase ethylation[26] was determined on 100 mL aliquots of lake water buffered to pH 5.0 with acetate. The samples were then ethylated for 25 min with sodium tetraethyl borate (10 mg/L in solution), purged onto Carbotrap™, and thermally desorbed into an isothermal (95° C) GC column packed with 1.6 m of 15% OV-3 on chromasorb WAW-DMSC.[27] The separated species (methylethyl-Hg for labile CH$_3$Hg and diethyl-Hg for labile Hg(II)) were then quantified by CVAFS, following pyrolytic decomposition to Hg^0. The detection limit for the method is approximately 0.005 ng/L as Hg. The relative precision of this and the other methods used in this study is approximately ±5 to 15% (Table 2).

Sample filtration was accomplished using disposable cellulose nitrate vacuum filtering units of the type designed for tissue culture sterilization. The unit funnel is first filled with 10% HCl, and allowed to drip under gravity for 2 to 6 h. The remainder of the acid is then pulled through by vacuum, and then two successive funnels of ultra-low mercury double-deionized water (DDW) are pulled through as rinses. When prepared in this manner in a clean room just prior to sample filtration, the blank associated with filtering is <0.05 ng/L Hg, using 1% HCl in DDW as the test media. Each filter unit is used only once.

III. RESULTS AND DISCUSSION

A. EFFECT OF ACIDIFIED STORAGE TIME

Presented in Figure 3 are the acid storage time results for acid labile mercury in three very different lakes: Russett (high color and DOC, low suspended matter). Max (very clear, low DOC), and Onondaga (eutrophic, high Hg, saturated with calcite). Storage experiments were carried out over a one-month period. Many of the points represent the mean of replicate analyses, which typically varies by less than 10% analytically (Table 2). Time is presented on a logarithmic scale, with the earliest time period (t = 10 min) representing half-way through the 20-min bubbling period of a sample purged immediately after acidification. Interestingly, all three of the lakes show dramatically different patterns in the Hg$_r$ concentration as a function of time. At this point, it would be pure speculation to explain these results, although the result for Onondaga lake may be due to desorption from or dissolution of the high inorganic (calcite) suspended load. The other patterns may be related to differing kinetics of humic matter coagulation and Hg^{++} displacement by H$^+$. In any case, the important observation is that the value obtained for Hg$_r$ in a given water sample is unpredictably a function of water type and time at which the analysis is made.

B. EFFECT OF pH

The effect of pH on the measured labile concentration of Russett lake is shown in Figure 4. The mercury determinations were made 2 to 4 h following adjustment of the solution pH. A pattern emerges where there is a broad region of low and stable Hg$_r$ values in the weakly

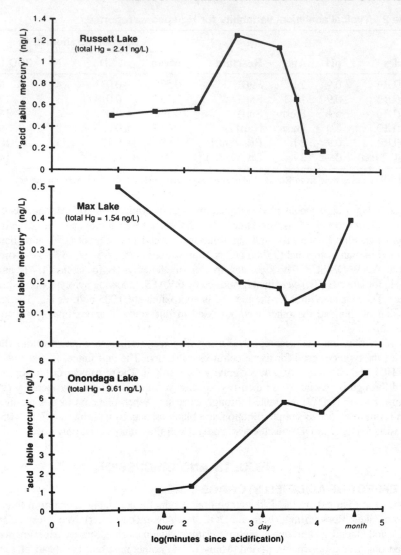

Figure 3 Effect of storage time with 1% HCl added on the recovery of Hg_r by Sn(II) reduction. The logarithmic time scale spans to 37 days.

acidic pH range (3.5 to 6.5), with rapidly increasing levels at either extreme in pH. This pattern has been observed, although for less complete data sets, for other lakes of this study. This result implies that the Hg_r results obtained by acid storage would be too high, compared to results collected at ambient pH. Interestingly, as shall be seen shortly, the opposite is often the case. These results do, however, indicate a region of stability near the ambient pH range of natural waters, where at least more reproducible measurements might be made.

A further examination of the pH effect was made by looking at the rate of release of Hg^0 from Sn(II)-reduced 0.45-μm-filtered natural waters spiked with Hg(II) to a level of 11 ng/L Hg_t, as a function of pH (Figure 5). When mercury in DDW is reduced with Sn(II), and purged out, it is removed at an exponential rate, largely a function of the flushing rate and scrubbing efficiency of the bubbler. Under the conditions used here, >98% of the Hg is removed within 10 min. In all cases where spiked natural waters were investigated, the rate

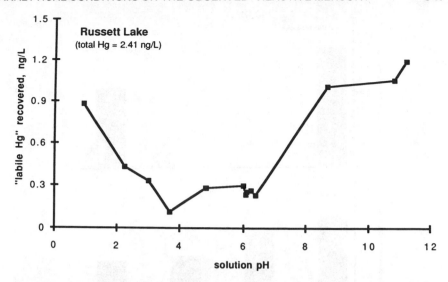

Figure 4 Effect of reduction pH on the recovery of Hg_r by Sn(II) in Russett Lake water. Water was equilibrated at buffered pH values for 2 to 4 h prior to reduction and analysis.

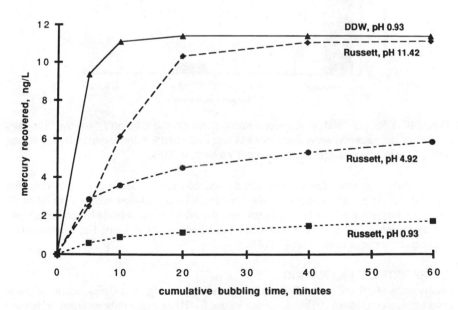

Figure 5 Effect of solution pH on the release rate of Hg^0 from Sn(II)-reduced spiked Russett Lake water. All samples were spiked to contain approximately 11.8 ng/L total Hg (ambient Hg plus spike) several hours before the beginning of the experiment.

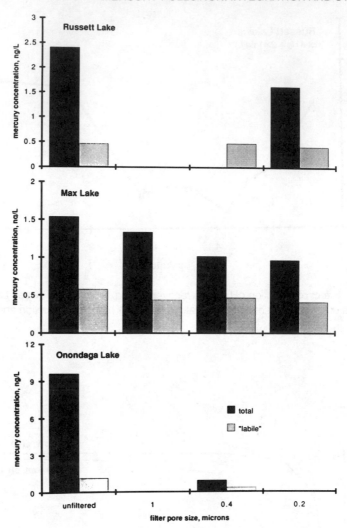

Figure 6 Effect of filtration at several spore sizes on the total and "reactive" mercury content of unpreserved water from these lakes. Each result is the mean of 2 to 4 points, which differed from their respective mean by less than 10%.

of conversion was much lower. However, the reaction rate is clearly pH dependent, with essentially complete release possible from strongly alkaline solution within a period of 40 min or so. For water acidified to the levels typically used with acid labile Hg procedure (pH 1 to 2), the rate of Hg release is slow and almost linear—meaning that Hg_r is arbitrarily a function of the purge time, as well as pH.

C. EFFECT OF FILTRATION

Observations of others have indicated that the majority of Hg_r in a sample seems to come from the dissolved phase.[19] That is, similar values for Hg_r are generally recorded on filtered and unfiltered samples, even if the Hg_t differs significantly. This has been generally verified for these lakes. For example, in Figure 6 is shown the effect of filtration at three different

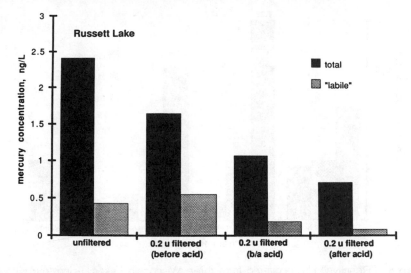

Figure 7 Effect of filtration (0.2 μm) both before and after acidification (pH 1.8) on the distribution of total and "reactive" Hg in Russett Lake water.

pore sizes on the Hg_t and Hg_r in several lakes. No meaningful differences in Hg_r were observed at any level of filtration, even as Hg_t became smaller in correspondence to smaller filter pore size. This trend has been generally true in our studies of remote lakes, with the lakes having the highest coloration (dissolved humic matter) showing the least effect of filtration. These results suggest that there is typically little exchangeable adsorbed Hg on lake water particulates. Instead, mercury seems to be strongly bound to (or inside) particulates or associated with dissolved complexes. An exception so far has been Onondaga Lake, which is particulate rich with precipitated calcite. In this lake, filtration reduces both the Hg_t and Hg_r substantially, which is consistent with the notion that labile Hg is adsorbed to the calcite, and released upon dissolution of that phase in acid.

Interestingly, when a lake water sample is filtered *after* acidification, a larger fraction is found in the particulate phase than when the same sample is filtered at ambient pH. This phenomenon is illustrated in Figure 7 for Russett lake, the most stained of the lakes reported here. The effect is considerably stronger in lakes with high levels of dissolved colored matter.

From the first two sets of bars, it is apparent that filtration of the fresh sample at 0.2 μm significantly reduces the Hg_t, while having little impact on the Hg_r. The slight increase in Hg_r has actually been observed on other occasions,[19] and is not a result of filtration contamination. There may be an inhibitory effect on the volatilization of Hg^0 by suspended matter,[28] which is removed by filtration. In the third set of bars is shown the result of taking 0.2 μm filtered fresh lake water (i.e., the water represented by the second set of bars), acidifying to pH 1.8 with HCl, and then filtering again. In the last set of bars, fresh unfiltered water was acidified to pH 1.8, and then filtered to 0.2 μm.

Clearly, acidification results in the formation of more suspended matter, which coprecipitates some of the Hg as well. The additional suspended matter likely comes from the coagulation of humic acids,[21] as the solutions become decidedly less colored after acidification and filtration. These surprising results support casual observations in the past of significant quantities of brown settled matter in filtered and acidified field samples. It also points out the importance of thoroughly homogenizing acidified samples prior to aliquoting and analysis, as a significant amount of the trace metals may reside on the precipitated organic matter.

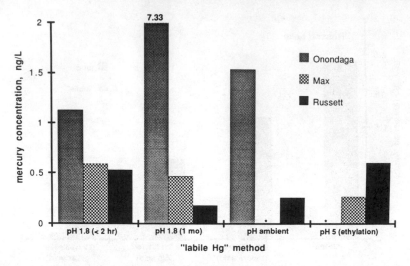

Figure 8 "Reactive" Hg recovered from three Wisconsin Lakes, using four different methods (* means not analyzed).

D. DIFFERENT MEASURES OF Hg, APPLIED TO DIFFERENT LAKES

To help obtain an initial insight to the eventual development of a more meaningful measure of the labile Hg(II) pool in aquatic samples, several chemical methods were applied to the lakes of this study (Figure 8). In addition to acid storage and reduction with Sn(II), the samples were acidified and immediately analyzed, reduced with the citrate-buffered Sn(II) at ambient pH, and analyzed by ethylation at pH 5. In general, the methods employing near-ambient pH analysis gave significantly higher results for Hg_r than did the ones which were acidified. In all cases, the lowest results were obtained where the samples were acidified and stored for 10 days prior to analysis. This is contrary to what might be expected from the pH experiments discussed earlier. The waters used, however, were collected in different seasons (July 1991 and April 1992), and so might have very different biogenic chemical constituents. Disappointingly, none of the methods consistently gave results similar to another over all lakes studied.

IV. CONCLUSION

To enable future modeling and understanding of the biogeochemical processes underlying the aquatic mercury cycle, a reliable measure of the biogeochemically reactive form of in-organic mercury is needed. It is assumed, however, that this parameter should encompass the mercuric ion (Hg^{++}), and those complexes of it which are labile at ambient conditions (i.e., chloro- and hydroxo-complexes, weak organic complexes, electrostatically adsorbed species, etc.).

As currently measured, however, the operatively defined parameter, "reactive mercury" seems to be analytically unreliable and biogeochemically uninterpretable. It is analytically unreliable because the ultimate result for a given sample is dependent on arbitrary choices for the analytical conditions such as pH, time of acidification, and purge time. The results obtained when using this method, while reproducible for the same sample analyzed in exactly the same way, give no coherent picture temporally with depth, between lakes, or when correlated with other important parameters such as Hg_t, CH_3Hg, DOC, etc. The results are also biogeochemically uninterpretable, as the process of reducing the pH by 4 to 6 orders of magnitude and storing for long periods has no theoretical or empirical relationship to the sought-after labile mercury speciation existing at ambient conditions.

It is possible to refine the current wet-chemical methods of estimating Hg_r by conducting the reaction on fresh samples and at ambient pH. Several such initial experiments were attempted, and the results presented in this volume. These other methods did show some interesting general differences from the acid-storage method—especially the tendency to yield much higher values for the Hg_r parameter. It is likely that further insights will be gained in further refining some of these methods, and comparing results obtained with them over a range of lakes, seasons, and depths. However, these methods also depend upon arbitrary selection of parameters such as purge time, reduction against reagent used, and its concentration.

A potentially more fruitful course of action, if the problem of detection limits can be overcome, will be the development of an in situ probe capable of responding to the concentration of Hg(II) without perturbing the system. This might be possible with UV-VIS spectrometry of Hg(II) complexes, selective ion electrodes, or biosensor technology. The ultimate test of the utility of any such method, however, will have to be in its ability to generate reliable data which produces a coherent picture of processes expected to be dependent upon the Hg_r concentration.

ACKNOWLEDGMENTS

I would like to thank Steve Claas for obtaining the samples necessary to continue research on this project, even when no other sampling was going on, and to Dr. Lian Liang for performing some of the chemical analyses described. This work was funded by the Electric Power Research Institute, Palo Alto, CA (RP 2020–10).

REFERENCES

1. Fitzgerald, W.F., Mason, R.P., and Vandal, G.M., Atmospheric cycling and air-water exchange of mercury over mid-continental Lacustrine regions, *Water Air Soil Pollut.*, 56:745, 1991.
2. Fitzgerald, W.F., Cycling of mercury between the atmosphere and oceans, in *The Role of Air-Sea Exchange in Geochemical Cycling,* Buat-Menard, P., Ed., D. Reidel, Dordrecht, 1986, 363.
3. Lindqvist, O., Mercury in the Swedish environment: recent research on causes, consequences and corrective methods, *Water Air Soil Pollut.*, 55:1, 1991.
4. Lindqvist, O., Jernelöv, A., Johansson, K., and Rhode, H., Mercury in the Swedish Environment, Global and Local Sources, SNV PM 1816, Swedish Environment Protection Agency, S-171–85 Solna, Sweden, 1984.
5. Ramlal, P.S., Rudd, J.W.M., Furutani, A., and Xun, L., The effect of pH on methyl mercury production and decomposition in lake sediments, *Can. J. Fish. Aquat. Sci.*, 42:685, 1985.
6. Xun, L., Campbell, N.E.R., and Rudd, J.W.M., Measurements of specific rates of methylmercury production in the watercolumn and surface sediments of acidified and circumneutral lakes, *Can. J. Fish. Aquat. Sci.*, 44:750, 1987.
7. Vandal, G.M., Mason, R.P., and Fitzgerald, W.F., Cycling of volatile mercury in temperate lakes, *Water Air Soil Pollut.*, 56:791, 1991.
8. Hurley, J.P., Watras, C.J., and Bloom, N.S., Mercury cycling in a northern Wisconsin seepage lake: the role of particulate matter in vertical transport, *Water Air Soil Pollut.*, 56:543, 1991.
9. Watras, C.J. and Bloom, N.S., Mercury and methylmercury in individual zooplankton: implications for bioaccumulation, *Limnol. Oceanogr.*, 7:1313, 1992.
10. Gherini, S., Personal communication, Lafayette, CA, 1991.

11. Matsunaga, K., Konishi, S., and Nishimura, M., Possible errors caused prior to measurement of mercury in natural waters with special reference to seawater. *Environ. Sci. Technol.*, 13:63, 1979.

12. Dalziel, J.A., and Yeats, P.A., Reactive mercury in the central North Atlantic Ocean, *Mar. Chem.*, 15:357, 1985.

13. Bloom, N.S., and Crecelius, E.A., Determination of mercury in seawater at subnanogram per liter levels, *Mar. Chem.*, 14:59, 1983.

14. Gill, G.A. and Fitzgerald, W.F., Picomolar mercury measurements in seawater and other materials using stannous chloride reduction and two-stage gold amalgamation with gas phase detection, *Mar. Chem.*, 20:227, 1987.

15. Fitzgerald, W.F. and Watras, C.J., Mercury in surficial waters of rural Wisconsin lakes, *Sci. Total Environ.*, 87/88:223, 1989.

16. Nelson, L.A., Mercury in the Thames Estuary, *Environ. Technol. Lett.*, 2:225, 1981.

17. Robertson, D.E., Sklarew, D.S., Olson, K.B., Bloom, N.S., Crecelius, E.A., and Apts, C.W., Measurement of Bioavailable Mercury Species in Freshwater and Sediments, EPRI EA-5197, Electric Power Research Institute, Palo Alto, CA, 1987.

18. Iverfeldt, Å, Mercury in the Norwegian fjord Framvaren, *Mar. Chem.*, 23:441, 1988.

19. Gill, G.A. and Bruland, K.W., Mercury speciation in surface freshwater systems in California and other areas, *Environ. Sci. Technol.*, 24:1392, 1990.

20. Schuster, E., The behaviour of mercury in the soil with special emphasis on complexation and adsorption processes—a review of the literature, *Water Air Soil Pollut.*, 56:667, 1991.

21. Aiken, G.R., McKnight, D.M., Wershaw, R.L., and MacCarthy, P., In: *Humic Substances in Soil, Sediment, and Water: Geochemistry, Isolation, and Characterization,* Aiken, G.R., McKnight, D.M., Wershaw, R.L., and MacCarthy, P., Eds., Wiley-Interscience, New York, 1985, 1–9.

22. Gill, G.A. and Fitzgerald, W.F., Mercury Sampling of open ocean waters at the picomolar level, *Deep Sea Res.*, 32:287, 1985.

23. Watras, C.J. and Frost, T.M., Little Rock Lake: perspectives on an experimental ecosystem approach to seepage lake acidification, *Arch. Environ. Contam. Toxicol.*, 18:157, 1989.

24. Bloom, N.S., Watras, C.J., and Hurley, J.P., Impact of acidification on the methylmercury cycling of remote seepage lakes, *Water Air Soil Pollut.*, 56:477, 1991.

25. Bloom, N.S. and Effler, S.W., Seasonal variability in the mercury speciation of Onondaga Lake (New York), *Water Air Soil Pollut.*, 53:251, 1990.

26. Bloom, N.S., Determination of picogram levels of methylmercury by aqueous phase ethylation, followed by cryogenic gas chromatography with cold vapor atomic fluorescence detection, *Can. J. Fish. Aquat. Sci.*, 46:1131, 1989.

27. Liang, L., Horvat, M., and Bloom, N.S., An improved speciation method for mercury by GC/CVAFS after aqueous phase ethylation and room temperature precollection, *Talanta*, 41:371, 1994.

28. Turner, R.R., Personal communication (Oak Ridge, TN), 1991.

A Passive Sampler for the Monitoring of Gaseous Mercury Amounts in the Atmosphere

Kestutis Kvietkus and Jonas Sakalys

CONTENTS

ABSTRACT: A new passive sampler for the monitoring of gaseous mercury amounts in the atmosphere is described. The accumulation rate and characteristics of mercury vapor on the sampler under different conditions are presented. It is established that the adsorbed mercury vapor amount on the sorbent depends directly on the exposure time, when mercury vapor concentration in the air is constant. This type of passive diffusional sampler may be useful for mercury vapor monitoring in the atmosphere and as a personal mercury dosimeter.

I. INTRODUCTION

At present, active diffusional methods for sampling air pollutants have been well developed by Ferm[1] for NO_2, by Kvietkus et al.[2] for mercury (Hg) vapor, and improved by Xiao et al.[3] for Hg vapor.

For determination of air pollutants at places where electricity is not available, small and light samplers are needed. The passive diffusional method of gaseous mercury sampling is efficient, convenient, and easy to use. The sampler does not require electric power or any equipment. Some data about passive samplers has been presented by Brown et al.,[4] Countant et al.,[5] and Lewis et al.[6] for volatile organic chemicals and by Ferm[7] for NO_2. Granular sorbents and impregnated filters were used to trap the gases. The use of passive samplers for the control of gaseous mercury amounts is very dependent on their unique characteristics. The sorbent presented by Kvietkus et al.[8] has the form of a plate coated with silver which adsorbs mercury vapor that can be detected by way of mercury vapor thermal desorption in a gas flow passing through the measuring device chamber. However, such a sampler has a major drawback: its useful surface is not reliably protected from mechanical damage and moisture.

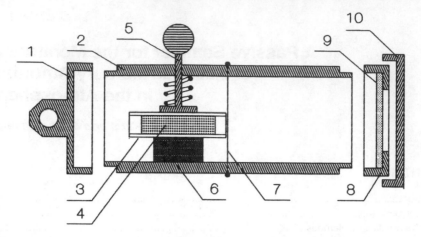

Figure 1   Storage container with passive sampler: (1) plastic cap; (2) storage container; (3) glass tube; (4) silver sorbent; (5 and 6) holders; (7) mark; (8) plastic holder with opening; (9) membrane filter; (10) bottom plastic cap.

II. MATERIALS AND METHODS

We have produced a new sampler which has no such drawback. The sampler is a thin-walled quartz tube, 0.8 cm inner diameter and 4 cm long. The inner surface of the tube is coated with a thin layer (about 30 μm) of silver or gold and its effective surface is 10 cm^2. In this case, a sampler coated with silver is studied. It is important to optimize the sampler sorbent area within the limits of sensitivity of the measuring device and the quantity of collected mercury from the air on the sorbent area during a sampling period. The analysis was performed by an atomic fluorescence analyzer, "Fluoran", produced in the Institute of Physics by Kvietkus et al.,[2] with an absolute detection limit of 0.2 pg (carrier gas—helium).

It is known from a study by Brown et al.[4] that the wind affects molecular diffusion when using a passive sampler. To eliminate this phenomenon, the mercury vapor sampler described here is placed into a plastic tube container (Figure 1). The container is closed at the bottom end by the membrane filter, and the sorbent in it is fixed at a definite place. Under such conditions, different samplers are equally stable during the exposure time. It is convenient to use the storage container together with the sampler at any place where mercury vapor concentration is to be measured. After desorption, the cleaned samples again may be used to collect gaseous mercury from the air. The passive sampler device described above is convenient to store and expose. The diffusional sampler can be recycled about 1000 times with a collection efficiency of 98% by using a thermal desorption technique in an inert gas atmosphere.

III. RESULTS AND DISCUSSION

An experiment was performed to understand the dynamics of mercury accumulation on silver sorbents when mercury vapor concentrations in the air in two rooms were different, i.e., 0.1 μg/m^3 and 0.15 μg/m^3. A small number of samplers (5 or 6) were exposed in each room for a period of time. Exposure time varied from 30 min to 3 h. The results obtained are presented in Figure 2. We may conclude that when the concentration is constant, the quantity of deposited mercury is in linear dependence with the time of exposure. The slope of the curve is in linear dependence with the concentration of mercury vapor in the air. The linear dependence of mercury on exposure time was also observed when the exposure time of the samplers was as long as several days under the conditions described (Figure 3). It should be

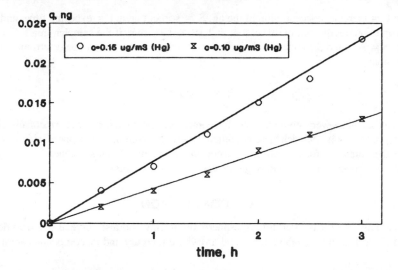

Figure 2 Sampling rate of the mercury concentration by the passive sampler as a function of time in two rooms with different concentrations. Each point represents at least five determinations.

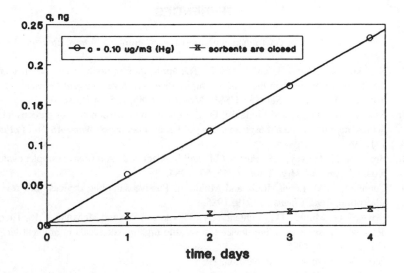

Figure 3 Sampling rate of the mercury concentration by the passive sampler as a function of time inside a room. Membrane filter (see Figure 1, No. 9) had been changed to an aerosol filter.

noted that the dispersion of mercury concentrations collected on different samplers in identical conditions increases after a longer exposure time. It is known from Fick's law and from a study by Kvietkus et al.[8] that the quantity of mercury collected on the sampler from the atmosphere in a definite time may be expressed as follows:

$$q = k \, c \, S \, t \qquad\qquad (1)$$

where c = mercury concentration in the air, S = sorbent area, t = exposure time, and k = the proportion coefficient. In this case k = 1.19×10^{-5} m/s. If the sampler area S = const, then k S = A^{-1} and the concentration of mercury vapor in the air may be determined by the expression:

$$c = A \, q \, / \, t \qquad (2)$$

Clean samplers were stored in a storage container. The container was hermetically closed with a plastic stopper. Under these conditions, mercury from the atmospheric air cannot get inside the container for several days, according to Figure 3. This manner of storing the samples is convenient for use during field measurement conditions.

IV. CONCLUSION

Passive samples may be used for atmosphere monitoring and also serve as personal dosimeters in industrial plants, laboratories, and anywhere mercury and its compounds are used.

ACKNOWLEDGMENTS

The authors are grateful to M. Kulakauskas for valuable technical assistance and to J. Krokys for correction of the manuscript.

REFERENCES

1. Ferm, M., Method for determination of atmospheric ammonia, *Atmos. Environ.*, 13, 1385, 1979.
2. Kvietkus, K., Sakalys, J., and Vebra, E., The application of fluorescence spectroscopy for the study of impurities in the atmosphere, in *Proc. Int. Symp. Urgent Problems of Spectroscopy,* Academy of Sciences USSR, Moscow, 1985, 235 (in Russian).
3. Xiao, Z., Munthe, J., and Lindqvist, O., Sampling and determination of gaseous and particulate mercury in the atmosphere using gold coated denuders. *Water Air Soil Pollut.,* 56, 141, 1991.
4. Brown, R.H., Harvey, R.P., Purnel, C.J., and Sanders, K.J., A diffusive sampler evaluation protocol, *Am. Ind. Hyg. Assoc. J.,* 45, 67, 1984.
5. Countant, R.W., Lewis, R.G., and Mulik, J., Passive sampling devices with reversible adsorption, *Anal. Chem.,* 57, 219, 1985.
6. Lewis, R.G., Mulik, J.D., Countant, R.V., Wooten, G.W., and McMillin, C.R., Thermally desorbable passive sampling device for volatile organic chemicals in ambient air, *Anal. Chem.,* 57, 214, 1985.
7. Ferm, M., Personal communication, 1988.
8. Kvietkus, K., Kulakauskas, M., and Sakalys, J., A passive sorbent characteristic for the monitoring of gaseous mercury amounts in the atmosphere. *Atmos. Phys.,* (Lithuania) 12, 188, 1988.

Experiences on Different Pretreatment Procedures in the Analysis of Low Methylmercury Levels in Environmental Samples by a GC-CVAFS Technique

Ying-Hua Lee, Åke Iverfeldt, and Elsemarie Lord

CONTENTS

I. INTRODUCTION

An increased interest in the dynamic behavior and fate of methylmercury (MeHg) in the biogeochemical mercury cycle has resulted in the development of new analytical tools for sensitive and accurate determinations of low MeHg levels in various environmental samples/materials. A newly developed gas chromatograph (GC) technique,[1,2] combined with aqueous phase ethylation and cold vapor atomic fluorescence spectrometry (CVAFS) detection, provides a subpicogram sensitivity for MeHg. However, the ethylation step in the analytical procedure requires, in general, a pretreatment of the environmental sample in order to isolate/separate MeHg from the sample matrix and to overcome the interference in the ethylation reaction process. The organic complexing forms, especially those containing ligands with an S donor atom (present in natural waters), interfere strongly with the ethylation, as do high chloride concentrations (>100 ppm) and low pH (<3). In this chapter we focus on two pretreatment procedures—extraction into an organic solvent and distillation with a nitrogen flow. We evaluate the accuracy and precision of the procedures using the extraction-GC-CVAFS method and compare the extraction and distillation techniques using the same samples. For the distillation procedure, a simplified model for simulating the process has been developed. The model calculations together with the experimental tests have provided important information on the distillation procedure. A certified reference material, some solid and several aqueous samples—including soil and surface waters as well as soils, litterfall, and biological material—are used in the present study.

1–56670–066–3/94/$0.00+$.50
© 1994 Lewis Publishers

II. EXPERIMENTAL

The extraction procedure using methylene chloride to extract MeHg chloride from the water sample matrix was performed according to the method described by Bloom.[1] An acidic and highly concentrated chloride medium is necessary for the extraction, in order to break down the bonds between MeHg and the complexing agents in the sample water.

The distillation procedure described by Horvat et al.[3,4] was used for water, soil, and litterfall samples. As the environmental water samples were already preserved with 0.5% Suprapure® HCl (9 M), less acid (0.3 mL 9 M H_2SO_4, instead of 0.5 mL for an unpreserved water sample) and 0.2 mL of 20% KCl were added to the 40 mL water sample. The experimental conditions for distillation, such as using HCl instead of H_2SO_4 and different concentrations of acid and chloride, were also examined.

MeHg was determined by GC-CVAFS using either cryogenic GC or isothermal GC after the isolation step. Prior to detection by CVAFS,[1,2] a volatile derivative of MeHg was generated from the sample by aqueous phase ethylation and then separated from other volatile alkymercury compounds.

III. DISTILLATION MODEL

A. DISTILLATION PROCESS

As the partial pressure of MeHgCl in aqueous solution increases with increasing temperature (as shown in Figure 1), the distillation process with nitrogen flow enables the MeHgCl to be purged within a short time (1 to 4 h) from a water phase (together with the water vapor) and to condense as a liquid. An acidic and chloride medium is necessary to strip MeHg from the sample matrix and to form a volatile chloride compound. However, the HCl will also be purged by the nitrogen flow and will condense in the receiving vessel. The concentration of chloride and pH in the distillate is dependent on the original concentrations of the acid and chloride in the distillation vessel, as well as the distillation time and temperature. Since high

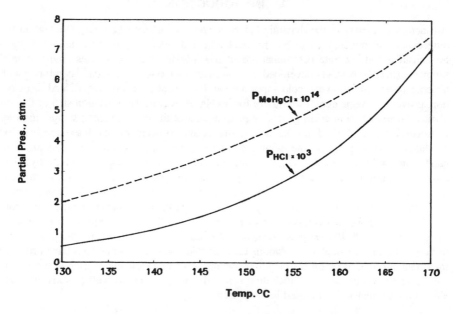

Figure 1   The variation of the partial pressure of MeHgCl in aqueous solution with increasing temperature.

chloride concentration (>100 ppm) and low pH (<3) will interfere with GC-CVAFS analysis,[1] the optimum conditions for distillation should minimize chloride and acid concentrations in the distillate while maximizing the yield of MeHgCl.

Experimental tests are generally used for finding the optimum conditions for an experimental procedure. However, a model calculation of the distillation process is a simplified way to provide information on the products of distillation, such as the pH and Cl$^-$ concentrations, which are important for the MeHg analysis using the aqueous phase ethylation/GC-CVAFS method. The basic consideration of the simplified model calculation is described as follows.

B. MODEL DESCRIPTION

The distillation model is based on physical/chemical principles. We assume that the whole distillation process with nitrogen flow can be divided into several short-period processes. Each process can be considered as a simplified dynamic flow system by assuming that the changes of the water phase volume and the activity coefficients of MeHgCl and HCl in the distillation vessel can be neglected.

For each short process an equation is used to calculate the fraction of the original MeHgCl and HCl left in the distillation vessel after sparging with a given nitrogen gas volume. Equation 1 is based on the law of conservation of mass and is derived from the fact that the decrease in the MeHgCl (or HCl) amount in the aqueous phase in the distillation vessel is equal to the increase in the MeHgCl (or HCl) amount in the receiving vessel. Equation 1 can be written as:

$$C_i/C_0 = \{\exp^{-V_g/(K*V_1(i)*RT)}\}*[C_{i-1}/C_0] \qquad (1)$$
$$= \{\exp^{-H*V_g/V_1(i)}\}*[C_{i-1}/C_0]$$

where C_0 is the original aqueous concentration of MeHgCl (or HCl) in distillation vessel, i is the series number of the short period process, C_i is the concentration after sparging with nitrogen gas volume $i*V_g$, and $V_1(i)$ is the volume of aqueous phase in the distillation vessel during the i period process. The dimensionless Henry's law constant, H = [MeHgCl(g)]/[MeHgCl(aq)] and the empirical distribution constant K = [H$^+$(aq)]*[Cl$^-$(aq)]/[HCl(g)] at various temperatures are estimated by an approach described below. For given values of V_g and $V_1(i)$, the fraction of the MeHgCl (or HCl) amount left in the distillation vessel, C_i/C_0, after i period process is calculated using Equation 1 and the estimated H and K values. The V_g is determined by the nitrogen purge rate (60 ml/min) applied, and V_1 is estimated from the distillation rate of 9 to 10 mL/h.

C. ESTIMATION OF H AND K AT VARIOUS TEMPERATURES

The dimensionless distribution constant H(MeHgCl) has previously been determined[5] using a constant ion medium, but only at a few different ionic strengths (I = 0.7, 1.0, and 1*10^{-4} M) and at a few temperatures (25, 15, and 10° C). The following approach has been used to estimate these constants at various other temperatures. Here, we consider the effects of the ionic strength and temperature on the activity of the MeHgCl in the water phase. The main equation used for estimating constant H(MeHgCl) at various temperatures is the Van't Hoff equation:

$$\ln H(MeHgCl) = -\Delta H^0/RT + constant \qquad (2)$$

where ΔH^0 is the standard enthalpy change for distribution of MeHgCl between the gas phase and the aqueous phase. In a first approximation, we assume that the temperature dependence of enthalpy can be neglected. In this calculation, the values of H(MeHgCl) and the standard

Table 1   **Comparison of the coefficient of variation (CV%) for the analysis of MeHg in various water samples using a cryogenic GC-CVAFS and an isothermal GC-CVAFS method, with and without an extraction procedure.**

Sample	GC method	Extrac. proc.	CV%	n
Stand. sol.	Cryogenic	Without	6.0–14.1	20
Stand. sol.	Isothermal	Without	1.0–6.3	37
Stand. sol. (0.25–4.09 ng/L)	Cryogenic	With	0.8–19	9
Surface water (0.3–0.6 ng/L)	Cryogenic	With	10.5–18.5	6
River water (0.12–0.37 ng/L)	Cryogenic	With	19–40	12
Environ. samp. (0.02–0.06 ng/L)	Isothermal	With	5.4–34.1	8
Environ. samp. (0.03–0.09 ng/L)	Cryogenic	With	1.5–59.6	8
Framvaren (0.5–1.6 ng/L)	Cryogenic	With	6–35	32
Framvaren (0.4–2.0 ng/L)	Isothermal	With	4.0–21	14

enthalpy change at 25° C are taken from previous studies.[5,6] The empirical distribution constant K for HCl is estimated by using the literature values of the equilibrium constant of HCl,[7] the partial pressure of HCl aqueous solutions at various temperatures,[8] and the activity coefficient at various ionic strengths.[9]

IV.   RESULTS AND DISCUSSION

In the present study, we compare the analytical results found for a MeHgCl standard solution, which were determined by the cryogenic GC-CVAFS and the isothermal GC-CVAFS method (see Table 1). We also compare the results from the determination of MeHg in different water samples, using these two different analytical methods in combination with the same extraction procedure (see Table 1 and Figure 2a). In Table 1, the CV% values (i.e., the relative standard deviation = standard deviation/mean value) are based on three to seven analytical values. The n value given in the table is the total number of analytical results used in evaluating the range of CV% for each type of sample, given in the first column of the table (Table 1). Also, the extraction and distillation isolation procedures are compared, using the same environmental water samples as previously discussed and the isothermal GC-CVAFS analytical technique (Figure 2b).

The isothermal GC-CVAFS method is statistically more reliable than the cryogenic GC-CVAFS method, which is also shown in Table 1 and Figure 2a when comparing the CV% of the standard solution. This is true both with and without the extraction procedure applied (Table 1). A decreased reliability in the MeHgCl determination in samples with low MeHg concentration is also found (Figure 2a). Using the isothermal GC-CVAFS method with alternative isolation techniques, a good agreement between duplicate analysis of the MeHg concentration is found (Figure 2b). However, when comparing the results obtained by the distillation and the extraction techniques in determining low level MeHg concentrations (<0.2 ng/L), lower MeHg concentrations were generally found by the extraction technique. Also, the standard deviation of duplicate analysis is most often higher using this latter technique (no such data are presented in this chapter). The main reason for these observations is

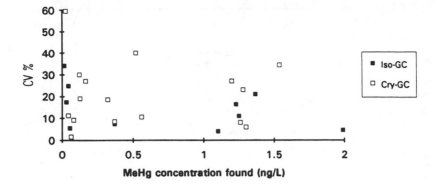

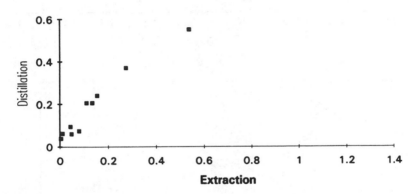

Figure 2   (a) Comparison of the coefficient of variation CV% (= relative standard deviation) for the analyses of the environmental samples, including samples with a low MeHg content and/or containing strong complexing ligands (Framvaren), using the isothermal GC-CVAFS method or the cryogenic GC-CVAFS method in combination with a methylene chloride extraction. (b) Comparison of MeHg concentrations in some environmental samples determined by two different isolation techniques (extraction or distillation) combined with the isothermal GC-CVAFS method.

a low extraction efficiency and a large variation (or spread) of this efficiency at low MeHg concentrations, even if the recovery of the standard addition to these samples is high (>75%). In humic-rich water (DOC >15 mg/L) the extraction technique could not be used to isolate MeHg from the water due to the formation of an emulsion. These latter samples are not included in the Figure 2b.

The simulation of the distillation process is illustrated in Figures 3a and 3b and Figures 4a to 4c. The ratios MeHg/MeHg-orig and HCl/HCl-orig used in Figure 3 and Figure 4 denote the fractions of MeHgCl (or HCl) left in the distillation vessel, which are negatively proportional to the MeHgCl and HCl in the receiving vessel. The concentrations of MeHgCl and HCl depend mainly on the distillation conditions (e.g., temperature, time, and the concentrations of the acid and chloride used in the distillation). After 50 min for a solid sample or 3 h for an aqueous sample and at a temperature greater than 140° C, more than 90% of the MeHg in the distillation vessel is distilled to the receiving vessel. The pH of the distillate decreases with increasing time of distillation and temperature. The model calculation (Figures

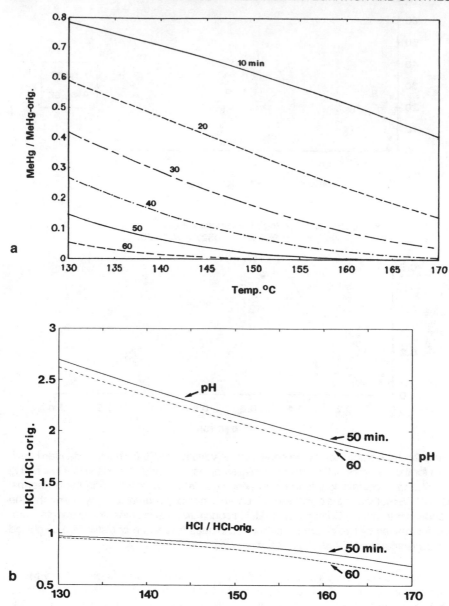

Figure 3 Model calculation of the distillation process for a mixture of a solid sample and a 10 mL solution containing 0.42 M H_2SO_4 and 0.05 M Cl^-, purged with a 60 mL/min nitrogen flow rate. Figure 3a: The temperature dependence of the MeHg/MeHg-orig (fraction of MeHgCl left in the distillation vessel) at various distillation times. Figure 3b: The temperature dependence of the HCl/HCl-orig (the fractions of HCl left in the distillation vessel) and the pH of the distillate at a distillation time of 50 or 60 min.

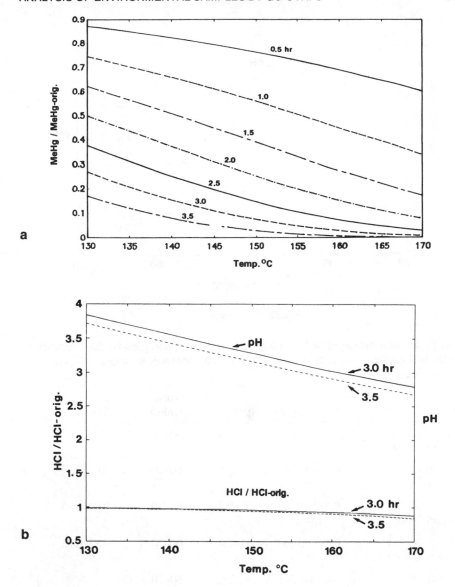

Figure 4   Model calculation of the distillation process for a water samples with a 40 mL original sample volume and using a 60 mL/min nitrogen flow rate. (4a) The temperature dependence of the MeHg/MeHg-orig (fraction of MeHgCl left in the distillation vessel) at various distillation times. The temperature dependence of the HCl/HCl-orig (fraction of HCl left in the distillation vessel) and the pH of the distillate at distillation time of 3 or 3.5 hours for a water sample containing 0.12 M H_2SO_4 and 0.014 M Cl^- (4b), and for a water sample containing 0.12 MHCl (4c).

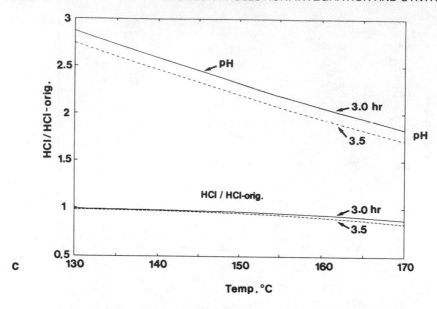

Figure 4 Continued.

Table 2   Determination of MeHg in biological and soil samples by the distillation/ GC-CVAFS method, using 0.42 M H_2SO_4 and 0.05 M chloride under different distillation conditions.

Sample	MeHg (ng/g)	(pH)$_{meas}$	Time (min.)	Vd/Vs
DORM-1(CRMs) (0.731 + 0.06)	0.738 ± 0.029 (n=5)	2.1–2.4	50–70	0.79–0.92
Soil (mor)	0.130 ± 0.004 (n=3)	2.1–2.5	60–70	0.91–0.98
Soil (upper B-hor.)	0.170 ± 0.001 (n=2)	1.9–2.3	60–70	0.90–0.93
Soil (lower B-hor.)	0.385 ± 0.005 (n=2)	2.1–2.4	60–70	0.87–0.93
Litterfall (various samples)	0.13–3.3 (n=16) (n=3) (n=3)	2.0–2.3 1.8–1.9 2.2–2.4	60–70	0.8–0.9 >0.9 0.7–0.8

Note: Vd/Vs is the ratio between the volume of the distillate and the aqueous phase in the distillation vessel.

4b and 4c) indicates that the same acid concentration of HCl, instead of H_2SO_4, results in a lower pH of the distillate due to the higher chloride concentration. This is in good agreement with the experimental results of the artificial water (Table 3). For a solid sample, Figure 3b shows that the pH in the distillate will be in the range 2.2 to 2.3 after 50 to 60 min distillation at 145° C, using H_2SO_4 and 0.05 M Cl⁻. The agreement between the actual pH in the distillate, presented in Table 2 and the model calculation is very good. Also, the pH in the distillate

Table 3   MeHg concentrations in artificial and surface water samples determined by the distillation/GC-CVAFS method under different distillation conditions.

Sample	MeHg (ng/L)	(pH)$_{meas}$	Time (hour)	Acid (+ Cl$^-$ M)
Artificial water (0.55 ng/L)	0.40–0.52 (n=3)	3.6–3.8	2.5–3.5	0.012 M H$_2$SO$_4$ (+0.014 M Cl$^-$)
	0.33–0.60 (n=3)	2.6–2.9	2.5–3.5	0.012 M HCl (+0.12 M Cl$^-$)
Surface water	0.09–0.18 (n=6)	1.3–2.6	2.5–3.5	0.012 M H$_2$SO$_4$ (+0.06 M Cl$^-$)
Surface water A	1.25	2.2(4.0)[a]	3.0	0.9 M H$_2$SO$_4$ (+0.18 M Cl$^-$)
Surface water B	2.10	2.8(3.3)[a]	3.0	0.9 M H$_2$SO$_4$ (+0.18 M Cl$^-$)
Surface water C	1.15	2.4(3.8)[a]	3.5	0.9 M H$_2$SO$_4$ (+0.18 M Cl$^-$)

[a] pH value calculated from the analyzed Cl$^-$ concentration.

decreases with increasing ratio between the volume of the distillate and the aqueous phase in the distillation vessel (i.e., the various litterfall samples). As the concentration of the HCl in the distillate determines the pH, the chloride concentration will be 282 ppm at pH = 2.1, which will interfere with the aqueous ethylation/GC-CVAFS analysis. Generally, since biological and soil samples as well as the other solid environmental samples contain high MeHg concentrations, a portion of the distillate (i.e., a low chloride amount) is in most cases enough for analysis. The results from the determination of MeHg in reference fish material and some soil samples (Table 2) show that the precision and accuracy of the distillation/GC-CVAFS method are good for solid environmental samples. The recovery of standard additions to soil samples are >90%, which indicates that the soil matrix does not influence the distillation procedure. For artificial water samples (see Table 3) the pH in the distillate when using H$_2$SO$_4$ (3.6 to 3.8) and when using HCl (2.6 to 2.9) is also in good agreement with the model simulation (3.3 to 3.4 and 2.3 to 2.4, respectively). However, the pH is surprisingly low for the surface water sample. This is probably due to the fact that naturally occurring organic substances which are present in the surface water sample may decompose at the end of the distillation. The product of the decomposition may be a volatile organic acid, which may be purged by the nitrogen gas together with water vapor, resulting in a lower pH than expected from the model calculation. Several humus-rich surface water samples (surface waters A, B, and C in Table 3) were also analyzed using a distillation procedure with higher concentrations of H$_2$SO$_4$ and Cl$^-$. The pH calculated from the analyzed chloride concentration in the distillate is much higher than the pH actually measured in the distillate. This suggests that acids other than HCl exist in the distillate, resulting in a lower pH but not higher Cl$^-$ concentration. The aqueous phase ethylation/GC-CVAFS analysis can therefore be used for water samples containing humic substances after neutralizing the low pH of the distillate.

V. CONCLUSION

The isothermal GC-CVAFS method is more reliable than the cryogenic GC-CVAFS method. The reliability of the MeHgCl determination is lower for samples with low MeHg concentrations (<0.1 ng/L), especially if the MeHg isolation is performed by the extraction technique. The results of the model calculation of optimal temperature and time interval are in good agreement with the distillation experiments performed. After 50 min for a solid sample and 3 h for an aqueous sample at a temperature greater than 140° C, more than 90% of the MeHg in the original aqueous solution in the distillation vessel is transferred to the receiving vessel. Adding the same acid concentration of HCl instead of H_2SO_4 to water samples, a lower pH of the distillate will be found. Preferably, H_2SO_4 and KCl should be used instead of HCl in the distillation procedure, which results in lower concentrations of chloride and hydrogen ions in the distillate.

For water samples containing humic substances, acids other than HCl may be transferred to the distillate, which results in a lower pH of the distillate than expected. The distillate should be neutralized prior to the aqueous phase ethylation/GC-CVAFS analysis.

The main advantages of the distillation preseparation procedure are (1) preparing blank samples free from MeHg (which is very important for water samples with a low MeHg content), and (2) the method is also suitable for humus-rich water samples. More work, however, needs to be done before we can accurately and precisely determine MeHg in all types of environmental samples.

ACKNOWLEDGMENTS

The study was performed as a part of the CEC STEP programme and was also supported by the Swedish State Power Board. The authors are indebted to Nicolas Bloom for valuable discussions and supplying the equipment. Thanks are also due to Jack Mowrer for preparing the GC column and revising the text, and to Milena Horvat for valuable discussions.

REFERENCES

1. Bloom, N. S., Determination of picogram levels of methylmercury by aqueous phase ethylation, followed by cryogenic gas chromatography with cold vapour atomic fluorescence detection, *Can. J. Fish Aquat. Sci.*, 46, 1131, 1989.
2. Liang, L., Horvat, M., and Bloom, N. S., An Improved Method for the Determination of Mercury by Aqueous Phase Ethylation, and Carbotrap Preconcentration, Followed by Isothermal Gas Chromotagraphy with Cold Vapour Atomic Fluorescence Detection (CVAFD), (submitted abstract), FACSS 1992 Conference.
3. Horvat, M., May, K., Stoeppler, M., and Byrne, A. R., Comparative studies of methylmercury determination in biological and environmental samples, *Appl. Organomet. Chem.*, 2, 515, 1988.
4. Horvat, M., Bloom, N. S., and Liang, L., Comparison of distillation with other current isolation methods for the determination of methyl mercury in low level environmental samples. I. Sediments, *Anal. Chim. Acta*, 281, 135, 1993.
5. Iverfeldt, Å. and Lindqvist, O., Distribution equilibrium of methyl mercury chloride between water and air, *Atmos. Environ.*, 16, 2197, 1982.
6. Iverfeldt, Å. and Persson, I., The solvation thermodynamics of methylmercury(II) species derived from measurements of the heat of solution and the Henry's Law Constant, *Inorg. Chim. Acta*, 103, 113, 1985.
7. Lewis, G. N. and Randall, M., *Thermodynamics and the Free Energy of Chemical Substances*, McGraw-Hill, New York, 1923, 470.
8. Peer, J. H., Ed. in chief, *Chemical Engineers' Handbook*, 4th ed., McGraw-Hill, Tokyo, 1963, 3–61.
9. Robinson, R. A., *Electrolyte Solutions*, 2nd ed., Butterworths, London, 1970, 560.

Determination of Low-Level Methylmercury Concentrations in Water and Complex Matrices by Different Analytical Methods: A Methodological Intercomparison

Susanne Padberg, Åke Iverfeldt, Ying-Hua Lee, Franco Baldi, Marco Filippelli, Karl May, and Markus Stoeppler

CONTENTS

ABSTRACT: As an interlaboratory comparison, MeHg was analyzed in various samples (rain, river, lake, and runoff water, as well as in soil and liquid bacterial cultures) using different separation procedures and detection techniques. Various combinations of different separation procedures were compared:

1. Anion exchange after acidification with HCl.
2. Water vapor distillation with NH_3 for liquid phases or with the addition of H_2SO_4 and NaCl for solid phases.
3. Cryogenic gas chromatography (GC) after aqueous phase ethylation.
4. Preconcentration of MeHg on a sulfhydryl cotton fiber, followed by HCl addition and extraction with benzene.
5. Acidification with a $CuSO_4/H_2SO_4/NaCl$ solution and toluene extraction, followed by a back extraction into $Na_2S_2O_3$ solution, and incubation with bacterial cells capable of enzymatically converting MeHg to CH_4.
6. HCl extraction into toluene and MeHg recovery with $Na_2S_2O_3$ solution. MeHg was subsequently back extracted into a toluene phase with $CuSO_4$.

Different detection techniques were also applied:

1. Cold vapor atomic absorption spectrometry (CVAAS).
2. Atomic fluorescence spectrometry (AFS).
3. Graphite furnace atomic absorption spectrometry (GFAAS).
4. Capillary column gas chromatography (GC-ECD).
5. Wide-bore column gas chromatography (GC-FID).

This intercomparison study ensures the reliability of anion exchange separation combined with a distillation procedure and detection by CVAFS for low-level methylmercury determination in water, soil, and liquid bacterial culture samples. The results were confirmed by different separation procedures and detection techniques, such as cryogenic gas chromatography separation after the ethylation step and detection with CVAFS, and benzene extraction after preconcentration of HCl-extracted MeHg on sulfhydryl cotton fiber and detection with GC-FID. Additionally, the MeHg concentration in liquid bacterial cultures, found by using toluene-thiosulfate separation and detection with GC-ECD after enzymatic derivatization of MeHg to methane, is comparable to the data obtained by the other techniques.

I. INTRODUCTION

The formation and fate of methylmercury (MeHg) as part of the biogeochemical cycle of mercury is of far-reaching importance. The occurrence of the ecotoxicologically significant low-level concentrations of methylmercury in the environment requires extremely accurate separation procedures and sensitive detection techniques.

The reliability of various methods for methylmercury (MeHg) identification in the environment is still very controversial, especially if the concentrations of MeHg are to be accurately quantified. Different laboratories reach different conclusions on the distribution of this organometallic compound in nature and its quantification is often affected by the methodology applied. The major problem in MeHg determination procedures frequently is the detection of extremely low concentrations of MeHg and, at the same time, accurate quantification of the concentration. Thus, the idea method would be the most sensitive one, but at the same time it should also be a compound/element-specific technique applicable to every kind of matrice. The continuous development of analytical methods (separation and detection) and their intercomparison provides the necessary prerequisites for accurate determination of low concentrations of MeHg in most environmental matrices.

The aim of this study is to intercompare results of MeHg determinations in various environmental and biological samples using different analytical methods performed by a group of laboratories involved in the European Community project, "Origin and Fate of Methylmercury."

The first step of the intercomparison procedure consisted of MeHg determination in aliquots from the same sample distributed to participating laboratories, where different separation procedures and detection techniques were applied (comparison of low-level MeHg concentrations).

As a second step in the exercise (methodological intercomparison), extracts from specific separation/extraction techniques applied by one laboratory were processed by another laboratory using their own separation procedure and detection technique.

The present intercomparison of analytical techniques for measuring low-level MeHg concentrations will increase the reliability of previously obtained data, and it will also identify suitable methods for MeHg quantification in different matrices in the future.

II. MATERIALS AND METHODS

The intercomparison of methylmercury determinations by different analytical methods was applied to various natural samples:

Material	Sampling location
Rainwater	Jülich, KFA (FRG)
River water	River Elbe, Magdeburg (FRG)
	River Elbe, Havelberg (FRG)

Lake water	Lake St. Lärebovattnet (S)
Runoff water	Catchment area of Lake Gårdsjön (S)
	(high and low humic content)
Soil	Bavarian Forest (FRG)
	(O_f layer of brown earth)
	Sample from the German Environmental
	Specimen Bank, Jülich
	Oldenburg (FRG)
	(industrial location)
Liquid bacterial cultures	Methylation experiments with pure culture of the sulfate-reducer strain ND-132 ($1.36 \pm 0.11 \times 10^9$ cells/ml) spiked with $HgCl_2$

The analytical methods applied for methylmercury determination are based on different separation procedures and detection techniques. They had been previously developed and details already have been described.[1-9] The procedures and techniques can be briefly characterized as follows:

1. Separation of MeHg by an anion exchange column after acidification with 6 M HCl and quantification by cold vapor atomic absorption spectrometry (CVAAS). This method was applied for water and soil samples.[1-4]

2. Separation of MeHg in water samples by an anion exchange column after acidification with 6 M HCl and, subsequently, separation by water vapor distillation with NH_3 for liquid phases.[3,4]

3. For solid phases, such as soil samples, separation of MeHg by water vapor distillation with the addition of H_2SO_4 and NaCl and subsequent separation with an anion exchange column after acidification with HCl.[3-5] A flowchart of these different extraction and separation steps for liquid and solid phases (1–3) is shown in Figure 1. CVAAS was used to quantify the amount of MeHg in both types of samples.[3-5] Additionally, CV atomic fluorescence spectrometry was used to detect MeHg in water samples. The detection limit of the modified cold vapor AAS is 5 pg; the determination limit of the whole procedure is about 50 to 100 pg. It is therefore possible to carry out determinations down to 0.2 ng MeHg/L water or 0.1 μg MeHg/kg of solid material.[2-4]

4. Separation of MeHg by cryogenic gas chromatography (GC) after extraction with CH_2Cl_2 and aqueous phase ethylation and quantification by atomic fluorescence spectrometry (AFS). The detection limit is about 0.6 pg Hg and the determination limit 0.003 ng/L for a 200 mL water sample.[6] This method was used for the determination of methylmercury in water samples, liquid bacterial cultures, and in the extracts of water and soil samples separated by the first, second, and third procedure.

5. Separation of MeHg by preconcentration on a sulfhydryl cotton fiber, followed by elution with 2 N HCl, extraction with benzene, and detection by a capillary column GC-ECD technique. The detection limit of the GC technique for MeHgCl/EtHgCl is about 2 to 3 pg using a 6-μl injection volume. Including the extraction and preconcentration procedure, the determination limit is about 0.04 ng/L for a 20 L water sample.[7,8] This method was applied in the intercomparison of the methylmercury concentrations in water samples.

6. Separation of MeHg by acidification with 0.5 M $CuSO_4$ in 11% H_2SO_4/3 M NaCl solution and toluene extraction, followed by back extraction with 0.01 N $Na_2S_2O_3$ solution, incubation with a dense culture of broad-spectrum Hg-resistant *Pseudomonas putida* (strain FB1) capable of enzymatically converting MeHg to CH_4 (biological derivatization of MeHg to methane), and detection by GC-FID. The detection limit of GC-FID is 15 ng MeHg, in view of the fact that the injected volume is 100 μL and the headspace of a 2-mL microreaction vessels is 0.7 mL. The determination limit is about 1.5 ng MeHg/mL extract.[9]

7. Separation of MeHg by 1 N HCl extraction into toluene and reextraction of MeHg by 0.01 N Na$_2$S$_2$O$_3$ solution and quantification by GFAAS. Additionally, MeHg was subsequently extracted back into a toluene phase with 0.5 M CuSO$_4$ followed by detection with GC-ECD. The detection limit of GFAAS is about 0.04 ng Hg with a 20-μL injection volume, and of GC-ECD it is roughly 0.003 ng MeHg with a 10-μL injection volume. The limit for MeHg determination with this separation procedure and detection by GFAAS is about 2 ng/mL. The limit is 0.15 ng/mL[10] for determination by applying additional toluene extractions and detection by GC-ECD.

These last two methods, based on the conventional (common) separation procedure of toluene extraction, were applied for the determination of methylmercury in liquid bacterial cultures from methylation experiments.

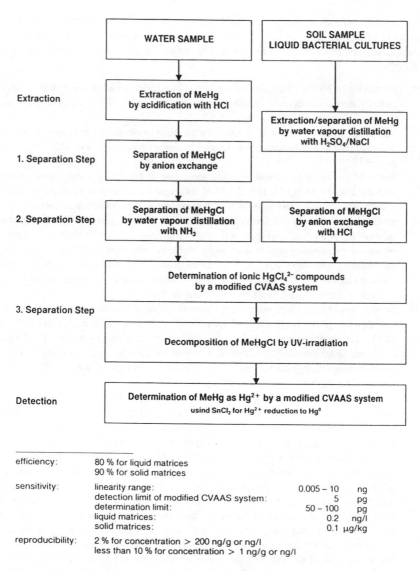

Figure 1   Methylmercury determination in water samples and complex matrices.

River water
Elbe / Magdeburg
MeHg – ng/l

Measurement Technique

Extraction/Separation Procedure	modified CVAAS	GFAAS	CVAFS	GC-ECD	GC-FID
anion exchange after acidification with HCl	10.17				
anion exchange after acidification with HCl / water vapour distillation with NH3	3.18		4.31		
water vapour distillation with H2SO4 and NaCl / anion exchange after acidification with HCl					
extraction with CH2Cl2/ cryogenic gas chromatography after ethylation			4.35 ± 0.04		
preconcentration on sulfhydryl cotton fibre / HCl addition and extraction with benzene				3.25	
acidification with CuSO4 in H2SO4 and NaCl/toluene extr./reextr. with Na2S2O3/derivatization with bacterial cells to CH4					
HCl extraction into toluene / MeHg reextr. with Na2S2O4 Subsequently back extr. into toluene phase with CuSO4					

Figure 2 Comparison of MeHg concentration in river water, sample I, (Elbe/Madgeburg) determined by different separation procedures and detection techniques.

III. DISCUSSION OF RESULTS

A. COMPARISON OF MeHg CONCENTRATION IN RIVER WATER, SAMPLE I, (ELBE/MADGEBURG) DETERMINED BY DIFFERENT SEPARATION PROCEDURES AND DETECTION TECHNIQUES (FIGURE 2)

Except for the Hg concentration determined with anion exchange separation and CVAAS detection, the results obtained by different methods are at same level of MeHg concentration. The MeHg concentrations determined by separation with anion exchange and distillation are comparable to GC separation after the ethylation step, and both were detected by CVAFS. The MeHg concentration determined by separation with anion exchange, distillation, and detection by CVAAS is comparable to the result obtained after preconcentration on sulfhydryl cotton fiber and extraction with benzene and GC-ECD detection.

B. COMPARISON OF MeHg CONCENTRATION IN RIVER WATER, SAMPLE II, (ELBE/HAVELBERG) DETERMINED BY DIFFERENT SEPARATION PROCEDURES AND DETECTION TECHNIQUES (FIGURE 3)

The variation in the MeHg concentrations of the second river water sample determined by different methods shows the same effect as observed in the first river water sample. Except for Hg concentration determined with anion exchange separation and CVAAS detection, the results from all separation procedures applied (anion exchange and distillation, GC after ethylation, benzene extraction after preconcentration on sulfhydryl cotton fiber) combined with different detection techniques (CVAAS, CVAFS, GC-ECD) are at the same level of MeHg concentration.

River water
Elbe / Havelberg
MeHg — ng/l

Measurement Technique

Extraction/Separation Procedure	modified CVAAS	GFAAS	CVAFS	GC–ECD	GC–FID
anion exchange after acidification with HCl	4.35				
anion exchange after acidification with HCl / water vapour distillation with NH3	2.55		1.57 ± 0.51		
water vapour distillation with H2SO4 and NaCl / anion exchanger after acidification with HCl					
extraction with CH2Cl2/ cryogenic gas chromatography after ethylation			2.25 ± 0.25		
preconcentration on sulfhydryl cotton fibre / HCl addition and extraction with benzene				2.29	
acidification with CuSO4 in H2SO4 and NaCl/toluene extr./reextr. with Na2S2O3/derivatization with bacterial cells to CH4					
HCl extraction into toluene / MeHg reextr. with Na2S2O4 Subsequently back extr. into toluene phase with CuSO4					

Figure 3 Comparison of MeHg concentration in river water, sample I, (Elbe/Havelberg) determined by different separation procedures and detection techniques.

C. COMPARISON OF MeHg CONCENTRATION IN RAINWATER (JÜLICH) DETERMINED BY DIFFERENT SEPARATION PROCEDURES AND DETECTION TECHNIQUES (FIGURE 4)

In the rainwater sample the MeHg concentration obtained only by the anion exchange separation procedure and detected by CVAAS is too high, as shown in the comparison of the results with those of river water.

Similar results were obtained with the three different separation procedures (anion exchange followed by distillation, GC separation after ethylation, benzene extraction after preconcentration on cotton fiber) combined with CVAAS, CVAFS, or GC-ECD detection, except for the concentration detected by CVAFS after separation with anion exchange followed by distillation. Using the same separation procedure, the value obtained by CVAFS detection is higher than that obtained by CVAAS detection.

D. COMPARISON OF MeHg CONCENTRATION IN LAKE WATER (LAKE STORA LÄREBOVATTNET) DETERMINED BY DIFFERENT SEPARATION PROCEDURES AND DETECTION TECHNIQUES (FIGURE 5)

In this case, a rather low MeHg concentration is confirmed by two methods. Comparing the results of the MeHg determination in lake water using different separation procedures, the same effect as in rainwater is observed. The concentration obtained by anion exchange separation followed by distillation and CVAFS detection is somewhat higher than the results obtained by GC separation after the ethylation step followed by CVAFS detection, and by separation with benzene extraction after preconcentration on sulfhydryl cotton fiber

Rain water
Jülich
MeHg — ng/l

Measurement Technique

		modified CVAAS	GFAAS	CVAFS	GC–ECD	GC–FID
Extraction/Separation Procedure	anion exchange after acidification with HCl	1.94				
	anion exchange after acidification with HCl / water vapour distillation with NH3	0.97		1.45 ± 0.21		
	water vapour distillation with H2SO4 and NaCl / anion exchange after acidification with HCl					
	extraction with CH2Cl2/ cryogenic gas chromatography after ethylation			0.70 ± 0.10		
	preconcentration on sulfhydryl cotton fibre / HCl addition and extraction with benzene				0.87	
	acidification with CuSO4 in H2SO4 and NaCl/toluene extr./reextr. with Na2S2O3/derivatization with bacterial cells to CH4					
	HCl extraction into toluene / MeHg reextr. with Na2S2O4 Subsequently back extr. into toluene phase with CuSO4					

Figure 4   Comparison of MeHg concentration in rain water (Jülich) determined by different separation procedures and detection techniques.

Runoff water
Lake Gårdsjön
MeHg — ng/l

Measurement Technique

		modified CVAAS	GFAAS	CVAFS	GC–ECD	GC–FID
Extraction/Separation Procedure	anion exchange after acidification with HCl					
	anion exchange after acidification with HCl / water vapour distillation with NH3			0.32 ± 0.02		
	water vapour distillation with H2SO4 and NaCl / anion exchange after acidification with HCl					
	extraction with CH2Cl2/ cryogenic gas chromatography after ethylation			0.05		
	preconcentration on sulfhydryl cotton fibre / HCl addition and extraction with benzene				0.19	
	acidification with CuSO4 in H2SO4 and NaCl/toluene extr./reextr. with Na2S2O3/derivatization with bacterial cells to CH4					
	HCl extraction into toluene / MeHg reextr. with Na2S2O4 Subsequently back extr. into toluene phase with CuSO4					

Figure 6   Comparison of MeHg concentration in runoff water (catchment area of Lake Gårdsjön) determined by different separation procedures and detection techniques.

Lake water
Stora Lärebovattnet
MeHg – ng/l

Measurement Technique

Extraction/Separation Procedure

Extraction/Separation Procedure	modified CVAAS	GFAAS	CVAFS	GC–ECD	GC–FID
anion exchange after acidification with HCl					
anion exchange after acidification with HCl / water vapour distillation with NH3			0.49 ± 0.03		
water vapour distillation with H2SO4 and NaCl / anion exchange after acidification with HCl					
extraction with CH2Cl2/ cryogenic gas chromatography after ethylation			0.16 ± 0.03		
preconcentration on sulfhydryl cotton fibre / HCl addition and extraction with benzene				0.28	
acidification with CuSO4 in H2SO4 and NaCl/toluene extr./reextr. with Na2S2O3/derivatization with bacterial cells to CH4					
HCl extraction into toluene / MeHg reextr. with Na2S2O4 Subsequently back extr. into toluene phase with CuSO4					

Figure 5 Comparison of MeHg concentration in lake water (Lake Stora Lärebovattnet) determined by different separation procedures and detection techniques.

and GC-ECD detection, which are both at about the same level (0.2 to 0.3 ng/L). This low-level concentration is close to the determination limit of the anion exchange separation procedure followed by distillation and detection by CVAAS.

Problems involved in a precise separation of such low-level concentrations near the determination limit could be the reason for the higher result. Until now, the determination limit of this separation procedure combined with CVAFS detection has not been well established but, most probably, is much better than for CVAAS. Thus, these types of problems probably were not previously revealed when using the CVAAS detector.

Additionally, as already observed for MeHg determination in rainwater, matrix effects may also disturb the ethylation step coupled to detection by CVAFS, since the water sample originated from Lake Stora Lärebovattnet, which is an acid, humic-rich lake.[11]

E. COMPARISON OF MeHg CONCENTRATION IN RUNOFF WATER (CATCHMENT AREA OF LAKE GÅRDSJÖN) DETERMINED BY DIFFERENT SEPARATION PROCEDURES AND DETECTION TECHNIQUES (FIGURE 6)

Similar effects (seen in Section III.D) may also explain the results observed in the comparative determination of MeHg in runoff water by using different methods. The concentration obtained with the combination of anion exchange separation followed by distillation and CVAFS detection may involve problems in precise separation (see above), but the differences between the concentration obtained with GC separation after the ethylation step and CVAFS detection, and benzene extraction after the preconcentration step and GC-ECD detection, is probably a result of the content of humic substances in runoff water. Specifically, the efficiency of the ethylation step is probably affected by very humic-rich waters.

F. DETERMINATION OF MeHg IN VARIOUS WATER SAMPLES BY DIFFERENT SEPARATION PROCEDURES AND DETECTION TECHNIQUES (TABLE 1)

The main factors causing different MeHg results in various water samples, depending on the different separation procedures and detection techniques used, can be summarized as follows (also see Table 1).

The higher concentrations in river water and rainwater samples determined by ion exchange separation, but without the additional separation step with water vapor distillation, are unreliable for the MeHg content of these samples. As already found in previous investigations,[3–5] the anion exchange resin is reliable if applied in the separation of MeHg from biological materials, but it does not only separate MeHgCl after the acidification of environmental samples like water. The unreliability of this separation procedure depends on the natural conditions of the sample, e.g., the concentration of organic material.

In contrast, the results for river water and rainwater obtained with ion exchange separation followed by the distillation procedure and detection with modified CVAAS are confirmed by the other separation procedures, such as cryogenic GC after aqueous phase ethylation, and extraction with benzene after preconcentration on a sulfhydryl cotton fiber.

In comparison to these latter methods, the different MeHg concentrations in rainwater, lake water, and runoff water obtained with ion exchange followed by distillation, but combined with CVAFS detection, may be explained by problems with determination at the level of the detection limit, and/or by matrix effects affecting the ethylation procedure, probably caused by humic substances.

The same effect may also be responsible for the differences in the MeHg concentrations in runoff water obtained by cryogenic GC after aqueous phase ethylation and CVAFS detection, and extraction with benzene after preconcentration on a sulfhydryl cotton fiber followed by GC-ECD detection.

Differences in the results are mainly due to matrix interferences in the separation steps rather than in the detection techniques. Therefore, the intercomparison study was expanded to also include the determination of MeHg in samples of runoff water with different contents of humic substances. After the first separation step with ion exchange followed by distillation, the separated MeHg was controlled by the use of a second separation step, i.e., the aqueous phase ethylation and cryogenic GC technique.

G. DETERMINATION OF METHYLMERCURY IN RUNOFF WATER AND SOIL SAMPLES BY DIFFERENT SEPARATION PROCEDURES AND DETECTION TECHNIQUES (TABLE 2)

Considering the MeHg concentrations in runoff water determined by the ion exchange/ distillation/CVAAS method and the MeHg concentrations determined by the ethylation/GC/ AFS method, the results obtained by the first separation procedure are confirmed by the second separation procedure.

The determination of MeHg in runoff water samples applying ion exchange separation combined with the distillation procedure and detection with CVAAS seems to be independent of the level of humic substances present. Based on the limited material present, no other conclusion is possible even for the ethylation/GC separation/CVAFS method. However, this contradicts the suggestion provided by previous findings. Furthermore the MeHg concentrations determined directly in a parallel sample by the ethylation/GC separation/CVAFS method are not significantly different compared to the above results. Therefore, we conclude for runoff water with low humic content that both separation procedures and detection techniques are accurate and reliable.

The reliability of the separation procedure has not yet been fully confirmed for the determination of MeHg in runoff water with a high humic content. After the first separation

Table 1 Determination of methylmercury in various water samples by different separation procedures and detection techniques.

Sample	Sampling location	Methylmercury concentration (ng/L)				
		Ion exchange[a] CVAAS	Ion exchange/distillation[b]		Ethylation/GC[c] CVAFS	Preconcentr.[d] GC-ECD
			CVAAS	CVAFS		
River water I	Elbe/Magdeburg (FRG)	10.17	3.18	4.31	4.35 ± 0.04	3.25
River water II	Elbe/Havelberg (FRG)	4.35	2.55	1.57 ± 0.51	2.25 ± 0.25	2.29
Rainwater	Jülich/KFA (FRG)	1.94	0.97	1.45 ± 0.21	0.70 ± 0.10	0.87
Lake water	St. Lärebovattnet (S)			0.49 ± 0.03	0.16 ± 0.03	0.28
Runoff water	Catchment area of Lake Gårdsjön (S)			0.32 ± 0.02	0.05	0.19

[a] Separation of MeHg by anion exchange after acidification with HCl, detection by CVAAS.[1,3,4]

[b] Separation of MeHg by anion exchange after acidification with HCl and subsequent separation by water vapor distillation with NH₃, detection by CVAAS[3,4] or with CVAFS.

[c] Separation of MeHg by GC after ethylation step and CVAFS detection.[6]

[d] Separation of MeHg by HCl extraction and preconcentration step with sulfhydryl cotton fiber adsorbent, detection by GC.[7,8]

Note: The values are expressed in mean ± S.D. of two measurements (n=2); if S.D. is not expressed, S.D. <0.01 ng/L.

Table 2 Determination of methylmercury in runoff water and soil samples by different separation procedures and detection techniques.

Sample	Sampling location	First MeHg separation		Second MeHg separation	
		Ion exchange[a] CVAAS MeHg separation	Ion exchange/distillation[b] CVAAS	Ethylation/GC[c] CVAFS MeHg separation	
		In sample	In sample	In extract A	In sample
Runoff water (ng/L)	Catchment area of Lake Gårdsjön (low humic content)		0.32 ± 0.13 (6)	0.26 ± 0.05 (7)	0.45 ± 0.05 (3)
Runoff water (ng/L)	Catchment area of Lake Gårdsjön (high humic content)		0.35 ± 0.06 (6)	0.23 ± 0.15 (6)	0.26 ± 0.03 (3)
Soil (ng/g)	Bavarian Forest (high conc. of organic matter)	6.59 (2) 8.78 (2)	1.07 ± 0.08 (2)	0.90 (2) / 1.04 (2) 0.77 (2) 1.16 (2)	
Soil (ng/g)	Oldenburg (industrial location)		925 (2) 195 (2)	1190 (1) 167 (2)	

[a] Separation of MeHg by anion exchange after acidification with HCl, detection as Hg by CVAAS.[1-4]

[b] Separation of MeHg by anion exchange after acidification with HCl and subsequent separation by water vapour distillation with NH_3, detection with CVAAS[3,4] or CVAFS (water). Separation of MeHg by water vapor distillation with H_2SO_4 and $NaCl$,[5] followed by anion exchange after acidification with HCl, detection with CVAAS (soil).[3,4]

[c] Separation of MeHg by GC after ethylation step and CVAFS detection.[6]

Note: The values are expressed in mean ± S.D. of several measurements (n); if S.D. is not expressed, S.D. ≤0.01 ng/L.

step with ion exchange/distillation, the MeHg concentration is certainly accurately separated by the ethylation/GC procedure. Without applying the distillation step, the humic substances may cause a lower recovery when using only the ethylation/GC separation procedure. Since the different content of humic substances does not significantly[5] influence the amount of methylmercury in the analyzed runoff water samples, these substances interfere with the recovery and the reliability of the separation procedures.

The influence of organic material on the determination of MeHg in soils has already been observed in previous studies.[3] Like the effect already mentioned for the determination of MeHg in river water and rainwater, results from ion exchange separation without an additional distillation procedure are always higher in comparison to the results derived using both separation procedures.

The ethylation/GC separation procedure was used to verify the results of MeHg in soils obtained by ion exchange separation combined with the distillation procedure. The MeHg concentration determined in a soil sample of the upper layer of a brown earth with a high content of organic materials, as well as in a soil sample from a mercury-contaminated industrial location, is certainly verified. In contrast, higher values are obtained if only the ion exchange separation procedure is applied after acidification with HCl. An additional HCl-extractable mercury compound is separated by the ion exchange resin.[3] Combined with the distillation procedure, ion exchange separation is reliable for the determination of low and high MeHg concentrations in soil samples.

H. DETERMINATION OF MeHg IN BACTERIAL CULTURE SAMPLES BY DIFFERENT SEPARATION PROCEDURES AND DETECTION TECHNIQUES (TABLE 3)

The results of the intercomparison of the MeHg concentration in liquid bacterial cultures determined by different separation procedures and detection techniques will also be discussed. Cultures of a sulfate-reducing bacteria, strain ND-132, were grown anaerobically in 50 mL liquid (Postgate's medium) with a low content of sulfates in the presence of 5 μg/mL of HgCl$_2$. After 2 days of incubation at 28° C, 5 mL of concentrated HCl was added to each

Table 3   **Determination of methylmercury in bacterial culture samples by different separation methods—methylation experiments by spiking liquid bacterial cultures with HgCl (University of Siena).**

Methylmercury concentration (Mean ± S.D.—ng/mL)			
Distillation/ion exchange[a] CVAAS	Ethylation/ GC[b] CVAFS	Toluene extr./ enzym. derivat.[c] GC/FID	Toluene extr.[d] GFAAS
12.6 ± 2.26 (9)	9.45 ± 2.5 (3)	19.9 ± 1.8 (6)	100 ± 9.3 (3)
0.98 ± 0.48 (4)[e]		1.5 ± 0.07 (2)[e]	

[a] Separation of MeHg by water vapor distillation with H$_2$SO$_4$ and NaCl[5], followed by anion exchange after acidification with HCl, detection with CVAAS.[3,4]

[b] Separation of MeHg by GC after ethylation step and CVAFS detection.[6]

[c] Separation of MeHg by acidification with CuSO$_4$ in H$_2$SO$_4$ and NaCl, toluene extraction, followed by a back extraction with Na$_2$S$_2$O$_3$, incubation with bacteria cells capable of enzymatically converting MeHg to CH$_4$ and detection by GC-FID.[9]

[d] Separation of MeHg by HCl extraction into toluene and MeHg reextraction by Na$_2$S$_2$O$_3$ and quantification by GFAAS. Additionally, subsequent back extraction of MeHg into toluene with CuSO$_4$ followed by detection with GC-ECD.[10]

[e] MeHg concentration in liquid bacterial cultures without HgCl$_2$ addition.

Note: Parenthesis indicate number of measurements.

serum bottle to stop growth and store the cultures for methylmercury determination. Triplicates of the samples, plus two control samples, were sent to each group participating in the intercalibration exercise.

The results obtained with the distillation/ion exchange separation procedure/CVAAS detection and with ethylation/GC separation and CVAFS detection are similar. The level is roughly confirmed by conventional toluene extraction coupled to the detection of MeHg based on its enzymatic derivatization after reextraction into an aqueous phase containing $Na_2S_2O_3$. The resulting CH_4 is subsequently detected by GC-FID.[9]

The method additionally applied,[10] also based on conventional toluene extraction and reextraction with $Na_2S_2O_3$, separated significantly higher MeHg concentrations in comparison to the levels of the other three methods. After this conventional separation procedure, MeHg is determined as total Hg in the extracted toluene-thiosulfate layer (GFAAS). Similar higher values were also obtained by Winfrey and Baldi,[12] using conventional toluene extraction. After incubation of the sample with $^{203}HgCl_2$, MeHg was separated by 0.5 M $CuSO_4$ and 3 M NaBr in 11% H_2SO_4 followed by 25 mL toluene extraction. Only 20 mL of the solvent was recovered after 15 min centrifugation at 2000 r.p.m. The toluene layer was then water-cleaned by two additions of anhydrous Na_2SO_4 aqueous solution in separate beakers. MeHg quantification was performed by ^{203}Hg detection with a liquid scintillation counter.[13] These extraction procedures, applied for samples of complex matrices, may have caused the scattered data.

IV. CONCLUSION

MeHg determination in liquid and solid environmental materials (various water samples and soil samples) by ion exchange separation combined with a distillation procedure and the CVAAS detection technique is verified by MeHg determination using cryogenic gas chromatography separation after an ethylation step and detection with CVAFS, and with MeHg determination by benzene extraction after preconcentration of the HCl-extracted MeHg on sulfhydryl cotton fiber and detection with GC-ECD.

Matrix effects, especially the content of humic substances, may cause a variation in the results depending on incomplete recovery in a derivatization step. These effects are also most probably responsible for the unreliable separation of MeHg in water and soil samples if the ion exchange procedure alone is used after acidification with HCl.

For the determination of MeHg in complex matrices such as liquid bacterial cultures, the reliability of the distillation/ion exchange/CVAAS method and the ethylation/GC/AFS method are additionally roughly confirmed by toluene-thiosulfate separation and detection with GC-ECD after enzymatic derivatization of MeHg to methane.

ACKNOWLEDGMENTS

We are grateful for financial support from the Commission of the European Communities under contract EV 4V-0138-D (BA) within the research project "Origin and Fate of Methylmercury." Thanks are due to Prof. Dr. Marko Branica from Rudjer Boskowic Institute, Zagreb and to Dr. Milena Horvat from Jazef Stefan Institute, Ljubljana for helpful discussions.

REFERENCES

1. Ahmed, R., May, K. Stoeppler, M., Ultratrace analysis of mercury and methylmercury in rain water using cold vapor atomic absorption spectrometry, *Fresenius Z. Anal. Chemie,* 326, 510, 1987.

2. May, K., Stoeppler, M., Reisinger, K., Studies of the ratio of total mercury/methylmercury in the aquatic food chain, *Toxicol. Environ. Chem.,* 13, 153, 1987.
3. Padberg, S., Mercury in a Terrestrial Ecosystem; Investigation of Transport and Transformation Mechanisms, Doctoral thesis, University of Tübingen, Report Jül-2534, KFA Jülich, Oct. 1991.
4. Padberg, S. and May, K., Occurrence and behaviour of mercury and methylmercury in the aquatic and terrestrial environment, In: *Specimen Banking, Environmental Monitoring and Analytical Approaches,* Rossbach, M., Schladot, J.D., Ostapczuk, P., Eds., Springer-Verlag, New York, 1992, 195.
5. Horvat, M., May, K., Stoeppler, M., Byrne, A.R., Comparative studies of methylmercury determination in biological and environmental samples, *Appl. Organomet. Chem.,* 2, 515, 1988.
6. Bloom, N., Determination of picogram levels of methylmercury by aqueous phase ethylation, followed by cryogenic gas chromatography with cold vapour atomic fluorescence detection, *Can. J. Fish. Aquat. Sci.,* 46, 1131, 1989.
7. Lee, Y.H., Determination of methyl- and ethylmercury in natural waters at sub-nanogram per liter using SCF-adsorbent preconcentration procedure, *Int. J. Environ. Anal. Chem.,* 29, 263, 1987.
8. Lee, Y.H. and Mowrer, J., Determination of methylmercury in natural waters at the sub-nanogram per litre level by capillary gas chromatography after adsorbent preconcentration, *Anal. Chim. Acta,* 221, 259, 1989.
9. Baldi, F. and Filippelli, M., New method for detecting methylmercury by its enzymatic conversion to methane, *Environ. Sci. Technol.,* 25, 302, 1991.
10. Filippelli, M., Determination of trace amounts of organic and inorganic mercury in biological materials by graphite furnace atomic absorption and organic mercury speciation by gas chromatography, *Anal. Chem.,* 59, 116, 1987.
11. Lee, Y.H. and Hultberg, H., Methylmercury in some Swedish surface waters, *Environ. Toxicol. Chem.,* 9, 833, 1990.
12. Winfrey, M.R. and Baldi, F., Unpublished data, 1991.
13. Furutani, A. and Rudd, J.W.M., Measurement of mercury methylation in lake water and sediment samples, *Appl. Environ. Microbiol.,* 40, 770, 1980.

Photochemical Behavior of Inorganic Mercury Compounds in Aqueous Solution

Z. F. Xiao, J. Munthe, D. Strömberg, and O. Lindqvist

CONTENTS

ABSTRACT: Laboratory experiments were carried out by irradiating several species of divalent mercury in aqueous solution with broadband light (wavelength >290 nm). $Hg(OH)_2$ was found to be the most photoreactive species and its photolysis first-order rate constant was determined to be 1.2×10^{-4} s^{-1} (n = 6). The reaction quantum yield was estimated to be 0.14. The photolysis rate constant and lifetime were calculated to be 3×10^{-7} s^{-1} and 600 h under summer conditions in the Northern Hemisphere (Stockholm's latitude). The photoreduction of HgS_2^{2-} results in the formation of Hg^0 and HgS(s). The potential environmental significance of these processes is discussed.

I. INTRODUCTION

The application of photochemistry in aqueous solutions to environmental transformation processes is a relatively new field.[1] Environmental photochemistry reflects the outdoor conditions and the sun's electromagnetic radiation spectrum. The shortest wavelength recorded on earth is 286 nm, thus the terrestrial environment is exposed to longer wavelengths than 286 nm.[2] In the last 10 years, there has been a growing interest in the photolysis of aqueous pollutants such as phenols, chlorophenols,[3] and chlorobenzenes,[4] since photochemical processes may play an important role in the environmental degradation of anthropogenic chemicals.[5] Some transition metal compounds and inorganic species such as NO_2^- and NO_3^- can also be photolyzed in sea water. This leads to a change in their oxidation states and to the generation of new radicals which undergo further reactions.[6]

Mercury, as a serious environmental pollutant, has attracted much attention during the last decade and research results from comprehensive investigations have been presented.[7]

High mercury content was found in fish from lakes which are far away from industrial areas. The mercury pollution problem is not simply due to point source discharges, but is the results of more widespread air pollution and long distance atmospheric transportation.[8] The form and complexation of Hg regulate the toxicity, transport pathways, and residence time of Hg in different compartments of the ecosystem. For example, the average residence time in air differs from hours for some oxidized Hg(II) compounds to perhaps a year or more for the elemental form.[9] Therefore, redox processes that results in change of the valence state of mercury are of great importance.

The photoreduction of some organic mercury compounds in solution has been studied. Photolysis of diphenylmercury resulted in a carbon-mercury bond cleavage and formation of elemental mercury.[10] This kind of reaction could lead to the transport of mercury from natural water bodies to the atmosphere via gas exchange. Atmospheric mobilization and air-water exchange are significant features of biogeochemical cycling of mercury in the environment.[11]

On-site measurement of volatile mercury emissions from lake surfaces using chamber technology demonstrated that the mercury flux over four lakes exhibited a very consistent diurnal pattern.[12] Emission values determined during daylight hours were always higher than during hours of darkness, suggesting that sunlight-induced photolysis may be one pathway for volatile mercury formation in natural waters.

Both laboratory experiments[13] and on-site observation[14] have indicated that divalent mercury compounds in aqueous solution can be reduced through photo-induced reactions. In the former case, wavelengths <290 nm were also used, and in the latter case Hg species are not identified, and thus further experiments under controlled conditions are needed to give us a clear picture of what might occur in natural systems.

In order to investigate the photochemical behavior of inorganic mercury species in aqueous solution and its environmental impacts, some experiments were conducted in this laboratory with simulated sunlight under controlled conditions. Preliminary results have been presented in Munthe and McElroy[15] and, hence, only a brief summary is provided here.

Among the mercury species investigated, mercuric dichloride was not reduced appreciably when irradiated. $Hg(OH)_2$ seemed to be the only photoactive form, with $Hg(CN)_2$ and $HgCl_4^{2-}$ being the most stable. Mercury (II) sulfite was decomposed to Hg^0 at the same rate, both in the dark and when irradiated. Addition of small pieces of quartz glass to the reaction bottle did not affect the photolysis rate, indicating that the primary process is not surface dependent. In these experiments, low mercury concentrations (1×10^{-8} to 1×10^{-9} M) were used.

Strömberg[16] estimated theoretically that $Hg(SH)_2$ and $HgS(SH)^-$ would probably absorb light with wavelengths longer than 290 nm and Dyrssen[17] reported that HgS_2^{2-}, $HgS(SH)^-$, and $Hg(SH)_2$ may be the predominant dissolved mercury species in water. Thus, photolysis of sulfur-containing mercury species seems to be of environmental interest.

In this chapter the photolytic behaviors of HgS_2^{2-} and especially that of $Hg(OH)_2$ are discussed in detail, together with the potential environmental significance of these processes.

II. EXPERIMENTAL

A. THEORETICAL BACKGROUND

A photochemical reactor and a sunlight-simulating lamp system were used together with the necessary analytical instrument to obtain kinetic data for the photolysis reactions and to determine reaction quantum yields. The latter can be used to calculate sunlight photolysis rate constants and environmental half-lives.

At low concentrations, the reactions are first order:[18]

$$\text{Ln} (C_0/C_t) = k_p t \qquad (1)$$

where C_0 and C_t are the concentrations of the reactive species at time zero and t, respectively, t is the reaction time, and k_p is the photolysis rate constant.

The reaction quantum yield Φ, is calculated using Equation 2:

$$\Phi = k_p/2.303 \ \Sigma \ I_\lambda \ E_\lambda \ L \qquad (2)$$

where L is the cell pathlength (cm), E_λ the molar extinction coefficient of the species of interest (L mol^{-1} cm^{-1}), and I_λ is the intensity of the incident light (Einstein's L^{-1} sec^{-1}) which can be estimated using a chemical actinometer with known photochemical properties. The actinometer, potassium ferrioxalate, was chosen in this investigation due to its convenience of use and its well established quantum yield at different wavelengths.[19] If the actinometer is exposed to light in the same way as the reactant solution, preferably at the same time and using the same or an identical reaction cell, then:

$$I_\lambda = [n_{Fe^{2+}}]/6.023 \times 10^{23} \times \Phi_{Fe} \times t \qquad (3)$$

where $[n_{Fe^{2+}}]$ is the amount (molecules L^{-1}) of photoproduct and Φ_{Fe} is the quantum yield of potassium ferrioxalate, t is the exposure time of the actinometer.

Using Φ, one can extrapolate laboratory kinetic results to the real environment. The sunlight photolysis rate constant is estimated using the following equation:

$$k_{sp} = \Phi \ k_a = \Phi \ \Sigma \ (2.303/j) \ E_\lambda \ Z_\lambda \qquad (4)$$

where k_a is the sunlight absorption rate summed over all the wavelengths that are absorbed by the pollutant, Z_λ is the sunlight intensity for a specific wavelength interval centered at a fixed wavelength (photons cm^{-2} sec^{-1} N nm^{-1}), and j = 6.023×10^{20} is a conversion constant that matches the units of E_λ and Z_λ.

The half-life, a convenient measure of the reaction rate, can be calculated by:

$$t_{1/2} = \ln 2/k_{sp} \qquad (5)$$

B. PREPARATION OF SOLUTIONS

All chemicals used were reagent grade, and Milli-Q® water was used throughout the investigation. Hg(OH)$_2$ solutions of 1×10^{-4} M were prepared at pH 7 by dissolving mercuric oxide (yellow, Merck) in 0.005 mol 1^{-1} H$_2$SO$_4$ (Titrisol, Merck) and then neutralizing with 0.01 mol 1^{-1} NaOH (Titrisol, Merck). At this pH the mercury was in the form of Hg(OH)$_2$.[20] As diluted Hg solutions undergo losses during storage, all Hg working solutions were freshly prepared prior to each experiment from this stock solution.

HgS$_2$$^{2-}$ solutions were prepared from solutions of Hg(OH)$_2$ by adding excess Na$_2$S. For example, a 1×10^{-4} M solution of HgS$_2$$^{2-}$ was prepared by adding 0.01 M Na$_2$S; a 1×10^{-6} M solution of HgS$_2$$^{2-}$ by adding 0.001 M Na$_2$S.

The chemical actinometer, potassium ferrioxalate, was synthesized by mixing solutions of 1.5 M K$_2$C$_2$O$_4$ and 1.5 M FeCl$_3$ (3:1 v/v). A 0.006 M working solution of the actinometer was prepared by dissolving recrystallized precipitation of K$_3$Fe(C$_2$O$_4$)$_3$·3H$_2$O in 0.1 N H$_2$SO$_4$ solution. All work concerning the preparation and manipulation of the ferrioxalate solution was performed in a dark room with a red photographic safelight. Ferrous ion, a photoproduct of the actinometer, was analyzed by formation of a 1,10-phenanthroline-Fe^{2+} complex which absorbs light strongly at 5100 Å. Hence, a standard curve of the Fe^{2+} complex is necessary. To do this, the following solutions were needed: 0.1 M Fe^{2+} solution in 0.1 N H$_2$SO$_4$; 0.1% (by weight) 1,10-phenanthroline solution in water; and 1 N buffer solution of HAc/NaAc in

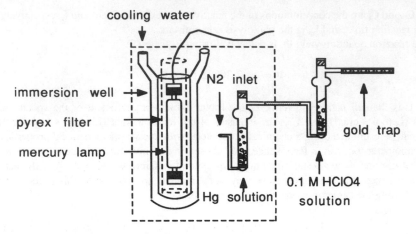

Figure 1   Setup of photochemical investigation with a medium-pressure mercury lamp as a light source.

1 N H_2SO_4. For detailed information on how to prepare these solutions and to analyze the Fe^{2+} complex, see Calvert and Pitts.[19]

C. EQUIPMENT AND SETUP

Two kinds of artificial sunlight sources were used in this investigation. One is a commercial xenon lamp (500 W) combined with a 6-mm Pyrex® glass filter which cuts off wavelengths ≤290 nm. The lamp irradiated the solution from the top of the reaction bottle at a distance of about 45 cm. It was noticed that an unknown portion of the incident light might be lost due to the reflection on the cap of the reaction bottle and the Hg solution was not evenly irradiated. Later, a photochemical turntable reactor (Model 7891, Ace Glass Incorporated, U.S.) was introduced with a Hanovia® 450 W medium-pressure Hg lamp as the light source. The distance between the Hg lamp and the reaction solution was about 8 cm. A Pyrex® filter was also used in this setup. The lamp was completely immersible in a quartz well, through which cooling water or other chemical solutions (for wavelength separation) can be circulated. This installation also gave a more uniform irradiation of the solution in the horizontal direction.

When performing experiments using the xenon lamp, mercury-free dry nitrogen was flushed through a 1000 mL quartz glass reaction bottle containing 250 mL reactant solution at a flow rate of 300 mL min^{-1}. When using the turntable reactor, the nitrogen flow rate was 75 mL and a 6-mL reaction bottle containing 3 mL solution was used.

A Shimadzu® UV-2100 UV-Visible Recording Spectrometer was employed in the establishment of a standard working curve of the Fe^{2+}-complex and the analyses of the exposed actinometer solutions. It was also used to the record of the absorption spectra of HgS_2^{2-} at the wavelengths of interest.

The elemental mercury formed in this light-induced reduction was collected in the gas phase on gold traps after a 0.1 M perchloric acid trap. The perchloric acid solution, which was placed between the reaction vessel and the sampling gold trap, is considered an ideal trap for any divalent Hg brought over from the reactant solution through physical processes. For a detailed setup see Munthe[21] for the xenon lamp system and Figure 1 for the mercury lamp installation.

The volatile mercury was sampled at regular time intervals and analyzed using a DC-plasma Atomic Emission Spectrometer (DC-AES).[22] Based on the characteristic absorption of elemental Hg^0 at 253.7 nm, the Hg^0 formed through photolysis was also positively

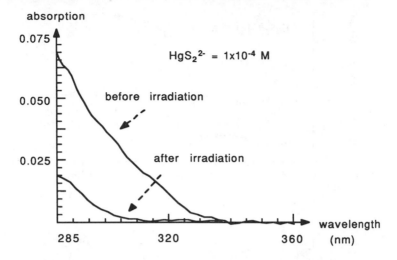

Figure 2 Absorbtion spectrum of HgS_2^{2-} solution before and after irradiation.

identified using a Atomic Absorption Spectrophotometer (AAS) which was directly connected to the reaction vessel.

All glassware used was soaked in aqua regia overnight and rinsed carefully and repeatedly with tap, distilled, and Milli-Q® water before use. All the experiments were conducted in a thermostatted room at temperatures of $20 \pm 0.5°$ C.

III. RESULTS AND DISCUSSION

A. HgS_2^{2-}

This compound was chosen as a model for other sulfur-containing species, such as $Hg(SH)_2$ and $HgS(SH)^-$, since these species are difficult to isolate in aqueous solution without the precipitation of HgS. The S-atoms of HgS_2^{2-} will probably be bound to the H-atoms of the surrounding water molecules, which means that the difference between the HgS_2^{2-} ion and the $HgS(SH)^-$ and $Hg(SH)_2$ complexes would be smaller in the aqueous solution than one would first expect by merely looking at the chemical formulas. Therefore, it is not unlikely that HgS_2^{2-}, $HgS(SH)^{-2}$, and $Hg(SH)_2$ have similar light absorption and photochemical behavior. The absorption spectrum of HgS_2^{2-} was recorded at a wavelength ≥ 290 nm (Figure 2).

Solutions of two different concentrations of HgS_2^{2-} were irradiated with the xenon lamp (Figure 3). Dark control experiments were always performed first until only small amounts of mercury were detected in the gas phase. The sudden increase of gas phase mercury when the lamp was turned on implied that a photo-induced reaction was occurring in the solution.

In order to evaluate the kinetics of this process, quartz test tubes containing 4 mL of 1×10^{-4} M HgS_2^{2-} solution were irradiated using the same lamp at a somewhat decreased distance between the lamp and the solution. A black precipitation, most likely HgS, was formed after 6 h irradiation. HgS was also produced when photolyzing a methyl mercuric sulfide solution.[23]

Analysis of the supernatant of the irradiated solution indicated that there was no dissolved mercury left in solution. Also, the absorption spectrum of the irradiated solution showed that there was no HgS_2^{2-} in the solution (Figure 2). When 1 mL 0.2 N BrCl was applied to the test tube, the precipitation was completely dissolved. Analysis for mercury in the resulting solution almost quantitatively accounted for the disappearance of mercury from the solution.

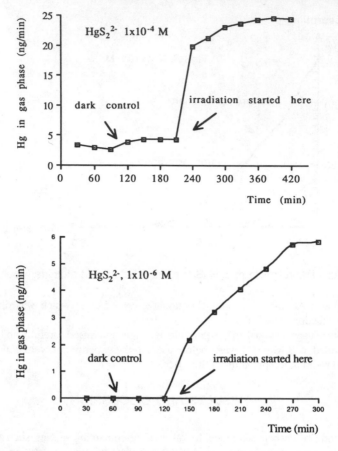

Figure 3 Results of irradiating HgS_2^{2-} solution at concentrations of 1×10^{-6} and 1×10^{-4} M (xenon lamp and a 1000 mL reaction bottle).

These results indicate that the photolysis of HgS_2^{2-} results in the production of both Hg^0 and HgS, with HgS being predominant. The formation of HgS is probably caused by photodegradation of S^{2-} or SH^-, which lowers the concentration of S^{2-} and shifts the equilibrium: $HgS_2^- \leftrightarrow HgS(s) + S^{2-}$ to the right. This notion was supported by an individual experiment. Yellow sulfur was precipitated out from a 0.01 M Na_2S solution after 5 h irradiation, suggesting that the photo-induced reactions cause the concentration of S^{2-} to decrease. It is not clear, however, which sulfur species was affected because both S^{2-} and SH^- exist in the solution at equilibrium: $S^{2-} + H_2O \leftrightarrow SH^- + OH^-$. A similar phenomenom was observed by Kern[24] when a solution of mercury dibenzyl mercaptide in benzene was irradiated. Both Hg^0 and HgS were formed by different mechanisms with different yields. The degradation from the $Hg^{2+}-S$ complex to HgS(s) is probably of minor interest under environmental conditions, because there is probably a steady-state concentration of S^{2-} in lake and sea water (see Dyssen[17] and references thereby).

The rate of formation of Hg^0 from HgS_2^{2-} photolysis is about 2 to 4 times lower than that from $Hg(OH)_2$, if we compare results from experiments performed at the same concentration levels (Figures 3 and 4). This was not expected, considering the strong UV absorption at wavelengths larger than 290 nm of HgS_2^{2-} (almost 500 times stronger than that for $Hg(OH)_2$). A possible explanation may be that the quantum yield Φ is very small, less than 1×10^{-4}.

According to a model calculation,[17] the most important dissolved species are $Hg(SH)_2$, $HgS(SH)^-$, and HgS_2^{2-} in low salinity water at pH 4 to 9, $[Cl^-] = 0.0002\ M$ and $[Hg^{2+}] = 2\ ng\ L^{-1}$. In order to observe light absorption clearly at the wavelengths of interest, solutions of high concentration are needed. At $[HgS_2^{2-}] = 1 \times 10^{-4}\ M$, the light absorption is only 0.075 (Figure 2). It is very difficult to prepare stable solutions of $HgS(SH)^-$ and $Hg(SH)_2$ by acidifying the solution of HgS_2^{2-} at this, or higher, concentration levels, since HgS will precipitate. Hence, no further effort has been made concerning the photolysis of sulfur-containing mercury compounds in solution.

B. $Hg(OH)_2$

The early investigations used the xenon lamp setup and a 1000 mL reaction bottle containing 250 mL solution. From the result presented in Figure 4, it is very difficult to evaluate the kinetic data. The only conclusion we could draw was that photo-induced reduction of Hg (II) occurred in the solution.

After irradiation, there was an initial sharp increase in gaseous mercury followed by an almost constant production rate for 3 h, although a slight decrease was expected due to the lowering of the concentration of $Hg(OH)_2$ in the solution. The rate of photolysis increased with increasing Hg concentration, but not in proportion to the increase in concentration of divalent mercury in the original solution. Probably, at high concentrations, the Hg^0 formed in solution cannot be effectively transported out of the solution.

Other possible explanations to the observed results could be that the reaction depends on other species, e.g., photosensitizers, which might exist in the solution in trace amounts or it could be a wall reaction. However, preliminary results with 150 mL of small pieces of quartz glass inside the 1 L reaction bottle showed no sign of increased photoreduction rate, which means that the wall reaction probably could be ruled out. The existence of a photosensitizer is also dubious based on the fact that Milli-Q® water and reagent grade (P.A.) chemicals were used throughout the investigation.

In this investigation, photochemical reactions occur in the aqueous phase and the product is transported to the gas phase by N_2 as a carrier gas. The transfer of elemental mercury is affected by the interfacial resistance, the concentration gradient, and the flow rate (300 mL min^{-1} in the xenon lamp investigation) of the carrier gas used. Hence, a reaction bottle with a large volume, an insufficient N_2 flow rate, and a defective bubbler frit in the reaction bottle may all contribute to less Hg^0 being transferred to the gas phase.

In order to investigate whether or not there was a mass transportation problem, a further series of experiments were performed using a small, 6-mL bubble reactor with a fritted bottom. A nitrogen stream (75 mL min^{-1}) passed upwards through the frit providing extensive agitation of the mercury solution (3 mL).

Kinetic data were evaluated by following the concentration decrease of divalent mercury in the solution. It was assumed that all the elemental mercury formed through photolysis was transferred from the solution to the gas phase and that there were no mercury losses in the system. In this case the following equation can be used:

$$[Hg^{2+}]_t = [Hg^{2+}]_0 - [Hg^0]_t \qquad (6)$$

where subscripts 0 and t refer to the reaction time at zero and t. By measuring Hg^0 in the gas phase and the known concentration of $[Hg^{2+}]_0$, $[Hg^{2+}]_t$ can be calculated. The corresponding first-order plots of the photolysis of $Hg(OH)_2$ solution at two different concentrations are presented in Figure 5. The special case $C_0 = C_t$ was not used in order to avoid forcing the plot through the origin.[25]

From the slope of these two plots, first-order rate constants were obtained, $(7.85 \pm 1.26) \times 10^{-3}\ min^{-1}$ (n = 3) and $(6.46 \pm 1.84) \times 10^{-3}\ min^{-1}$ (n = 3) for mercury at starting concentrations of $1 \times 10^{-6}\ M$ and $1 \times 10^{-7}\ M$, respectively, with a total average value of

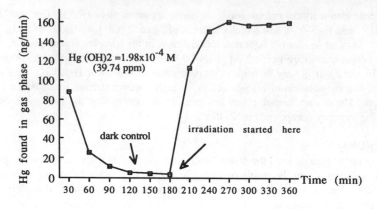

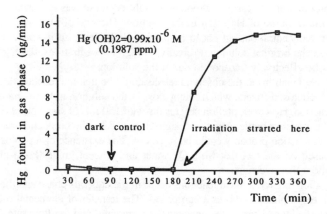

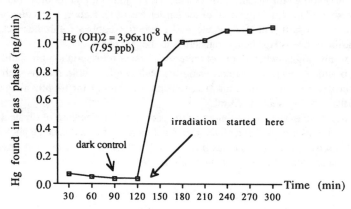

Figure 4 Results of irradiating $Hg(OH)_2$ solution at concentrations of 1×10^{-8}, 1×10^{-6}, and 1×10^{-4} M (xenon lamp and a 1000 mL reaction bottle).

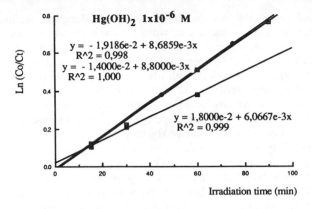

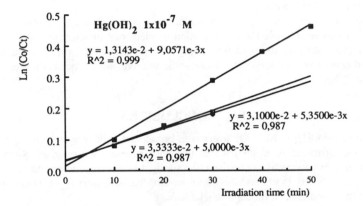

Figure 5   Plots of first-order photolysis of Hg(OH)$_2$ solution at concentrations of 1×10^{-6} and 1×10^{-7} M (mercury lamp and a 6 mL reaction bottle).

$(7.2 \pm 1.6) \times 10^{-3}$ min^{-1}, or $(1.20 \pm 0.27) \times 10^{-4}$ s^{-1}. The fact that the difference between the two first-order rate constants is minor for the two different Hg concentrations with more effective nitrogen streams supports our assumption that there was a mass transportation problem in the xenon lamp experiments.

There is only one estimation of the molar extinction coefficient for Hg(OH)$_2$ at a wavelength longer than 290 nm in the literature,[14] due to the rather low solubility of HgO in water (maximum 2.25×10^{-4} M) combined with the small absorption in this region. The molar extinction coefficient in this range was estimated by Strömberg[16] to be around 1 or less (L mol^{-1} cm^{-1}).

Due to the difficulties in measuring molar extinction coefficiencies in this region we have to make an assumption: the molar extinction coefficients decrease from 1 at 290 nm, passing through 0.12 at 360 nm, down to 0 at 370 nm. The quantum yield can then be estimated by using Equation 2, to approximately 0.14 for the mercury lamp experiment.

The actinic flux of 60° N (Stockholm's latitude) was taken from the literature.[26] Using Equation 4 and a quantum yield of 0.14, the k$_{sp}$ is calculated to be 3×10^{-7}s^{-1}. This k$_{sp}$ corresponds to a half-life of 600 sunlight hours. Such a long half-life means that photoreduction is too slow to have any significance for the atmospheric processes in clouds. However, if we consider a lake surface, the situation may be different. Taking an unpolluted natural lake in Sweden as an example, consider a 1 m³ water volume at the lake surface (1 m² × 1 m), which contains approximately 4 µg of mercury.[8] We further assume that the light intensity is constant throughout the whole cubic meter and that $Hg(OH)_2$ is the dominant species (a rather crude assumption) together with the half-life of mercury of 600 sunlight hours. An Hg^0 emission rate of approximately 5 ng $h^{-1}m^{-2}$ was obtained. This is likely to be an overestimation of the rate of mercury emission in this volume, since the light would be absorbed and its intensity be attenuated with depth. However, this decrease could be compensated for by an additional photoreduction of Hg(II) at depths greater than 1 m.

An evasion rate of 5 ng $h^{-1}m^{-2}$ is of the same order of magnitude as those estimated from field measurements (3 to 20 ng $h^{-1}m^{-2}$; Xiao et al.[12]). It is only in pristine natural waters that $Hg(OH)_2$ would be the predominant mercury species.[21] However, in many lakes, a large fraction of the Hg^{2+} is most probably bound to humic matter.[27] Therefore, further investigations concerning Hg photolysis in the presence of humic matter is needed to ascertain their influence on Hg^0 formation.

Dark control experiments suggest that in addition to photolysis, there exists other processes that may also be responsible for the formation of elemental mercury in aqueous solution. Possible explanations are the reduction of Hg^{2+} to its elemental form by trace reductants introduced either from the reagents used or some microorganisms normally present in laboratory air.[28-30]

IV. CONCLUSIONS

Irradiation of $Hg(OH)_2$ in Milli-Q® water with broadband light (wavelength >290 nm) resulted in the formation of elemental mercury with an estimated quantum yield of 0.14. Extrapolated to environmental conditions, the sunlight photolysis rate constant, k_{sp}, was estimated to be 3×10^{-17} s^{-1} and the corresponding half-life time ($t_{1/2}$) 600 sunlight hours under summer conditions at Stochholm's latitude.

Although elemental mercury was obtained when solutions of HgS_2^{2-} were irradiated under the employed experimental conditions, the main reaction product, however, was probably HgS due to the decrease in S^{2-} concentration by photolysis. The photoreduction of mercury in the form of Hg^{2+}-S complexes is probably of minor importance in the atmospheric clouds ($t_{1/2}$ too large). However, the degradation pathway (from Hg^{2+} to Hg^0) could perhaps be of great importance for lake and sea water where a steady-state concentration of S^{2-} occurs and the total mass of Hg is very large.

Speciation of divalent mercury in natural water is not possible to examine experimentally in field experiments. This is due to the extremely low concentrations of mercury found in the natural waters and the presence of a multitude of ligands that are capable of forming strong bonds with Hg(II). $Hg(OH)_2$ is the predominant form of divalent mercury in aqueous solution in the absence of ligands and in the pH range 4 to 8. It is also difficult to theoretically predict the speciation of Hg(II) in natural waters since the concentrations of all relevant species are not always known.

The results presented in this chapter are only relevant to the photo-induced reaction of aqueous $Hg(OH)_2$ and HgS_2^{2-}. It is not possible to identify these species in natural waters and thus it is not possible to assess the environmental significance of these processes with certainty. The results do, however, indicate that the field observations of processes involving the sunlight-induced reduction of Hg can be explained by the photo-reduction of simple inorganic compounds. The photolysis of $Hg(OH)_2$ is potentially important only in very clean waters such as clouds, precipitation, and nutrient poor lakes with low concentrations of humic matter.

REFERENCES

1. Zika, R. G. and Cooper, W. J., Eds., *Photochemistry of Environmental Aquatic System*, *ACS Symposium Series* 327, American Chemical Society, Washington D.C., 1987, preface VII.

2. Roof, A. A. M., Basic principles of environment photochemistry, In: *The Handbook of Environment Chemistry*, Vol. 2, Part B., Hulzinger, O., Ed., Springer-Verlag, Berlin, 1982.

3. Boule, P., Guyon, C., Tissot, A., and Lemaire, J., Specific phototransformation of xeno-biotic compounds: chlorobenzenes and halophenols, In: *Photochemistry of Environmental Aquatic System*, *ACS Symposium Series* 327, American Chemical Society, Washington, D.C., 1987, chap. 2, 10.

4. Huang, H., Hodson, R. E., and Lee, R. F., Degradation of phenol and chlorophenols by sunlight and microbes in estuarine water, *Environ. Sci. Technol.*, 20, 1002, 1986.

5. Zafiriou, O. C., Joussot-Dubin, J., Zepp, R. G., and Zika, R. G., Photochemistry of natural water, *Environ. Sci. Technol.*, 18, 359, 1984.

6. Zafiriou, O. C., Marine organic photochemistry previewed, *Mar. Chem.*, 5, 497, 1976.

7. Lindqvist, O., Ed., Mercury as an environmental pollutant, *Water Air Soil Pollut.*, 56, 1, 1991.

8. Lindqvist, O., Johansson, K., Austrap, M., Andersson, A., Bringmark, L., Hovsenias, G., Håkanson, L., Iverfeldt, Å., Meili, M., and Timm, R. Mercury in the Swedish environ-ment—recent research on causes, consequences and corrective methods, *Water Air Soil Pollut.*, 55, 1, 1991.

9. Lindqvist, O. and Rodhe, H., Atmospheric mercury—a review, *Tellus*, 37B, 136, 1985.

10. Zepp, R. G., Wolf, N. L., and Gordon, J. A., Photodecomposition of phenylmercury com-pounds in sunlight, *Chemosphere*, 3, 93, 1973.

11. Fitzgerald, W. F., Manson, R. P., and Vandal, G. M., Atmospheric cycling and air-water exchange of mercury over mid-continental lacustrine regions, *Water Air Soil Pollut.*, 56, 745, 1991.

12. Xiao, Z. F., Munthe, J., Schroeder, W. H., and Lindqvist, O., Vertical fluxes of volatile mercury over forest soil and lake surfaces in Sweden, *Tellus*, 43B, 267, 1991.

13. Iverfeldt, Å., Structure, Thermodynamic and Kinetic Studies of Mercury Compounds. Application Within the Environmental Mercury Cycle, Thesis, Department of inorganic Chemistry, University of Göteborg, S-412 96, Göteborg, Sweden, 1984.

14. Brosset, C., The behavior of mercury in the physical environment, *Water Air Soil Pollut.*, 34, 145, 1987.

15. Munthe, J. and McElroy, W. J., Some aqueous reactions of potential importance in the atmospheric chemistry of mercury, *Atmos. Environ.*, 26A, 553, 1992.

16. Strömberg, D., Some Mercury Compounds Studied by Relativistic Quantum Chemical Methods, Thesis, Department of Inorganic Chemistry, University of Göteborg, S-412 96, Göteborg, Sweden, 1990.

17. Dyrssen, D. and Wedberg, M., The sulphur-mercury (II) system in natural waters, *Water Air Soil Pollut.*, 56, 507, 1991.

18. Zepp, R. G., Experimental approaches to environmental photochemistry, In: *The Hand-book of Environmental Chemistry*, Hutzinger, O., Ed., Springer-Verlag, Berlin, Vol. 2, Part B, 1982.

19. Calvert, J. G. and Pitts, J. N., *Photochemistry*, John Wiley & Sons, New York, 1966.

20. Benes, P. and Havlik, B., Specification of mercury in natural waters, In: *The Biogeochem-istry of Mercury in the Environment*, Nriagu, J. O., Ed., Elsevier/North-Holland, Am-sterdam, 1979, chap. 5.

21. Munthe, J., The Redox Cycling of Mercury in the Atmosphere, Thesis, Department of Inorganic Chemistry, University of Göteborg, S-412 96, Göteborg, Sweden, 1991.

22. Iverfeldt, Å. and Lindqvist, O., Distribution equilibrium of methyl mercury chloride between water and air, *Atmos. Environ.*, 16, 2917, 1982.

23. Baughman, G. L., Gordon, J. A., Wolfe, H. L., and Zepp, R. G., Chemistry of Organomercurials in Aqueous Systems, U.S. EPA Report, EPA-660/3–73–012, September, 1973.

24. Kern, R. J., The photolysis of some mercury dimercaptides, *J. Am. Chem. Soc.*, 75, 1865, 1953.

25. Dulin, D. and Mill, T., Development and evaluation of sunlight actinometers, *Environ. Sci. Technol.*, 16, 815, 1982.

26. Svenson, A. and Björndal, H., Test av fotokemisk omvandling i akvatisk miljö, *Swedish Environ. Res. Inst., Publication*, 887, 1988, in Swedish.

27. Allard, B. and Arsenie, I., Abiotic reduction of mercury by humic substances in aquatic system—an important process for the mercury cycle, *Water Air Soil Pollut.*, 56, 457, 1991.

28. Toribara, T. Y., Shields, C. P., and Koval, L., Behaviour of dilute solution of mercury, *Talanta*, 17, 1025, 1970.

29. Tokunaga, S., Some observations on the loss of mercury from aqueous solutions, *Anal. Sci.*, 2, 89, 1986.

30. Krivan, V. and Haas, H. F., Prevention of loss of mercury(II) during storage of dilute solutions in various containers, *Fresenius Z. Anal. Chem.*, 332, 1, 1988.

SECTION VII

Pollution and Remediation

Natural Gas Industry Sites Contaminated with Elemental Mercury: An Interdisciplinary Research Approach

David S. Charlton, John A. Harju, Daniel J. Stepan,
Vit Kühnel, Craig R. Schmit, Raymond D. Butler,
Kevin R. Henke, Frank W. Beaver, and James M. Evans

CONTENTS

I. INTRODUCTION

The past use of elemental mercury in monitoring and control instrumentation at natural gas industry sites has resulted in soils, sediments, and the possibility of groundwater becoming contaminated with mercury. The most common types of mercury-filled instrumentation in the gas industry are manometers that are used to measure the volume of gas flowing through metering stations.

The Energy and Environmental Research Center (EERC) has been contracted by the Gas Research Institute (GRI), with support from the U.S. Department of Energy (USDOE), to investigate potential issues related to, and remediation technologies for, elemental mercury spills. The investigation of elemental mercury spills is an outgrowth of a broader research program designed to address issues related to a variety of potential contaminants in gas industry wastes and products. A goal of this research program is to provide the gas industry with sufficient information about the presence and behavior of mercury in the subsurface to be able to deal with contamination in the most efficient and cost-effective manner, while protecting human health and the environment.

II. SOURCES OF MERCURY CONTAMINATION

Mercury-filled flowmeters (manometers) have been the traditional instrumentation used in the gas industry for the past half century. Flowmeters may contain 8 to 10 lb (3.6 to 4.5 kg) of mercury. Soils in the immediate area of some mercury flowmeters have become contaminated with mercury due to leakage, spills, equipment failure, pressure surges, vandalism,

1–56670–066–3/94/$0.00+$.50
© 1994 Lewis Publishers

and operator error. A better understanding of these causes of mercury contamination has resulted in more effective containment systems and improved management practices, resulting in a drastic reduction in mercury reaching the soil surface.

Gas-metering sites still exist that were contaminated in the past with elemental mercury. Because of the small amounts of elemental mercury involved, most of these sites have relatively small volumes of mercury-contaminated soil, approximately 1 to 2 yd^3 (0.75 to 1.5 m^3) per site. Nevertheless, the gas industry is interested in the most efficient and cost-effective technologies for remediating these sites.

III. ENVIRONMENTAL REGULATIONS

Mercury first became a concern within the gas industry as a worker-safety issue. Gas companies voluntarily began cleaning up their sites because of this issue. Site cleanups resulted in mercury-contaminated soils that were sent to hazardous waste landfills for disposal. The land disposal of hazardous waste is regulated under the Resource Conservation and Recovery Act (RCRA). Of particular concern, therefore, to the gas industry was the May 8, 1992 deadline under the Land Disposal Restrictions (LDR), also called "Land Ban".[1] After that deadline, certain types of mercury-contaminated wastes could no longer be sent to hazardous waste landfills without meeting specific treatment requirements.

Mercury-contaminated soil typical of many gas-metering sites, if excavated, would probably be classified as a characteristic waste, D009, under RCRA. A D009 waste is one which has leachable mercury above a specific regulated limit. These D009 soils are subject to specific treatment requirements under the LDR. For D009 wastes with total mercury concentrations greater than 260 mg/kg, the required Best Demonstrated Available Technology (BDAT) is roasting or retorting, with condensation and recovery of the volatilized mercury. Within the gas industry, there are concerns that roasting or retorting may not be the most economical and efficient technology for dealing with mercury-contaminated soils. Another concern is that the national capacity does not exist to deal with the potential volume of contaminated soils that could require treatment.

The regulations do provide mechanisms for addressing these two concerns. One available mechanism is to apply for a treatability variance when the BDAT is inappropriate for the waste. This variance would allow the use of another technology besides the BDAT.

Another available mechanism is to apply for a national capacity variance when it can be demonstrated that there is insufficient capacity nationwide to handle the volume requiring treatment. This variance would allow an extension to the date when all D009 wastes are subject to the restrictions under the LDR. However, both of these mechanisms can cause substantial delays in initiating a remediation effort.

An additional regulatory consideration is that individual states can impose corrective action levels, the levels that trigger a cleanup and levels that must be achieved after the cleanup. As an example, a number of states currently require cleanup to background levels, but states are often willing to negotiate the cleanup level on a case-by-case basis.

The regulatory issues discussed above are primarily those affected by RCRA and its amendments. However, there are other federal programs and regulations that can also affect the cleanup of mercury-contaminated soils, including those related to the Occupational Safety and Health Administration (OSHA); the comprehensive Environmental Response, Compensation, and Liability Act (CERCLA, "Superfund"); and the Clean Air Act (CAA) amendments.

IV. MERCURY RESEARCH PROGRAM

In 1989, GRI saw the need for an interdisciplinary research program to address the potential environmental impacts of contaminants at gas industry sites, in the context of a changing regulatory framework. The EERC, with the support of both GRI and the USDOE, subsequently designed and implemented a research program to develop a better understanding of complex interactions between mercury and a range of variables at gas industry sites. The primary elements of this mercury research program include:

- A mercury workshop that was organized and conducted for the gas industry at GRI headquarters in Chicago, Illinois on February 10 and 11, 1992. Representatives of the gas industry, the research community, and remediation companies participated in the workshop. A proceedings volume is available from GRI.[2]
- A literature review that was conducted to identify relevant mercury-related literature, and a bibliographic database was constructed for use by gas industry personnel and researchers. A summary report addressing key mercury issues was written based upon the literature review.[3]
- Computer models that were assessed for their applicability to mercury migration in the subsurface.
- A risk assessment model that was developed for mercury-contaminated sites. Risk assessment models are expected to be useful as mechanisms for setting site-specific cleanup levels or for ranking sites for remediation.
- Sampling, preservation, and analytical protocols for mercury-contaminated solids and liquids that were evaluated and developed as necessary.
- Laboratory experiments that evaluated (1) leaching techniques applied to mercury-contaminated materials and (2) mercury speciation in subsurface environments.
- Mercury-contaminated gas industry field sites that were characterized and instrumented in several parts of the country. Sites have been selected to represent a range of site-specific variables.
- Existing and developing remedial technologies with the potential to deal with mercury in soils that have been reviewed. In addition, one remedial technology has actually been demonstrated, and three others were tested in response to a recently circulated Request for Proposal (RFP).

The following sections include details about some of the major research program elements listed above.

V. COMPUTER MODELING AND RISK ASSESSMENT

Models can serve a wide variety of purposes for investigating mercury at gas industry sites. In investigating the movement of a contaminant in the subsurface, transport and fate models can help elucidate the dynamics of movement. In toxicological and risk assessment studies, the models can be used to predict health threats and to prioritize sites for remediation.

The subsurface-related models that are used in the gas industry estimate the transport and fate of contaminants in groundwater, sediment, and soil. Specifically, these models approximate the extent of contamination at a site or predict future contaminant transport and fate. These transport and fate models can be very useful in remediation efforts.

Transport models for the liquid flow of elemental mercury in the subsurface have been investigated as a research tool in increasing the understanding of differences in behavior in different geologic materials, in the depth of penetration, and in the initial source volume. A one-dimensional computer model for nonaqueous phase liquids (NAPLs) was modified for the infiltration and redistribution of elemental mercury because it showed promise for this application. The physical parameters of elemental mercury that dominate this flow model are density (13.5 g/cm^3) and surface tension (420 dyne/cm), both of which are high compared to other liquids.

Several models exist that simulate vapor movement above a contaminant source. Modeling the flow of mercury in the vapor phase is extremely complex, because the rates of volatilization are a function of many factors, including, but not limited to, moisture content and air permeability of the soil, temperature, relative humidity, barometric pressure, and size of the mercury droplets. As an example, one of those variables, temperature, which can range from -20 to $70°$ C within enclosed metering stations, can change the saturation vapor pressure of mercury by three orders of magnitude. The complexity of mercury vapor movement may be so great that modeling may not be possible or, at best, the results may be highly suspect.

In aqueous solutions, chemical properties dominate the transport and fate of mercury species. The key chemical characteristics are solubility and the partitioning coefficients. Because the solubility of elemental mercury is very low and the partitioning coefficients are very high, the expected distance that dissolved mercury will travel is small.

A risk assessment model is currently under development for gas industry sites contaminated by mercury. Key parameters that will be entered into a risk model are being measured at mercury-contaminated sites during the field investigation phase of research activities. Risk assessment models are expected to be useful as mechanisms for setting site-specific cleanup levels and/or ranking sites for remediation.

VI. FIELD RESEARCH SITE INSTRUMENTATION

Field research sites were selected and instrumented in various regions of the U.S. to represent a combination of key climatologic, hydrologic, and geologic variables. In this way, the results will have application beyond the study sites themselves. The purpose of this task is to understand the distribution of mercury at gas industry sites and the interaction of mercury with key site components. This understanding is critical in making decisions concerning the assessment and remediation of mercury-contaminated sites.

The field sites were instrumented with stainless steel monitoring wells (piezometers). Each piece of instrumentation undergoes thorough decontamination procedures prior to installation, including steam cleaning, detergent wash, and a final rinse with deionized water to enable the collection of truly representative groundwater samples.

To further ensure the utmost reliability in sample integrity, dedicated pumps constructed of stainless steel and Teflon® were installed in each of the monitoring wells. In conjunction with the drilling at each site, continuous cores were collected, and both geologic and chemical analyses were conducted on selected representative portions of each core.

VII. COUPLED LABORATORY RESEARCH

Selected sediment cores were subjected to batch sorption experiments to determine site-specific distribution coefficients for various chemical species of mercury. Additionally, TCLP, the Toxicity Characteristic Leaching Procedure (EPA Method 1311), and SGLP, the Synthetic Groundwater Leaching Procedure[4] tests were performed on selected cores whose total mercury analyses indicated the presence of elemental mercury or mercury compounds. The determination of mercury species present allows the direct comparisons of data obtained from various leaching tests with total mercury analyses. In addition, the leachability of various chemical forms of mercury can be determined.

Batch sorption experiments utilizing various mercury compounds were conducted on sampling equipment, such as pumps, tubing, and containers that are or might be used at mercury-contaminated sites. Similar experiments were conducted on well construction materials such as stainless steel, silica sand, and bentonitic seals. These experiments will serve as a basis for the interpretation of data obtained from chemical analyses of groundwater samples, as well as the determination of the suitability of construction materials for use in groundwater instrumentation for mercury.

The majority of samples were analyzed by cold vapor atomic absorption spectrometry (CVAAS), with the use of a gold trap and to achieve lower detection limits. Where the detection limits of CVAAS were not adequate, cold vapor atomic fluorescence spectrometry (CVAFS) was used. Standard protocols and techniques were assessed and, as appropriate, additional analytical protocols were developed.

VIII. REMEDIATION OPTIONS

A major goal of this research program was to develop an understanding of mercury contamination at gas industry sites so that the impacted materials can be remediated in an efficient and cost-effective manner. To evaluate the range of remediation options available to the gas industry, information was solicited from companies and researchers that have available or are developing remedial technologies for mercury-contaminated soils. The remediation options were grouped into six major categories: (1) physical treatment, (2) chemical treatment, (3) immobilization, (4) thermal treatment, (5) electrolytic treatment, and (6) biological treatment. A summary report was written based on this remedial technology review.[5]

A technology based on one of these six remediation options, physical separation, was tested using native soil from the southwestern U.S. spiked with elemental mercury. This particular remedial technology was selected for testing because a unit was already under development as a mobile prototype. Also, preliminary information suggested that the unit, based upon established mining technologies, might be effective in separating elemental mercury from soil.

In order to increase the number of remediation options available to the gas industry, other mercury remediation technologies were selected for development and testing based upon the evaluation of proposals solicited through the circulation of an RFP. These technologies included a pilot-scale thermal treatment process, a pilot-scale chlorine leaching process combined with physical separation, and a bench-scale oxidative chemical leaching process.[6]

IX. SUMMARY

This ongoing research program has been designed to address a range of key issues regarding elemental mercury contamination at gas industry sites. In addition, the results of this program may be useful for addressing mercury contamination problems in other industries. An integrated, multidisciplinary research approach is being used to develop a better understanding of the complex interactions that can occur between mercury and a range of variables at the sites. A better understanding of these complex interactions is crucial to the design and evaluation of technologies to remediate mercury-contaminated materials at gas industry sites.

REFERENCES

1. Federal Register, Rules and regulations, *Fed. Reg.,* 55:(106), 22520, 1990.
2. Charlton, D.S. and Harju, J.A., Eds., *Proceedings of the Workshop on Mercury Contamination at Natural Gas Industry Sites,* Gas Research Institute, Chicago, 1992, GRI-92/0214, 235 p.
3. Henke, K.R., Kühnel, V., Stepan, D.J., Fraley, R.H., Robinson, C.M., Charlton, D.S., Gust, H., and Bloom, N.S., Critical Review of Mercury-Contamination Issues Relevant to Nanometers at Natural Gas Industry Sites, Gas Research Institute, Chicago, GRI-93/0117, 1993, 92 p.

4. Hassett, D.J., A generic test of leachability: the synthetic groundwater leaching procedure, In: *Proceedings of the Waste Management for the Energy Industries Conference*, University of North Dakota, Grand Forks, 1987, 30.

5. Stepan, D.J., Fraley, R.H., Henke, K.R., Gust, H., Hassett, D.J., Charlton, D.S., and Schmit, C.R., A Review of Remediation Technologies Applicable to Mercury Contamination at Natural Gas Industry Sites, Gas Research Institute, Chicago, GRI-93/0099, 1993, 43 p.

6. Stepan, D.J., Fraley, R.H., and Charlton, D.S., The Demonstration and Testing of Technologies for the Remediation of Mercury-Contaminated Soils, Gas Research Institute, Chicago, (in preparation).

Could the Geothermal Power Plant at Mt. Amiata (Italy) be a Source of Mercury Contamination?

R. Ferrara, B.E. Maserti, A. De Liso, H. Edner,
P. Ragnarson, S. Svanberg, and E. Wallinder

CONTENTS

ABSTRACT: The results of a first direct determination, obtained with the lidar technique, of the mercury flux emitted from a geothermal power plant are reported in this chapter. The value determined for the Bellavista power plant (20 MWe) (Mt. Amiata, Italy) proved to be 158 to 210 kg/year, or one third the value measured with the same technique for a large chlor-alkali plant (120,000 Mg/year of chlorine). The measured flux value is particularly high, probably because the power plant is located in a highly mineralized zone. The determinations of the mercury concentration in the air and soil reveal that the contamination of the environment surrounding the geothermal power plant is limited to a restricted zone (500 to 600 m from the plant) with an asymmetry linked to the direction of the prevailing winds.

I. INTRODUCTION

Geothermal energy is the energy extractable from the naturally or artificially cycling water in areas where the thermal gradient of the ground (temperature behavior as a function of depth) is higher than the usual. It is generally used both for electrical energy production and for domestic space heating and horticulture.

Only the "high-enthalpy" geothermal fields (with a fluid temperature greater than 200° C) are well suited for direct electricity production, while those with a low enthalpy (fluid temperature lower than 200° C) are at present used exclusively for heating.

Regions with geothermal potential are widely scattered throughout the world (Central America, Confederation of Independent States, Hawaii, Iceland, Italy, Japan, New Zealand, and the Philippines). The first example of electrical power generation from geothermal sources dates back to 1904 at Larderello (Italy). Today 40 countries are interested in geothermal fields to produce electricity. Barbier[1] reported that production was 936 MWe in the U.S. and 580 MWe in the Philippines.

In Europe, Italian fields presently supply electrical energy for about 3% (600 MWe) of the country's total production; this figure is about 10% of the corresponding worldwide capacity. Besides Italy, only Iceland (39 MWe), Turkey (20 MWe), Greece (2 MWe), and Portugal (3 MWe) have installed practical power plants.[2]

In Italy the geothermal activities for electric energy production are found mainly in the region of Tuscany (Larderello-Travale and Mt. Amiata areas). These two fields differ in that the Larderello geothermal field, one of the two deposits of dry steam known in the world, is situated in a nonmineralized region, while the Mt. Amiata hydrothermal system is situated in a strongly mineralized region rich in cinnabar deposits (Ferrara et al.[3]). Despite the fact that geothermal power is considered "clean energy", since 1973 from studies conducted in New Zealand[4,5] it has been ascertained that significant effects on the environment can be produced by geothermal power plants as a result of mercury and arsenic contamination.

In 1975, in the U.S. following the development of research in the field of geothermics for the exploitation of endogenous steam in the production of electric energy, Siegel and Siegel[6] observed that environmental contamination from mercury could have both short- and long-term consequences. Dall'Aglio and Ferrara[7] confirmed in 1986 that geothermal emissions could also constitute a long-term environmental hazard with regard to radon.

Subsequent research on mercury contamination[8-10] has shown that the metal is emitted in mineralized and nonmineralized geothermal areas. The authors report an estimate of the global continental mercury flux from geothermal areas of 60 Mg/year. Phelps and Buseck[9] affirm that the mercury anomalies in soils of Yellowstone National Park, although very high (up to 5000 times the background value), were restricted to the immediate vicinity of the upwelling steam. Analogous results were reported by Varekamp and Buseck[11] in soils from several geothermal areas in the western U.S. The mercury concentration present in the condensed steam from geothermal emissions is quite modest (5 to 100 ng/L).[3] In agreement with the findings of White,[12] much of the mercury escapes with the vapor phase.

The data reported in the literature on the distribution of the mercury concentration in the atmosphere in the vicinity of geothermal power plants were obtained with point monitors and therefore do not express a spatial distribution of the mercury. Only recently, thanks to the use of lidar techniques, have Edner et al.[13] been able to map atmospheric mercury in some Italian geothermal fields.

Taking into account the high degree of mineralization of the Mt. Amiata area, a study has been carried out to verify whether the extraction of geothermal energy is connected with a substantial emission of mercury into the atmosphere.

The determinations were made both with point monitors and a remote sensing technique, which for the first time allowed direct estimation of the mercury flux emitted by a geothermal power plant.

II. STUDY AREA

The Mt. Amiata volcanic area, located in the region of Tuscany (central Italy) and known for its extensive cinnabar deposits and mining activity, covers approximately 400 km^2. The maximum elevation of the mountain is 1738 m. At Bagnore and Piancastagnaio, steam jets used for electric power production are present. Sulfurous thermal springs at temperatures of 50 to 60° C are widely distributed in this area.

The geothermal gradient exceeds 1° C/10 m throughout most of the region, and heat flows average greater than 3 and often 5 μcal/cm^2/sec. Temperatures at the top of the carbonate complex are commonly above 100° C, with temperatures of 150 to 200° C recorded in the Bagnore and Piancastagnaio fields.

The geothermal region covers more than 200 km^2, but the more intense heat flow anomalies correspond to several rather small geologic structures of only 5 to 10 km^2 each. From

1959 through the present, only three power plants have been built in this area. The geothermal power plant examined in this work is that of "Bellavista", located in the Piancastagnaio zone, with an installed capacity of 20 MWe, realized in 1980. The cooling tower, of the mechanical-draft type, has an air flow of 6 to 7 Mm^3/h. This power plant is the prototype of a series of automated modular plants now being constructed in this area.

III. EXPERIMENTAL

Atmospheric mercury was sampled using the point monitors described elsewhere.[14] Air was sucked through two parallel gold traps at a flow rate of 1 L/min by means of a membrane pump. Each trap is made up of a quartz tube containing a thin plate of pure gold (0.8 g) with a helicoidal shape. A timer makes it possible to perform the sampling at set times. The sampling time ranges from 10 to 120 min, depending on the mercury concentration in the air. Electrothermally desorbed mercury was determined with a modified atomic absorption spectrometer. The detection limit was 0.01 ng of mercury.

The lidar system, extensively described by Edner et al.,[15] is housed in a Volvo F610 truck towing a 20 kW diesel generator. The laser radar equipment is capable of generating pulse energies of up to 5 mJ with a linewidth of 0.001 nm at the mercury resonance line (253.7 nm), with a repetition rate of 10 Hz.

The laser beam is directed out into the atmosphere via a large computer-controlled mirror on top of the lidar van. Back-scattered light from the atmosphere is reflected via the same mirror down into a vertical Newtonian telescope, and the lidar signal is detected by a photomultiplier tube. The laser is tuned on and off the resonance line of mercury every second laser shot, allowing differential absorption measurements.

The detection limit is of the order of 2 ng/m^3 of elemental mercury (the only form of mercury detectable with the lidar system). The theoretical spatial resolution is about 1 m. This remote sensing technique, in addition to the determination of the mercury concentration in the atmosphere, also permits an estimate of the flux emitted from a punctiform source once the wind direction and speed are known.

Surface soil samples were collected all around the geothermal power plant, taking into account the direction of the prevailing winds. The soil was dried at 50° C in an electric oven for 72 h, then was sieved with a 250 μm net size sieve. Then 500 mg of the sample was mineralized with 10 mL of a mixture of nitric, sulfuric, and hydrochloric acid (5:3:2) at 120° C under reflux for 60 min. Mercury determinations were performed using a Coleman MAS 50B mercury analyzer.

The results of the analyses on the reference standard material (CRM 277) "Estuarine Sediment" issued by the Commission of the European Communities[16] was 1.75 ± 0.08 μg/g d.w. with respect to the certified value of 1.77 ± 0.06 μg/g.

IV. RESULTS AND DISCUSSION

The determinations made during the month of September 1990 with the lidar near the Bellavista geothermal power plant (Mt. Amiata) yielded an emitted mercury flux value of the order of 18 to 24 g/h. This value must be considered high; it corresponds to an annual atmospheric mercury discharge of 158 to 210 kg, equal to one third the emission of a large chlor-alkali plant (R. Solvay, Livorno, Italy).[17]

Weissberg and Rhode[18] and Robertson et al.[19] estimate mercury release from geothermal power plants at 1 to 2 g/MW/day; this value is much lower than the one we measured directly with the lidar (7 to 9 g/MW/day). This may be due to an erroneous estimate by the cited authors or, much more likely, to the fact that the power plant considered by us is located in a highly mineralized area.

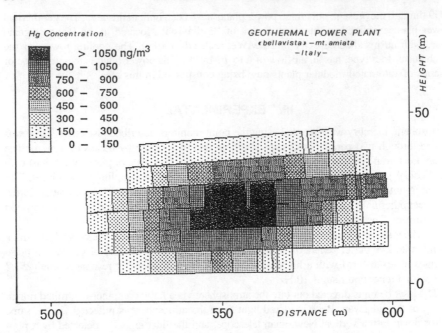

Figure 1 Mercury concentrations determined with the lidar system in a vertical scan through the plume of the "Bellavista" geothermal power plant.

Figure 1 gives an example of the concentrations observed in one section of the plume. These determinations were made approximately 100 m downwind from the plant and about 500 to 600 m from the position of the lidar. Analysis of Figure 1 reveals that the highest mercury concentration is found between 20 and 30 m from the ground, and that the plume is found at a height of less the 50 m above ground level. This rapid fallout of the mercury vapor is due to the rather low height of the cooling tower and to the presence in the plume of numerous small water droplets dragged from the cooling tower by the aspiration system of the air.

To evaluate the effect of this mercury emission on the environment surrounding the power plant, mercury determinations were performed in the soil and air by means of point monitors.

Figure 2 reports the average values of the concentrations observed during the sampling done in the months of September and May.

The mercury concentrations in the air are very high only near the cooling tower where the presence of the plume was detected and around the openings of the tower where air is taken in to cool the water. The concentrations rapidly decrease with increasing distance from the power plant and display the highest values along the direction of the wind. Background values for the area in question are reached at a distance of 200 to 300 m from the plant.

Determinations made with the lidar and point monitors in the zone beneath the plume yielded comparable mercury concentration values; considering the response characteristics of these instruments to the different chemicophysical forms in which mercury may be present in the atmosphere, the obtained results demonstrate that the mercury emitted from the power plant is chiefly found in elemental form. A similar trend was found for the mercury present in the surface soil; from the distribution of the concentrations the influence of the prevailing winds is evident.

The highest values (80 µg/g d.w.) were measured about 20 m downwind from the cooling tower. The soils around the power plant displays high concentrations, due to the fallout of the mercury vapor emitted from the plant, for a radius of some 500 to 600 m. The value of

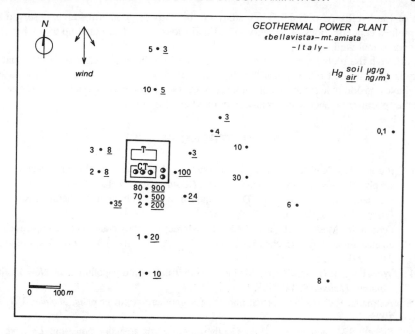

Figure 2 Mercury concentrations in air and surface soil around the geothermal power plant.

0.1 µg/g d.w., found about 1 to 2 km from the power plant, was particularly low considering that the study area is located in a highly mineralized zone. The sampling done in the month of September supplied mercury concentration values comparable to those of the month of May.

The same behavior of the mercury concentration in the soil as a function of the distance from the power plant was observed by Baldi[20] near well Travale 22, which feeds a geothermal power plant in southern Tuscany. However, since the Travale plant is located in a nonmineralized area the mercury concentration in the soil varies from a minimum of 0.35 to a maximum value of 0.5 µg/g—values much lower than those observed at the Bellavista power plant. The mercury enrichment of the soil of the latter geothermal area is much higher since the steam jet that feeds the power plant evidently transports a larger amount of mercury.

V. CONCLUSION

The mercury flux emitted by the geothermal power plant of Bellavista (Mt. Amiata), determined with lidar techniques, proved to be 158 to 210 kg/year. Considering that this power plant has a capacity of just 20 MWe and that in the study area other power plants are in an advanced state of construction, we feel that the exploitation of the geothermal energy of this highly mineralized area may cause considerable mercury enrichment in the surrounding environment.

The mercury determinations made in the soil, in agreement with what has been reported in the literature for other geothermal areas, indicate that the area affected by the contamination is restricted to a belt 500 to 600 m around the power plant. Analogous results on the extent of the area affected by mercury enrichment of the environment have been observed by Bargagli and Barghigiani[21] by means of the use of epiphyte lichens. The authors observed that the mercury concentrations in the lichens sampled around the power plants of Mt. Amiata are 3 to 10 times higher than those measured near the goethermal plants in the unmineralized area of Larderello. Also, Baldi,[20] in a study on the Travale geothermal area, observed that the highest concentrations of mercury in soil and mosses are distributed up to 0.6 km from the geothermal well.

Although the global amount of mercury emitted into the atmosphere by geothermal activity is quite modest (60 Mg/year)[10] with respect to other natural sources, in some cases it may have considerable environmental impact at the local level, making it a legitimate question whether geothermal energy can really be defined as clean energy.

REFERENCES

1. Barbier, E., Geothermal energy in the context of energy in general and electric power supply. National and international aspects, *Geothermics,* 14, 131, 1984.
2. Batchelor, A.S. and Garnish, J.D., The industrial exploitation of geothermal resources in Europe, *Tectonophysics,* 178, 269, 1990.
3. Ferrara, R., Maserti, B.E., and Breder, R., Mercury in abiotic and biotic compartments of an area affected by a geochemical anomaly (Mt. Amiata, Italy), *Water Air Soil Pollut.,* 56, 219, 1991.
4. Weissberg, B.G. and Zobel, M.G.R., Geothermal mercury pollution in New Zealand, *Contam. Toxicol.,* 9, 148, 1973.
5. Axtmann, R.C., Environmental impact of a geothermal power plant, *Science,* 187, 795, 1975.
6. Siegel, S.M. and Siegel, B.Z., Geothermal hazards mercury emission, *Environ. Sci. Technol.,* 9, 473, 1975.
7. Dall'Aglio, M. and Ferrara, G.C., Impatto ambientale dell'energia geotermica, *Acqua e Aria,* 10, 1091, 1986.
8. Klusman, R.W., Cowling, S.S., Culvey, B., Roberts, C., and Schwab, A.P., Preliminary evaluation of secondary controls on Hg in soil of geothermal districts, *Geothermics,* 6, 1, 1977.
9. Phelps, D. and Buseck, P.R., Distribution of soil mercury and the development of soil mercury anomalies in the Yellowstone geothermal area, WY, *Econ. Geol.,* 75, 730, 1980.
10. Varekamp, J.C. and Buseck, P.R., Global mercury flux from volcanic and geothermal sources, *Appl. Geochem.,* 1, 65, 1986.
11. Varekamp, J.C. and Buseck, P.R., Hg anomalies in soils: a geochemical exploration method for geothermal areas, *Geothermics,* 12, 29, 1983.
12. White, D.E., Active geothermal systems and hydrothermal ore deposits, 75th Anniv. Vol., *Econ. Geol.,* 392, 1981.
13. Edner, H., Ragnarson, P., Svanberg, S., Wallinder, E., De Liso, A., Ferrara, R., and Maserti, B.E., Differential absorption mapping of atmospheric atomic mercury in Italian geothermal fields, *J. Geophys. Res.,* 97, 3779, 1992.
14. Ferrara, R., Maserti, B.E., Edner, H., Ragnarson, P., Svanberg, S., and Wallinder, E., Atmospheric mercury determination by Lidar and Point Monitors in environmental studies, *Chemical Speciation and Bioavailability,* (Special suppl.), 29, 1991.
15. Edner, H., Fredriksson, K., Sunesson, S., Uneus, L., and Wendt, W., Mobile remote sensing system for atmospheric monitoring, *Appl. Opt.,* 26, 4330, 1987.
16. Griepink, B. and Muntau, H., BCR Information—Reference Material. Commission of the European Communities. Report EUR 11850 EN, 1988.
17. Ferrara, R., Maserti, B.E., Edner, H., Ragnarson, P., Svanberg, S., and Wallinder, E., Mercury emissions into the atmosphere from a chlor-alkali complex measured with the lidar technique, *Atmos. Environ.,* 26A, 1253, 1992.
18. Weissberg, B.G. and Rhode, A.G., Mercury in some New Zealand geothermal discharges, *N.Z.J. Sci.,* 21, 365, 1978.

19. Robertson, D.E., Crecelius, E.A., Fruchter, J.S., and Ludwick, J.D., Mercury emissions from geothermal powerplants, *Science,* 196, 1094, 1977.

20. Baldi, F., Mercury pollution in the soil and mosses around a geothermal power plant, *Water Air Soil Pollut.,* 38, 111, 1988.

21. Bargagli, R. and Barghigiani, C., Lichen biomonitoring of mercury emission and deposition in mining, geothermal and volcanic areas of Italy, *Environ. Monit. Asses.,* 16, 265, 1991.

Mercury in Soils and Plants in the Florida Everglades Sugarcane Area

W. H. Patrick, Jr., R. P. Gambrell,
P. Parkpian, and F. Tan

CONTENTS

ABSTRACT: Mercury emission from preharvest burning of sugarcane in the Florida Everglades Agricultural Area has been suggested as an important source of atmospheric mercury. A study was carried out utilizing 17 soil locations for soil samples and 9 locations for sugarcane plant samples to determine the amount of mercury subject to emission upon burning. Based on results of this study, emission from both plants and muck soils as a result of the preharvest burning was small, with an average emission of approximately 35 kg for the entire 174,000 ha sugarcane crop in the Everglades.

I. INTRODUCTION

Serious mercury contamination of freshwater fish in the Florida Everglades was discovered in 1989. The body burden of mercury is high enough in some top predator fish such as the largemouth bass (ranging up to 4 ppm fresh weight in extreme cases) that over 400,000 ha of the Everglades is under a health advisory to completely avoid consumption of several species of fish. In an additional 400,000 ha limited consumption of largemouth bass is recommended. Terrestrial animals also appear to be seriously impacted, with raccoons and panthers showing high levels of mercury.[1]

Mercury contamination in south Florida was unexpected, since the area is nonindustrial and very sparsely settled except for the eastern seaboard. Mercurial compounds have been banned for agricultural use for many years and apparently were never used in amounts large enough to cause the levels of contamination seen in the affected animals. A number of sources of the mercury found in fish have been proposed. These include regional emissions from municipal incinerators and other industrial sources, regional drainage, long range transport of mercury as a result of a general increase in Northern Hemisphere atmospheric levels, burning of bagasse at sugarmills, and preharvest burning in the field that is carried out to lower the trash content of the cane going to the mill. This last source of mercury has received considerable attention and the present study was designed to determine the significance of this atmospheric source.

1–56670–066–3/94/$0.00+$.50
© 1994 Lewis Publishers

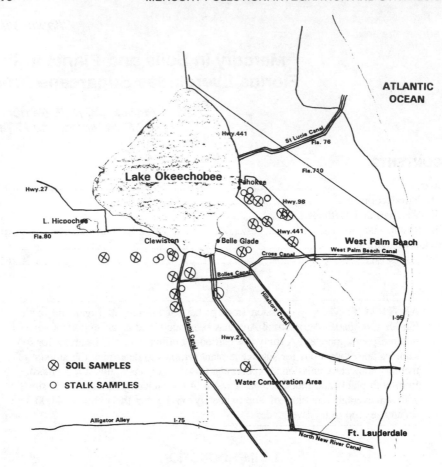

Figure 1 Location of 17 sites for soil samples and 9 sites for sugarcane samples.

II. MATERIALS AND METHODS

Seventeen locations in the Everglades Agricultural Area (EAA) were sampled in October and December 1991. The sites were selected throughout the EAA in an attempt to obtain soil samples that were representative of the area. At each site approximately 5 subsamples were collected down to a depth of 15 to 20 cm and mixed together to provide a composite sample of approximately 2 kg of soil. Thirteen of the sites were muck soils and the other four were mineral soils. Composite samples of green leaves and dead leaves were also collected at the same locations. Most of the dead leaves undergo combustion while the green leaves and tops usually are not burned. The location of the 17 sample sites is shown in Figure 1.

In addition to the soil and leaves, samples of millable cane were collected from nine fields ready for harvest in February 1992. Nine short sections of representative stalks were collected at each site that represented the top one third of the stalk, the middle one third of the stalk, and the lower one third of the stalk. The locations of these nine sugarcane samples are also shown in Figure 1.

The soil samples were thoroughly mixed and a dry weight equivalent of a 1.0 to 1.5 g aliquot taken, weighed to 0.01 g, and placed in a BOD bottle. A separate sample was collected and oven dried at 105° C for 24 hours to determine dry weight of the samples. Green and

Table 1 **Analyses of mercury in soils and plants in Florida Everglades sugarcane area.**

	Mercury concentration in ppm on dry wt. basis	Number of samples
Soils		
Sandy soils	0.036 ± 0.017	4
Muck soils	0.154 ± 0.066	13
Leaves		
Green leaves	0.085 ± 0.043	17
Dead leaves	0.068 ± 0.025	17
Stalks	0.086 ± 0.030	9
Standard reference materials		
NBS 1645 river sediment. True value: 1.1 ppm	1.2 ± 0.05	4
NBS orchard leaves. True value: 0.155 ppm	0.146 ± 0.014	11

dead leaf samples and sugarcane stalks were handled in a similar manner, with the plant samples being ground with a Wiley mill before analysis.

The samples were extracted for mercury using the procedure described by Plumb[2] and analyzed by the cold vapor atomic absorption method. Standard reference materials were used as quality control checks. These were National Bureau of Standards River Sediment (NBS 1645), NBS Orchard Leaves (NBS 1571), and Environmental Protection Agency Municipal Digested Sludge (EPA 2111).

In addition to the sugarcane leaves and trash, another possible source of mercury to the atmosphere during preharvest burning is the muck soil. During the burning operation mercury might be volatilized directed from the organic soil. A study was carried out to determine if mercury is emitted from the soil during burning. Three surface soil samples were collected on March 11, 1992 from three fields that were ready for harvest. The individual soil samples were thoroughly mixed and packed into half pint glass jars to approximately the same density as in the field. Care was taken to minimize any drying of the soil during preparation. Four replications of the jars containing each of the three soils were placed in a field that was scheduled for burning. The jars were buried so that the surface of the soil in the jar was level with the outside soil. Four replications of each soil were retained without burning to provide a check on the burning operation. After burning of the field the jars were removed and along with the unburned samples were analyzed for mercury as described above.

III. RESULTS AND DISCUSSION

Total mercury contents of the soils, green and dead leaves, and stalks are shown in Table 1. The sandy soils were considerably lower in mercury than the organic soils, as expected, since mercury in soils is usually associated with the organic matter. The dead leaves were slightly lower in mercury content than the green leaves, probably reflecting the loss of mercury associated with soluble organic material leaching from the dead leaves. The mercury content of the stalks was approximately the same as the green leaves. The average mercury content of the Everglades agricultural muck soils was 0.15 ppm. This value is similar to the 0.2 ppm mercury content of muck soils in Sweden reported by Hakanson et al.[3]

The mercury content of the various components of the above-ground portion of the mature sugarcane crop is shown in Table 2. For these calculations a 30 t/ha dry matter yield based on an 80 t/ha crop of millable cane was assumed. The 30 t of dry matter consisted of 24 t

Table 2   **Mercury content of various components of above-ground sugarcane crop (all values based on dry weight).**

	Metric t/ha	Hg-ppm	Kg Hg/ha	Kg Hg in EAA[a]
Millable cane	24.0			
Stalks	21.6	0.086	0.00183	318
Trash	2.3	0.068	0.00016	28
Burned leaves	3.0	0.068	0.00020	35
Tops and green leaves	2.7	0.085	0.00022	38
Total above-ground dry matter	30.0	0.083	0.00241	419

[a] Based on 174,000 hectares.

Table 3   **Mercury emissions from preharvest burning of Everglades sugarcane.**

Average above-ground field weight (moist weight) of millable cane	80 t/ha
Total above-ground dry matter represented by 35 t crop (stalks + leaves)	30 t/ha[a]
Percent of above-ground biomass burned at harvest (mostly dead leaves)	10%[b]
Dry weight of burned material	2910 kg/ha
Concentration of mercury in burned material	0.068 ppm (17 replications)
Amount of mercury in burned material	0.00020 kg/ha
Area burned	174,000 ha
Total amount of mercury emitted from entire Everglades sugarcane area by preharvest burning	35 kg

[a] Based on study by Dr. Laron Golden, Louisiana Agricultural Experiment Station. No comparable data available for Florida sugarcane crop.
[b] Data from Dr. John Dunckleman, Florida Sugar Cane League.

of millable cane (21.6 t of stalks and 2.3 t of trash), 3 t of burned leaves, and 2.7 t of green leaves and tops that usually do not burn. The total mercury content of these various components was calculated on a per hectare basis, and on the basis of the whole Everglades Agricultural Area sugarcane crop (174,000 hectares). The total mercury content for the whole EAA sugarcane crop is approximately 410 kg, most of which is in the millable cane that is taken to the mill for processing.

The burned portion of the crop produces approximately 0.0002 kg of Hg/ha or approximately 35 kg emitted into the atmosphere from the whole EAA. This value was calculated using the assumptions shown in Table 3. A larger amount of mercury is represented by the millable cane.

The results of the muck soil burning experiment showed no soil mercury lost from the preharvest cane burning (Table 4). None of the muck soil became ignited during the burning operation. It is also likely that no direct volatilization of mercury occurred as a result of the high temperature caused by the burning, at least under the conditions of this experiment. The surface of the soil was still moist after the burning operation. The high moisture content of the muck soil apparently provided an effective insulation that prevented a high temperature buildup in the soil.

This study showed that direct emission of mercury from the sugarcane fields during preharvest burning is a minor source of atmospheric mercury. A comparison of this source

Table 4 **Mercury lost from sugarcane field by emission from muck soil during preharvest burning.**

	Soil no.	Hg content of soil-Kg/ha[a]	Overall average[b]
Before burning	1	0.094 ± 0.004	0.094
	2	0.095 ± 0.009	
	3	0.093 ± 0.005	
After burning	1	0.108 ± 0.005	0.097
	2	0.091 ± 0.003	
	3	0.093 ± 0.003	

[a] Average values and standard deviations shown. Based on soil depth of 9.4 cm.
[b] Number of replications = 12.
Note: No statistically significant difference from burning at 5% level of probability.

Table 5 **Comparison of mercury emissions from sugarcane burning with other sources in South Florida and with normal atmospheric fallout.**

Annual total amount of mercury emitted from preharvest sugarcane burning for entire Everglades area	35 kg
Annual amount of mercury removed from field in millable cane	345 kg
Annual total amount of mercury emitted from industrial and municipal sources in south Florida[4]	3060 kg
Annual amount of mercury falling on sugarcane area from normal atmospheric fallout (based on average fallout of 15 $\mu g/M^2$/year on 174,000 ha)[5]	30 kg

of mercury with that emitted from municipal incinerators in south Florida[4] shows that less than 2% of the mercury known to go into the atmosphere in south Florida is from preharvest burning (Table 5). This amount is little more than the normal annual background deposition of mercury, using the atmospheric deposition rate of 15 $\mu g/m^2$/year reported for the upper Midwest.[5] This value was used because no comparable data on dry and wet deposition of mercury is available for the Southeast. A potentially larger source of atmospheric mercury is from the burning of bagasse. However, this source represents less than 10% of total atmospheric emissions.

ACKNOWLEDGMENT

This study was supported by the Louisiana State University Wetland Biogeochemistry Institute and the Florida Sugar Cane League. Appreciation is expressed to Dr. John Dunckelman of the Florida Sugar Cane League for his assistance in the location and collection of samples.

REFERENCES

1. Lambou, V. W. et al., Mercury Technical Committee Interim Report, Center for Biomedical & Toxicological Research and Waste Management, Florida State University, Tallahassee, FL, 1991.
2. Plumb, R. H., Jr., Procedure for handling and chemical analysis of sediment and water samples, Technical Report EPA/CE 81-1, U.S. Army Engineer Waterways Experiment Station, CE, Vicksburg, MS, 1981.
3. Hakanson, L., A. Nilsson, and T. Andersson, Mercury in the Swedish mor layer—linkages to mercury deposition and sources of emission, *Water Air Soil Pollut.,* 50:311, 1990.

4. Hunt, B., Mercury point source mass balance analysis: Dade, Broward and Palm Beach Counties, Florida, 8 pages. The Florida Mercury Conference, Miami, FL, Nov. 14, 1991.

5. Anonymous, Mercury in the Environment. Technical Brief, Electric Power Research Institute, Palo Alto, CA, 1990.

Mercury Pollution: The Impact of U.S. Government Stockpile Releases

Michael Rieber and DeVerle P. Harris

CONTENTS

I. PROBLEM DIMENSIONS

U.S. Government stockpiles of mercury include the National Defense Stockpile (NDS) and the Department of Energy (DOE) stockpile; both are administered by the Defense Logistics Agency (DLA). While mercury from the NDS is prime virgin material, that from DOE is recycled secondary material, which is said to be guaranteed at 99.9% but which may be contaminated with lithium, iron from old flasks, dust, and other debris. Preparation for auction of DOE material may require filtration, reflasking, and rehandling. Sale of stockpiled material exceeding U.S. Department of Defense needs is predicated not only on desired reductions, but also on the costs of holding the stocks, which involves upkeep of the six separate facilities, periodic replacement and inspection of flasks, and the market risk of stockholding in a time of secularly declining demand and prices.

The dimensions of the problem are easily stated.[1] At the end of October 1989, a year in which U.S. mercury consumption totaled 36,800 flasks (76 pounds/flask), the DOE stockpile totaled 17,592 flasks. The NDS stockpile was 148,603 flasks in excess of the strategic goal of 10,500 flasks.[2,3] In brief, at the then current rate of consumption, DLA-administered mercury equalled 4.5 years of total U.S. consumption. According to some mercury dealers, the DLA stockpile, particularly the DOE portion, is significantly larger than that which was claimed, but this has been hard to verify.

By 1990, U.S. demand had fallen to 20,880 flasks, with preliminary estimates for 1991 totalling 11,600 flasks.[4] In 1990, DLA sold 4,855 flasks of DOE material, but none from the NDS stockpile because of industry pressures to reduce its market influence.[5] It appears that in 1991 the DLA sold 6,750 flasks of DOE mercury and 3,000 of NDS material. As a result, the DLA administered stockpile at the beginning of 1992 totaled 151,590 flasks or 13.1 years of total U.S. consumption at the 1991 rate. In future years, U.S. consumption is expected to continue to decline.

Alternatively, if the current rate of stockpile release, 750 flasks/month each for DOE and NDS, is not increased the DOE stockpile will run out by August 1992, assuming allowable sales are consummated monthly; but at the current allowable rate, the NDS stockpile will last 16.2 years. Finally, though there are no figures compiled for world annual consumption of mercury, U.S. Government holdings are more than 1.5 times the estimated 1991 world rate of less than 100,000 flasks/year.[6]

The disposal problem does not become easier on either the supply or the demand side. The changes in both consumption and the structure of demand in the U.S., the only country

1–56670–066–3/94/$0.00+$.50
© 1994 Lewis Publishers

Table 1   **U.S. industrial consumption of refined mercury metal.**

	(1000 flasks)[a]			(%)[b]		
	1988	1989	1990	1988	1989	1990
Chlorine and caustic soda	12.9	11.0	7.2	27.9	31.3	34.4
Laboratory uses	0.8	0.5	0.9	1.7	1.4	4.3
Paint	5.7	5.6	0.6	12.3	16.0	2.9
Other chemical and allied products	2.5	1.2	1.0	5.4	3.4	4.8
Electric lighting	0.9	0.9	1.0	1.9	2.6	4.8
Wiring devices and switches	5.1	4.1	2.0	11.0	11.7	9.6
Batteries	13.0	7.2	3.1	28.1	20.5	14.8
Measuring and control instruments	2.2	2.5	3.1	4.8	7.1	14.8
Dental equipment and supplies	1.5	1.1	1.3	3.2	3.1	6.2
Other uses	1.6	1.0	0.7	3.5	2.8	3.3
Total	46.2	35.1	20.9	99.8	99.9	99.9

[a] 29 flasks/metric ton.
[b] Derived from volumes.
From the U.S. Bureau of Mines, Mercury in 1990, *Min. Ind. Surv.*, Dec. 13, 1991, Table 3.

for which the data are available, are shown in Table 1. The decline in mercury consumption for paint, batteries, and the production of chlorine and caustic soda from brines is particularly relevant. Virtually all of the reduction can be ascribed to environmental actions related to toxicity. There is simply no other way to explain such radical movements in the short time span. The rest of the Western industrialized world is likely to follow these movements relatively quickly. It may be noted, however, that while there are substitutes for mercury in paint, the decline in domestic consumption in batteries is largely offset by the significant increase in the importation of batteries; another example of the export of pollution. Most of the remaining uses will decline, but slowly, as mercury substitutes or new technology become available. Of more importance here is mercury use and consumption in the chlor-alkali industry.

Producers of chlorine-caustic soda (NaOH) from brines are not only mercury consumers, but also holders of a very large stockpile ready for release. In process terms, theoretically no mercury is consumed; it is used as a moving cathode to separate sodium and chlorine. Consumption occurs as mercury is lost in waste-water treatment, recaptured, reprocessed, and sent to landfills. EPA regulations effective May 1992 will further limit such disposal, accelerating the shift to a nonmercury-using membrane cell technology. By the end of 1989, only 17% of U.S. plants were still using mercury cells, in Western Europe, 25% remained. Japan completed its transition in 1986. In all cases, however, the quantity of mercury represented by the floating cathode became an involuntary stockpile to be disposed on the market. The disposal takes time if adverse price effects are to be avoided. The Japanese stockpile was finally eliminated in 1990.

During the 1980s, the U.S. chlor-alkali industry released about 32,000 flasks into the market, about half of all recorded U.S. secondary supply. But at the end of 1989, it was still holding at least 76,500 flasks. If, as has been claimed, some 20% of existing U.S. plants recently closed to meet the May 1992 deadline,[7] some 15,300 flasks are currently available (more than 1 year of total U.S. consumption), with the rest to be available in the relatively near future. As noted above, the storage of this mercury is costly and hazardous. Industry is less able to bear these costs than is the U.S. Government. For the market economy countries, it has been estimated that in 1990 some 250,000 flasks of mercury were still in use as cathodes,[8] to become stockpiles as plants closed or shifted technology. Eastern European and

Chinese conditions are unknown, but as their industrial structure is generally antiquated, it is likely that the older mercury cell technology is dominant.

The remaining mercury sources are secondary mercury (by-product and recycled) and mined prime virgin material. In 1989, U.S. production alone of by-product mercury from gold mining was 2,700 flasks/year. Zinc metallurgy yielded over 3,000 flasks/year from operations in Finland, Spain, and elsewhere.

Recycled mercury production in the U.S. amounted to 8071 flasks in 1988. In Spain, Almaden's MAYASA plant has an operating capacity of 15,000 flasks/year. The company has a stockpile of about 8,000 metric tons of residues to reprocess at an expected rate of 1,000 flasks/year.[9] Additional recycled mercury is available from industrial waste, mercury-containing instruments and equipment, as well as batteries.

The major producers of primary mercury are the former U.S.S.R., China, Turkey, and Algeria. The mines are state owned and may be operated for "social" rather than profit objectives. The U.S. and Spain both ceased mine production of primary mercury in 1990. Output in the U.S.S.R. in 1990 was estimated at 72,450 flasks, China's output was 20,212 flasks, that of Algeria was 20,286 flasks. The rest of the world mine production of prime virgin mercury was estimated at 13,892 flasks.[10] This mine capacity can be further contracted, or expanded, relatively quickly. There is no mercury shortage. At current rates of world consumption the mercury resource exceeds 150 years coverage without further exploration. In the short-run, the ratio of mine productive capacity to mine production is at least 2 to 1.

U.S. mercury imports are negligible, 435 flasks in 1990, but our exports are significant, 9,019 flasks in that year. Of these exports, 22.1% went to Latin America (primarily Brazil and Colombia) and 26.4% went to Southeast Asia (primarily India). A further 34.7% of the exported mercury was sent to the Netherlands.[11] Rotterdam is a re-export and world market center for both scrap mercury and noncontract spot or barter disposals. The remaining 16.8% was shipped to OECD nations.

As the import and export figures do not capture mercury embodied in finished products (e.g., switchgear, instruments, and batteries), the U.S. mercury trade is understated. Our export of mercury-containing products, however, is small while our imports are significantly larger. An estimate could be made for batteries, the remainder would be more difficult.

II. DISPOSAL IMPACTS

As mercury moves from the status of a commodity towards that of a toxic waste, markets dry up, distribution organizations dissolve, prices plunge. In 1988 there were six mercury dealers in the U.S.; one remains. In 1988 the average price for prime virgin material with prompt delivery in the New York market was $335.52/flask. It dropped steadily to $181.25/flask in December 1990 and $175/flask in February 1992. European (Rotterdam) prices generally ranged from $25 to $75/flask lower over the period.

The decrease in price and the increase in environmentally mandated handling and storage costs will have some fairly predictable effects. Where mercury is a primary product, mines operating for profit must close. Mines run on a statist basis (some 85% of total output in the market economy nations and all of the output in the former U.S.S.R., Eastern Europe, and China) may actually increase output as prices fall and they seek to maintain labor income, hard currency revenues, or both. Spain no longer mines mercury and, under some pressure, Algeria has apparently begun contracting with consumers rather than shipping to the open market. But during 1990 and 1991, Russia was reportedly shipping 2,000 flasks/month to the open market, and China was an aggressive seller. Their shipments might have been more had there not been problems in the Chinese mines and recurring distributional difficulties with Russian shipments. While the developed market economy countries, including those with state-owned mines, are likely to curtail output over time, there is little reason to expect similar action from developing or the former socialist nations.

The producers of by-product mercury (gold, silver, and zinc producers) will recover the material as long as they can profitably mine the primary product. In the U.S., this may mean closure for some of the eight by-product producing gold mines, if the mercury handling/disposal costs increase while the market price falls, and the gold price does not provide a compensatory increase. The same behavior need not be followed by state-owned mines. Zinc mines in Spain now provide the bulk of the material required to be processed by the Spanish company Almeden's MAYASA plant.

It is the processors and reprocessors of scrap mercury who may be most affected. Processing involves a cost. As environmental regulations become stricter, these costs will rise or recycling will move to less discriminating places. As disposal costs rise and landfill options are reduced, the quantity of scrap available to the recyclers will increase and prices will fall. In any event mercury scrap prices on the London market are correlated with the Rotterdam price for prime virgin material. On the other hand, the price received for reprocessed and recycled mercury is limited by the going price for virgin material. As that price falls toward processors' costs, fewer non-state-owned processors will remain; recycling mercury will become more difficult. Additionally, at low prices consumers prefer virgin material to reprocessed scrap. At the end of 1991, only five companies remained in the U.S. If landfills continue to close and if mercury shipments to existing landfills are further restricted, more residues may be exported.

III. STOCKPILE IMPACTS

The impact of stockpiles on commodity prices has been known since Exodus. The mere existence of stockpiles serves as a price depressant; their disposal, or even the threat of their release, is sufficient to quell price increases and enrage producers. The stockpiles need not even be large relative to annual consumption to affect market prices. For example, the Strategic Petroleum Reserve is a small fraction of U.S. annual consumption. Agricultural stocks are disposed through school aid and special export aid programs to avoid spoiling the domestic market. Efforts to corner the silver market failed as unanticipated, largely misunderstood stocks were released from private hoards.

Both U.S. Government and private stocks of mercury are very large relative to domestic or even annual world consumption. The private stocks will be released in any case. Presumably the market can and is adjusting price, in part due to a correct anticipation of this imminent surge in supply. Other producers will make compensating changes in their output and stock holdings. Consumers will also anticipate the new supply source and its probable timing. U.S. Government stockpile releases differ because the government is not compelled by costs as are the private firms. These releases are subject to administrative fiat rather than to market forces.

Given the conditions prevailing in the mercury market in 1989 it was estimated that the price effect alone of U.S. Government mercury stockpile releases (i.e., excluding the reduced output effect on producers) was about \$20/flask per 1,500 flasks released. At the allowable rate of joint NDS and DOE releases, 4,500 flasks per quarter, the expected price reduction was about \$60/flask. The validity of the projection was borne out by events during 1990.

The statistically significant variables best explaining New York dealer mercury prices were (1) the auto-correlated mercury prices of the five preceding quarters, (2) the auto-correlated changes in producers' stocks three and four quarters earlier, (3) changes in total stocks in the prior quarter, (4) changes in primary mercury production in the prior quarter, and (5) releases of DLA-administered stocks two quarters earlier. The squared coefficient of multiple correlation is 0.83, which implies that 83% of the variation in mercury price is explained by these five variables. This value 0.83 is statistically significant at the 0.01% level, meaning that the probability for this result when there really is no relation between mercury price and the five variables is 0.0001 or a 1 in 10,000 chance.

Conditions have changed, and the equation has not been reestimated. It is possible, however, to anticipate some of the results of stockpile release under present conditions. As prices have fallen by half their level in 1989, their impact will be less in absolute terms. Producers' stocks are now limited to those mines in the U.S. yielding by-product mercury; consequently, this variable is likely now to be nonsignificant. Total stocks, which do not include the chlor-alkali industry, are now limited, except for the government holdings. What remains as the major variable unambiguously affecting prices is the releases from the government stockpile. For this reason it seems likely that the impact of such releases is likely to have increased in importance and to be felt within one quarter (3 months) rather than being delayed for two quarters.

The results of the original study were limited to the U.S. and predicated on the partial but significant separation of the U.S., Western European, and Japanese mercury markets. The proof depended on the lack of sufficient net mercury imports by the U.S. to offset a persistent price premium in the New York market that exceeded the cost of shipping and handling.

Today, that price differential is as high as it has ever been. Yet imports are negligible and exports are significant. The market is becoming global, and as demand declines, the U.S. is becoming a supplier, primarily from stocks. From the trade data, we are not particular as to destination. On past evidence, as demand declines worldwide, other suppliers will be even less discriminating.

It is generally supposed that the rest of the world consumes mercury in roughly the same manner as does the U.S. In fact, this is the basis for estimates of world demand. While probably true for OECD nations, this assumption is probably weak when applied to Eastern Europe, and unlikely when applied to China, India, and Latin America. In pollution terms, the evidence from gold mining in Brazil is simply the most notorious.

It is impossible to require private companies permanently to store and monitor mercury. It would be difficult to ask the U.S. Government to do so. Yet, as disposal sites are foreclosed, as consumer demand is further proscribed, the choice must be storage or export. If other suppliers do not hold their hand, prices must fall as the U.S. Government reduces its stockpile. Those consumers remaining will reap the benefits. They may be the wrong type of consumer.

The study of mercury impacts, their measurement, chemistry, and transfer are of unquestioned importance. It is possible that attention paid to long-term storage might be as useful. Further work on environmental reclamation could be an alternative.

REFERENCES

1. Unless otherwise noted, this paper is based on Rieber, M. and Harris, D.P., U.S. Mercury Marketing and the Impact of Government Stockpile Releases, unpublished report, Placer Dome U.S., Inc., 1990; presented to U.S. House of Representatives, Committee on Interior and Insular Affairs, Oversight Hearing, Market Factors Affecting the Mercury Mining Industry, June 28, 1990.
2. U.S. Bureau of Mines, Mercury in 1990, *Min. Indust. Sur.,* December 13, 1991, p. 2.
3. Lawrence, B.J., Mercury prices adrift, *Eng. Min. J.,* 191(3), 25, 1990.
4. Lawrence, B.J., Mercury adjusts to a new market, *Eng. Min. J.,* 193(3), 21, 1992.
5. Lawrence, B.J., Mercury historical low for 1990, *Eng. Min. J.,* 192(3), 23, 1991.
6. Lawrence, B.J., Mercury adjusts to a new market, *Eng. Min. J.,* 193(3), 21, 1992.
7. Lawrence, B.J., Mercury adjusts to a new market, *Eng. Min. J.,* 193(3), 21, 1992.
8. Masters, H. and Neve, C., Mercury, metals and minerals—annual review, *Min. J.,* 43(6), 89, 1991.

9. Masters, H. and Neve, C., Mercury, metals and minerals—annual review, *Min. J.,* 43(6), 89, 1991.

10. Masters, H. and Neve, C., Mercury, metals and minerals—annual review, *Min. J.,* 43(6), 89, 1991.

11. U.S. Bureau of Mines, Mercury in 1990, *Min. Indust. Sur.,* Dec. 13, 1991, p. 7, Tables 6–7.

The Retention of Gaseous Mercury on Flyashes

Peter Schager, Björn Hall, and Oliver Lindqvist

CONTENTS

ABSTRACT: The main objective of the present investigation was to study the adsorption of mercury on solid materials, such as flyash and calcium compounds, and thereby assess which parameters are relevant in influencing mercury retention. It is known that elemental and oxidized mercury are adsorbed by activated carbon and flyash. Oxidized mercury is more easily adsorbed than elemental mercury. The degree of adsorption and the influence of different trace gas components, however, are unknown. It is therefore of interest to study the influence of trace gases, such as SO_2 and HCl, on these adsorption processes.

The retention of mercury by solid materials captured on fabric filters was investigated using a flue gas generator. The solid materials, activated carbon, flyashes, silica sand, lime, and gypsum were exposed to mercury-containing flue gas in a 1 to 3 mm thick layer supported by a standard fabric filter at 150° C. The mercury retention capacity of the solid material was judged from measurements of the mercury concentration both upstream and downstream of the fiber. The effect on the retention of mercury by addition of HCl and SO_2 was studied. The results indicate that it is possible to retain mercury on flyashes in general, and activated carbon in particular, while sand and calcium compounds showed negligible retention capacity. The carbon content in the flyashes is the crucial factor which determines mercury retention. High residual carbon content is reflected in a high retention of mercury on the flyashes. The results showed that mercury adsorption on flyashes and activated carbon was enhanced by oxidizing trace gases such as HCl and decreased by reducing species like SO_2. This is in agreement with the hypothesis that oxidized mercury is more easily captured on the solids than elemental mercury.

1–56670–066–3/94/$0.00+$.50
© 1994 Lewis Publishers

I. INTRODUCTION

In contrast to most other heavy metals, mercury is emitted to the atmosphere in the gas phase. Coal combustion and waste incineration are two of the most important anthropogenic sources of mercury pollution. There is a significant difference in the resulting distribution of elemental mercury (Hg^0) and oxidized forms (Hg^{2+} or possibly Hg^+) in the emission from these processes. While coal combustion mainly emits elemental mercury (approximately 80%), waste incineration emits almost all mercury in oxidized form.[1] This effect is often explained by the high concentration of hydrochloric acid in waste incineration conditions. Combustion of coal of marine origin produces a relatively high concentration of HCl (150 mg m^{-3} NTP) and this can be reflected in a large fraction of oxidized mercury.[2] An important reaction is HCl combining with mercury to form $HgCl_2$ or possibly Hg_2Cl_2, depending on temperature and the Hg to HCl ratio. The reaction is believed to proceed according to Reactions (1) and (2).[3] The oxidation of mercury in coal combustion could be explained by Reaction (3), where HgO(s,g) is formed. This reaction can proceed at temperatures below 400° C, if an active surface such as soot or flyash is present in the flue gas.[4]

$$Hg(g) + 2\ HCl(g) + 1/2\ O_2(g) \rightarrow HgCl_2(g) + H_2O(g) \qquad (1)$$
$$2\ Hg(g) + 2\ HCl(g) + 1/2\ O_2(g) \rightarrow Hg_2Cl_2(s) + H_2O \qquad (2)$$
$$Hg(g,ads) + 1/2\ O_2(g) \rightarrow HgO(s,ads,g) \qquad (3)$$

The abatement of mercury emissions by dry filters, from an environmental point of view, is an important subject. Elemental and oxidized mercury is adsorbed on activated carbon and flyash—oxidized forms being more easily retained than elemental mercury.[5,6] The chemical form of mercury developed in anthropogenic combustion processes is obviously a crucial factor which needs to be resolved before flue gas cleaning schemes can be effectively operated. If solid materials are trapped on the fabric filter, it is likely that some mercury in the flue gas also will be adsorbed. The main objective of the present investigation was to study the influence of SO_2 and HCl on the mercury retention capacity of some different solid flue gas components.

II. EXPERIMENTAL

The experiments were performed with a 17-kW, propane-fired flue gas generator consisting of a combustion chamber connected to a 12-m long flue gas duct.[3,4,7] A filter chamber was situated at the end of the duct, where the solid materials were exposed in a 1- to 3-mm thick layer supported by a standard fabric filter. The flue gas flow rate was 0.6 m^3/min (NTP) in all experiments and the oxygen concentration was 10%. The CO_2 and CO concentrations were not measured but the CO_2 concentration was calculated to be 5.4% (dry flue gas). The temperature of the filter was 150 ± 5° C. The residence time of the flue gas in the duct before the filter was 4.5 s and the residence time for the flue gas in the filter material was approximately 5 to 30 ms. The mercury retention capacity of the different materials was measured as the difference between the concentration of mercury in the flue gas before and after the filter. Measurement of the gaseous forms of mercury was performed with the continuous CVAA method.[3,8] The error limits for the values given in Tables 2 to 5, according to this measuring model, is approximately ±5 to 10%. Only a few replicate tests were made and these were in the range of this error limit. Trace gas components ([SO_2] = 100 ppm and [HCl] = 100 ppm) together with mercury were fed into the flue gas generator near the propane burner and the effect of these additions on mercury retention was studied. The following solids were tested:

1. Pulverized activated carbon from Kebo Lab (Sweden) with a specific surface of 850 ± 50 m^{2}g^{-1}; 55% (by weight) of the particles had a size of less than 40 μm.

2. Flyash A and B (baghouse) originate from experiments with the 16 MW Atmospheric Fluidized Bed Combustor (AFBC) at Chalmers University of Technology, Göteborg, Sweden, during operation on coal. Flyash A was collected from experiments without addition of limestone, while flyash B was collected from experiments with additions of limestone to the bed corresponding to an amount of 350 kg of limestone h^{-1}.
3. Flyash C (ESP) was collected from the Sävenäs municipal waste incineration plant, Göteborg, Sweden.
4. Silica sand used as bed material in the AFBC at Chalmers University of Technology, Göteborg, Sweden. The sand was fresh and had not previously been used in the AFB combustor.
5. CaO pro analysi.
6. $CaSO_4 \cdot 2H_2O$ pro analysi.

Table 1 shows the composition of the flyashes used.[9] The combustible material consists of a large fraction of char and has characteristics similar to activated carbon. The fraction of combustible material is crucial to the mercury retention capacity of the flyashes, as will be discussed later. Note that the residual unburnt carbon content is higher in flyash A and B than in C, and that a higher calcium concentration occurs in flyash B.

III. RESULTS AND DISCUSSION

A. RETENTION ON ACTIVATED CARBON

The mercury retention capacity of activated carbon was investigated in four experiments presented in Table 2. The flue gas flow was 0.6 m^3 min^{-1} (NTP) in the experiments and the filter temperature was $150 \pm 5°$ C. The flue gas was passed through the filter for approximately 60 min and 15 g activated carbon was exposed on the filter (20 g when SO_2 was added). Table 2 shows the amount of total, oxidized, and elemental mercury as a mean value for the time period during which the experiments took place.

The retention of mercury on activated carbon is 95% without any additions of trace gas. It is increased to 99% if HCl is present, but decreases considerably when SO_2 is added to

Table 1 **Composition of the flyashes given in weight %.[9]**

	Content in %		
Flyash	**A**	**B**	**C**
Combustible	41.1	34.9	12.5
Ash	58.9	58.5	87.5
Ca	3.0	17.1	10.8
Mg	0.7	0.4	1.6
Na	0.2	0.1	3.4
K	0.8	0.5	3.6
Fe	4.9	2.3	2.4
Si	20.2	13.9	12.8
Al	8.2	3.8	8.4
$CaCO_3$ (maximum)	0.4	15.0	0.8
$CaSO_4$ (maximum)	3.8	8.0	10.3
CaO	0	5.1	0.0

Note: The amount of $CaCO_3$ and $CaSO_4$ are calculated from determinations of CO_2 and sulfur, respectively, and assuming that these species originated only from $CaCO_3$ and $CaSO_4$ in the sample. The amount of unburned material is calculated from the ash content, where [ash] + [combustible] + [moisture] = 100%. The moisture content is 0%.

Table 2   **Mercury retention capacity of a 1-mm thick layer of activated carbon at 150° C.**

	No addition	$[SO_2]$ = 100 ppm	$[HCl]$ = 100 ppm	$[SO_2]$ = 100 ppm $[HCl]$ = 100 ppm
Before filter				
$[Hg(tot)]$	160	310	147	170
$[Hg^0]$	122	250	68	137
$[Hg(ox)]$	38	60	79	33
After filter				
$[Hg(tot)]$	8	160	2	30
$[Hg^0]$	8	150	2	30
$[Hg(ox)]$	0	10	0	0
Retention (in %)				
$[Hg(tot)]$	95	47	99	82
$[Hg^0]$	93	40	97	78
$[Hg(ox)]$	100	80	100	100

Note: Mercury concentrations are given in μg Hg m^{-3} (NTP).

the flue gas. A possible explanation is that SO_2 reduces oxidized mercury compounds to $Hg^0(g)$. Since $Hg^0(g)$ has a greater tendency of escaping the filter, this would be reflected in a lower mercury retention. This reaction could proceed homogeneously in the gas phase with gaseous HgO and SO_2, but more likely is proceeding through a heterogenous reaction where adsorbed HgO reacts with SO_2. The reaction of mercury with sulfur dioxide has been studied by Zacharewski et al.[10] The authors suggested that HgO(s) reacts with $SO_2(g)$ forming $HgSO_4$ and Hg_2SO_4. This reaction will not explain the decreased adsorption of mercury with SO_2 added to the flue gas. Another possibility is the overall Reaction (4), where the reaction product is $HgSO_4(s)$ and $Hg^0(g)$.

$$2HgO(g, s) + SO_2(g, ads) \rightarrow Hg^0 + HgSO_4(s) \qquad (4)$$

In preliminary results from experiments performed in a newly built fixed bed reactor, it was found that orthorhombic mercuric oxide is partly reduced to elemental mercury by SO_2 at 150° C. In the same experimental setup, 0.5 g of activated carbon was exposed to a flow of 300 μg Hg(g) m^{-3} in nitrogen and 5% oxygen at a temperature of 150° C. The oxygen increased the adsorption of mercury, which confirms previous studies where it was found that the adsorption of mercury on activated carbon is enhanced by oxygen.[12] The reaction exhibits a half-order oxygen dependence and a first-order dependence with respect to mercury.[12] This could be due to formation of HgO on the surface of the activated carbon. When the activated carbon is exposed to 100 ppm SO_2 in the fixed bed reactor, the previously adsorbed mercury desorbs. This phenomenon could explain the decreased adsorption when SO_2 was added to the flue gas. An alternative explanation could be that SO_2 is adsorbed on the surface of the activated carbon and subsequently inhibits the adsorption of mercury, by covering the active sites otherwise available for mercury sorption.

The positive effect of HCl addition on mercury retention is probably caused by the increased fraction of oxidized mercury in the flue gas, i.e., oxidized mercury is more easily adsorbed on solid materials than elemental mercury. In some experiments the trace gas composition was changed rapidly. HCl was fed into the flue gas generator for 2 h and mercury adsorption was almost constant (>95%) throughout this period. The HCl addition was then interrupted and the mercury adsorption decreased from 95 to 55% over a period of 5 min.

Table 3 **Mercury retention of a 1-mm thick layer of flyash A at 150° C.**

	No addition	[SO$_2$] = 100 ppm	[HCl] = 100 ppm
Before filter			
[Hg(tot)]	200	285	210
[Hg^0]	120	260	34
[Hg(ox)]	80	25	176
After filter			
[Hg(tot)]	110	260	14
[Hg^0]	110	260	14
[Hg(ox)]	0	0	0
Retention (in %)			
[Hg(tot)]	45	9	93
[Hg^0]	8	0	59
[Hg(ox)]	100	100	100

Note: Mercury concentrations are given in μg Hg m^{-3} (NTP).

This implies that the mercury retention is significantly enhanced by the presence of HCl in flue gases. The effect is presumably due to the formation of HgCl$_2$, according to Reaction (1). HgCl$_2$ is more easily adsorbed on activated carbon than elemental mercury.[11]

When both SO$_2$ and HCl are added, the effect of HCl seems to dominate the adsorption capacity of the solid. The retention for Hg(tot), Hg^0, and Hg(ox) was 83, 78, and 100%, respectively. The decrease of the adsorption, due to SO$_2$, is apparently not so significant when HCl is present in the flue gas. However, it is still not known in which concentration range of SO$_2$ and HCl the inhibition of mercury adsorption will occur.

B. RETENTION ON FLYASH

A series of experiments was performed with flyashes (A, B, C) as adsorbent. The flue gas flow was approximately 0.6 m^3 min^{-1} and the flyashes were exposed to the mercury-containing flue gas during approximately 1 hr. In each experiment 30 g of the flyash was exposed in a 1-mm thick layer on the fabric filter.

The results from the experiments with flyashes A and B, which originated from an AFB combustor, are shown in Tables 3 and 4. The tables show the amount of total, oxidized, and elemental mercury as a mean value for the time period during which the experiments took place. Flyash A has a low calcium content while flyash B has a high calcium content (17%). The carbon content is 40 and 35%, respectively, for the two flyashes. The results are similar to those from the experiments with activated carbon (see Table 2). The mercury retention is enhanced by addition of HCl but decreased by SO$_2$. Also, in these materials the trapping of oxidized mercury is more efficient than that of elemental mercury.

Flyash B has a higher retention ability than flyash A when no trace gas addition is made. The main difference between the two ashes is the higher concentration of Ca in flyash B. How this affects the adsorption of mercury is not fully understood. As will be discussed later, the adsorption of mercury onto CaO and CaSO$_4$ is negligible. Thus, the difference between the two flyashes cannot be explained simply in terms of adsorption to the calcium compounds. The synergistic effect could possibly be the result of the lime controlling the chemical environment in sample B.

The results from experiments with flyash C, which originated from waste incineration, are shown in Table 5. The mercury retention capacity is lower for this flyash in comparison with the other flyashes, both with and without addition of trace gases. The trends seen in the other experiments, where an increase in the retention occurred when HCl was added and a

Table 4   **Mercury retention capacity of a 1-mm-thick layer of flyash B (calcium content 17%) at 150° C.**

	No addition	$[SO_2] = 100$ ppm	$[HCl] = 100$ ppm
Before filter			
[Hg(tot)]	180	280	174
[Hg^0]	125	250	74
[Hg(ox)]	55	30	100
After filter			
[Hg(tot)]	35	260	26
[Hg^0]	35	250	9
[Hg(ox)]	0	10	17
Retention (in %)			
[Hg(tot)]	81	7	85
[Hg^0]	72	0	88
[Hg(ox)]	100	67	83

Note: Mercury concentrations are given in μg Hg m^{-3} (NTP).

Table 5   **Mercury retention capacity of a 1-mm thick layer of flyash C at 150° C.**

	No addition	$[SO_2] = 100$ ppm	$[HCl] = 100$ ppm
Before filter			
[Hg(tot)]	170	198	209
[Hg^0]	110	157	53
[Hg(ox)]	60	41	156
After filter			
[Hg(tot)]	130	150	129
[Hg^0]	87	108	42
[Hg(ox)]	43	42	87
Retention (in %)			
[Hg(tot)]	24	24	38
[Hg^0]	21	31	21
[Hg(ox)]	28	0	44

Note: Mercury concentrations are given in μg Hg m^{-3} (NTP).

decrease when SO_2 was added, does not occur in the experiments with flyash C. An additional difference is that oxidized mercury passes through flyash C. The explanation is likely to be found in the low content of residual carbon (13%). In contrast to the experiments with activated carbon, the retention of mercury had dropped to 0 to 20% at the end of the experiments with the flyashes. It is important to notice this breakthrough effect, since it is not obvious from the figures in Tables 3 to 5. The retention capacity of the flyashes is therefore only a fraction of that of the activated carbon.

    Attempts were made to do mass balance calculations for mercury. Mercury was determined in the samples by closed digestion at 150° C with concentrated nitric acid followed by the conventional Cool Vapor Atomic Absorption technique.[12] The concentration of mercury in the samples was determined both before and after exposure to the mercury-containing flue gas. The mercury concentration was low in the flyashes before exposure. The adsorbed amount was at least five times greater than the initial concentration (flyash C), and for flyash A and B it was between 50 and 100 times greater. However, it was difficult to close the mass balance and the results differed between 50 and 147%. The reason for this is not known, but possible errors are inhomogeneous samples and losses to ambient air when transferring the samples from the fabric filters to the glass flasks. There is also a suspicion, especially in the experiments with HCl additions, that mercuric chloride did adsorb on the stainless steel surface of the filter chamber. The recovery was conformly low (approximately 50%) in the HCl experiments. Furthermore, after opening the filter chamber, at the end of an experiment with HCl, it was possible to detect up to 50 μg Hg m^{-3} desorbing from the inner stainless steel surface of the filter chamber. The desorption of mercury, however, ended after a few minutes, when the filter had cooled down.

C. RETENTION ON SAND AND Ca COMPOUNDS

The mercury retention capacity of silica sand, used as a bed material in a fluidized bed boiler, is negligible. Mercury is only to a small extent trapped in the presence of HCl. This was also the case with $CaSO_4 \cdot 2H_2O$ and CaO. The retention was negligible in all experiments, except when HCl was present as the only trace gas. Even in this case the retention was small, and only approximately 10% of the total amount of mercury was adsorbed.

IV. SUMMARY AND CONCLUSIONS

Activated carbon adsorbs mercury most effectively. The flyashes also adsorb mercury but to a lesser extent. The flyash from a waste incinerator, flyash C, has a lower retention capacity than the flyashes from AFB combustion. This is believed to be due to the comparatively low carbon content (13%) in flyash C. Flyash with a low content of unburned material (carbon) together with a high Hg concentration in the flue gas, which is often the case in waste incineration, will be saturated with adsorbed mercury rather quickly.

Concerning the effects of the flue gas composition, the most important result is that the presence of SO_2 in the flue gas diminishes mercury retention. This could be due to SO_2 competing with mercury for adsorption sites on the surface of the solid material, but it is more probable that SO_2 reduces oxidized forms of mercury to $Hg^0(g)$. Since $Hg^0(g)$ has a greater tendency of escaping the solid filter, a lower mercury retention would then be expected. On the other hand, HCl increases the retention of mercury. The effect is caused by the increased amount of oxidized mercury through oxidation of $Hg^0(g)$ by HCl—100% of the oxidized mercury is retained in most experiments where activated carbon or flyash A and B are used. The combined effect of HCl and SO_2 resembles the case with only HCl added, i.e., considerable mercury retention is achieved. However, further studies are necessary to find out how higher levels of SO_2 would affect the adsorption of mercury on flyash and activated carbon.

ACKNOWLEDGMENT

For the financial support of this project the authors are in debt to the Swedish National Energy Administration.

REFERENCES

1. Levander, T., Emissions of Mercury to Air in Sweden during the Period 1860–1987, National Swedish Environment Protection Board, S-171 25 Solna, Sweden, SNV Report 890414, 1989, (In Swedish).
2. Meij, R., The fate of mercury in coal-fired power plants and the influence of wet flue-gas desulphurisation, *Water Air Soil Polut.,* 56, 21, 1991.
3. Hall, B., Lindqvist, O., and Ljungström, E., Mercury chemistry in simulated flue gases related to waste incineration conditions, *Environ. Sci. Technol.,* 24, 108, 1990.

4. Hall, B., Schager, P., and Lindqvist, O., Chemical reactions of mercury in combustion flue gases, *Water Air Soil Pollut.,* 56, 3, 1991.

5. Vogg, H., Braun, H., Metzger, M., and Schneider, J., *In situ* mercury speciation in flue gas by liquid and solid sorption systems, *Chemosphere,* 16, 821, 1987.

6. Matsumura, Y., Adsorption of mercury vapour on the surface of activated carbon modified by oxidation iodization, *Atmos. Environ.,* 8, 1321, 1974.

7. Schager, P., Hall, B., and Lindqvist, O., The Retention of Mercury from Flue Gases, Report No. II to The National Energy Administration, ISSN 0283–8575 Report OOK 88:7, 1988 (In Swedish, Abstract in English).

8. Lindqvist, O. and Schager, P., Continuous measurements of mercury in flue gases from waste incinerators and combustion plants, *VDI Berichte,* 838, 401, 1990.

9. Puromäki, K., Analytical Report, No. 105, Department of Inorganic Chemistry, GU/CTH, Kemigården 3, 412 96 Göteborg, Sweden, 1990.

10. Zacharewski, T. R., Cherniak, E. A., and Schroeder, W. H., FTIR investigations of the heterogeneous reaction of HgO(s) with SO_2(g) at ambient temperature, *Atmos. Environ.,* 21, 2327, 1987.

11. Vogg, H., Braun, H., Metzger, M., and Schneider, J., The specific role of cadmium and mercury in municipal waste incineration, *Waste Manag. Res.,* 4, 65, 1986.

12. Schager, P., The Behaviour of Mercury in Flue Gases, Department of Inorganic Chemistry, Chalmers University of Technology and University of Göteborg, ISSN 0283–8575, Report OOK 90:07, Göteborg, Sweden, 1990.

SECTION VIII

Human Health
and
Public Policy

The Toxicology of Mercury and its Compounds

Thomas W. Clarkson

CONTENTS

ABSTRACT: As an element present in the earth's crust, in water, and in the atmosphere, living cells have been exposed to mercury throughout evolution of life on this planet. Elemental mercury vapor was probably present even in Archean times and may have been oxidized to the highly toxic divalent cation, Hg^{++}, when oxygen first appeared in concentrations approaching present-day levels. As a result, cells have developed mechanisms to protect themselves from its toxic effects. Some of these mechanisms are discussed. On the other hand, methylmercury compounds, although also present at the earliest times, may not have presented a toxic hazard until the evolution of an advanced nervous system. This may be one reason why we lack the same degree of cellular protection as we enjoy for inorganic mercury. The production of methylmercury by biomethylation in the environment, its bioaccumulation in aquatic food chains, and its potency to produce irreversible brain damage makes it the species of mercury of greatest public health concern.

I. INTRODUCTION

Today, human exposure is mainly to two forms of mercury: the vapor of metallic mercury, Hg^0, and monomethylmercury compounds, MeHg. The former is present in the working environment in certain workplaces where metallic mercury is used, is emitted by dental amalgam tooth fillings, and is present in the ambient atmosphere.[1] The latter is found principally in fish and tissues of marine mammals.[2] However, divalent inorganic mercury, Hg^{++}, is a metabolic product of both Hg^0 and MeHg and may be the proximate toxic agent for Hg^0 if not for MeHg. Thus this chapter will address these three forms of mercury.

Occupational exposure to Hg^0 first started about 2000 to 3000 years ago.[3] The records of mercury production in the Almaden mines in Spain go back at least 2000 years. It is still the largest producer today. Over the past 2000 years, mercury has found many uses:[4] in agriculture as fungicides, in the electrical and chemical industries as an electrode and catalyst, in war as a detonator, in medicine as an emetic and diuretic, and in the dental profession in the form of mercury amalgam tooth fillings. The use of cinnabar as the pigment in red ink in China perhaps credits mercury with the birth of bureaucracy almost 3000 years ago.

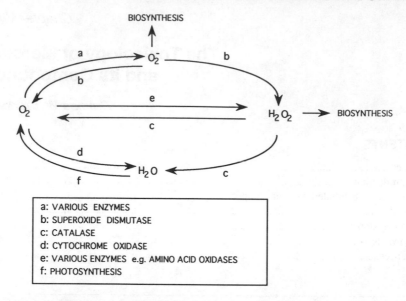

Figure 1   The Oxygen Cycle. (Adapted from Hill, H.O.A., *New Trends in Bio-Inorganic Chemistry*, Williams, R.J.P. and DaSilva, J.R.R.F., Eds., Academic Press, New York, 1978, 173. With permission.)

It has also contributed to major scientific discoveries.[4] Its use in temperature and pressure measurements was important to the discovery of many of the early laws of physics and its special chemical properties were involved in the discovery of at least 28 elements. As an ubiquitous element, traces of mercury are always present in our food, water, and air. Presumably, we are all tolerant to these background levels. When mercury levels rise above a toxic threshold, adverse health effects appear or, to quote from Paracelsus, "Dose makes the poison."[5] Thresholds differ from one individual to another, and may do so by factors of 10 or more.

What are the mechanisms of tolerance or defense against mercury? Under what circumstances do these defenses fail? What makes some individuals more resistant than others? The answer to these questions is key to present and future health-risk assessments. The answer may go back to events not 2000, but 2 to 4 billion years ago! The point is that at that time the atmospheric concentrations of mercury vapor were far higher than today's levels.

In the earliest days, when life was first forming on this planet, emission of mercury vapor from land surfaces was probably far higher than what it is today due to higher temperatures and greater volcanic activity. The earth's atmosphere contained no oxygen so the conditions existed to support high levels of Hg^0 in the primitive atmosphere. The major gases in the atmosphere were CH_4, CO_2, H_2, H_2O, H_2S, N_2, and NH_3, but no oxygen. Debate continues on the relative atmospheric abundance of these gases, but there is general agreement that oxygen levels were virtually zero.[6] In this atmosphere, no possibility existed to oxidize Hg^0 to the water-soluble Hg^{++} and thereby remove it from the atmosphere.

At this time, so long as mercury stayed in the form of Hg^0, there was no threat to the developing life on the planet as Hg^0 is itself chemically unreactive and nontoxic. This idyllic existence changed dramatically when readily available hydrogen supplies started to run out. One group of microorganisms, in their desperate search for hydrogen, used sunlight to split the water molecule and released oxygen to the atmosphere.[7]

Not only was oxygen itself toxic, but its metabolic products were even more so (Figure 1). Thus, the superoxide anion and hydrogen peroxide were produced—probably other active species such as hydroxyl radicals were produced as well.

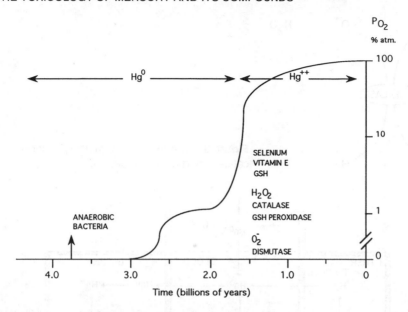

Figure 2 Atmospheric oxygen tension throughout evolutionary history (for details, see text).

The oxygen tension 4 billion years ago was at least 10 orders of magnitude below today's levels.[9] At 2 billion years, the rise in oxygen started (Figure 2). Most microorganisms were defenseless against the onslaught of this very toxic gas. In the meantime, the surviving organisms developed a variety of biochemical defenses against oxygen as well as the respiratory enzymes to utilize this toxic gas for energy production.[10] Some of these defense factors were oxygen scavenging agents like vitamin E and selenium compounds. Reduced glutathione was accumulated to high levels inside the cells to maintain a reducing atmosphere. Enzymes appeared that maintained levels of GSH, and destroyed active oxygen species.

The thesis here is that these cells were hit with a second highly toxic chemical, divalent ionic mercury. The rising levels of atmospheric oxygen would allow oxidation of mercury vapor to ionic mercury that would be returned to the earth's surface in rain water. How did these early cells survive this second attack?

A plausible answer is they did this by using virtually the same metabolic machinery used to protect themselves from oxygen. Apparently, we have inherited many of the biochemical defenses. Let us see to what extent these Archean defenses still play a vital role in determining our tolerance to mercury and its compounds. In particular, we will compare and contrast the defenses against inhaled mercury vapor with those against methylmercury in our food supply.

II. TOLERANCE TO INHALED MERCURY VAPOR

The oxidation of mercury vapor to divalent ionic mercury is conducted exclusively by the enzyme catalase (Figure 3). One molecule of hydrogen peroxide combines with catalase to form compound I. In the next step, compound I reacts with another molecule of hydrogen peroxide to release oxygen and water and the enzyme returns to its native state. In this way, catalase removes a highly toxic product of oxygen and recycles oxygen for the respiratory pathways.

However, compound I can also accept an atom of mercury and oxidize it to divalent ionic mercury. At first thought, it would appear that the oxygen defense system is doing the very

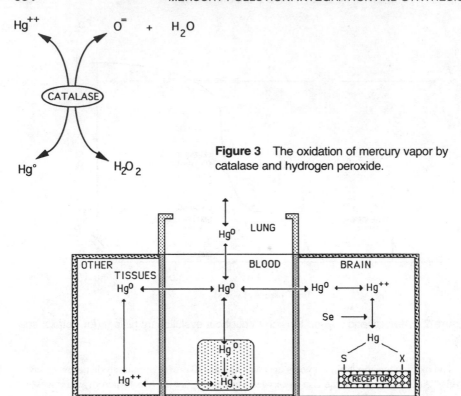

Figure 3   The oxidation of mercury vapor by catalase and hydrogen peroxide.

Figure 4   The inhalation, tissue deposition, and metabolism of inhaled mercury vapor in humans.

opposite to what was postulated—it is actually increasing the toxic potential of mercury vapor by producing a more toxic species of mercury. However, the catalase reaction also plays a protective role.

Consider the example of the transport of inhaled vapor from blood to brain (Figure 4). Mercury vapor is a monatomic gas that is nonpolar and soluble in lipids.[11] It crosses cell membranes with ease. It is cleared from the alveolar spaces of the lung and dissolves in plasma. The dissolved vapor is carried to all tissues in the body including the major target tissue, the brain.

Once inside the cells it is oxidized to divalent mercury. The latter is depicted as attaching to a receptor in brain tissue, but the actual biochemical disturbances leading to clinical symptoms are not yet known. These disturbances include erethism, a bizarre syndrome typified by the behavior of the Mad Hatter in *Alice in Wonderland.*

Despite the fact that divalent mercury is the proximate toxic species, the oxidation reaction can play a protective role because the oxidation first takes place in the red blood cells before the vapor reaches the brain. Divalent mercury is trapped in the cells. Moreover the divalent form penetrates the blood-brain barrier much more slowly than mercury vapor. This allows time for other defense mechanisms to come into play. For example, selenium can form an insoluble complex with ionic mercury. Indeed, all the brain mercury in autopsy samples from mercury miners has been found to be present as a 1:1 complex with selenium.[12]

Also, other enzymes in the oxygen defense system are capable of reducing ionic to mercury vapor (Figure 5). For example, superoxide dismutase that normally destroys the

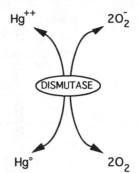

Figure 5 A biochemical pathway of reduction of divalent mercury.

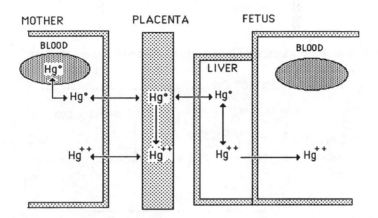

Figure 6 The role of catalase in protecting the fetal brain from inhaled mercury vapor.

superoxide anion can reduce divalent mercury.[13] The reduction of ionic mercury is known to take place in the liver.[14] Given the ubiquity of the biochemical pathways that can reduce divalent mercury, it is likely that reduction occurs in most tissues in the body. Since the lung is interposed between the liver and brain, the vapor released from the liver is carried first to the lung. This organ has been shown to efficiently clear dissolved vapor from blood into expired air.[15] Thus mercury vapor produced in other tissues never reaches the brain.

The catalase system plays a clear protective role in the case of prenatal exposures (Figure 6). As in the case of the blood-brain barrier, mercury vapor readily crosses the placental barrier and enters the fetal bloodstream. However, a substantial fraction of the portal circulation first passes through the fetal liver before entering the general circulation. It is in the fetal liver that catalase plays its protective role. Here the mercury vapor is oxidized to ionic mercury which cannot pass easily into the fetal brain. This may explain why so few cases, if any, of prenatal poisoning have been reported despite the long history of human exposure to this form of mercury.

The oxygen defense system also assists in the elimination of mercury from the body (Figure 7). Mercury vapor enters the liver where catalase converts it to divalent mercury. This, in turn, reacts with reduced glutathione and the complex is secreted into bile on the glutathione carrier across the canicular border of the liver cell and into bile.[16,17] Divalent mercury is carried in bile to the intestines and finally excreted in the feces. Only a small fraction, probably no more than 10% of the amount secreted in bile, is reabsorbed back into the bloodstream.

The reason that the mercury-glutathione complex travels on the glutathione carrier is the structural similarity to oxidized glutathione that is normally transported on this carrier

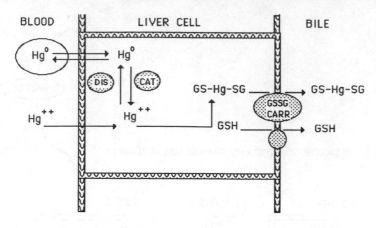

Figure 7 The role of glutathione in the fecal excretion of divalent mercury.

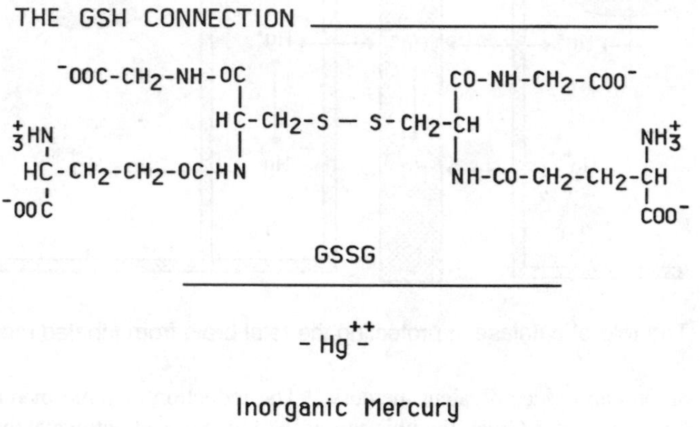

THE GSH CONNECTION

GSSG

$$- Hg^{++} -$$

Inorganic Mercury

Figure 8 The structural similarity of the complex of divalent mercury with glutathione and of oxidized glutathione.

(Figure 8). Note that the two sulfur mercury bonds are at 180°, giving the complex an almost identical structure to the endogenous oxidized form of glutathione.

III. METHYLMERCURY

Some of the oxygen defense mechanisms may be operative to protect us from methylmercury, but nothing like to the extent they function with mercury vapor. Also, other defense barriers such as the blood-brain barrier and the placenta fail to impede the transport of methylmercury. Of all living species, humans appear to have the weakest defenses against methylmercury.

A remarkable aspect of the toxicology of methylmercury is that it is amazingly selective to the central nervous system and to the higher and evolutionary more recent structures of the brain such as the cortical areas.[18] It has been reported that some organisms that methylate inorganic mercury are among the oldest recorded life forms.[19] Perhaps methylation was evolved to protect these organisms against divalent inorganic mercury. Indeed, vulnerability to methylmercury may not have been a problem throughout most of evolution, not until recent times when the brain developed in mammals and especially in primates. One might

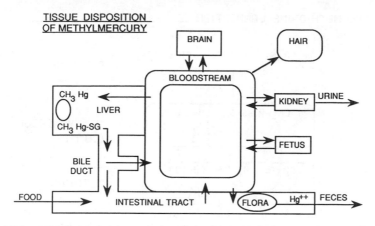

Figure 9 The disposition of methylmercury in humans.

speculate that the recent nature of this threat may not yet have allowed sufficient time to develop the needed defenses.

The disposition of methylmercury in humans in depicted in Figure 9. The most striking aspect is the ease with which methylmercury crosses the body's main diffusion barriers. Thus, 95% of an oral dose is absorbed across the intestine into the bloodstream. Methylmercury crosses into the brain, crosses the placental barrier, and just as easily passes into the fetal brain. The reason for the rapid mobility across membrane is not obvious. All analytical data to date indicate that methylmercury forms water-soluble compounds with thiol-containing amino acids and proteins. We cannot invoke lipid solubility as in the case of mercury vapor. Recently, we have obtained some fascinating clues to the mechanisms of transport which will be discussed in a moment.

To continue with the disposition picture, scalp hair accumulates methylmercury avidly. The average ratio of hair-to-blood levels is 250 to 1 in humans.[1] This phenomenon is useful insomuch that we can use scalp hair to monitor people's exposure to methylmercury, but we do not have enough hair for this uptake process to remove significant amounts from the body. With animals, it is a different story. Even in primates, half the body burden can be removed by the hair.[20] The uptake by newly grown fur in suckling rodents is so effective that over half the methylmercury can be removed in 24 hours.[21] The mechanisms of hair uptake is not understood, but it is suspected that the high level of thiol compounds in the hair follicle plays a role.

The body's method of excreting methylmercury involves a series of steps starting in the liver. At first it follows the exact same pathways as inorganic mercury. It forms a complex with glutathione which, in turn, is excreted into bile on the glutathione carrier.[22] However, unlike inorganic mercury, methylmercury is readily resorbed both in the biliary tree and in the intestine. In fact we would never get rid of methylmercury except to the extent it is broken down to inorganic mercury. One site of breakdown is the lower intestine, where the resident microflora split the carbon-mercury bond and the liberated inorganic mercury is excreted in the feces.[23] The kidney excretes ony small amounts of methylmercury since it is reabsorbed in the renal tubules.[24,25]

Recent evidence has indicated that some breakdown of methylmercury may take place in certain cells in the spleen.[26] The liberated inorganic mercury is secreted in bile and subsequently in feces. This breakdown step, whether in microflora or in mammalian tissues, is so important to the excretion process that virtually all the mercury in feces and urine is in the inorganic form.

THE METHIONINE CONNECTION _____

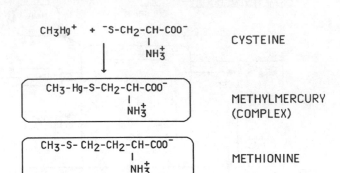

Figure 10   The structural mimicry between the methylmercury-cysteine complex and the large neutral amino acid, methionine.

The fact that methylmercury is transported from liver to bile as a glutathione complex gave the first clue as to how methylmercury could so easily cross cell membranes. Thus, in the case of the liver cell, it formed a complex whose structure mimicked that of an endogenous substrate. We now have evidence that methylmercury can cross the cell membranes of the blood-brain barrier by forming a compound that structurally mimics amino acids (Figure 10). Thus, methylmercury reacts with the thiol-containing amino acid, cysteine, in plasma. This complex is also formed from the hydrolysis of glutathione as cysteine is a constituent of this peptide. The structure of this cysteine complex is very close to that of the large neutral amino acid, methionine. It thus travels across the cell membrane on the large neutral amino acid carrier.[27]

The blood-brain barrier consists primarily of the tightly packed endothelial cells lining the walls of the blood capillaries. These cells are highly active in amino acid transport. Methylmercury enters from the bloodstream as the cysteine complex (Figure 11). Due to the high concentrations of glutathione inside these cells, it probably exchanges a cysteine molecule for glutathione and the latter is then pumped out of the cell on the glutathione carrier as described for the liver cell. Indeed, recent evidence suggests that these two pathways are important in all mammalian cells, the cysteine entering the cell and the glutathione complex leaving the cell.[25]

In any event, reduced glutathione present in the cell as a protection against oxygen, and effective against inorganic mercury, actually plays a negative role with regard to methylmercury! Once inside the brain, the glutathione complex is broken down to the cysteine complex which can now enter the nerve cells in the brain.

Why the brain cells are selectively damaged by methylmercury is still a mystery. Recent evidence indicates that glutathione levels are low in certain brain neurons.[28] Thus, the methylmercury-cysteine complex may persist inside these cells and interfere with protein synthesis. It is of interest that methionine is always the first amino acid to start the peptide chain. Modification of protein synthesis is one of the earliest biochemical effects of methylmercury on the brain, but the mechanisms underlying these effects are not fully elucidated.[29]

The fetal brain is even more vulnerable. Amino acid transport is two or more times faster than in the adult, a fact that explains why methylmercury levels in fetal brain are double those in the maternal brain.[30]

The lowered glutathione levels in brain cells will allow methylmercury to bind to other SH groups, particularly those proteins key to brain development. Thus, methylmercury

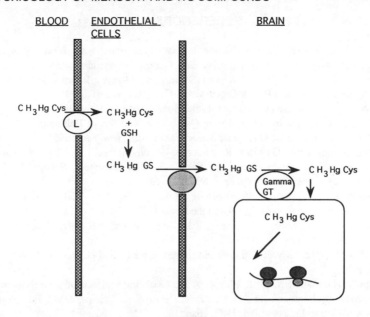

Figure 11 The transport of methylmercury across the blood-brain barrier to its target sites in neuronal cells.

depolymerizes the microtubules of the cytoskeleton, a structure essential for cells division and migration.[31]

The microtubules are constantly formed and disassembled in a treadmilling process. Methylmercury reacts with the SH groups of the tubulin monomers and blocks the assembly. The disassembly continues and the microtubules disappear. Thus, two processes unique to brain growth, neuronal division and migration, are inhibited.

Cell migration is an especially interesting process whereby the neuronal cells literally climb up a support cell to travel to their final destination. The larger the brain, the further the distance and the more vulnerable to methylmercury.

IV. CONCLUSION

Protective mechanisms developed eons ago against oxygen and its active species seemed to function against inorganic mercury including mercury vapor. Unfortunately, there is little evidence that such defenses work against methylmercury. Indeed, glutathione, built up in cells to protect against oxygen, appears to be of dubious value against methylmercury.

In terms of human risk assessment and the limited research resources we have available, it makes more sense to focus on that form of mercury against which we have the weakest defense. The thesis of this presentation is that this form of mercury is methylmercury.

ACKNOWLEDGMENT

This work was supported in part by NIEHS Center Grant #5 P30 ESO1247 and in part by NIEHS Program Project Grant #1 P01 ES05197.

REFERENCES

1. WHO, *Environmental Health Criteria 101, Methylmercury,* International Program on Chemical Safety, World Health Organization, Geneva, Switzerland, 1990.
2. WHO, *Environmental Health Criteria 118, Inorganic Mercury,* International Program on Chemical Safety, World Health Organization, Geneva, Switzerland, 1991.
3. Hunter, D., *Diseases of Occupations,* Little, Brown, Boston, 1969, 290.
4. Goldwater, L. J., *Mercury: A History of Quicksilver,* York Press, Baltimore, MD, 1972.
5. Pachter, M., *Paracelsus: Magic into Science,* Schuman, New York, 1951, 86.
6. Chang, S., DesMarais, D., Mack, R., Miller, S. L., and Strathearn, G. E., Prebiotic organic syntheses and the origin of life, In: *Earth's Earliest Biosphere,* Schopf, J. W., Ed., Princeton University Press, Princeton, NJ, 1983, chap. 4.
7. Margulis, L. and Sagan, D., *Microcosmos,* Summit Books, New York, 1986, chap. 6.
8. Hill, H. O. A., The superoxide ion and the toxicity of molecular oxygen, In: *New Trends in Bio-Inorganic Chemistry,* Williams, R.J.P. and DaSilva, J. R. R. F., Academic Press, New York, 1978, 173.
9. Knoll, A. H., The early evolution of eukaryotes: a geological perspective, *Science,* 256, 622, 1992.
10. Chapman, D. J. and Schopf, J., Biological and biochemical effects of the development of an aerobic environment, In: *Earth's Earliest Biosphere,* Schopf, J. W., Ed., Princeton University Press, Princeton, NJ, 1983, chap. 13.
11. Hursh, J. B., Partition coefficients of mercury (^{203}Hg) vapor between air and biological fluids, *J. Appl. Toxicol.,* 5, 327, 1985.
12. Kosta, L., Byrnne, A. R., and Selenko, V., Correlation between selenium and mercury in man following exposure to inorganic mercury, *Nature,* 254, 238, 1975.
13. Ogata, M., Kenmotsu, K., Hirota, N., Meguro, T., and Aikoh, H., Reduction of mercuric ion and exhalation of mercury in acatalasemic and normal mice, *Arch. Environ. Health,* 42, 26, 1987.
14. Dunn, J. D., Clarkson, T. W., and Magos, L., Ethanol reveals novel mercury detoxification step in tissues, *Science,* 213, 1123, 1981.
15. Magos, L., Clarkson, T. W., and Hudson, A. R., The effects of dose of elemental mercury and first-pass circulation time on exhalation and organ distribution of inorganic mercury in rats, *Biochim. Biophys. Acta,* 991, 85, 1989.
16. Ballatori, N. and Clarkson, T. W., Inorganic mercury secretion into bile as a low molecular weight complex, *Biochem. Pharmacol.,* 33(7), 1087, 1984.
17. Ballatori, N. and Clarkson, T. W., Dependence on biliary secretion of inorganic mercury on the biliary transport of glutathione, *Biochem. Pharmcol.,* 33(7), 1093, 1984.
18. Clarkson, T. W., Hursh, J. B., Sager, P. R., and Syversen, T. L. M., Mercury, In: *Biological Monitoring of Toxic Metals,* Clarkson, T. W., Friberg, L., Nordberg, G. F., and Sager, P. R., Eds., Plenum Press, New York, 1988, 199.
19. Wood, J. M., Kennedy, F. S., and Rosen, C. G., Synthesis of methylmercury compounds by extracts of a methanogenic bacterium, *Nature,* 220, 173, 1986.
20. Nordberg, G. F., Berlin, M. H., and Grant, C. A., Methylmercury in the monkey: autoradiographical distribution and nephrotoxicity, In: *Proceedings of the 16th International Congress on Occupational Health,* Tokyo, 1971, 234.
21. Shi, C.-Y. and Clarkson, T. W., Uptake of mercury by the hair of methylmercury-treated newborn mice, *Environ. Res.,* 51, 170, 1990.
22. Ballatori, N. and Clarkson, T. W., Biliary transport of glutathione and methylmercury, *Am. J. Physiol.,* 244, G435, 1983.
23. Rowland, I. R., Davies, M. J., and Grasso, P., Metabolism of methylmercuric chloride by the gastrointestinal flora of the rat, *Xenobiotica,* 8, 37, 1978.

24. Bakir, F., Damluji, S. F., Amin-Zaki, L., Murtadha, M., Khalidi, A., Al-Rawi, N. Y., Tikriti, S., Dhahir, H. I., Clarkson, T. W., Smith, J. C., and Doherty, R. A., Methylmercury poisoning in Iraq, *Science,* 181, 230, 1973.

25. Hirayama, K., Yasutake, A., and Adachi, T., Mechanism for renal handling of methyl-mercury, In: *Advances in Mercury Toxicology,* Suzuki, T., Imura, N., and Clarkson, T. W., Eds., Plenum Press, New York, 1991, 121.

26. Suda, I. and Takahashi, H., Enhanced and inhibited biotransformation of methylmercury in the rat spleen, *Toxicol. Appl. Pharmacol.,* 82, 45, 1986.

27. Kerper, L, Ballatori, N., and Clarkson, T. W., Methylmercury transport across the blood-brain barrier by the amino acid transport system, *Am. J. Physiol.,* 262, R761, 1992.

28. Omata, S., Teru, Y., Kasama, H., Ichimura, T., Horigome, T., and Sugano, H., Alterations in gene expression due to methylmercury in central and peripheral nervous tissues of the rat, In: *Advances in Mercury Toxicology,* Suzuki, T., Imura, N., and Clarkson, T.W., Eds., Plenum Press, New York, 1991, 223.

29. Philbert, M. A., Beiswanger, C. M., Waters, D. K., Reuhl, K. R., and Lowndes, H. E., Cellular and regional distribution of reduced glutathione in the nervous system of the rat: histochemical localization by mercury orange and o-phthaldialdehyde-induced histoflu-orescence, *Toxicol. Appl. Pharmacol.,* 107, 215, 1991.

30. Aschner, M. and Clarkson, T. W., Uptake of methylmercury in the rat brain: effects of amino acids, *Brain Res.,* 462, 31, 1988.

31. Miura, K. and Imura, N., Microtubules: a susceptible target of methylmercury cytotoxicity, In: *Advances in Mercury Toxicology,* Suzuki, T., Imura, N., and Clarkson, T. W., Eds., Plenum Press, New York, 1989, 241.

Policy Implications of Communicating Health Risks from Fish Consumption

Barbara A. Knuth

CONTENTS

ABSTRACT: Fish consumption health advisories are commonly used to protect anglers and fish consumers from the risks associated with local fisheries. Development of advisories typically involves state health, environmental quality, and fishery management agencies, each with its own policy objectives to be achieved through advisories. These objectives may include enabling people to make their own informed decision about fish consumption, reducing public health risks, helping people select a variety of risk-reducing behaviors, encouraging public support for toxic cleanup programs, and promoting enjoyment of sport-fishery resources. The process of risk communication includes identification of these objectives, assessing the needs of various target audiences, preparing and disseminating the advisory, and evaluating the effectiveness of the advisory, at least in part relative to the original objectives.

I. INTRODUCTION

Other authors in this volume demonstrate the environmental and human health impacts associated with mercury contamination, particularly in fish tissue. Following identification of a potentially harmful environmental contaminant, risk management decisions must be made by health professionals and natural resource managers. One component of risk management involves communicating the potential health risks to those who may be affected. In this chapter, we discuss the policy implications associated with communicating health risks from mercury (and other environmental contaminants), focusing specifically on issues associated with noncommercial (sport or subsistence) fisheries.

Responsibilities for managing inland and near-shore sport-fisheries in the U.S., including the health risks associated with these fisheries, rests largely with the individual states.[1] Health advisories (or fish consumption advisories) and bans on fishing are the most common mechanisms used to protect anglers and fish consumers from the risks associated with these local fisheries. Health advisories have been issued by at least 37 states,[2] including at least 26 states that have issued health advisories in response to mercury contamination.[3] Health advisories are the focus of this chapter, and are used to illustrate how policy objectives associated with risk management and communication are articulated, implemented, and evaluated.

II. RISK MANAGEMENT AND RISK COMMUNICATION

The process of risk management for environmental contaminants involves the integration of health effects and exposure data with social, economic, and political information to decide how to reduce or eliminate potential human health risks. Risk communication through health advisories is one result of the decisions made about managing environmental contaminants in sport-fisheries.

Risk communication is a process of sharing information about perceived and potential dangers associated with a risk, e.g., mercury contamination in fish tissue. The concept of sharing information is important. Good risk communication is not a one-way process, but rather a constant, evolving exchange of information among all parties concerned, designed to influence perceptions and behaviors associated with the risk. This definition of risk communication implies that many parties should be involved, particularly those with information and with needs for information, that objectives of risk communication programs often revolve around influencing human perceptions and behaviors, and that the evolving exchange of information should be evaluated, particularly in relation to the associated objectives.

The following sections of this chapter address policy objectives related to health advisories, considerations that should be made by those participating in the health advisory communication process, and a discussion of several health advisory program evaluations. The chapter concludes with a set of recommendations for those involved in health advisory risk communication—scientists, health and resource managers, communication specialists, and health advisory audiences.

III. HEALTH ADVISORY POLICY OBJECTIVES

Developing and issuing health advisories typically involves staff from three types of state agencies: health, environmental quality, and fishery management. Each of these agencies has its own mandates, professional cultures, constituents, and roles within the health advisory process.

In 1991, the policy objectives held for health advisories by U.S. state agencies within the Great Lakes basin were assessed.[4] Health, environmental quality, and fishery management agencies shared some common objectives, but differed on others and on the priority they placed on particular objectives. In each of the state jurisdictions, the health agency is designated as the lead agency for developing health advisory criteria,[5] but the fishery management and environmental quality agencies are involved in fish tissue sampling, monitoring, and advisory communication. Differences among these agencies in objectives held for health advisories contribute to differences in approaches to health advisory risk communication and other elements of the risk management process. The results from that study, presented in detail in Knuth and Connelly,[4] are summarized below to provide ideas for other agencies and risk communicators involved in the health advisory process.

A. HEALTH AGENCIES

Great Lakes state health agencies identified four priority policy objectives. First, health advisories should reduce public health risks, especially for at-risk groups (e.g., high fish consumers, women of childbearing age, children), subsistence fishers, and licensed anglers. Second, health advisories should enable people to make their own, informed decision about eating Great Lakes fish. Third, advisories should help people select a variety of risk-reducing behaviors such as fish cleaning and cooking procedures, fishing for lesser-contaminated species, and fishing in less-contaminated locations. Fourth, health advisories should inform people about the health benefits of eating fish.

B. FISHERY AGENCIES

Great Lakes state fishery management agencies shared the top three objectives with state health agencies noted above. Fishery agencies, however, placed much less priority on communicating the health benefits of eating fish and rather added several distinct high-priority objectives. These included: (1) encouraging public support for programs to reduce or clean up toxic contamination in Great Lakes waters; (2) encouraging the enjoyment and beneficial uses of sport-fishery resources; and (3) encouraging public support for Great Lakes fishery management.

C. ENVIRONMENTAL QUALITY AGENCIES

Environmental quality agencies shared the top three objectives of health and fishery management agencies, but placed a clear top priority on promoting risk-reducing behavior. Environmental quality agencies assigned considerably lower priority to some of the "positive" objectives of health and fishery management agencies, such as promoting the health benefits of eating fish and encouraging the use and enjoyment of fishery resources.

Although these three types of agencies could likely agree on several of the main health-protection objectives of health advisories, the different emphasis the agencies place on promoting the benefits of angling or fish consumption relative to the hazards of eating fish could produce conflict in a risk communication program. Such potential conflicts should be addressed explicitly as health advisories are being developed and communication strategies planned.

IV. HEALTH ADVISORY RISK COMMUNICATION PROCESS

Identifying the objectives of health advisory communication is the first step in the risk communication process. The entire risk communication process involves articulating the objectives and intentions of risk messengers (e.g., state agencies), the preparation and dissemination through appropriate channels of the health advisory message, and the attitudes and behaviors of the receivers of the message (e.g., anglers, fish consumers).[6]

A. THE AGENCIES: RISK MESSENGERS

As noted earlier, the objectives of agencies involved with health advisory risk communication may differ, creating conflicts over what information to include in an advisory, which audiences to target with advisory information, and how to present that information. Other decisions and responsibilities of state agencies may influence their abilities to develop and disseminate health advisories effectively. Sims and Baumann[7] emphasized the importance of the communicating agencies being perceived as credible by the audiences they target with information. For example, a fishery management agency can create confusion or produce credibility questions by stocking fish in contaminated waters, or allowing high limits on the numbers of fish taken from contaminated waters while at the same time issuing restrictive health advisories for those same fish. Doubts about agency credibility may be reduced if the agency explains why it is taking such an approach to fishery management.

B. PREPARING AND DISSEMINATING THE HEALTH ADVISORY

The types of information that should be included in a health advisory depend on the originating agency's objectives and the information needs and characteristics of the target audiences. For example, agencies that seek to allow fish consumers to make their own informed choice about sport-caught fish consumption must have an understanding of what constitutes an informed choice. Several studies have demonstrated that target audiences desire information about the relative safety of different fish species and sizes and of different fishing locations; how to reduce health risks by adopting proper fish cleaning and cooking techniques rather than eliminating fish consumption altogether; and an explanation of the particular potential health risks associated with levels of fish consumption above the recommended limits.[7–10] Similarly, if agency objectives relate to promoting the health benefits of fish consumption or the associated benefits of fishing, health advisory information should be constructed to help achieve these objectives.

Methods of disseminating advisory information differ in their effectiveness relative to specific target audiences.[8,11] Printing health advisories in fishing license regulations booklets has proven to be an effective and efficient method for disseminating information to a majority of licensed anglers[10] (Knuth, unpublished data). At least as important, and sometimes more important as a health advisory information source, however, are mass media channels such as newspapers and television. These information dissemination mechanisms are likely to reach not only licensed anglers, but unlicensed anglers and other fish consumers. Health and fishery professionals charged with disseminating health advisories, however, have traditionally had minimal formal contact with these mass media sources, and therefore little influence over the kinds of information portrayed via these methods. Several studies have shown that personal professional contacts (e.g., with health agency or fishery agency personnel, doctors, or health clinic staff) are the most effective method for providing credible, correct, and influential information to potential fish consumers[10] (Knuth, unpublished data). Personal contacts, although sometimes costly, have the added benefit of reaching those fish consumers who may not be licensed anglers.

Information dissemination methods should be geared to the information needs of the target audiences and should consider the accessibility of the information to the audiences of most concern. Accessibility includes ensuring people receive the information as well as ensuring it is understandable considering the abilities of the specific audience.[9]

C. HEALTH ADVISORY AUDIENCES

Health advisory-related agencies may differ in their definition of the audiences of concern. In our analysis of agency objectives, we identified several agencies that viewed licensed anglers as their primary constituency and concern, whereas others were most concerned with potential high-risk groups such as women of childbearing age, children, and subsistence anglers.[4] The characteristics and needs of these various audiences will determine the appropriate health advisory content and dissemination methods as discussed earlier. In addition to understanding what information sources are used by these audiences to obtain health advisory-related information, advisory communicators should understand the attitudes and behaviors of the audiences.

Most health advisory recommendations are geared toward influencing fish consumption behaviors. Behaviors, however, are intertwined with complex attitudes such as those about fish consumption, the role of fishing in an individual's lifestyle, and proclivity to accept or reject risk.[10] Often, the initial perceptions agencies hold about the information needs (and associated attitudes and behaviors of audiences) are at least in part incorrect.[9] Communicators seeking to influence audiences with which the agency has not traditionally worked may need to begin their communication programs with assessments of the initial attitudes and behaviors of target audiences to be able to design communication strategies that can effectively influence ultimate fish consumption behaviors.

V. HOW SUCCESSFUL IS HEALTH ADVISORY RISK COMMUNICATION?

A. EVALUATION CRITERIA

Evaluating the success of health advisory risk communication programs can occur on two levels. First, an evaluation can focus on the communication process itself. Was information disseminated as planned (e.g., consider content, timeline, dissemination methods anticipated)? Ultimately, however, evaluation should focus on the original policy objectives and whether or not those objectives were achieved. Did people feel they were provided the proper information to enable them to make their own informed decision? Were anglers aware of a variety of behavioral changes possible to reduce risk, such as reducing fish consumption, choosing safer locations or species, or using proper fish-cleaning and cooking procedures when organic contaminants are involved? Did fish consumers keep their consumption within the limits recommended in the advisory?

Several studies have evaluated the success of health advisories in a variety of states. Most of these were based on mail surveys of licensed anglers, and so could not produce a complete evaluation of advisory effectiveness relative to all audiences of concern. Evaluation criteria used in these studies included health advisory awareness, angler behavior, and attainment of agency objectives.

B. HEALTH ADVISORY AWARENESS

Most evaluations in recent years have found relatively high levels of health advisory awareness among licensed anglers, ranging from 60 to 90% of licensed anglers aware of health advisories[10,12,13] (Knuth, unpublished data). Awareness of advisories among other audiences, for example, migrant farmworkers, may be considerably lower (e.g., 10% reported by Springer[9]).

Most measures of awareness have established only that anglers know the advisories exist. Connelly et al.[10] found that although anglers may be aware of advisories, they were weak in specific areas of knowledge such as reliable outward indicators of fish contamination (e.g., odd taste or behavior not indicative of contamination), types of negative health effects associated with contaminants in fish, benefits of fish consumption, and specific (more restrictive) advisory recommendations for women of child-bearing age and children. Such detailed measures of awareness are necessary to understand the actual effectiveness of advisories in allowing people to make their own informed choice about fish consumption, a major priority of most health advisory-related agencies.

C. ANGLER BEHAVIOR

Determining desired angler behavior relative to health advisories is complicated. Measures of success may include the level of fish consumption relative to the recommended levels, or the extent to which other risk-reducing practices (e.g., selecting safer species or locations) are used. Dar et al.[14] concluded that widespread awareness of health advisories and contaminants in the Green Bay, Wisconsin area apparently led to a decrease in sport fish consumption by pregnant women. Other studies have shown that 35 to 65% of licensed anglers have made changes in their fishing or fish consumption behavior[9,10,13] (Knuth, unpublished data). In New York, 80% of licensed anglers statewide keep their fish consumption within the recommended levels, including those who have ceased fishing altogether because of health advisories.[10]

How close fish consumption should be to advisory recommendations depends on the objectives of the agency. For example, is it acceptable that people reduce their fish consumption far below the levels recommended in an advisory, or is it more desirable that people keep their consumption at or slightly below the recommended levels? Those agencies putting a high priority on the health benefits from fish, or those who view enjoyment and use of fishery resources as important, might view fish consumption driven far below the recommended levels as being undesirable. People may be unnecessarily avoiding a food resource that provides a beneficial source of protein or associated recreational and economic benefits. Agencies putting a priority solely on reduction of human exposure to contaminants (via fish consumption) may have no concerns that consumption moves far below the recommended levels. This difference in objectives influences not only evaluation conclusions about advisory effectiveness, but also willingness to include diverse information (e.g., health benefits) within health advisory communications.

VI. RECOMMENDATIONS FOR RISK COMMUNICATION

Agencies and constituents concerned with health advisory development and dissemination should recognize that a variety of policy objectives are associated with advisories. Understanding what other groups or agencies hope to achieve through health advisories is the first step toward reaching agreement on appropriate policy objectives and health advisory communication strategies. Scientists must understand that information they produce relative to the health advisory monitoring and contaminant sampling programs is just one of the considerations entering into risk management and risk communication program decisions.

Any public policy program, particularly one with substantial implications for public health and local economies, should include an evaluation component. Evaluations should be twofold, including an assessment of the process and of the program outcomes, especially relative to the original policy objectives.

Finally, health advisory communicators should realize that the public is not monolithic. Many audiences involved in fishing or fish consumption can be identified, and respond differently to health advisory information. Their information needs must be considered in developing successful advisory communication programs. Scientists and others involved in health advisory risk communication sometimes believe there is one right way to communicate fish consumption information. Health advisory evaluation research, however, indicates that a single message designed for diverse audiences and multiple agency objectives will not meet each of their needs, at least not very well.

ACKNOWLEDGMENTS

Much of the health advisory research described has been supported by the New York Sea Grant Institute, the New York Agricultural Experiment Station, the New York State Department of Environmental Conservation, the Great Lakes Protection Fund, and the U.S. Environmental Protection Agency.

REFERENCES

1. Reinert, R. E., B. A. Knuth, M. A. Kamrin, and Q. J. Stober, Risk assessment, risk management, and fish consumption advisories in the United States, *Fisheries,* 16(6):5–12, 1991.
2. Cunningham, P. A., J. M. McCarthy, and D. Zeitlin, Results of the 1989 Census of State Fish/Shellfish Consumption Advisory Programs, Research Institute Report, Research Triangle Park, North Carolina, 1990.

3. Cole, H. S., A. L. Hitchcock, and R. Collins, Mercury Warning: The Fish You Catch May Be Unsafe to Eat, Clean Water Fund and Clean Water Action, Washington, D.C., 1992.
4. Knuth, B. A. and N. A. Connelly, *Objectives and Evaluation Criteria for Great Lakes Health Advisories: Perspectives From Fishery, Health, and Environmental Quality Agencies,* HDRU Series No. 91–11, Department of Natural Resources, Cornell University, Ithaca, New York, 1991.
5. Hesse, J. L., Summary and Analysis of Existing Sportfish Consumption Advisory Programs in the Great Lakes Basin, The Great Lakes Fish Consumption Advisory Task Force, State of Wisconsin, 1990.
6. Knuth, B. A., Risk communication: a new dimension in sport-fisheries management, *N. Am. J. Fisheries Management,* 10:374–381, 1990.
7. Sims, J. H. and D. D. Baumann, Educational programs and human response to natural hazards, *Environ. Behavior,* 15:165–189, 1983.
8. Wendt, M. E., Low Income Families' Consumption of Freshwater Fish Caught From New York State Waters, Master's Thesis, Cornell University, Ithaca, New York, 1986.
9. Springer, C. M., Risk Management and Risk Communication Perspectives Regarding Lake Ontario's Chemically Contaminated Sport Fishery, Master's Thesis, Cornell University, Ithaca, New York, 1990.
10. Connelly, N. A., B. A. Knuth, and C. A. Bisogni, *Effects of the Health Advisory and Advisory Changes on Fishing Habits and Fish Consumption in New York Sport Fisheries,* HDRU Series Report 92–9, Department of Natural Resources, Cornell University, Ithaca, New York, 1992.
11. Smith, B. F. and E. E. Enger, A Survey of Attitudes and Fish Consumption of Anglers on the Lower Tittabawasee River, Michigan, Michigan Department of Public Health, Center for Environmental Health Science, Lansing, 1988.
12. Fiore, B. J., H. A. Anderson, L. P. Hanrahan, L. J. Olson, and W. C. Sonzogni, Sport fish consumption and body burden levels of chlorinated hydrocarbons: a study of Wisconsin anglers, *Arch. Environ. Health,* 44:82–88, 1989.
13. Connelly, N. A., T. L. Brown, and B. A. Knuth, New York Statewide Angler Survey, 1988, New York State Department of Environmental Conservation, Albany, 1990.
14. Dar, E., M. S. Kanarek, H. A. Anderson, and W. C. Sonzogni, Fish consumption and reproductive outcomes in Green Bay, Wisconsin, *Environ. Res.,* 59:189–201, 1992.

Mercury Imbalances in Patients with Neurodegenerative Diseases

W. D. Ehmann, E. J. Kasarskis, and W. R. Markesbery

CONTENTS

ABSTRACT: Mercury (Hg) is one of several elements that have been implicated in the etiology of a number of age-related neurological diseases. In our studies, Hg has been determined by instrumental neutron activation analysis (INAA) along with up to 20 other elements in samples of brain, hair, and nail from subjects with Alzheimer's disease (AD), in samples of brain, spinal cord, nail, blood cells, and blood serum from subjects with amyotrophic lateral sclerosis (ALS), and in corresponding samples from age-matched, neurologically normal control subjects. Control and AD brain temporal lobes were subjected to subcellular fractionation by ultracentrifugation and fractions enriched in nuclei, mitochondria, and microsomes were then also analyzed by INAA. Our results from these studies, together with observations from recent related in vitro and animal model studies, are summarized in this chapter. From the data we have obtained, it is clear that significant Hg imbalances relative to control subjects do exist in both AD and ALS subjects. However, additional studies are required to define the principal pathways for Hg entry into the central nervous system and the role of the observed imbalances in the etiology and pathogenesis of these diseases.

I. INTRODUCTION

Age-related dementing disorders in adults are a major national health concern. Alzheimer's disease (AD), affecting 4 million persons, is the most common cause of dementia in the U.S.[1] Alzheimer's disease is characterized clinically by progressive decline of intellectual function. Pathologically, the principal findings are neurofibrillary tangles (NFT), senile plaques (SP) containing amyloid, dystrophic neurites, and neuron and synapse loss.[2] Suggested etiological hypotheses for AD include: slow virus infection, genetic predisposition, generalized membrane or metabolic disorders, autoimmune defect, and trace element neurotoxicity. Amyotrophic lateral sclerosis (ALS) is a less common age-related neurodegenerative disorder characterized clinically by progressive weakness and muscle atrophy with preserved cognitive function, and pathologically by degeneration of motor neurons in the brain stem and spinal cord. Speculations that the etiology of ALS may be linked to trace element toxicity can be found in the literature as early as 1850.[3] Summaries of reported elemental imbalances and reviews of hypotheses relating to trace element neurotoxicity in AD and ALS can be found in recent articles by Markesbery and Ehmann[4] and Kasarskis et al.[5,6]

Mercury (Hg) is one of several elements that have been implicated in the etiology and/ or pathogenesis of AD and ALS.[7,8] In our laboratory we have used instrumental neutron activation analysis (INAA) to determine Hg concentrations in a variety of tissues from AD and ALS patients and age-matched control subjects. Results of these studies are reviewed below.

II. NEUTRON ACTIVATION ANALYSIS FOR MERCURY IN HUMAN TISSUE

A major advantage of INAA is its relative freedom from problems associated with reagent and laboratory contamination. Because Hg is often used in research facilities, minimizing the opportunities for contamination is critical for analyses at the ng/g concentration level. In our INAA procedure, autopsy samples of brain, spinal cord, and extra-neural tissues (heart, liver, kidney, and spleen) are taken with quartz or high purity titanium knives and stored in virgin polyethylene vials at $-70°$ C prior to analysis. Brain autopsy samples are obtained through the University of Kentucky Alzheimer's Disease Research Center. All AD patients must meet the established clinical[9] and pathological[10] diagnostic criteria for AD. Clinical diagnostic and histological criteria used for ALS patients are described by Khare et al.[8] Blood is collected from living patients via an indwelling Teflon® catheter directly into the acid-washed poly-ethylene vials. The specimen is allowed to clot, and then is centrifuged to separate the blood clot (blood cells) from the serum. Samples of blood cells and serum are lyophilized separately for analysis. Fingernails and hair are collected by use of surgical scissors, and washed prior to analysis using the procedure recommended by the International Atomic Energy Agency.[11] Hair is taken from the nape of the neck, 5 cm proximal to the scalp.

Tissue samples are freeze-dried to constant weight and the freeze-dried to wet weight ratio (FD/WET) is computed for each sample. For INAA, representative 10 to 20 mg samples of freeze-dried tissue are placed in Suprasil® synthetic quartz vials and heat-sealed. Up to 50 samples and comparator standards are irradiated simultaneously at the University of Missouri Research Reactor. Laboratory sample handling, and hence opportunities for laboratory contamination, are minimal. After sampling, freeze-drying and encapsulation in quartz vials, no further opportunities for contamination exist because the Hg content is determined solely by the induced radioactivity from ^{203}Hg produced in the irradiation.

A major advantage of INAA is that it is a multielement analytical technique. Although obtaining Hg concentrations in the tissues may be a primary objective, INAA simultaneously provides data for some 15 to 20 other elements in the same sample. Of particular interest are other elements which may act to counter Hg toxicity. Table 1 lists the experimental parameters for the determination of Hg by INAA in our laboratory. Data obtained in our laboratory for Hg in international biological standards are presented in Table 2 and additional information on our INAA procedures may be found in Ehmann et al.[12,13]

III. MERCURY IN ALZHEIMER'S DISEASE

Adult (age >20 years) control brain Hg levels[13] and brain Hg concentrations as a function of age from premature infants to age 85[14] have been determined in our laboratory. A summary of our bulk brain (sample mass ≥50 mg, wet) Hg data for control subjects is given in Table 3. In our control populations, females had significantly higher brain Hg levels than males. No significant trend in brain Hg levels with age was observed for adult controls. The absence

Table 1   **INAA determination of mercury.**

Irradiation	40 h at a neutron flux density of $\sim$5.5 $\times$ 10^{13} n cm^{-2} s^{-1}
Nuclear reaction	^{202}Hg, 29.86% (n,γ) ^{203}Hg
Decay mode	^{203}Hg, t$_{1/2}$ = 46.612 d, γ ray = 0.2792 MeV
Interferences	^{75}Se, t$_{1/2}$ = 119.779 d, γ rays = 0.2795, 0.2647 MeV, + others
Detection	Shorter-lived interferences eliminated by 30 d delay prior to counting. Count for 4 h with a 40+% relative efficiency, 1.9 keV (at ^{60}Co) resolution, HPGe detector.
Procedure	Measure (Se-0.2795 γ)/(Se-0.2647 γ) ratio in Se standard under normal counting conditions. Multiply this ratio by the 0.2647 MeV γ peak in the unknown to obtain the Se-0.2795 γ count to be subtracted from the composite 0.279 MeV peak to obtain the Hg-0.2792 γ count.
Comparator standards	NIST biological standard reference materials

Table 2   **Mercury by neutron activation analysis in some international reference standards.**

ng Hg/g $\pm$ σ_m (No. of analyses)	
This laboratory	**Other laboratories**
Bowen's kale	
153. $\pm$ 4.5 (11)[13]	171 $\pm$ 4.4, Bowen, recommended[18]
155. $\pm$ 7.2 (22)[15]	150, Lievens et al.[19]
150. $\pm$ 20 (5)[16]	180, Muramatsu et al.[20]
164. $\pm$ 7.6 (19)[17]	
NIES #5 human hair, Japan	
4560 $\pm$ 118 (3)[17]	4400 $\pm$ 400(σ), certified value
	4200, Muramatsu et al.[20]
	4470, Zhuang et al.[21]

of a significant difference in adult and infant brain Hg levels agrees with observations of Massaro et al.[22] Mercury in brain correlates positively with mercury in fingernails for subjects in the same control population.[23]

Mercury imbalances in bulk AD brain compared to neurologically normal control brain were first reported by our laboratory in 1986.[7] In that study $\sim$75% of the samples were from the cerebral cortex. Bulk brain Hg concentrations in AD and a population of age-matched controls are presented in Table 4. AD brain is significantly (p <0.01) enriched in Hg compared to the age-matched control brain. In the comparison of gray matter and white matter, Hg in AD gray matter is significantly elevated compared to AD white matter, but in an equivalent comparison for controls a smaller apparent elevation in gray matter is not significant (p >0.05). AD-control gray matter and AD-control white matter differences in Hg concentrations were not significant.

In a later study, Hg and Se levels were determined in several brain regions most strongly affected by AD; e.g., amygdala, hippocampus, and nucleus basalis of Meynert (nbM).[24] Biological interactions of Se and Hg have been recently reviewed by Cuvinaralar and Furness.[25] Our results are presented in Table 5. The elevation of Hg in the nbM of AD patients (p <0.01) is the largest bulk brain trace-element imbalance observed to date. Selenium appears to be elevated along with Hg in each of these three regions. However, as was the case for Hg, the elevation is significant at the p <0.05 level only in the nbM.

Table 3 **INAA data for mercury in bulk control brain.**

Age factor	Infants (<1 year) 14.8 ± 1.3 ng Hg/g wet wt. (sam. = 30, pat. = 6)
	Adults (>20 years) 18.1 ± 1.5 ng Hg/g wet wt. (sam. = 140, pat. = 15)
	Adult/infant difference not significant, (p >0.05), in agreement with Massaro, et al.[22]
	No significant trend with age in adults. Hg levels are relatively constant during the first 80 years of life, then appear to decline.
Sex	Male controls 13.2 (92)
	Female controls 22.1 (48)
	Male/female difference is significant (p <0.05). Se is also significantly higher in female brain (0.209 μg/g, n=22), than in male brain (0.180 μg/g, n = 112).
Concentration ranges	Infants ≤6.0 to 33.1 ng Hg/g wet wt.
	Adults ≤6.0 to 158 ng Hg/g wet wt.
	75% of samples from cerebral cortex
Regional variation	Highest Middle frontal lobe (28.7 ng/g, wet)
	Cerebellar vermis (25.3 ng/g, wet)
	Lowest Cerebellar hemisphere (12.4 ng/g, wet)
	Globus pallidus (12.8 ng/g, wet)
Freeze dry to wet weight ratio	Infants 0.124 ± 0.005
	Adults 0.211 ± 0.002
Interelement correlations	Hg correlates positively with Ag & P, negatively with Cr in control brain (p <0.05).
Brain/fingernail correlation	Hg in control brain correlates positively with Hg in fingernails of the same subjects (p <0.05).[23]

Data from References 13 and 14.

Concentrations of Hg and several other elements in isolated enriched subcellular fractions of AD and control brains have also been determined in our laboratory.[26] In that study, sections of the temporal neocortex were minced with titanium knives and homogenized in a chilled solution of 0.25 M sucrose and 5 × 10^{-3} M imidazole. The homogenate was subjected to ultracentrifugation to separate fractions enriched in nuclei, mitochondria, and microsomes. A summary of the subcellular Hg data on a freeze-dried sample weight basis is presented in Table 6. In addition, Hg/Se and Hg/Zn mass ratios are given for whole brain and the several subcellular fractions. In each fraction, the concentration of Hg was higher in AD than in the corresponding controls, but the difference was significant in the p <0.05 level only for the microsomal fraction. In this group of AD and control patients, Hg levels in AD whole brain were again much higher than in controls, but the difference was not significant due to the high variability of the data in the AD temporal lobe of the small population sampled. Significant increases in Hg/Se (nuclear and microsomal fractions) and Hg/Zn (microsomal fraction) mass ratios were present in AD. In comparing wet weight Hg concentrations with freeze-dried concentrations it should be noted that the typical FD/WET weight ratio for adult brain is approximately 0.21.

Several extra neural tissues have been examined for elemental imbalances in AD. A summary of Hg data in hair and finger nails of AD patients and age-matched controls is

Table 4 **Comparison of bulk brain mercury concentrations in AD and controls.**

		ng Hg/g sample, wet wt. basis (No. samples)
Bulk brain	Controls, both sexes, age-matched	17.5 ± 1.3 (107)
	AD, both sexes	31.4 ± 3.7 (67)
	AD/control diff. = sig., $p < 0.01$.	
Separated tissue	Control gray matter	29.0 ± 7.2 (14)
	Control white matter	20.5 ± 4.9 (13)
	Control gray/white diff. = not sig.	
	AD gray matter	42.7 ± 6.8 (25)
	AD white matter	14.7 ± 2.9 (12)
	AD gray/white diff. = sig., $p < 0.05$.	
	AD/control diffs. are not significant for either gray or white matter.	
Interelement correlations	No significant interelement correlations were found for Hg in AD brain.	

Data from Reference 7.

Table 5 **Mercury and selenium in brain regions strongly affected by AD.**

Region	AD ng Hg/g wet wt. (No. patients and samples)	Control ng Hg/g wet wt. (No. patients and samples)	Sig. (p)
Amygdala	24.3 ×/÷ 1.23 (14)	14.7 ×/÷ 1.20 (15)	<0.10
Hippocampus	15.2 ×/÷ 1.14 (27)	12.4 ×/÷ 1.15 (23)	>0.10
nbM	39.3 ×/÷ 1.17 (11)	8.9 ×/÷ 1.43 (11)	<0.01
	μ Se/g wet wt. (No. patients and samples)		
Amygdala	0.193 ± 0.014 (14)	0.177 ± 0.010 (15)	>0.10
Hippocampus	0.161 ± 0.013 (27)	0.157 ± 0.005 (23)	>0.10
nbM	0.221 ± 0.022 (11)	0.174 ± 0.007 (11)	<0.05

Note: Distributions of Hg are log-normal, so geometric means ×/÷ SEM are reported. Groups are age-matched.
Data from Reference 24.

presented in Table 7.[27] The decrease of Hg concentration in AD nails compared to controls could be due to a lower environmental exposure rate of AD patients who are often confined in long-term care facilities. Mercury levels in AD nails also decreased with increasing age of the patient and with increasing duration and severity of the dementia.[28] Hair and nails may reflect recent exposure to environmental Hg intake, in contrast to brain. An alternate explanation would be that AD somehow alters the distribution of Hg in the body. It has been demonstrated that Hg ions can penetrate the blood-brain barrier and impair its normal function,[30] and that this barrier is altered in AD.[31] If this is the case, Hg could selectively accumulate in the AD brain and be depleted in other tissues that are reflective of more recent

Table 6   **Mercury concentrations and Hg/Se and Hg/Zn mass ratios in AD and control enriched sub-cellular brain fractions.**

		AD (n = 10)	Control (n = 12)	Sig. (p)
Whole brain	Hg, ng/g[a]	176 ± 129	69.6 ± 19.7	NS
	Hg/Se[b]	0.213 ± 0.90	0.081 ± 0.005	NS
	Hg/Zn[c]	3.21 ± 1.92	1.28 ± 0.21	NS
Nuclei	Hg, ng/g	107 ± 37	75.3 ± 40.0	NS
	Hg/Se	0.166 ± 0.016	0.080 ± 0.011	$p < 0.01$
	Hg/Zn	4.16 ± 1.79	2.39 ± 1.52	NS
Mitochondria	Hg, ng/g	174 ± 50	127 ± 30	NS
	Hg/Se	0.349 ± 0.68	0.202 ± 0.041	NS
	Zn/Hg	4.99 ± 1.46	3.52 ± 0.64	NS
Microsomes	Hg, ng/g	59.8 ± 29.1	40.0 ± 8.7	$p < 0.05$
	Hg/Se	0.331 ± 0.053	0.134 ± 0.033	$p < 0.05$
	Zn/Hg	311 ± 0.76	1.49 ± 0.20	$p < 0.05$

[a]   Freeze-dry weight basis.
[b]   Mercury to selenium mass ratio.
[c]   Mercury to zinc mass ratio times 10^3.
Note: NS = not significant, $p > 0.05$.
Data from Reference 26.

Table 7   **Mercury in AD and control hair and fingernails.**

		ng Hg/g fresh wt. basis (No. samples and patients)	
		AD	Control, age >45
Hair	Male	355 ×/÷1.19 (23)	439 ×/÷ 1.09 (32)
	Female	557 ×/÷ 1.25 (20)	525 ×/÷ 1.11 (58)
		Male/Female diff. = sig., $p < 0.05$. AD/Control diff. = not sig.	
		Only untreated hair samples were used for comparison.	
Nails	M + F	132 ×/÷ 1.14 (63)	170 ×/÷ 1.07 (117)
		Male/Female diff. = not sig. AD/Control diff. = sig., $p < 0.05$.	
		Hg tends to decrease in nail of AD patients with increasing age, and increasing duration and severity of the dementia.[28]	

Data from References 27–29.

exposure. This is consistent with the findings of Chaudhary et al.[23] that levels of Hg in brain and fingernail of controls are positively correlated, but not in AD patients.

IV.   MERCURY IN AMYOTROPHIC LATERAL SCLEROSIS (ALS; "LOU GEHRIG'S DISEASE")

We have investigated ALS utilizing the exact methodology as in the AD studies.[8] This strategy was developed for several reasons. Alterations in toxic and essential metals have been proposed as important factors in the etiopathogenesis of ALS.[5,6] Furthermore, in select Western Pacific populations, ALS and dementia coexist in the same individuals with Parkinson's disease,[32] and more recently a frontal lobe type of dementia associated with clinical features of ALS has been described.[33] Therefore, it is possible that trace element alterations

Table 8 **Mercury in ALS and control CNS tissue.**

| | ng Hg/g, wet wt. basis (No. samples) | | |
	ALS	Control	Sig. (*p*)
Spinal cord	15.5 ×/÷ 1.14 (21)	23.5 ×/÷1.25 (20)	NS
Bulk brain	28.0 ×/÷ 1.10 (29)	11.7 ×/÷ 1.11 (71)	0.001
Brain motor region	23.4 ×/÷ 1.18 (12)	16.1 ×/÷ 1.35 (20)	NS

Note: NS = not significant.
Data from Reference 8.

Table 9 **Mercury in ALS and control non-CNS tissue.**

| | | ng Hg/g, wet wt. (No. samples and patients) | | Sig. (*p*) |
		ALS	Control	ALS/control
Blood cells	Males	38.5 ×/÷ 1.10 (28)	21.0 ×/÷ 1.22 (12)	0.001
	Females	51.0 ×/÷ 1.14 (9)	32.1 ×/÷ 1.14 (7)	0.010
		Male/female diff. = sig., both ALS and control, *p* <0.05		
Serum	Males	5.06 ×/÷ 1.17 (27)	2.38 ×/÷ 1.19 (21)	0.009
	Females	6.20 ×/÷ 1.30 (12)	6.86 ×/÷ 1.27 (7)	NS
		Male/female diff. = sig., both ALS and control, *p* <0.05		
Nails	M + F	101 ×/÷ 1.12 (39)	147 ×/÷ 1.15 (40)	NS
		Male/female diff. = NS		

Note: Age-matched controls for nails are from same pool as the AD study.[27,29] NS = not significant.
Data from Reference 8.

Table 10 **Hg/Se molar ratios in ALS and controls.**

	ALS	Controls	Sig. (*p*)
Bulk brain	0.066	0.029	0.001
Motor region	0.053	0.040	NS
Spinal cord	0.059	0.066	NS
Blood cells	0.131	0.077	0.004
Serum male	0.037	0.013	0.001
Serum female	0.077	0.030	0.005
Nail	0.052	0.070	NS

Note: For AD bulk brain the Hg/Se molar ratio is 0.131, as compared to 0.038 for an adult control population.[7] This difference is highly significant, *p* <0.001. NS = not significant.
Data from Reference 8.

are a common signature for both ALS and AD, which may represent a necessary precondition for these phenotypes to develop.

Although we found alterations in the elemental composition of several ALS neural and extra-neural tissues, the most consistent finding was an elevation in Hg.[8] These alterations were found in samples of bulk brain, blood cells, and serum (Tables 8 and 9). Finding a reduction of Se in ALS serum and blood cells (females) and cognizant of the biological interactions between Se and Hg, we computed the molar ratios of Hg/Se for each sample. This analysis revealed large increases in Hg relative to Se in these same tissues (Table 10).

We anticipated finding alterations of Hg in ALS spinal cord and motor cortex. However, the apparently lower Hg levels in ALS spinal cord and higher Hg levels in ALS motor cortex as compared to controls did not reach the significance level of p <0.05, due principally to the broad distribution of values in the control group. Spinal cord is profoundly altered in ALS as indicated by a reduction in percent dry matter (lowered FD/WET ratio). This may be attributed to the loss of motor neurons, axons, and myelin. Other intracellular elements such as zinc (Zn) and potassium (K) were also reduced in the spinal cord from patients dying with advanced ALS. Therefore, it is uncertain whether or not Hg may be present to excess in ALS cord early in the course of the disease before the significant tissue alterations occur.

V. IMPLICATIONS OF IMBALANCES, IN VITRO STUDIES, AND ANIMAL MODELS

Our INAA studies have shown a consistent elevation of Hg in AD and ALS bulk brain and subcellular brain fractions in AD. In AD bulk brain, Hg is significantly elevated in the cerebral cortex, but the most striking increase is in the nbM, the major cholinergic nucleus projecting to the cerebral cortex. The most consistent neurochemical change in the brain in AD is the loss of cholinergic markers in the cerebral cortex[34] which is thought to relate to the severe neuron loss and degeneration in the nbM.[35] Perhaps the neurotoxic effects of Hg play a role in the degeneration of the nbM. In subcellular fractions of the temporal lobe, although there are elevations of Hg in all fractions, the most prominent is in the microsomal fraction, the site of localization in rats intoxicated with Hg.[36] Selenium and Zn have a protective function against Hg in biological tissue.[25,37] The elevation in the Hg/Se and Hg/Zn ratios linked to the elevation of Hg and decrease in Se and Zn found in some subcellular fractions in AD leads us to speculate that Se and Zn may be biologically active in detoxifying Hg in this disorder.

There are several potential mechanisms of Hg toxicity of neurons in AD. Previous studies have shown that Hg binds to tubulin.[38] Other studies have shown that defective microtubule assembly occurs in the AD brain.[39] Abnormal phosphorylation of the microtubule-associated protein tau is present in paired helical filaments of NFT in AD.[40,41] A recent study by Khatoon et al.,[42] using a nucleotide photoaffinity probe of guanosine triphosphate (GPT), revealed that the exchangeable GTP site of the beta subunit of tubulin is not available to add guanine nucleotide in AD brain homogenates. Similar changes were not found in autopsied brains of normal controls or patients with Parkinson's disease, Pick's disease, Creutzfeldt-Jakob disease, or in chronic Al-intoxicated rabbits. Subsequent studies by this group revealed that low μM levels of EDTA-chelated Hg specifically blocks the GTP photolabeling of beta tubulin in normal brain homogenate, like that observed in AD brains.[43] EDTA complexes of Al, Pb, Zn, or Mg did not block GTP binding. Homogenate of brains of rats fed mercury chloride chronically displayed an absence of photolabeled GTP-tubulin interaction similar to that found in the AD brain.[44]

The elevation of Hg in the brain in AD, the potential autoprotection by Se and Zn, plus the finding of Hg-EDTA-induced defective microtubule assembly similar to that observed in AD suggests that Hg plays a role in the pathogenesis of AD. It is possible that Hg could cause alterations in cytoskeletal proteins and relate to NFT or dystrophic neurite formation in AD. Whether Hg deposition in the brain is a primary or secondary event in AD is not known. If it is a secondary event superimposed on degenerating neurons, the toxic effect of Hg could accelerate the rate of neuron degeneration and death.

Elevated brain Hg is a feature common to both ALS and AD. Because all tissues from experimental subjects and controls were obtained and analyzed in a uniform manner, systematic errors and different analytical techniques are not responsible for these elevations. Also at issue is whether or not the changes in the concentration of other elements such as bromine (Br), cesium (Cs), K, and rubidium (Rb) are a consequence of the increased Hg in tissue.

Table 11   **Elemental concentration ratios relative to controls in AD, ALS, and in Sprague-Dawley rats fed mercuric chloride.**

		Bulk brain	Spinal cord
Br	AD	1.42[a]	ND[b]
	ALS	NS[c]	1.32
	Hg rat[d]	NS	1.03
Cs	AD	0.84	ND
	ALS	0.65	NS
	Hg rat	3.2	NS
Hg	AD	1.8	ND
	ALS	2.4	0.66
	Hg rat	10.1	11.1
K	AD	0.92	ND
	ALS	1.12	0.77
	Hg rat	1.04	0.95
Rb	AD	0.77	ND
	ALS	0.46	NS
	Hg rat	0.82	NS
Zn	AD	NS	ND
	ALS	NS	0.88
	Hg rat	NS	1.04
Co, Fe,	ALS	NS	NS
Na, Se	Hg rat	NS	NS

[a] Underline = significant difference relative to control.

[b] ND = not determined.

[c] NS = not significantly different from control.

[d] Hg rat = $HgCl_2$-treated rat.

Data from Reference 45.

In order to determine which of the elemental imbalances could potentially be secondary to chronic Hg toxicity, we exposed rats to oral mercuric chloride for 8 months and analyzed the brain and spinal cord utilizing INAA as in our human studies.[45] Briefly, adult male Sprague-Dawley rats weighing 209 ± 18 g were fed commercial Rodent Chow and distilled water *ad libitum*. Mercury-exposed rats were treated identically except that they consumed water containing 20 mg $HgCl_2$/L. At the time of sacrifice, the entire brain and spinal cord were removed using titanium knives and submitted in toto for INAA.

Mercury was increased 10.1- and 11.1-fold, respectively, in rat brain and spinal cord due to chronic Hg ingestion (Table 11). In contrast, Hg was increased 1.8- and 2.4-fold, respectively, in AD and ALS samples of bulk brain compared to age-matched controls. Therefore, the Hg levels in brain and spinal cord achieved in our chronic Hg-ingestion paradigm exceeded the endogenous levels of Hg in AD and ALS.

Several elements (cobalt, iron, sodium, Se, and Zn) were not perturbed by chronic Hg ingestion and were not altered in ALS brain.[8] Selenium is found to be imbalanced in the nbM[24] and microsomal subcellular fraction of AD brain.[26] Iron is imbalanced in AD neocortical gray matter,[7] amygdala, and hippocampus.[24] Of these, two elements merit further comment. Burnet has speculated that Zn may be important in dementia,[46] although we found no alteration in Zn concentration in AD bulk brain.[7] Muto et al.[47] reported acute changes in Zn concentration in the brains of rats receiving near-lethal doses of $HgCl_2$ or methylmercury chloride, but we observed no changes as a consequence of chronic Hg exposure (Table 11).

Based upon the data of Nylander and Weiner,[48] we anticipated that Se might be elevated in concert with Hg in ALS and AD brain. We found no alterations in Se in ALS bulk brain,[8] but Se was significantly elevated in AD nbM and significantly depleted in AD microsomes, compared to respective controls.[24,26] These observations cannot be simply explained. Moreover, we were not able to induce secondary changes in Se concentrations in rat brain as a consequence of Hg ingestion (Table 11). In a separate experiment however, Se was elevated in a group of rats which were withdrawn from Hg after an 8 month exposure period. Therefore, it appears possible that the (steady-state) concentration of Se is not altered during a period of continual chronic Hg exposure, whereas it may increase during periods of net Hg loss from the brain as in the case of human dental personnel exposed to Hg in the remote past.[48]

Based upon our data (Table 11), it appears that the alterations in Br and Cs cannot be attributed solely to Hg exposure. In AD brain, Br was elevated 1.42-fold over control. No change in Br was detected in either ALS, or as a consequence of Hg ingestion in the rat study. Cesium was decreased in both AD and ALS, whereas controlled Hg exposure caused a 3.2-fold increase in rat brain Cs. Changes in K were minor in AD, ALS, and Hg-exposed rats. The Rb concentration was reduced in all three conditions, but to a greater degree in ALS and AD brains.

Observations on the elemental composition of spinal cord are restricted to Hg-exposed rats and to ALS. The interpretation of any change in elemental composition of ALS cord is problematic because the microscopic structure of the spinal cord itself is altered in end-tate ALS. In addition to the loss of motor neurons in the ventral spinal cord, axons and myelin sheaths are lost in the white matter tracts as well. This pathological change is reflected in a reduced FD/WET ratio in ALS cord, whereas no change was detected as a result of chronic Hg ingestion. Therefore, the reduced concentrations of Hg, K, and Zn in ALS spinal cord probably reflect a reduction of intracellular protein. Bromine, on the other hand, was increased in ALS cord, although not as a consequence of chronic Hg exposure in rats (Table 11).

If Hg is important in the pathogenesis of AD or ALS, the source of Hg in the brain must be sought. All living organisms are exposed to Hg in varying degrees. Numerous environmental sources of Hg are available including seafood, medications, paints, and coal.[38] Dental amalgam has received considerable recent attention as a source of Hg vapor in humans. In individuals with a large number of dental amalgam fillings, Hg vapors from the fillings are the source of a large fraction of the Hg distributed in the body.[49] Animal studies using radioactive Hg in dental amalgams have shown the isotope in various organs, including the brain, within four weeks.[50] Hg in the brain has been found to correlate with the number of amalgam restorations.[51] It is not known whether there is a relationship between dental amalgam fillings, brain Hg, and AD. We are presently performing a prospective and retrospective study of AD and control individuals to answer this question.

The analysis of post-mortem tissues in these disorders can define the chemical architecture which may be characteristic of these diseases, and serve to shape hypotheses which may be tested in animal models of toxicity. Our study of chronic Hg exposure is a specific example of this approach. However, the complexities of even a "simple", controlled exposure to a recognized toxin such as Hg are clear. Factors such as species, dose and route of administration, chemical form, duration of exposure, timing of sacrifice vis a vis the toxin exposure, attention to limiting sample contamination, and the analytical method all require meticulous control. However, it is only through this approach that insights into the potential pathophysiological mechanisms of human diseases can be studied.

ACKNOWLEDGMENTS

This work has been supported in part by NIH grants AG05119, AG05144, NS25165, and a grant from the Muscular Dystrophy Association.

REFERENCES

1. Katzman, R., The prevalence and malignancy of Alzheimer disease: a major killer, *Arch. Neurol.,* 33, 217, 1976.
2. Markesbery, W. R., Alzheimer's disease, In: *Diseases of the Nervous System—Clinical Neurobiology,* Asbury, A. K., McKhann, G. M., and McDonald W. I., Eds., W.B. Saunders, Philadelphia, 1992, chap. 59, 797.
3. Aran, F. A., Researches Sur une Maladie Non Encore de'-Crite de Systeme Musculaire, *Arch. Gen. Med.,* 24, 15, 1850.
4. Markesbery, W. R. and Ehmann, W. D., Trace elements in dementing disorders, In: *Nutritional Modulation of Neural Function,* Morley, J. E., Walsh, J. H., and Sterman, M., Eds., Academic Press, Orlando, 1988, 179.
5. Kasarskis, E. J., Ehmann, W. D., Markesbery, W. R., Khare, S., and Henson, J. C., Toxic and essential elements in amyotrophic lateral sclerosis, In: *ALS. New Advances in Toxicology and Epidemiology,* Rose, F. C. and Norris, F. H., Eds., Smith-Gordon, London, 1990, chap. 22, 181.
6. Kasarskis, E. J., Neurotoxicology: heavy metals, In: *Handbook of Amyotrophic Lateral Sclerosis,* Smith, R. A., Ed., Marcel Dekker, New York, 1992, chap. 25, 559.
7. Ehmann, W. D., Markesbery, W. R., Alauddin, A., Hossain, T. I. M., and Brubaker, E. H., Brain trace elements in Alzheimer's disease, *NeuroToxicology,* 7, 197, 1986.
8. Khare, S. S., Ehmann, W. D., Kasarskis, E. J., and Markesbery, W. R., Trace element imbalances in amyotrophic lateral sclerosis, *NeuroToxicology,* 11, 521, 1990.
9. McKhann G., Drachman, D., Folstein, M. F., Price, D., and Stadlin, E. M., Clinical diagnosis of Alzheimer's disease: report of the NINCDS-ADRDA Work Group, *Neurology (NY),* 34, 939, 1984.
10. Khachaturian, Z. S., Diagnosis of Alzheimer's disease, *Arch. Neurol.,* 42, 1097, 1985.
11. International Atomic Energy Agency, Quality Assurance in Biomedical Neutron Activation Analysis, IAEA, Vienna, Austria, 1984, 323.
12. Ehmann, W. D., Markesbery, W. R., Kasarskis, E. J., Vance, D. E., Khare, S. S., Hord, J. D., and Thompson, C. M., Applications of neutron activation analysis to the study of age-related neurological diseases, *Biol. Trace Element Res.,* 13, 19, 1987.
13. Ehmann, W. D., Markesbery, W. R., Hossain, T. I. M., Alauddin, M., and Goodin, D. T., Trace elements in human brain tissue by INAA, *J. Radioanal. Chem.,* 70, 57, 1982.
14. Markesbery, W. R., Ehmann, W. D., Alauddin, M., and Hossain, T. I. M., Brain trace element concentrations in aging, *Neurobiol. Aging,* 5, 19, 1984.
15. Khare, S. S., INAA Studies of Trace Element Imbalances Associated with Amyotrophic Lateral Sclerosis, Ph.D. dissertation, University of Kentucky, Lexington, 1987.
16. Wenstrup, D., Trace Elements at the Subcellular Level in Alzheimer's Disease, Ph.D. dissertation, University of Kentucky, Lexington, 1988.
17. Vance, D. E., Trace Element Relationships in Alzheimer's Disease: Hair and Nail Analysis by INAA, Ph.D. dissertation, University of Kentucky, Lexington, 1986.
18. Bowen, H. J. M., Kale as a reference material, In: *Biological Reference Materials: Availability, Uses, and Need for Validation of Nutrient Measurement,* Wolf, W. R., Ed., John Wiley & Sons, New York, 1985, chap. 1, 3.
19. Lievens, P., Cornells, R., and Hoste, J., A separation scheme for the determination of trace elements in biological materials by neutron activation analysis, *Anal. Chim. Acta,* 80, 97, 1975.

20. Muramatsu, Y., Ogris, R., Relchel, F., and Parr, R. M., Simple destructive neutron activation analysis of mercury and selenium in biological materials using activated charcoal, *J. Radioanal. Nucl. Chem.*, 125, 175, 1988.

21. Zhuang, G., Wang, Y., Zhi, M., Zhou, W., Yin, J., Tan, M., and Cheng, Y., Determination of arsenic, cadmium, mercury, copper, and zinc in biological samples by radiochemical neutron activation analysis, *Radioanal. Nucl. Chem.*, 129, 459, 1989.

22. Massaro, E. J., Yaffee, S. J., and Thomas, C. C., Mercury levels in human brain, skeletal muscle and body fluids, *Life Sci.*, 14, 1939, 1975.

23. Chaudhary, K., Ehmann, W. D., Rengan, K., and Markesbery, W.R., Trace element correlations between human brain and fingernails, *J. Trace Microprobe Techn.*, 10, 225, 1992.

24. Thompson, C. M., Markesbery, W. R., Ehmann, W. D., Mao, Y.-X., and Vance, D. E., Regional brain trace-element studies in Alzheimer's disease, *NeuroToxicology*, 9, 1, 1988.

25. Cuvinaralar, M. L. A. and Furness, R. W., Mercury and selenium interaction—a review, *Ecotoxicol. Environ. Safety*, 21, 348, 1991.

26. Wenstrup, D., Ehmann, W. D., and Markesbery, W. R., Trace element imbalances in isolated subcellular fractions of Alzheimer's disease brains, *Brain Res.*, 533, 125, 1990.

27. Vance, D. E., Ehmann, W. D., Markesbery, W. R., Trace element imbalances in hair and nails of Alzheimer's disease patients, *NeuroToxicology*, 9, 197, 1988.

28. Vance, D. E., Ehmann, W. D., and Markesbery, W. R., A search for longitudinal variations in trace element levels in nails of Alzheimer's disease patients, *Biol. Trace Element Res.*, 26/27, 461, 1990.

29. Vance, D. E., Ehmann, W. D., and Markesbery, W. R., Trace element content in fingernails and hair of a nonindustrialized US control population, *Biol. Trace Element Res.*, 17, 109, 1988.

30. Ware, R. A., Chang, L. W., and Burkholder, P. M., An ultrastructural study on the blood-brain barrier dysfunction following mercury intoxication, *Acta Neurophathol. (Berlin)*, 30, 211, 1974.

31. Wisniewski, H. M. and Kollowski, P. B., Evidence for blood-brain barrier changes in senile dementia of the Alzheimer type (SDAT), *Ann. N.Y. Acad. Sci.*, 396, 119, 1982.

32. Elizan, T. S., Hirano, A., Abrams, B. M., Need, R. L., Van Nuls, C., and Kurland, L. T., Amyotrophic lateral sclerosis and Parkinson-dementia complex of Guam. Neurological reevaluation, *Arch. Neurol.*, 14, 356, 1966.

33. Neary, D., Non Alzheimer's disease forms of cerebral atrophy, *J. Neurol. Neurosurg. Psychiat.*, 53, 929, 1990.

34. Bartus, R. T., Dean, R. L., Beer, B., and Lippa, A. S., The cholinergic hypothesis of geriatric memory dysfunction, *Science*, 217, 408, 1982.

35. Whitehouse, P. J., Price, D. L., Struble, R. G., Clark, A. W., Coyle, J. T., and Delong, M. R., Alzheimer's disease and senile dementia: loss of neurons in the basal forebrain, *Science*, 215, 1237, 1982.

36. Yoshino, Y., Mozai, T., and Nakao, K., Biochemical changes in the brain in rats poisoned with alkyl mercuric compound, with special reference to the inhibition of protein synthesis in brain cortex slices, *J. Neurochem.*, 13, 1223, 1966.

37. Yoshikawa, H. and Ohta, H., Interaction of metals and metallothionein, *Dev. Toxicol. Environ. Sci.*, 9, 11, 1982.

38. Clarkson, T. W., Mercury, In *Trace Element in Human and Animal Nutrition*, Mertz, W., Ed., Vol. 1, Academic Press, San Diego, CA, 1987, 417.

39. Iqbal, K., Grundke-Iqbal, I., Zaidi, T., Merz, P. A., Wen, G. Y., Shaikh, S. S., and Wisniewski, H. M., Defective brain microtubule assembly in Alzheimer's disease, *Lancet*, 2, 421, 1986.

40. Kosik, K. S., Joachim, C. L., and Selkoe, D. J., Microtubule-associated protein tau (τ) is a major antigenic component of paired helical filaments in Alzheimer disease, *Proc. Natl. Acad. Sci. U.S.A.*, 83, 4044, 1986.

41. Grundke-Iqbal, I., Iqbal, K., Tung, Y. C., Quinlan, M., Wisniewski, H. M., and Binder, L. I., Abnormal phosphorylation of the microtubule-associated protein tau (τ) in Alzheimer cytoskeletal pathology, *Proc. Natl. Acad. Sci. U.S.A.* 83, 4913, 1986.

42. Khatoon, S., Campbell, S. R., Haley, B. E., and Slevin, J. T., Aberrant guanosine triphosphate-beta tubulin interaction in Alzheimer's disease, *Ann. Neurol.,* 26, 210, 1989.

43. Slevin, J. T., Gunnersen, D. J., Duhr, E., and Haley, B. E., Implication for mercury in the alteration of β-tubulin observed in Alzheimer disease, *Ann. Neurol.,* 28, 230, 1990.

44. Duhr, E., Pendergrass, C., Kasarskis, E., Slevin, J., and Haley, B., Hg^{2+} induces GTP-tubulin interactions in rat brain similar to those observed in Alzheimer's disease, *FASEB J.,* 5(4), A456, 1991.

45. Tandon, L., Kasarskis, E. J., and Ehmann, W. D., INAA for interelement correlations in rats after mercuric chloride exposure, *J. Radioanal. Nucl. Chem.,* 161, 39, 1992.

46. Burnet, F. M., A possible role of zinc in the pathology of dementia, *Lancet,* 1, 186, 1981.

47. Muto, H., Shinada, M., Tokuta, K., and Takizawa, Y., Rapid changes in concentrations of essential elements in organs of rats exposed to methylmercury chloride and mercuric chloride as shown by simultaneous multielemental analysis, *Br. J. Ind. Med.,* 48, 382, 1991.

48. Nylander, M. and Weiner, J., Mercury and selenium concentrations and their interrelations in organs from dental staff and the general population, *Br. J. Ind. Med.,* 48, 729, 1991.

49. Clarkson, T. W., Hursch, J. B., and Nylander, M., The prediction of intake of mercury vapor from amalgams, In: *Biological Monitoring of Toxic Metals,* Clarkson, T. W., Friberg, L., Nordberg, G. F., and Sager P. R., Eds., Plenum Press, New York, 1988, 247.

50. Hahn, L. J., Kloiber, R., Vimy, M. J., Takahashi, Y., and Lorscheider, F. L., Dental "silver" tooth fillings: a source of mercury exposure revealed by whole-body image scan and tissue analysis, *FASEB J.,* 3, 2641, 1989.

51. Eggleston, D. W. and Nylander, M., Correlation of dental amalgam with mercury in brain tissue, *J. Prosthet. Dent.,* 58, 704, 1987.

Trends in Mercury Concentrations in the Hair of Women of Nome, Alaska: Evidence of Seafood Consumption or Abiotic Absorption?

Brenda Lasorsa

CONTENTS

ABSTRACT: Eighty samples of hair from women of childbearing age from Nome, Alaska and 7 control samples from women living in Sequim, Washington were analyzed for mercury concentration by segmental analysis in an effort to determine whether seasonal fluctuations in mercury concentration in the hair samples can be correlated to seasonal seafood consumption. Full-length hair strands were analyzed in 1.1-cm segments representing 1 month of growth using a strong acid digestion and cold vapor atomic fluorescence analysis. It was assumed that the concentration of mercury in each segment is an indicator of the mercury body burden during the month in which the segment emerged from the scalp.

When mercury concentration vs. growth month is plotted for each participant, a number of trends are seen. Of the hair samples, 40 including 1 control are either too short to show any particular trend or have steady concentrations between 0.2 and 4.5 ppm for all segments; 18 of the samples show seasonal variability, with 5 of the controls and 1 Nome resident showing winter highs while the remainder, all Nome residents, show summer highs. There were 26 samples that showed an increase in mercury concentration toward the distal end of the strand regardless of month of growth—14 of the 26 distally increasing samples, including 1 control, have a maximum of less than 3 ppm, while the remainder, all Nome residents, have maximums as high as 16 ppm. The remaining three samples show a distal increase with a superimposed seasonal variation.

The 12 individuals with maximums over 3 ppm are of interest. These 12 individuals exceed normal levels for people consuming fish 1 to 4 times per month and in some cases 1 to 4 times per week, and some also exceed the commonly accepted levels of concern for fetal effects of mercury poisoning. However, the trend of increasing mercury concentrations toward the distal end of the hair strand regardless of month of emergence, and the documented presence of elevated levels of elemental mercury in the Nome area, suggest that these elevated levels may

1-56670-066-3/94/$0.00+$.50

actually be due to external contamination of the hair strands by adsorption and not due to ingestion of contaminated foodstuffs such as seafood.

I. INTRODUCTION

In the autumn of 1989, 200 samples of human hair from women of childbearing age residing in Nome, Alaska were analyzed for total mercury. The mercury analyses were conducted at Battelle Marine Sciences Laboratory (MSL) as part of a baseline monitoring study undertaken by Minerals Management Service (MMS) during the preparation of an environmental impact statement evaluating the feasibility of off-shore gold dredging leases. There was concern that off-shore dredging could release elemental mercury, which is often associated with gold deposits, to the waters of Norton Sound. This mercury could then be accumulated by marine mammals and fish of the region that, in turn, are consumed by the population of Nome.

The results of the 1989 study[1] prompted MMS to pursue a more thorough investigation of the mercury levels. To that end, 80 full-length hair samples were collected in the autumn of 1990 from 27 participants of the original study, including 10 of 16 with relatively high mercury levels, plus 53 additional heavy users of subsistence foods. The goal of this study was to analyze the full-length hair samples in segments equivalent to one month of growth to ascertain whether variations occur in the levels of mercury in the hair as a function of dietary habits, such as seasonal consumption of certain forms of marine life.

Samples were collected by personnel from Norton Sound Health Corporation using methods and equipment supplied by MSL. Samples were taken as close to the scalp as possible by a gloved staff member using clean scissors. The sample was carefully bound with tape within 2 cm of the scalp end to maintain the hair in a bundle and placed in labeled polyethylene bags for shipment to MSL. Upon arrival, each sample was carefully removed from the bag and sectioned into 1.1-cm lengths. This length has been determined to be equivalent to 1-month growth on average.[2] Each segment was placed in a labeled, pre-weighed, acid-cleaned glass vial. The portion of the sample that was in contact with the tape was discarded and the amount discarded was recorded. To reduce sample loss from static electricity, samples were wetted with distilled water during segmentation and dried prior to weighing. The number of segments generated from each sample has varied from 2 to 26, recording between 2 and 26 months of mercury exposure.

II. ANALYTICAL METHODS

The samples were received and logged in at Battelle MSL on November 1, 1991. Prior to beginning analysis of the samples, preliminary experiments were performed to be certain the procedure was appropriate and would work as expected. An experiment to ascertain whether any significant contamination to the hair sample would occur during storage in polyethylene bags indicated an insignificant contamination level of 0.013 ng of mercury. Using samples of the Japanese certified hair standard, NIES-5, a series of digestions were performed to optimize the digestion method and time while still assuring complete digestion. This resulted in the HNO_3/H_2SO_4 digestion at 350°F for 3 h as described below.

Each 1.1 cm hair segment was placed in an acid-cleaned, preweighed glass scintillation vial and 5.0 mL of a 70% HNO_3/30% H_2SO_4 solution was pipetted into the vial and swirled to mix. An acid-cleaned glass sphere was placed over the mouth of the vial, and the samples were predigested at room temperature for about an hour. Samples were then placed on a hot plate, and brought up to a refluxing boil in small temperature increments to avoid excessive foaming, which is especially common with tissue samples. The samples were refluxed (hot plate temperature about 300° C) for 2 to 3 h or until all organic matter had dissolved. The solution looked almost colorless or light yellow, and the brown gas above the liquid had almost disappeared. The samples were allowed to cool on the hot plate. Each sample was then diluted to the neck of the vial with 1% BrCl, capped, and thoroughly mixed prior to analysis. The final volume was determined to be 21.378 ± 0.183 mL for the lot of vials used for this study. Samples were digested in groups of 24 to 30 segments. This usually included two or three segmented samples (depending on the length), a blank, a NIES hair standard, a spiked NIES hair standard, and another tissue standard (usually DORM-1 dogfish muscle). Digestion batches were visually separated from each other using different colored labels so that the appropriate batch blank could be applied during analysis. Individual samples were identified by their participant number, and the segments were identified alphabetically, beginning with "a" at the scalp.

Samples were analyzed by a cold vapor atomic fluorescence technique, based upon the emission of 254 nm radiation by excited Hg^0 atoms in an inert gas stream. Mercuric ions in the oxidized sample are reduced to Hg^0 with $SnCl_2$, and then purged onto gold-coated sand traps as a means of preconcentration and interference removal. Mercury vapor is thermally desorbed to a second "analytical" gold trap, and from that into the fluorescence cell. Fluorescence (peak area) is proportional to the quantity of mercury collected, which is quantified using a standard curve as a function of the quantity of sample purged. Due to the strong oxidation step, followed by dual gold amalgamation, there are no observed interferences with the method.[3]

The instrumentation was calibrated daily using a four-point linear regression and a calibration check standard NBS-164lb. The average of the daily calibration checks was 1.51 ± 0.07 µg/mL, which compares very well with the certified value of 1.52 ± 0.04 µg/mL. Two tissue standards (NIES-5 human hair and DORM-1 dogfish muscle) were digested with each set and analyzed several times daily. A spiked NIES-5 hair sample was analyzed for matrix spike recovery as well.

A total of 828 hair segments were analyzed, not including duplicates, control samples, and standards. Seven samples from women of childbearing age living in the Sequim, Washington area were segmented and analyzed as controls. Two of the samples were split prior to segmentation and analyzed as duplicates: control sample #7 was duplicated at MSL, and sample #62 was duplicated at another laboratory. The mean deviation between mercury concentrations in each segment of the duplicated sample was 6.5% for the sample duplicated inhouse and 19% for the sample duplicated at another laboratory.

Initially, a problem was encountered with the mercury values in the tissue standards consistently running 10 to 15% high. This problem was finally resolved when it was discovered that when the sample vials were warmed to dry the samples after segmentation, the labels were actually losing weight as some of the adhesive evaporated. When the vials were reweighed following sample addition, the calculated weight difference, therefore, was too small, resulting in calculated mercury concentrations being too high. Because the weights of the segmented hair samples were very small, this weight difference is significant. This problem was rectified by heating the labeled vials briefly prior to the initial weighing. Because the "blank" vial was always weighed and treated exactly like the samples, its weight difference after heating was used to correct the concentrations of the samples analyzed prior to identification of the cause of the problem. Once this was done, the tissue standards once again fell into their certified ranges.

Another problem, related to the very low sample weights, was that a small percentage of the segments at the distal end of the samples (where there were fewer strands than at the scalp) were so light that we were often working near the limits of the balance, resulting in a potentially larger margin of error in the sample weight and therefore in the final concentration. Samples exhibiting this problem are flagged on the final graph. Two segments were lost in the course of the study: one caused by a vial rupture during digestion and one because of an apparent weighing error.

Throughout the course of analysis standard reference materials (NIES-5 hair and DORM-1) and matrix spikes of NIES-5 were analyzed. The mean value determined for NIES-5 (certified value = 4.4 ± 0.4 ppm) was 4.51 ± 0.44 (n = 81) and the mean value determined for DORM-1 (certified value = 0.798 ± 0.074 ppm) was 0.864 ± 0.091 ppm (n = 72). Spike samples of NIES-5 yielded a mean recovery of $96.6 \pm 12.7\%$ (n = 47).

III. RESULTS AND DISCUSSION

A summary of minimum, maximum, and average concentrations for each participant, as well as the concentration determined in the previous study when applicable, is presented in Table 1. When two participant numbers are given, the first is for the present study and the second is for the 1989 study. Trend abbreviations are as follows: SVWH = seasonal variability summer high, SVWH = seasonal variability winter high, SSDI = superimposed seasonal and distally increasing, NT = not apparent trend, DIL = distal increase low concentration, and DIH = distal increase high concentration. No statistically significant correlation was found between mercury concentration and chemical hair treatments as indicated in the perm/color column. When mercury concentration is plotted vs. growth month, the data reveal several interesting trends.

Of the 87 hair samples which were segmented for analysis (80 Nome residents and 7 controls), 40 exhibited no statistically significant trend, mainly because the samples were too short (5 segments or fewer). None of these trendless samples had an average mercury concentration greater than 3 ppm. In the remaining 47 samples, all but 12 samples had maximum concentrations less than 3 ppm. In the low level group (maximum [Hg] <3 ppm) several trends stand out: (1) seasonal increases, (2) distal increases, or (3) some combination thereof. In the 12 higher level samples (maximum [Hg] >3 ppm) only the latter two trends were seen.

Idealized examples of the common trends are presented in Figure 1. The first two examples illustrate seasonal variability with both summer and winter highs. Samples with summer highs tended to have overall higher concentrations as well. The next two examples illustrate both large and small distal increases. The fifth example illustrates a superimposed distal increase and seasonal fluctuation and the final example illustrates a sample showing no trend, most often due to lack of length.

The most common trend in the "less than 3 ppm" group appears to be seasonal increases with or without a superimposed distal increase. Note that six of the seven control subjects fall into this category; however, the control subjects show winter increases while almost all of the Nome participants show seasonal variation peak during the summer months. None of the controls analyzed has an average concentration over 1 ppm.

In the remaining 12 samples, those with maximums in the 3 to 16 ppm range, all of the participants except participant #3 exhibit a nearly constant concentration for the first 3 to 5 months of emergence followed by a steady increase toward the distal end regardless of month of emergence. Participant #3 showed this general trend, but the values fluctuate somewhat because of low segment weights.

This trend suggests that the participants showing the distal increase (particularly those greater than 3 ppm) are exposed to some source of mercury that results in hair strand uptake by adsorption rather than ingestion. In order for this trend to be related to the ingestion of contaminated foodstuffs, the contamination would have to be continuously decreasing with an onset sometime prior to the sampling and an end coinciding with the sampling (because all of the participants exhibiting this trend had normal levels [<3 ppm] at the scalp). This is highly unlikely. This type of trend would also be expected to some degree from exposure of the hair strand to airborne contaminants. The longer the hair strand has been exposed to the environment, the greater the degree of external contamination. Most of the hair samples that were long enough to exhibit any trend at all exhibit this distal increase, but the overall Hg concentrations are still at or below normal levels of 1.9 ± 0.9 ppm (derived from an average of 559 samples from 13 industrialized countries from individuals consuming fish 1 to 4 times

Table 1 Tabulation of data for all 80 participants of the study.

Participant ID	Number of segments	Minimum [Hg] PPM	Maximum [Hg] PPM	Average [Hg] PPM	Factor	Trend type	1989 Study average [Hg] PPM	Length	Perm/color
Control 2	15	0.590	1.572	0.963		SVWH			
Control 4	14	0.427	1.341	0.874		SVWH			
26/67	22	0.727	1.278	1.056		SVWH	0.49		
Control 1	19	0.403	0.726	0.605		SVWH			
Control 6	14	0.249	0.597	0.475		SVWH			P
Control 5	10	0.493	1.272	0.764		SVWH			
17	12	0.531	1.433	0.861		SVSH			P
27/27	13	0.413	1.026	0.537		SVSH	0.59		P
68	12	0.565	1.402	0.828		SVSH			P
64	21	0.712	1.232	0.973		SVSH			
41	22	0.992	2.996	1.961		SVSH			P
44/150	16	0.463	0.960	0.669		SVSH	0.85		P
52	13	0.508	1.613	0.850		SVSH			P
32	12	0.802	1.790	1.166		SVSH			
53	13	0.209	0.380	0.270		SVSH			P
13	12	0.423	1.139	0.637		SVSH			P
80	17	0.516	1.192	0.895		SVSH			
78	20	0.852	1.537	1.096		SVSH			
54	24	0.429	1.242	0.830	0.036	SSDI			
25/20	16	1.049	2.90	1.610	0.069	SSDI	2.15		P
34	26	0.821	2.859	1.781	0.090	SSDI			
57	12	1.195	2.324	1.543		NT			
59	4	2.025	2.123	2.068		NT		Short	
55	8	1.176	2.387	1.755		NT			P
56	4	1.394	2.628	2.073		NT		Short	P
Control 7DUP	11	0.392	1.149	0.848		NT			
49	4	1.348	1.710	1.532		NT		Short	P
48	6	1.457	4.532	2.611		NT			
50	12	1.402	1.902	1.626		NT			
79/93	2	1.381	1.604	1.493			0.54	Short	P

Table 1 Continued.

Participant ID	Number of segments	Minimum [Hg] PPM	Maximum [Hg] PPM	Average [Hg] PPM	Factor	Trend type	1989 Study average [Hg] PPM	Length	Perm/color
Control 7	11	0.430	1.178	0.872		NT			
51	13	1.197	1.644	1.403		NT			
67	3	1.037	1.790	1.490		NT		Short	
66	8	2.051	3.113	2.306		NT			
65	7	0.954	1.764	1.196		NT		Short	P
74	4	1.028	1.358	1.207		NT			P
72	6	1.700	2.583	1.993		NT		Short	P
73	3	0.420	0.794	0.612		NT		Short	P
70	5	0.221	0.770	0.455		NT			
62	6	0.836	1.215	1.031		NT		Short	
61	5	0.647	2.533	1.366		NT			
60	6	0.564	1.064	0.748		NT			
62 DUP	6	0.660	1.240	0.973		NT		Short	P
63	5	1.664	2.284	1.990		NT		Short	P
75	2	0.309	0.322	0.316		NT			
76	6	0.530	0.895	0.712		NT			
21	7	0.891	1.566	1.129		NT		Short	P
20	3	0.487	0.611	0.545		NT		Short	P
22	3	1.520	2.788	2.063		NT		Short	
33	3	1.947	3.333	2.564		NT		Short	
26	4	0.661	1.366	0.980		NT		Short	P
18/168	10	0.252	0.661	0.507		NT	3.80	Short	C
7/166	4	0.653	0.742	0.697		NT	0.39	Short	
4/146	5	1.712	1.917	1.783		NT	3.16	Short	P
8	2	0.599	0.842	0.720		NT		Short	
11	6	0.532	0.866	0.657		NT		Short	P
10	5	0.430	0.842	0.573		NT		Short	P

ID	n					Group			
48/116	8	0.457	0.814	0.611		NT	0.37		P
38/185	6	0.667	1.591	0.992		NT	0.70		P
40	4	0.727	0.994	0.853		NT		Short	P
42	17	0.365	0.807	0.578		NT			P
45/66	2	0.715	0.836	0.775		NT	1.38	Short	P
39/32	16	0.628	0.924	0.717		NT	0.58		P
69/155	8	0.563	1.224	0.872	0.105	DIL	0.20		
71/108	9	0.332	1.459	0.642	0.223	DIL	0.66		
77	12	0.254	1.767	0.577	0.154	DIL			P
43	18	0.817	2.525	1.365	0.123	DIL			
Control 3	13	0.164	0.805	0.359	0.074	DIL	3.82		P
6/84	16	0.755	2.527	1.625	0.125	DIL	1.12		P
29/70	13	0.709	2.596	1.150	0.167	DIL			
36	6	0.290	0.793	0.440	0.086	DIL			P
58	10	0.315	0.704	0.417	0.067	DIL			
1	11	1.078	2.744	1.384	0.195	DIL			
47	10	0.800	1.806	1.255	0.146	DIL			
37/193	11	0.388	1.091	0.590	0.084	DIL	0.53		P
23/30	16	0.272	1.106	0.659	0.063	DIL	0.57		
19/194	18	0.382	1.680	0.759	0.094	DIL	2.05		P
2/38	22	1.290	15.194	4.979	0.585	DIH	0.89		
5/152	16	1.335	6.575	2.664	0.445	DIH	3.75		P
3/99	16	2.901	12.743	7.423	0.650	DIH	2.15		P
30	9	0.789	3.396	1.643	0.483	DIH			
24/85	18	1.027	3.896	1.931	0.206	DIH	3.70		P
35	13	0.559	3.153	1.336	0.293	DIH			P
31	15	1.259	4.386	1.944	0.224	DIH			P
16/117	5	0.945	3.283	1.565	0.797	DIH	6.22	Short	
129	6	1.061	6.198	2.898	1.850	DIH	1.96		
9	11	0.729	7.535	2.853	0.860	DIH			
15/170	19	1.129	6.424	3.450	0.313	DIH	3.01		
14/145	14	0.762	3.975	1.838	0.312	DIH	0.80		P

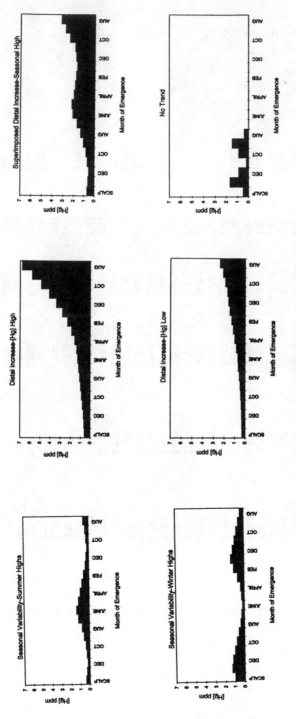

Figure 1 Idealized examples of the trends common in the segmented hair samples.

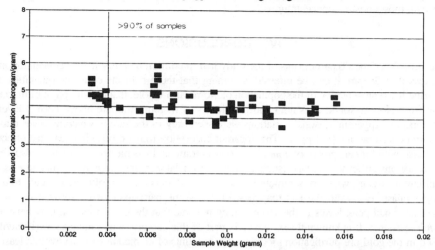

Figure 2 Correlation of sample weight vs. measured mercury concentration.

per month).[4] However, 12 of the samples exhibit this trend to a greater extent, with distal end concentrations approaching 16 ppm. Because the level of concern is generally considered to be 10 ppm,[4] it is important to determine whether these high concentrations are truly representative of body burden in these 12 individuals.

An "adsorption factor" was calculated based on the slope of a least-squares regression of concentration vs. month of growth (excluding the first four constant level segments) to examine the rate of mercury concentration increase. Assuming the increase can be entirely attributed to external adsorption, the rate of adsorption varies from 0.034 to 0.090 ppm/month (mean = 0.072 ppm/month) for the samples exhibiting superimposed seasonal variation and distal increase. The rate varies from 0.063 ppm/month to 0.223 ppm/month (mean = 0.123 ppm/month) for the distally increasing samples with maximum concentrations less than 3 ppm. Finally, for the samples with maximum concentrations greater than 3 ppm, the rate varies from 0.206 ppm/month to 1.850 ppm/month (mean = 0.583 ppm/month).

There was initial suspicion that the distal increases could be an analytical artifact caused by processing contamination or weighing error in the distal ends of the hair strands. These segments were sometimes up to 50% lighter than segments near the scalp from the same subject because of layered haircuts or breakage. However, examination of the numerous replicate analyses of the 31 digestions of the NIES-5 hair standard varying in weight from 0.0032 to 0.0156 g (a total of 81 analyses) shows that this is not probable (see Figure 2). This graph shows the correlation between sample weights of 31 digestions of the NIES hair standard, ranging from 0.003 and 0.0156 g, and the corresponding analytical result. The range of NIES-5 digestion weights bracket the sample weights with the exception of a few which were flagged as somewhat unreliable because of low weights. The graph shows that there is little or no correlation between sample size and analyzed concentration, except for sample weights less than 0.004 g. However, even the slight correlation seen at weights less than 0.004 g does not account for the order-of-magnitude increases seen in many of the samples.

Similar trends were seen by Clarkson in the hair of infants and their families exposed to diapers containing phenylmercury[6] and by Wilson et al.[5] in a family using a shampoo containing an unusually high concentration of mercury. In these cases, it is probably the first few segments (those most recently emerged from the scalp) that are indicative of the true body burden of the participant. These are the segments which, in this study, exhibited the lowest, most constant concentrations.

IV. CONCLUSIONS

Because 65 of the 80 participants of the study had maximum mercury in hair concentrations of less than 3 ppm, it can be inferred, assuming that these subjects consume quantities of marine life representative of the population of Nome as a whole, that consumption of marine life from Norton Sound does not contribute levels of mercury that are above normal levels of concern. Apparent seasonal variations were seen only in the samples having lower maximum concentrations (<3 ppm). The apparent absorption may be masking seasonal effects in those with higher, probably nondietary, concentrations. Possible sources of this adsorbed mercury are airborne mercury (such as vapor from latex paints containing mercury as a mildew retardant), water, or sediment tracked into buildings and contributing mercury to the vapor phase. However, the fact that most of the Nome participants who showed seasonal variations had peak levels in the summer may indicate that the contamination may have an outdoor source. It is known that in the early part of this century, when mercury was heavily used in the gold ore purification process, large amounts of elemental mercury were released to the environment in the vicinity of Nome and soil levels in the range of 350 to 1000 ppm have been measured within the city limits.[2] The relatively high mercury concentrations measured in the hair of the individuals exhibiting this trend may be indicative of this rather large source of mercury contamination in the Nome area.

If these data are to used as a reliable indicator for exposure assessment, it is important to determine whether the steady distal increase seen in many of the samples is indeed due to abiotic adsorption from the environment. If that is the case, there are implications with respect to the interpretation of this and other mercury in hair data, both segmental and total hair. If adsorption is taking place, it would be in the form of elemental mercury, as opposed to methylmercury, which is the predominant species of mercury found in marine mammals and fish. It would be possible to resolve the question of the origin of the mercury increase by analyzing segmented hair from the same subject for both methylmercury and total mercury (methylmercury + elemental mercury). If the increase is an accurate reflection of the mercury ingested by the subject, the trend should be the same for both types of mercury. If the increase is due to abiotic adsorption of airborne elemental mercury, the methylmercury levels should remain relatively constant while the total mercury level increases toward the distal end of the hair. By analyzing hair in this way, the results would be indicative of both body burden as a result of ingestion of organomercury compounds common in fish and marine mammals and abiotically adsorbed mercury.

ACKNOWLEDGMENT

This work was supported by the Minerals Management Service under a Related Services Agreement with the U.S. Department of Energy. The Battelle Marine Sciences Laboratory is part of the Pacific Northwest Laboratory, which is operated for the U.S. Department of Energy by Battelle Memorial Institute under Contract DE-AC06–76RLO 1830.

REFERENCES

1.   Crecelius, E.A., C.W. Apts, and B.K. Lasorsa, 1990. *Concentrations of Metals in Norton Sound Seawater Samples and Human Hair Samples,* OCS Study MMS 90–0010, Minerals Management Service, Anchorage, AK.

2. Marsh, D.O., 1989. Methylmercury poisoning in Iraq, In: *Mercury in the Marine Environment,* Workshop Proceedings, OCS Study MMS 89–0049, Minerals Management Service, Anchorage, AK.
3. Bloom, N.S. and W.F. Fitzgerald, 1988. Determination of volatile mercury species at the picogram level by low-temperature gas chromatography with cold-vapor atomic fluorescence detection, *Anal. Chim. Acta,* 208:151.
4. Mitra, S., 1986. *Mercury in the Ecosystem,* Trans Tech Publications, Brookfield, VT, p. 142.
5. Wilson, K., D. Stark, T. Kolotyluk, and E. E. Daniel, 1974. A technical problem discovered in testing for mercury in human hair, *Arch. Environ. Health,* 28:18.
6. Clarkson, T., Personal communication, May 1991.

Inhibition of Progression Through the S Phase of the Cell Cycle: A Mechanism of Cytoxicity of Methylmercury*

Edward J. Massaro

CONTENTS

ABSTRACT: The effect of methylmercury (MeHg) on cell cycle progression was investigated by flow cytometry (FCM) employing the murine erythroleukemic cell as a model of the proliferating cellular system. Exposure in vitro to 5 to 10 μM (μmol/L) MeHg for 6 h results in a dose-dependent decrease in the rate of cell replication. Investigation of nuclear binding of propidium iodide (PI) indicates that MeHg inhibits the rate of DNA synthesis (rate of passage through the S phase of the cell cycle). Following exposure to 5 μM MeHg, only a modest accumulation of cells with a G_2M (4n) DNA content is observed, indicating that the primary effect of MeHg is not colchicine-like. Light microscopy reveals progressive chromosomal damage as a function of MeHg dose, ranging from condensation to pulverization at concentrations <10 μM to the formation of wreath-like ring structures at/above 10 μM. Ring formation appears to involve fusion of the entire chromosomal complement. The data indicate DNA synthesis, not mitosis, to be the primary target of MeHg toxicity in proliferating cells.

*DISCLAIMER: This manuscript has been reviewed in accordance with the policy of the Health Effects Research Laboratory, U.S. Environmental Protection Agency, and approved for publication. Approval does not signify that the contents necessarily reflect the views and policies of the Agency, nor does mention of trade names or commercial products constitute endorsement or recommendation for use.

I. INTRODUCTION

Methylmercury (MeHg), a potent developmental neurotoxin and teratogen,[1-4] is an ubiquitous contaminant of the aqueous environment.[5] Apparently, the mechanism of MeHg cytotoxicity is complex. It has been observed that MeHg inhibits DNA, RNA, and protein synthesis,[1,2,6] interacts with the cytoskeleton,[7,8] alters the properties of biomembranes,[1,2,9-11] and disrupts axoplasmic transport.[12] In the proliferating cell, MeHg inhibits mitosis and/or decreases the rate of the cell cycle. It has been reported that mitotic arrest results from inhibition of microtubule assembly[7,13-15] while a decreased cycling rate has been attributed to lengthening of the duration of the G_1 phase of the cell cycle as a consequence of inhibition of protein synthesis.[16] Whether the duration of other premitotic phases of the cell cycle is altered is not clear.

Flow cytometric analysis (FCM) reveals that MeHg perturbs the kinetics of the murine erythroleukemic (MEL) cell cycle. At relatively low levels (5 μM for 6 h), MeHg predominately inhibits DNA synthesis (i.e., progression through the S phase of the cell cycle). Only a modest accumulation of cells with a 4n DNA content (i.e., in the G_2/M phase of the cycle as defined by FCM) is observed. At higher concentrations ≥ 10 μM), progression through all phases of the cell cycle is blocked. Light microscopy reveals a dose-dependent increase in the incidence of chromosomal aberrations. Chromosomal condensation is observed at doses ≤ 10 μM. At 10 μM MeHg, both condensation and pulverization are observed and more than 50% of the aberrant chromosomal figures appear in the form of wreath-like ring structures apparently comprised of the entire chromosomal complement. At higher dose levels ≥ 25 μM), all (100%) of the resolvable chromosomal figures appear as ring structures. In addition, at dose levels >5 μM, progressive perturbation of the cell membrane/cytoplasm complex is observed: the cells exhibit increased 90° light scatter (refractive index,[17] protein content),[18] decreased axial light loss (apparent cell volume, cell size),[19] simultaneous propidium iodide (PI) and carboxyfluorescein (CF) fluorescence, and resistance to detergent (NP-40)-mediated cytolysis similar to that resulting from tributyltin exposure.[20]

In summary, our observations indicate that DNA synthesis is the primary target of MeHg cytotoxicity and that apparent targets and degree of cytotoxicity are a complex function of dose.

II. MATERIALS AND METHODS

A. CELLS

Friend murine erythroleukemic (MEL) cells (T3CL2: obtained from Dr. Clyde Hutchinson, University of North Carolina, Chapel Hill, NC) were grown in suspension culture in RPMI 1640 (GIBCO, Grand Island, NY) supplemented with 10% fetal bovine serum (FBS) and 25 mM HEPES (Sigma, St. Louis, MO #H3375). Cell density was monitored by a Coulter Counter (Model ZBI: Coulter Electronics, Inc., Hialeah, FL) and the cells were subcultured every 2 to 3 days to maintain exponential growth.

B. VIABILITY ASSAY

Viability was estimated by FCM employing the carboxyfluorescein diacetate (CFDA: Molecular Probes, Eugene, OR)/propidium iodide (PI: Sigma #P5264) assay.[17,20,21]

C. STANDARD PREPARATION OF NUCLEI FOR CELL CYCLE ANALYSIS

Exponentially growing cells were harvested and washed as described previously.[20] Nuclei were isolated by nonionic detergent [Nonidet P-40 (NP-40): Sigma #N6507]-mediated solubilization of the plasma membrane/cytoplasm complex and stained with fluorescein isothiocyanate (FITC: Sigma #F7250) for protein content and PI for DNA content.[20]

D. FLOW CYTOMETRY

Cytometric analyses were accomplished as described previously.[18,20,22]

E. MeHg EXPOSURE PROTOCOL

Methylmercury (II) chloride (Alfa Inorganics, Danvers, MA, #37123) in methanol was added to exponentially growing MEL cells to final concentrations (C) of 0.1, 0.25, 0.5, 1, 2.5, 5, 7.5, 10, 25, or 50 μM. The final methanol concentration of the medium was 0.1% which had no effect on viability (CFDA/PI assay) or growth rate. Duration of exposure was 1, 2, 4, or 6 h (T). To investigate recoverability from the effects of MeHg exposure, cells were exposed to MeHg for 6 h, washed twice with prewarmed (37° C) MeHg-free FBS-supplemented medium, and reincubated for 18 h.

F. PROGRESSION ASSAY

The relative rate of movement of cells through the compartments of the cell cycle (G, S, G_2/M as defined by FCM) was estimated by a modification of the stathmokinesis assay of Darzynkiewicz et al.[23] which is based on the rate of accumulation of cells in the G_2/M phase of the cell cycle following treatment with Colcemid. Following 4 h of exposure to MeHg, aliquots of cell cultures received either 0.2 µg/mL Colcemid (Demecolcine: Sigma #D7385) from a stock solution of 0.1 mg/mL 95% ethanol or (control) an equivalent amount of 95% ethanol. After 2 h incubation (6 h total), the cells were harvested and nuclei prepared for cell cycle analysis. Quantification of the DNA distribution in the G_0/G_1, S, and G_2/M compartments of the cell cycle, as a function of time, allows estimation of the rate of progression of cells through the cycle (progression assay). In our experiments, exposure of control cells to Colcemid for 2 h assured accumulation of cells in the G_2/M phase of the cycle without total depletion of the G_0/G_1 population, permitting quantification of the cell cycle phase distribution of cells by Multicycle®, a cell cycle analysis PV software package (Phoenix Flow Systems, San Diego, CA).

G. QUANTIFICATION OF THE MITOTIC FRACTION
OF G₂/M NUCLEI

To estimate the percentage of cells in the M phase of the cell cycle, nuclei were prepared (from 1×10^6 cells per sample) according to the method of Pollack et al.,[24] which allows flow cytometric discrimination of the M subpopulation.[18]

H. CHROMOSOME MORPHOLOGY

Cells (1×10^6 per sample) were washed twice with phosphate-buffered saline (PBS: Sigma #4417) by centrifugation ($120 \times$ g, 5 min), resuspended in 10 ml of 75 mM KCl, and fixed in methanol-acetic acid (3:1). Chromosome spreads were prepared by centrifugation (2×10^4 fixed cells, $500 \times$ g for 10 min at room temperature) onto glass slides in Leif Cytobuckets (Coulter kit #322, Coulter Electronics, Inc., Hialeah, FL), drying at 56° C, and staining with 6% Giemsa (Sigma #G5637). The percentages of normal and abnormal chromosomal complements were obtained from 200 mitotic figures. The mitotic index was obtained from 500 cells. The data represent the mean ± standard deviation of three experiments.

I. DATA ANALYSIS

The data reported are from representative experiments. The experiments were repeated at least 3 times. The data within and among replicate experiments were consistent within 10%. For each cytometric parameter investigated (PI or FITC fluorescence, 90° light scatter, axial light loss), the distribution or mean of 10^4 events (cells or nuclei) per condition (dose, duration of exposure) or combination of conditions was determined. Data derived from cells exposed to MeHg concentrations below 5 µM (i.e., 2.5 µM did not differ from the control condition and are not included.

Table 1   **Cytogenetic effects of methylmercury.**

	Viability (%)	Mitotics (%)	Chromosomal aberrations			
			Normal (%)	Condensed (%)	Pulverized (%)	Rings (%)
Control	98	3.1 ± 1.0	92 ± 6	5 ± 2	—	3 ± 4
2.5 μM	98	4.2 ± 2.0	84 ± 8	13 ± 8	2 ± 2	2 ± 2
5.0 μM	95	5.0 ± 0.9	74 ± 22	22 ± 15	4 ± 4	—
10.0 μM	14	4.8 ± 3.0	3 ± 4	30 ± 34	12 ± 9	55 ± 41
25.0 μM	3	4.5 ± 0.8	—	—	—	100 ± 0
50.0 μM	5	3.9 ± 1.8	—	—	—	100 ± 0
Colc.	97	41 ± 0	90 ± 4	10 ± 4	—	—

Note: Viability (estimated by the CFDA/PI assay[20]) the mitotic index and percentage of chromosomal abberations occurring in MEL cells following 6 h exposure to 0 to 50 μM MeHg or 0.2 µg/mL Colcemid (colc.).

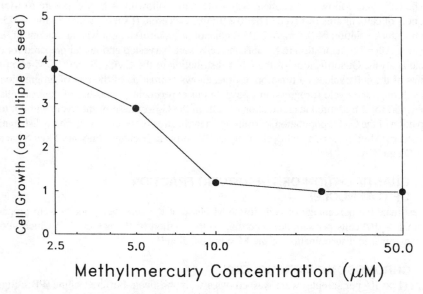

Figure 1   Rate of MEL cell replication following MeHg exposure (6 h), washout and reincubation (18 h) in MeHg-free medium. MEL cell doubling time after exposure to 2.5 μM MeHg was essentially equal to that of control cells. Multiple of seed = $\frac{\text{No. cells } T_{18}}{\text{No. cell } T_0} \times T_0$ = time of inoculation into MeHg-free medium. T_{18} = 18 h postinoculation.

III. RESULTS

It was observed that the maximal C × T exposure of MEL cells to MeHg that has no significant effect on viability (estimated by the CFDA/PI assay, Table 1), 90° light scatter (a measure of protein content,[18] Figure 7), or axial light loss (cell size[17]) is 5 μM for 6 h. However, exposure at this level results in a decreased rate of cell replication (Figure 1) following MeHg washout and reincubation for 18 h in fresh, MeHg-free medium. Exposure to higher doses results in significant loss of viability (Table 1). At/above 5 μM, MeHg alters the percentage of distribution of cells across the cell cycle (Figures 2 through 6). Following exposure to 5 μM MeHg, DNA histogram analysis (Figure 3: no Colcemid) reveals depletion

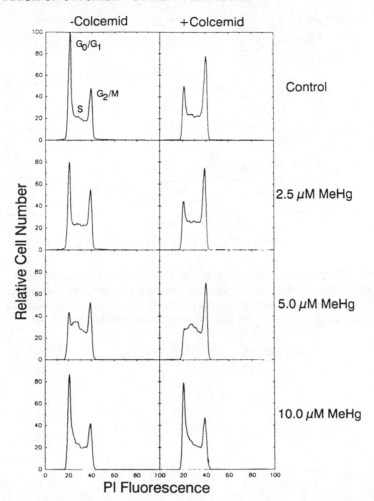

Figure 2 Representative DNA histograms of nuclei of MEL cells exposed to MeHg for 6 h with or without Colcemid for the last 2 h of exposure (progression assay). G_0/G_1 represents the pre-DNA synthetic phase of the cell cycle; S, the phase of DNA synthesis; G_2, the postsynthetic phase preceding mitosis; and M, mitosis. Following exposure to 5 μM MeHg, movement of cells through the S phase of the cycle appears to be retarded. At 10 μM MeHg, there is complete cessation of cycling.

of the G_0/G_1 compartment, an increase in the percentage of cells in S phase (to a relatively steady state), and no increase in G_2/M compared to control cells, suggesting little movement of cells out of the S-phase.

Compared to control cells treated with Colcemid, addition of Colcemid to the 5 μM MeHg-treated cells has little effect on the percentage of G_2M cells. If MeHg primarily inhibits microtubule assembly (a Colcemid-like effect) and S phase progression is normal, cells would accumulate in the G_2/M phase of the cycle at the expense of the other phases (Figure 5). Apparently, cells treated with MeHg can enter S phase; but the rate at which they traverse this compartment is retarded, resulting in reduction of the rate of influx of cells into G_2/M. Compared to control cells, exposure to 10 μM MeHg results in an increased percentage of cells in the S phase of the cycle, a slightly decreased percentage in G_0/G_1, and little effect

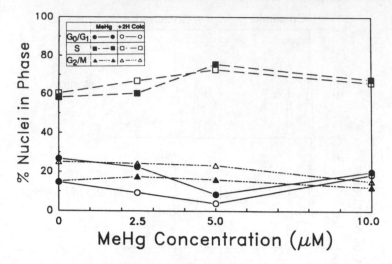

Figure 3 The percentage of cells (in terms of nuclear DNA content) in each cell cycle phase was determined by computerized mathematical analysis of the histograms of Figure 2. The nuclear DNA distribution across the phases of the cell cycle in the absence of MeHg depicts the control condition. +2H colc. = exposure to Colcemid for 2 h (see Figure 2).

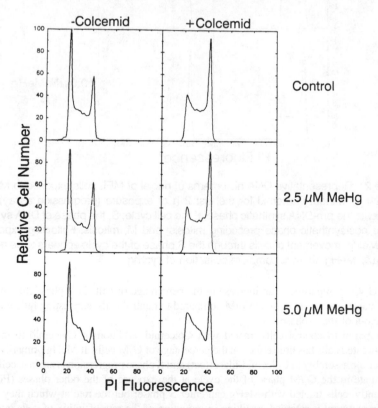

Figure 4 Representative DNA histograms of nuclei obtained from MeHg-exposed (6 h) MEL cells reincubated in MeHg-free medium for 18 h. The cell cycle distribution of cells recovering from exposure to 5 μM MeHg approaches normality, but still indicates retardation of progression into, through, and out of the S phase.

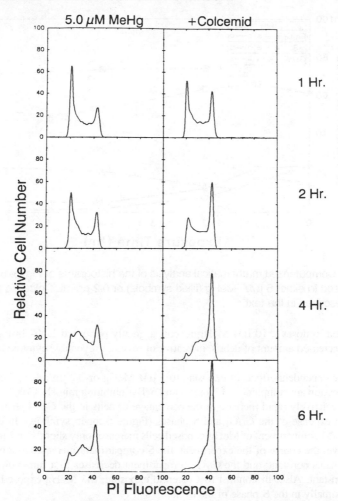

Figure 5 Exposure to Colcemid (0.2 µg/mL) results in time-dependent accumulation of MEL cells in the G$_2$/M compartment and a sequential depletion of the G$_0$/G$_1$ and S compartments. In contrast, exposure to 5 µM MeHg results in accumulation of cells in the S phase and retardation of the rate of efflux out of this compartment. As a result, the G$_0$/G$_1$ compartment becomes depleted and accumulation of cells in the G$_2$/M compartment is inhibited.

on the percentage of G$_2$/M cells. Exposure of such cells to Colcemid has minimal effect on phase distribution, indicating essentially complete cessation of cycling.

At 18 h after MeHg washout and reincubation under standard conditions, the DNA histogram of nuclei obtained from MEL cells exposed to 2.5 µM MeHg appears identical to that of control cells (Figure 4). The DNA histogram obtained from MEL cells exposed to 5 µM MeHg indicates recovery toward a pattern of DNA distribution similar to that of exponentially growing cells (compare Figure 4 with Figure 2). However, there is persistent retardation of the rate of S phase traverse as evidenced by the accumulation of cells in early and mid S phase and depletion of cells in late S and the G$_2$/M phase. Colcemid treatment confirms recovery of normal cell cycle kinetics by cells exposed to 2.5 µM MeHg and the persistence of S phase retardation in cells exposed to 5 µM MeHg. Reincubation of MEL

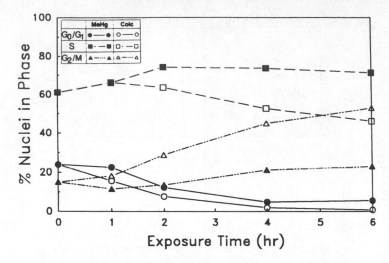

Figure 6 Computerized mathematical analysis of the histograms of Figure 5. MEL cells were exposed to either 5 μM MeHg (filled symbols) or 0.2 µg/mL Colcemid (open symbols) as described in the text.

cells exposed to doses ≥10 µM MeHg reveals a greatly perturbed DNA histogram manifesting an increased amount of debris, indicative of severe, irreversible cytotoxicity (data not shown).

The time-dependent effects of exposure to 5 µM MeHg or 0.2 µg/mL Colcemid on cell cycle progression are compared in Figures 5 and 6. By inhibiting mitosis, Colcemid exposure results in a relatively rapid increase in the percentage of cells in the G_2/M phase of the cell cycle at the expense of the G_0/G_1 and S phases (Figure 6, open symbols). In contrast, the apparent G_2/M compartment of MeHg-exposed cells increases only slightly and at a relatively slow rate over the course of the experiment; the S compartment increases to a maximum at 2 h and remains constant; and the G_0/G_1 compartment decreases to a minimum at 4 h and remains constant. Also, in contrast to Colcemid exposure, the MeHg-exposed cells accumulate maximally in the S phase of the cycle.

To gain insight into the effect of MeHg on the G_2/M phase of the cell cycle, nuclei were prepared from MEL cells by an isolation procedure that allows flow cytometric discrimination of mitotic nuclei.[18,24] On a contour cytogram of 90° scatter vs. PI fluorescence (Figure 7), M phase nuclei appear as a distinct subpopulation exhibiting decreased 90° scatter and PI fluorescence compared to G_2 nuclei. Following 6 h of exposure to concentrations of MeHg ≤7.5 µM, the percentage of G_2/M cells remains relatively constant over dose (control, 16.4%; 2.5 µM MeHg, 14.7%; 5 µM MeHg, 18.2%; 7.5 µM MeHg, 19.4%). However, the contribution of the M subcompartment appears to increase considerably. This suggests that although progression from S phase into G_2 is retarded, subsequent progression from G_2 into M occurs. However, the percentage of cells exhibiting recognizable chromosomes does not increase substantially as a function of dose (Table 1), suggesting that cells leaving G_2 become arrested in a premitotic phase in which their nuclei exhibit the same biophysical properties as those of M phase cells, but chromosome morphology is disrupted.

Chromosome analysis reveals an increase in the percentage of condensed and pulverized chromosomes (Table 1) and, at best, a modest increase in the mitotic index (considerably less than that caused by Colcemid) as a function of MeHg dose up to 10 µM. At 10 µM MeHg, chromosome spreading is inhibited and the chromosomes of more than half of the mitotic cells appear in the form of wreath-like ring structures (Table 1). Following exposure

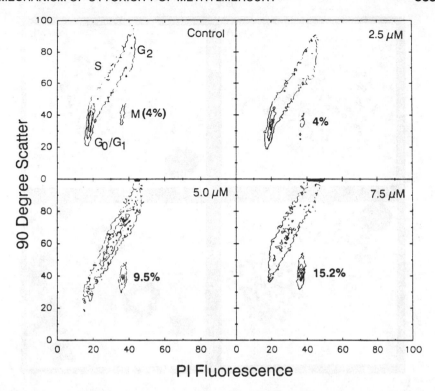

Figure 7   Contour cytograms of 90° scatter vs. PI fluorescence of nuclei isolated from MeHg-exposed (6 h) MEL cells by the method of Pollack et al.[24] Mitotic nuclei appear as a distinct subpopulation exhibiting decreased 90° light scatter and PI fluorescence compared to the G_2 subpopulation. The relative percentage of M cells appears to increase as a function of MeHg dose.

to 25 or 50 μM MeHg for as little as 1 h (the shortest time period investigated), all resolvable spreads appear in the form of wreath-like ring structures (Figure 8).

During the preparation of nuclei it was observed that following exposure to concentrations of MeHg $\geq$10 μM, nuclear FITC fluorescence (i.e., protein content) was increased. Fluorescence microscopy revealed proteinaceous tags adherent to the nuclei indicative of incomplete detergent solubilization (i.e., resistance to solubilization) of the plasma membrane/cytoplasm complex.

IV. DISCUSSION

The mechanism through which MeHg exerts its cytotoxicity has been postulated to involve binding to sulfhydryl groups and disruption of disulfide bonds. Indeed, inhibition of mitosis has been attributed to the binding of MeHg to sulfhydryl groups resulting in inhibition of microtubule assembly and/or disruption of assembled microtubules.[7,13–15] It is expected that sulfhydryl binding also would affect other cellular functions, including the structure and function of biomembranes and the cytoskeleton; synthesis, repair, and structure of DNA and RNA (and, therefore, the cell cycle); protein synthesis, turnover, structure and function; and chromosome structure and function.[2,6–12]

Cell cycle analysis (by FCM) reveals that compared to the control condition, MeHg induces a dose-dependent increase in the percentage of cells in S phase, little change in the

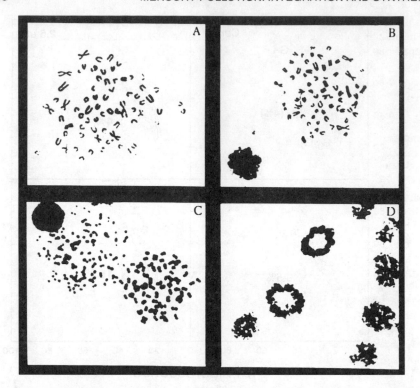

Figure 8   Representative photomicrographs (630×) of chromosomal aberrations induced by exposure (6 h) of MEL cells to MeHg. The chromosomes of cells exposed to 5 µ*M* MeHg (B) appear to be of normal morphology but of smaller size than those of control cells (A), and both the rate of cell duplication (Figure 1) and S phase progression (Figures 2 through 6) are inhibited. Following exposure to 10 µ*M* MeHg, approximately 45% of mitotic figures were comprised of condensed pulverized chromosomes while approximately 55% were in the form of wreath-like ring structures (C). At/above 25 µ*M* MeHg, the chromosomes of resolvable mitotic figures appear in the form of wreath-like ring structures apparently containing the entire chromosomal complement (D).

percentage of cells in G_2/M, and depletion of the G_0/G_1 compartment (Figures 2, 3, 5, and 6). Depletion of the G_0/G_1 compartment indicates retardation/inhibition of mitosis reducing the supply of G_0/G_1 cells. If MeHg primarily retards/inhibits mitosis, then an increase in the percentage of cells in the G_2/M compartment (a colchicine-like effect) should be observed. That no increase is observed indicates retardation of S phase traverse. This is supported by the increase in the size of the S compartment, as a result of the emptying of the G_0/G_1 compartment (Figures 2, 3, 5, and 6). Apparently, recovery of cycling following MeHg exposure can occur. Cells exposed to concentrations of MeHg up to 5 µ*M* exhibit partial, if not complete, recovery following reincubation (for 18 h) in MeHg-free medium (Figure 4).

To obtain more precise information on cell cycle effects, the time course of interaction of MeHg (5 µ*M*) with exponentially growing MEL cells was investigated (Figures 5 and 6). Computerized mathematical analysis of the DNA histogram indicates that the percentage of cells in S phase increases with time, reaching a maximum after 2 h of exposure, and remaining constant thereafter (Figures 5 and 6). Synchronously, the percentage of cells in G_0/G_1 decreases and continues to decrease as a function of exposure up to 4 h, indicating reduction of the rate of influx of cells into this compartment. The percentage of cells in G_2/M changes little as a function of time. This indicates (in light of the increased percentage of cells in S phase) retardation of influx into as well as efflux out of this compartment (the size of the G_2/M compartment is limited by the relative rates of S phase efflux and mitosis).

Increasing the MeHg dose to 10 μM does not increase the size of the G_2/M compartment (Figures 2 and 3). If the primary effect of MeHg is disruption of the microtubule assembly, increasing the MeHg dose below the cytotoxic level would be expected to progressively inhibit mitosis and, thereby, increase the percentage of cells in the G_2/M compartment and the mitotic index. However, this is not the case (Figures 2 and 3; Table 1). Indeed, cells accumulate in the S compartment.

Quantification of the relative contribution of M phase cells to the G_2/M compartment (Figure 7) reveals that although the percentage of cells in G_2/M changes little as a function of dose (Figure 3), the M subpopulation appears to increase following exposure to MeHg concentrations greater than 2.5 μM. Thus, it would appear that at concentrations of MeHg (>2.5 μM) above which influx into G_2 is retarded, efflux out of M (i.e., mitosis) is retarded to a greater extent. However, morphologic analysis reveals that the mitotic index changes little as a function of MeHg dose (Table 1). This suggests that the apparent increase in the percentage of M phase nuclei results from the cells leaving G_2 becoming arrested in a pre-mitotic phase in which the nucleus exhibits biophysical properties similar to those of M phase nuclei; but chromosomal condensation apparently is inhibited. This argues against the hypothesis that the mitotic spindle (i.e., the microtubule) is the primary target of MeHg, as does our observation of only limited accumulation of cells in the G_2/M phase of the cell cycle; far less than that seen after treatment with Colcemid, an agent that specifically blocks microtubule assembly (Table 1).

Following exposure to concentrations of MeHg $\geq$10 μM, viability decreases (Table 1), growth is completely inhibited (Figure 1), and traverse through all phases of the cell cycle is blocked (Figures 2 and 3). In addition, exposure to such concentrations results in a fixation-like (protein denaturation, cross-linking, etc.) alteration of the plasma membrane/cytoplasm complex. This phenomenon is manifested as resistance to detergent-mediated solubilization of the plasma membrane/cytoplasm complex, resulting in increased nuclear FITC fluorescence due to the presence of proteinaceous material adherent to the nuclei.

Cytologic examination reveals that MeHg exposure perturbs chromosome structure. At 10 μM, chromosomal condensation, pulverization and wreath-like chromosomal ring structures, apparently involving fusion of the entire chromosomal complement, are observed. Above 10 μM only the chromosomal ring structures are observed.

Although the mechanism of ring structure formation is unknown, direct interaction of MeHg with chromatin, as well as perturbation of the plasma membrane/cytoplasm complex resulting in alteration of the intracellular environment, may be involved. It has been observed repeatedly that MeHg disrupts the structure/function of biomembranes.[2,9-11] Indeed, in cultured mouse neuroblastoma cells, Koerker[9] reported that exposure to 1 μM MeHg for 24 to 72 h at 37° C in Ham's F-12 medium supplemented with serum resulted in perturbation of the function of the plasma membrane, lysosomes, mitochondria, and endoplasmic reticulum.

REFERENCES

1. Harada, Y., Congenital Minimata Disease, In: *Minimata Disease, Methylmercury Poisoning in Minimata and Niigata, Japan,* Tsubaki, T. and Irukayama, K., Eds., Elsevier/North-Holland, Amsterdam, (1977), 209.

2. Olson, F.C. and Massaro, E.J., Effects of methylmercury on murine fetal amino acid uptake, protein synthesis and palate closure, *Teratology,* 16:187–194, (1977).

3. Amin-Zaki, L., Majeed, M.A., Elhassani, S.B., Clarkson, T.W., Greenwood, M.R., and Doherty, R.A., Prenatal methylmercury poisoning, *Am. J. Dis. Child.,* 133:172–177, (1979).

4. Geelen, J.A., Dormans, J.A., and Verhoef, A., The early effects of methylmercury on the developing rat brain, *Acta Neuropathol. (Berl.)*, 80:432–438, (1990).

5. Mason, R.P. and Fitzgerald W.F., Alkylmercury species in the equatorial Pacific, *Nature*, 347:457–459, (1990).

6. Gruenwedel, D.W. and Cruikshank, M.K., Effect of methylmercury (II) on the synthesis of deoxyribonucleic acid, ribonucleic acid and protein in Hela S3 cells, *Biochem. Pharmacol.*, 28:651–655, (1979).

7. Sager, P.R., Selectivity of methylmercury effects on cytoskeleton and mitotic progression in cultured cells, *Toxicol. Appl. Pharmacol.*, 94:473–486 (1988).

8. Wasteneys, G.O., Cadrin, M., Reuhl, K.R., and Brown, D.L., The effect of methylmercury on the cytoskeleton of murine embryonal carcinoma cells, *Cell Biol. Toxicol.*, 4:41–60, (1988).

9. Koerker, R.L., The cytotoxicity of methymercuric hydroxide and colchicine in cultural mouse neuroblastoma cells, *Toxicol. Appl. Pharmacol.*, 53:458–469, (1980).

10. Goodman, D.R., Fant, M.E., and Harbison, K.D., Perturbation of aminoisobutyric acid transport in human placental membranes: direct effects by $HgCl_2$, CH_3HgCl, and $CdCl_2$, *Teratogen. Carcinogen. Mutagen*, 3:89–100, (1983).

11. Peckham, N.H. and Choi, B.H., Surface charge alterations in mouse fetal astrocytes due to methyl mercury: an ultrastructural study with cationized ferritin, *Exp. Mol. Pathol.*, 44:230–234, (1986).

12. Abe, T., Haga, T., and Kurokawa, M., Blockage of axoplasmic transport and depolymerization of reassembled microtubules by methylmercury, *Brain Res.*, 86:504–508, (1975).

13. Ramel, C., Methylmercury as a mitosis disturbing agent, *J. Jpn. Med. Assoc.*, 61:1072–1081, (1969).

14. Miura, K., Suzuki, K., and Imura, N., Effects of methylmercury on mitotic mouse glioma cells, *Environ. Res.*, 17:453–471, (1978).

15. Vogel, D.G., Margolis, R.L., and Mottet, N.K., The effects of methylmercury binding to microtubules, *Toxicol. Appl. Pharmacol.*, 80:473–486, (1985).

16. Vogel, D.G., Rabinovitch, P.S., and Mottet, N.K., Methylmercury effects on cell cycle kinetics, *Cell Tissue Kinet.*, 19:227–242, (1986).

17. Shapiro, H., *Practical Flow Cytometry*, 2nd ed., Alan R. Liss, New York, (1988).

18. Zucker, R.M., Elstein, K.H., Easterling, R.E., and Massaro, E.J., Flow cytometric discrimination of mitotic nuclei by right-angle light scatter, *Cytometry*, 9:226–231, (1988).

19. Cambier, J.C. and Monroe, J.G., Flow cytometry as an analytical tool for studies of neuroendocrine function, In: *Methods in Enzymology*, Conn, P.M., Ed., Vol. 103, Academic Press, New York, (1983), 227.

20. Zucker, R.M., Elstein, K.H., Easterling, R.E., Ting-Beall, H.P., Allis, J.W., and Massaro, E.J., Effects of tributyltin on biomembranes: alteration of flow cytometric parameters and inhibition of Na+, K+—ATPase two-dimensional crystallization, *Toxicol. Appl. Pharmacol.*, 96:393–403, (1989).

21. Rotman, B. and Papermaster, B.W., Membrane properties of living cells as studied by enzymatic hydrolysis of fluorogenic esters, *Proc. Natl. Acad. Sci. U.S.A.*, 55:766–771, (1966).

22. Crissman, H.A., Darzynkiewicz, Z., Tobey, R.A., and Steinkamp, J.A., Correlated measurements of DNA, RNA and protein in individual cells by flow cytometry, *Science*, 228:1321–1324, (1986).

23. Darzynkiewicz, Z., Traganos, F., and Kimmel, M., Assay of cell cycle kinetics by multivariate flow cytometry using the principle of stathmokinesis, In: *Techniques of Cell Cycle Analysis*, Gray, J.W., and Darzynkiewicz, Z., Eds., Humana Press, Clifton, NJ, (1987), 291.

24. Pollack, A., Moulis, H., Block, N.L., and Irvin, G.L., III, Quantitation of cell kinetic responses using flow cytometric measurements of correlated nuclear DNA and protein, *Cytometry*, 5:473–481, (1984).

Mercury Levels in Fisherman Groups of the North Adriatic Sea

G. Moretti, A. Bortoli, V. Marin, and E. Ravazzolo

CONTENTS

I. INTRODUCTION

A number of studies on Hg levels in sea organisms, carried out in various areas of the Mediterranean Sea since the early 1970s, indicate that the concentrations of Hg in this region are generally higher than those recorded in other seas. The same conclusion was reached with regard to other matrices in the Mediterranean marine environment.

The potential health effects on the population by excessive intake of Hg (particularly methylmercury) have led to considerable international activities (WHO, World Health Organization; FAO, Food and Agricultural Organization; IAEA, International Atomic Energy Agency) in an effort to identify and quantify the problem.

These studies show that most of the general Mediterranean population seems to have a low Hg intake through the consumption of seafood and could be considered not at risk.[1] On the other hand, it appears equally evident that some population groups in the Mediterranean Region could have an intake of methylmercury through seafood in excess of permissible levels.

The north Adriatic Sea receives great amounts of water from rivers that have run through open country and many industrial areas of north Italy. Moreover, it should be emphasized that Italy is one of the most important producers of Hg and that east of the Alps there is a Hg mine (Idrjia) that may influence the waters of small rivers that flow into the north Adriatic sea.

It is known that selenium can play the role of an antagonist in mercury toxicity: seafood is one of the main sources of selenium intake.[2] An inhibition in the mechanism of mercury chelation by the —SH groups in the cell can be suggested in relationship with the antioxidant properties of selenium.[2]

Besides, as selenium is both essential and toxic, there is an optimum intake to prevent both seleniosis and selenium deficiency: no correlation is demonstrated between high selenium intake and increased risk of cancer, on the contrary, it seems that selenium may contribute to prevention of cancer.[3]

II. MATERIALS AND METHODS

In order to evaluate the possible health risk of the presence of Hg in particular groups of the coastal population of the north Adriatic sea, hair samples of fishermen from different zones were collected to measure the levels of Hg, methylmercury, and selenium. Fishermen are known, on the average, to have a greater intake of seafood than other people, nevertheless, each participant was requested to answer a questionnaire containing questions about his/her daily intake of seafood.

Chioggia and Grado (two small coastal towns on the northern side of the Adriatic sea) were selected, because of their tradition of fishing activity, for hair sampling. Hairs were cut with stainless steel scissors, according to the standard method.[1]

In Chioggia 77 hair samples were collected from fishermen and their relatives and, subsequently, another 26 from fishermen only. In Grado 46 hair samples were collected from individuals of a fisherman group.

The control population was represented by 92 inhabitants of Sarntal, a mountain district of the Dolomites where food habits do not provide for significant seafood consumption.

Every month, Hg and methylmercury levels were determined in migratory and nonmigratory fish, which represent the most important vehicle for mercury intake. Fish were collected at the fish markets and almost all the edible species were represented: *Thunnus thynnus, Squalus fernandinus, Mustelus asterias, Lamna nasus, Solea solea, Mullus barbatus, Sparus auratus, Platichthys flesus, Mugil* sp., *Anguilla anguilla,* and *Scomber scombrus.*

The determination of total Hg was carried out by flameless atomic absorption spectrometry,[4] while the determination of methylmercury was carried out by gas chromatography, using the partly modified Westoo method.[5]

Selenium analyses were performed using electrothermal atomic spectroscopy with S.T.P.F. conditions, Zeeman background correction, and palladium as the matrix modifier.[6]

III. RESULTS

The mean values of the mercury (Hg), methylmercury (MeHg), and selenium (Se) concentrations found in the three considered groups (fishermen and their relatives, fishermen of Chioggia, fishermen of Grado) are represented in Table 1. The standard deviations and the minimum and maximum values are also reported.

Table 2 shows the mean values, standard deviations, and minimum and maximum values of Hg and Se of the inhabitants of Sarntal (the control population) and also the Hg and MeHg values of the inhabitants of Chioggia and of Caorle (another small coastal town of the north Adriatic sea) that have been investigated in a previous study.[7]

Table 3 reports the Hg and MeHg mean values found in 112 nonmigratory and 64 migratory fish of the north Adriatic sea.

In Figures 1, 2, and 3 the Hg and MeHg distributions in the three groups are represented, while Figures 4, 5, and 6 show the correlation between Hg and MeHg in the same groups. Figure 7 reports the Se distribution in the three groups.

The distributions of Hg and Se in the control population (Sarntal's inhabitants) are represented in Figures 8 and 9, respectively.

IV. DISCUSSION

The mean values of the Hg and MeHg levels (Table 1) found in the considered fisherman groups when compared with those found in other fisherman groups or populations[8,9] are not considered high, even if some values seem quite elevated.

Each group shows significantly higher Hg and MeHg levels, not only in comparison with those of the control population, but also with those of the inhabitants of Chioggia and Caorle (Table 2): it means that fishermen and their relatives are more exposed to Hg than other people.

The Hg and MeHg values that have been found were not homogeneously distributed in the considered population: fisherman who live in the south side of the Venetian Region

Table 1 **Mean, standard deviation, and minimum and maximum values of Hg, MeHg, and Se levels of some fisherman groups of the north Adriatic sea.**

mg/Kg	Fishermen of Chioggia and their relatives (1989, 77 specimens)		
	Hg	MeHg	Se
Mean	3.30	1.95	0.70
Standard deviation	2.05	1.48	0.27
Min.-max values	0.69–10.83	0.28–6.99	0.30–1.86

mg/Kg	Fishermen of Chioggia (1990, 26 specimens)		
	Hg	MeHg	Se
Mean	3.28	2.29	0.79
Standard deviation	1.49	1.44	0.26
Min.-max values	1.40–7.48	0.46–6.99	0.35–1.28

mg/Kg	Fishermen of Grado (1990–1991, 46 specimens)		
	Hg	MeHg	Se
Mean	7.45	6.31	0.47
Standard deviation	4.54	3.97	0.09
Min.-max values	1.03–19.87	0.71–18.70	0.10–1.04

Table 2 **Mean, standard deviation, and minimum and maximum values of Hg, MeHg, and Se levels of the inhabitants of Sarntal, Chioggia, and Caorle.**

mg/Kg	Inhabitants of Sarntal (1989, 92 specimens)	
	Hg	Se
Mean	1.40	1.19
Standard deviation	1.91	0.48
Min.-max values	0.03–12.50	0.50–2.25

mg/Kg	Inhabitants of Chioggia (1988, 143 specimens)	
	Hg	MeHg
Mean	2.44	1.45
Standard deviation	2.60	1.97
Min.-max values	0.10–24.14	0.00–17.28

mg/Kg	Inhabitants of Caorle (1989, 296 specimens)	
	Hg	MeHg
Mean	3.27	1.36
Standard deviation	2.93	1.78
Min.-max values	0.20–16.10	0.00–11.82

Table 3 **Mean, standard deviation, and minimum and maximum values of Hg and MeHg levels of nonmigratory and migratory fish of the north Adriatic sea.**

	Nonmigratory fish (1989–90, 112 specimens)	
mg/Kg	Hg	MeHg
Mean	0.34	0.21
Standard deviation	0.28	0.22
Min.-max values	0.01–1.83	0.00–0.88

	Migratory fish (1989–1990, 64 specimens)	
mg/Hg	Hg	MeHg
Mean	1.32	0.81
Standard deviation	0.48	0.34
Min.-max values	0.33–1.92	0.33–1.22

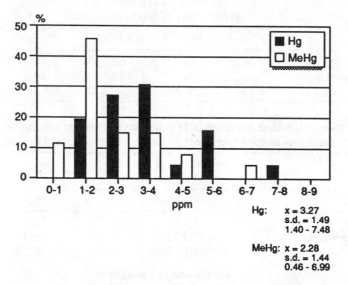

Figure 1 Mercury and methylmercury distribution in fishermen of Chioggia and their relatives.

(Chioggia) were shown to have significantly lower levels of Hg than those who live in the northern side (Grado). In fact, the mean values of these two fisherman groups were, for total Hg, respectively, 3.30 mg/kg and 7.45 mg/kg (Table 1).

If these Hg and MeHg data are compared with those of the inhabitants of Chioggia and Caorle (Table 2), it also appears that the mean values increase gradually from the south towards the north and that fishermen have higher Hg levels than other inhabitants of the same zone.

This could likely be caused by the fact that the northern side of the Region is more influenced by the nearness of the Hg mine of Idrjia, so that the Hg levels of the corresponding waters and of the local seafood might be higher, as has been pointed out by investigations carried out by other authors.[10]

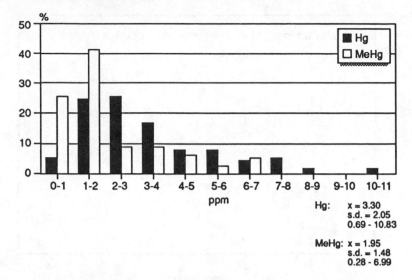

Figure 2   Mercury and methylmercury distribution in fishermen of Chioggia.

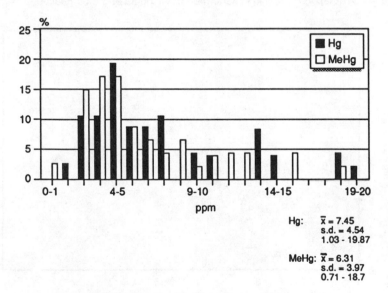

Figure 3   Mercury and methylmercury distribution in fishermen of Grado.

On the other hand, the Hg levels shown in the fisherman groups of Chioggia and Grado are significantly lower if compared with those found in some fisherman groups of south Italy;[11] this difference depends on the fact that in south Italy fishermen eat large-size fish such as swordfish and scabbard fish, which usually contain higher amounts of Hg than the smaller fish.

Meals based on fish were, on the average, 3.5 per week and a good relationship was found between the intake of fish and the Hg levels, while no significant difference was observed by sex or age.

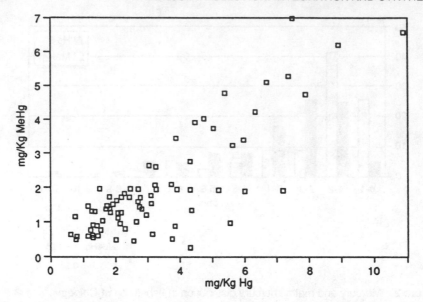

Figure 4   Correlation Hg/MeHg in fishermen of Chioggia and their relatives.

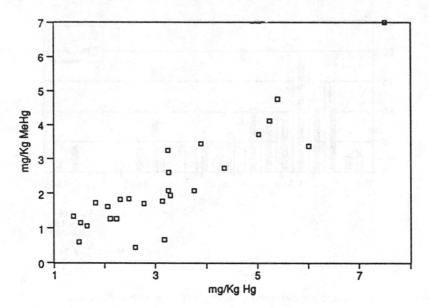

Figure 5   Correlation Hg/MeHg in fishermen of Chioggia.

The Hg and MeHg levels found in the fish of the north Adriatic sea differ according to size and age. The results show that small and young fish have lower Hg and MeHg concentrations than bigger and older ones and that food habits play a significant role, with higher Hg concentrations in those fish that feed on crustaceans and fish.

Nonmigratory fish do not show remarkable Hg levels, but in several species of migratory fish (as *Thunnus thynnus*, *Mustelus asterias*, and *Lamna nasus*) the mean Hg concentrations substantially exceed the value of 0.7 mg/kg—the limit value permitted by Italian law (Table 3).

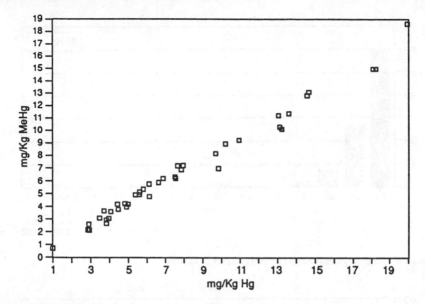

Figure 6   Correlation Hg/MeHg in fishermen of Grado.

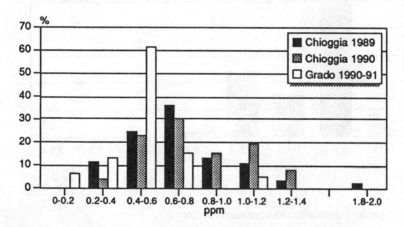

Figure 7   Selenium distribution in fishermen of Chioggia and Grado.

The three considered fisherman groups do not differ very much in Hg and McHg distribution (Figures 1, 2, and 3). In the three groups, the greatest percentage of individuals shows values higher than 1 ppm Hg, while in the control population (Figure 8) almost all values are between 0 and 1 ppm Hg. This indicates that Hg is preferentially transmitted by fish or seafood.

More than the 20% of the tested fisherman population have a MeHg hair concentration higher than 6 ppm, and the correlation between Hg and MeHg is stronger in the fisherman group of Grado, with a ratio MeHg/Hg of 84.7%, than in the two groups of Chioggia (Figures 4, 5, and 6).

The Se values, even if in the normal range, seem to be rather low, maybe because of the presence of Hg. In fact, a correlation with Hg is suspected, in that the fisherman group with higher Hg concentrations have Se values that are lower, and in the group where Hg is low, Se is higher (Table 1 and Figure 7).

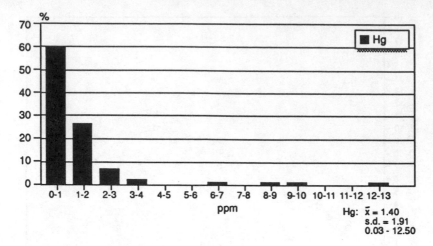

Hg: x̄ = 1.40
s.d. = 1.91
0.03 - 12.50

Figure 8 Mercury distribution in inhabitants of Sarntal.

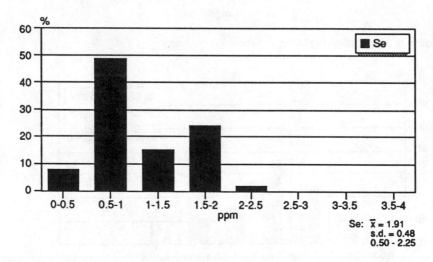

Se: x̄ = 1.91
s.d. = 0.48
0.50 - 2.25

Figure 9 Selenium distribution in inhabitants of Sarntal.

A comparison with the Se levels of the control population (Table 2 and Figure 9) seems to confirm that there is an inverse proportion between Se and Hg.

In conclusion, even if it seems that at the moment there are no great problems for Hg levels for the coastal populations of the north Adriatic sea, the biological indicators confirm the critical position of those groups that consume large-size fish and of other groups situated near the areas at risk.

ACKNOWLEDGMENT

This research was supported by the Health Department of the Veneto Region.

REFERENCES

1. UNEP-FAO, Assessment of the state of pollution of the Mediterranean sea by mercury and mercury compounds, *UNEP,* Athens, 1987.
2. Magos L. and Berg G.G., Selenium, In: *Biological Monitoring of Toxic Metals,* Clarkson T.W., Friberg L., Nordberg G.F., and Sager P.R., Eds., Plenum Press, New York, 1988, 383.
3. Alabaster, O., *Alimentazione e Cancro,* Ferro Edizioni, Milano, 1986, chap. 9.
4. Decreto Ministeriale, 14.12.1971, G.U. Rep. Ital., 328, 28, 12, 1971.
5. Westoo, G., Determination of methylmercury compounds in foodstuffs, *Acta Chim. Scand.,* 22, 2277, 1968.
6. Eckerlin, R.H., Hoult, D.M., and Carnick, G.R., Selenium determination in animal whole blood using stabilized temperature platform furnace and Zeeman background correction, *Act. Spectr.,* 8, 64, 1987.
7. Moretti, G., Bortoli, A., and Marin, V., Indagine sulla presenza di mercurio e di metil-mercurio in popolazioni costiere venete ed in organismi eduli dell'Adriatico settentrionale, *L'Igiene Moderna,* 93, 403, 1990.
8. Kyle, J.H. and Ghani, N., Methylmercury in human hair: a study of a Papua New Guinean population exposed to methylmercury through fish consumption, *Arch. Environ. Health,* 37(5), 255, 1982.
9. Sherlock, J.C. and Lindsay, D.J., Duplication diet study on mercury intake by fish consumers in the United Kingdom, *Arch. Environ. Health,* 37(5), 271, 1982.
10. Majori, L., Nedoclan, G., and Modonutti, G.B., Inquinamento da Hg nell'alto Adriatico, *Riv. Acqua Aria,* 3, 22, 1976.
11. Paccagnella, B., Prati L., and Bigoni A., Studio epidemiologico sul mercurio nei pesci e la salute umana in un'isola italiana del Mediterraneo, *L'Igiene Moderna,* 66, 479, 1973.

Average Mercury Intake in an Italian Diet

Torquato Ristori and Corrado Barghigiani

CONTENTS

ABSTRACT: Data are reported on total Hg concentration in 101 foodstuffs which often enter the Italian diet. Using these data, the possible average Hg intake per week with typical Italian diets was calculated. The introduction was also considered of particular foodstuffs, such as edible marine species caught in the Italian seas and in particular off the southern Tuscan coast, an area that has long been known to be influenced by the Hg anomaly of Mount Amiata.

I. INTRODUCTION

Mercury, which is commonly known as a particularly toxic element, is released into the environment as a consequence of both natural and anthropogenic activities.[1,2,3] Therefore this metal is distributed in all the environmental compartments, enters aquatic and terrestrial food chains, and reaches man. Data are available on the presence of this contaminant in many foodstuffs.[4-8]

For these reasons, diet has been considered one of the most important nonoccupational ways of mercury intake for humans, especially when the diet is mainly based on fish consumption.[9-13] The aim of this study is to evaluate the average mercury intake per week with the foodstuffs which most commonly enter the Italian diet and whether it is hazardous for human health according to the WHO Provisional Weekly Tolerable Intake (PWTI).[14]

II. EXPERIMENT

Diets for people of different ages were chosen, and foodstuffs were collected and analyzed for mercury.

A. DIET AND COLLECTION OF FOODSTUFFS

The diet for adults was chosen on the basis of logical considerations regarding the alimentary behavior of Italian people. Taking into account that this varies greatly depending on geographic regions, age, social classes, and job types, we took as a landmark a diet used in the canteen of a factory in Pisa,[15] where the quantities and types of the food served are a result of inquiries into the needs, likings, and alimentary habits of the workers and clerks who are distributed over a wide range of age classes and who come from different parts of Italy.

As a diet for children and adolescents (age between 4 and 14 years), we chose the one prescribed by the Children's Hospital of the University of Pisa for the meals served in nursery, elementary, and secondary schools.[16]

Both diets include two meals per day. Each meal consists of a first dish made of pasta, rice, or vegetable soup; a second dish consisting of different kinds of meat, cheese, an omelette, or fish; bread, vegetables, fruit, and beverage (water and/or wine). Fish was included in the diet 1 day per week for adults and 1 day per week, 3 times a month, for students. Moreover, the menu changes each day of the week, for 4 weeks. A continental breakfast was also considered, based on bread and a beverage such as milk and/or coffee, tea, or chocolate.

The collection of foodstuffs was made in Pisa (Tuscany), and were bought at supermarkets. The most commonly used brands were chosen. Fresh, frozen, and canned foodstuffs were analyzed.

B. Hg ANALYSIS

Before the analyses, the samples were mineralized with nitric acid at 120° C for 6 h in a closed system under pressure. Hg determination was performed by atomic absorption spectrometry on cold vapor. The analytical procedures were tested using Certified Reference Materials 1572 (citrus leaves: 0.08 ± 0.02 µg g^{-1}Hg) of the National Bureau of Standards (U.S.) B.C.R. (olea europea: 0.28 ± 0.02 µg g^{-1}Hg) from the Community Bureau of Reference of Brussels (Belgium); DORM-1 (dogfish muscle: 0.798 ± 0.074 µg g^{-1}Hg) and DOLT-1 (dogfish liver: 0.225 ± 0.057 µg g^{-1}Hg) of the National Research Council of Canada.

III. RESULTS AND DISCUSSION

Referring to the EC standards, which provide maximum Hg limits for seafood and land-grown products of 0.7 and 0.2 µg g^{-1} fresh weight, respectively, our results showed that there were few foodstuffs which presented relatively high metal contents (Table 1). In particular, among these the most contaminated (Hg concentrations are given as ng g^{-1} fresh weight ± S.D.) canned foods were salmon, 612 ± 80; anchovies, 135 ± 8.5; pilchard, 117 ± 4; a brand of tuna, 435 ± 5; small clams, 591 ± 66; and tripe, 441 ± 15. The most Hg contaminated frozen foods were swordfish, 736 ± 36 and smooth hound, 339 ± 15; and among fresh foods were dory, 341 ± 33.5; smooth hound, 327 ± 6.8; swordfish, 1082 ± 95; and dry boletes of mushrooms, 3420 ± 585.

Except for the above-reported values, Hg concentrations in fish ranged between 7.5 and 100 ng g^{-1}, while the other foodstuffs displayed a metal content below 56 ng g^{-1}. All terrestrial products of animal origin except tripe had a mercury concentration below 5 ng g^{-1}, results very similar to those found by other authors.[5,6,17,18]

With the diet for adults taken as a landmark, the mercury intake proved to be quite low, 80.9 µg per week, as reported in Table 2—very close to that found for adults with diets of other countries.[19,20] It must be pointed out that pasta and bread played the most important role in the intake, with 58 µg of Hg per week.

As concerns the weekly mercury intake for children and adolescents (Table 2), our results show that it increases with age. This is due to the fact that although the diet is the same for all ages, the amounts of food change.

Unlike that observed for adults, in the diet for students it is fish that plays the most important role in the intake of mercury, with about 100 µg per week, even if it is contained in the diet only 1 day per week.

Table 1 **Average mercury concentration (ng g⁻¹ fresh weight) ± S.D. in different foodstuffs; n=3.**

	Hg		Hg
Canned fish		**Starchy foods/legumes**	
Anchovy	135 ± 8.5	Bean	<2
Clam	591 ± 66	Brown bread	17 ± 5.9
Mackerel	25 ± 4.5	Cracker	12.8 ± 3
Pilchard	117 ± 4	Pasta[a]	33.4 ± 5.2
Salmon	612 ± 80	Potato	2.2 ± 0.6
Tuna[a]	233 ± 117	Rice	<2
		White bread	11.7 ± 3.1
Frozen fish		**Fruits/vegetables**	
Cod	79 ± 18	Apple	1.4 ± 0.2
Common shrimp	12.4 ± 3.2	Artichoke	3.5 ± 1.2
Smooth hound	339 ± 15	Carrot	2.6 ± 0.5
Eledone cirrhosa[b]	11.3 ± 2.8	Olive oil	<2
Hake	28.3 ± 3	Orange	2.2 ± 0.4
Mullet	15.2 ± 3.8	Pear	5.6 ± 1.3
Norway lobster	16.6 ± 7.2	Pickled vegetables	2.0 ± 0.3
Plaice	30 ± 2.3	Radish	2.4 ± 0.6
Prawn	11.5 ± 3	Salad	1.8 ± 0.5
Ray	7.5 ± 2.3	Strawberry	2 ± 0.4
Swordfish	736 ± 36	Tomato sauce	1.3 ± 0.5
Fresh fish		**Meat/game/egg**	
Anchovy	99 ± 4.2	Beefsteak	<2
Angler	74.8 ± 8.4	Turkey breast	<2
Clam	10.6 ± 6.9	Canned meat	<2
Common squid	13 ± 6.7	Canned tripe	441 ± 15
Cuttlefish	78 ± 5.2	Chicken	<2
Smooth hound	327.6 ± 6.8	Chicken liver	<2
Dory	341 ± 33.5	Deer	2.6 ± 0.5
Eel	98 ± 2.8	Egg	<2
Eledone cirrhosa[b]	9.7 ± 3	Hare	3 ± 1.5
Grouper	40.1 ± 11.3	Horse	<2
Mantis shrimp	33 ± 6.5	Horse liver	<2
Mussel	19.5 ± 6.6	Lamb	2.5 ± 0.8
Octopus	9.7 ± 3	Pork	<2
Prawn	14 ± 3.7	Rabbit	2 ± 0.1
Salmon	48 ± 2.8	Bouillon cube	56.0 ± 14
Sole	42.6 ± 2.3	Sparrow	2.3 ± 0.4
Strandsnile	100 ± 43	Veal	2.8 ± 0.6
Swordfish[e]	1082 ± 95	Veal liver	3.2 ± 1.3
Trout	30.5 ± 13	Veal kidney	6 ± 0.4
Cheeses		**Nuts/dried fruits/seeds**	
Caciotta	3.4 ± 1.1	Almond	4.8 ± 0.5
Gruyère	22.6 ± 4	Date	2.9 ± 0.2
Mozzarella	14.1 ± 8.5	Dried fig	<2
Parmesan	<2	Pine seed	19 ± 7
Taleggio	<2	Pistachio	3.6 ± 0.8
		Powdered chestnuts	<2
		Walnut	3 ± 1

Table 1   **Continued.**

	Hg		Hg
Cold meats		**Beverages**	
Boiled ham	2.5 ± 0.1	Cocoa[d]	<2
Frankfurter	3.8 ± 0.4	Coffee[d]	5.3 ± 3.3
Dry-cured ham	<2	Milk	<2
Salami	<2	Red wine	1 ± 0.3
		Tea[d]	<2
Mushrooms		White wine	1.1 ± 0.3
Armillariella mellea	45 ± 8.6		
Boletus edulis[c]	3420 ± 585	**Sweeteners**	
Psalliota bispora	22.1 ± 1.8	Honey	<2
		Sugar	3.2 ± 1.5
		Chocolate-cream	6.8 ± 2.3

[a] n=15.
[b] Small ocotpus-like cephalopod.
[c] Dried.
[d] Ground or leaves.
[e] n=6.

Table 2   **Weekly mercury intake, range, and average ± S.D. (µg).**

	Range	Average ± SD
First class		
4–6 years	35–174.8	109.3 ± 75.5
Nursery School		
Second class		
6–11 years	47.9–235.4	147.4 ± 101
Elementary school		
Third class		
11–14 years	58–243.9	157 ± 100.3
Secondary school		
Fourth class		
Adults	79.4–83.6	80.9 ± 3.7
Industry		

Paradoxically, it is students and not adults that ingest a greater amount of this toxic metal. However, the reason for this is merely due to the fact that in the factory canteen it is possible to choose dishes other than fish, and in the school a choice is not provided.

It must be emphasized that the above mercury amounts may be greatly enhanced by frequent introduction of other particularly contaminated foodstuffs, such as most fish from the Northern Tyrrhenian sea.[7] Table 3 reports on the most commercially important edible fishes caught off the southern coast of Tuscany, which have long been known to be affected by the geological anomaly of Mount Amiata.[1,7,21] In comparison to the Hg values found in fish bought at the market (Table 1), those reported in Table 3 for the same species are much higher. The introduction in the diet of these fish increased the average weekly mercury intake to 205.6 ± 115.6 µg for the first class, 275.9 ± 154.3 µg for the second class, 288.7 ± 157.5 µg for the third class, and 120.7 ± 1.1 µg for the fourth class. Since, according to WHO,[14] a weekly intake of 300 µg of Hg is the maximum tolerable for an individual weighing 70

Table 3 **Hg content in fish caught off the coast of southern Tuscany.**

Fish	Hg	Length (cm)	No.
Merluccius merluccius	249.2 ± 269.6 (8.2–1504)	6–62	84
Parapeneus longirostris	641.4 ± 475.2 (75.4–1361)	0.9–3.6	77
Eledone cirrhosa	651.9 ± 263.7 (142–1784)	2–14.5	230

Note: Average values of Hg (ng g^{-1} w.w.) ± S.D. and range are reported.

kg, from our data and calculation based on the chosen diets it turns out that young people of school age living where these fish are commonly consumed could be exposed to a health risk. It must be added that fishermen and their families living in southern Tuscany could be even more subject to risk since, as is known, they are accustomed to eating fish often. The same problem could exist for fishermen living in Sicily, who often include swordfish—a fish highly contaminated by mercury—in their diet (see Table 1).

IV. CONCLUSIONS

For most Italians, diet is not a very important source of mercury intake. This is due to the fact that the main Italian basic foods such as pasta, bread, and most of the other land-grown products have very low Hg concentrations. Only a frequent introduction into the diet of a particular species of fish such as swordfish, or marine organisms from Hg-contaminated areas such as southern Tuscany, could represent a risk for human health.

REFERENCES

1. UNEP/FAO/WHO, Assessment of the state of pollution of the Mediterranean sea by mercury and mercury compounds, *MAP Tech. Rep. Series,* No. 18, 1987.
2. Andren, A.W. and Nriagu, J.O., The global cycle of mercury, In: *The Biochemistry of Mercury in the Environment,* Nriagu, J.O. Eds., Elsevier/North-Holland, Amsterdam, 1979, 1.
3. Lindberg, S.E., Mercury, In: *Lead, Mercury, Cadmium and Arsenic in the Environment,* Scope 31, Hutchinson, T.C. and Meema, K.M., Eds., John Wiley & Sons, Chichester, 1987, chap. 2.
4. Westoo, G., Determination of methylmercury compounds in foodstuffs, *Acta Chem. Scand.,* 21, 1790, 1967.
5. May, K. and Stoeppler, M., Studies on the biochemical cycle of mercury. Mercury in sea and inland water and food products, In: *Proc. Int. Conf. on Heavy Metals in the Environment,* Vol. 1, CEP Consultants Ltd., Edinburgh, U.K., 1983, 241.
6. Kambamanoli-Dimou, A., Kilikidis, S., and Kamarianos, A., Methyl mercury concentration in broiler's meat and hen's meat and eggs, *Bull. Environ. Contam. Toxicol.,* 42, 728, 1989.
7. Barghigiani, C. and De Ranieri, S., Mercury content in different size classes of important edible species of the Northern Tyrrhenian sea, *Mar. Pollut. Bull.,* 17, 424, 1992.
8. Buzina, R., Suboticanec, K., Vucusic, J., Sapunar, J., Antonic, K., and Zorica, M., Effect of industrial pollution on seafood content and dietary intake of total and methylmercury, *Sci. Total Environ.,* 78, 45, 1989.

9. Turner, M.D., Marsh, D.O., Rubio, C.E., Chiriboga, J., Collazos Chiriboga, C., Crispin Smith, J., and Clarkson, T.W., Methylmercury in population eating large quantities of marine fish. I. Northern Perù, in Proc. of the 1st Congreso International del Mercurio, Tomo I, Fabrica National de Moneda y Timbre, Madrid, 1974, 229.

10. Marsh, D.O., Turner, M.D., Smith, J.C., Wun Choi, J., and Clarkson, T.W., Methyl mercury (MeHg) in human populations eating large quantities of marine fish. II. American Samoa; cannery workers and fishermen, in Proc. of the 1st Congreso International del Mercurio, Tomo I, Fabrica National de Moneda y Timbre, Madrid, 1974, 235.

11. Suzuki, T., Kashiwazaki, S.S., Igata, A., and Niina, K., Hair mercury value and fish-eating habit, *Ecol. Food. Nutr.*, 8, 117, 1979.

12. Nauen, C.E., Tomassi, G., Sartorini, C.P., and Josupeit H., Results of the first pilot study on the chance of Italian seafood consumers exceeding their individual allowable daily mercury intake. VI Journées Etud. Pollutions, Cannes, C.I.E.S.M., 1982, 571.

13. Skerfving, S., Mercury in women exposed to methylmercury through fish consumption, and in their newborn babies and breast milk, *Bull. Environ. Contam. Toxicol.*, 41, 475, 1988.

14. WHO, Report on Consultation to Re-examine the WHO Environmental Health Criteria for Mercury, Geneva, April 21–25, 1980.

15. Biasi, G., Personal communication, 1991.

16. Vignoli, R., Personal communication, 1991.

17. Falandysz, J., Manganese, copper, zinc, iron, cadmium, mercury and lead in muscle meat, liver and kidneys of poultry, rabbit and sheep slaughtered in the northern part of Poland, 1987, *Food Addit. Contam.*, 8, 71, 1991.

18. Jorhem, L., Slorach, S., Sundström, B., and Ohlin, B., Lead, cadmium, arsenic and mercury in meat, liver and kidney of Swedish pigs and cattle in 1984–88, *Food Addit. Contam.*, 8, 201, 1991.

19. Menendez, R., Beltrán, G., and Torres, O., Estimation of the weekly intake of mercury which may be consumed by university students from the City of Havana, *Die Nahrung*, 34, 95, 1990.

20. Palusová, O., Ursínyová, M., and Uhnák, J., Mercury levels in the components of the environment and diets, *Sci. Total Environ.*, 101, 79, 1991.

21. Barghigiani, C., Pellegrini, D., D'Ulivo, A., and De Ranieri, S., Mercury assessment and its relation to selenium levels in edible species of the Northern Tyrrehenian Sea, *Mar. Pollut. Bull.*, 22, 406, 1991.

Changes in Human Dietary Intake of Mercury in Polluted Areas in Finland Between 1967 and 1990

Kimmo Louekari, Arun B. Mukherjee, and Matti Verta

CONTENTS

I. INTRODUCTION

In recent decades, much attention and public debate has been focused on mercury (Hg) as an environmental pollutant. There were outbreaks of methylmercury (MeHg) poisoning (known as Minamata disease) in Japan in the 1950s caused by eating contaminated fish, and in Iraq at the beginning of 1970s. These incidences led to the awareness of the toxic effects of MeHg.[1-3] Due to environmental concerns emission of Hg has been reduced.[4] In Finland, anthropogenic Hg emission has decreased from 14 tons (t) in 1967 to 3.7 t in 1987.[5] Still, it has been estimated by Nriagu[6] in 1989 that the amount of global anthropogenic Hg emission is considerable (3600 t year^{-1}).

The accumulation of Hg in fish in Finnish freshwater and brackish water has been studied for more than 20 years. In the 1960s, high Hg concentrations of up to 5 to 6 mg kg^{-1} (wet wt.) in northern pike were reported due to industrial discharge to receiving waters.[7] Since then, there has been a decrease of Hg concentrations in fish in these heavily polluted areas: firstly, due to the ban on the use of Hg as a slimicide in the pulp and paper industry, and secondly, due to the considerable reduction of Hg discharges from chloralkali plants.[8-9,20]

Several studies on dietary intake of Hg in Finland have been published in recent years. Past studies indicated that the average dietary intake of Hg was 6 µg d^{-1} of which 60% stemmed from the consumption of fish.[13] Studies of different socioeconomic groups based on food consumption data from a 1985 household survey indicated that Hg intake of farmers was 13 µg d^{-1} whereas that of single people was 6 µg d^{-1}.[13] In that study, however, data on Hg concentration in fish was obtained from a publication of Nuurtamo et al.[19] in 1980, which does not reflect the present situation. The dietary intake of total Hg in different countries is given in Table 1.

In Finland, between 1967 and 1990 there has been a change of food habits as well as a decline in the Hg concentration in fish. Therefore, the objective of our study was to find:

1. The trend of dietary intake of Hg in areas contaminated by the pulp and paper industry, covering 10 to 15% of the lakes and coastal waters in Finland
2. The differences between socioeconomic groups in respect to exposure to Hg in our study areas.

1-56670-066-3/94/$0.00+$.50
© 1994 Lewis Publishers

Table 1   **The intake of total Hg for adult populations in selected countries.**

Country	Sampling method	Hg-intake $\mu g\ d^{-1}$	Ref.
Germany	C[a]	23	24
U.S.	Not known	3	10
Australia	Not known	3	10
Finland	MB[b]	2.3	12
Farmers	C	12.9	13
White-collar workers	C	7.3	13
Sweden	MB	1.6	14
Yugoslavia			
Polluted areas	ASF[c]	9.2–25	15
Belgium	DP[d]	14	23
Netherlands	MB	5	22
U.K.	Not known	3	21
Polluted areas	DP	15	11

[a] C: Calculation method (food items are analysed separately, not combined to food groups as in market basket method).

[b] MB: Market basket method.

[c] ASF: Analysis of selected foods (only sea foods were analysed in this study).

[d] DP: Duplicate portion method.

Note: PTWI established by WHO is 29 µg/day for MeHg and 43 µg/day for total Hg.

II. MATERIALS AND METHODS

A. CONCENTRATION OF MERCURY IN FISH AND OTHER FOOD ITEMS

The first reports of aquatic Hg pollution were published in the late 1960s.[7,16] These data consisted of 460 fish samples of different species. The other data source used in this study for concentration of Hg in fish was an extensive monitoring study of northern pike (*Esox lucius*) (n = 920) in contaminated fresh and brackish water areas conducted by the National Board of Waters and the Environment in 1970 through 1990. Since monitoring of other fish species has not been as extensive, change in Hg concentrations in other fish was assumed to be relative to the change in Hg concentration of pike. This assumption was substantiated by analysis for Hg concentration of perch (*Perca fluviatitis*), roach (*Rutilus rutilus*), and vendice (*Coregonus albula*) from one of the largest polluted areas, lake Päijänne.[17]

The Hg content in food items other than fish has been measured in a Finnish trace element study by Koivistoinen[18] and the results of that investigation were utilized in the present study. In that study, sampling covered the whole country and was directed to large food processors, dairies, slaughterhouses, etc., and seasonal variation was taken into account for some food groups. The Hg concentrations of imported and processed fish (n = 24) were obtained from Nuurtamo et al.[19]

B. FOOD CONSUMPTION DATA

The food consumption data were obtained from household surveys made by the Central Statistical Office of Finland, which recorded the food purchases in a nationally representative sample of households. The households were randomly sampled from files of the Central Population Register. The number of participating households varied between 3253 and 8640 in the years 1966 to 1990.

Recording of food consumption was made in households, usually by the housewife, who listed all food purchases in a record book. The survey provides data on different socioeconomic groups. In addition to the average food consumption data, data of farmers (small farms

Table 2   Consumption of fish (kg year^{-1}) by farmers and white-collar workers, 1967 to 1990.

Year	Farmers	W-C workers	Average
Baltic Herring			
1967	2.60	3.00	3.10
1971	2.60	2.40	2.90
1981	2.61	1.76	3.07
1990	1.82	0.82	2.49
Pike			
1967	1.00	0.50	0.60
1971	2.85	0.30	1.00
1981	4.10	0.47	0.91
1990	3.00	0.25	0.68
Salmon and trout			
1967	0.00	0.00	0.00
1971	0.05	0.10	0.10
1981	0.48	0.75	0.87
1990	1.30	3.40	2.83
Other fish			
1967	8.06	4.90	5.30
1971	9.00	7.20	9.60
1981	13.58	5.22	6.77
1990	6.93	4.32	6.84
Total			
1967	11.66	8.40	9.00
1971	14.50	7.20	9.60
1981	20.76	8.21	11.61
1990	13.04	8.78	12.84

Note: "Other fish" include perch, vendice, whitefish, bream, imported frozen fish like cod and redfish, etc.

of 2 to 10 hectares cultivated; 215 households in 1981) and white-collar workers (2123 households in 1981) were selected for comparison. Farmers are known to consume local, and often, self-caught fish, whereas white-collar workers eat less pike and perch. The food consumption data indicated a tenfold difference in the consumption of pike (Table 2). Since the Hg concentration in pike is high, it was reasonable to study the difference in total intake between these socioeconomic groups.

On an average, fish consumption has slightly increased during the whole study period and most of this is due to consumption of rainbow trout (Table 2), which is now available fresh in practically every larger retailer. Other fresh fish are sometimes available in retailers, but are often bought in fish markets in Finland. The Hg concentration in rainbow trout is low compared to pike and perch.

Contribution of other foods to the Hg intake was assumed to remain at the same level through the study period, since major changes in Hg concentration in other foods have not been reported.

C. ESTIMATION OF DIETARY INTAKE OF MERCURY

The database used for calculation of the intake of total-Hg, contained data of the food consumption (kg year^{-1} person^{-1}) and mean concentration of total-Hg (μg kg^{-1} wet wt.) in different fish species, processed foods of fish, and other food items. Intakes (μg d^{-1}) were

Table 3 **The area weighed Hg-concentration of pike (1 kg), perch and vendice in contaminated areas, 1967 to 1990.**

Year	Total Hg ($\mu g\ kg^{-1}$, fresh weight)		
	Pike	**Perch**	**Vendice**
1967/1968[7]	1520	930	750
1970/1971	1150	600	420
1980	850	420	310
1990	600	330	220

Table 4 **Dietary intake of total Hg ($\mu g\ d^{-1}$) in different socioeconomic groups.**

	1966	**1970**	**1980**	**1990**
Average	14.3	11.9	11.2	8.6
Farmers	22.4	23.2	25.2	15.1
White-collar workers	12.6	7.4	8.5	7.9

calculated by multiplying the total-Hg concentration of a food item with the consumption of the respective food item and adding up the results.

III. RESULTS AND DISCUSSION

Monitoring revealed a pronounced change of Hg concentration in fish in the polluted areas during the study period. The average pike Hg concentration decreased from 1.52 mg kg^{-1} in 1967 to 1968 to 0.60 mg kg^{-1} in 1990 (Table 3). At the same time, the areas where pikes have a Hg concentration exceeding 0.5 mg kg^{-1} diminished from 4000 to 3000 km^2. Most of these lakes and coastal areas are located in Southern Finland and are the sites of pulp and paper and chloralkali plants. In 1967 to 1968 there were several lakes and coastal areas where the Hg concentration of pikes exceeded 1 mg kg^{-1}, but no such large areas were found in 1990.

In 1971 the National Board of Health recommended that fish containing 0.5 to 1 mg kg^{-1} (wet wt.) should not be eaten more than once a week and that fish containing more than 1 mg Hg kg^{-1} should not be used for human consumption. With new data in 1980, updated recommendations were published in 1981. In these recommendations, the clearing of certain polluted areas was taken into account. In addition, restriction for fish use was also established for several artificial reservoirs.

It is interesting to note that in spite of the awareness and remarkable public concern of high Hg concentration in fish in the early 1970s, farmers, who have the highest dietary intake of Hg, actually increased consumption of pike and fish in general during the period 1967 to 1980 (Table 2). Either farmers were not convinced of the existence of a health risk or fish was too important to their diet. Surprisingly, white-collar workers also increased their consumption of pike in the 1970s after a clear decrease at the end of the 1960s. It is likely that white-collar workers consumed pike from brackish waters, which were much less contaminated than pike from the lakes. In the 1980s, consumption of pike decreased in all groups included in the present study.

The average dietary intake of total Hg decreased from 14 to 9 $\mu g\ d^{-1}$ during the period of 1966 to 1990 (Table 4). One explanation of this trend is that the Hg content of fish in contaminated lakes decreased by about 60 to 70% (Table 3). The other explanation is that consumers prefer rainbow trout, which have a low concentration of Hg (130 $\mu g\ kg^{-1}$), whereas the fishes with highest Hg concentration, namely pike and perch (300 to 600 $\mu g/\ kg^{-1}$ in the year 1990) are consumed less than previously.

The decrease in the total-Hg intake of farmers has been faster in the 1980s than that of the average consumer (Table 4). This is explained by the fact that farmers consume less fish than 10 years ago, whereas the total fish consumption of an average person and that of white-collar workers has increased slightly (Table 2). This increase is presumably due to the nutritional education emphasizing the favorable qualities of fish as compared to meat in terms of cardiovascular disease risk factors.

Intake of total-Hg by white-collar workers is smaller than the average intake (Table 4). In this socioeconomic group sport/recreational fishing is not common, and therefore consumption of the predator fishes of lakes has remained at a low level (Table 2). White-collar workers especially prefer rainbow trout and imported frozen fish instead of pike and perch. Rainbow trout and convenience foods containing imported sea fish are available in most retail shops, whereas other fishes are usually sold in fish markets or are self-caught.

The contribution of foods other than fish to the total dietary intake of Hg in 1990 was 37% for white-collar workers and only 15% for farmers.

Our results cover the areas contaminated by emissions from the pulp and paper industry in Finland and it is obvious from the data that the intake of Hg in the Finnish population is lower on an average. The decreasing trend of Hg intake observed in the study probably applies to the whole Finnish population, because fish with low Hg concentrations have replaced pike, perch, and other predator fish in the Finnish diet.

The method used for estimating Hg intake in the present study was not optimal. Household surveys provide information about the food consumption of households and not individuals, and furthermore, loss of food in household is not taken into account. Therefore these estimates are not accurate. More reliable estimates of intake would require use of the duplicate portion method or analysis of market basket samples which cannot be collected afterwards. Since our aim was to study the change of Hg intake over a relatively long period, we have had to use statistical data from household surveys, applying the same method used during the years 1967 to 1990, which is also the case for the monitoring study of Hg concentration in fish.

IV. SUMMARY

This study identifies the change in Hg intake of people of different socioeconomic groups during the period 1967 to 1990 in polluted areas covering 10 to 15% of the lakes and coastal waters in Finland. In the study areas, Hg loads stemmed mainly from the pulp and paper industry and also from chloralkali plants. The pulp and paper industry discarded the use of Hg as a slimicide in 1968 and anthropogenic Hg emissions (via air and water) have decreased from 14 t in 1967 to 3.7 t in 1987. It is well documented that MeHg enters into the food chain through fish. Because of this, an extensive Hg survey of fish populations, mainly northern pike (*Esox lucius* L.), has been carried out by the national Board of Waters and the Environment in 60 areas over 10-year intervals (e.g., 1970 to 1971; 1980 to 1981, and 1990 to 1991). It was observed that the Hg concentration in fish has remarkably decreased in polluted areas. The weighed average pike Hg concentration decreased from 1.52 mg kg^{-1} (in 1967 to 1968) to 0.60 mg kg^{-1} (in 1990–1991). Dietary intake of Hg among farmers and white-collar workers living in the study areas was estimated by combining the data on Hg concentrations in food with the data on food consumption habits during the period 1967 to 1990. It was observed that total dietary Hg intake for farmers was 22, 23, 25, and 15 µg d^{-1} in the years 1967, 1970, 1980, and 1990, respectively. On the other hand, the total dietary Hg intake for white-collar workers was 13, 7, 9, and 8 µg d^{-1} in the same years. Our study suggests that although the consumption of fish has slightly increased in the Finnish population, except for farmers, the intake of Hg has decreased remarkably (by 39% on the average). This is due to a rapid decline of aquatic Hg discharge, especially from the pulp and paper industry, and a changed preference of fish species, for example, to rainbow trout, containing much less Hg than pike and perch.

REFERENCES

1. WHO, Environmental Health Criteria 1: Mercury, World Health Organization, Geneva, 1–32, 1976.
2. WHO, Conference on intoxication due to alkylmercury treated seed, *Bull. WHO*, 53 (Suppl.), 1–138, 1976.
3. ILO, Mercury, In: *Encyclopaedia of Occupational Health and Safety*, 3rd ed., International Labour Organization, Geneva, 1332, 1983.
4. Lindqvist, O., Johansson, K., Aastrup, M., Andersson, A., Bringmark, L., Hovsenius, G., Håkanson, L., Iverfeldt, Å., Meili, M., and Timm, B., Mercury in the Swedish environment—recent research on causes, consequences and corrective methods, *Water Air Soil Pollut.*, 55:23, 1991.
5. Mukherjee, A.B., Industrial emissions of mercury in Finland between 1967 and 1987, *Water Air Soil Pollut.*, 56, 35, 1991.
6. Nriagu, J.O., A global assessment of natural sources of atmospheric trace metals, *Nature*, 338, 47, 1989.
7. Häsänen, E. and Sjöblom, V., Mercury concentration in fish in Finland in 1967 *Suomen Kalatalous (Finl. Fisk.)*, 36, 1, 1968 (in Finnish).
8. Nuorteva, P., Lodenius, M., and Nuorteva, S.-L., Decrease in the mercury levels of *Esox lucius* L. and *Abramis farenus* L. (Teleostei) in the Hämeenkyrö watercourse after the phenylmercury ban in Finland, *Aquilo Ser. Zool.*, 19, 97, 1979.
9. Lodenius, M., Mercury content of fish from Hämeenkyrö, SW Finland in 1971–1986, *Suomen Kalatalous (Finl. Fisk.)* 53, 14, 1988 (in Finnish).
10. UNEP/FAO/WHO, Assessment of Chemical Contaminants in Food. Report on the Results of the UNEP/FAO/WHO Programme on Health-Related Environmental Monitoring, Monitoring and Assessment Research Centre, London, 1988.
11. Sherlock, J.S., Lindsay, D.G., Hislop, J.E., Evans, W.H., and Collier, T.R., Duplication diet study on mercury intake by fish consumers in the United Kingdom, *Arch. Environ. Health*, 37, 271, 1982.
12. Kumpulainen, J. and Tahvonen, R., Report on the Activities of the Sub-network on Trace Elements Status in Food, Presented in: 1989 Consultation of the European Cooperative Research Network on Trace Element, Lausanne, Switzerland, 5–8 Sept., 1989.
13. Louekari, K., Estimation of heavy metal intakes based on household survey, *Näringsforskning*, 34, 107, 1990.
14. Kumpulainen, J., Intake of heavy metals, comparisons of methods, In: *Monitoring of Dietary Intakes*, McDonald, J., Ed., Springer-Verlag, Berlin, 1991, 61.
15. Buzina, R., Suboticanec, K., Vukusic, J., Sapunar, J., Antonoc, K., and Zorica, M., Effect of industrial pollution on seafood content and dietary intake of total and methylmercury, *Sci. Total Environ.*, 78, 45, 1989.
16. Nuorteva, P., Metylkvicksilver i naturens näringskedjor—ett aktuellt problem, *Nordenskiöldsamfunderts Tidskr.*, 29, 6, 1969.
17. Hattula, M.L., Särkka, J., Janatuinen, J., Paasivirta, J., and Roos, A., Total mercury and methylmercury contents in fish from lake Päijänne, *Environ. Pollut.*, 17, 19, 1978.
18. Koivistoinen, P., Ed., Mineral element composition of Finnish foods: N, K, Ca, Mg, P, S, Fe, Cu, Mn, Zn, Mo, Co, Ni, Cr, F, Se, Si, Rb, Al, B, Br, Hg, As, Cd, Pb and ash, *Acta Agric. Scand.*, 22, 1972, 1980.
19. Nuurtamo, M., Varo, P., Saari, E., and Koivistoinen, P., Fish and fish products, *Acta Agric. Scand.*, 22, 77, 1980.
20. Verta, M., Mercury in Finnish Forest Lakes and Reservoirs: Anthropogenic Contribution to the Load and Accumulation in Fish, Thesis, Dept. of Limnology, University of Helsinki, 1–34, 1990.

21. Ministry of Agriculture, Fisheries and Food, Survey of Mercury in Food: Second Supplementary Report, London, 1987.

22. de Vos, R.H., van Dokkum, W., Olthof, P.D.A., Quirijns, J.K., Muys, T., and van der Poll, J.M., Pesticides and other chemical residues in Dutch total diet samples (June 1976–July 1978), *Food Chem. Toxicol.*, 22 11, 1984.

23. Buchet, P. and Lauwerys, R., Oral daily intake of cadmium, lead, manganese, copper, chromium, mercury, calcium, zinc and arsenic in Belgium: a duplicate meal study, *Food Chem. Toxicol.*, 21, 19, 1983.

24. Weigert, P., Muller, J., Klein, H., Zufelde, K., and Hillebrandt, J., Arsen, Blei, Cadmium und Quecksilber in und auf Lebensmitteln. Zentrale Erfassungs- und Bewertungsstelle fur Umweltchemikalien des Bundesgesundheitsamtes, ZEBS-Hefte, 1, 1984.

INDEX

INDEX

A

Accumulation, 9
Acid extraction, mercury, 26, 27
Acidification
  boreal lake, 100, 103
  Little Rock Lake Project, 206
  terrestrial ecosystem, 343–354
Adirondack drainage lakes, perch, 457–469
Adriatic Sea, fishermen, 689–697
Aerosol
  atmospheric measurement, 283
  fraction
    dry deposition of mercury, 267
    throughfall, 267
  southern Ontario, Canada, 283
Air sampling unit, southern Ontario, Canada, 282
Air-water cycling
  dimethyl-Hg, 212
  evasional flux, 213
  flux in deposition, 211
  gaseous mercury, 208
  humics, sequestering with, 217
  hypolimnion, 216
  interlake variation, 216
  lake, 203–220
  Little Rock Lake, 209
  Max Lake, 209
  oxidation, mercury, 211
  ozone, 211
  Pallette Lake, 213
  photoplankton productivity, 216
  photosynthetic activity, 216
  precipitation, 207, 208
  saturation, 212
  scavenging, 204
  scavenging ratio, 210
  soot, 217
  sulfite, 211
  University of Connecticut, 207
  "unreactive" mercury, 217
  water column methylation rate, 213
Aluminum, ombrotrophic peat bog, 194
Alzheimer's disease, mercury imbalance, 651–662
Amazon, Tucuruí reservoir, 21–40
Amytrophic lateral sclerosis, mercury imbalance, 651–662

Anoxia
  flooded vegetation/soil, 355–365
  meromictic lake, 58, 60
  plankton layer, 140
  vertical distribution of mercury, 140
Antecedent period
  dry deposition of mercury, 264
  throughfall, 264
Anthropogenic emission, 5, 6, 183
  southern Ontario, Canada, 281
Aquatic macrophytes, 24
Aqueous mercury
  atmospheric chemistry, 274
  boreal lake, 99–106
  photochemical behavior, 581–594
  redox reaction, 274
Atmosphere, 6
  aqueous mercury, 274
  bulk deposition
    atomic fluorescence spectrometry, 223
    coastal forest station, 222
    Denmark, 222, 223
    forest station, 222
    plasma emission spectrometry, 223
    precipitation, 225
    radiochemical neutron activation analysis, 222
    southern Baltic Sea area, 221–228
    Sweden, 223
  chloride, 276
  cycling
    mercury, 181–185
    Wisconsin lake, 153
  deposition, 4
    ombrotrophic peat bog, 187
  dry deposition, 273
  exchange, 7
  gas phase oxidation, 274
  gaseous mercury, 553–556
  Henry's law, 275
  hydroxide, 276
  input, boreal lake, 99
  kinetic constant, 274
  ligand, 276
  measurement
    aerosol, 283
    commercial uses, 281
    dimethyl mercury, 286
    dispersal, 281